BIODIVERSITY OF THE WESTERN GHATS AND SRI LANKA

BIODIVERSITY HOTSPOT
OF THE
WESTERN GHATS AND SRI LANKA

Edited by
T. Pullaiah, PhD

First edition published 2024

Apple Academic Press Inc.
1265 Goldenrod Circle, NE,
Palm Bay, FL 32905 USA

760 Laurentian Drive, Unit 19,
Burlington, ON L7N 0A4, CANADA

CRC Press
2385 NW Executive Center Drive,
Suite 320, Boca Raton FL 33431

4 Park Square, Milton Park,
Abingdon, Oxon, OX14 4RN UK

© 2024 by Apple Academic Press, Inc.

Apple Academic Press exclusively co-publishes with CRC Press, an imprint of Taylor & Francis Group, LLC

Library and Archives Canada Cataloguing in Publication

Title: Biodiversity hotspot of the Western Ghats and Sri Lanka / edited by T. Pullaiah, PhD.
Names: Pullaiah, T., editor.
Description: First edition. | Series statement: Biodiversity hotspots of the world | Includes bibliographical references and index.
Identifiers: Canadiana (print) 20230506437 | Canadiana (ebook) 20230506445 | ISBN 9781774913758 (hardcover) |
    ISBN 9781774913765 (softcover) | ISBN 9781003408758 (ebook)
Subjects: LCSH: Biodiversity—India—Western Ghats. | LCSH: Biodiversity—Sri Lanka.
Classification: LCC QH183 .B56 2024 | DDC 578.0954/7—dc23

Library of Congress Cataloging-in-Publication Data

Names: Pullaiah, T., editor.

Title: Biodiversity hotspot of the Western Ghats and Sri Lanka / edited by T. Pullaiah, PhD.

Description: First edition. | Palm Bay, FL, USA : AAP/Apple Academic Press, 2024. | Series: Biodiversity hotspots of the world | Includes bibliographical references and index. | Summary: "Biodiversity is declining at an alarming rate due to anthropogenic activities around the world. This book is the first volume in the new series Biodiversity Hotspots of the World, which will highlight the 36 hotspot regions of the world, regions that were designated as reaping maximum benefit from preservation efforts. This series is our humble attempt to document these hotspots as a conservation and preservation measure. The concise volumes in this series focus on the most interesting and important properties of these hotspots, covering physiography and climatology, vegetation and forest types, amphibian and reptile biodiversity, genetic diversity of crops and plants, fishes, butterflies and insects, birds, mammals, angiosperms and gymnosperms, and much more. And of course, the unique threats and conservation efforts for the area are addressed as well. This first volume focuses on the Western Ghats and Sri Lanka, construed as forming a community of species because of their shared biogeographical history. The Western Ghats and Sri Lanka Hotspot is extraordinarily rich in species, especially plants that are found nowhere else. However, its forests face tremendous population pressure and have been dramatically impacted by demands for timber and agricultural land. The Western Ghats and Sri Lanka hotspot are rich with over 5,000 flowering plants, 139 mammals, 508 birds, and 179 amphibian species. And over 300 of these species are globally threatened. These volumes will be essential resources for researchers and practitioners in the fields of conservation biology, ecology, and evolution as the series concisely records the existing biodiversity of these biodiversity hotspots of the world"-- Provided by publisher.

Identifiers: LCCN 2023035720 (print) | LCCN 2023035721 (ebook) | ISBN 9781774913758 (hbk) | ISBN 9781774913765 (pbk) | ISBN 9781003408758 (ebook)

Subjects: LCSH: Biodiversity--India--Western Ghats. | Biodiversity--Sri Lanka. | Biodiversity conservation--India--Western Ghats. | Biodiversity conservation--Sri Lanka.

Classification: LCC QH183 .B54336 2024 (print) | LCC QH183 (ebook) | DDC 577.095493--dc23/eng/20231003

LC record available at https://lccn.loc.gov/2023035720
LC ebook record available at https://lccn.loc.gov/2023035721

ISBN: 978-1-77491-375-8 (hbk)
ISBN: 978-1-77491-376-5 (pbk)
ISBN: 978-1-00340-875-8 (ebk)

ABOUT THE BOOK SERIES
BIODIVERSITY HOTSPOTS OF THE WORLD

Series Editor: **T. Pullaiah, PhD**
Former Professor, Department of Botany, Sri Krishnadevaraya University,
Andhra Pradesh, India

The term 'biodiversity' came into common usage after the 1986 National Forum on Biodiversity, held in Washington, DC, and the publication of selected papers from that event, titled *Biodiversity,* edited by Wilson (1988). Wilson credits Walter G. Rosen for coining the term. In 1988, British ecologist Norman Myers published a seminal paper identifying 10 tropical forest "hotspots." Myers described ten tropical forest "hotspots" on the basis of extraordinary plant endemism and high levels of habitat loss, albeit without quantitative criteria for the designation of "hotspot" status. A subsequent analysis added eight additional hotspots, including four from Mediterranean-type ecosystems (Myers 1990).

A biodiversity hotspot is a biogeographic region that is both a significant reservoir of biodiversity and is threatened with destruction. Conservation International introduced two strict quantitative criteria to qualify as a hotspot: a region had to contain at least 1,500 vascular plants as endemics (>0.5% of the world's total), and it had to have 30% or less of its original vegetation (extent of historical habitat cover) remaining. These efforts culminated in an extensive global review (Mittermeier et al. 1999) and scientific publication (Myers et al. 2000) that introduced seven new hotspots on the basis of both the better-defined criteria and new data. In 2005, an additional analysis brought the total number of biodiversity hotspots to 34, based on the work of nearly 400 specialists. In 2011, the Forests of East Australia was identified as the 35th hotspot by a team of researchers from the Commonwealth Scientific and Industrial Research Organisation (CSIRO) working with Conservation International. In February 2016, the North American Coastal Plain was recognized as meeting the criteria and became the Earth's 36th hotspot.

Very few books have been written/edited to document the biodiversity of these hotspots. Hence, these new volumes which aim to be a reference source for all foresters, wildlife biologists, conservationists, botanists, zoologists, policymakers, and others concerned with the conservation of the biodiversity of our planet.

REFERENCES

Mittermeier, R.A., Myers, N., Robles-Gil, P., Mittermeier, C.G., **1999**. Hotspots: Earth's Biologically Richest and Most Endangered Terrestrial Ecoregions. CEMEX/Agrupación Sierra Madre, Mexico City.
Myers, N. The Biodiversity Challenge: Expanded Hot-Spots Analysis. *Environmentalist* **1990**, *10*, 243–256.
Myers, N., Mittermeier, R., Mittermeier, C. *et al.* Biodiversity hotspots for conservation priorities. *Nature* 403, 853–858 (**2000**). https://doi.org/10.1038/35002501
Wilson, E. O.; Peter, F. M. (eds). National Academies of Sciences, Engineering, and Medicine. **1988**. Biodiversity. Washington, DC: The National Academies Press. https://doi.org/10.17226/989.

BIODIVERSITY HOTSPOTS OF THE WORLD

North and Central America
- California Floristic Province
- Madren Pine-Oak woodlands
- Mesoamerica
- North American Coastal Plain

The Carribbean
- Caribbean Islands

South America
- Atlantic Forest
- Cerrado
- Chilean Winter Rainfall-Valdivian Forests
- Tumbes-Choco-Magdalena
- Tropical Andes

Europe
- Mediterranean Basin

Africa
- Cape Floristic Region
- Coastal Forests of Eastern Africa
- Eastern Afromontane
- Guinean Forests of West Africa
- Horn of Africa
- Madagascar and Indian Ocean Islands
- Maputaland-Pondoland-Albany
- Succulent Kaoo

Central Asia
- Mountains of Central Asia

South Asia
- Himalayas/Eastern Himalaya
- Indo-Burma
- Western Ghats–Sri Lanka

Southeast Asia and Asia-Pacific
- East Melanesian Islands
- New Caledonia
- New Zealand
- Philippines
- Polynesia–Micronesia
- Eastern Australian Temperate Forests
- Southwest Australia
- Sundland and Nicobar Islands
- Wallacea

East Asia

- Japan
- Mountains of Southwest China

West Asia

- Caucasus
- Irano-Anatolian

ABOUT THE EDITOR

T. Pullaiah, PhD
Former Professor, Department of Botany,
Sri Krishnadevaraya University,
Andhra Pradesh, India

T. Pullaiah, PhD, is a former Professor in the Department of Botany at Sri Krishnadevaraya University in Andhra Pradesh, India, where he has taught for more than 35 years. He has held several positions at the university, including Dean, Faculty of Biosciences, Head of the Department of Botany, Head of the Department of Biotechnology, and Member of Academic Senate. He was the President of Indian Botanical Society (2014), President of the Indian Association for Angiosperm Taxonomy (2013), and Fellow of Andhra Pradesh Akademi of Sciences. Under his guidance, 54 students have obtained their doctoral degrees. He has authored 70 books, edited 40 books, and published over 340 research papers, including reviews and book chapters. His books include *Redsanders: Silviculture and Conservation* (Springer), *Genetically Modified Crops* (Springer), *Sandalwood: Silviculture, Conservation and Applications* (Springer), *Advances in Cell and Molecular Diagnostics* (Elsevier), *Camptothecin and Camptothecin Producing Plants* (Elsevier), *Paclitaxel* (Elsevier), *Monograph on Brachystelma and Ceropegia in India* (CRC Press), *Ethnobotany of India* (five volumes, Apple Academic Press), *Global Biodiversity* (four volumes, Apple Academic Press), and *Invasive Alien Species* (four volumes, Wiley Blackwell). He was also a member of Species Survival Commission of the International Union for Conservation of Nature (IUCN). Professor Pullaiah received his PhD from Andhra University, India, attended Moscow State University, Russia, and worked as postdoctoral fellow from 1976–1978.

CONTENTS

CONTRIBUTORS

M. V. Anju
Institute of Forest Genetics and Tree Breeding, Forest Campus, R. S. Puram, Coimbatore, India

R. Balaji
Institute of Forest Genetics and Tree Breeding, Forest Campus, R. S. Puram, Coimbatore, India

Nithya Basith
Department of Botany, Madras Christian College (Autonomous), Tambaram, Chennai, Tamil Nadu, India

V. K. Chandini
Department of Botany, St. Teresa's College (Autonomous), Ernakulam, Kerala, India
PG and Research Department of Botany, The Zamorin's Guruvayurappan College (Affiliated to the University of Calicut), Kerala, India

Lucas Dauner
Kunming Institute of Botany, Chinese Academy of Sciences, Kunming, Yunnan, China

Salindra K. Dayananda
Guangxi Key Laboratory of Forest Ecology and Conservation, College of Forestry, Guangxi University, Nanning, China
Field Ornithology Group of Sri Lanka, Department of Zoology and Environment Science, Faculty of Science, University of Colombo, Colombo, Sri Lanka

P. Deepak
Mount Carmel College, Vasanth Nagar, Bengaluru, Karnataka, India
St. Carmels College, Bangalore, India

K. P. Dinesh
Zoological Survey of India (ZSI), Western Regional Centre (WRC), Vidyanagar, Pune, India

Mahesh C. A. Galappaththi
Postgraduate Institute of Science, University of Peradeniya, Sri Lanka

Smpath De Alwis Goonatilake
International Union for Conservation of Nature, Pelawatte, Battaramulla, Sri Lanka
IUCN Freshwater Fish Specialist Group

Biju Haridas
Microbiology Division, KSCSTE-Jawaharlal Nehru Tropical Botanic Garden and Research Institute, Palode, Thiruvananthapuram, Kerala, India

Peter Janzen
Justus-von-Liebig-Schule, Duisburg, Germany

Udeni Jayalal
Department of Natural Resources, Sabaragamuwa University of Sri Lanka, Belihuloya

Samantha C. Karunarathna
Kunming Institute of Botany, Chinese Academy of Sciences, Kunming, Yunnan, China

S. Karuppusamy
Department of Botany, The Madura College, Madurai, Tamil Nadu, India

Sarath W. Kotagama
Field Ornithology Group of Sri Lanka, Department of Zoology and Environment Science, Faculty of Science, University of Colombo, Colombo, Sri Lanka

S. Arun Kumar
Institute of Forest Genetics and Tree Breeding, Forest Campus, R. S. Puram, Coimbatore, India

C. Kunhikannan
Institute of Forest Genetics and Tree Breeding, Forest Campus, R. S. Puram, Coimbatore, India

R. Kuralarasan
Institute of Forest Genetics and Tree Breeding, Forest Campus, R. S. Puram, Coimbatore, India

Alagu Manickavelu
Department of Genomic Science, Central University of Kerala, Kasaragod, Kerala, India

C. N. Manju
Bryology Laboratory, Department of Botany, University of Calicut, Malappuram, Kerala, India

K. M. Manjula
PG and Research Department of Botany, The Zamorin's Guruvayurappan College (Affiliated to the University of Calicut), Kerala, India
KSCST - Malabar Botanical Garden and Institute for Plant Science, Kozhikode, Kerala, India

B. Mufeed
Bryology Laboratory, Department of Botany, University of Calicut, Malappuram, Kerala

Kesavan Madhavan Nampoothiri
Microbial Processes and Technology Division (MPTD), CSIR- National Institute for Interdisciplinary Science and Technology (NIIST), Trivandrum, Kerala, India

Chandrika M. Nanayakkara
Department of Plant Sciences, University of Colombo, Colombo, Sri Lanka

Sanjeeva Nayaka
Lichenology Laboratory, CSIR-National Botanical Research Institute, Rana Pratap Marg, Lucknow, India

Hopeland P.
Arihant Heirlooms, Thalambur, Navalur Post Office, Chengalpattu District, Tamil Nadu, India

M. Arun Padmakumar
Microbial Processes and Technology Division (MPTD), CSIR- National Institute for Interdisciplinary Science and Technology (NIIST), Trivandrum, Kerala, India

S. Vazeed Pasha
Forest Biodiversity and Ecology Division, National Remote Sensing Centre, Indian Space Research Organisation, Balanagar, Hyderabad, Telangana, India

Umesh Pavukandy
Pavukandy House, Narayamkulam, Kozhikode, Kerala, India

Sandun J. Perera
Department of Natural Resources, Faculty of Applied Sciences, Sabaragamauwa University of Sri Lanka, Sri Lanka
Field Ornithology Group of Sri Lanka, Department of Zoology and Environment Science, Faculty of Science, University of Colombo, Colombo, Sri Lanka

Maya Peringottillam
Department of Genomic Science, Central University of Kerala, Kasaragod, Kerala, India

R. Prasanna
Institute of Forest Genetics and Tree Breeding, Forest Campus, R. S. Puram, Coimbatore, India

T. Pullaiah
Department of Botany, Sri Krishnadevaraya University, Anantapur, Andhra Pradesh, India

R. W. K. Punchihewa
ThiSaBi Institute (Bees for Sustainable Development), Homagama, Sri Lanka

K. P. Rajesh
Department of Botany, the Zamorin's Guruvayurappan College (Affiliated to the University of Calicut), Kozhikode, Kerala, India

Ramakrishna
Zoological Survey of India, Bangalore, Karnataka, India

Sudheera M. W. Ranwala
Department of Plant Sciences, Faculty of Science, University of Colombo, Sri Lanka

C. Sudhakar Reddy
Forest Biodiversity and Ecology Division, National Remote Sensing Centre, Indian Space Research Organisation, Balanagar, Hyderabad, Telangana, India

Paulraj Selva Singh Richard
Department of Botany, Madras Christian College (Autonomous), Tambaram, Chennai, Tamil Nadu, India

K. V. Satish
Forest Biodiversity and Ecology Division, National Remote Sensing Centre, Indian Space Research Organisation, Balanagar, Hyderabad, Telangana, India

Bharath Setturu
Energy and Wetlands Research Group [CES TE 15], Centre for Ecological Sciences, Indian Institute of Science, Bangalore, Karnataka, India

Sampath S. Senevirathne
Avian Evolution Node, Department of Zoology and Environment Sciences, Faculty of Science, University of Colombo, Colombo, Sri Lanka

Kodandoor Sharathchandra
Department of Biosciences, Mangalore University, Mangalagangotri, Karnataka, India

Kandikere R. Sridhar
Department of Biosciences, Mangalore University, Mangalagangotri, Karnataka, India

Steven L. Stephenson
Department of Biological Sciences, University of Arkansas, Fayetteville, Arkansas, USA

Amila P. Sumanapala
Department of Zoology and Environment Sciences, University of Colombo, Sri Lanka

R. H. S. Suranjanfernando
Centre for Applied Biodiversity Research and Education, Doragamuwa, Kandy, Sri Lanka

Seetharamaiah Thippeswamy
Department of Environmental Science, Mangalore University, Mangalagangothri, Mangalore, Karnataka, India

Ramachandra T. V.
Energy and Wetlands Research Group [CES TE 15], Centre for Ecological Sciences, Indian Institute of Science, Bangalore, Karnataka, India
Centre for infrastructure, Sustainable Transportation and Urban Planning [CiSTUP], Bangalore, India
Centre for Sustainable Technologies (astra), Indian Institute of Science, Karnataka, India

Nalin Wijayawardene
Center for Yunnan Plateau Biological Resources Protection and Utilization, College of Biological Resource and Food Engineering, Qujing Normal University, Qujing, Yunnan, P.R. China

ABBREVIATIONS

AFLP	amplified fragment length polymorphism
BCAP	Biodiversity Conservation Action Plan
BD	biotic disturbance
BES	Biodiversity and Ecosystem Services
BMNH	British Museum of Natural History
BOW	birds of the world
BR	biological richness
BRs	biosphere reserves
BSI	Botanical Survey of India
Bt	*Bacillus thuringiensis*
BV	biodiversity value
BWLS	Bhadra Wildlife Sanctuary
BZ	buffer zone
CCB	Central College Bangalore
CCD	Coast Conservation Department
CI	conservation international
COI	cytochrome *c* oxidase subunit I
CSR	corporate social responsibility
CR	critically endangered
DCA	drone congregation area
DD	data deficient
DEM	Digital Elevation Model
DI	disturbance index
DOS	department of space
DSG	Dharwar Super Group
DSS	decision support systems
EBA	Endemic Bird Area
ED	Evolutionarily distinct
EDGE	Evolutionarily Distinct and Globally Endangered
EG	Eastern Ghats
EN	endangered
EU	ecosystem uniqueness
FARA	Fisheries and Aquatic Resources Act
FFPO	Fauna and Flora Protection Ordinance
FI	fragmentation index
FOGSL	Field Ornithology Group of Sri Lanka
FOSS	Free and Open-Source Software
GMOs	genetically modified organisms

GnRH	gonadotropin-releasing hormone
GPS	Global Positioning System
GtH	gonadotropin
HADP	Hill Area Development Programme
HCG	human chorionic gonadotropin
IAA	indole acetic acid
IAS	invasive alien species
IBA	important bird area
InDels	insertion–deletions
IRAP	inter retrotransposons
IRS	Indian Remote Sensing
ISSR	inter-simple sequence repeat
IUCN	International Union for Conservation of Nature
KDN	Kanneliya–Dediyagala–Nakiyadeniys
LC	least concern
LULC	land use and land cover
LUT	look-up table
MAB	man and biosphere
MCA	Markov cellular automata
MCE	multi-criterion evaluation
MCZ	Museum of Comparative Zoology
MNHNP	Museum National Histoire Naturelle
NISSG	National Invasive Species Specialist Group
MPCA	Medicinal Plants Conservation Area
m.y.a.	million years ago
NARF	net annual rainfall
NBR	Nilgiri Biosphere Reserve
NP	national parks
NSFC	National Science Foundation of China
NT	near threatened
OGC	Open Geospatial Consortium
OGUs	operational geographic units
PA	protected areas
PGR	plant genetic resources
PIFI	President's International Fellowship Initiative
PWS	Periyar Wildlife Sanctuary
QDGC	Quarter-Degree Grid Cell
RAPD	random amplified polymorphic DNA
RET	rare, endangered, and threatened
SCA	special central assistance
SDSS	spatial decision support systems
SNP	single nucleotide polymorphism
SOI	Survey of India
SPLAM	Spatial Landscape Analysis and Modelling

SRTM	Shuttle Radar Topography Mission
SSR	simple sequence repeats
STMS	sequence-tagged microsatellite sites
SVNP	Silent Valley National Park
SWLS	Someshwara Wildlife Sanctuary
SWWZ	South-Western Wet Zone
TC	terrain complexity
TCP	tri-calcium phosphate
TEEB	The Economics of Ecosystems and Biodiversity
TN	Tamil Nadu
UNDP	United Nations Development Programme
UNFCC	UN Framework Convention on Climate Change
UPGMA	unweighted pair-group method with arithmetic averages
VU	vulnerable
WCMC	World Conservation Monitoring Centre
WCS	web coverage service
WFS	web feature service
WG	Western Ghats
WHT	Wildlife Heritage Trust
WLS	wild life sanctuaries
WMS	standards, web map service
YTB	yellow-throated bulbul
ZMB	Zoologisches Museum
ZSI	Zoological Survey of India

PREFACE

Forests and vegetation are being degraded at an alarming proportion today, resulting in loss of biodiversity. A lot of concern on the loss of this precious gift of nature is being raised. Myers et al. (2000) proposed the concept of biodiversity hotspots, where there is a high rate of endemism and high rate of habitat destruction. There is, hence, an urgent need to take stock of the biodiversity hotspots around the world for taking necessary action for its conservation. Very few books have been written/edited to document the biodiversity of these hotspots. There is no book documenting the complete biodiversity of all the biodiversity hotspots. The present series is a humble attempt to give a summary of the biodiversity in some of these hotspots. I should have started the series from east or west, but I took hotspots of the Indian region first, since I am from this region and have expertise in it. For volumes on regions other than India, I am taking the help of experts from that region as coeditors.

Initially, we thought to cover biodiversity of both the Western Ghats and Sri Lanka together. But our expert on a particular group of organisms of the Western Ghats had less knowledge on Sri Lanka and vice versa. Consequently, we had to go for separate accounts on the Western Ghats and Sri Lanka. Some aspects could not be covered due to two reasons. First reason was expertise on a topic on the region was not available, while the second reason was the experts on that topic were too busy with their professional activities. I thank all the contributors for their critical review of the biodiversity of the Western Ghats and Sri Lanka.

Publishers Apple Academic Press were very strict regarding plagiarism. They insisted to bring down the plagiarism to 5% and below. It was extremely difficult to bring down plagiarism in a book providing an account on biodiversity with repeated scientific names and concepts.

I thank all the contributors for keeping the similarity to 5% or less. I thank Rakesh Kumar and Sandra Sickels for their support and encouragement. Their patience and perseverance are very much appreciated.

I request the readers to bring to our notice any omissions/mistakes so that the same can be rectified in our further volumes and future editions.

—**T. Pullaiah, PhD**

CHAPTER 1

BIODIVERSITY HOTSPOTS

T. PULLAIAH

Department of Botany, Sri Krishnadevaraya University,
Anantapur, Andhra Pradesh, India

ABSTRACT

Biodiversity hotspot is a region which had to contain at least 1500 vascular plants as endemics (>0.5% of the world's total), and it had to have 30% or less of its original vegetation remaining. There are 36 biodiversity hotspots on the planet earth. These include (1) California Floristic Province, (2) Madrean Pine-Oak Woodlands, (3) Mesoamerica, (4) North American Coastal Plain, (5) Caribbean Islands, (6) Atlantic Forest, (7) Cerrado, (8) Chilean Winter Rainfall-Valdivian Forests, (9) Tumbes-Choco-Magdalena, (10) Tropical Andes, (11) Mediterranean Basin, (12) Cape Floristic Region, (13) Coastal Forests of Eastern Africa, (14) Eastern Afromontane, (15) Guinean Forests of West Africa, (16) Horn of Africa, (17) Madagascar and Indian Ocean Islands, (18) Maputaland-Pondoland- Albany, (19) Succulent Karoo, (20) Mountains of Central Asia, (21) Himalayas, (22) Indo-Burma, (23) Western Ghats–Sri Lanka, (24) East Melanesian Islands, (25) New Caledonia, (26) New Zealand, (27) Philippines, (28) Polynesia–Micronesia, (29) Eastern Australian Temperate forests, (30) Southwest Australia, (31) Sundaland and Nicobar Islands, (32) Wallacea, (33) Japan, (34) Mountains of Southwest China, (35) Caucasus, and (36) Irano-Anatolian.

1.1 INTRODUCTION

A biodiversity hotspot is referred to as a biogeographic region with significant levels of biodiversity under threat due to natural or anthropogenic causes. Norman Myers, a British biologist, was the first to introduce the concept of "biodiversity hotspots" in 1988 based on plant endemism and levels of habitat loss. Afterwards, in 1989 Conservation International (CI), USA adopted Myer's hotspots as its institutional blueprint. In order to qualify as a biodiversity hotspot, a region must meet two strict criteria: (1) it must have > 0.5% of the world's total flora (at least 1500 vascular plant species) as endemics, and (2) it must have lost at least 70% of its primary vegetation. In other words, the region must contain only 30% or less of its original natural vegetation. Myers in 1988 identified 10 hotspots of biodiversity and a subsequent analysis added eight additional hotspots (Myers, 1990). Efforts by Conservation International culminated in an extensive global review (Mittermeier et al., 1999) and scientific publication (Myers et al., 2000) that introduced seven new hotspots bringing their total to 25. An additional analysis brought the total number of biodiversity hotspots to 34. In 2011, Forests of East Australia was recognized as the 35th hotspot, and in 2016, the North American Coastal Plain was identified as the 36th hotspot.

1.2   LIST OF HOTSPOTS OF BIODIVERSITY

North and Central America
1.   California Floristic Province
2.   Madrean Pine-Oak woodlands
3.   Mesoamerica
4.   North American coastal plain

The Caribbean
5.   Caribbean Islands

South America
6.   Atlantic Forest
7.   Cerrado
8.   Chilean Winter Rainfall-Valdivian Forests
9.   Tumbes-Choco-Magdalena
10. Tropical Andes

Europe
11. Mediterranean Basin

Africa
12. Cape Floristic Region
13. Coastal Forests of Eastern Africa
14. Eastern Afromontane
15. Guinean Forests of West Africa
16. Horn of Africa
17. Madagascar and Indian Ocean Islands
18. Maputaland- Pondoland- Albany
19. Succulent Karoo

Central Asia
20. Mountains of Central Asia

South Asia
21. Himalayas
22. Indo-Burma
23. Western Ghats – Sri Lanka

South East Asia and Asia-Pacific

24. East Melanesian Islands
25. New Caledonia
26. New Zealand
27. Philippines
28. Polynesia–Micronesia
29. Eastern Australian Temperate forests
30. Southwest Australia
31. Sundaland and Nicobar Islands
32. Wallacea

East Asia

33. Japan
34. Mountains of Southwest China

West Asia

35. Caucasus
36. Irano-Anatolian

Of the 36 biodiversity hotspots, Tropical Andes is on top with 20,000 endemic plant species, Sundaland is second with 15,000 endemics, followed by the Mediterranean basin with 13,000 species, and Madagascar with 9704 species. Myers team (2000) identified 25 hotspots for endemic tropical forest plants that together contain 133,149 endemics, which represent 44% of the total number of world species recorded. Thirty six biodiversity hotspots cover an area of just 2.4% of the Earth's land surface but contain more than 50% of the world's plant species and 42% of vertebrate species are endemic to these hotspots.

WWF (2020) says that biodiversity continues to decline. Habitat destruction, invasive alien species, climate change, pollution, and overexploitation are the main factors responsible for the decline of biodiversity. Hence, it is necessary to take stock of the biodiversity of hotspots. This book series is an attempt in this direction.

KEYWORDS

- **biodiversity**
- **hotspots**
- **Western Ghats**
- **Sri Lanka**
- **Himalayas**
- **East Melanesian Islands**
- **Wallacea**
- **Madagascar**

REFERENCES

Mittermeier, R. A.; Myers, N.; Robles-Gil, P.; Mittermeier, C. G. Hotspots: Earth's Biologically Richest and Most Endangered Terrestrial Ecoregions. CEMEX/Agrupacio´n Sierra Madre, Mexico City, **1999**.

Myers, N. Threatened Biotas: 'Hotspots' in Tropical Forests. *Environmentalist* **1988**, *8*, 187–208.

Myers, N. The Biodiversity Challenge: Expanded Hotspots Analysis. *Environmentalist* **1990**, *10*, 243–256.

Myers, N.; Mittermeier, R. A.; Mittermeier, C. G.; da Fonseka, G. A. B.; Kent, J. Biodiversity Hotspots for Conservation Priorities. *Nature* **2000**, *403*, 853–858. https://doi.org/10.1038/35002501

WWF. *Living Planet Report 2020-Bending the Curve of Biodiversity Loss*. Almond, R. E. A., Grooten, M., Petersen, T. Eds.; WWF, Gland, Switzerland, **2020**.

CHAPTER 2

PHYSIOGRAPHY AND CLIMATOLOGY OF THE WESTERN GHATS

S. KARUPPUSAMY

Department of Botany, The Madura College, Madurai, Tamil Nadu, India

ABSTRACT

Western Ghats is a continuous chain of hill peaks situated on the Western escarpment of Peninsular India ranging from 8 to 21°N latitude. It runs to a length of around 1490 km parallel to the west coast of Peninsular India, and it is interrupted at 11°N by a 30 km wide pass at Palghat Gap. These Western Ghats ranges are known for their rich repository of floral and faunal diversity. It has been recognized as one of the major global biodiversity hotspots in India. The southwest monsoon facilitates the major proportion of rainfall of southern Western Ghats and withdraws in the reverse direction in south to north gradient. The length of the dry period is also increased from south to north. Altitude in the Western Ghats varies from 50 m to 2300 m above mean sea level (msl). The highest peak of Western Ghats range is Anaimudi (2695 m) in the Annamalai hills. As toward the north, the average elevation is declined around 1000 m. The temperature is also parallelly declined lower to higher elevations. The climatic variables like rainfall, dry period length, and temperature, which are responsible for heterogeneous vegetation types of the Western Ghats with a strong north-south and east-west gradient. Vegetation types vary from tropical wet evergreen, montane sholas, moist deciduous, dry deciduous to scrub forests. This chapter highlights the geology, formation, topography, rock types, soil types, rainfall, temperature range, moisture, rivers and water bodies, biogeography, and general vegetation of Western Ghats.

2.1 INTRODUCTION

The Western Ghats is one of the global biodiversity hotspots, with a rich repository of biodiversity and ecologically important region of the world that originated form Gondwanaland. The region is locally called the Sahyadris and runs parallel to the west coast of Peninsular India about 1490 Km, starting from the Tapti river in Gujarat downwards to Tamil Nadu through Maharashtra, Goa, Karnataka, and Kerala. Usually, the Western Ghats are continuous mountain ranges located between 08° 19' 8" to 21° 16' 24" N and

Biodiversity Hotspot of the Western Ghats and Sri Lanka. T. Pullaiah, PhD (Ed.)

72 56' 24" to 78° 19'40" E enclaving an area of 129, 037 km², which is 5% of the total area of India. The average elevation of Ghat peaks reached about 1200 m ranging from 300 to 2695 m with some important ranges at Kalsubai (1646 m) and Salher (1567 m) in northern Western Ghats, Dodabetta (2637 m), Mukurti (2554 m), and Anamudi (2695 m) in southern Western Ghats. There are number of natural passes in the Western Ghats ranging from 3 km to 24 km wide as known one larger gap is Palakkad (Palghat) gap. The Western Ghats is one of the UNESCO World Heritage Site and is one among the eight "hottest hotspots" of world biological diversity (Myers et al., 2000; UNESCO, 2012). The range runs parallel to north-south along the western edge of the Deccan Plateau known as Great Escarpment of India and separates the Deccan plateau from a narrow coastal plain, called Konkan, along the Arabian Sea coast. A total of 39 legally protected areas are there for biodiversity conservation including national parks, wildlife sanctuaries, and reserve forests. They were represented twenty in Kerala, ten in Karnataka, five in Tamil Nadu, and four in Maharashtra.

2.2 GEOLOGY

Geological age of the Western Ghats ranges is estimated to be older than the Himalayan mountains. The geological history of the Western Ghats dates back to the origin of the earth's crust.. The oldest part of the Western Ghats region was named as Gondwanaland by an Austrian geologist Edward Suess. The Gondwanaland lost its haze of a distant past by the subsequent geological events. The *Gonds* a tribal community inhabited in Madhya Pradesh, Odisha, and northern Andhra Pradesh of present day are probable ancestors, and those who have custodians of the mother land of Indian Peninsula and a part of a super-continent later it called as Gondwanaland. The oldest fossil plant *Glossopteris* was found in Gondwanaland firstly and subsequently it was recorded all over India, Southern Africa, South America, Australia, and Antarctica. Continental drift theory explains the initial landmass breaking apart due to various changes on the mantle of the earth crust during the formation of the earth. The Pangaea ("all lands") primarily made up of Gondwana in the south and Laurasia in the north. Laurasia consists of North America, Europe, and Asia. The remaining part of the earth crust broke away about 200–150 million years ago (m.y.a.) and formed Gondwanaland, including continents of Australia, India, South America, Africa, and Antarctica. During the Carboniferous and the Permian period (300–260 m.y.a), most part of the Gondwanaland was covered by ice. In those times, very little evidence of animals and plants actually living on the Gondwana landmass is available. The ice melted in the late Permian period (240 m.y.a.) and Triassic period (225 m.y.a.). Slowly animals were able to colonize and continued warm until 40 m.y.a. The part of Antarctica now froze over the period again. The drift of Gondwanaland began at the time when Dinosaurs were the dominant on land (Cretaceous—120 m.y.a.), though it was a spontaneous process, but they were unlikely to be affected by it.

The moving of continents was very slow and a large chunk from Gondwanaland broke up from Antarctica in the south. This chunk fissured to form South America and Africa to its west and Australia to its east. The main central portion moving toward northern and about

70 m.y.a. Madagascar was separated from the Indian Plate and continued its migration in a north-east direction. The primary Gondwanaland was a triangular stable "shield" area known as the Deccan Plateau that covers the Peninsula of India (and which includes Sri Lanka). Most geologists supported this view as having formed Indian Peninsula as a part of Gondwanaland. The Indian Plate consisted of a deep seated volcanic hot spot of the united Gondwana islands. The underneath heat beneath evolved basaltic magma, which gives rise to lithosphere of the crust. All these events caused the Indian Plate move in the eastward direction. This tilt rose the altitude of western portion and the major rivers of the peninsular India flow eastward.

Around 45 m.y.a., the Peninsular shield moved toward the Asian landmass after having undergone many uplifts and modifications. This uplift caused the Asian Plate to form the elevated Tibetan plateau. The Gondwana landmass and the Asian Plate was obliterated and the Tethys sea caught up between the two moving landmasses subsequently by its sediments raised up to form the Himalayas. These landmasses not only define the subcontinent and also separate it from the adjacent plate. Usually, the mountains are formed in the folded type, consisting of raised ridges and deep valleys, all peaks parallel to one another, generally stretching in long chains parallel to the colliding areas.

2.3   THE DECCAN VOLCANIC EPISODE

The events of turbulence, upheavals, and extinctions occurred from late Cretaceous to the early Tertiary period known as the Deccan Volcanic Episode. The huge fissures erupted the enormous Lava basalt that has flowed over the Deccan region between 60 and 65 m.y.a. Palaeoecologists considered the recent palaeo-magnetic and isotopic data indicated the possibility of magma outbursts in a short duration of about 2 million years around 65 m.y.a. It is evidenced from the absence of typical cone-like peaks in the southern Peninsula formed by the vast spread of basaltic rock points to a series of outbursts confined to magma spread out layer after layer. It is cited by Birbal Sahni, the renowned palaeo-botanist as "this terrible drama of fire and thunder." About 90% of the fauna and flora went extinct in the Cretaceous—Tertiary period due to this remarkable episode. The huge range of extinction includes large reptiles (Dinosaurs) and ammonites at the end of the Cretaceous period. It is believed that the magnitude and suddenness of volcanic activity and abrupt changes in environment caused the mega-extinction of plant and animal species. The Deccan basalts lava might have played a crucial role in the extinction of the Dinosaurs by releasing eruption fumes that led to a great change in the global climate and the demise of the large mammals, including Dinosaurs.

The Deccan Plateau basalts ranked as the fourth largest (volumetrically) in the world subcontinental outburst of lava extrudes during active tectonic plate movements. The Deccan Traps covers one-sixth of Indian landmass of over 500,000 sq. km. in west-central India with more than 2000 m deep flat lying basalt layer. The original basalt covered by the lava flows as high as 1.5 million sq. km. The total volume of the basalt is estimated to be approximately 512,000 cubic km. The term "Deccan Trap" was first introduced by Col. W. H. Sykes for explaining lithological rock basalts.

An entire Western Ghat landscape characterized under the Deccan Volcanic Province of Peninsular India. The upland region of Deccan Trap was not preserved by any sedimentological records during the stratigraphic period from post-Eocene to pre-Pleistocene, which modified lateritic covers extensively. Hence, it has reconstructed the Tertiary history of Deccan Plateau landscape due to the absence of any sedimentological records. The major known geological events in the Gondwanaland constituted the northward drift of Indian subcontinent and the formation of Western Ghats landscape. All these episodes lead to the establishment of monsoon system and, in response, rise of the Himalayan altitude. The remarkable geotectonic events in the course of time on the landforms are not yet properly understood (Elizen et al., 2013).

2.4   FORMATION OF THE WESTERN GHATS

The Indian Peninsula was separated from the Gondwanaland by the Deccan volcanic episode around 150 m.y.a. In this event, the landmass started moving toward the northern direction. The continuous northward drift ended at 100 million years, finally colliding the Peninsula with the Asian mainland around 45 m.y.a. It was a major geological transformation as the Peninsula drifted over the Asian plate and acted as localized volcanic center on the earth's lithosphere 200–300 km across and remained active for several million years. Further, it generated heat beneath the basaltic magma, which raised the lithosphere and by crustal arching around 120–130 m.y.a. This event resulted in the upliftment of the Western Ghats in the easterly direction. Subsequently, there was a series of volcanic eruptions around 65 m.y.a., which transformed the extensive Deccan Traps. These volcanic episodes added multitude of magma layers largely to the northern Western Ghats (Subrahmanya, 1987).

The Western Ghats were formed as a result of a domal uplift and the underlying rocks were aged around 2000 million years old. The oldest rocks of Gondwana crust are found in the Nilgiris and the high altitudes of the southern Western Ghats ranges. The Indian Peninsula broke along its axis of weakness and the western range drifted westward toward the Arabian sea, and this region rose to the present-day hill ranges of the Western Ghats. The exposed face of the eastern unsubmerged plate was lifted during the Eocene period (between 45 and 65 m.y.a.) form the scarp of the Western Ghats. At this time, the Indian Peninsula ended its northern drift and conjoined with the Asian mainland and its well-marked eastward tilt permanently changed. The western faulting processes led to "river capture" and easterly diversion of drainage to the west in many instances. The rivers of Sharavathy and Kali in Karnataka are some classical examples of westerly drainage due to uplift and faulting. The Western Ghats metamorphosed with tectonically active center, fast rates of uplift, high summit altitudes, steep slopes, deep gorges, and potential for erosion with correspondingly high sedimentation yields.

The uplift of the western part of the Indian Peninsula led to several effects. There was an orientation and direction of the big rivers toward the east. This event started a chain of regressive erosion on the steep and abruptly in western slopes. By this orientation, the Western Ghats chiseled into steps following the sedimentation of basaltic beds and carved by deep valleys. The Sharavathy river was cleaved by vertical cliffs toward Agumbe

(Karnataka). The Western Ghats escarpment was separated from the Seychelles, and it was closely joined with the late Cretaceous by Deccan flood basalt eruption. It oriented the low-lying seaward coastal plains in the west and elevated the Deccan plateau to the east. The Western Ghat escarpment is one of the largest volcanic continental margins with over 1500 km stretch of paralleling seacoast and elevation, with immediate start of sea level ranging over 1 km. Therefore, Western Ghats is a classic example to understand the relationship between volcanic episodes, uplift and extension of plateaus and high ranges (Radhrakrishna et al., 2019).

The northern Western Ghats from Tapti valley down to Goa formed horizontal beds of massive Deccan Trap. The western face of Western Ghats appeared as a steeply-cut valley toward the Arabian Sea. The Deccan scarp ranges in this region presents a vertical profile of over 1000 m elevation of successive sedimentation layers. From Goa to Mahabaleshwar known as "Arthur's Seat," the horizontal traps gave place to steeply dipping gneisses and schists beds. In this section, the Ghats lose their abrupt steep and an average height fell to lesser than 2000 ft. The western flowing rivers were affected by cutting deep gorges and canyons on sedimented rocks. The water flows separated the easterly and westerly Gangavali drainage by a nearly 100 km inland near Hubli. The scarp again lying back on the Arabian coast of south Honnavar formed the significant peaks of Kodachadri (1343 m) and Kudremukh (1892 m).

Charnokite sediments formed the Kodagu and southern Mysore Ghats. In this Charnokitic region, some of the hill peaks formed an average elevation of over 6000 ft., whereas the highest point of Dodabetta was in Nilgiris (2637 m.; 8650 ft). The Western Ghats hill masses oriented with the continuous Ghat ranges, known as tableland, was lifted abruptly to elevations of over 2440 m. (8000 ft). The southern part of the Western Ghats continued to the south of Nilgiris, where it formed a remarkable cleft called the "Palghat Gap", which bifurcated the Nilgiri massif of north and Anamalai massif to the south. The Palghat gap was separated by a valley which was 20 miles wide and with a maximum height of about 300 m. above the sea level. South of the Palghat Gap, the Western Ghats formed complex group of hills such as Anaimalais, Palnis, Varushanadu hills, and Cardamom hills, mainly composed of Charnokite rocks. The course of the river flow in the western side was straight and the drainage pattern was usually rectilinear, but in eastern side its flow receded inland, leaving a fertile coastal plain.

2.5 TOPOGRAPHY

The Western Ghats are metamorphosed basaltic rock mountains formed with faulted edge of an elevated plateau. The northern part of Ghats made up of Deccan basalts is about 650 km north of the Krishna basin, and the southern granitoid gneiss basement rocks that stretch for about 150 km along the coastal strips are considered as more recent sediments. The Volcano-Sedimentary archaean rock shield is mainly composed of crystalline, mostly gneisses, granites, and charnockites alternating with Schist-belts. The Schist-belts are mainly sedimented in "Dharwar–Schists" (Bourgeon, 1987). The Indian Peninsula formed a great east-west dissymmetry and was composed of a fragment of the archaean

Gondwanaland. The Western Ghats may have raised during the Jurassic period, and further episode of tectonic drifts combined with segregating river flows, may have constituted the present day topography.

On the Western side, the Ghats appear abruptly rising from the Arabian Sea, whereas they merge through a raised hill peaks with the Deccan Plateau in the east. They show escarpment formed by coalesced amphitheater shaped valley heads. The Ghat is composed of steep valleys, narrow gorges, and high waterfalls. In the southern section, the range lies close to the coast till it joins the Nilgiri mountains. Northern Nilgiri plateau of eastern flank merged with somewhat raised into the elevated plateau of Karnataka, but the western slopes suddenly elevated from the lower coast. Western Ghats occupies a major position in the humid tropical belt and is situated to the southwest in Karnataka state. The southern section of the Ghats, the Brahmagiris (1360 m), forms the southern boundary with the Wyanad plateau of Kerala, and there is another Brahmagiri peak near Bhagamandala, which is the source of Cauvery river. The highest peak of Brahmagiri district is Tadiandamol (1734 m). Madikeri plateau (Mercara) at an average elevation of about 1050 m extends northwards as far as Somwarpet, a distance of 30 km, but on the east slopes down to the Cauvery.

The Nilgiris is the meeting ground of three mountain systems of Peninsular India viz, the Western Ghats, the Southern Ghats, and the Eastern Ghats. The abrupt rise of the Nilgiri from the surrounding area is very striking. The southwestern part is more hilly and is traversed by bold ranges and intersected by deep valleys. The rest of the Nilgiri is extremely undulating. The highest peak in this range is Dodabetta (2637 m). The Palghat Gap, 24 km wide, is the most prominent break in the continuity of the range. In the south of Palghat gap, the Ghats are very steep and rugged on both the slopes. The Ghats raised prominent hill peaks in the Anamallais-Palnis block, which formed the hill ranges in the Nellaiampathi plateau to the west, the Anamalais plateau of the north, and the Palni hills plateau in the northeast. The higher ranges of the Anamallais consist of a series of plateaus of high elevations. These are covered with rolling downs and dark evergreen forests and are cut off by deep valleys. All these hill ranges branch off from the southern highest peak of Anaimudi about 2695 m. To the south of Anamallais plateau, west-east oriented hills ranges, the Ghats reappear further changes. The elevated hill blocks are slanting toward the west of the Periyar plateau, formed by the flow of Periyar river. The Cardamom range adjoined the eastern part of the Periyar, the plateau is known for cardamom plantations. This middle hill ranges attain a peak at Devar Malai (1922 m) and ended the hill range in the east by a steady cliff around 1000 m high. The southwest-northeast orientation of Varushanad massif is detached from Periyar-Anamalais plateaus and continued to the Andipatti hills, which together with the Palani hills embrace the Cumbam valley.

The usual orientation of the Ghats is northwest-southeast directions and there are three more or less arcuate projections extending eastward toward the south Madras plains. The Nilgiri range oriented northeast between the Coimbatore plains through Gudalur Ghat and the Mysore plateau and continued northward to Satyamangalam–Kollegal hills and Biligirirangan hills of Karnataka, the Palni range (Kodaikanal) running also northeast and continued east of southeast, after a narrow pass at Dindigul in the east of Sirumalai–Amma-yanayakkanur-Ayyalur hills, and southwest of the low Varshanad–Antipatti range running northeast toward the Vaigai valley. Lying between arcuate projections of the Nilgiri range and the Anamalai-Palni range is the Palghat Gap, with an elevation of 250–1000 ft, the highest point in the gap being a little over 1000 ft at Pollachi (Jacob and Narayanaswami, 1953).

In the southernmost direction of the Devar Malai cliff, at about 9°N, the Ghats are once again cleaved by the narrow pass at Shenkottah (alt. 160 m). From there, elevated land stretch continues with a narrow ridge and steep slopes to the west-east, near about 20 km away from Cape Comarin (Kannyakumari). This is a last bit and very rugged by its highest peak in the Agastyamalai at 1869 m. In this region, three major hill peaks, such as Agasthiyamalai proper, Mahendragiri, and eastern Tirunelveli hills of Kalakadu-Mudnathurai hill slopes are found.

In Kerala, the range adjoins the laterite terraces, which gradually slope down toward the coast. This coastal zone (30–50 m) constituting Kerala is made up of converge shaped hills with rounded summits and extends up to Kanniyakumari. Here, the tabular reliefs are not found. Kerala Ghats are generally isolated from the Deccan plateau by rocky mountainous belts. The mountainous belts consist of a series of hilly ranges clothed with dense forest vegetation. The eastern border of Kerala mountains bounded by an almost unbroken stretch of mountainous wall. The continuity of the Ghats is broken by a prominent depression, the Palghat Saddle. On the lower southwestern side of the Nilgiri plateau lies the Verdant Silent Valley. The Ghats area receives both southwest and northeast monsoons. A major proportion of the annual precipitation is received from the southwest monsoon and northeast monsoon brings light showers.

2.6 LITHOLOGY

The Western Ghats can be conveniently divided into three distinct geographic regions based on lithology and structure.

2.6.1 NORTHERN KONKAN REGION

Northern Konkan Region mainly is composed of Deccan Basalt/trap rocks belong to Tertiary age. The northern part of Western Ghats mountain range covering Maharashtra and Goa is better known as "*Konkan.*" It consists of horizontal to nearly horizontal sheets of basalts, cut into a steep scarp on the western side and eastward gentle slope. The highest peak of the northern range is Mahabaleswar, with 4540 ft above mean sea level. It is a flat-topped and slightly elevated hilly terrain compared to Deccan plateau. The steep western margin is usually terraced and resembles the *ghat* or landing–a *stair* from which the name is derived for the region. The landward eastern slope is gentle but also terraced. Krishna, Bhima, and Godavari rivers take their birth in this part of the Western Ghats. There are a few passes in the Ghats which provide lines of communication between the coastal plains and the interior, and have also had great strategic significance in the past phenomena of the Peninsula.

2.6.2 CENTRAL SAHYADRI OR MALNAD REGION

Central Sahyadri or Malnad Region represented by Metamorphosed Volcano-sedimentary sequence of Archaean and Proterozoics of Dharwar Super Group (DSG). This hill range forms the central part of the Western Ghats covering Karnataka state starting from the south of Konkan (south of about 16°N latitude) up to Kasaragodu in Kerala. Nearly horizontally disposed Deccan Basalts of Konkan gives way to the ancient meta volcano-sedimentary sequence of Dharwar Super Group (DSG). Structurally, this part of the Western Ghat shows regional syncline overturned to west and possess low angle plunge toward northeast. The general trend of this hill range coincides with the regional trend of Western Ghats. The highest peak of middle Western Ghats region is Mullayanagiri with an elevation of 1892 m above mean sea level. The major rivers like Cauveri, Sharavathi, Chandranadi, Varahi, Tunga, Bhadra, Kalinadi, Gangavalli, Swamanadi, Sita Nadi, and Netravathi originate from this Sahyadri hill range. Besides biodiversity, this part of the Western Ghats ranges possess rich mineral resources viz., iron and manganese ores, asbestos, corundum resources, clay, and laterite, etc.

2.6.3 SOUTHERN MALABAR OR NILGIRI REGION

Southern Malabar or Nilgiri Region comprises high grade granulite terrain of Archaean age. The southernmost part of the Western Ghats clothed over the Kerala and Tamil Nadu states, it extends from south of Kasaragodu down to Kanyakumari (Cape Comorin), which is known as Malabar hill ranges. These hill ranges are essentially made up of high-grade granulite rocks of Archaean age. The general foliation of these formations coincides with the Western Ghats, and its ranges vary from 40° to sub vertical toward the northeast direction. The highest peak of the Anaimalai is a major portion of this region at 2695 m above mean sea level. Periyar and Idamalayar rivers form the main drainage system in this region. A part of central Kerala in India Peninsula is a basaltic product of prolonged late Mesozoic and Cenozoic formation of overprinted on Archaeo-Proterozoic basement. The Gondwana supercontinent is a formation of post-late Mesozoic breakup episodes (Radhakrishna et al., 1999), subsequently by the migration of Gondwana plate northwards, and finally collided with Eurasian plates and resulted the raising of the Himalayan highlands (Müller et al., 2012; Chatterjee et al., 2013; Yatheesh et al., 2013; Ramkumar et al., 2016).

2.7 ROCK COMPOSITION

The Western Ghats are constituted of the following litho units (1) gneisses (2) chloride schist (3) phyllites.

2.7.1 GNEISSES

A northern and eastern part of the hill range is comprised of gneisses/migmatites of PGC represented by gray gneisses of granodiorite to trondhjemite composition. Gneisses generally occur as bouldery outcrops in low-lying mounds. The failed slopes near Anegundi, Hariharpura, Thirthahalli, and Megaravalli are represented by this lithounit.

2.7.2 CHLORITE SCHISTS

It is the second major lithouint of the Western Ghats. The rock is light green in color and shows well-developed schistosity with slaty cleavages, often shows pinching and swelling structure. The general trend of the schistosity plane is N50°W -SSE, with varying dip of 35° to 50°E. It is essentially comprised of quartz, chlorite, and biotite mica. The rock is highly friable in nature and susceptible for weathering and denudation. Chlorite schist is a dominant litho unit in the famous Agumbe landslide area, which is variously weathered and fragmented rock units.

2.7.3 PHYLLITE

Phyllite forms the major litho unit in Agumbe, Hosur, and Begar slide areas.

2.7.4 CRYSTALLINE ROCKS

The Western Ghats are recognized with three major Archean rock types, such as Archaean sediment (oldest rocks), tertiary bed rocks (upper Miocene to Pliocene), and quaternary sedimented rocks (recent rock deposits).

2.7.4.1 DHARWAR QUARTIZITE FORMATION

This kind of rock is composed of garnetiferous-ferruginous quartzite, mica, talc, and schist. Mostly they are found in the exposed rocks in southeast of Wayanad and northwest of Gudalur and south of Dharwar. They may be formed in Malabar ghats only during the Deccan volcanic episode.

2.7.4.2 CHAMPION GNEISSIC ROCKS

They are recognized in south and southeast Wayanad blocks and have formed with elemental in nature. They appear to be of post Peninsular deposits.

2.7.4.3   PENINSULAR GNEISSIC ROCKS

This is a major widespread sedimented rock types found in Kerala and Tamil Nadu. They constitute the minerals of quartz, feldspars, biotite, and granite. Southern Malabar is formed of these mineralized rock beds extensively in Cochin. They usually made up of biotite and hornblende gneissic.

2.7.4.4   CHARNOCKITE

A major portion of the Western Ghats constitutes of this rock type. The southern part of Travancore areas are well foliated charnokite sediments characterized with peninsular gneissic. They are highly composed of garnetiferous in southern Kerala and Tamil Nadu, whereas in northern Kerala charnokite is absent.

2.7.4.5   CLOSEPET GRANITE

The Malabar region is found in the post charnockite rocks of Archaean intrusions. There are two kinds of intrusions recognized, such as biotite granite in Kalpetta hills and dorneg-neissic granite of Hazaribagh.

2.7.5   PRECAMBRIAN ROCK SYSTEM

2.7.5.1   BASIC DYKES

The oldest rocks are made up of mylonitized crusts. They mainly found in southern Western Ghats, which composed of dolerite. The basic dykes generally formed from fine to medium grained, free form of olivine in Cochin area. The coarse-grained crystalline forms are made by gabbros sediments. This kind of exposed rocks are also found in the Cochin area.

2.7.5.2   RESIDUAL LATERITES

The eastern boundary of the Kerala state constitutes a linear zone of lateritic crystalline rocks to the west direction. These are Varkalai formation exposed on the surface of laterite, which exhibits the characteristics, variations from the typical lateritic. These lateritic formations preserve the ancient features of the parent rock with less compact. The Western Ghats part of Maharashtra showed extensive development of "laterite." The high altitudes plateau corresponding to the lateritic crest of the Western Ghats and the low-lying coastal margins over the west coast of India formed the widespread development of "laterites" (Pascoe, 1962; Krishnan, 1982). The lateritic features of Maharashtra have been pointed out by Patil (1992), where the "laterites" are nothing but the "red beds," and it has been considered as the secondary laterites (Pascoe, 1962; Sahasrabudhe and Rajguru, 1990).

2.7.5.3 THE VARKALI FORMATION

It is represented by the most conspicuous characteristic sedimentary bed rock composition found in Varkala. The cliffs near the Varkala seashore consist of clayey sandstone, white, and variegated clay with thin layer of carbonaceous lignite.

2.7.5.4 RECENT DEPOSITS

After the post Deccan episode, mainly developed sand and silt layers in northern parts of Quilon in southern Western Ghats. It extends the coast of Alleppey and all along the seashore of Kerala which deposits the lacustrine mud banks. It contains valuable mineral sands for the economic point of view.

2.8 RIVERS AND WATER BODIES

The Western Ghats ranges oriented parallel to the Arabian coast of its western side at an average elevation of 1200 m. The Western Ghats is the source of origin of the several important perennial rivers of India namely the Godavari, Bhima, Koyna, and Krishna originate. Northern Ghats of Maharashtra has originated from more than 11 west flowing rivers, namely Damanganga, Surya, Vaitarna, Ulhas, Savitri, Kundalika, Patalganga, Vashisti, Shastri, Karli, and Terekhol. There are several smaller water flowing course joining the creeks in the Ghats (WGEEP, 2011). The Western Ghats are regarded as the main water source of south India, and it supplies more than 60% of surface water to Karnataka (Malhotra and Prasad, 1984), 70% to Kerala, 60% to Tamil Nadu, 50% to Maharashtra, 25% to Goa, and 10% to Gujarat (Putty and Madhusoodhanan, 2014).

Important other rivers originate from the Western Ghats are the Godavari, Tungabhadra, Krishna, Thamiraparani, and Kaveri, and they flow into the Bay of Bengal. The west flowing rivers include Periyar, Bharathappuzha, Netravati, Sharavathi, Mandovi, and Zuari drain into the Arabian Sea and the Laccadive Sea. A number of these rivers feed the backwaters to Kerala and Maharashtra. Kaveri and Krishna rivers are comparatively slow moving ones due to larger rivers. Several larger tributaries originate from the Ghats and merge with some other rivers including the Tunga river, Bhadra river, Bhima river, Malaprabha river, Ghataprabha river, Hemavathi river, Kabini river. There is some smaller riverian course flow from the Ghats namely Chittar river, Gadananathi river, Manimuthar river, Kallayi river, Kundali river, and the Pachaiyar river (Jacob and Narayanaswami, 1953).

Several streams originating from the Western Ghats join the rivers Krishna and Kaveri in monsoon months only. Across the many rivers, dams have been constructed for the production of electricity and irrigation purposes. The major reservoirs constructed in Western Ghats region include Lonavala and Walwahn in Maharashtra; V.V. Sagar, K.R. Sagar, Bhadra, Lingamakki in Karnataka; Mettur, Upper Bhavani, Lower Bhavani, Mukurthi, Parson's Valley, Porthumund, Avalanche, Emerald, Pykara, Sandynulla, Karaiyar, Servalar,

Kodaiyar, Gadananathi river, Manimuthar, and Glenmorgan in Tamil Nadu; and Kundallay and Maddupatty in Kerala. Of these Lonavla, Walwahn, Upper Bhavani, Mukurthi, Parson's Valley, Porthumund, Avalanche, Emerald, Pykara, Sandynulla, Glenmorgan, Kundally, and Madupatty reservoirs are used for fisheries and irrigation purposes.

About 50 major reservoirs constructed along the length of the Western Ghats, among them most notable projects are the Koyna Hydroelectric Project in Maharashtra, the Parambikulam dam in Kerala, and the Linganmakki dam in Karnataka. Another major hydroelectric dam is Idukki dam of Kerala. The well-known water falls of the Western Ghats are Jog Falls, Kunchikal Falls, Dudhsagar Falls, Sivasamudram Falls, and Unchalli Falls. Talakaveri wildlife sanctuary is a spectacular watershed body from the source of river Kaveri. In Shimoga, Sharavathi, and Someshvara Wildlife sanctuaries are the main source of the Tungabhadra river.

The several manmade reservoirs in the Western Ghats created for tourism, fisheries, and small scale hydroelectric power stations. The known reservoirs are in Ooty (34.0 ha) in Nilgiris and Kodaikanal (26 ha) and Berijam in Palani Hills. The Pookode lake of Wayanad in Kerala is a beautiful scenic with boating and garden establishment for tourism. Two smaller lakes in Kerala are Devikulam (6.0 ha) and Letchmi (2.0 ha) in the Munnar.

2.9   SOIL TYPES

With a range of parent rock materials occurring in the Peninsula and a wide range of climatic factors, a wide variety of soils is developed. The development of soils is greatly influenced by the climate, which consists of precipitation, humidity, evaporation, temperature, wind, and biological agencies, which include vegetation and other biological components. The works of Bourgeon (1989) is extensively analyzed the edaphic features of Western Ghats. The Ghats appear to be homogeneous but can be distinguished into three major subdivisions for the understanding the major compartments: (1) the escarpment, (2) the coastal hinter land, and (3) the residual reliefs in the coastal area, which is the retreat of the escarpment. The escarpment of the Ghats is made up of a succession of sub-vertical rocky crags and zones, where huge rocky boulders measuring several cubic meters alternate with deep soils. The whole escarpment is carved by numerous V-shaped sub-parallel ravines. The slope is very steep and projected on a horizontal plane, the total width of the Ghats rarely exceeds 10 km in the east-west direction. The upper horizon of the soil is reddish brown, humiferous clayey, and about 20 cm thick. This horizon is largely porous with plenty of plant root system and abundant of microbial activity. The deeper horizons are of variable thickness (50 cm to several meters). It is red in color with clayey texture. The empty space is generally filled with micro-aggregates. The humiferous zone is slightly acidic, highly saturated with water and rich in exchangeable mineral elements. The deeper layers are with coarse elements, acidic, and with medium saturation. The foot of the escarpment shows two types of soils viz, soils on debris and soils on regolith. Soils on debris contains high proportion of alluvial with coarse elemental compositions. It shows two horizons, upper dark brown humiferous layer and lower yellow brown horizon. These soils are acidic and contains about 5% of organic matter with sufficient amount of exchangeable bases.

The soils on regolith are more weathered and show a humiferous horizon of about 10 cm thick, which is dark brown with sandy loam. It is followed by a yellowish brown alluvial horizon of about 20 cm thick with sandy loam. Deeper horizons show red and yellow mottles in the form of vertical streaks. These soils are acidic to very acidic. With a wide range of climatic factors, a wide range of soil types are seen.

Raychaudury et al. (1963) categorized the south Indian hill soils into four main types viz, red soils, black soils, lateritic soils, and alluvial soils. Red soils vary considerably in texture, some are sandy soils, a great many are barns of different kinds and a few are clay soils. The red barns and the red sandy soils are two prominent classes in the region. Laterite is generally reddish in color and often has a vermicular structure. These soil types are formed due to high rainfall with alternating wet and dry seasons. The subsoil contains hydrated oxides of aluminum and iron. These soils classes are well organized on the historic summits of Ghats. Alluvial soils are formed by transportation by streams and rivers and deposited over plains, valleys, or along coastal belts. These soils are light in texture and well drained.

2.10   CLIMATE

Climate of the Western Ghats is highly influenced by altitudinal gradation and latitudinal distance from the equator. Usually, the climate is tropical humid in the lower part and reaches temperate in high altitudes above 2000 m mean sea level. An altitude of 1500 m (4921 ft) and above in the northern Ghats and 2000 m (6562 ft) and above in the southern Ghats have attained temperate climate. Average annual temperature noted around 15°C, but in high altitudes of Ghat area touch the freezing point in winter months. Mean annual temperature ranges from 20°C in the south to 24°C in the north. There have also been recorded the coldest periods in the southern Western Ghats with wettest in the same period. The works of Pascal (1988) and Pascal et al. (2004) have been extensively used to explain the climate in the region. The broad picture of the climate of Western Ghats is determined by its geographical situation. The southern Peninsular hill ranges is a meteorological unit in itself, and has a great diversity of climates with contrasting ombrothermic patterns. The Western Ghats has a tropical-warm and humid climate. The weather systems of the Ghats have vast variation due to heterogeneity of the region, largely depending on geographical positions and physiographical features. The distinctive monsoon pattern with alternation of seasons is well known factor for maintenance of climatic cycle of the Western Ghats.

The Western Ghats chain acts as a barrier to moisture bounding clouds during the monsoon season between June and September. The most precipitation of rain on the windward side of the Ghats by force of heavy clouds moving eastward. The Western Ghats, situated in the wind ward direction, receives a heavy rainfall of more than 500 cm per year. The increase of temperature in the month of May over the mainland of Asia plate, the air pressure decreases in that area, and the air circulation in the Indian area and the neighboring seas becomes more and more intense. Almost entire India is quickly overrun by cool and moisture laden air known as the "southwest monsoon." The southwest monsoon from the Arabian sea and the northeast monsoon from the Bay of Bengal are two sets in different seasons prevails the more precipitation of the Western Ghats. The southwest monsoon strikes the west coast receiving heavy rain to the coastal districts and mountains of the Western Ghats. During the southwest monsoon approximately 80–85% of total rainfall occurs in the Western Ghats. In June, rainfall exceeds 760 mm along the west coast of the peninsula. In July also, most of the west coast gets about 760 mm of rainfall. It fell down to the eastern side of the Ghats, which is less than 125 mm. The distribution of rain fall in August follows the same general pattern as in July, but the amounts are generally less. The distribution of September rainfall over the area is similar to that of August but amounts are still less. Subsequently, the monsoon withdraws and retreats by October, by this time the winter season begins in the northern part of the Ghats, and the northeast monsoon begins where it prevails in a dry climate in the southern Ghats. The dry winter season is followed by the pre-monsoon transition period of a hot summer extending generally from March to the beginning of June. Next to rainfall, temperature is perhaps the most important climatic factor.

2.10.1   RAINFALL

The heterogeneity in the physical features of Western Ghats has a profound influence on the climate of the region (Nair, 1991). The Western Ghats is lying in the tropical monsoon tract characterized with well distinct wet summers and dry winters (Putty and Madhu-soodhanan, 2014). The normal annual rainfall pattern is recorded in the Western Ghats, where there is a systematic variation in rainfall across the region. The western side of the Western Ghats receives threshold rainfall of the southwest annual monsoon about 203–254 cm, and the eastern side is a rain shadow area with lesser annual rainfall (Subramanyam and Nayar, 1974). The Ghat area below 15°N is influenced by both southwest and north-east monsoons, with the northeast monsoon being more effective deeper south, where tropical cyclones originating in the Bay of Bengal bring substantial rains. The influence of southwest monsoon increases gradually toward north, parts of Kerala, and the rest of Ghats getting more than 90% of their rains during this season. The pre-monsoon rainfall that occurs in the months of April and May due to local convection is never widespread. Throughout the length of Ghats and coast, daily intensities are very high. In the areas where Net Annual Rainfall (NARF) is above 5000 mm, daily rainfall exceeding 250 mm is quite common feature, and even where NARF is less than 2000 mm, rainfalls exceeding 100 mm contribute about 25% of the annual rainfall. During very heavy rainy days (≥150 mm) in the region, it rains 20–23 h and the contribution of low intensity rainfall is always much greater (Thipperudrapa, 2009). Even in very heavy rainfall area (NARF ≥ 5000 mm) rainfalls less than 6 mm/15 min (24 mm/h) last 95% of the time and contribute 75% of the total rain (Putty and Madhusoodhanan, 2014).

2.10.2   TEMPERATURE

The temperatures in the Western Ghats are highly variable depending on the seasons and other physiographic factors. In winter season, the temperature decreases but in summer months increases. Temperature, humidity, and rainfall are largely determining the features of extent vegetation. Naturally March, April, and May months are hot months. But, these months are mellowed by the effects of temperature and humidity, which formed cloud and rain of the southwest monsoon. The cold weather months from November to February prevails dry continental winds over the Ghats. December and January are the coldest months in high altitudes of the Ghats and the mean temperature ranges from 5 to 15°C. The mean temperature is somewhat highest in the months of March to May in the eastern side of the Ghats due to dry winds prevailing to coastal side. The monsoon period is prevailing almost uniform temperature in all over the Ghats with very little variation. The autumn period follows the rainy days in the Ghats during November to January to maintain uniform low temperature.

2.10.3 MOISTURE

The atmospheric moisture is an important factor for vegetation concerned. The characteristic wind flow from the sea is the main factor, which determines the hydro vapor content of air at any place. Generally, humidity increases toward the Arabian sea in the west border of the Western Ghats. In the hot summer season, dried air is blown at the foot hills, especially in the border of the hills. The commencement of oceanic winds on the coast begin to increase the humidity. In cold weather seasons, temperature as well as humidity pressure decreases along the coastal region toward the interior. In summer months, during the hot afternoons, usually relative humidity is very low. Humidity is varying in monsoon seasons from sea level to mountain summits. Hence, humidity variability is of general characteristic features of the Ghats in different months. During the rainy season usual weather is saturated, but alternatively in summer hot winds blow and humidity becomes low.

2.11 BIOGEOGRAPHY OF WESTERN GHATS

The Western Ghats served as refugia for wet evergreen elements since the periods of volcanism, the post reformation of volcanic period, might have led the evergreen species dispersed from southern Western Ghats to central and northern Western Ghats. The detailed phytogeographic analysis of Western Ghats has been recorded by Subramanyam and Nayar (1974). Mani (1974) has characterized the Western Ghats as one of the oldest region of evolutionary center of flora and fauna thereby, highlighted the significance of the Western Ghats formation in oriental biogeography. Biogeography explicit the origin, evolution, and distribution pattern of biological species through the time course (Parthasarathy, 1988). This concept can be examined in a historical and biogeographical framework that could be compared in modern molecular phylogenetics with conjunction molecular prospecting and event-based biogeographical explorations. The modern molecular approach to the phylogeny of Western Ghats endemics, nested within well-separated clades in the central Western Ghats and northern Western Ghats, differentiating from southern Western Ghats species. Additionally, the experiments proved the ancestral areas of the deeper center of origin divergent in the phylogeny that could be assessed in the southern Western Ghats and dispersals from southern to northern in the post-volcanic period. The Peninsular Indian plate probably departed from ancient Gondwana elements and collided with Asian plate after breaking of the Gondwana supercontinent. It has been clearly addressed in the "Biotic ferry" model or "Out-of-India" hypothesis (Joshi and Karanth, 2013). The biodiversity of Western Ghats received many relatively young invaded intrusive elements from other adjoining parts, such as South China, Malayan archipelago, Ethiopia, Mediterranean, and Himalayan Pleistocene relicts. Therefore, it is interesting to recognize the biodiversity components of the Western Ghats that has a close relationship with Gondwana affinities. A few phylogenetic studies have been explored from the complex evolutionary origins of selected Western Ghats biota (Joshi and Karanth, 2011a).

Recent molecular phylogenetic studies proved the ancient lineages of the amphibians and centipedes in the southern Western Ghats (Biju and Bossuyt, 2003; Bossuyt et al., 2004; Roelants et al., 2007; Joshi and Karanth, 2011a; Vijayakumar et al., 2016). Possibly, physiographical or ecological barriers might have played a crucial role in the present distribution of biotic groups. Distribution pattern of plants, birds, dragonflies, arboreal frogs (*Philautus*), and fishes of the Western Ghats suggested that Palghat gap might have a potential geographical barrier for dispersal (Fraser, 1936; Subramanyam and Nayar, 1974; Ali and Ripley, 1987; Dahanukar et al., 2004; Biju et al., 2005). A few molecular-based biogeographic approaches have confirmed the role of the Palghat gap in orientation of current distribution of flora and fauna. It is strongly supported by microsatellite data on the distribution of Asian Elephant population across the Palghat gap (Vidya et al., 2005).

Plant populations of *Eurya nitida* and *Gaultheria fragrantissima* are structuring their distribution patterns across Palghat gap, and this gap act as an important geographical barrier for population distribution even within the species (Bahulikar et al., 2004; Apte et al., 2006). Hill peaks and deep valleys are appearing to be a significant barrier for high-elevation endemic bird White-bellied Shortwing (*Brachypteryx major*) (Robin et al., 2010). Thus, these studies have concluded that Palghat gap is serving as a spatial geographical barrier to gene flow of plants and animal species distributed across the gap. All these studies have narrowly focused on intraspecific variations in the species and elucidated the impact of Palghat gap in distribution of interspecific populations. More interestingly, a study on a group of caecilians showed that the gap did not function as a spatial barrier for the species and may have traditionally considered as limited dispersal ability (Gower et al., 2007). Nevertheless, experimental conclusion from a molecular phylogenetic approach, which included multiple taxonomic groups of Indian and Sri Lankan species, has been pointed toward the impact of Palghat gap in shaping species assemblages (Bossuyt et al., 2004).

The four other biogeographic subdivisions of the Western Ghats have also been restricted species distribution patterns, and they are serving as ecological barriers in addition to the Palghat gap. These subdivisions are spatially separated and differ by their climatic conditions and vegetal composition (Subramanyam and Nayar, 1974). Species in every subdivision might also be more closely related to each other than the species from other subdivisions. This phenomenon would be reflected in the molecular phylogenetic clades formed by species from each subdivision, forming distinct divergent clades. Up to date, there is no published molecular work that addressed the ecological barrier for species distribution in the Western Ghats.

A molecular phylogenetic investigation on scolopendrid centipedes established the monophyly of Indian *Digitipes* species (Joshi and Karanth, 2011a), and another investigation on species delimitation by a comprehensive phylogenetic hypothesis with revised species distribution maps (Joshi and Karanth, 2011b). An investigation revealed the distribution patterns of *Digitipes* species in the Western Ghats and showed that the southern Western Ghats is having more diversity with many more endemic species than the central and northern Western Ghats. It has been reported that out of five species, three are endemic in Nilgiris, no endemics exist in the central Western Ghats and one endemic species exist in the northern Western Ghats. One species is nonendemic reported elsewhere (Joshi and Karanth, 2011b). Furthermore, these species are distributed along an elevational gradient at lower (>500 mean sea level), mid-elevation forests (500–1500 msl), and in the high-elevation forests of the Western Ghats (<1500 msl) (Joshi and Karanth, 2011b).

The evolution and diversification of *Digitipes* lineage occurred in the same age when the Peninsular Indian plate was drifting on its northward movement during the late Cretaceous period and it had been processed by long periods of isolation (Joshi and Karanth, 2011a). The *Digitipes* species of the Western Ghats is suitable for testing the biogeographical hypothesis mentioned above as they represent an ancient endemic lineage and also exhibit a decreasing trend in species richness with endemicity when increasing latitude and altitude. A recent report on *Digitipes* phylogeny estimate the divergence lineages by the Bayesian approach. It clearly spells out the historical and biogeographic relationships among the *Digitipes* species in terms of their origin and distribution, and the experiments help to reconstruct the ancestral areas and test the biogeographic hypothesis (Vijayakumar et al., 2016).

KEYWORDS

- **Western Ghats**
- **physiography**
- **geology**
- **climatology**
- **river system**
- **lithology**
- **topography**

REFERENCES

Ali, S.; Ripley, S. D. In *Handbook of the Birds of India and Pakistan*; Oxford University Press, **1987**.

Apte, G. S.; Bahulikar, R. A.; Kulkarni, R. S.; Lagu, M. D.; Kulkarni, B. G.; Suresh, H. S. Genetic Diversity Analysis in *Gaultheria fragrantissima* Wall. (Ericaceae) from the Two Biodiversity Hotspots in India using ISSR Markers. *Curr. Sci.* **2006**, *91*, 1634–1640.

Bahulikar, R. A.; Lagu, M. D.; Kulkarni, B. G.; Pandit, S. S.; Suresh, H. S.; Rao, M. K. V. Genetic Diversity Among Spatially Isolated Populations of *Eurya nitida* Korth. (Theaceae) Based on Inter-Simple Sequence Repeats. *Curr. Sci.* **2004**, *86*, 824–831.

Biju, S. D.; Bossuyt, F. New Frog Family from India Reveals an Ancient Biogeographical Link with the Seychelles. *Nature* **2003**, *425*, 711–714.

Biju, S. D.; Bossuyt, F.; Lannoo, M. M. Two New *Philautus* (Anura: Ranidae: Rhacophorinae) from Ponmudi Hill in the Western Ghats of India. *Copeia* **2005**, *2005*, 29–37.

Bossuyt, F.; Meegaskumbura, M.; Beenaerts, N.; Gower, D. J.; Pethiyagoda, R.; Roelants, K. Local Endemism within the Western Ghats-Sri Lanka Biodiversity Hotspot. *Science* **2004**, *306*, 479–481.

Bourgeon, G. Les "sols rouges" oles Regions Semi-andes du sud de linde I & III. Proprietes of classification pedologique L. *Agron. Trop.* **1987**, *42* (3), 153–170.

Bourgeon, G. *Reconnaissance Soil Map of Forest Area Western Karnataka and Goa;* Institut Francais de Pondicherry: India, **1989**.

Chatterjee, S.; Goswami, A.; Scotese; Christopher, R. The Longest Voyage: Tectonic, Magmatic, and Paleo-climatic Evolution of the Indian Plate During its Northward Flight from Gondwana to Asia. *Gondwana Res.* **2013**, *23*, 238–267.

Dahanukar, N.; Raut, R.; Bhat, A. Distribution, Endemism and Threat Status of Freshwater Fishes in the Western Ghats of India. *J. Biogeogr.* **2004**, *31*, 123–136.

Elizen, S. M.; Patil, D. N.; Al-Imam, O. A. O. Genesis of the Khanapur 'Red beds' Maharashtra, India. *Intern. J. Eng. Sci. Res. Technol.* **2013**, *2* (6), 1422–1437.

Fraser, F. C. Odonata. In *The Fauna of British India, Including Ceylon and Burma.* Taylor and Francis: London, **1936**.

Gower, D. J.; Dharne, M.; Bhatta, G.; Giri, V.; Vyas, R.; Govindappa, V. Remarkable Genetic Homogeneity in Unstriped, Long-Tailed *Ichthyophis* Along 1500 km of the Western Ghats, India. *J. Zool.* **2007**, *272*, 266–275.

Jacob, K.; Narayaswami, S. *The Structural and Drainage Patterns of the Western Ghats in the Vicinity of the Palghat Gap;* Geological Survey of India: Calcutta, **1953**.

Joshi, J.; Karanth, K. P. Cretaceous-Tertiary Diversification among Select Colopendrid Centipedes of South India. *Mol. Phylogenet. Evol.* **2011a**, *60*, 287–294.

Joshi, J.; Karanth, K. P. Coalescent Approach in Conjunction with Niche Modeling Reveals Cryptic Diversity Among Centipedes in the Western Ghats Biodiversity Hotspot of South India. *PLoS ONE* **2011b**, *7*, e42225.

Joshi, J.; Karanth, P. Did Southern Western Ghats of Peninsular India Serve as Refugia for its Endemic Biota During the Cretaceous Volcanism? *Ecol. Evolut.* **2013**, *3*, 3275–3282.

Krishnan, M. S. *Geology of India and Burma;* CBS Publishers and Distributors: New Delhi, **1982**.

Mani, M. S. *Ecology and Biogeography of India;* Junk, Dr. W.; B.V. Publihsers: The Hague, **1974**.

Malhotra, K.; Prasad, R. *Some Important Aspects of Water Resources in Karnataka, with Particular Reference to Western Ghats.* PhD. Thesis, The National Institute of Engineering, Mysuru, India, **1984**.

Müller, R. D.; Yatheesh, V.; Shuhail, M. The Tectonic Stress Field Evolution of India since the Oligocene. *Gondwana Res.* **2012**, *28*, 612–624.

Myers, M.; Mittermeier, R. A.; Mittermeier, C. G.; da Fonseca, G. A. B.; Kent, J. Biodiversity Hotspots for Conservation Priorities. *Nature* **2000**, *403*, 853–858.

Nair, S. C. *Southern Western Ghats.* INTACH: New Delhi, **1991**.

Parthasarathy, N. A Phytogeographic Analysis of the Flora of Kalakkad Reserve Forest, Western Ghats. *J. Indian Bot. Soc.* **1998**, *67*, 342–346.

Pascal, J. P. *Wet Evergreen Forests of the Western Ghats of India.* Institut Francais de Pondicherry Tra y Sec Sci tech de l'Institut Francaise de Pondichery, **1988**.

Pascal, J. P.; Ramesh, B. R.; De Franceschp, D. Wet Evergreen Forest Types of the Southern Western Ghats, India. *Trop. Ecol.* **2004**, *45*, 281–292.

Pascoe, E. H. *A Manual of the Geology of India and Burma,* vol. III; Govt. of India Publication, **1962**.

Patil, D. N. Laterites from the Semiarid Regions of Deccan Volcanic Province, India: A Reappraisal. *Tropical Geomorphology Newsletter,* **1992**, *13*, 8–10.

Putty, Y. M. R.; Madhusoodhanan, C. G. Water Resources and Hydrology of Western Ghats: Their Role and Significance in South India. *NNRMS Bull.* **2014**. http://www.reserchgate.net/publication/268147276.

Radhakrishna, T.; Maluski, H.; Mitchell, J. G.; Joseph, M. 40Ar/39Ar and K/Ar Geochronology of the Dykes from the South Indian Granulite Terrain. *Tectonophysics* **1999**, *304*, 109–129.

Ramkumar, M.; Menier, D.; Mathew, M.; Santosh, M.; Siddiqui, N. A. Early Cenozoic Rapid Flight Enigma of the Indian Subcontinent Resolved: Roles of Topographic Top Loading and Subcrustal Erosion. *Geosci. Front.* **2016**, *8* (1), 15–23. http:// dx.doi.org/10.1016/j.gsf.2016.05.004.

Raychuduri, S. P.; Aarwal, R. R.; Datta Biswas, N. R.; Gupta, S. P.; Thomas, P. K. *Soils of India;* Indian Council of Agricultural Research: New Delhi, **1963**.

Robin, V. V.; Sinha, A.; Ramakrishnan, U. Ancient Geographical Gaps and Paleo-Climate Shape the Phylogeography of an Endemic Bird in the Sky Islands of Southern India. *PLoS ONE* **2010**, *5*, e13321.

Roelants, K.; Gower, D. J.; Wilkinson, M.; Loader, S. P.; Biju, S. D.; Guillaume, K. Global Patterns of Diversification in the History of Modern Amphibians. *Proc. Natl. Acad. Sci. USA* **2007**, *104*, 887–892.

Sahasrabudhe, Y. S.; Rajaguru, S. N. The Laterites of the Maharashtra State. *Bull. Deccan College Research Institute* **1990**, *49*, 357–370.

Subrahmanya, K. R. Evolution of the Western Ghats in India–A Simple Model. *J. Geol. Soc. India* **1987**, *29*, 446–449.

Subramanyam, K.; Nayar, M. P. Vegetation and Phytogeography of the Western Ghats. *Ecol. Biogeogr. India* **1974**, *23*, 178–196.

Thipperudrapa, S. *Some Studies on the Intensity Pattern of Rainfall in Karnataka, with Particular Reference to Western Ghats.* PhD Thesis, The National Institute of Engineering, Mysuru, **2009**.

UNESCO. Decision Adopted by the World Heritage Committee at its 36th Session, [Online] **2012**. http://whc.unesco.org/en/sessions/36COM.

Vidya, T. N. C.; Fernando, P.; Melnick, D. J.; Sukumar, R. Population Differentiation Within and Among Asian Elephant (*Elephas maximus*) Populations in Southern India. *Heredity* **2005**, *94*, 71–80.

Vijayakumar, S. P.; Menezes, R. C.; Jayarajan, A.; Shanker, K. Glaciations, Gradients, and Geography: Multiple Drivers of Diversification of Bush Frogs in the Western Ghats Escarpment. *Proc. Roy. Soc. B* **2016**, *283*, 20161011. http://dx.doi.org/10.1098/rspb.2016.1011.

WGEEP. *Report on the Western Ghats Ecology Expert Panel;* Submitted to the Ministry of Environment and Forests, Government of India: New Delhi, **2011**.

Yatheesh, V.; Dyment, J.; Bhattacharya, G. C.; Muller, R. D. Deciphering Detailed Plate Kinematics of the Indian Ocean and Developing a Unified Model for East Gondwana Land Reconstruction: An Indian Australian-French Initiative. *DCS-DST News Lett.* **2013**, *23*, 1–9.

CHAPTER 3

VEGETATION AND FOREST TYPES OF THE WESTERN GHATS

S. KARUPPUSAMY

Department of Botany, The Madura College (Autonomous), Madurai, India

ABSTRACT

The Western Ghats, parallel to the west coast of India, occupied an area of 129,037 km² comprised a major portion of the Western Ghats and Sri Lankan Hotspot for conservation in the Indian subcontinent. It is one among the 36 global hotspots of the world with its unique assemblages of plant and animal communities and rich endemic species of well-known conservation endemic and heritage centres of the world. All these biological species assemblages over the various forest areas of the Ghats were determined by its climatic, edaphic, and topographic gradients. This chapter reveals the forest types and their characteristic floristic compositions existing the Western Ghats hotspot. The Western Ghats has four major categories of the forests such as moist tropical forests, dry tropical forests, montane subtropical forests, and montane temperate forests with 39 subtypes of forests.

3.1 INTRODUCTION

The Western Ghats, otherwise called the Sahyadri Hills, represented with extraordinarily rich in biodiversity of flora and fauna. The Western Ghats is one among the 36 biodiversity hotspots of the world recognized by Norman Myers (Marchese, 2015). It has been identified rich number of endemic species diversity and a major endemic center in the Indian subcontinent (Nayar et al., 2014). The Western Ghats may be classified geologically into two segments that are northern basaltic and southern gneissic. The south of Krishna seldom beyond 1500 m is the region of archean crystalline Precambrian rocks with major composition of granites, schists, gneisses, and quartzites which are approximately 2000 million years old. In the montane high rainfall belts, soils varied from rich peat humus to laterite in the lower elevation. Generally, soils are acidic due to high humus contents. The climatic conditions of the Western Ghats are highly influenced by the altitudinal gradients

Biodiversity Hotspot of the Western Ghats and Sri Lanka. T. Pullaiah, PhD (Ed.)

and physical proximity of the Arabian Sea. The Western Ghats experienced a tropical climate and being a warm and humid during most of the months in a year with an average temperature ranging from 20°C in the southern to 24°C in the northern Ghats, and in the higher elevations experienced from subtropical climates to temperate and rare occasions of the frost. Further, it has been recorded that the coldest months in the southern Western Ghats coincide with the wettest climate. These enriched amidst weather condition prevails the vast assemblage of floristic and faunastic communities of Western Ghats.

The varied climatic and topographic heterogeneity from 2695 m of the highest peak to sea level and a distinct monsoon pattern with rainfall gradient of annual precipitation <50 cm in sheltered high altitude valleys in the east to >700 cm over the west-facing hill slopes, the combined effect of climate gave rise to a tremendous diversity of varied life forms and vegetation types. The forest types include tropical wet evergreen forest, montane stunted evergreen hill peak forest and grasslands, lateritic plateaus, moist deciduous and dry deciduous forests, dry thorn forests, and low level savanna grassland. All these forest types are considered as critical habitats for many plants and animal species. The typical sholas and grasslands of the southern Western Ghats are unique as well as highly vulnerable to future climate change. The riparian forests along the rivers and streams of the Ghats sheltered high levels of plant and animal diversity in addition to acting as relict ecotone forests, where confined the lowland *Dipterocarp* forests and *Mysristica* swamps on the western slopes are highly threatened. Around 4000 and odd species of flowering plants with about 27% of the total plant species of the country are known to occur in the Western Ghats. Out of a total 645 species of evergreen trees, about 56% are endemic to the Ghats. The diversity of lower plant groups especially bryophytes is highly impressive in the Ghats with around 1224 species, of which 824 species are mosses with 28% of endemism and 379 species are liverworts with 43% endemism.

The Western Ghats consists of more than 30% of all plants, fish, herpetofauna, bird, and mammal species with less than 6% of the total land area of India (Daniels, 2002). The Ghat region has a spectacular assemblage of wild flora and fauna and the area is serving for several globally significant wildlife sanctuaries, tiger reserves, and national parks. A total of 58 protected areas are there along the Ghats consisting of 3 biosphere reserves, 14 national parks, and 44 wildlife sanctuaries within its boundaries. The legally protected total area cover has 13,595 sq km contributing 9.06% of the Western Ghats (Rodgers et al., 2002). The Western Ghats has one of the highest levels of the protected cover area in India and another one is being the Andaman and Nicobar Islands (Rodgers and Panwar, 1988).

3.2   EARLIER EXPLORATIONS OF VEGETATIONS IN WESTERN GHATS

The Western Ghats, especially the Malabar Coast is well known in world history and commerce as an important and perhaps the sole centre of spice trade, especially pepper, ginger, and cardamom. Greek, Arab, and later European traders, lured by the spices, found their way to the Western Coast of India. The Portuguese settlement at Goa and the Dutch possessions of Malabar, interested in the exploration of flora of this region, contributed for the first time to the scientific study of the plants of the region. In 1565, Garcia de Orta enumerated a list of medicinal plants of India. Heinrich von Rheede has published *Hortus Malabaricus,* a monumental work on the plants of Malabar between 1678 and 1703. Robert Wight, in the middle of the nineteenth century, published a series of books with illustrations on South Indian plants. Between 1872 and 1894, Hooker, assisted by several botanists, published 7 volumes of Flora of the British India and gave the first comprehensive report on the Flora of Indian subcontinent. This flora accelerated the publication of provincial floras like the Flora of the Madras Presidency by Gamble and subsequently completed by C. E. C. Fischer (1915–1936), and the Flora of the Presidency of Bombay by Cooke (1901–1908). The remarkable contribution to the biodiversity survey of southern India was made by Richard Henry Beddome. Rama Rao (1914) enumerated the flowering plants of Travancore. Other workers like Blatter, Bourdillon, Cleghorn, Dalzell, Fischer, Fyson, Gibson, Lawson, Santapau, and Talbot have made the greatest contributions for understanding the flora of the Western Ghats. The geological antiquity, evolutionary history, and biogeographic patterns, with special emphasis on the endemism of the flora and fauna of the Western Ghats, have been accounted by several authors in different times (Govindarajulu and Swamy, 1958; Blasco, 1970; Subramanyam and Nayar, 1974; Nayar, 1977; Ahmedullah and Nayar, 1986; Rao, 1978; Pascal, 1988; Nair and Daniel, 1986; Nair, 1991). The forests of the Western Ghats represent a dynamic repository of a wide array of invaluable human resources.

Botanical explorations as carried out by different workers in the recent past in various parts of Western Ghats, the details of plant explorers are given in Chapter 9. In fact, these publications have made worthwhile contributions to the existing account of plant wealth of the Western Ghats At present the Botanical Survey of India (BSI), some State and Central Universities, Forest and Environment Departments, conservational organizations are actively engaged in floristic revisions, vegetation structure, ecological interactions, rate of deforestation and afforestation, and conservation programmes of Western Ghats.

3.3 VEGETATION

The vegetation component of the Western Ghats is one of the most complex and diverse types found in India. A part of megadiversity region, the flora and vegetation of this tract were under constant surveillance of botanists since the 19th century, more active among earlier ones being Hooker and Thomson (1855), Beddome (1869–1874), Bourdillon (1908), Rama Rao (1914), Gamble and Fischer (1915–1936), and others. Of the later contributions of Champion (1936), and Champion and Seth (1968), seem to be the most comprehensive and outstanding. They classified the forests of India into six major types with several subdivisions and edaphic or seral variations. The forests of Western Ghats belongs to 4 major types with more than 40 subtypes in various proportions of the forested areas from the north Maharashtra to south of Kanyakumari. The entire Western Ghats areas records about 7402 flowering plant species in 1480 genera under 210 families. Form which 5588 species are indigenous, 376 species are exotics naturalized and 1438 species are cultivated or planted. Of the 5588 indigenous species of plants recorded in India, 2253 plant species are endemic and 1273 plant species are endemic to the Western Ghats. Apart from the above, there are 593 taxa with subspecies and vaiety status recorded for the Western Ghats (Nayar et al., 2014). The original forest vegetation of Western Ghats and natural biodiversity can be seen in pockets of legally protected areas or within locally semiprotected regions like sacred groves. The rest of the Western Ghats landscapes are degraded in various degrees of stages depending on the grading of disturbance and it resulted in six major vegetation groups namely rocky outcrops, open grasslands with scrub, dense shrubbery, dwarf canopy forests, riparian forests, and tall woodland forests. The composition of the biological community pertaining to each of these vegetation classes is more or less definite with some overlaps.

The Western Ghats holds rich repository of biodiversity with indigenous vegetation. There is no comprehensive account on endemic plant diversity of the Western Ghats but sporadic publications are available for literature. Chatterjee (1940) enumerated only 34 endemic genera of dicotyledons from Indian Peninsula. Rao (1972) lists about 64 plant genera are confined to Indian floristic region, from which nearly 60 plant genera are narrowly distributed to Peninsular India and Sri Lanka. Subramanyam and Nayar (1974) accounted endemic plants of the Western Ghats. Nayar (1977) enumerated that about 2100 endemic flowering plants occur in Peninsular India including the Western Ghats, which contributed about 32% of its flora. However, Ahmedullah and Nayar (1986) gave the first exhaustive list on the endemic plants of Peninsular India. They have accounted 1940 endemic plant species including infraspecific taxa of Peninsular India. Three volumes of Red Data Books of Indian Plants (Nayar and Sastry, 1987, 1990), some 90 endemic plant taxa were included from Northern Western Ghats alone. Nayar (1996) has recorded 2150 endemic plants in Peninsular India. Tetali et al. (2000) have recorded 439 endemic taxa of plants for Maharashtra state alone. Mishra and Singh (2001) have given elaborate account of 215 endemic and threatened taxa from Maharashtra. Irwin and Narasimhan (2011) have reviewed endemic genera of India with about 49 endemic genera, of which 40 are confined to Peninsular India. A total of 159 plant species (including infraspecific taxa) are confined to the northern Western Ghats of India. Most of them are confined their distribution in small geographical area and they are facing a high risk of extinction (Gaikwad et al., 2014).

The Western Ghats, being one of the 36 global biodiversity hotspots (Marchese, 2015), is very rich in plant diversity and endemism (Nayar, 1996). Most of the arboreal taxa (63%) of the Western Ghats are narrowly confined to the region and characteristically the evergreen forests have distributed a very high percentage of endemic plant species (Ramesh, 2001). The mega centers of Western Ghats have been further classified into eight microcenters of endemism (Nayar, 1996). The Nilgiris forms a part of one of the micro-endemic center along with its contiguous regions such as Silent valley–Wyanad and Koadgu. This whole region ranks second in terms of total number of endemic species next to Agasthyarnalai (Narasimhan and Irwin, 2009).

3.4 FOREST TYPES OF WESTERN GHATS

Champion and Seth (1968) have classified the Indian forests into 5 major forest groups and 16 minor forest groups based on temperature and rainfall (climatic types) and further categorized more than 200 subforest types. Champion and Seth's forest classification was widely accepted in India but a large number of forest subtypes is not feasible for managing forest classification at a time. Forest survey of India, FSI (2012) has proposed the Indian forest atlas with digitized forest classification of India for the first time. The official Indian forest maps constructed according to the Champion and Seth's classification including States and Union Territories have been prepared on 1:50,000 scale. In order to prepare further upscale of forest distribution map done by FSI, and has been attempted to reclassification forest types in India and to rework the forest types based on ground survey. According to Champion and Seth (1968), the forest types of the Western Ghats have listed out in Table 3.1.

Tropical forests of Western Ghats have been divided into four series of climatic forest types including dense and multilayered moist tropical forests due to high temperature and rainfall. These areas comprise characteristically with wet evergreen forests composed of dense growth of tall trees, rich in climbers, lianas, epiphytes, and shrub communities but poor in ground vegetation and grasses. The semievergreen forests are made by dense growth of intermixed tall deciduous and evergreen trees with rich layer of ground herbs, grasses, and ferns. The moist deciduous forests constituted by the dominance of deciduous trees with a lower storeied evergreen trees and shrubs. The littoral and swamp forests near the sea coast are characterized by the dominance of halophytic evergreen trees, shrubs, and herbaceous vegetation (Singh and Chaturvedi, 2017).

Dry tropical vegetation is comprised of dry-deciduous forests, where the abundance of shruby vegetation and open canopy of small trees, the climate experienced around 6 months of dry period in the annual cycle. The thorn forests vegetation in lower elevational gradient experienced more than 6 months of dry period every year, which are characterized by very sparse distribution of small thorny trees with abundance of shrubs. Ground floor vegetation appears during only in rainy season when trees and shrubs have flushed leaves. The dry-evergreen forests prevailed in high temperature and low rainfall experienced during the summer season. Bamboos are usually absent from these dry forests but grasses and small thorny trees are abundant. The dry deciduous forests located on the leeward side of the Western Ghats ranges within an elevational gradient of 300–900 m where mean annual rainfall is around 900–2000 mm. They were distributed across the southern Ghats of Karnataka and Tamil Nadu. The high-altitude Western Ghat ranges increased the moisture content from the southwest monsoon; hence, the eastern slopes of the Deccan Plateau receive relatively little rainfall approximately 900 to 1500 mm. The undulating lower-range valleys have very shallow soils. Thorny vegetation becomes more common in the buffer zone of foothills where grazing pressure is usually high.

The moist deciduous forests occupied the largest areas of the Western Ghats with an elevational range of 500–900 m, where they experienced mean annual rainfall of 2500–3500 mm. The narrow strip of moist deciduous forests is on the windward side of the Ghat range, where the southwest monsoon promotes to develop dense wet evergreen forests. The leeward side of the Ghats has more dry weather with less steep, which caused the rain shadow area to develop in a broader and uneven strip of moist deciduous forests. This forest further extends into the Deccan Plateau of the northern side. The rainfall pattern of the leeward side of Ghats is influenced by complex climatic conditions, with some areas

TABLE 3.1 Forest Types Existing Western Ghats of India (Champion and Seth, 1968).

Forest categories	Major forest group	Subgroup	Subtypes
I. Moist tropical forests	Tropical wet evergreen forests	Southern tropical wet evergreen forests	Southern hill top evergreen tropical forests (1A/C3)
			West coast tropical evergreen forests (1A/C4)
		Edaphic and seral wet evergreen forest types	Pioneer Euphorbiaceous scrub (1A/2S1)
			Cane bakes (1A/E1)
			Wet bamboo bakes (1A/E2)
	Tropical semievergreen forests	Southern tropical semievergreen forests	West coast semievergreen forests (2A/C2)
			Tirunelveli semievergreen forests (2A/C3)
			West coast semievergreen Diptercarp forests (2A/2S1)
		Edaphic and seral semievergreen forests types	Cane brakes (2B/E1)
			Wet bamboo brakes (2B/E2)
			Moist bamboo brakes (2B/E3)
			Leteritic semievergreen forests (2B/E4)
			Secondary moist bamboo brakes (2B/2S1)
	Tropical moist deciduous forests	South Indian moist deciduous forests	Slightly moist and moist teak forests (3B/C1c)
			Southern moist mixed deciduous forests (3B/C2)
			Southern secondary moist mixed deciduous forests (3B/2S1)
	Littoral and swamp forests	Littoral forests	Littoral forests (4A/L1)
		Tidal swamp forests	Mangrove scrub and mangrove forests (4B/TS1)
		Tropical fresh water swamp forests	Submontane hill valley swamp forests (4C/FS2)
		Tropical riparian fringing forests	Riparian fringing forests (4E/RS1)
II. Dry tropical forests	Tropical dry deciduous forests	Southern tropical dry deciduous forests	Dry teak forests (5A/C1b)
			Very dry teak forests (5A/C1a)
			Southern dry mixed deciduous forests (5A/C3)
		Degradation stage of tropical dry deciduous forests	Dry deciduous scrub (5/DS1)

TABLE 3.1 *(Continued)*

Forest categories	Major forest group	Subgroup	Subtypes
		General edaphic types of deciduous forests	Dry savannah forests (5/DS2)
			Euphorbia scrub (5/DS3)
			Dry grassland (5/DS4)
			Boswellia forests (5/E2)
			Hardwickia forests (5/E4)
			Butea forest (5/E5)
			Dry bamboo brakes (5/E9)
		General seral types of deciduous forests	Dry tropical riverain forests (5/1S1)
			Secondary dry deciduous forests (5/2S1)
	Tropical thorn forests	Southern tropical thorn forests	Southern thorn forests (6A/C1)
			Karnatak umbrella thorn forests (6A/C2)
			Southern thorn scrub (6A/DS1)
			Southern *Euphorbia* scrub (6A/DS2)
		General edaphic, degraded and seral types of thorn forests	*Cassia auriculata* scrub (6B/DS1)
	Tropical dry evergreen forests	Dry evergreen forests	Tropical dry evergreen forests (7/C1)
			Tropical dry evergreen scrub (7/DS1)
III. Montane subtropical forests	Subtropical broad-leaved hill forests	Southern tropical broad-leaved hill forests	Nilgiri subtropical hill forests (8A/C1)
			Ochlandra reed brakes (8A/E1)
			South India subtropical hill savannah (8A/DS1)
IV. Montane temperate forests	Montane wet temperate forests	Southern montane wet temperate forests	Southern montane wet temperate forests (11A/C1)
			Southern montane wet temperate scrub (11A/DS1)
			Southern montane wet grass land (11A/DS2)

Source: Adapted from Champion and Seth, 1968.

experiencing less than one-fifth of the 3000 mm or more of average annual precipitation usually at higher elevation of the mountains.

Montane sub-tropical forests are occupied between the altitudes of 1000 and 2000 m where experienced cooler climate than tropical but warmer than temperate regions. Semi-xerophytic evergreen vegetations are predominantly colonized and the forests are classified into broad-leaved hill forests, which prevail the abundance of climbers and epiphytic ferns, orchids, and dense growth of evergreen broad-leaved trees. The pine forests comprising the open formations of pine trees which are usually cultivated and dry-evergreen forests are mainly characterized by small-leaved evergreen trees and thorny xerophytes.

Montane temperate forests are occupied in areas of low temperature with comparatively high humidity in high altitudes of the Ghats. The temperate forests comprised with dense tall coniferous trees or evergreen angiosperm trees and their branches clothed with epiphytic mosses, lichens, and ferns. These forests are further classified into three groups: montane wet temperate forests which are characterized in northern and southern vegetation subtype (Singh and Chaturvedi, 2017). The habitat types of the southern Western Ghats typical tropical evergreen forests include the wet montane evergreen forests and *shola*-grassland complexes in the higher altitudes (1900–2200 m). In the montane evergreen forests, vegetations are highly diverse, multistoried with rich epiphytes, at a stunted low canopy at 15–20 m (Ganesh et al., 1996). The montane evergreen forests hold rich diversity of endemism usually more than half of the tree species are confined only in this forest type, especially major plant families of Dipterocarpaceae and Ebenaceae. The majority of monotypic species are found in this region. The distribution and richness of endemic plants are not uniform within this forest type, where some areas have higher accumulation of endemic plants. For the convenient of descriptions of the forest types of the Western Ghats, the regions are divided into two ranges are such as northern Western Ghats and southern Western Ghats. Northern Western Ghats covered the states of Gujarat, Maharashtra, and Goa, whereas in southern Western Ghats occupied in the states of Karnataka, Kerala, and Tamil Nadu (Figure 3.1).

3.4.1 NORTHERN WESTERN GHATS

3.4.1.1 TROPICAL WET EVERGREEN FORESTS

This forest type consists of two storied canopies with ground vegetation. The top canopy is composed the tree species mainly of *Beilschmiedia dalzellii, Carallia brachiata, Diospyros malabarica, D. montana, D. sylvatica, Mangifera indica, Myristica malabarica, Olea dioica, Syzygium cumini,* and *Symplocos racemosa.* The lower canopy is represented by *Actinodaphnae gullavara, Cinnamomum nitidum, Dimorphocalyx glabellus, Ixora brachiata, I. nigricans, Litsea josephii, Mallotus aureopunctatus, M. resinosus,* and *Memecylon umbellatum.* A number of climbing species among the trees are *Piper hookeri, P. trichostachyon, Stephania japonica, Ancistrocladus heyneanus,* and *Premna obtusifolia* var. *pubescens*. The ground floor of the forest is clothed only in the openings of the canopy, along the sides of streams, which is grown by *Achyranthes coynei, Asystasia dalzelliana, Canscora diffusa* var. *diffusa, Curcuma pseudomontana, Chlorophytum tuberosum, Cyathocline purpurea, Desmodium ritchiei, Ecbolium ligustrinum, Habenaria foliosa, Leucas deodikarii, Impatiens dalzellii, I. pulcherrima, Rhinacanthus nasutus, Paracaryopsis coelestina, Senecio bombayensis, Rubia cordifolia, Smithia bigemina, S. setulosa, S. purpurea, Vigna dalzelliana, Sida rhombifolia, Peristylus lawii, Zingiber neesanum,* and *Arisaema murrayi.*

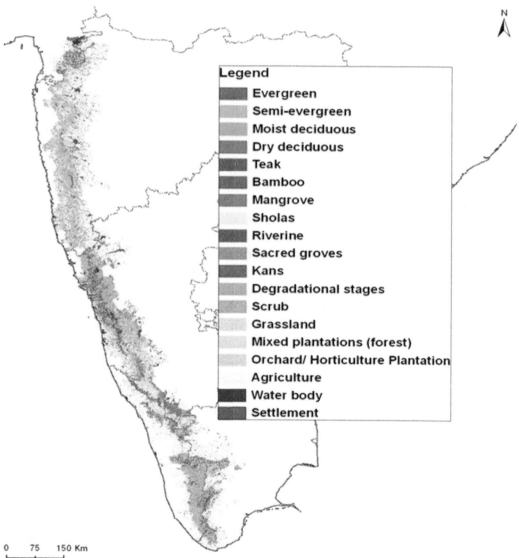

FIGURE 3.1 Forest type map of Western Ghats, India.

Source: Adapted from Google Maps

3.4.1.2   TROPICAL SEMIEVERGREEN FORESTS

This forest type occurs mostly on upper hills slope from 450 to 1050 m above the MSL in the Western Ghats. The major tree species are *Actinodaphne hookeri, Olea dioica, Mangifera indica, Memecylon umbellatum, Syzigium cumini, Terminalia paniculata,* and *T. chebula,* etc. Foothills of the Western Ghats toward Konkan are covered by semievergreen forest. The common trees that form the first storey of this forest are *Albizia lebbeck, Atalantia racemosa, Bombax ceiba, Calophylum inophyllum, Carallia brachiata, Ficus racemosa, Mangifera indica, Syzygium cumini, Terminalia bellirica,* and *T. chebula.* The dominant endemics found in this evergreen class were *Knema attenuata, Hopea ponga, Reinwardtiodendron anamallayanam, Holigarna grahamii, Diospyros candolleana, Holigarna arnottiana, Ixora brachiata,* and *Flacourtia montana.* The climax evergreen forests of the Western Ghats are composed of the characteristic tree species such as *Palaquium ellipticum, Vateria indica, Dipterocarpus indicus, Poeciloneuron indicum,* and *Dysoxylum malabaricum* are found only in this very high evergreen class. Major endemics observed in the forest types are *Holigarna grahamii, H. beddomei,* and *Polyalthia fragrans.* Other dominant endemics recorded were *Garcinia indica, Blachia denudata, Dimorphocalyx lawianus, Sageraea laurifolia, Gordonia obtusa, Hydnocarpus laurifolia, Drypetes elata,* and *Diospyros saldanhae.* These forest vegetations developed in the high rainfall areas of Ghats usually seen in northern Sahydri Plateaus like Mahabaleshwar, Matheran, Bhimashankar, Amboli Ghat, etc.

Tropical semievergreen forests are primarily distributed in the states of Maharashtra, Goa, and Karnataka in the Western Ghats, within a range of 300–900 m elevational gradients. This forest type includes secondary evergreen dipterocarp forests, lateritic semievergreen forests, bamboo brakes, and riparian forests (Champion and Seth, 1968). These forests were established with varied composition and structure from north to south and especially from eastern to western slopes of the Ghats. The predominant tree species include *Aporusa lindleyana, Careya arborea, Celtis timorensis, Elaeocarpus tuberculatus, Hopea parviflora, Holigarna arnottiana, Hydnocarpus laurina, Mesua ferrea, Lagerstroemia microcarpa, Memecylon umbellatum, Olea dioica, Terminalia paniculata, Syzygium* spp., and *Vateria indica.* These forest types also noted high levels of tree diversity and endemism.

In Bhimashankar wildlife sanctuary of Maharashtra, the top canopy of trees consists of *Caryota urens, Ficus racemosa, Firmiana colorata, Garcinia indica, Mangifera indica, Mallotus philippensis, Memecylon umbellatum, Olea dioica, Sterculia guttata, Syzygium cumini,* and *Xantolis tomentosa.* The lower canopy is composed by *Atalantia racemosa, Callicarpa tomentosa, Dimorphocalyx glabellus, Litsea deccanensis, L. ghatica, L. josephii, Hymenodictyon obovatum, Ixora brachiata, Mallotus stenanthus,* and *Rubus ellipticus.* Common climbers are included *Cyclea peltata, Diploclisia glaucescens, Gnetum ula, Oxyceros rugulosus, Piper hookeri, P. trichostachyon, Gymnema cuspidata, Stephania japonica, Salacia macrosperma,* and *Tinospora sinensis.* The ground floor is occupied by *Curcuma angustifolia, C. decipiens, C. pseudomontana, Malaxis rheedii, Nervilia aragoana, N. plicata, Peristylus lawii, Rubia cordifolia, Zingiber neesanum,* and *Z. nimmonii.* This forest vegetation supports to grow more diversity of epiphytes when compared to other forest types. The common epiphytes are *Aerides crispa, A. maculosa, Bulbophyllum fimbriatum, Conchidium braccatum, C. microchilos, Dendrobium aqueum, D. barbatulum, Hoya wightii, Remusatia vivipara, Oberonia recurva, Smithsonia viridiflora,* and *Thunia alba* var. *bracteata* as well as other ferns and mosses (Savita and Sanjaykumar, 2017).

A decrease in the altitude in the south of Maharashtra as well as disturbance resulted in the appearance of deciduous elements and the evergreens vegetation makes way for the semievergreen type of vegetation. These forests resemble evergreen forests in their structure except for the profuse under growth. The tree species, which dominate in these forests, are *Schleichera* spp., *Memecylon* spp., *Madhuca* spp., *Michelia* spp., *Semecarpus* spp., *Olea* spp., *Syzygium* spp. *Persa* spp., *Hopea* spp., *Hydnocarpus* spp., *Ceiba* spp., *Mappia* spp., *Polyalthia* spp., *Stereospermum* spp., *Pterocarpus* spp., and *Terminalia* spp. The shrubs like *Ixora* spp., *Allophylus cobbe. Calamus* spp., *Ardisia solanacea,* and *Alangium salvifolium* dominate the ground flora. The opening of the canopy has resulted in the establishment of a large number of herbs, including bryophytes, pteridophytes, and orchids.

3.4.2 *TROPICAL MOIST DECIDUOUS FOREST*

This type of forest is found all along the eastern side of Sahyadri Ghats where the average annual rainfall ranges from 130 to 180 cm and the average annual temperature from 24 to 27°C. The forests of Allapalli subdivision in Gadchiroli district also fall under this vegetation. Usually, the top of northern Sahyadri ranges has constituted moist deciduous forests, which become drier toward east. The dominant tree species throughout the Ghats occupy the upper canopy such as *Tectona grandis,* and other deciduous species like *Acacia chundra, A. ferruginea, Albizia lebbeck, Bombax ceiba, Dalbergia lanceolaria, D. latifolia, Mangifera indica, Tamarindus indica, Terminalia bellirica,* and *T. chebula.* These tree species compositions are found in mid-elevation of Bhimashankar wildlife sanctuary, especially along the hill slopes surrounding rocky areas and on the western side. The part of the sanctuary in the southern side of the Konkan region has constituted the dry deciduous nature of vegetation. The top canopy trees are composed by small trees of *Anogeissus latifolia, Bridelia retusa, Firmiana simplex, Erythrina stricta, Ficus racemosa, Gnidia glauca, Heterophragma quadriloculare, Mallotus philippensis, Terminalia chebula, T. bellirica, T. cuneata, Tectona grandis,* and *Xantolis tomentosa.* This vegetation is intermixed with *Butea monosperma, Boswellia serrata,* and *Lannea coromandelica.* The understoried shrubby vegetation comprises of *Atalantia racemosa, Breynia retusa, Carissa spinarum, Helicteres isora, Holarrhena pubescens, Murraya koenigii, Pavetta crassicaulis,* and *Strobilanthes callosa.* The lower elevation of dry zones species like, *Balanites aegyptiaca* was also noticed at a few places in the Konkan region. The tree vegetation supports diverse species of climbers and ground cover. The common climbers are *Asparagus racemosus, Dioscorea pentaphylla* var. *pentaphylla, Hemidesmus indicus* var. *indicus, Gymnema sylvestre,* and *Jasminum malabaricum.* The ground vegetation covered *Curculigo orchioides, Drimia indica, Euphorbia fusiformis, Hypoxis aurea, Pancratium parvum, Zingiber neesanum,* and grasses (Savita and Sanjaykumar, 2017).

These vegetation types are found in Chandrapur, Gadchiroli, Bhandara, Gondia, and Amaravati districts and also on the slopes of northern part of the Western Ghats spreading from Nandurbar, Dhule, Nashik, Thane to Kolhapur districts. There are two sub-types represented in this group.

3.4.2.1   MOIST TEAK BEARING FORESTS

The important and valuable forests in the commercial viewpoint of the Maharashtra state, are restricted to the Project Tiger area in the Melghat region with major tree of *Tectona grandis*, and associate with *Haldina cardifolia, Dalbergia latifolia, Mitragyna parviflora, Madhuca indica, Terminalia tomentosa, Pterocarpus marsupium, Salmalia malabaricum,* and *Dendrocalamus strictus,* etc.

3.4.2.2   MOIST MIXED DECIDUOUS FORESTS

Teak is present occasionally and the characteristic evergreen component is occasionally intermixed with teak and a major component of semievergreen elements. The major tree species are *Pterocarpus marsupium, Salmalia malabaricum, Terminalaia bellarica, Dalbergia latifolia, Syzigium cumini, Terminalia tomentosa, Lagerstroemia parviflora,* etc.

3.4.3   TROPICAL DRY DECIDUOUS FORESTS

These forests are situated on the eastern slopes of the Western Ghats where the mean annual rainfall ranges from 50 to 150 cm and the mean annual temperature from 26 to 27.3°C. The forests of the districts of Kolhapur, Sangli, Satara, Pune, Ahmednagar, Nasik, Dhule, Jalgoan, Aurangabad, Jalna, Parbhani, Nanded, Yavatmal, Buldhana, Akola, Amravati, Wardha, Nagpur, Chandrapur, Bhandara, and Gadchiroli are fully and partially covered by this type of vegetation. The common trees are *Acacia chundra, A. ferruginea, Albizia lebbeck, Azadirachta indica, Dalbergia latifolia, Euphorbia ligularia, Ficus amplissima, Mangifera indica,* and *Tectona grandis,* etc. These forest types occupied about 62% of total forested area of Maharashtra state. The Ghat region of Nagpur, Amravati, and Uttar Maharashtra and their eastern hill slopes represented a major portion of the tropical dry deciduous forests. The dominant trees are Teak *(Tectona grandis),* Tiwas *(Ougeinia dalbergioides),* Khair *(Acacia catechu),* Shivan *(Gmelina arborea),* Dhavda *(Anogoissus latifolia)* in this type of forests. In this region, two types of following subforests vegetations are met with.

3.4.3.1   DRY TEAK BEARING FORESTS

Principal predominant species is *Tectona grandis* and their associate trees are *Acacia catechu* (Khair), *Anogeissus latifolia* (Dhawada), *Gmelania arborea* (Shivan), *Ougeinia dalbergiaoides* (Tiwas), etc.

3.4.3.2   DRY MIXED FORESTS

This forest type occurs mostly on elevational hill slopes and valleys from 450 to 1050 m m.s.l. in the Western Ghats. The main tree species are *Actinodaphne hookeri* (Pisa), *Mangifera indica* (mango), *Memocylon umbellatum* (Anjani), *Olea dioica* (Parjamun), *Syzigium cumini* (Jambul), *Terminalia paniculata* (Kinjal), *T. chebula* (Hirda), etc.

3.4.4   TROPICAL THORN FORESTS

The forest types occupied in drier parts of Desh, Khandesh, Vidarbha, and Marathwada districts, where the average annual rainfall ranges from 35 to 80 cm and mean temperature from 26 to 27°C. Major portion of these forests are heavily degraded by low soil fertility associated with lower annual rainfall. The thorn forests are characterically blank, scattered, restricted to shallow soils, and buffered by cultivable lands on all sides. These are also subjected to heavy grazing, lopping, and illicit felling. The trees usually have short boles and the usual height is 6–9 m. There is an ill-defined lower storey of smaller trees and large shrubs, mostly spiny, and often with other xerophytic adaptations. During rainy seasons, various herbaceous plants are also found in *Acacia arabica, A. leucophloea, Azadirachta indica, Balanites roxburghii, Bauhinia racemosa, Butea monosperma, Cassia auriculata, Dolichanrone falcata, Euphorbia ligularia, Ixora arborea, Lantana camara, Mimosa hamata,* and *Zizyphiis mauritiana*. These portions of the forests are predominantly occupied by *Euphorbia* and *Cassia* scrub. A few places where the scrub vegetation are dominated by grasslands mostly on gradual hill slopes near the foothills of the northern Ghats. Intermittent areas are seen in coppiced trees, as an indicator of remenants or earlier canopied forests over these areas. The zones are most exploited and highly disturbed whch presents hardy species indicating heavy degradation. The ground cover is clothed with dominant herbs and grasses include *Apluda mutica, Heteropogon contortus, Themeda cymbaria* along with shrubs of *Carrisa congesta, Gnidia gluaca, Meyna laxiflora,* and also sparse trees of *Mangifera indica, Syzygium cumini, Terminalia elliptica,* etc. (Ghate, 2014).

3.4.5   MONTANE SUBTROPICAL BROAD LEAVES HILL FORESTS

These forests are found in the higher ghats of Sahyadri, restricted to patches or narrow strips usually over 900 m altitude. The mean annual rainfall of this region is above 300 cm and the mean annual temperature is around 20°C. The soil is usually rich in humic contents and the climate is humid. Actually, there is no sharp distinction between the moist mixed deciduous type and montane braod leaved hill type, the latter probably represents the post-climax of the former type. Most of the forests of this type have fundamentally been altered by shifting cultivation or lopping, but where properly developed, it has dense evergreen elements mixed with the deciduous elements. The trees are usually of low to medium average height (5–15 m) with comparatively smaller girth (up to 2 m) and spreading canopies. The top storey consists of trees like *Acacia ferruginea, Albizia procera, Bombax ceiba. Celtis timorensis. Cinnamomum macrocarpum, Dalbergia lanceolaria, D. latifolia, Terminalia bellerica,* and *T. chebula.* The members of second storey are *Mallotus philippensis, Mangifera indica, Memecylon umbellatum,* and *Bambusa arundinacea,* etc.

3.4.6 LITTORAL AND SWAMP FORESTS

This forest type occurs along the creeks and littoral areas of Sindhudurg and Thane districts. Coastal areas have mostly sandy rocky beaches do not possess muddy flats and evergreen mangrove forests. There are few patches of mangrove forests along coastal areas of Sindhudurg and Goa. The mangrove forest of soft tidal mud flats in lower elevation near the sea level supports to develop evergreen shrubs or small trees. The northernmost distribution of *Myristica* swamps has been reported in the Western Ghats areas of "Nirankarachi Rai," in Goa. The predominant tree species were reported here, such as *Myristica fatua, Holigarna arnottiana,*and *Stereospermum colais* (Sreedharan and Indulkar, 2018).

3.5 SOUTHERN WESTERN GHATS

The forests of southern Western Ghats, according to Champion and Seth (1968) fall under four major categories viz., (1) moist tropical forests, (2) dry tropical forests, (3) moist subtropical forests, and (4) montane temperate forests, with about 30 subdivisions. Chandrasekaran (1962, 1973) classified little modification from Champion and Seth, they classified the forests of southern Western Ghats into five categories: (1) tropical wet evergreen forests with three climax forest types viz., (a) low-level evergreen forests, (b) high-level evergreen forests, and (c) low-tropical ghat forests; seven secondary types viz., (a) semievergreen forests, (b) secondary evergreen forests, (c) moist deciduous forests, (d) open deciduous forests, (e) wet bamboo (reed) brakes, (f) moist bamboo brakes, and (g) low-level grasslands, and six edaphic forest types such as: (a1) *Myristica* swamps, (b1) Tropical valley freshwater swamps, (c1) tropical riverine forests, (d1) cane brakes, (e1) *Xylia* mixed forests, and (f1) laeterite scrubs; (2) tropical deciduous forests of their two climax types as (a) Wynad Plateau deciduous forests and (b) moist mixed deciduous forest; two secondary forms as: (a1) moist savannah and (b1) secondary bamboo forests and single edaphic type namely post climax evergreen forests; (3) tropical dry deciduous forests having only climax and secondary forest types. The climax types are (a) tropical dry deciduous forests and (b) mixed dry deciduous forests with sandal; (4) montane subtropical forests having only one climax formation viz., subtropical wet hill forest, and (5) montane temperate forests with one climax type namely wet temperate forests and one secondary forest as high level (montane) grasslands.

3.5.1 MOIST TROPICAL FORESTS

They occur in warm humid areas characterized by high rainfall, temperature, humidity, and uniform photoperiods throughout the year. Winter in such forests will be of a short period.

The occurrence of frost or fog is uncommon. These forests are mostly confined to altitude between 200 and 1500 m in areas with several broken hills, long spurs, and extensive ravines. Trees are generally tall and broad leaved.

3.5.1.1   TROPICAL WET EVERGREEN FORESTS

This forest type is confined to southern the Western Ghats, which are occupied at an altitude of 1500 m or above. The vegetation is characterized by three-layered structure, that is, forest floor, with rich herbaceous ground cover; under storey layer with small or medium-sized trees, dense epiphytic growth of orchids, and pteridophytes; upper tall trees canopy layer continuous layer to attain a height more than 45 m. Some tree species also emerge out of the canopy. Species of *Artocarpus, Durio, Dysoxylum, Elaeocarpus,* and *Lophopetalum* are common in these forests.

Eighteen tree species are indicators of wet evergreen forests in the southern Western Ghats such as *Bhesa indica, Calophyllum austro-indicum, Cinnamomum keralense, Cullenia exarillata, Diospyros atrata, D. bourdillonii, Dipterocarpus bourdillonii, D. indicus, Ficus beddomei, Gluta travancorica, Hopea parviflora, Kingiodendron pinnatum, Lophopetalum wightianum, Mangifera indica, Nageia wallichiana, Palaquium ellipticum, Strombosia ceylanica,* and *Vateria indica.* Some of the tree species are considered as better indicators of the ecological conditions among the above associates, and their ecological amplitude is very narrow such species of *Bhesa indica, Cullenia exarillata, Dipterocarpus bourdillonii, D. indicus, Gluta travancorica,* and *Nageia wallichiana.*

The two types of tropical wet evergreen forests met with are (1) west coast tropical evergreen forests and (2) southern hilltop tropical evergreen forests.

3.5.1.2   WEST COAST TROPICAL EVERGREEN FORESTS

This type is generally found in valleys of high altitudes between 200 and 1500 m. *Cullenia–Mesua–Palaquium* series is the major association here. This association is considered biologically the richest. The top canopy is formed of trees like *Acrocarpus fraxinifolius, Aglaia elaeagnoidea, Ailanthus triphysa, Antiaris toxicaria, Artocarpus heterophyllus, A. hirsutus, Bischofia javanica, Chrysophyllum roxburghii, Calophyllum calaba, Canarium strictum, Cullenia exarillata, Dillenia bracteata, Diospyros bourdillonii, D. candolleana, Dimocarpus longan, Dipterocarpus indicus, Dyrpetes elata, D. macrophylla, Dysoxylum malabaricum, Elaeocarpus tuberculatus, Epiprinus mallotiformis, Garcinia gummi-gutta, Holigarna arnottiana, H. beddomei, Hopea parviflora, H. racophloea, Litsea oleoides, Lophopetalum wightianum, Mangifera indica, Mastixia arborea, Mesua ferrea, Myristica dactyloides, Ostodes zeylanicus, Palaquium ellipticum, Paracroton pendulus* subsp. *zeylanicus, Poeciloneron indicum, Reinwardtiodendron anamallaiense, Syzygium chavaran, S. gardneri, S. parameswaranii, Terminalia travancorensis, Tetramelos nudiflora,* and *Vateria indica.*

The second storey has medium-sized trees adapted to partial shady conditions such as *Actinodaphne bourdillonii, A. tadulingamii, Agalaia elaeagnoidea, A. lawii, A. malabarica,*

Aphanamixis polystachya, Apodytes dimidiata, Aporusa lindleyana, Beilschmiedia bourdillonii, Canthium dicoccum, Carallia brachiata, Cryptocarya beddomei, Dendrocnide sinuata, Diospyros paniculata, D. pruriens, Elaeocarpus glandulosus, E. serratus, Flacourtia montana, Glochidion ellipticum, Gordonia obtusa, Goniothalamus cardiopetalus, Gymnacranthera farquhariana, Glyptopetalum zeylanicum, Hydnocarpus pentandra, Knema attanuata, Litsea coriacea, L. floribunda, Macaranga peltata, Mallotus philippensis, Melicope lunuankenda, Mimusops elengi, Myristica malabarica, Nageia wallichiana, Nothapodytes nimmoniana, Polyalthia fragrans, Symplocos cochinchnensis, Turpinia malabarica, and *Xanthophyllum flavescens.*

The third storey possesses small trees and large shrubs like *Acronychia pedunculata, Agrostistachys indica, Antidesma menasu, Atalantia wightii, Callicarpa tomentosa, Canthium angustifolium, C. travancoricum, Celtis timorensis, Chassalia curviflora, Clausena indica, Chionanthus malabaricus, Elaeocarpus munoroii, Euonymus indicus, Flacourtia montana, Grewia flavescens, Humboldtia trijuga, Ixora elongata, I. malabarica, I. nigrescens, Isonandra lanceolata, Lasianthus jackianus, Leea indica, Leptonychia caudata, Nothopegia beddomei, Pandanus thwaitesii, Pavetta hispidula, Psychotria nudiflora, Sarcococca coriacea, Tabernaemonatana gamblei, Tournefortia heyneana,* etc.

The fourth storey is formed by shrubs, undershrubs, and herbs like *Acranthera grandiflora, Alpinia galanga, A. malaccensis, Amomum muricatum, Agrostemma courtallense, Asplenium indicum, Begonia malabarica, Boesenbergia pulcherrima, Barleria courtallica, Blachia denudata, Costus speciosus, Curcuma calcarata, C. neilgherrensis, Calamus gamblei, C. travancorensis, Elatostemma acuminatum, E. lineolatum, Elettaria cardamomum, Murdannia trichocarpa, Munronia pinnata, Mycetia acuminata, Neanotis monosperma, Neurocalyx calycinus, Ophiorrhiza brunonis, O. eriantha, O. hirsutula, O. mungos, Orophea unniflora, O. zeylanica, Pavetta calophylla, Psychotria congesta, Phyrnium rheedei, Saprosma corymbosum, Sarcandra chloranthoides, Sonerila rheedei, Indianthus virgatus, Thottea siliquosa, Tarenna monosperma, Zingiber macrostachyum, Z. roseum,* etc.

3.5.1.3 *SOUTHERN HILLTOP TROPICAL WET EVERGREEN FORESTS*

These types of forests generally are seen above 1200 m in valleys surrounding ridges. The trees are usually shorter compared to the previous type and are with fuller and rounded crowns. This forest area is considered a transitional stage from tropical to subtropical climax. The top storey is characterized by trees like *Bischofia javanica, Calophyllum austroindicum, Cullenia exarillata, Elaeocarpus serratus, E. tuberculatus, Ficus arnottiana, Garcinia morella, Glochidion ellipticum, Gluta travancorica, Gordonia obtusa, Heritiera papilla, Holigarna arnottiana, Mesua ferrea, Persea macrantha, Prunus zeylanica, Semecarpus anacardium, Syzygium zeylanicum,* etc.

The second storey consists of medium-sized trees of younger forms of the upper storey along with *Alstonia venenata, Alseodaphne semecarpifolia* var. *parvifolia, Baccauera courtallensis, Beilschmiedia bourdillonii, Cinnamomum filipedicellatum, Deberegeasia ceylanica, Elaeocarpus glandulosus, Harpullia arborea, Humboldtia brunonis, Litsea bourdillonii, L. floribunda, L. oleoides, Memecylon angustifolium, Meliosma pinnata, Mallotus simplicifolia, M. tetracoccus, Otonephelium stipulaceum, Toona ciliata, Heynea trijuga, Tupinia malabarica,* etc.

The third-storey vegetation composed of large shrubs and small trees merges with the lower ground layer formed of herbaceous species. The herbs, shrubs, and smaller trees of this storey are *Aglaia tomentosa, Aporusa acuminata, Ardisia pauciflora, Asystasia dalzelliana, Casearia ovata, Clausena austroindica, Gomphandra tetrandra, Lepisanthes erecta, Meiogyne pannosa, Phlogacanthus grandis, Strobilanthes foliosus, S. heyneanus, S. warrensis, Oreocnide integrifolia, Orophea uniflora, Pavetta oblanceolata, Phaulopsis imbricata, Plectranthus wightii, Pogostemon paniculatus, P. travancoricus, Psychotria anamalayana, P. flavida, Rauvolfia densiflora, Rostellularia quinquangularis, Sarcandra chloranthoides, Sarococca coriacea, Tarenna alpestris, T. caranrica, Thottea siliquosa,* and *Wendlandia bicuspidata.* The common climbers are *Aganosma cymosa, Calamus gamblei, Elaeagnus conferta, Embelia ribes, Jasminum cordifolium, Milletia rubiginosa, Molluva spicata, Mucuna hirsuta, Salacia beddomei, Shuteria vestita,* and *Vigna dalzelliana.* The trunks of trees are more carpeted with mosses and lichens compared to the earlier type.

Pascal et al. (2004) categorized the wet evergreen forests into three main groups such as A, B, and C, based on elevational ranges, which are low, medium, and high elevational level wet evergreen forest types, respectively.

3.5.1.3.1 Low Elevation Forests (to 800 m)

In the Western Ghats, the low-elevation forests are characteristically composed by the presence of *Dipterocarpus indicus* and *Strombosia ceylanica,* occurring above 800 m. Further north of the Ghats, *Dipterocarpus indicus* is a characteristic canopy tree species in the other low-elevation evergreen forests. However, it has been found more sporadic patches because of over exploitation of the tree species. *Strombosia ceylanica* is distributed widely throughout the Western Ghats, but it is a common canopy tree in the low-elevation types. Hence, these two tree species were selected for naming of the low elevation types. Some other tree species were contributing to the low-elevation forests such as *Diospyros bourdillonii, Dipterocarpus bourdillonii, Ficus beddomei, Hopea parviflora, Kingiodendron pinnatum,* and *Lophopetalum wightianum.* Among these trees, *Dipterocarpus bourdillonii* has markedly separated the locations to the north of the Ariankavu Pass and to the south which are distributed more prominent presence among the *Kingiodendron pinnatum.* This tree species shows a peculiar distribution pattern in the north of the Ariankavu Pass, but it has become prominent again in the dry evergreen forest in the east and south. Further, it has been classified into two low-elevation forest types based on the composition of tree species (*sensu* Oldeman, 1974) with a canopy height reaching about 35–45 m.

Dipterocarpus indicus–Kingiodendron pinnatum–Strombosia ceylanica type: this forest type occurs to south of the Ariankavu Pass (8°20′N–9°00′N), where the period of the dry climates varies from 2 to 3 months. This forest is less disturbed now mainly represented above 8°40′N. At present, it is mostly in the fragmented form or in a highly degraded where in *Dipterocarpus indicus* that has become very rare due to overexploitation. In the eastern side of the southern Western Ghats, the forest type is restricted to moist valleys in Tambraparni river (east of Agastyamalai), at an elevation 450 to 750 m. This region is more or less as a transitional stage to medium elevation type. The characteristic endemic species are represented in this region namely *Palaquium bourdillonii, Diospyros barberi,* and *Memecylon subramanii* in the lower strata of the forest. These forest communities have been identified under *Dipterocarpus–Kingiodendron–Strombosia* type are both occurring in humid areas. One group is characterized by the abundance of local endemics (e.g., *Hopea racophloea*), and the other forest corresponds to a specific ecosystem adapted usually to water-logged areas.

a) *Hopea racophloea–Humboldtia decurrens facies:* This forest community is characterized by *Hopea racophloea* and *Humboldtia decurrens.* Other common trees are found in intermixed canopy: *Artocarpus gomezianus, Cynometra* sp., *Ficus beddomei, Holigarna nigra, Otonephelium stipulaceum,* and *Vateria indica. Poeciloneuron indicum* occupied rarely in patches. This forest community is confined only in humid valleys of Kallar and Shendurni rivers, located west of Agastryamalai zone.

b) *Myristica swamps*: The swamps are water-logged muddy flat terrains throughout the year. The soil types are generally sandy loam in this region. The unique feature of this ecosystem found the abundance of Myristicaceae, especially with two species such as *Gymnacranthera canarica* and *Myristica fatua* var. *magnifica.* They produce numerous stilt roots and pneumatophores as in mangroves. Only a few such swamps in the entire Western Ghats are met with near Kulathupuzha in Travancore, where rare patches of relatively well-preserved *Myristica* swamps can be observed. Along with Myristicaceae members, other riparian woody species such as *Caralia brachiata, Elaeocarpus tuberculatus, Humboldtia vahliana, Lophopetalum wightianum, Lagerstroemia flos-reginae, Syzygium montana,* and *Tetrameles nudiflora* are met with.

c) *Dipterocarpus indicus-Diptercarpus bourdillonii–Strombosia ceylanica type*: *Dipterocarpus indicus* and *Stromosia ceylanica* are characteristically found in this forest. Here, *Kingiodendron pinnatum* has been replaced by the dipterocarp *Dipterocarpus bourdillonii*. The dipterocarps are also again sporadic patchy due to overexploitation. They are generally distributed at lower elevations (<600 m) and it is the most threatened species in the southern Western Ghats. Usually, this forest community is now encountered only along streams sides and in some inaccessible valleys.

These forests were encountered for many species of hardwoods and softwoods for several decades. Several of their important species are *Vateria indica, Palaquium ellipticum, Calophyllum polyanthum, Otonephelium stipulaceum, Chrysophyllum lanceolatum, Dipterocarpus indicus, Semecarpus auriculata,* and *Poeciloneuron indicum.* However, several other species in the foersts are having wider ecological amplitude viz. *Artocarpus gomezianus, Antiaris toxicaria, Bombax ceiba, Polyalthia fragrans,* and *Pterygota alata.* This forest covers a wide area between 9°N and 11°N in the Western Ghats. This forest community is found in north of Ariankavu Pass, and it is highly discontinuous.

The continuous stretch of *Dipterocarpus indicus-Diptercarpus bourdillonii–Strombosia ceylanica* forest between 9°10′N and 9°35′N, and other smaller fragments of secondary moist deciduous forest and in degraded stages enclaved by either thickets or savannas, where intermixed with several other deciduous tree species.

3.5.1.3.2 *Medium Elevation Types (800–1450 m)*

The medium-elevation forests may be comprised of between 800 and 1450 m, or lower elevation (650 m) depending on local climatic variations and exposure. These forests are structurally similar to low-elevation ones but vary with canopy reaching 35–45 m with four structural layers. In the upper elevation limit, these forests have become stunted with two to three layers canopy (<18 m). They varied floristically from the low elevation types by the distribution of certain indicator tree species or the variation in relative abundance of species. The floristic composition of the zone showed that *Cullenia exarillata* is a good indicator of the elevation and its range usually from 700 to 1400 m, very rarely below to 550 m in the occurrence of moist valleys. *Mesua ferrea* and *Palaquium ellipticum* are well representing this forest and also occur at lower elevation but with a fewer frequency. In the south of the Ariankavu Pass, the original floristic composition is replaced by locally colonized tree species: *Gluta travancorica* and *Calophyllum austro-indicum*. In other areas of the southern Ghats have been identified specifically with the presence of *Nageia wallichiana* and *Diospyros atrata* in medium elevation.

a) *Cullenia exarillata – Mesua ferrea – Palaquium ellipticum – Gluta travancorica* type: This unique forest type is located only in Agastyamalai and Mahendragiri regions and south of the Ariankavu Pass between 8°20′N and 8°50′N. Some tree species were rarely found in the medium elevation type of further north of the region. The major tree species are *Atuna travancorica, Calophyllum austroindicum, Diospyros barberi, Garcinia rubro-echinata, G. imbertii,* and *G. travancorica* at top canopy and subcanopy and lower canopy with *Phlogacanthus grandis, Goniothalamus rhyncantherus, Memecylon subramanii, M. gracile, Octotropis travancorica, Popowia beddomeana,* and *Vernonia travancorica*.

b) *Nageia wallichiana* facies: This composition is found by the abundance of some indigenous Gymnosperm tree species in South India, specifically *Nageia wallichiana*. This species distribution is confined only in eastern side of the Western Ghats between 8°20′N and 9°30′N and above 1000 m. Other Angiosperm tree species were characteristically found in this type: *Actinodaphne campanulata, Aglaia bourdillonii, Diospyros atrata, Elaeocarpus venustus, Eugenia floccosa,* and *Syzygium microphyllum*. An endemic palm, *Bentinckia condapanna* is usually found along the margins of the vallyes near the cliffs.

c) *Cullenia exarillata–Mesua ferrea–Palaquium ellipticum* type: This forest type is found along the southern Western Ghats between the Ariankavu Pass and the Brahmagiri ghat, and north of the Palghat Gap (Pascal, 1982). The lower limit of the forest type locally varies from 550 to 700 m elevation and in the eastern side established only above 800 m. This tree species composition has been found along the southern part of the Periyar plateau and Elamalai region. Most part of the forest areas have been converted into cardamom plantations. The surrounding of Idukki dam of the northern part of Periyar plateau, these forests are highly fragmented and established secondary forests like thickets and savannas. Other tree species are common as canopy trees or emergents that are *Aglaia jainii*, *Cinnamomum keralense*, *Dimocarpus longan*, *Diospyros sylvatica*, *Drypetes elata*, *Litsea oleoides*, and *Syzygium gardneri*.

There are species with second and third layers with wide distribution such species like *Agrostistachys meeboldii*, *Diospyros paniculata*, *Homalium travancoricum*, *Myristica dactyloides*, *Symphilia mallotiformis*, and *Tricalysia apiocarpa*. Other common tree species found in the elevation are *Aglaia exstipulata*, *Bhesa indica*, *Diospyros nilagirica*, *Drypetes venusta*, *Litsea bourdillonii*, *L. keralana*, and *Semecarpus travancorica*. This forest composition may extend up to the Palghat gap. Beyond the Palghat gap, the vegetation changed somewhat with other canopy trees and understoried shrubs. The understorey vegetation constitutes *Ardisia pauciflora*, *Goniothalamus wightiana*, *Tabernaemontana gamblei*, *Psychotria anamalayana*, *Lasianthus jackianus*. *Nageia wallichiana* is very rarely distributed between Kottai Malai and Devar Malai.

3.5.1.3.3   *High Elevation Type (1400–1800 m)*

Bhesa indica–Gomphandra coriacea–Litsea spp. Type: This forest type is characteristically formed by *Bhesa indica* above 1350 m, especially on exposed terrain. The abundant lower storied tree species is *Gomphandra coriacea* because it was used to name the forest type, some part of the forest dominates the lower storey by different other species of *Litsea*.

This type is established along the high-elevation eastern ridges of the Ghats in between Kottai Malai and Devar Malai and further, extends to Megamalai and Cardamom hills. Several species in the lower elevation forests disappear, for example, *Cullenia exarillata*, *Palaquium ellipticum*, *Diospyros* spp., and *Agrostistachys borneensis*. The dominant family is Annonaceae at lower strata in both low and medium elevations, but they are completely absent at high elevations. Some other tree species like *Gomphandra coriacea*, *Hydnocarpus alpina* are less important at lower elevations, but they are much more significant at higher elevations. Several other tree species are *Schefflera capitata*, *Mastixia arborea*, *Archidendron clypearia*, *Cocculus laurifolius*, *Acronychia pedunculata*, *Isonandra* spp. *Meliosma* spp., *Symplocos* spp. in this forest type. Structurally, these forests constituted two storied stunted canopy layers which seldom exceeds 15 m.

Champion and Seth (1968) revised forest classification of India concerned on the whole part of the Western Ghats which belongs to the Southern tropical wet evergreen forests (sub-group 1A) and they subdivided into two groups depending on the exposure to wind, edaphic, and climatic conditions. The West coast tropical evergreen forests (1A/C4) and the Southern hilltop tropical evergreen forests (1A/C3). These forest group 1A/C4 extends from the extreme southern part of India up to in Maharastra. The lower plains and foothills at the sea level established the two medium-elevational forest types. All these forests are growing under the more unfavorable corresponds rather than the group 1A/C3. At higher altitude, these are divided into two major categories around 1500 m: Montane subtropical forests (sub-group 8A Southern subtropical broad-leaved hill forests) and Montane temperate forests (sub-group 11A Southern montane wet temperate forests). Below the elevation of 2000 m, the sub-group 11A is found, therefore, limited and it has not been individualized in this classification due to specific indicator tree species replaced by other local population.

Chandrasekharan (1962b, 1962c, 1962d) has proposed a detailed classification based on plant communities of the forest types in Kerala. He separated the low-level evergreen forests from the high-level forests at about 450 m msl. It extends from 450 to 1050 m and above 900 m the low tropical ghats; evergreen forests start to appear on the upper slopes to the top of ghat ridges. In between 1050–1200 m and 1500–1650 m established the montane sub-tropical forests, and above 1500 m elevation occurs the montane temperate forests. He has identified that there was no marked zonal variation between the forest types, as they were overlapping directly with one another.

3.5.2 TROPICAL SEMIEVERGREEN FORESTS

These forests form a bridge between wet evergreen and moist deciduous forests; usually occur as a narrow strip between 450 and 1200 m elevations in southern Western Ghats and greatly altered due to biotic influences. This forest is composed of evergreen species as lower storey and deciduous components as dominants. Trees such as *Haldina cordifolia, Albizzia odoratissima, Artocarpus hirsutus, Emblica officinalis, Hopea parviflora, Holoptelea integrifolia, Hydnocarpus alpina*; shrubs and stragglers or climbers including *Clerodendrum, Glycosmis, Strobilanthes, Entada, Strychnos,* and *Dioscorea* are commonly occurring in these forests.

Of the three main semievergreen forests recognized in the southern Western Ghats such as West coast semievergreen forests which are found along the banks and rivers and streams adjoining evergreen forests in Kerala. These are confined to the area rich in alluvial soils. This type has a mixture of both evergreen and deciduous trees as it is intermediate between these two types. The stratification in this forest type is not as marked as in the evergreen types. However, two storey can be easily discernible. Champion and Seth (1968) considered it to be a climatic climax while Chandrasekaran (1962a, 1962b, 1962c) and others considered it to be a transitional stage with much biotic interference.

The dominant elements of the first storey are *Ailanthus triphysa, Alstonia scholaris, Antiaris toxicaria, Artocarpus heterophylla, A. hirsutus, Baccaurea courtallensis, Bombax ceiba, Buchanania lanceolata, Calophyllum calaba, C. inophyllum, Carallia brachiata, Chukrasia tabularis, Dimocarpus longan, Diospyros buxifolia, Elaeocarpus tuberculatus, Grewia tiliifolia, Holigarna arnottiana, Hopea parviflora, H. ponga, Hydnocarpus alpina, H. pentandra, Humboldtia vahliana, Ixora brachiata, I. nigricans, Knema attenuata, Lagerstroemia microcarpa, L. speciosa, Madhuca neriifolia, Mangifera indica, Mastixia arborea, Nothopegia travancorica, Otonephelium stipulaceum, Pterocarpus marsupium, Schelichera oleosa, Sterculia guttata, Syzygium gardneri, Tamilnadia uliginosa, Terminalia bellirica, T. paniculata, Trewia nudiflora, Vateria indica, Vitex altissima, V. pinnata,* and *Xylia xylocarpa.*

The lower storey is formed by smaller trees, shrubs, and undershrubs like *Aglaia barberi, A. lawii, Agrostistachys borneensis, Aporusa lindleyana, Arenga wightii, Bischofia javanica, Cinnamomum verum, Clerodendrum infortunatum, Diospyros montana, Eleaocarpus serratus, Flacourtia montana, Flemingia semialata, F. strobilifera, Ficus nervosa, Homonoia riparia, Mallotus philippensis, Saraca asoca, Xanthophyllum flavescens,* etc. Common climbers are *Abrus pulchellus, Adenia hondala, Acacia concinna, A. pinnata, A. sinuata, Anamirta cocculus, Anodendron paniculatum, Aristolochia indica, Butea parviflora, Caesalpinia bonduc, Calycopteris floribunda, Combretum latifolium, Elaeagnus indica, Entada pursaetha, Gouania microcarpa, Mezoneuron cucullatum, Cissus repanda, Olax imbricata, Salacia oblonga, Strychnos colubrina, S. minor, Tiliacora acuminata, Tinospora cordifolia, Uvaria narum,* etc., along with some of those found in the evergreen type. More common herbs found in the ground cover are *Centotheca latifolia, Costus speciosus, Geophila repens, Gymnostachyum febrifugum, Mussaenda frondosa, Ophiorrhiza mungos,* and *Spodiopogon rhizophorus.*

3.5.3 TROPICAL MOIST DECIDUOUS FORESTS

These forests are established between semievergreen and dry deciduous forests at an elevation of 500 m altitude. Usually, these are located in damp valleys or in mosit habitats with shallow or porous soil. Deciduous trees along with some evergreen ground vegetation including climbers and canes are common characteristic features yet the proportion of teak is very limited in these forests. *Anogeissus latifolia, Hardwickia binata, Pterocarpus marsupium, Schleichera oleosa,* and *Terminalia bellerica* are some of the common species occurring in these forests. Commonly found in Papanasam hills, Kalakad, Periyakulam, Gundar valley, and lower elevations of Kodaikanal.

The moist deciduous forests of Western Ghats particularly bearing teak are variously classified as moist, very moist, and slightly moist teak forests. The other two types are the southern moist mixed deciduous forests and southern secondary moist mixed deciduous forests. These two forests are considered together as there is little difference in the floristic composition. The main differences are the degree of degradation.

3.5.3.1 SOUTHERN MOIST MIXED DECIDUOUS FORESTS AND SOUTHERN SECONDARY MOIST MIXED DECIDUOUS FORESTS

These forests occur at lower elevation (500–900 m) receiving less rainfall (250–350 cm). They have a mixed composition with a few evergreen trees as well. The characteristic composition found is *Tectona–Dillenia–Lagerstroemia–Terminalia* series. The canopy is rather open and defoliation begins during December. These forests have the following dominant trees: *Achronychia pedunculata, Albizia amara, A. lebbeck, A. odoratissima, A. procera, Alstonia scholaris, Anogeissus latifolia, Bombax ceiba, B. insigne, Bridelia retusa, Buchanania lanzan, Careya arborea, Chukrasia tabularis, Dalbergia latifolia, D.*

sissoides, Dillenia pentagyna, Haldina cordifolia, Hymenodictyon obovatum, H. orixense, Lagerstroemia lanceolata, L. microcarpa, L. speciosa, Lannea coromandelica, Macaranga peltata, Mastixia arborea, Melia dubia, Mitragyna parviflora, M. tubulosa, Olea dioica, Pongamia pinnata, Pterocarpus marsupium, Pterosepermum reticulatum, Radermachera xylocarpa, Scleropyrum pentandrum, Stereospermum colais, Tectona grandis, Terminalia auriculata, T. bellerica, Tetrameles nudiflora, Vitex altissima, Xylia xylocarpa, etc. Scattered amidst dense grasses with *Phyllanthus emblica* are found.

The middle layer of the forests constitutes of *Bauhinia malabarica, B. racemosa, Bridelia retusa, Buchanania lanzan, Butea monosperma, Careya arborea, Cipadessa baccifera, Cassia fistula, Clausena austroindica, Catunaregum spinosa, Cleistanthus collinus, Cycas circinalis, Dalbergia sissoo, Eriolaena quinquelocularis, Ficus callosa, F. exasperata, F. hispida, F. racemosa, Garuga pinnata, Gmelina arborea, Grewia tiliifolia, Macaranga peltata, Miliusa tomentosa, Kydia calycina, Morinda pubescens, Nothapodytes nimmoniana, Schleichera oleosa, Spondias pinnata, Sterculia foetida, S. guttata, S. villosa, Trema orientalis, Trewia nudiflora, Wrightia arborea, Zanthoxylum rhetsa, Ziziphus xylopyrus,* etc.

The lower layer forming the understorey has *Ablemoschus moschatus, Abutilon indicum, Allophylus cobbe, Baliospermum montanum, Canthium rheedei, Clerodendrum serratum, C. infortunatum, Cynoglossum zeylanicaum, Ecbolium viride, Embelia tsjerium-cottam, Flemingia strobilifra, Gomphostemma heyneana, Gardenia gummifera, Glycosmis pentaphylla, Helicters isora,* and several members of Acanthaceae, Orchidaceae, and ferns. The frequent climbers are *Acacia pennata, A. torta, Asparagus gonocladus, Butea parviflora, Caesalpinia bonduc, Calycopteris floribunda, Cynanchum callialatum, Capparis rheedei, Combretum albidum, Connarus wightii, Dalbergia horrida, D. volubilis, Dioscorea pentaphylla, D. tomentosa, D. wallichii, Diploclisia glaucescens, Erythropalum scandens, Jasminum azoricum, Naravelia zeylanica, Tiliacora acuminata,* etc. The ground floor is covered with *Apluda mutica, Bothriochloa pseudoischaemum, Brachiaria distachya, Centotheca lappacea, Ottochloa nodosa, Panicum trypheron,* and *Rottboellia exaltata.* The under storey is of evergreen type in nature giving the forest as a whole a more or less evergreen appearance in most part of the year. Species like *Tectona grandis, Terminalia* spp., *Garcinia* spp., *Mallotus* spp., *Macaranga peltata, Macaranga indica, Artocarpus heterophyllus, Hopea parviflora, Ixora* spp., *Pavetta indica, Xylia xylocarpa,* and *Tabernaemontana heyneana* dominate the stand. These forests are rich in ground flora. The stand is dominated by members of Combretaceae such as *Terminalia tomentosa* and *T. paniculata,* especially in Ghats of Karnataka.

3.5.3.2   *MOIST TEAK FORESTS*

These are found in lower elevations in the Nilambur and Attapadi valleys of Kerala, where the rainfall is 2000–3000 mm. Besides teak, these forests may also have *Albizia odoratissima, Anogeissus latifolia, Bombax ceiba, Butea monosperma, Careya aborea, Cassia fistula, Catunaregum spinosa, Cleistanthus collinus, Dalbergia latifolia, D. sissodes,*

Grewia tiliifolia, Holarrhena pubescens, Hydnocarpus pentandra, Kydia calycina, Lagerstroemia parviflora, Mallotus philippensis, M. tetracoccus, Mirtagyna parviflora, Pterocarpus marsupium, Radermachera xylocarpa, etc. The floor is covered by gregarious growth of grasses, the dominant ones being species of *Cymbopogon* and *Themeda.* Amidst grasses may be present some herbs and subshrubs like *Alpinia malaccesnsis, Cipadessa baccifera, Clerodendrum infortunatum, Curcuma angustifolia, Helicteres isora, Leea indica, Solanum erianthum,* etc. Large portions of these forests have been occupied by exotic weeds like *Lantana camara* and *Chromolaena odorata.*

3.5.3.3 *XYLIA XYLOCARPA DOMINANT LATERITIC DECIDUOUS FORESTS*

This type of forests is seen in lower ranges and lateritic soil in sufficient depth and is an edaphic climax type dominated by *Xylia xylocarpa.* Other species frequently seen also include *Canthium parviflorum, Cassia fistula, Catunaregum spinosa, Strychnos nux-vomica,* and *Terminalia paniculata.*

3.5.3.4 *LOW-LYING MARSHY DECIDUOUS FORESTS*

These types of vegetation are seen in northern parts of Kerala (Sivarajan and Mathew, 1997), such areas usually dry up during summer leaving the floor with the top soil contains a large proportion of clay. These areas are unsuitable for farming, hence were dominated by grasses and weeds forming grazing grounds for cattle. A few trees of *Cordia wallichii, Haldina cordifolia, Tamilnadia uliginosa,* and *Terminalia alata* may be seen along with hydrophilous species in the undergrowth. These are *Ardisia solanacea, Artanema longifolia, Crotalaria quinquefolia, Limnophila repens, Ludwigia perennis, Rotala indica,* species of *Cyperus* and *Fimbristylis,* grasses like *Brachiaria repens, Cyrtococcum trigonum, Digitaria longiflora, Echinochloa crus-galli, Eragrostis atrovirens, Isachne globosa, I. miliacea, Ischaemum rugosum, Panicum paludosum, P. psilopodium, Paspalidium flavidum, Rottboellia exaltata, Sacciolepis indica,* etc.

3.5.4 *SWAMP FORESTS*

The characteristic forest types occur in low-level-hill valleys of the southern Western Ghats as *Myristica* swamps. The *Myristica* swamps were first reported by Krishnamoorthy (1960) from the Travancore of the southern Western Ghats. This forest type is restricted to the valleys of Shendurney, Kulathupuzha, and Anchal forest ranges in Kerala. Champion and Seth (1968) categorized the forest vegetation as tropical freshwater swamp forests (4C/FS1). Pascal (1988) and Rodgers and Panwar (1992) have described the special vegetation as most critically needed for conservation. In the northern Ghats, swamps have also been reported from Uttara Kannada of Karnataka (Chandran et al., 2010) and Satari region in Goa (Santhakumaran et al., 1995).

3.5.4.1   MYRISTICA SWAMP FORESTS

The characteristic forest with the chief composition of *Myrisitica magnifica, Gymnacranthera farquhariana,* and other *Myristica* species occasionally found that are *Myristica malabarica* and *Knema attenuata.* They formed characteristically pneumatophores or breathing roots, which are specialized survival features of trees in waterlogged conditions. They also produced superficial lateral roots that emerged out the air and loop back into the muddy soil which are called breathing roots. The intermixed undergrowth of the forests are spiny plants such as *Pandanus* sp. and *Calamus* sp. and along the herbs like *Alpinia malaccensis, Lagenandra ovata, Phrynium* sp., *Codariocalyx motorius, Selaginella* spp., *Ochlandra* spp., *Gnetum* spp., *Helicteres* spp., *Bauhinia* spp., and rare occurrence of creepers as *Chilocarpus* spp., *Kunstleria* spp., and *Piper* spp. (Senthilkumar et al., 2014).

Myristica swamps are highly restricted, vulnerable, and fragmented ecosystems in the Western Ghats. These swamps cover up less than 0.01% of the total forest area of the Western Ghats. The biodiversity of the unique ecosystem is solely depending on the existence of the forest type. The plant communities varied in degrees of environmental factors and species composition differed in specific niches. Therefore, this forest type is very important for conservation because the vegetation composition of the special ecosystem is composed of Myristicaceae trees along with narrow endemic tree species like *Lophopetalum wightianum, Vateria indica, Holigarna beddommi, Semecarpus auriculata, Syzygium travancoricum,* etc. On the margin of the forests are seen *Humboldtia vahliana, Hopea parviflora, Persea macrantha, Vateria indica,* etc., with sometimes gregarious growth of *Strobilanthes* spp.

3.5.4.2   TROPICAL RIVERINE FORESTS

Several rivers and their tributaries together form a specialized habitat for a plant community remarkably similar in appearance but with a totally dissimilar taxonomic affinity. Trees common to the habitat are *Agrostistachys borneensis, Calophyllum apetalum, Cleidion javanicum, Ficus heterophylla, Garcinia wightii, Helicia nilagirica, Holigarna arnottiana, Homonoia riparia, Hydnocarpus alpina, H. pentandra, Lagerstroemia microcarpa, L. reginae, L. speciosa, Lophopetalum wightianum, Madhuca neriifolia, Neolamarckia cadamba, Ochreinauclea missionis, Ochlandra* spp., *Rapanea wightiana, Syzygium hemisphericum, S. heynanum, S. occidentale, Tetrameles nudiflora, Vateria indica, Vitex alata,* and *V. leucoxylon.*

3.5.5   WET BAMBOO BRAKES

This type occurs in places wherever there is a break in the canopy such as slides of streams, higher slopes or gaps created by felling. Being very aggressive, reeds colonize very rapidly, and form large populations especially in gaps that are partially shaded. Common such reeds are *Ochlandra ebracteata, O. scriptoria, O. travancorica,* and *O. wightii.*

3.5.6   LOW LEVEL GRASSLANDS

The grasslands found below 1000 m are usually classified as low-level grasslands. These include secondary grasslands of the lower elevations also. They are scattered amidst moist deciduous and dry deciduous forests based on which they are divided into (1) dry and (2) wet grasslands.

3.5.6.1   DRY GRASSLANDS

They are limited in occurrence. The northern part particularly Kasargod, Kozhikode, parts of Palakaad and Wynad districts have dry rocky areas which are covered by grasses intermixed with scrub jungles and dry deciduous forests. Grasses frequent here are *Arundinella cannanorica, A. mesophylla, Bhidea brunsiana, Danthonidium gammiei, Dimeria bialata, Heteropogon contortus, Ischaemum indicum,* and *Pseudanthistiria heteroclita*.

Vast areas in the dry and moist deciduous forests are covered by grasses which are formed consequent to biotic interference like fire, grazing, and indiscriminate felling of trees. In such areas are found *Arundinella ciliata, A. mesophylla, Chrysopogon orientalis, C. zeylanicus, Cymbopogon flexuosus, Digitaria ciliaris, Ergarostis tenuifolia, Eulalia trispicata, Heteropogon contortus, Ischaemum timorense,* etc. Gregarious amidst these grasses occur the fire-resistant *Phoenix loureiri*. More common shrubs associated with these grasses are *Hedyotis purpurascens, Uraria rufescens,* and herbaceous species of *Biophytum, Crotalaria, Cyanotis, Fimbristylis, Leucas, Phyllanthus, Rhynchosia, Striga,* and *Vernonia*.

3.5.6.2   WET GRASSLANDS

These are characteristics of low-lying wetlands of the coastal regions, particularly in Alappuzha district of Kerala. The common grasses adapted to this situation are *Brachiaria mutica, Imperata cylindrica, Eriochloa procera, Paspalum conjugatum, P. geminatum, Panicum repens,* etc. Grasses found in areas with tidal action adapted to saline conditions are *Sporobolus virginicus, Paspalum scrobiculatum, Zoysia matrella,* etc.

3.5.7   TROPICAL DRY DECIDUOUS FORESTS

Dry deciduous forests are restricted to well-drained hill slopes or undulating terrain between 600 and 1000 m elevations. Some of the characteristic species include *Tectona grandis, Bridelia retusa, Pterocarpus marsupium, Chloroxylon swietenia, Erythroxylon monogynum, Gardenia gummifera, Buchanania lanzan, Grewia* spp., *Albizia amara, A. lebbeck, Wrightia tinctoria, Morinda coreja, Phoenix sylvestris,* etc. These forests are distributed widely in Tamil Nadu especially in lower elevations of Kodaikanal and Kerala of Chinnar, Palakkad, and Kannur.

3.5.7.1 SOUTHERN DRY MIXED DECIDUOUS FORESTS

They occur in drier areas of southern Western Ghats and it is characterized by the light understorey canopies, these forests occur at low elevations between 300 and 900 m and receive less rainfall, 100–200 cm. the undergrowth is dense as light penetrates the canopy sufficiently but epiphytes and ferns are rare. The common association here is *Terminalia – Anogeissus latifolia – Tectona grandis* series. Dominant trees are *Albizia amara, Anogeissus latifolia, Bauhinia* spp., *Buchanania lanzan, Butea monosperma, Commiphora caudata, Dalbergia latifolia, Dillenia pentagyna, Ehretia laevis, Garuga pinnata, Gmelina arborea, Haldina cordifolia, Phyllanthus emblica, Pterocarpus marsupium, Semecarpus anacardium, Terminalia chebula,* etc.

The smaller trees and shrubs are *Acacia pennata, Benkara malabarica, Carissa congesta, Chassalia curviflora, Desmodium gangeticum, D. laxiflorum, D. triangulare, Erythroxylum monogynum, Holarrhena pubescens, Leea indica, Phyllanthus polyphyllus, Securinega leucopyrus, Solanum americanum, S. torvum, Writghtia tinctoria,* and *Zingiber* spp. The herbaceous flora after the monsoon may have *Corchorus aestuans, Crotalaria evolvuloides, C. heyneana, C. nana, Desmodium heterocarpum, D. triflorum, Indigofera trifoliata, Mimosa pudica, Sida acuta, S. alnifolia, S. rhombifolia, Rhychosia densiflora* and grasses like *Apluda mutica, Eragrostis japonica, E. tenella, Oplismenus compositus,* and *Themeda triandra.* Gregarious growth of the bamboo *Dendrocalamus strictus* is a distinguishing feature in the forest type.

The dominant species in the Ghat area of Karnataka shows teak-bearing forests comprising *Tectona grandis, Anogeissus latifolia,* and *Terminalia* spp., and in the non-teak forests, *A. latifolia* and *Terminalia* spp., *Diospyros* spp., and *Sterculia* spp. are dominant. Important timber species like *T. grandis, Lagerstroemia microcarpa, Terminalia* spp., *Dalbergia latifolia,* and *Pterocarpus marsupium* are abundant in these forests. Bamboo *(Dendrocalamus strictus* and *Bambusa arundinacea)* generally present along the streams and watercourses are more toward the southern region and less in northern region of the Ghats. These dry deciduous forests are abundant with shrubs like *Gardenia* spp., *Randia dumetorum, Grewia microcos, Flacourtia indica, Ziziphus* spp., *Capparis* spp., *Canthium* spp., and *Carissa* spp. The rainy season results in profuse ground flora. Grasses mainly constitute these annuals. Climbers like *Hemidesmus indicus, Ichnocarpus frutescens, Tinospora cordifolia, Tylophora asthmatica, Cocculus hirsutus, Aristolochia* spp., *Decalepis hamiltonii,* and *Cryptolepis buchanani* are common in these forests. Sandal wood trees are found in the dry deciduous and scrub forests of Mysore, Hassan, and Chikmagalur districts.

3.5.7.2 TROPICAL THORN FORESTS

These forests are also known as scrub jungles occurring in the dry plains and low elevations or foothills of the Western Ghats, particularly in Madurai, Tirunelveli, and Tiruchirappalli. The annual rainfall ranges from 500 to 700 mm. Tropical thorn forests in Tamil Nadu are the most exposed forests to human disturbances. Some of the dominant species of these forests include *Acacia chundra, Albizia amara, Anogeissus latifolia, Azadirachta indica,*

Butea monosperma, Cassia fistula, Cordia dichotoma, Diospyros candolleana, Holar-rhena pubescens, Limonia acidissima, Strychnoa nux-vomica, Tectona grandis, Terminalia bellerica, Ziziphus mauritiana, Z. xylopyrus. Z. oenoplia, etc. The common stragglers include *Capparis zeylanica, Carissa spinarum, Flacourtia indica, Maytenus emarginata,* and *Toddalia asiatica.* More common climbers are *Cardiospermum halicacabum, Ceropegia candelabrum, C. juncea, Cissus quadrangularis, C. repanda, Cyclea peltata, Hemidesmus indicus, Ichnocarpus frutescens, Jasminum angustifolium, Vicetoxicum indicum,* etc. The area being dry for the most part of the year, the ground is almost barren.

After the monsoon, it is covered by herbs like *Achyranthes aspera, Allmania nodi-flora, Apluda mutica, Aristida setacea, Blumea mollis* and species of *Barleria, Blepharis, Cleome, Cymbopogon, Glinus, Indigofera, Leucas, Mollugo,* and *Hedyotis* and a few shrubs like *Canthium coromandelicum, Catunaregum spinosa, Pavetta indica, Ixora coccinea, Pseudaidia speciosa,* etc. The other dominant species of these scrub lands in Karnataka are the species of *Albizia, Pterolobium, Canthium, Capparis, Cadaba,* and *Ziziphus.* Shrubs like *Scutia myrtina, Maytenus emarginata, M. heyneana, Dodonaea viscosa, Cippadessa baccifera, Tarenna asiatica,* and *Gmelina asiatica* are also common.

3.5.7.3 *TROPICAL DRY EVERGREEN FORESTS*

Tropical dry evergreen forests are usually confined to coastal regions of Tamil Nadu and patches forests in Kerala coasts. These forests occur along sandy beaches, coastal plains with isolated hillocks of nearby coastal area usually contain red lateritic soil. The forests are dense toward the west and these thickets thin out as scrub land mixed with open grasslands toward the coast. Notable trees include *Ficus benghalensis, Syzygium cumini,* and *Pongamia pinnata.* However, the typical dry evergreen forests are categorized under sub-type 7/C1, which are constituted by common trees like *Manilkara hexandra, Memecylon umbellatum, Diospyros ferrea, Chloroxylon swietenia, Albizia amara,* and other evergreen species. Further, the sub-type 7/DS1 was described under tropical dry evergreen scrub, where dominated trees are *Memecylon edule, Ziziphus xylopyrus, Dichrostachys cinerea, Psydrax dicoccos, Carissa spinarum, Albizia amara, Buchanania axillaris, Dodonaea viscosa,* and other associate species. The phyto-sociological composition of the tropical dry evergreen forest vegetation were also mentioned by several other authors as *Manilkara hexandra* series, generally *Manilkara hexandra – Drypetes sepiaria – Chloroxylon swietenia – Memecylon umbellatum* series, the *Manilkara hexandra – Memecylon umbellatum – Drypetes sepiaria – Pterospermum suberifolium – Carmona microphylla* facies of the *Albizia amara* community, the *Manilkara hexandra – Chloroxylon swietenia* vegetation type within the *Albizia amara* zone and the *Memecylon edule – Atalantia monophylla* series. This forest is floristically varied by a good representation of unique and preferential species, exclusively confined to the vegetation type. The preferential woody species are *Atalantia monophylla, Lannea coromandelica, Lepisanthes tetraphylla, Manilkara hexandra, Memecylon umbellatum, Psydrax dicoccos, Pamburus missionis, Sapindus emarginatus, Putranjiva roxburghii, Dolichandrone falcata, Buchanania axillaris,* etc. The common climbers and woody liana are *Combretum albidum, Ventilago maderaspatana, Grewia orientalis,* and *Hugonia mystax.*

Southernmost tip of Western Ghats with some remnant patches of tropical dry ever-green forest occurs with core species of *Albizia amara, A. lebbeck, Atalantia monophylla, Azadirachta indica, Cassia fistula, Chionanthus mala-elengi, Crateva magna, Dalbergia lanceolaria, Diospyros ebenum, D. ferrea, Drypetes sepiaria, Lannea coromandelica, Lepisanthes tetraphylla, Manilkara hexandra, Psydrax dicoccos, Pterospermum cane-scens, Sapindus emarginata,* and *Syzygium cumini.* The common shrubs are *Benkara malabarica, Cadaba fruticosa, Canthium parviflorum, Capparis brevispina, Carissa spinarum, Carmona retusa, Ecbolium viride, Flacourtia indica, Glycosmis mauritiana, Gmelina asiatica, Ixora pavetta, Memecylon umbellatum, Opuntia dillenii, Phyllanthus reticulatus, Premna alstoni, Catunaregum spinosa, Securinega leucopyrus,* and *Tarenna asiatica.* Climbers and stragglers are *Acacia caesia, Adenia wightiana, Allophylus cobbe, Asparagus racemosus, Capparis zeylanica, Cassytha filiformis, Cayratia pedata, Cissus vitiginea, C. quadrangularis, Cocculus hirsutus, Combretum ovalifolium, Dioscorea oppositifolia, Grewia carpinifolia, Gymnema sylvestre, Hugonia mystax, Jasminum angus-tifolium, Passiflora foetida, Reissantia indica, Rivea hypocrateriformis, Sarcostemma intermedium, Scutia myrtina, Solanum trilobatum, Solena amplexicaulis, Strychnos minor, Symphorema involucratum, Toddalia asiatica, Tylophora indica, Ventilago maderaspatana,* and *Ziziphus oenoplia.* In between found the twiners like *Abrus precatorius, Aristolochia indica, Canavalia cathartica, Ichnocarpus frutescens,* and *Ipomoea sepiaria.* Occasionally found *Borassus flabellifer* in the sandy vegetation. Some common perennial succulent and tuberous plants are *Caralluma adscendens, Curculigo orchioides, Gloriosa superba, Sansevieria roxburghiana,* and *Theriophonum fischeri.*

3.5.8   SUBTROPICAL BROAD LEAVED HILL FORESTS

These forests are also referred as "stunted rain forests" found between 1100 and 1800 m elevations of Nilgiris and Kanyakumari. Though the attributes resemble tropical rain forests the components are not as dense and the trees are smaller with less shapely trunks, often with epiphytes. Species such as *Syzygium arnottianum, S. montanum, Microtropis ovalifolia, Dillenia* sp., *Antidesma* sp., etc., commonly occur in these forests.

3.5.9   SUBTROPICAL MONTANE FORESTS

These are the southern subtropical hill forests. They occur mostly on plateaus in the upper reaches of slopes in higher altitudes (1000–1800 m). A transitional stage between tropical wet evergreen and temperate shola forests, they are miniature sholas with a floristic compo-sition of both these types. Though they may not be as luxuriant as the wet evergreen forests, where the trees are taller above 20–30 m when compared to wet temperate forests. The low stature of the trees is attributed to the high velocity of winds and less favorable conditions of the soil. The stratification is not clearly marked. The dominant elements in the top storey are *Actinodaphne bourdillonii, A. bourneae, A. lanata, A. salicina, Bischofia javanica, Callicarpa tomentosa, Chionanthus courtallensis, C. mala-elengi, Celtis tetrandra,*

Cinnamomum macrocarpum, C. sulphuratum, C. verum, C. wightii, Cryptocarya lawsonii, C. neilgherrensis, Daphniphyllum neilgherrense, Dimocarpus longan, Elaeocarpus serratus, E. tuberculatus, Eurya nitida, Fagraea ceylanica, Ficus arnottiana, Glochidion neilgherrense, Gordonia obtusa, Heritiera papilio, Ligustrum robustum, Maesa indica, Neolitsea cassia, Persea macrantha, Pittosporum neelgherrense, Syzygium cumini, S. densiflorum, Toona ciliata, and *Turpinia malabarica.*

The second storey may have *Actinodaphne campanulata, Byrsophyllum tetrandrum, Canthium dicoccum, Cinnamomum filipedicellatum, C. sulphuratum, C. travancoricum, Gomphandra tetrandra, Hydnocarpus pentandra, Memecylon angustifolium, Pavetta brevifolia, Pittosporum napaulense, Rapanea wightiana, Schefflera wallichiana, Symplocos cochinchinensis* subsp. *laurina, Ternstroemia gymnanthera, Vernonia monosis, Wendlandia notoniana,* etc. The third storey is formed of younger forms of the upper two storeys along with shrubs and undershrubs like *Canthium neilgherrense, Hedyotis purpurascens, H. ramarowii, H. swertioides, Knoxia heyneana, Pavetta hispidula, Psychotria nigra, Sparosma fragrans,* etc.

The ground is often covered by *Arundinella tuberculata, Elatostemma lineolatum, Strobilanthes heyneanus, S. wightianus, Pogostemon purpurascens, Polygala arillata, Viola pilosa* in addition to pteridophyte species of *Asplenium, Osmunda, Selaginella,* etc. The common epiphytes are orchids like *Aerides crispa, A. ringens, Bulbophyllum fischeri, B. tremulum, Coelogyne nervosa, Dendrobium herbaceum, Eria nana, E. polystachya, E. reticosa, Liparis elliptica, Oberonia santapaui, Trichoglottis tenera,* etc. *Impatiens auriculata* and certain *Peperomia* spp. are also common. More common ground orchids are *Calanthe masuca, Habenaria longicorniculata, H. plantaginea, Peristylus aristatus, P. densus, Pectilis gigantea, Phaius luridus,* etc. *Lagerstroemia lanceolata* was the most dominant tree in the southern region of Karnataka and the other dominant trees are *Elaeocarpus tuberculatus, Dimocarpus longan, Canarium strictum,* and *Hopea glabra.* It is interesting to note that a deciduous tree is the most dominant one indicating that the forests have been subjected to disturbance resulting in modification of forest structure and composition (Sathish et al., 2013).

The Western Ghats of Karnataka with heavy rainfall supported tropical evergreen forests. The characteristic climatic conditions provided the prolonged wet periods, and established storied canopy with poor under growth. Thick layer of decaying biomass on the ground is another feature of these forests. Trees attained heights greater than 30 m and girths exceeding 300 cm. The canopy is extremely dense and the differentiation of definite canopy layers probably does not exist. These forests are confined to the higher altitudes of the slopes and in the regions of low disturbance of the Ghats. The windward sides of the Ghats harbor more of this type of vegetation as they received rain in most part of the year. The characteristic vegetation composition made by the tree species of *Myristica, Cinnamomum, Bischofia, Dipterocarpus, Vateria, Holigarna, Hopea, Litsea, Persea, Artocarpus, Kingiodendron, Poeciloneuron, Dysoxylum,* and *Aglaia* as the top canopy trees. The middle canopy is formed by species of *Diospyros, Olea, Humboldtia,* and *Knema,* while the lower canopy is formed by species of *Garcinia, Mappia, Memecylon, Canthium,* and *Euodia.* The ground is devoid of any vegetation. The tree trunks are abounding with epiphytes. Many of the endemic orchids occur in these forests. The species like *Poeciloneuron indicum,*

Kingiodendron pinnatum, Knema attenuata., Myristica fatua, Hopea canarensis, Diptero-carpus indicus, Vateria indica, and *Holigarna* species are endemic to the Western Ghats and are found distributed in these primary forests.

3.5.10   MONTANE TEMPERATE FORESTS

These forests are also known as "Hill shola forests" and occur above 1500 m altitude. The height of the forest is relatively low. Trees are short and their branches are densely covered with lichens, mosses, ferns, and epiphytes. Montane forests are usually found in patches in more sheltered areas like valleys interspersed with grasslands. These forests occur in Anamalais, Nilgiris, Palnis, and Tirunelveli hills. Some of the notable species include *Brysophyllum tetrandrum, Hedyotis leschnaultiana, Rhododendron arboretum* subsp. *nilagiricum, Eugenia floccosa, Euphorbia santapaui, Olea polygama, Cryptocarya stocksii,* etc.

These are two types (1) Southern montane wet temperate (sholas) and (2) southern montane wet grasslands (high altitude grasslands). Both are closely associated with each other.

3.5.10.1   SOUTHERN MONTANE WET TEMPERATE FORESTS (SHOLAS)

These types of forests are confined to depressions and moist sheltered wet grasslands, the sholas are unique types of vegetation distinct from other types by various features. They are small, isolated, compact, and sharply defined evergreen woodlands scattered amidst vast stretches of grasslands at higher altitudes above 1500 m. The sholas in their best vegetation composition can be observed in the Silent valley adjacent to the Nilgiris and in the high Ranges. A fast-receding community (Vishnu Mittre and Gupta, 1968), the sholas are characterized by stunted trees with short boles reaching less than 15 m high. The crowns of these trees are usually dense and rounded with entire and coriaceous leaves usually reddish when young. The trunks are not buttressed and branches are crooked and densely clothed with lichens, mosses, ferns, and epiphytic orchids. The stunted growth and slow spread of this vegetation are generally attributed to the high velocity of winds and unsuitable soil nearby (Adriel, 1964). A mixture of both temperate and tropical elements is found here that is another remarkable feature. The occurrence of sino-Himalayan *Rhododendron* in such forests in the high ranges is a notable feature. Unlike tropical evergreen forests, this type has only two well-marked storeys. The dominant trees in the top canopy are *Actinodaphne bourneae, Cinnamomum wightii, C. perrottetii, Callicarpa tomentosa, Daphniphyllum neilgherrense, Elaeocarpus tuberrculatus, Euonymus indicus, Glochidion neilgherrense, Gordonia obtusa, Litsea wightiana, Melicope lunuankenda, Meliosma simplicifolia, Microtropis ramiflora, Neolitsea cassia, Pheobe wightii, Syzygium calophyllifolium, S. densiflorum,* etc. Along the fringes are found light-loving *Elaeocarpus munroii, E. recurvatus, Eurya nitida, Prunus zeylanica, Rhododendron arboreum* var. *nilagiricum, Rhodomyrtus tomentosa, Symplocos cochinchinensis* subsp. *laurina,* and *Trenstroemia gymnanthera.*

The second storey is with smaller trees like *Actinodaphne salicina*, *Appollonias arnottii*, *Cryptocarya stocksii*, *Gomphandra coriacea*, *Hydnocarpys alpina*, *Ixora notoniana*, *Ligustrum roxburghii*, *Memecylon malabaricum*, *Pittosporum tetaspermum*, *Rapanea wightiana*, *Schefflera capitata*, *S. racemosa*, *Symplocos foliosa*, *S. pendula*, *Syzygium gardneri*, *Staphylea cochinchinensis*, *Vaccinium leschenaultii*, *V. neilgherrense*, *Viburnum erubescens*, *Vernonia monosis,* and *Wendlandia notoniana.*

The ground layer may have shrubs, subshrubs, and herbs such as *Anotis monosperma*, *Canthium neilgherrense*, *Elatostemma lineolatum*, *Gnidia glauca*, *Hedyotis stylosa*, *Hypericum mysorense*, *Hypochaeris glabra*, *Laportea bulbifera*, *Lasianthus cinereus*, *L. jackianus*, *Linocera ligustrina*, *Maesa indica*, *Memecylon flavescens*, *Micromeria biflora*, *Polygala arillata*, *Psychotria congesta*, *Sarcococca saligna*, *Senecio lavendulaefolius*, *Rotala rotundifolia*, *Rungia elatior*, *Viola pilosa*, etc. Lianas are not common. The common climbers are *Clematis wightiana*, *Elaeagnus indica*, *E. kologa*, *Linocera leschenaultia*, *Piper* spp., *Pentapanax leschenaultia*, *Rosa leschenaultiana*, *Rubia cordifolia*, *Rubus racemosus*, *Schefflera venulosa*, *S. wallichiana*, *Senecio candicans*, *S. corymbosa*, *S. intermedius*, *Vincetoxicum pauciflora*, etc. Parasites are rare but epiphytes are abundant. More common epiphytes are *Aerides ringens*, *Bulbophyllum tremulum*, *B. sterile*, *Chilochista pusilla*, *Coelogyne mossiae*, *C. odoratissima*, *Dendrobium barbatulum*, *D. herbaceum*, *D. heterocarpum*, *D. heyneanum*, *Eria dalzellii*, *E. polystachya*, *Liparis viridiflora*, *Oberonia proudlockii*, *O. santapaui*, *Sirhookera latifolia*, *Trias bonaccordensis*, *T. stocksii*, and *Vanda spathulata*. *Hoya ovalifolia*, *Peperomia tetraphylla,* and ferns are also common. Mosses and lichens completely cover the branches and trunks of trees. On the fringes of water courses, *Ochlandra beddomei* may be found.

The status of shola forests has been a subject of dispute. While Champion and Seth (1968) and Chandrasekharan (1962a, 1962b, 1962c) considered them as temperate vegetation, but Meher-Homji (1965) treated them as subtropical montane type.

3.5.10.2 SOUTHERN MONTANE WET GRASSLANDS (HIGH ALTITUDE GRASSLANDS)

These forests associated with and dominating the shola woodlands are vast stretches of grasslands, a characteristic feature of the high mountains above 1500 m. The Eravikulam National Park, Munnar, and Devikulam in the High Ranges of Idukki, Silent valley of Kerala, Avalanch of Nilgiris, Thirukurungudi range of Tirunelveli hills, and Muthukuzhivayal of Kalakadu-Mundanthurai Tiger Reserve in Tamil Nadu are such extensive grasslands in southern Western Ghats. These are mixed with herbs and shrubs in varying proportions and hence called shrub savannahs. These montane grasslands have superficial resemblance to the "patenas" of Sri Lanka (Sreekumar and Nair, 1991). The chief communities of grasses are (1) *Sehima – Dichanthium* types (Gupta et al., 1967), wherein *Heteropogon contortus*, *Sehima nervosum*, and *Dichanthium* species are the dominant elements. (2) *Cymbopogon – Themeda* type, the dominant species are *Cymbopogon flexuosus* and *Themeda cymbaria*, (3) *Saccharum – Imperata – Phragmites* association consisting of grasses like *Phragmites karka*, *Saccharum spontaneum*,

Imperata cylindrica, Paspalum conjugatum, etc. A number of temperate grasses such as *Agrostis schmidii, A. zenkeri, Brachypodium sylvaticum, Bromus diandrus, Poa gamblei* along with temperate species of *Exacum, Gentiana, Bupleurum, Dipsacus, Ranunculus* are also found. Some other common grass species in the areas are *Agrostis peninsularis, A. setosa, Andropogon polyptychus, Arthraxon castratus, Bromus ramosus, Chrysopogon zeylanicus, Cymbopogon flexuosus, Eragrostis nigra, E. tenuifolia, Indochloa oligantha, Isachne bourneorum,* and *Tripogon bromoides*. The sedges found amidst these grasses are *Killinga cylindrica, Mariscus cyperinus, Pycreus angulatus*, etc. *Anaphalis neilgerriana, Conyza bonariensis, Crotalaria barbata, C. formosa, C. ovalifolia, Hedyotis lechenaultiana, Helichrysum* spp., *Impatiens* spp., *Memecylon lawsonii, Osbeckia* spp., *Ranunculus peninsularis,* and *Senecio* spp. are common. The ecological status of the shola grasslands has been worked on extensively by various workers (Bor, 1938; De Rosaryo, 1945; Gupta, 1960; Legris, 1960; Chandrasekharan, 1962a, 1962b, 1962c; Noble, 1967).

The manifestation of these grassy vegetation cover is primarily constituted by climatic, edaphic, and chiefly by latitudinal influence. Whereas in *Sehima–Dichanthium* composition is tropical, and *Dichanthium–Cenchrus–Lasiurus, Phragmites–Saccharum–Imperata,* and *Themeda–Arundinella* compositions are sub-tropical. Usually, the temperate alpine grass vegetation is varied from the above types. The altitudinal gradients separate *Themeda-Arundinella* grass cover which is confined to the northern hill ranges from *Phragmites–Saccharum–Imperata* and from *Dichanthium–Cenchrus–Lasiurus* covers, generally occurs in plains. In low altitudes, moisture level separates *Phragmites–Saccharum–Imperata* type from the *Dichanthium–Cenchrus–Lasiurus* type (relatively dry habitats).

Earlier investigations described the vegetation of these grassland types by the most ubiquitous plant species composition (Ranganathan, 1938; Meher-Homji, 1965). Generally, botanical exploreres have categorized the shola grasslands based on the dominance of certain grass communities (Gupta et al., 1967; Blasco, 1970). An investigation on the ecology of Nilgiri tahr described, the tahr habitat is dominated by two grass species such as *Eulalia phaeothrix* and *Andropogon polyptychos* (Rice, 1988). There is no clear and exhaustive inventory of the plant species composition of the montane grasslands of the Western Ghats (Shetty and Vivekanadan, 1971). However, the grasslands with high rates of endemism are associated with the island-like phenomena of mountain tops (Karunakaran et al., 1998). The grassland communities are considered undoubtedly a high conservation priority. These grasslands are considered to be frost-related climactic climax (Meher-Homji, 1965, 1967), or considerable amount of disturbance controlled by sub-climax (Bor, 1938; Noble 1967), or existing dual climax (Ranganathan, 1938). Some other workers considered these grassland forests are to be edaphic climax vegetation caused by differential soil moisture (Ranganathan, 1938), and there exist distinct edaphic microclimatic variations between the forests and grasslands (Jose et al., 1994).

Meher-Homji (1965, 1967) made an exhaustive study on the phytogeography of the Western Ghats shola species, which showed distinctly two principal floristic elements, that is tropical and extra-tropical (or temperate). These typical shola species are found mostly in tropical regions of the Western Ghats and Sri Lanka, or Indo-Malayan region, while woody species established along the fringes of the shola forests and in adjacent open high-altitude grasslands supported subtropical to temperate origins (Meher-Homji, 1967). The high-altitude grasslands or montane grasslans may explain the frost-induced distributional pattern. The grassland community in order to adapt for climate-specific vegetation is controlled by environmental niche and temperature (Meher-Homji, 1967). Biogeographic similarities showed that high-altitude montane grasslands of the Western Ghats are more related to the Western Himalayan vegetation than to the tropical montane grasslands of Sri Lanka (Karunakaran et al., 1998).

3.6 CONCLUSION

The Western Ghats is known for biodiversity hotspots which are threatened by varieties of human influences and natural factors. Approximately 180,000 km^2 area in the Western Ghats region, only one-third of the land cover is clothed with natural vegetation. Moreover, the existing forested areas are overexploited and highly fragmented, which is facing the increase of threat by fast degradation of its natural resources. The vegetation types and the distribution of forest profiles provided an overview of the causes and loss of biodiversity. It could be ascertained the current institutional frameworks and investments for the conservation of biodiversity in the hotspot area. The forests of the Western Ghats are unique representatives of non-equatorial tropical evergreen forests in the world. They may have evolved into one of the richest centers of endemism owing to their isolation from other moist areas. A fundamental part of conservation action is identifying endangered species, communities, or ecosystems followed by the creation of preserved areas or sanctuaries for the conservation of specific plant or animal communities to be conserved *in situ*. It has been created of as many as 578 protected areas (national parks and sanctuaries) around 4.7% of the total land area of the Ghats for serving the habitat to rich number of endemic species of the Western Ghats landscape. Therefore, these baseline data strengthened the need of goal-oriented management of this wilderness and such management systems depend largely on the ecological knowledge of the species, communities, or ecosystems with scientific principles.

KEYWORDS

- **Western Ghats**
- **vegetation**
- **forest types**
- **montane forests**
- **shoals**
- **swamps**
- **grasslands**

REFERENCES

Ahmedullah, M.; Nayar, M. P. *Endemic Plants of the Indian Region*, vol. 1; Botanical Survey of India: Calcutta, **1986**.

Beddome, R. H. *Flora Sylvatica for Southern India*, vol. 2; Madras, **1869–1874**.

Blasco, F. Aspects of the Flora and Ecology of Savannas of South Indian Hills. *J. Bombay Nat. Hist. Soc.* **1970**, *67*, 522–534.

Bourdillon, T. F. *The Forest Trees of Travancore*; The Travancore Govt. Press: Thiruvananthpuram, **1908**.

Champion, H. G. A Preliminary Survey of Forests Types of Indian and Burma. *Indian Forest Rec. Bot. (n.s) Silva.* **1936**, *2* (1), 1–286.

Champion, H. G.; Seth, S. K. *A Revised Survey of the Forest Types of India*; Government of India Press: Nasik, India, **1968**.

Chandran, M. D. S.; Rao, G. R.; Gururaja, K. V.; Ramachandra, T. V. Ecology of the Swampy Relic Forests of Kathalekan from Central Western Ghats, India. Bioremediation, Biodiversity and Bioavailability. *Global Sci. Books* **2010**, *4* (1), 54–68.

Chandrasekharan, C. Forest Types of Kerala State (1,2,3). *Indian Forester* **1962a,b,c**, *88*, 660–674, 731–747, 837–847.

Chatterjee, D. Studies of the Endemic Flora of India and Burma. *J. Asiatic Soc. Bengal Sci.* **1940**, *5* (1), 19–68.

Cooke, T. *Flora of Presidency of Bombay*, Govt. of India, 1901–1908.

Daniels, R. J. *National Biodiversity Strategy and Action Plan: Western Ghats Eco-region*; Care Earth: Chennai, 2002.

De Rosaryo, R. A. The Montane Grasslands of Ceylon. An Ecological Study with Reference to Afforestation. *Trop. Agriculture* **1945**, *101*, 208–213.

Gaikwad, S.; Gore, R.; Garad, K.; Gaikwad, S. Endemic Floering Plants of Northern Western Ghats (Sahaydri ranges) of India: A Check List. *Check List* **2014**, *10* (3), 461–472.

Gamble, J. S.; Fischer, C. E. C. *Flora of the Presidency of Madras*, 3 vol.; Adlard & Sons Ltd.: London, **1915–1936**.

Ganesh, T.; Ganeshan, R.; Soubadra Devy, M.; Davidar, P.; Bawa, K. Assessment of Plant Biodiversity at a Mid-Elevation Evergreen Forest of Kalakad–Mudanthurai Tiger Reserve, Western Ghats, India. *Curr. Sci.* **1996**, *71*, 379–392.

Ghate, K. Management of Forests in Northern Western Ghats. *J. Ecol. Soc.* **2014**, *26* & *27*, 7–71.

Govindarajulu, E.; Swamy, B. G. L. Enumeration of Plants Collected in Mundanthurai and its Neighbourhood. *J. Madras University* **1958**, *28B*, 167–177.

Gupta, R. K. Ecological Notes on the Vegetation of Kodaikanal in South India. *J. Indian Bot. Soc.* **1960**, *39*, 601–607.

Gupta, S. C.; Chinnamani, S.; Rege, N. D. Ecological Relationship between High Altitude Grasslands in the Nilgiris. *Indian For.* **1967**, *9*, 164–168.

Hooker, J. D.; Thomson, T. *Flora Indica: Being a Systematic Account of the Plants of British India, Together with Observations on the Structure and Affinities of their Natural Orders and Genera*; London, W. Pamplin, **1855**.

Irwin, S. J.; Narasimhan, D. Endemic Genera of Angiosperms in India: A review. *Rheedea* **2011**, *21* (1), 87–105.

Jose, S.; Sreepathy, A.; Kumar, B. M.; Venugopal, V. K. Structural, Floristic and Edaphic Attributes of the Grassland-Shola Forests of Eravikulam in Peninsular India. *Forest Ecol. Manage.* **1994**, *65*, 279–291.

Karunakaran, P. V.; Rawat, G. S.; Uniyal, V. K. *Ecology and Conservation of the Grasslands of Eravikulam National Park, Western Ghats;* Wildlife Institute of India: Dehra Dun, India, **1998**.

Krishnamoorthy, K. *Myristica* Swamps in Evergreen Forests of Travancore. *Tropical Moist Evergr. Symposium*: FRI, Dehradun, 1960.

Legris, P. *La vegetation de l'Inde*. Inst. fr. Pondichéry, trav. sec.sci. tech. Tome 6, **1960**.

Marchese, C. Biodiversity Hotspots: A Shortcut for a more Complicated Concept. *Glob. Ecol. Conserv.* **2015**, *3*, 297–309.

Meher-Homji, V. M. Ecological Status of the Montane Grasslands of the South Indian Hills: A Phytogeographic Reassessment. *Indian For.* **1965**, *91*, 210–215.

Meher-Homji, V. M. Phytogeography of the South Indian Hill Stations. *Bull. Torrey Bot. Club* **1967**, *94*, 230–242.

Mishra, D. K.; Singh, N. P. *Endemic and Threatened Flowering Plants of Maharashtra*; Botanical Survey of India, Kolkata, **2001**.

Nair, S. C. *The Southern Western Ghats: A Biodiversity Conservation Plan*. INTACH: New Delhi, India, **1991**.

Nair, N. C.; Daniel, J. C. The Floristic Diversity of Western Ghats and its Conservation: A Review. *Proc. Indian Acd. Sci. Suppl.* **1986**, 127–163.

Narasimhan, D.; Irwin, S. Endemic Angiosperms of Tamil Nadu. *Envis News Lett.* **2009**, *5* (4), 26–34.

Nayar, M. P. Changing Patterns of the Indian Floras. *Bull. Bot. Surv. India* **1977**, *19*, 145–155.

Nayar, M. P. '*Hot Spots' of Endemic Plants of India, Nepal and Bhutan*; Tropical Botanical Research Institute: Thiruvanathapuram, **1996**.

Nayar, T. S.; Beegam, A. R.; Sibi, M. *Flowering Plants of the Western Ghats, India, Vol. 1 & 2;* Jawaharlal Nehru Tropical Botanic Garden and Research Institute: Palode, Thiruvananthapuram, Kerala, India, **2014**.

Nayar, M. P.; Sastry, A. R. K. *Red Data Book of Indian Plants,* vol. 1; Botanical Survey of India, Calcutta, **1987**.

Nayar, M. P.; Sastry, A. R. K. *Red Data Book of Indian Plants,* vol.3; Botanical Survey of India, Calcutta, **1990**.

Noble, W. A. The Shifting Balance of Grasslands, Shola Forests, and Planted Trees in the Upper Nilgiris, Southern India. *Indian For.* **1967**, *93*, 691–693.

Oldeman, R. A. A. *L'architecture de la Forêt Guyanaise*. Mémoire ORSTOM No. 73, **1974**.

Pascal, J. P. Bioclomates of the Western Ghats at 1/250,000 (2 sheets). *Inst. fr. Pondichéry, trav. sec. sci. tech. Hors série* 17, **1982**.

Pascal, J. P. *Wet Evergreen Forests of the Western Ghats of India: Ecology, Structure, Floristic Composition and Succession*; French Institute of Pondicherry: India, **1988**.

Pascal, J.; Ramesh, B. R.; Franceschi, D. D. Wet Evergreen Forest Types of Southern Western Ghats, India. *Tropical Ecol.* **2004**, *45* (2), 281–292.

Rama Rao, M. *Flowering Plants of Travancore*; Government Press: Trivandrum, **1914**.

Ramesh, B. R. Patterns of Richness and Endemism of Arborescent Species in the Evergreen Forests of the Western Ghats, India, *Proceedings of International Conference on Tropical Ecosystems: Structure, Diversity and Human Welfare,* Ganeshaiah, K. N., Shaanker, R. U., Bawa, K. S., Eds.; Oxford-IBH: New Delhi, **2001**; pp. 539–544.

Ranganathan, C. R. Studies in the Ecology of the Shola Grassland Vegetation of the Nilgiri Plateau. *Indian For.* **1938**, *64*, 523–541.

Rao, C. K. Angiosperm Genera Endemic to the Indian Floristic Region and its Neighboring Areas. *Indian For.* **1972**, *98* (9), 560–566.

Rao, C. K. Angiosperm Genera Endemic to the Indian Floristic Region and its Neighboring Areas. *Indian For.* **1978**, *105*, 335–341.

Rice, C. G. Habitat, Population Dynamics and Conservation of the Nilgiri Tahr, *Hemitragus hylocrius. Biol. Conserv.* **1988**, *44*, 137–156.

Rodgers, W. A.; Panwar, H. S. *Planning a Protected Area Network in India*; Wildlife Institute of India: Dehra Dun, **1988**.

Rodgers, W. A.; Panwar, H. S. *Planninga Wildlife Protected Area - An Ecological History of India;* Oxford University Press: New Delhi, **1992**.

Rodgers, W. A.; Panwar, H. S.; Mathur, V. B. *Wildlife Protected Area Network in India: A Review*; Wildlife Institute of India: Dehradun, **2002**.

Santhakumaran, L. N.; Singh, A.; Thomas, V. T. Description of a Sacred Grove in Goa (India), with Notes on the Unusual Aerial Roots Produced by its Vegetation. In *Wood (Octobe-December)*; **1995**, pp. 24–28.

Sathish, B. N.; Viswanath, S.; Kushalappa, C. G.; Jagadish, M. R.; Ganeshaiah, K. N. Comparative Assessment of Floristic Structure, Diversity and Regeneration Status of Tropical Rain Forests of Western Ghats of Karnataka, India. *J. Appl. Nat. Sci.* **2013**, *5* (1), 157–164.

Savita, S. R.; Sanjaykumar, R. R. Floristic Diversity of Bhimashankar Wildlife Sanctuary, Northern Western Ghats, Maharashtra, India. *J. Threat. Taxa* **2017**, *9* (8), 10493–10527.

Senthilkumar, N.; Prakash, S.; Rajesh Kannan, C.; Arun Prasath, A.; Krishnakumar, N. Revisiting Forest Types of India (Champion and Seth, 1968): A Case Study on *Myristica* Swamp Forests in Kerala. *Int. Nat. J. Adv. Res.* **2014,** *2* (2), 492–501.

Shetty, B. V.; Vivekanadan, K. Studies on the Vascular Floras of Anaimudi and Surrounding Regions; Kottayam District, Kerala. *Bull. Bot. Surv. India* **1971,** *13,* 16–42.

Singh, J. S.; Chaturvedi, R. K. Diversity of Ecosystem Types in India: A Review. *Proc. Indian Natn. Sci. Acad.* **2017,** *83* (3), 569–594.

Sivarajan, V. V.; Mathew, P. *Flora of Nilambur (Western Ghats, Kerala)*; Dehradun, **1997.**

Sreedharan, G.; Indulkar, M. New Distributional Record of the Northmost *Myristica* Swamp from the Western Ghats of Maharashtra. *Curr. Sci.* **2018,** *115* (8), 1434–1436.

Sreekumar, P. V.; Nair, V. J. In *Flora of Kerala – Grasses*; Botanical Survey of India: Calcutta, **1991.**

Subramanyam, K.; Nayar, M. P. Vegetation and Phytogeography of the Western Ghats; In *Ecology and Biogeography in India*, vol. 23; Mani, M. S., Ed The Hague, Netherlands, **1974**; pp. 178–196.

Tetali, P.; Tetali, S.; Kulkarni, B. G.; Prasanna, P. V.; Lakshminarasimhan, P.; Lale, M.; Kumbhojkar, M. S.; Kulkarni, D. K.; Jagtap, A. P. *Endemic Plants of India (A Status Report of Maharashtra State)*; Naoroji Godrej Centre for Research: Shindewadi, Satara, India, **2000**; pp. 87.

Vishnu-Mittre; Gupta, H. P. A Living Fossil Plant Community in South Indian Hills. *Cur. Sci.* **1968,** *37,* 671–672.

CHAPTER 4

BACTERIAL DIVERSITY OF THE WESTERN GHATS AS AN AFFLUENT SOURCE FOR BIOTECHNOLOGICAL APPLICATIONS

M. ARUN PADMAKUMAR and KESAVAN MADHAVAN NAMPOOTHIRI

Microbial Processes and Technology Division (MPTD), CSIR-National Institute for Interdisciplinary Science and Technology (NIIST), Trivandrum, Kerala, India

ABSTRACT

The diversity and phylogeny of bacterial populations in soil samples collected from various locations of Western Ghats were studied by using both cultivation-dependent and cultivation-independent methods by various research groups. Based on their sequence similarity, these isolates were found to be distributed in different genera belonging to Proteobacteria, Firmicutes, Actinobacteria, and Bacteroidetes etc. Most of these microbes are of unique nature and could have been exploited for the production of value-added products, such as industrial enzymes, bioactive molecules, and platform chemicals. This chapter addresses the commercial exploitation of the bacterial diversity of Western Ghats.

4.1 INTRODUCTION

Biodiversity, the variability among living organisms on the Earth includes variations in plants, animals, fungi, and bacteria. Earth's biodiversity is so rich that many species are yet to be discovered. Biodiversity helps in promoting stability in the ecosystem by broader utilization of resources, various biotic interactions, and energy exchange. Biodiversity can be studied by understanding the genetic diversity, species diversity, and ecosystem diversity. All the species that are alive today are a result of evolution which makes them distinct from other species. These organisms have evolved to be so different that they can no longer reproduce with each other. All organisms that reproduce with each other come under a species. All these different species make up biodiversity. Scientists have been eager to study different species to understand the patterns of energy flow within the ecosystem and the potential uses of these organisms to mankind.

Biodiversity Hotspot of the Western Ghats and Sri Lanka. T. Pullaiah, PhD (Ed.)
© 2024 Apple Academic Press, Inc. Co-published with CRC Press (Taylor & Francis)

Microbial biodiversity also plays a major role in the survival of human being and economic stability. Microorganisms play an important role in biogeochemical cycles and food chains, thus establishing elegant relationships between themselves and higher organisms (Devi et al., 2014). This chapter deals with the diversity of microorganisms seen in the Western Ghats and how nicely it is exploited in Industrial biotechnology.

4.2 THE WESTERN GHATS

The Himalayas and the Western Ghats are the two major mega biodiversity hotspots in India. Among these, the Western Ghats of India is considered to be one of the oldest biodiversity and the eighth hottest biological hotspots in the world. The Western Ghats region covers about 25% of the India's biodiversity. According to UNESCO, the Western Ghats is considered to be a world heritage site. The Western Ghats is an ideal site for collecting samples to study about the microbes as it occupies the world's most biodiversity that runs along the western part of India through Gujarat and traverses the states of Maharashtra, Goa, Karnataka, and ends at Cape Comorin, a rocky headland on the Indian Ocean in Tamil Nadu. The Western Ghats is the home for many species because of the plenteous vegetation and the forest ecosystem in the Western Ghats influences the Indian monsoon weather pattern that results in a warm tropical climate thereby playing an indispensable role in regulating the climate in the country. Kudremukh, Agasthyamalai hills, the Anaimalai hills, the Nilgiri Mountains, Periyar national park, Talakaveri wild life sanctuary are some of the hotspots in the Western Ghats. Kaveri, Krishna, Thamirabarani, Godavari, and Tungabhadra are some of the main rivers that originate from the Western Ghats.

Being a gene pool, it is not only flourished with flora and fauna but also contains a high resource of unexploited microbial biodiversity (Nampoothiri et al., 2013). Efficacious knowledge of distribution of the microrganisms is required for the effective exploitation of microbial diversity (Bohannan and Huges, 2003), and there is a great interplay between the microbial community and the other species in ecological balancing (Ruckmani et al., 2011). For instance, some rhizosphere microorganisms interact with the roots of plants and make symbiotic association that would be beneficial for both species. Not only rhizosphere microorganisms but also other microorganisms indulge in these kinds of associations and get the benefit. Maybe this is one reason for the abundant growth of flora in the Western Ghats. So it is crucial to understand the distribution of microbial diversity in the area. A large number of microorganisms are characterized and isolated from the Western Ghats to understand the distribution of microbial diversity and the use of the microorganisms for the well-being of humans (Ruckmani et al., 2011; Balachandran et al., 2012; Vasudevan et al., 2015). It is becoming an authentic hotspot of microorganisms attracting more researchers who are in search of novel microorganisms. Developments in molecular techniques have revealed an astonishing diversity of many uncharacterized microorganisms (Bohannan and Huges, 2003).

4.3 BACTERIAL DIVERSITY WESTERN GHATS

Bacteria that can survive almost everywhere on the earth are found even in extreme conditions, such as hot springs, salty, and marshy areas. Many bacteria can also survive without the presence of oxygen and are commonly called as anaerobic bacteria (Dunlap, 2001). They have the ability to colonize and to support many life forms including both animals and plants. Commonly, bacteria are classified into two categories, namely, Gram-positive and Gram-negative bacteria.

4.3.1 GRAM-POSITIVE BACTERIA

4.3.1.1 ACTINOMYCETES AND BACILLI CULTURES

Actinomycetes, the Gram-positive mycelia bacteria are widely used for the production of industrially and medically relevant compounds. Actinomycete is one of the prevalent species studied in the Western Ghats. A predominant number of actinomycete species had been isolated from the different regions of Western Ghats. About 127 actinomycetes strains were isolated from the various regions of Western Ghats in Kerala and they were screened for β-glucanase enzymes (Edison and Pradeep, 2020) and 367 Actinomycetes strains were isolated from the regions of the Western Ghats of Kanyakumari, Thirunelveli, Dindigul, and Nilgiri districts of Tamil Nadu (Arasu et al., 2012). Similarly, from the soil samples collected from the above-mentioned regions of Western Ghats, about 150 actinomycetes strains were isolated. Among these, ERI-15 and ERI-17 demonstrated a favorable inhibitory activity against *E. faecalis* and *C. albicans* (Devadass et al., 2016) and 32 isolates were obtained from the soil samples of different forest locations of Bisle Ghats and Virjapet in the Western Ghats of Karnataka which included *Streptomyces* and rare actinomycetes, such as *Nocardiopsis* and *Nocardioides* (Siddharth et al., 2020). Similarly, 230 Actinomycetes were isolated from the coffee plantation of the Western Ghats including *Streptomyces*, along with rare actinomycetes, such as *Actinomadura, Spirillospora, Actinocorallia, Arthrobacter, Saccharopolyspora,* and *Nonomuraea.* This was the first report on the species *Nonomuraea antimicrobica,* identified as soil actinomycetes from the Western Ghats (Sameera et al., 2018). Newly isolated Actinomycetes and *Bacillus* species from the soil samples of Western Ghats of Kerala exhibited remarkable antagonistic activity (Ramkumar et al., 2015). Five actinomycetes, namely, TGH30, FMS-20, TGH-31, TGH 31-1, and IS-4 from 106 actinomycetes isolated from Nilgiri district of Tamil Nadu showed potential antimicrobial activity against various pathogens (Ganesan et al., 2016). A novel actinomycetes isolate ERI-26 isolated from Nilgiri forest soil of Western Ghats exhibited a great antimicrobial activity against pathogenic bacteria and fungi (Arasu et al., 2008) and SH7, a *Streptomyces* species isolated from the cardamom field of the Western Ghats revealed characteristic antibacterial activity against Gram-positive and Gram-negative bacteria (Ramani et al., 2012). Likewise, ERI-3, a *Streptomyces* species with prominent antimicrobial activity was isolated from the rocky soil of Western Ghats (Arasu et al., 2009). A *Streptomyces* strain, *Streptomyces fradiae,* GOS1 isolated from the Western Ghats of Agumbe, Karnataka (Annappa et al., 2013). Six *Streptomycetes* isolates KSRO1, KSRO2, KSRO3, KSRO4, KSRO5, and KSRO6 isolated from the Kodachathiri soil showed antifungal activity against pathogenic yeasts *C. albicans, C. neoformens, C. lipolytica, S. cerevisiae* (Shoba and Onkarappa, 2011) and another actinomycete bacteria *RAMPP-065* from the soil of Kudremukh, Karnataka showed antimicrobial activity against 8 test bacteria and 2 fungi (Manasa et al., 2012). A study was conducted with 525 soil samples collected from 14 different locations of the Western Ghats in Tamil Nadu from which about 316 new strains of *Bacillus thuringiensis* (Bt) that produces seven different types of crystalline inclusions were isolated. About 26.9% of *Bacillus thuringiensis* isolates showed cuboidal crystalline inclusions, and further characterization of 70 isolates from the 316 strains revealed six different types of crystalline inclusions with different sizes. This variation in the mass of the crystalline proteins is conveying proof of the molecular diversity of *Bacillus thuringiensis* in the Western Ghats region (Ramalakshmi and Udaya-suriyan, 2009). The characterization of 44 *Bacillus thuringiensis* species isolated from the region of Western Ghats of Karnataka showed that all 44 isolates were found to be effective insecticidal agents (Goudar et al., 2012). An endophytic bacterium PVMX4, a *Bacillus* species was isolated from the root of *Phyllanthus amarus* from the Kalakad region of Western Ghats in Tamil Nadu, showed good salt tolerance and phosphate solubilization (Joe et al., 2016). A cellulose-producing *Bacillus subtilis* was isolated from reserve forest of Western Ghats of Nilgiri district in Tamil Nadu (Karthi et al., 2019). WGB1 strains of *Bacillus cereus* isolated from the Western Ghats have the potential to degrade methylene blue, a dye used in the textile industries (Mary et al., 2020).

4.3.2   GRAM-NEGATIVE BACTERIA

Like Actinomycetes, *Methylobacterium* is a natural source of industrially important compounds and an alternative to bacterial expression systems for the production of recombinant proteins (Balachandran et al., 2012), which can also be potentially utilized as a plant growth promoting bacteria (Yim et al., 2010). ERI-135, a *Methylobacterium* species isolated from the Western Ghats showed greater antibacterial, antifungal, and cytotoxicity activity. The isolates of ERI-135 could also produce the enzyme protease (Balachandran et al., 2012). Three novel bacteria *Pontibacter niistensis* NII-0905 (Dastager et al., 2010a), *Bacillus thioparus* NII-0902 (Deepa et al., 2010), and *Paracoccus niistensis* NII-0918 (Dastager et al., 2010b) were isolated from the Western Ghats. Similarly, *Pantoea* NII-186, a bacterium was also isolated which can be used as bioinoculants for promoting growth in agriculture (Dastager et al., 2009). A *Pseudomonas* sp. AF1 isolated from the western Ghats of Tamil Nadu (Durai et al., 2012), five strains isolated from the Western Ghats of Kerala (Mohandas et al., 2012), and 52 fluorescent pseudomonads isolated from the forest soil of Western Ghats of Karnataka (Megha et al., 2007) rendered as potential bioactive compounds. Several other significant microbes included *Serratia marcescens* strain SN5gR, isolated from the scat of endangered monkey living in the forest of Nilgiri biosphere

(Gupta et al., 2013). A flavor-producing species *Chryseobacterium indologenes* strain WG4 from the water samples of Western Ghats (Kumar et al., 2011) and a green phototrophic bacterium, *Rhodobacter viridis* sp. novel Strain JA737 of the genus *Rhodobacter* from the mud of a stream in the Western Ghats were also isolated (Raj et al., 2013). From the 150 rhizosphere soil samples collected from six different locations of Maharashtra, 93 *Azospirillum* species (Murumkar et al., 2013), 65 fluorescent pseudomonads (Sonawane et al., 2014) and 3 sulfur-oxidizing bacteria, *Acidithiobacillus* (Sonawane et al., 2017) were isolated. One of these six different sites was the Western Ghats. From the four strains isolated from the *Curcuma* rhizosphere soil of Western Ghats, A nitrogen-fixing bacterium *Rhizobium pusense* PR II showed a good nitrogen-fixing capability (Chandran, 2018). From the 94 species of cyanobacteria, a photosynthetic nitrogen-fixing bacteria belonging to 38 genera, 14 families, and five orders were isolated from the Western Ghats region of Maharashtra (Nikam et al., 2013). Newly isolated *Pseudomonas* species from the soil of Western Ghats showed good antagonistic activities (Ramkumar et al., 2015). The Western Ghats is endowed with a wide variety of bacterial endophytes, the symbionts of bacteria were found inside the tissue of the host plants which promote and help the host plants in many ways. These plants behold medicinal properties. About 75 bacterial endophytes that are capable of producing bioactives and therapeutic metabolites were isolated from 24 plant species of the Western Ghats (Webster et al., 2020). A salt-tolerant endophytic and phosphate-solubilizing bacterium, ACMS25, a *Acinetobacter* species was isolated from the roots of *Phyllanthus amarus* from the Kalakad region of Western Ghats in Tamil Nadu (Joe et al., 2016).

4.3.3 METAGENOMICS ANALYSIS OF BACTERIA ISOLATED FROM THE WESTERN GHATS

Metagenomics analysis of bacteria isolated from the soil samples collected from different regions of Thattekad bird sanctuary in the Western Ghats of Kerala identified 20,544 bacteria. The identified bacteria belonged to 24 Phyla, 47 Classes, 97 Orders, 202 Families, 478 Genera, and 1275 Species (Soman et al., 2018). Similarly, metagenomics analysis of the samples collected from the sediments of the Periyar river that flows through the reserve forest of Western Ghats reveals that from the discovered 4674 operational taxonomic units, Proteobacteria (33.12%), Actinobacteria (14.58%), Acidobacteria (12.81%), and Bacteroidetes (9.89%) were dominant phyla (Rajeev et al., 2018). A novel α-amylase enzyme was isolated and characterized from a metagenomics library of Western Ghats of Kerala (Vidya et al., 2011).

4.4 BIOTECHNOLOGICAL APPLICATIONS

Soils are the greater source of microbes. All the studies underwent in Western Ghats revealed information that the Western Ghats contains potential microbial diversity. Isolation and characterizations of these microorganisms paved a way for the potential application of these organisms in biotechnology for developing some of the industrial and pharmaceutical products. Some of the application of the isolated microbes especially bacteria are depicted below.

4.4.1 MICROBIAL BIOACTIVES

Diverse number of bacterial strains with various antagonistic activities were isolated from the Western Ghats. The soils from the Western Ghats have a rich source of uncharacterized bacteria that are able to produce potential antagonistic compounds against various pathogens. From the soil samples collected from the Western Ghats of Kerala, Ramkumar et al. (2015) isolated some strains of *Pseudomonas, Bacillus, Actinomycetes,* and *Trichoderma* which showed a prominent antagonistic activity against the pathogens, such as *Phytophthora capsica* which causes foot rot disease in pepper plant and *Rhizoctonia solani* which causes collar root disease in chick pea. A *Methylobacterium* species ERI 135 isolated from the Nilgiri forest soil of Western Ghats showed antimicrobial activity and cytotoxic (A549) activity against various Gram-negative and Gram-positive bacterial species, such as *Bacillus subtilis, Klebsiella pneumoniae, Pseudomonas aeruginosa, Salmonella typhimurium, Shigella flexneri, Enterobacter aerogenes, Staphylococcus aureus,* and *Staphylococcus epidermidis* and fungi, such as *Candida albicans* and *Trichophyton rubrum* (Balachandran et al., 2012). From the 32 isolates, five *Streptomyces* isolates S1A, SS4, SS5, SS6, SCA35, showed antimicrobial activity against MRSA strains. Among these isolates, SCA35, SCA11, S1A, and SS5 showed antifungal activity against *Fusarium moniliforme*. Shoba et al. (2011) isolated six potent strains from the Kodachathiri soil of Western Ghats showed antifungal activity against pathogenic yeasts *C. albicans, C. neoformens C. lipolytica*, and *S. cerevisiae*. RAMPP-065, an actinomycetes isolate showed antibacterial and antioxidant activity (Manasa et al., 2012). Five strains NII713, NII716, NII167, NII714, and NII1054 isolated from the Western Ghats showed extended *β*-lactamase inhibitory activity. These strains belong to *Streptomyces* and *Bacillus* family. Among these two strains, NII 167 and NII 1054 showed potent antimicrobial activity against *β*-lactamase resistant strains (Mohan Das et al., 2012).

Bacillus thuringiensis having the capability to produce parasporal crystalline inclusions which is basically a toxin that acts as insecticides (Schnepf et al., 1998). About 316 *Bacillus thuringiensis* species were isolated from the 512 samples collected from different region of Western Ghats. All the 316 strains showed the parasporal crystal formation (Ramalakshmi and Udayasuriyan, 2009) and Goudar et al. (2012) isolated 44 *Bacillus thuringiensis* species from the Uttara Kannada district of Karnataka which also showed parasporal crystal formation. About 52 fluorescent pseudomonads from the same region were isolated by Megha et al. (2007), and about 44 isolates showed biocontrol activity against plant pathogens and 14 isolates showed activity against *Ralstonia solanacearum*, 10 against *Xanthomonas campestris*, 32 against *Fusarium* and 24 against *Rhizoctonia bataticola*. ERI-26 and ERI-3, are new *Streptomyces* sp. isolated from the Nilgiri forest soil showed a potent activity against streptomycin and antimicrobial activity against various bacterial strains and several fungi species. A *Pseudomonas* strain AF1, which showed great antibiotic resistance against various Gram-positive and Gram-negative bacteria. It showed about 8.5 µg/mL MIC against *Bacillus* species. Similarly, a *Streptomyces* sp. SH 7 isolated from the cardamom field of western Kerala (Ramani and Kumar, 2012) and ACTK2 isolated from Kodagu (Dezfully and Ramanayaka, 2015) are *Streptomyces* sp. and showed great antibacterial activity against various Gram-positive and Gram-negative bacteria and several fungi. *Pseudomonas fluorescens* species, namely, P-31, P-51, P-56, P-121, P-134, P-136, and P-145 showed high antagonistic activity against plant pathogens than the commercial *Pseudomonas* species (Sonawane et al., 2014).

Ganesan et al. (2016) isolated 106 actinomycetes from the Nilgiri district of Tamil Nadu. Among these, 44 strains showed antimicrobial activity and they selected five strains (TGH30, FMS-20, TGH-31, TGH 31-1, and IS-4) for the secondary screening using filtrate. Among these, FMS-20 showed inhibition against tested pathogens and the intracellular and extracellular extract of FMS 20 showed maximum zone of inhibition against *A. brasiliensis* (22 mm) and against *B. subtilis* (25 mm). Devadass et al. (2016) isolated 150 strains from soil samples collected from Kanyakumari, Thirunelveli, Dindigul, Coimbatore and the Nilgiri districts of Western Ghats. About 26 strains were potentially active against tested pathogens and about five strains (ERI-11, ERI-14, ERI-15, ERI-17, ERI-28) showed sensitivity to vancomycin and ERI-15 and ERI-17 showed a good inhibitory activity against *E. faecalis* and *C. albicans*. Similarly, Arasu et al. (2012) isolated 367 actinomycetes from the above-mentioned regions and these isolates showed great antimicrobial activity about 140 isolates showed activity against *B. subtilis*, 128 isolates against *S. aureus*, 123 against *S. epidermidis*, 105 against *P. aeruginosa*, 88 against *K. pneumoniae* and 62 against *Xanthomonas* sp. and some of them showed activity against *S. thyphi*, *V. fishcheri*, and *P. vulgaris* and 103 isolates showed activity against fungi, such as *B. cinerea*, *A. niger*, and 25 isolates against *T. simii*.

4.4.2 PLANT GROWTH PROMOTING BACTERIA

Plant growth promoting bacteria are organisms which promote the growth of the plants by forming a symbiotic or mutualistic association with plants. Some bacteria colonize in rhizosphere (rhizo bacteria), epiphyte, and some bacteria called endophyte can colonize inside the plants (Glick, 2012). Plant growth promoting microorganisms can promote the plant growth directly and indirectly by regulating plant growth hormone or by inhibiting the effects of pathogens on plant development and growth (Glick, 1995). *Azotobacter*, bacteria like cynobacteria have the capability to fix the atmospheric nitrogen. Western Ghats soils have vast diversity of microorganisms. Some of the soil bacteria help the plants by its unique properties, such as phosphate solubilization activity, Indole acetic acid (IAA) production, Siderophore production, and HCN production. *Pontoea* NII-186 and *Bacillus thioparus* sp. NII-0902 isolated from the west coast of Western Ghats showed the characteristics of PGPB. Both strains are endowed with phosphate solubilization activity, Indole acetic acid (IAA) production, siderophore production and HCN production and both can tolerate different environmental stresses also. So both strains can be used as bioinoculant to attain desired plant growth promoting activity. Nitrogen-fixing bacteria have great importance in plant growth promotion. The application of bacteria with nitrogenous activity may reduce the environment pollution of excess use of fertilizers in the agriculture field. An endophytic bacterium *Gluconacetobacter diazotrophicus* isolated from the Western Ghats by Madhaiyan et al. (2004) showed enhanced capability to nitrogen fixing and this endophyte involved in phytohormone Indole acetic acid (IAA) production and also phosphate and zinc solubilization activity. Two *Azospirillum* species *A. lipoferum* and *A. brasilense* strains isolated from the 150 rhizosphere soil samples collected from six different locations of Maharashtra. 93 *Azospirillum* species were isolated from the 150 soil samples. Among 93 *Azospirillum*, 83 isolates were identified as *A. lipoferum* and 10 isolates were identified as *A. brasilense*. Comparatively both strains showed good nitrogenase activity. However, the results showed that *A. lipoferum* showed good nitrogenase activity than the *A. brasilense*. About 94 cynobacterial species were isolated from the Western Ghats region of Maharashtra. A nitrogen-fixing bacterium, *Rhizobium pusense* PRII isolated from the *Curcuma* rhizosphere soil. PRII exhibited good nitrogenase activity, proved it as a good inoculant and showed good result in pot trial studies as well. Joe et al. (2016) isolated two endophytic phosphate-solubilizing strains an *Acinetobacter* sp. ACMS25 and *Bacillus* sp. PVMX4 from *Phyllanthus amarus*. About 52 *Pseudomonas* strains having the capability to produce IAA and GA were isolated from the Western Ghats region of Uttara Kannada, Karnataka by Megha et al. (2007) and these strains are also capable of solubilizing tricalcium phosphate (TCP).

4.4.3 INDUSTRIAL ENZYMES

Enzymes are widely used in many industries for the processing of food, beverage, paper, and textile, etc. We can use enzyme even in small quantities. Microorganisms play a vital role in the production of enzymes in biological ways like fermentation. We can easily manipulate microbes for the production of desired enzymes. Saravanan et al. (2015) isolated 32 strains from the shoal forest soils of the Western Ghats and showed a wide range enzyme production, including amylase, glutaminase, urease, agarase, pectinase, lipase, and invertase. These isolates also showed some plant growth promoting properties. Gupta et al. (2013) isolated *Serratia marcescens* strain SN5gR from the scat of lion-tailed macaque and showed a potentiality on the production of lipase enzyme. Karthi et al. (2019) isolated *Bacillus subtilis* strain from the Western Ghats which exhibited potential cellulase production through submerged and solid-state fermentation. They achieved cellulase production of 5.02 IU/mL by solid-state fermentation and 2.21 U/mL by submerged fermentation. About 127 strains are isolated from the Western Ghats and were screened for β-Glucanase enzyme production by Edison et al. (2020). About 106 strains produced exo-β-1,4-glucanase enzyme and 79 strains produced endo-β-1,3-glucanase enzyme. By screening, they concluded that the strains TBG-MR17 and TBG-AL13 are the dominant producers of the enzymes exo-β-1,4-glucanase and endo-β-1,3-glucanase. Among the 230 actinomycetes isolated from the coffee plantation of Western Ghats region of Karnataka by Sameera et al. (2018), the strain MH470335 belongs to the genus *Streptomyces,* exhibited a potential production of cellulase, pectinase, xylanase, amylase, and protease enzymes, respectively.

4.4.4 OTHER INDUSTRIAL APPLICATIONS

Mary et al. (2020) isolated a ligninolytic bacterium *Bacillus cereus* WGB1 from the Western Ghats; they have the potential to degrade methylene blue which is used in the textile industry. A fruity aroma-producing strain *Chryseobacterium indologenes* WG4 has been isolated from the Western Ghats by Kumar et al. (2011). Due to the presence of aromatic compounds such as ethyl-2-methylbutyrate and ethyl-3-methylbutyrate, this strain has the ability to produce the fruity aroma. A *Streptomyces* species *Streptomyces fradiae* GOS1 isolated from the Western Ghats of Agumbe, Karnataka by Annappa et al. (2013) showed potential production of metabolite having pharmacological activities, such as anti-inflammatory, analgesic, and antipyretic activity.

4.5 CONCLUSION

Biodiversity is earth's greatest fortune. Microorganisms, regardless of their smaller size, occur everywhere and have potential applications. They are considered to be the pioneer species to populate the planet, now spread vastly even in extreme climate conditions. They are important to maintain a balanced ecosystem by ensuring adequate energy flow via food chain, biogeochemical cycles. Numerous applications of microbes are utilized by humans. Even though microbes were not identified in earlier times, they were used for the preparation of different kinds of food products, such as cheese, vinegar, and yogurt. As they are a reservoir of enzymes, they can be used in pulp processing, degradation of organic matter, cleaning up toxic wastes, waste water treatment etc. This is why there is an urgent requirement for conservation of biodiversity especially microbial diversity which is damaged by human activities. This chapter contains details of microbial diversity at different location of the Western Ghats and their potential application which can be helpful for understanding the impact of the same, also to conserve them in a sustainable manner. Through this chapter, we can understand that the Western Ghats is blessed not only with flora and fauna but also with a vast variety of microbial biodiversity. If we can use this bacterial biodiversity, then that could be a great blessing to mankind.

ACKNOWLEDGMENT

The corresponding author acknowledges CSIR, New Delhi, for various research fundings and facilities.

KEYWORDS

- **bacterial diversity**
- **bioactives**
- **industrial enzymes**
- **metagenomics**
- **Western Ghats**

REFERENCES

Annappa, G.; Onkarappa, R.; Kuppast, I. J. Pharmacological Activities of Metabolite from *Streptomyces fradiae* Strain GOS 1. *Int. J. Chem. Sci.* **2013**, *11* (1), 583–590.

Arasu, M. V.; Duraipandiyan, V.; Agastian, P.; Ignacimuthu, S. Antimicrobial Activity of *Streptomyces* sp. ERI-26 Recovered from Western Ghats of Tamil Nadu. *J. Med. Mycol.* **2008**, *18* (3), 147–153.

Arasu, M. V.; Duraipandiyan, V.; Agastian, P.; Ignacimuthu, S. In vitro Antimicrobial Activity of *Streptomyces* sp. ERI-3 Isolated from Western Ghats Rock Soil. *J. Med. Mycol.* **2009**, *19* (1), 22–28.

Arasu, M. V.; Ignacimuthu, S.; Agastian, P. *Actinomycetes* from Western Ghats of Tamil Nadu with Its Antimicrobial Properties. *Asian Pac. J. Trop. Biomed.* **2012**, *2* (2), 830–837.

Balachandran, C.; Duraipandiyan, V.; Ignacimuthu, S. Cytotoxic (A549) and Antimicrobial Effects of *Methylobacterium* sp. Isolate (ERI-135) from Nilgiris Forest Soil, India. *Asian Pac. J. Trop. Biomed.* **2012**, *2* (9), 712–716.

Bohannan, B. J. M.; Hughes, J. New Approaches to Analyzing Microbial Biodiversity Data. *Curr. Opin. Microbiol.* **2003**, *6* (3), 282–287.

Chandran, S. J. G. Isolation and Characterization of N_2 Fixing Bacteria from Curcuma Rhizosphere Soil of Western Ghats. *Int. J. Res. Anal. Rev.* **2018**, *6* (1), 487–491.

Daniels, R. J. R. Taxonomic Uncertainties and Conservation Assessment of the Western Ghats. *Curr. Sci.* **1997**, *73* (2), 169–170.

Dastager, S. G.; Deepa, K.; Puneet, C.; Nautiyal, C. S.; Pandey, A. Isolation and Characterization of Plant Growth-Promoting Strain *pantoea* NII-186 from Western Ghat Forest Soil, India. *Lett. Appl. Microbiol.* **2009**, *49* (1), 20–25.

Dastager, S. G.; Raziuddin, Q. S.; Deepa, C. K.; Li, W. J.; Pandey, A. *Pontibacter niistensis* sp. *nov.*, Isolated from Forest Soil. *Int. J. Syst. Evol. Microbiol.* **2010a**, *61* (3), 2867–2870.

Dastager, S.; Kumaran, D.; Li, W.; Tank,S.; Pandey, A. *Paracoccus niistensis* sp. nov., Isolated from Forest Soil, India. *Antonie van Leeuwenhoek,* **2010b**, *99* (3), 501–506.

Deepa, C. K.; Dastager, G. S.; Pandey, A. Plant Growth-Promoting Activity in Newly Isolated *Bacillus thioparus* (NII-0902) from Western Ghat Forest, India. *World J. Microbiol. Biotechnol.* **2010**, *26* (12), 2277–2283.

Devadass, B. J.; Paulraj, M. G.; Ignacimuthu, S.; Theoder, A. S.; Al Dhabi, N. A. Antimicrobial Activity of Soil *Actinomycetes* Isolated from Western Ghats in Tamil Nadu, India. *J. Bacteriol. Mycol.* **2016**, *3* (2), 224–232.

Devi, N. S. K.; Padmavathy, S.; Sangeetha, K. Microbial Diversity in Leaf Litter and Sediments of Selected Streams of Palani hills, Southern Western Ghats, India. *Issues Biol. Sci. Pharm. Res.* **2014**, *2* (6), 54–61.

Dezfully, N. K.; Ramanayaka, J. G. Isolation, Identification and Evaluation of Antimicrobial Activity of *Streptomyces flavogriseus*, Strain ACTK2 from Soil Sample of Kodagu, Karnataka State (India). *Jundishapur J. Microbiol.* **2015**, *8* (2), 1–8.

Dunlap, P. V. Microbial Diversity. In *Encyclopedia of Biodiversity*; Levin, S. A., Ed.; Academic Press: Waltham, MA, **2001**; pp 280–291.

Durai, S.; Selvaraj, B.; Radhakrishnan, M.; Balagurunathan, R. Exploitation of Bacteria from Forest Ecosystem for Antimicrobial Compounds. *J. Appl. Pharm. Sci.* **2012**, *2* (3), 120–123.

Edison, L. K.; Pradeep, N. S. Functional Screening of β-Glucanase Producing Actinomycetes strains from Western Ghats Ecosystems of Kerala, India. *BioRxiv* **2020**, 036731.

Ganesan, P.; Daniel, A.; Host, R.; David, A.; Rajiv, M.; Gabriel, M.; Al-Dhabi, N. A.; Ignacimuthu,S. Antimicrobial Activity of Some Actinomycetes from Western Ghats of Tamil Nadu, India. *Alexandria J. Med.* **2016**, *53* (2), 101–110.

Glick, B. R. Plant Growth-Promoting Bacteria: Mechanisms and Applications. *Scientifica* **2012**, *2012*, 1–15.

Glick, B. R. The Enhancement of Plant Growth by Free-Living Bacteria. *Can. J. Microbiol.* **1995**, *41* (2), 109–117.

Goudar, G.; Alagawadi, A.; Krishnaraj, P. U. Characterization of *Bacillus thuringiensis* Isolates of Western Ghats and Their Insecticidal Activity Against Diamond Back Moth (*Plutella xylostella* L). *Karnataka J. Agric. Sci.* **2012**, *25* (2), 199–202.

Gupta, B.; Gupta, K.; Mukherji, S. Lipase Production by *Serratia marcescens* Strain SN5gR Isolated from the Scat of Lion-Tailed Macaque (*Macaca silenus*) in Silent Valley National Park, A Biodiversity Hotspot in India. *Ann. Microbiol.* **2013**, *63*, 649–659.

Joe, M. M.; Devaraj, S.; Benson, A.;Tongmin, sa. Isolation of Phosphate Solubilizing Endophytic Bacteria from *Phyllanthus amarus* Schum & Thonn: Evaluation of Plant Growth Promotion and Antioxidant Activity Under Salt Stress. *J. Appl. Res. Med. Aromat. Plants.* **2016**, *3* (2), 71–77.

Karthi, V.; Kumar, A. R.; Ramar, R.; Vidyapith, R. M.; Stephen, A. Isolation, Identification and Optimization of Cellulase Producing Bacteria from Forests of Western Ghat. *J. Emerg. Technol. Innov. Res.* **2019**, *6* (4), 240–247.

Kumar, P. A.; Srinivas, T. N.; Prasad, A. R.; Shivaji, S. Identification of Fruity Aroma-Producing Compounds from *Chryseobacterium* sp. Isolated from the Western Ghats, India. *Curr. Microbiol.* **2011**, *63* (2), 193–197.

Madhaiyan, M.; Saravanan, V. S.; Jovi, B. S. S.; Lee, H.; Thenmozhi, R.; Hari, K.; Sa, T. Occurrence of *Gluconacetobacter diazotrophicus* in Tropical and Subtropical Plants of Western Ghats, India. *Microbiol. Res.* **2004**, *159* (3), 233–243.

Manasa, M.; Poornima, G.; Abhipsa, V.; Rekha, C.; PrashithKekuda, T. R.; Onkarappa, R.; Mukunda, S. Antimicrobial and Antioxidant Potential of *Streptomyces* sp. RAMPP 065 Isolated from Kudremukh Soil, Karnataka, India. *Sci. Technol. Arts Res. J.* **2012**, *1* (3), 39–44.

Mary, J. E.; Krithika, T.; Kavitha, R. Biodegradation of Textile Dye by Ligninolytic Bacteria Isolated from Western Ghats. *Int. J. Res. Rev.* **2020**, *7* (4), 22–29.

Megha, Y.; Alagawadi, A.; Krishnaraj, P. U. Multiple Beneficial Functions of Fluorescent *pseudomonads* of Western Ghats of Uttara Kannada District. *Karnataka J. Agric. Sci.* **2007**, *20* (2), 305–309.

Mohandas, S. P.; Ravikumar, S.; Menachery,S. J.; Suseelan, G.; Narayanan, S. S.; Nandanwar, H.; Nampoothiri, K. M. Bioactives of Microbes Isolated from Western Ghat Belt of Kerala Show β-Lactamase Inhibition Along with Wide Spectrum Antimicrobial Activity. *Appl. Biochem. Biotechnol.* **2012**, *167* (6), 1753–1762.

Murumkar, D.; Borkar, S. G.; Chimote, V. P. Diversity for Cell Morphology, Nitrogenase Activity and DNA Profile of *Azospirillum* Species Present In Rhizosphere Soils of Six Different Physiographic Regions of Maharashtra. *Res. J. Biotechnol.* **2013**, *8* (4), 16–25.

Nampoothiri, K. M.; Ramkumar, B.; Pandey, A. Western Ghats of India : Rich Source of Microbial Biodiversity. *J. Sci. Ind. Res.* **2013**, *72*, 617–623.

Nikam, T. D.; Nehul, J. N.; Gahile, Y. R.; Auti, B. K.; Ahire, M. L.; Nitnaware, K. M.; Joshi, B. N.; Jawali, N. Cyanobacterial Diversity in Western Ghats Region of Maharashtra, India. *Bioremediat. Biodivers. Bioavailab.* **2013**, *7*, 70–80.

Raj, P. S.; Ramaprasad, E. V. V.; Vaseef, S.; Sasikala, C.; Ramana, C. V. *Rhodobacter viridis* sp. nov., A Phototrophic Bacterium Isolated from Mud of a Stream. *Int. J. Syst. Evol. Microbiol.* **2013**, *63* (1), 181–186.

Rajeev, A. C.; Sahu, N.; Deori, M.; Dev, S. A.; Yadav, V. P.; Ghosh, I. Metagenomic Exploration of Microbial Signatures on Periyar River Sediments from the Periyar Tiger Reserve in the Western Ghats. *Genome Announc.* **2018**, *6* (11), 17–18.

Ramalakshmi, A.; Udayasuriyan, V. Diversity of *Bacillus thuringiensis* Isolated from Western Ghats of Tamil Nadu State, India. *Curr. Microbiol.* **2009**, *61* (1), 13–18.

Ramani, D. G.; Kumar, T. V. Antibacterial Activity of *Streptomyces* sp. SH7 Isolated from Cardamon Fields of Western Ghats in South India. *Int. J. Pharma Bio. Sci.* **2012**, *3* (4), 957–968.

Ramkumar, B. N.; Nampoothiri, K. M.; Sheeba, U.; Jayachandran, P.; Sreeshma, N. S.; Sneha, S. M.; Meena Kumari, K. S.; Sivaprasad, P. Exploring Western Ghats Microbial Diversity for Antagonistic Microorganisms Against Fungal Phytopathogens of Pepper and Chickpea. *J. BioSci. Biotechnol.* **2015**, *4* (2), 207–218.

Ruckmani, A.; Chakrabarti, T. Analysis of Bacterial Community Composition of a Spring Water from the Western Ghats, India Using Culture Dependent and Molecular Approaches. *Curr. Microbiol.* **2011**, *62* (1), 7–15.

Sameera, B.; Prakash, H. S.; Nalini, M. S. *Actinomycetes* from the Coffee Plantation Soils of Western Ghats: Diversity and Enzymatic Potentials. *Int. J. Curr. Microbiol. App. Sci.* **2018**, *7* (8), 3599–3611.

Saravanan, D.; Radhakrishnan, M.; Balagurunathan, R. Bioprospecting of Bacteria from Less Explored Ecosystem. *J. Chem. Pharm. Res.* **2015**, *7* (3), 852–857.

Schnepf, E.; Crickmore, N.; Van Rie, J.; Lereclus, D.; Baum, J.; Feitelson, J.; Zeigler, D. R.; Dean, D. H. *Bacillus thuringiensis* and Its Pesticidal Crystal Proteins. *Microbiol. Mol. Biol. Rev.* **1998**, *62* (3), 775–806.

Shoba, K. S.; Onkarappa, R. In vitro Susceptibility of *C. albicans* and *C. neoformens* to Potential Metabolites from *Streptomycetes*. *Indian J. Microbiol.* **2011**, *51* (4), 445–449.

Siddharth, S.; Vittal, R. R.; Wink, J.; Steinert, M. Diversity and Bioactive Potential of *Actinobacteria* from Unexplored Regions of Western Ghats, India. *Microorganisms* **2020**, *8* (225), 1–13.

Soman, R.; Thomas, A.; Sarma, M.; Jagadeesh, V.; Sankarshanan, M. Bacterial Diversity in Soil from Peripheral Areas of Thattekad Bird Sanctuary. *Indian J. Sci. Res.* **2018**, *18* (2), 80–87.

Sonawane, R. B.; Deokar, C. D.; Chimote, V. P. Isolation and Characterization of Sulphur Oxidising Bacteria from the Soils of Maharashtra and Molecular Diversity. *Bioinfolet* **2017**, *14* (2), 138–141.

Sonawane, R. B.; Deokar, C. D.; Chimote, V. P. Isolation, Characterization, Functional Potential and Molecular Diversity of *Pseudomonas fluorescens* Isolated from the Soils of Maharashtra. *Res. J. Biotechnol.* **2014**, *9* (11), 95–103.

Vasudevan, G.; Siddarthan, V.; Ramatchandirane, P. S. Predominance of *Bacillus* sp. in Soil Samples of the Southern Regions of Western Ghats, India. *Ann. Microbiol.* **2015**, *65*, 431–441.

Vidya, J.; Swaroop, S.; Singh, S.; Alex, D.; Sukumaran, R.; Pandey, A. Isolation and Characterization of a Novel α-Amylase from a Metagenomic Library of Western Ghats of Kerala, India. *Biologia* **2011**, *66* (6), 939–944.

Webster, G.; Mullins, A. J.; Cunningham-oakes, E.; Renganathan, A.; Aswathanarayan, J. B.; Mahenthiralingam, E.; Vittal, R. R.; Webster, G. Culturable Diversity of Bacterial Endophytes Associated with Medicinal Plants of the Western Ghats, India. *FEMS Microbiol. Ecol.* **2020**, *96* (9), 1–17.

Yim, W.; Poonguzhali, S.; Boruah, H. P. D.; Palaniappan, P.; Tong-Min, S. In *Colonization Pattern of gfp Tagged Methylobacterium suomiens on Rice and Tomato Plant Root and Leaf Surfaces*, 19th World Congress of Soil Science, Soil Solutions for a Changing World, Brisbane, Australia, Aug 1–6, **2010**; pp 31–34.

CHAPTER 5

DIVERSITY OF CYANOBACTERIA IN THE WESTERN GHATS

KODANDOOR SHARATHCHANDRA and KANDIKERE R. SRIDHAR

Department of Biosciences, Mangalore University, Mangalagangotri, Karnataka, India

ABSTRACT

The Western Ghats being one of the hotspots of biodiversity, support numerous flora, fauna, and microbes. Cyanobacteria are the important components of terrestrial and aquatic ecosystems; they play a major role in nutrient cycling. The habitats studied in the Western Ghats include soils, rocks, reservoirs, rivers, ponds, lakes, thermal springs, paddy fields, and polluted habitats spread out in five political states (Maharashtra, Goa, Karnataka, Kerala, and Tamil Nadu). Studies carried out in the Western Ghats revealed the highest species richness of cyanobacteria in the paddy fields followed by soils and polluted habitats. Polluted habitats possess higher cyanobacterial richness than other habitats probably due to nutrient enrichment or the capability of cyanobacteria to degrade pollutants. Cyanobacterial richness (160 spp.) in the Western Ghats although accounts for only 6% of the global estimate, they are up to 57% of the species reported from the Asian region (240 spp.). Because of the importance of cyanobacteria in nutrition, metabolites, bioactive compounds, and bioremediation, precise and explicit studies are warranted on the diversity and richness of cyanobacteria in the Western Ghats.

5.1 INTRODUCTION

Cyanobacteria are versatile prokaryotes with wide geographic distribution in freshwater, marine water, and terrestrial ecosystems. They are also called blue-green algae due to their characteristic pigments. Being photosynthetic, cyanobacteria meet their organic carbon requirement using carbondioxide, light, and water. Many species are capable of fixing atmospheric nitrogen; hence, they are of great importance to maintain the nitrogen balance in various ecosystems. Usually, the nutrient-deficient natural waters harbor lower number of cyanobacteria, while the nutrient-enriched water bodies possess a high population. Their morphology and diversity range from unicellular to multicellular, coccoid to branched

Biodiversity Hotspot of the Western Ghats and Sri Lanka. T. Pullaiah, PhD (Ed.)

filaments, autotrophic to heterotrophic, psychrophilic to thermophilic, acidophilic to alka-liphilic, planktonic to barophilic, and freshwater to marine (including salt pans and estuaries) (Dvořák et al., 2017). They are widely distributed from the Polar Regions through temperate to tropical waters and they are more abundant in nutrient-rich aquatic habitats (Deep et al., 2013; Sorichetti et al., 2013). Cyanobacteria grow as epiphytes on tree bark, as epiliths on rocks/stones, and are also capable of establishing on the surfaces of buildings (concrete roofs and other artificial surfaces leading to decoloration and deterioration) (Crispim et al., 2006; Barberousse et al., 2007). Such growths occur commonly in humid places on uneven surfaces (e.g., holes, crevices, and also on damp building walls). Some species of cyanobacteria are also reported from stromatolitic mats, calcareous rocks, and also from the lime encrusted caves (Abdelhad, 1989; Abdelhad and Bazzichelli, 1991; Subramanian and Uma, 2001).

The success of cyanobacteria in a wide range of habitats has been attributed to their unique physiological characteristics and high adaptability to a varied range of environmental conditions. They are the major source for a variety of compounds, such as polysaccharides, fatty acids, proteins, vitamins, sterols, enzymes, pharmaceuticals, pigments, and other compounds. In addition, various biochemical compounds of cyanobacteria and their significance in biotechnology are highly appreciable (Abed et al., 2009). Many habitats consist of toxin-producing cyanobacteria, depending on the nutritional qualities of waters or habitats. Besides biotic factors (e.g., eutrophication), cyanobacterial diversity, distribution, and biochemical composition are dependent on the physicochemical as well as abiotic factors. Cyanobacteria are known for many broad-spectrum antimicrobial as well as antiviral activities. Nontoxic *Arthrospira* and *Spirulina* are the richest sources of proteins and amino acids; hence, the WHO recommended them as nutritional sources for humans as well as livestock. Cyanobacteria are suitable candidates to assess the state of pollution in aquatic habitats. They have great potential for bioremediation to degrade petroleum products, treat industrial effluents, treat wastewaters, and are capable of adsorbing heavy metals. This mini-review addresses the occurrence, distribution, diversity, ecology, and significance of cyanobacteria in the Western Ghats.

5.2   HABITATS AND DISTRIBUTION

Cyanobacteria being primary producers, they have strategically positioned in the first trophic level in the food chain in aquatic, semi-aquatic, and terrestrial habitats. The various habitats in five states of the Western Ghats studied for the occurrence and distribution of cyanobacteria include terrestrial (soils and surfaces of structures), aquatic (ponds, lakes, paddy fields, rivers, reservoirs, thermal springs, estuaries, and mangroves), and polluted habitats (domestic and industrial) (Table 5.1). Hosmani and Bharathi (1981) erected a new species from the freshwaters of Haliyal region of the Western Ghats-based morphological characteristics and named *Scenecocystis karnatakensis*. Another new species of cyanobacteria *Iningainema sahyadrensis* has been described from the Sahyadri mountain range based on morphological, cultural, and molecular analysis (Maltsev et al., 2021). Hossain et al. (2021) obtained 74 monocultures of cyanobacteria from the freshwater bodies of Sri

Lanka and Jaffna. Three species were considered rare and novel isolates include *Alkalinema pantanalense*, *Geitlerinema* sp., and *Westiellopsis prolifica*, while another novel species *Cephalothrix komarekiana* was identified for the first time in Sri Lanka. Another study was carried out for cyanobacteria occurring in two water samples in Sri Lanka, 13 salt marshes, 18 mangroves, 4 thermal springs, and 4 lagoons (Bowange et al., 2020). A total of 150 monocultures were obtained consisting of 15 genera with the dominance of *Leptolyngbya* in marine habitats (salt marshes and mangroves), whereas Chroococcales and *Pseudanabaena* were frequent in four thermal springs.

5.2.1 TERRESTRIAL HABITATS

Soil samples were studied for cyanobacteria in the Western Ghats region of Maharashtra, Kerala, and Tamil Nadu (Nikam et al., 2011; Suresh et al., 2012; Philip and Krishnan, 2020). The surface of rocks provides potential habitat for the growth and activity of cyanobacteria. Rock surfaces in Goa, Karnataka, Kerala, and Tamil Nadu attracted several researchers (Pereira and Almeida, 2012; Suresh et al., 2012; Singh and Singh, 2019; Philip and Krishnan, 2020). The surface of bark, pots, pipes, rocks, and concrete structures was also subjected to the study of cyanobacteria (Pereira and Almeida, 2012; Singh and Singh, 2019).

5.2.2 AQUATIC HABITATS

Ponds (Tamil Nadu) and artificial tanks (Karnataka) are also attractive to study about the cyanobacterial population (Balasingh et al., 2015; Sharathchandra and Rajashekhar, 2016). In addition to water, scrapings of plant detritus and tank walls were assessed by Sharathchandra and Rajashekhar (2016). Reservoirs in Kaiga (Karnataka), Peringalkuthu, and Pothundi (Kerala) were studied by many researchers (Sharathchandra and Rajashekhar, 2013b; Nasser and Sureshkumar, 2013; Philip et al., 2016). The seasonal occurrence of anoxygenic photosynthesis by cyanobacteria in Tillari and Selaulim reservoirs of Goa was investigated by Kurian et al. (2012). Pingale and Deshmukh (2005) identified 87 cyanobacterial species belonging to 43 genera from the Wilson Dam (Maharashtra). Many rivers in Karnataka have attracted the interest of researchers to study the cyanobacteria: Nethravathi, Gurupura, Kumaradhara, Shambhavi, and Sita (Rajeshwari and Rajashekhar, 2012; Joishi, 2014; Severes et al., 2018). Paddy fields in Maharashtra and Kerala have been studied (Ghadage et al. 2019; Vijayan and Ray, 2015). Estuary (Karnataka) and mangroves (Kerala) were also studied by Severes et al. (2018) and Ram and Shamina (2017), respectively.

5.2.3 THERMAL SPRINGS

Panekal thermal sulfur spring and Kaiga thermal nuclear power station discharge canal of Karnataka were evaluated for cyanobacteria (Rajeshwari and Rajashekhar, 2012; Sharathchandra and Rajashekhar, 2013b; Sharathchandra et al., 2021). A few reports on cyanobacterial composition and physicochemical properties of thermal springs of India are available (e.g., Rajeshwari and Rajashekhar, 2012; Sharathchandra et al., 2021). This clearly depicts the temperature specificity of cyanobacteria and probably temperature-induced metabolites will be of biotechnological significance. In addition, Panekal thermal spring is also known for its sulfate (range, 55.9−79 mg/l) and sulfide (3.1−4.5 mg/l) contents (Rajeshwari and Rajashekhar, 2012). Another study has compared the cyanobacterial population of Panekal sulfur spring with that of the thermal water discharge canal of a nuclear power station in Kaiga (Sharathchandra et al., 2021).

5.2.4 POLLUTED HABITATS

Many polluted habitats in the Mangalore region (fertilizer, dairy, petrochemical refinery, plywood, and sewage drains) were studied by Rajeshwari and Rajashekhar (2012). Among the five polluted sites, the sewage drain possess the highest number of cyanobacteria (43 spp.) followed by dairy effluents (32 spp.), fertilizer factory effluent (15 spp.), and they were least (10−11 spp.) in the plywood industry effluent. The dominance of cyanobacteria in the wastewaters from the rubber industry near Mysore, Karnataka was recorded by Ramaswamy et al. (1982). *Microcoleus* sp. was recorded for the first time in the wastewaters of this industry. Mahadev and Hosmani (2005) studied the community structure of cyanobacteria in two polluted lakes of Mysore city. Raghavendra and Hosmani (2002) observed the dominance of four species of cyanobacteria, among them *Microcystis aeruginosa* occurred as a bloom followed by *Phormidium fragile* in Mandakally lake, which receives sewage from Mysore city.

5.3 RICHNESS AND DIVERSITY

Richness and diversity (qualitative and quantitative) studies on cyanobacteria aid to assess the quality of water (Shekhar et al., 2008). Benthic forms of cyanobacteria are also useful as indicators of water quality, their diversity responds rapidly to changes in the nutrients in the aquatic environment (Anand et al., 1995; Selvakumar and Sundararaman, 2007). The development of cyanobacterial blooms favored under conditions of high rates of nutrient loading, low rates of vertical mixing, and warm water temperatures. According to Kamble (2017), the heterogeneous environmental conditions that prevail in the estuaries are responsible for the high biodiversity of cyanobacteria. Their abundance and community composition vary seasonally as a result of changes in water temperature, solar irradiance, meteorological conditions, hydrology, and nutrient supply. These conditions also apply to the polluted aquatic bodies receiving high nutrient loads. Such blooms are characterized either by single-species or multiple-species dominance.

5.3.1 RICHNESS

According to Dvořák et al. (2017), cyanobacteria are morphologically distinct and represent up to 23.4% of the prokaryotes so far known. About 2698 species of cyanobacteria are described with the addition of 15 new species per annum (Nabout et al., 2013). Based on the predictive models, it has been speculated that the range of cyanobacteria lies between 4484 and 8000 species (Guiry, 2012; Guiry and Guiry, 2015). About 160 cyanobacterial species in the Western Ghats account only for 6% of the global estimate, while 57% of the species are reported from Asia (240 spp.) (Hauer et al., 2015). It is likely that conditions responsible for cyanobacterial blooms decrease the richness and diversity owing to the dominance of one or a few species.

Table 5.1 presents the details of the richness of cyanobacteria in different habitats in five states of the Western Ghats region. Maharashtra showed the highest number as well as the average number of cyanobacteria followed by Karnataka, Kerala, Tamil Nadu, and Goa (Figure 5.1).

5.3.2 DIVERSITY

Cyanobacterial diversity in the soils of Pune and Satara of the Western Ghats region of Maharashtra was carried out by Nikam et al. (2011). In their study, among the 627 soil samples screened from different locations, 94 cyanobacterial species were identified (in 38 genera, 14 families, and 5 orders). However, the soil samples of forest locations of Kerala consist of lower richness (11 spp.) (Philip and Krishnan, 2020). The surfaces of different materials (bark, pots, pipes, rocks, and concrete walls) were harbored by 59 species of cyanobacteria (Singh and Singh, 2019), while the surface of rocks and macroalgae showed only 12 species (Pereira and Almeida, 2012). Ponds in Tamil Nadu and tanks of Karnataka yielded 32 and 43 species, respectively (Balasingh et al., 2015; Sharathchandra and Rajashekhar, 2016).

Freshwater cyanobacteria from the Kaiga reservoir region of the Western Ghats of Karnataka were studied by Sharathchandra and Rajashekhar (2013b) who reported the occurrence of 59 species. Peringalkuthu Reservoir in Kerala yielded only nine species of cyanobacteria (Nasser and Sureshkumar, 2013), while Pothundi Dam harbored up to 15 species (Philip et al., 2016). Four rivers in Karnataka (Nethravathi, Kumaradhara, Shambhavi, and Sita) harbored up to 41 species. Up to 87 cyanobacterial species were detected in the Wilson Dam of Maharashtra (Pingale and Deshmukh, 2005).

Panekal thermal sulfur spring yielded up to 29 species of cyanobacteria (Rajeshwari and Rajashekhar, 2012), while the same spot showed up to 35 species as reported by Sharathchandra et al. (2021). Kaiga thermal nuclear power discharge waters possess 29 species (Sharathchandra et al., 2021). Up to 43 species of cyanobacteria (in 23 genera and 14 families) were recorded with 50% of species common in these thermal bodies. Total species, species richness, and diversity were higher in thermal spring compared with the thermal canal. Twenty-one species were common to water bodies, while 14 and 8 species were confined to the thermal spring and thermal canal, respectively. It is of immense interest to study those species found exclusively in thermal springs or thermal canals for their biotechnological applications.

Many researchers studied the distribution of cyanobacteria in the paddy fields of Western Ghats with the highest of 137 species in Maharashtra followed by 64 species

TABLE 5.1 Selected Studies Carried Out on Cyanobacteria in the Five States of the Western Ghats Region.

State	Location	Habitat	Number of species	References
Maharashtra	Ahmednagar, Pune and Satara	Soils	94	Nikam et al. (2011)
	Karad and Patan	Paddy fields	137	Ghadage et al. (2019)
Goa	Anjuna, Baga, Betul, Bogmalo, Cabo-de-Rama, Chapora, Dona Paula, Holant, Mormugoa, Palolem, Polem, Reis Magos, San Jacinto, Siridao, Talpona, Terekhol and Vagator	Macroalgae and rocks	12	Pereira and Almeida (2012)
Karnataka	Rivers: Nethravathi and Gurupura	Chemical fertilizer factory, Mangalore dairy, petrochemical refinery, plywood industry, and sewage drain	84	Rajeshwari and Rakashekhar (2012)
	Uppinangadi	Panekal thermal sulfur spring	29	Rajeshwari and Rakashekhar (2012)
	Kaiga	Hartuga, Kadra reservoir, thermal water discharge channel, labor colony, and Virje	59	Sharathchandra and Rajashekhar (2013a)
	Dharmasthala, Karkala, Sanoor and Subrahmanya	Rivers (Nethravathi, Kumaradhara, Shambhavi, and Sita)	41	Joishi (2014)
	Belur, Gajanoor, Kowarkolli and Lakkavalli	Artificial tanks: water, surface of moist detritus, and surface of tank wall	43	Sharathchandra and Rajashekhar (2016)
	Gurupura and Mangalore	Rivers, estuary, and lake	5	Severes et al. (2018)
	Sirsi	Surface of bark, pots, pipes, rocks, and walls	59	Singh and Singh (2019)
Kerala	Thrissur	Peringalkuthu Reservoir	9	Nasser and Sureshkumar (2013)
	Kuttanadu	Paddy fields	64	Vijayan and Ray (2015)
	Nelliyampathy	Pothundi Dam	15	Philip et al. (2016)
	Calicut, Kottayam, Malappuram and Ernakulam	Mangroves	31	Ram and Shamina (2017)
	Forests (Calicut, Idukki, Palghat and Thrissur)	Bark, rock, soil, and streams	11	Philip and Krishnan (2020)
Tamil Nadu	Kanyakumari	Coastal wetlands	17	Sivakumar et al. (2012)
	Agasthiyar falls, Gudalur, Kodaikanal and Kolli hills	Rocks, gravel, moist soil, and water	41	Suresh et al. (2012)
	Kanyakumari	Ponds	32	Balasingh et al. (2015)

in Kerala (Vijayan and Ray, 2015; Ghadage et al., 2019). Seven mangroves of Kerala (Vallikunnu, Kadalundi, Kallayi, Mangalavanam, Kumbalam, Kumarakom, and Mekkara) possess 31 species of cyanobacteria (Ram and Shamina, 2017). Polluted habitats of the Mangalore region (fertilizer, dairy, petrochemical refinery, plywood, and sewage drains) revealed the occurrence of 84 species (Rajeshwari and Rajashekhar, 2012), which is the third highest number after paddy fields and soils (Nikam et al., 2011; Ghadage et al., 2019).

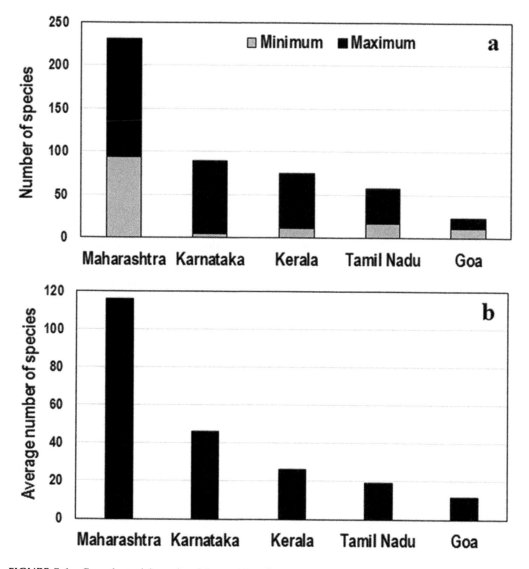

FIGURE 5.1 Cyanobacterial species richness (a) and average number of species (b) in the Western Ghats region.

5.3.3 BLOOM

Cyanobacterial blooms also occur in oligotrophic (nutrient-poor and low-productivity) systems, although not as frequently as in eutrophic systems. Blooms in oligotrophic systems are often associated with benthic cyanobacteria or favorable nutrient and light conditions at the depth in a stably stratified water column (Nayak et al., 2001; Kamble, 2017). Cyanobacterial blooms are frequent in places where there are intensive agricultural activities, resulting in nutrient loading and eutrophication (Smith, 2003). The diversity of freshwater cyanobacteria from some unexplored water bodies of a rapidly developing industrial region in the Western Ghats was studied by Severes et al. (2018). Such conditions are also applicable to the polluted aquatic bodies, which receive high nutrient loads. Four species of cyanobacteria were dominated in the Mandakally Lake (which receives sewage from Mysore city) with the bloom of *Microcystis aeruginosa* and *Phormidium fragile* (Raghavendra and Hosmani, 2002). Hosmani and Anita (1998) correlated the free carbon dioxide level in the Kukkarahalli Lake in Mysore when cyanobacterial bloom occurred with high amount of *Microcystis aeruginosa*.

5.3.4 SEASONAL OCCURRENCE

Studies on annual variation in species diversity of cyanobacteria in four rivers of the Western Ghats (Nethravathi, Kumaradhara, Shambhavi, and Sita) were carried out by Joishi (2014). A total of 41 species (in 16 genera) were observed in these rivers. *Oscillatoria limosa* was the most dominant species, which was prevalent in all these rivers. Species diversity and annual and seasonal variation of the cyanobacterial assemblage were investigated in different habitats, especially in temple ponds (Joishi, 2009). The distribution pattern of cyanobacteria in Kaiga region of the Western Ghats of Karnataka was carried out by Sharathchandra and Rajashekhar (2013b). Fifty-nine species (27 genera) were recorded with the dominance of the members of Chroococcaceae followed by Phormidaceae. Seasonal occurrence of cyanobacteria in four artificial tanks of the Western Ghats of Karnataka was reported by Sharathchandra and Rajashekhar (2016). A total of 43 species (in 19 genera) were recorded with the dominance of non-heterocystous filamentous forms. The species richness and diversity were higher during the monsoon season than in other seasons in these tanks.

5.4 INFLUENCE OF FACTORS

Cyanobacteria in the Western Ghats are influenced by various biotic and abiotic factors in different aquatic, semi-aquatic, and terrestrial conditions. The main abiotic factors that influence the cyanobacterial occurrence, richness, and their diversity include pH, temperature, conductivity, chemical/biochemical oxygen demand, dissolved oxygen, salinity, and alkalinity (Rajeshwari and Rajashekhar, 2012; Sharathchandra and Rajashekhar, 2016).

The effect of pH on freshwater cyanobacteria isolated from different habitats of Southern Karnataka was studied by Sharathchandra and Rajashekhar (2016). This study revealed that cyanobacteria under acidic to neutral pH (5.5–7) can synthesize higher carotenoids rather than under alkaline pH. This study also indicated that cyanobacteria can produce higher phycobilin pigments, such as phycocyanin, allo-phycocyanin, and phycoerythrin at alkaline pH (8–8.5). In addition, cyanobacteria produce higher phycoerythrin pigment at alkaline pH (between 7.5 and 8) except for *Microcystis aeruginosa*, which showed higher phycoerythrin content at pH 9.

Hosmani (2010) observed that the diversity of cyanobacterial species was high at a temperature range between 30 and 34.0°C prior to bloom formation. The effects of temperature and sulfide on cyanobacteria isolated from a sulfur spring in the Western Ghat forests of Karnataka were studied by Sharathchandra and Rajashekhar (2017). Furthermore, the effects of salinity on six species of freshwater cyanobacteria isolated from different habitats of Southern Karnataka were studied by Sharathchandra and Rajashekhar (2016).

Hosmani (2010) observed that the species diversity indices decreased when the dissolved organic matter was high and it favors the bloom formation. Positive correlation between the dissolved oxygen and population of cyanobacteria has been established by Hosmani and Anita (1998). These authors also found that the free carbon dioxide content of the Kukkarahalli Lake in Mysore was higher during the bloom of *Microcystis aeruginosa* compared with other lakes.

5.5   APPLICATIONS OF CYANOBACTERIA

Cyanobacteria are well known for their nutritional constituents (e.g., proteins and polyunsaturated fatty acids) (e.g., *Arthrospira* and *Spirulina*). Many species possess substantial bioactive potential (antioxidant capacity, antimicrobial activity, and pharmaceutical values) (e.g., *Calothrix, Lyngbya,* and *Scytonema*) (Abed et al., 2009). Besides nutrition and bioactive compounds, cyanobacteria are used in biotechnological, biopesticidal, bioenergy, biofertilizer, and bioremediation purposes. Many species possess the capacity to adsorb heavy metals, degrade pesticides, and are helpful in the treatment of wastewaters.

5.5.1   NUTRITION

Species of *Anabaena, Spirulina,* and *Nostoc* are used in the human diet in many countries (e.g., Chile, Mexico, Peru, and Philippines) (Abed et al., 2009). Nontoxic *Arthrospira platensis* and *Spirulina platensis* are grown in large scale either in outdoor ponds or in bioreactors for marketing (as powder, flakes, tablets, and capsules) as nutritional boosters owing to their protein content (60–70%) (Ishimi et al., 2006; Moreira et al., 2013). Total lipid and fatty acid composition in freshwater cyanobacteria from the Western Ghats was studied by Sharathchandra and Rajashekhar (2011). Among the fatty acids, over 60% consists of polyunsaturated fatty acids. Palmitic acid ($C_{16:0}$) was common in many isolates followed by linoleic acid ($C_{18:2}$). In some species, long-chain fatty acids ($C_{20:1}$ and $C_{24:0}$) were also found in low concentration. According to Murray and Mitsui (1982), 22 species belonging to the genera *Anabaena, Calothrix,* and *Nostoc* showed increased growth as well as body weight of *Tilapia*, which is equivalent to those fishes fed with excellent food conversion ratios. *Spirulina* can be called a protein supplement; it can replace up to 40% of the protein content in the diets of *Tilapia* (Olvera-Novoa et al., 1990). Marine nitrogen-fixing cyanobacteria have also been tested as fish feed in aquacultures and *Tilapia* showed high growth rates when fed with marine cyanobacteria in indoor and outdoor cultures (Mitsui et al., 1983). Among them, *Phormidium valderianum* has been used in India to serve as a complete aquaculture feed source based on its nutritional value and nontoxic nature. A wide variety and strains of the above cyanobacteria are well known from different habitats of the Western Ghats and the west coast of India provides ample scope to adapt in human and livestock nutrition.

5.5.2 BIOACTIVE COMPONENTS

Besides proteins and fatty acids, cyanobacteria are composed of beta-carotene, iron, vitamins, phenolic compounds, and pigments (Belay et al., 1993; Colla et al., 2007; Sajilata et al., 2008; Abed et al., 2009). Cyanobacteria are also known for various pharmacological attributes in addition to nutritional novelties (Nuhu, 2013). Biochemical evaluation of freshwater cyanobacteria isolated from different freshwater habitats of the Western Ghats showed value-added components useful in pharmaceutical industries (Rajeshwari and Rajashekhar, 2011; Sharathchandra and Rajashekhar, 2017). Similarly, the estuarine *Synechocystis aquatilis* possesses appreciable quantities of amino acids (Shashikumar and Madhyastha, 2002). Antioxidant activity in four species of cyanobacteria isolated from the Panekal thermal sulfur spring was assessed by Sharathchandra and Rajashekhar (2013a). This study showed significant antioxidant potential in *Calothrix fusca, Lyngbya limnetica,* and *Scytonema bohnerii*.

5.5.3 BIOREMEDIATION

In addition to the cyanobacterial potential to serve as indicators of pollution of aquatic bodies, many species are useful in the abatement of pollution (Abed et al., 2009). Biodegradation of pesticides by cyanobacteria has been reported by a few investigators (e.g., Subramanian et al., 1994). Uma and Subramanian (1990) treated Ossein factory effluent with selected locally isolated species of *Aphanocapsa* and *Oscillatoria*, which resulted in reduced calcium and chloride contents in the effluents. Manoharan and Subramanian (1993) and Kalavathi et al. (2001) have shown the ability of these cyanobacteria in the treatment of various effluents from paper mill, domestic effluents, and distillery effluents. *Anabaena cylindrica, Calothrix parietina, Gloeocapsa* sp., *Nostoc* sp., *Oscillatoria* sp., *Phormidium* sp., *Scytonema schmidlei,* and *Tolypothrix tenuis* are used successfully to adsorb several heavy metals (e.g., mercury, lead, cobalt, chromium, and cadmium) (Nagase et al., 2005; Ruangsomboon et al., 2007; Shukla et al., 2012). Interestingly, Rajeshwari and Rajashekhar (2012) reported the third highest number of cyanobacteria (84 species in 30 genera) in polluted habitats (fertilizer factory, dairy, petrochemical refinery, plywood industry, and sewage drain) after paddy fields and soils (Nikam et al., 2011; Ghadage et al., 2019). *Microcytis aeruginosa* was common to all polluted locations (93.8%) followed by *Phormedium foreaui* (44%), and *Phormidium limosum* (42%) seems to be resistant to pollution or has the potential to degrade pollutants. Since a substantial number of cyanobacteria occur in the polluted sites, it is interesting to study the capability of consortia of cyanobacteria as dominant core-group species (frequency of occurrence, > 10%).

5.6 CONCLUSION

Cyanobacteria, being Archaean origin (2.7 bya), are genetically diverse with wide geographic distribution in aquatic and terrestrial ecosystems. Their carbon- and nitrogen-fixing abilities are responsible for worldwide distribution. They possess value-added components, such as proteins, carbohydrates, fatty acids, vitamins, pigments, toxins, and secondary metabolites. *Arthrospira* and *Spirulina* are nontoxic, rich in proteins and essential amino acids, and considered suitable for human and livestock nutrition. They are also known for having broad-spectrum antimicrobial properties and authentic indicators to assess the extent of aquatic pollution. Industrial and sewage drains showed the third highest richness of cyanobacteria in the Western Ghat region and provide hopes for their potential to degrade recalcitrant compounds or rehabilitation/restoration potential. Cyanobacteria are used to degrade petroleum products, treat industrial effluents, process wastewaters and biosorption of heavy metals. Studies on the cyanobacteria in the Western Ghats mainly concentrated on the aquatic bodies, while the terrestrial ecosystems are neglected. Buildings, monuments, rocks, tree surfaces, surfaces of plant debris, and symbiotic associations (plants, lichens, and algae) are some of the neglected habitats.

KEYWORDS

- **biodiversity**
- **species richness**
- **habitats**
- **bloom**
- **bioactive compounds**
- **bioremediation**

REFERENCES

Abdelhad, N. On Four *Myxosarcina* Like Species (Cyanophyta) Living in the Inferniglio Cave (Italy). *Arch. Hydrobiol.* **1989**, *82*, 3–13.

Abdelhad, N.; Bazzichelli, G. The Genus *Gloeocapsa* Kutz. (Cyanophyta) on Calcareous Rock Surfaces in the Upper Valley of the River Aniene (Latium, Italy). *Cryptogam. Bot.* **1991**, *2* (2/3), 155–160.

Abed, R. M. M.; Doberetsov, S.; Sudesh, K. Applications of Cyanobacteria in Biotechnology. *J. Appl. Microbiol.* **2009**, *106*, 1–12.

Anand, N.; Hooper, R. S.; Kumar, S. Distribution of Blue-Green Algae in Rice Fields of Kerala State, India. *Phykos* **1995**, *35*, 55–64.

Balasingh, G. S. R.; Jeeva, S.; Vincy, W. Species Composition of Phytoplankton in the Two Migratory Birds Visited Ponds of Kanyakumari District. *Int. J. Bot.* **2015**, *5*, 1–8.

Barberousse, H.; Brayner, R.; Do Rego, A. M. B.; Castaing, J.; Beurdeley-Saudou, P.; Colombet, J. Adhesion of Facade Coating Colonisers, as Mediated by Physico-Chemical Properties. *Biofouling* **2007**, *23*, 15–24.

Belay, A.; Ota, Y.; Miyakawa, K.; Shimamatsu, H. Current Knowledge on Potential Health Benefits of *Spirulina*. *J. Appl. Phycol.* **1993**, *5*, 235–241.

Bowange, R. W. T. M. R. T. K.; Jayasinghe, M. M. P. M.; Yakandawala, D. M. D.; Kumara, K. L. W.; Abeynayake, S.; Ratnayake, R. R. Morphological Characterization of Cyanobacteria in Extreme Ecosystems of Sri Lanka, *Proceedings of the Postgraduate Institute of Science Research Congress*, Sri Lanka, Nov 26–28, 2020; p 155. http://www.pgis.pdn.ac.lk/rescon2020/doc/abstracts/LS/185.pdf.

Colla, L. M.; Reinehr, C. O.; Reichert, C.; Costa, J. A. W. Production of Biomass and Nutraceutical Compounds by *Spirulina platensis* Under Different Temperature and Nitrogen Regime. *Biores. Technol.* **2007**, *98*, 1489–1493.

Crispim, C. A.; Gaylarde, P. M.; Gaylarde, C. C.; Neilan, B. A. Deteriogenic Cyanobacteria on Historic Buildings in Brazil Detected by Culture and Molecular Techniques. *Int. Biodeterior. Biodegrad.* **2006**, *57*, 239–243.

Deep, P. R.; Bhattacharyya, S.; Nayak, B. Cyanobacteria in Wetlands of the Industrialized Sambalpur District of India. *Aqua. Biosyst.* **2013**, *1*, 9–14.

Dvořák, P.; Casamatta, D. A.; Haler, P.; Jahodářová, E.; Norwich, A. R.; Poulíčková, A. Diversity of Cyanobacteria. In *Modern Topics in the Phototrophic Prokaryotes*, Hallenbeck, P. C., Ed.; Springer International Publishing: Switzerland, **2017**; pp 346.

Ghadage, S. J.; Karande, V. C. The Distribution of Blue-Green Algae (Cyanobacteria) from the Paddy Fields of Patan and Karad Tehsils of Satara District, Maharashtra, India. *J. Threat. Taxa.* **2019**, *11*, 14862–14869.

Guiry, M. D. How Many Species of Algae Are There? *J. Phycol.* **2012**, *48*, 1057–1063.

Guiry, M. D.; Guiry, G. M. AlgaeBase. World-Wide Electronic Publication, National University of Ireland, Galway, **2015**, http://www.algaebase.org (accessed Sept 02, 2015).

Hauer, T.; Mühlsteinová, R.; Bohunická, M.; Kaštovská, J.; Mareš, J. Diversity of Cyanobacteria on Rock Surfaces. *Biodivers. Conserv.* **2015**, *24*, 759–779.

Hosmani, S. P. Phytoplankton Diversity in Lakes of Mysore District, Karnataka State, India. *Ecoscan* **2010**, *4*, 53–57.

Hosmani, S. P.; Anita, M. C. Biochemical Study of *Microcystis aeruginosa* Kutz. *Ecol. Environ. Conser.* **1998**, *4*, 255–257.

Hosmani, S. P.; Bharati, S. G. A New Genus and Species of Alga from Karnataka (India). *J. Bombay Nat. Hist. Soc.* **1981**, *78*, 579580.

Hossain, M. F.; Bowange, R. W. T. M. R. T. K.; Kumara, K. L. W.; Magana-Arachchi, D. N.; Ratnayake, R. R. First Record of Cyanobacteria Species: *Cephalothrix komarekiana*, from Tropical Asia. *Environ. Eng. Res.* **2021**, *26*, 200040. DOI: 10.4491/eer.2020.040

Ishimi, Y.; Sugiyama, F.; Ezaki, J.; Fujioka, M.; Wu, J. Effects of *Spirulina*, A Blue-Green Alga, on Bone Metabolism in Overiectomized Rats and Hindlimb-Unloaded Mice. *Biosci. Biotechnol. Biochem.* **2006**, *70*, 363–368.

Joishi, M. S. Studies on Annual Variation in Species Diversity of Cyanobacteria in Four Rivers of Western Ghats Region, Karnataka, India. *Int. Res. J. Plant Sci.* **2014**, *5*, 43–52.

Joishi, M. S. Studies on the Cyanobacteria in Some of the Aquatic Systems of the Western Ghats with Special Reference to Coastal Regions of Dakshina Kannada and Udupi District. PhD Dissertation, Mangalore University, Mangalore, India, **2009**; p 161.

Kalavathi, F. D.; Uma, L.; Subramanian, G. Degradation and Metabolization of the Pigment Melanoidin in Distillery Effluent by the Marine Cyanobacterium *Oscillatoria boryana* BDU 92181. *Eng. Microb. Technol.* **2001**, *29*, 249–251.

Kamble, P. B. Studies on Filamentous Blue Green Algae from Satara District Maharashtra. PhD Dissertation, Shivaji University, Kolhapur, India, **2017**; p 278.

Kurian, S.; Roy, R.; Repeta, D. J.; Gauns, M.; Shenoy, D. M.; et al. Seasonal Occurrence of Anoxygenic Photosynthesis in Tillari and Selaulim Reservoirs, Western India. *Biogeosciences* **2012**, *9*, 2485–2495.

Mahadev, J.; Hosmani, S. P. Algae for Biomonitoring of Organic Pollution in Two Lakes of Mysore City. *Nat. Environ. Poll. Technol.* **2005**, *4*, 97–99.

Maltsev, Y.; Kezlya, E.; Maltseva, S.; Karthick, B.; Kociolek, P. D.; Kulikovskiy, M. A. New Species of the Previously Monotypic Genus *Iningainema* (Cyanobacteria, Scytonemataceae) from the Western Ghats, India. *Euro. J. Phycol.* **2021**, *56* (3), 348–358. DOI: 10.1080/09670262.2020.1834147

Manoharan, C.; Subramanian, G. Influence of Effluents on Fatty Acid Content of a Cyanobacterium. *Curr. Sci.* **1993**, *65*, 353–355.

Mitsui, A.; Phlips, E. J.; Kumazawa, S.; Reddy, K. J.; Ramachandran, S.; et al. Progress in Research Toward Outdoor Biological Hydrogen Production Using Solar Energy, Sea Water, and Marine Photosynthetic Microorganisms. *Ann. NY. Acad. Sci.* **1983**, *413*, 514–530.

Moreira, L. M.; Ribeiro, A. C.; Duarte, F. A.; Morais, M. G. D.; Soares, L. A. D. *Spirulina platensis* Biomass Cultivated in Southern Brazil as a Source of Essential Minerals and Other Nutrients. *Afr. J. Food Sci.* **2013**, *7*, 451–455.

Murray, R. L.; Mitsui, A. Growth of Hybrid *Talipia* Fry Fed Nitrogen Fixing Marine Blue Green Algae in Sea Water. *J. World Maricult. Soc.* **1982**, *13*, 198–209.

Nabout, J. C.; da Silva, R. B., Carneiro, F. M.; Sant'Anna, C. L. How Many Species of Cyanobacteria Are There? Using a Discovery Curve to Predict the Species Number. *Biodivers. Conserv.* **2013**, *22*, 2907–2918.

Nagase, H.; Inthorn, D.; Isaji, Y.; Oda, A.; Kajiwara, Y.; et al. Improvement of Selective Removal of Heavy Metals in Cyanobacteria by NaOH Treatment. *J. Biosci. Bioeng.* **2005**, *99*, 372–377.

Nasser, K. M. M.; Sureshkumar, S. Interaction Between Microalgal Species Richness and Environmental Variables in Peringalkuthu Reservoir, Western Ghats, Kerala. *J. Enviorn. Biol.* **2013**, *34*, 1001–1005.

Nayak, S.; Prassana, R.; Dominic, T. K.; Singh, P. K. Floristic Abundance and Relative Distribution of Different Cyanobacterial Genera in Rice Field Soil at Different Crop Growth Stages. *Phykos* **2001**, *40*, 14–21.

Neustupa, J.; Skaloud, P. Diversity of Subaerial Algae and Cyanobacteria Growing on Bark and Wood in the Lowland Tropical Forests of Singapore. *Plant Ecol. Evol.* **2010**, *143*, 51–62.

Nikam, T. D.; Nehul, J. N.; Gahile, Y. R.; Auti, B. K.; Ahire, M. L.; et al. Cyanobacterial Diversity in the Western Ghats Region of Maharashtra, India. *Biorem. Biodiver. Bioavail.* **2011**, *7*, 70–80.

Nuhu, A. A. *Spirulina* (*Arthrospira*): An Important Source of Nutritional and Medicinal Compounds. *J. Mar. Biol.* **2013**, *2*, 1–8.

Olvera-Novoa, M. A.; Campos, S.; Sabido, M.; Martínez-Palacios, C. A. The Use of Alfalfa Leaf Protein Concentrates as a Protein Source in Diets for Tilapia *Oreochromis mossambicus*. *Aquaculture* **1990**, *90*, 291–302.

Pereira, N.; Almeida, M. R. New Records of Blue-Green Algae from Goa. *J. Algal Biomass Util.* **2012**, *3*, 27–29.

Philip, M.; Farhad, V. P.; Shamina, M. Diversity of Cyanobacterial Flora at Nelliyampathy, Kerala. *South Indian J. Biol. Sci.* **2016**, *1*, 198–202.

Philip, M.; Krishnan, V. V. R. Diversity of the Genus *Chroococcus nageli* (Chroococcaceae, cyanobacteria) from Southern Western Ghat regions of Kerala, India. *Int. J. Adv. Res.* **2020**, *8*, 817–820.

Pingale, S. D.; Deshmukh, B. S. Some Freshwater Algae from Amphitheatre of Wilson Dam. *Indian J. Hydrobiol.* **2005**, *7*, 97–100.

Raghavendra; Hosmani, S. P. Hydrobiological Study of Mandakally Lake, A Polluted Water Body at Mysore. *Nat. Environ. Pollut. Technol.* **2002**, *1*, 291–293.

Rajeshwari, K. R.; Rajashekhar, M. Biochemical Composition of Seven Species of Cyanobacteria Isolated from Different Aquatic Habitats of Western Ghats, Southern India. *Braz. Arch. Biol. Technol.* **2011**, *54*, 849–857.

Rajeshwari, K. R.; Rajashekhar, M. Diversity of Cyanobacteria in Some Polluted Aquatic Habitats and a Sulfur Spring in the Western Ghats of India. *Algol. Stud.* **2012**, *138*, 37–56.

Ram, A. T.; Shamina, M. Cyanobacterial Diversity from Seven Mangrove Environments of Kerala, India. *World News Nat. Sci.* **2017**, *9*, 91–97.

Ramaswamy, S. N.; Somasekhar, R. K.; Arekal, G. D. Ecological Studies on Algae in Waste Waters from Rubber Tyre Factory. *Indian J. Environ. Health.* **1982**, *24*, 1–7.

Ruangsomboon, S.; Chidthaisong, A.; Bunnag, B.; Inthorn, D.; Harvey, N. W. Lead (Pb^{2+}) Adsorption Characteristics and Sugar Composition of Capsular Polysaccharides of Cyanobacterium *Calothrix marchica. Songklanakarian. J. Sci. Technol.* **2007**, *29*, 529–541.

Sajilata, M. G.; Singhal, R. S.; Kamat, M. Y. Fractionation of Lipids and Purification of γ-Linolenicacid (GLA) from *Spirulina platensis. Food Chem.* **2008**, *109*, 580–586.

Selvakumar, G.; Sundararaman, M. Seasonal Variations of Cyanobacterial Flora in the Back Water. *J. Sci. Trans. Environ. Technol.* **2007**, *3*, 69–73.

Severes, A.; Nivas, S.; D'Souza, L.; Hegde, S. Diversity Study of Freshwater Microalgae of Some Unexplored Water Bodies of a Rapidly Developing Industrial Region in India. *J. Algal Biomass Util.* **2018**, *9*, 31–40.

Sharathchandra, K.; Rajashekhar, M. Total Lipid and Fatty Acid Composition in Some Freshwater Cyanobacteria. *J. Algal Biomass Util.* **2011**, *2*, 83–97.

Sharathchandra, K.; Rajashekhar, M. Antioxidant Activity in the Four Species of Cyanobacteria Isolated from a Sulfur Spring in the Western Ghats of Karnataka. *Int. J. Pharm. Bio Sci.* **2013a**, *4*, 275–285.

Sharathchandra, K.; Rajashekhar, M. Freshwater Cyanobacteria from Kaiga, Uttara Kannada District in the Western Ghats of Karnataka. *Phykos* **2013b**, *43*, 51–66.

Sharathchandra, K.; Rajashekhar, M. Seasonal Occurrence of Freshwater Cyanobacteria in the Four Artificial Tanks of Western Ghats of Karnataka. *J. Algal Biomass Util.* **2016**, *7*, 1–17.

Sharathchandra, K.; Rajashekhar, M. Effects of Temperature and Sulfide on Freshwater Cyanobacteria Isolated from a Sulfur Spring in the Western Ghat forests of Karnataka. *Int. J. Res. Biosci.* **2017**, *6*, 52–65.

Sharathchandra, K.; Sridhar, K. R.; Rajashekhar, M. Diversity of Cyanobacteria in Thermal Water Bodies of Southwest India. In *Biodiversity, Conservation and Sustainability in Asia*: *Prospects and Challenges in South and Middle Asia*; Öztürk, M., Khan, S. M., Altay, V., Efe, R., Egamberdieva, D., Khassanov, F. O., Eds.; Springer International Publishing: Switzerland, **2021** (in press).

Shashikumar, K. C.; Madhyastha, M. N. Amino Acid Profile of *Synechocystis aquatilis* Isolated from an Estuary Near Mangalore Coast. *Indian J. Microbiol.* **2002**, *42*, 81–82.

Shekhar, R. T.; Kiran, B. R.; Puttiah, E. T.; Shivaraj, Y.; Mahadeven, K. M. Phytoplankton as Index of Water Quality with Reference to Industrial pollution. *J. Environ. Biol.* **2008**, *29*, 233–236.

Shukla, D.; Vankar, P. S.; Srivastava, S. K. Bioremediation of Hexavalent Chromium by a Cyanobacterial Mat. *Appl. Water Sci.* **2012**, *2*, 245–251.

Singh, P.; Singh, D. M. Subaerial Nonheterocytous and Heterocytous Cyanobacteria from Sirsi Taluk, Uttara Kannada, Karnataka, India. *Int. J. Sci. Res. Rev.* **2019**, *8*, 1248–1274.

Sivakumar, N.; Viji, V.; Satheesh, S.; Varalakshmi, P.; Ashokkumar, B.; Pandi, M. Cyanobacterial Abundance and Diversity in Coastal Wetlands of Kanyakumari District, Tamil Nadu (India). *Afr. J. Microbiol. Res.* **2012**, *6*, 4409–4416.

Smith, V. H. Eutrophication of Freshwater and Coastal Marine Ecosystems–A Global Problem. *Environ. Sci. Pollut. Res.* **2003**, *10*, 126–130.

Sorichetti, R. J.; Creed, I. F.; Trick, C. G. Evidence for Iron-Regulated Cyanobacterial Predominance in Oligotrophic Lakes. *Freshwat. Biol.* **2013**, *1*, 1–13.

Subramanian, G.; Sekar, S.; Sampoornam, S. Biodegradation and Utilization of Organophosphorus Pesticides by Cyanobacteria. *Int. Biodeter. Biodegr.* **1994**, *33*, 129–143.

Subramanian, G.; Uma, L. Potential Applications of Cyanobacteria in Environmental Biotechnology. In *Photosynthetic Microorganisms in Environmental Biotechnology*; Kojima, H., Lee, Y. K., Eds.; Springer: Singapore, **2001**; pp 41–49.

Suresh, A.; Kumar R. P.; Dhanasekaran, D.; Thajuddin, N. Biodiversity of Microalgae in Western and Eastern Ghats, India. *Pak. J. Biol. Sci.* **2012**, *15*, 919–928.

Uma, L.; Subramanian, G. In *Effective Use of Cyanobacteria in Effluent Treatment*, Proceedings of the National Symposium on Cyanobacterial Nitrogen Fixation, Indian Agriculture Research Institute, New Delhi, **1990**; pp 437–444.

Vijayan, D.; Ray, J. G. Ecology and Diversity of Cyanobacteria in Kuttanadu Paddy Wetlands, Kerala, India. *Am. J. Plant Sci.* **2015**, *6*, 2924-2938.

CHAPTER 6

LICHEN-FORMING AND LICHENICOLOUS FUNGI OF THE WESTERN GHATS, INDIA

SANJEEVA NAYAKA[1] and BIJU HARIDAS[2]

[1]*Lichenology Laboratory, CSIR-National Botanical Research Institute, Rana Pratap Marg, Lucknow, India*

[2]*Microbiology Division, KSCSTE-Jawaharlal Nehru Tropical Botanic Garden and Research Institute, Palode, Thiruvananthapuram, Kerala, India*

ABSTRACT

The biota of lichen-forming and lichenicolous fungi occurring in the Western Ghats were analyzed utilizing recent publications. The study revealed the occurrence of 1617 taxa of lichen-forming fungi (or lichens) with 1597 species, 19 varieties, 2 subspecies, 1 forma. The lichenicolous fungi were represented by 28 species. As the lichenicolous fungi are fewer in number and unexplored groups, the emphasis is given to lichens in the present communication. A total of 251 lichen taxa are endemic, of which, 129 are restricted in their distribution to the Western Ghats only. The lichen biota of Western Ghats is dominated by crustose lichens with 1117 taxa while foliose and fruticose forms represent 393 and 107 taxa. All the lichens taxa reported belonged to phylum Ascomycota and are dominated by Graphidales and pyrenocarpous lichens with 404 and 224 taxa, respectively. *Graphis*, Graphidaceae, Lecanorales and Lecanoromycetes are the most speciose genus, family, order and class in the region with 78, 210, 421, and 1288 taxa, respectively. As many as 1231 lichens in the region preferred to grow on bark only while 50 taxa shared substrates such as bark, rock, and soil. This region also represented a good diversity of cyanolichens (117) which are known indicators of moistness in the forests. Similarly, a study indicated that some parts of the Western Ghats are healthy and undisturbed by the presence of 88 foliicolous lichens. Several species of lichens are rare in the Western Ghats, and among them, at least five of them are not collected for a long time and are suitable for inclusion in the Red Data Book. Among different states, Tamil Nadu, Kerala, and Karnataka represent maximum number of lichens represented by 963, 783, and 658 taxa, respectively. The lichens of Western Ghats show their affinities to the lichen biota of north-eastern India and

---

Biodiversity Hotspot of the Western Ghats and Sri Lanka. T. Pullaiah, PhD (Ed.)
© 2024 Apple Academic Press, Inc. Co-published with CRC Press (Taylor & Francis)

tropical regions of the world. Although large quantities of lichens are being harvested from the Western Ghats, there is no authentic published account of the same. Also, there are no serious efforts for conservation of lichens in the Western Ghats.

6.1 INTRODUCTION

Fungi are a diverse group of organisms that originated during 760 to 1.06 billion years ago (Money, 2016). During the course of their evolution, they adapted wide range of lifestyles ranging from free-living to symbiosis. Lichenization is one such lifestyle of fungi where they develop a symbiotic association with cyanobacteria or algae, which are otherwise known as photobionts. It is now well known that in a lichen, apart from a fungus and photosynthetic partner, there are several other microbes that are closely associated. These microbes also include secondary fungi and algae. Therefore, a lichen is redefined as "a self-sustaining ecosystem formed by the interaction of an exhabitant fungus and an extracellular arrangement of one or more photosynthetic partners and an indeterminate number of other microscopic organisms" (Hawksworth and Grube, 2020). Further, the classification and nomenclature of lichens or lichen-forming fungi follow fungal systematics and all the associated organisms (including photobiont) have their independent names. Among the associated microbes in lichens, endolichenic (living inside) and lichenicolous (growing over lichens) fungi are interesting groups with regard to their taxonomy, physiology, ecology, co-evolution, and unique metabolites (Chakarwarti et al., 2020).

It is estimated that approximately 8% of the earth's terrestrial surface has lichens as its most dominant life-form (Ahmadjian, 1995). Various authors proposed different numbers of lichens occurring in the world viz., 22,000 (Zahlbruckner, 1922–1940); 13,500 (Hawksworth, 1995); 20,000 (Sipman and Aptroot, 2001), and 17,322 (Feuerer and Hawksworth, 2007). As per the latest records in the world, a total of 19,409 species of lichens are known under 1,002 genera, 119 Families, and 40 Orders (Lücking et al., 2017). However, at any given point in time, the total number of lichens in the world can be summed to 20,000 species. It is interesting to note that the maximum species richness of lichens in the world is found in the tropical rain forests and more than 600 species can be found in a km^2 area (Lücking et al., 2009). There are more lichen species than a tree or bird species in any given area of a tropical jungle. Unfortunately, 50% of the tropical lichen biota is unexplored (Aptroot and Sipman, 1997).

India is one of the megadiverse countries with rich lichen biota. The list of lichens in India grew from 960 species in 1965 (Awasthi, 1965), 2305 in 2010 (Singh and Sinha, 2010) to 2714 in 2018 (Sinha et al., 2018). At present, India is represented by about 2902 species under 407 genera and 79 families. It is about 14.8% of the world's known lichens, and approximately, 18% of it is endemic to India (Sinha and Jagadeesh Ram, 2020). The exploratory and revisionary studies in recent years have added a number of novel taxa to the Indian lichen biota. Several unexplored and ecologically interesting areas were surveyed resulting in an increase in the number of species at the regional level. Among different regions, the Western Himalaya, north-eastern India (including eastern Himalaya), the Western Ghats, and Andaman Nicobar Island can be considered as hotspots for lichen diversity with approximately 1200, 1580, 1475, and 525 species, respectively (personal comm. Dr. G. P. Sinha, based on Sinha and Jagadeesh Ram, 2020).

The lichenicolous fungi represent a highly specialized and successful group of organisms that live exclusively on lichens, most commonly as host-specific parasites but also as broad-spectrum pathogens, saprotrophs, and commensals (Diederich et al., 2018). They are also a relatively less explored group of organisms in the world. The current accepted number of lichenicolous fungi in the world is 2319 species, of which, 2000 are obligate associates. Out of the total, about 96% belongs to the ascomycetes group and only 4% belongs to the basidiomycetes. The experts' opinion is that about 3000–4000 species of lichenicolous fungi may be present in the world (Lawrey and Diederich, 2018); however, their exploration is far from completion. The studies of lichenicolous fungi in India have started recently. The number of currently known lichenicolous fungi in India is 179, which is very less compared with vast diversity of lichens in the country (Joshi, 2018). So far, there is no separate list of lichenicolous fungi available for the Western Ghats.

6.2   STUDIES ON LICHEN-FORMING FUNGI OF WESTERN GHATS

The Western Ghats being a biodiversity-rich region, lichens from various localities have been collected and included in monographs, regional, and local-level inventories. It can be noted that there has been a steady increase in the number of lichens reported from the Western Ghats (Figure 6.1). Awasthi (1965) listed around 250 taxa of lichens from the Western Ghats and about 225 species were added to list during 1966–1977 (Singh 1980). Patwardhan (1983) and Singh and Sinha (1977) estimated about 800 lichen species to occur in Western Ghats while Kumar and Sequiera (1999) could list 771 species. Nayaka and Upreti (2005) in their attempt to update the checklist of lichens from Western Ghats enumerated a total of 949 lichen taxa belonging to 929 species 20 varieties. Nayaka and Upreti (2011) further updated the numbers with 1114 lichen taxa with 1096 species and 16 intra-specific taxa. Later, according to Nayaka and Asthana (2014) the number increased to 1138 species under 193 genera and 53 families. As per the recent calculation total number of lichens known to occur in Western Ghats is 1472 species (personal comm. Dr. G. P. Sinha, based on Sinha and Jagadeesh Ram, 2020).

6.3   CURRENT STATUS OF LICHEN-FORMING AND LICHENICOLOUS FUNGI IN WESTERN GHATS

In the present chapter, numbers of lichen-forming and lichenicolous fungi from the Western Ghats are updated with recent publications. The annotated checklist of lichens of India by Singh and Sinha (2010) was taken as a base for counting the species. Although only the name of the states is given in Singh and Sinha (2010), the references cited in the book indicated that these lichens are reported mostly from the Western Ghats. The

Western Ghats range starts from the Sonagadh town of Gujarat and traverses toward south up to Marunthuwamalai in the southern tip of India. In between the Western Ghats states, such as Maharashtra, Goa, Karnataka, Kerala, and Tamil Nadu are present. More than 80 species of lichens are known to occur in Gujarat (Nayaka et al., 2017); however, only two species (*Arthonia dispersula* Nyl. and *A. medusula* (Pers.) Nyl.) reported from the Valsad district falling within the Western Ghats are included here (Nayaka et al., 2013). Similarly, some of the lichens reports from islands, mangroves, and the Eastern Ghats parts in Karnataka and Tamil Nadu states are excluded. Some of the important and recent publications referred include Randive et al. (2017a,b, 2018, 2019) for Goa; Chitale and Makhija (2012), Chitale et al. (2009, 2011), Makhija et al. (2014), Mishra et al. (2017) for Maharashtra; Krishnamurthy and Subramanya (2017), Rashmi and Rajkumar (2015a,b), Singh (website 4), Singh and Singh (2017), Subramanya and Krishnamurthy (2015a,b, 2016a,b, 2017), Subramanya et al. (2010), Vinayaka (2016), Vinayaka and Krishnamurthy (2010a,b), Vinayaka et al. (2010, 2011, 2012, 2016) for Karnataka; Biju et al. (2010, 2012, 2014 a,b, 2021), Kumar (2000), Kumar and Sequiera (2002), Easa (2003), Sequiera (2012), Zachariah et al. (2018, 2019, 2020) for Kerala; and Balaji and Hariharan (2013a,b), Hariharan and Balaji (website 5), Ingle et al. (2016), Joseph et al. (2011) for Tamil Nadu. It is observed that several taxa reported by Kumar (2000), Kumar and Sequiera (2001), Easa (2003), Balaji and Hariharan (2013a,b), and Rashmi and Rajkumar (2015a,b) are doubtful in occurrence in the Western Ghats. However, they are considered for the present study and such species need to be verified by studying the specimens in the future.

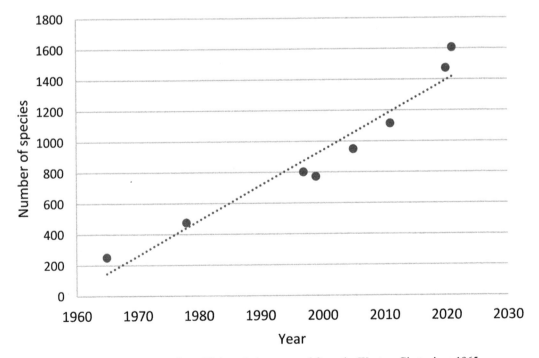

FIGURE 6.1   Increase in the number of lichens being reported from the Western Ghats since 1965.

In the recent years, there were dramatic changes in lichen taxonomy which resulted in several nomenclature changes, especially in pyrenocarpous, graphidaceous, the lotremataceous, and cyanolichens. The databases, such as indexfungorum.org; gbif.org and mycobank.org are consulted for nomenclatural changes. Also, related literature is referred to confirm such name changes. The endemism of lichens is checked by referring to relevant literature and occurrence record in gbif.org. The recent fungal classification proposed by Wijayawardene et al. (2020) is followed for the hierarchical arrangement of lichen taxa under various families, order, class, and phylum. However, some taxa not included in Wijayawardene et al. (2020) were referred from indexfungorum.org.

6.3.1 DIVERSITY

The compilation revealed the occurrence of 1617 taxa of lichens with 1597 species 19 varieties 2 subspecies 1 forma in Western Ghats. Whereas region represented 28 species of lichenicolous fungi. The lichen taxa are distributed under 304 genera, 71 families, 29 orders, 7 classes (Table 6.1) and only one phylum Ascomycota. The lichenicolous fungi belong to 22 genera and 18 families, 14 orders, 7 classes, and 2 phyla (Table 6.2). The lichen biota of Western Ghats is dominated by crustose forms with 1117 taxa followed by foliose and fruticose forms with 393 and 107 taxa, respectively. The Western Ghats has got at least 25 genera with more than 15 species; however, *Graphis* (Figure 6.3c) is the most speciose genus with 78 taxa followed by *Pyrenula* (Figure 6.3g) with 57 taxa, while *Parmotrema* (Figure 6.3f) and *Usnea* have 51 taxa each (Table 6.3). Among various families, Graphidaceae is the most speciose 210 taxa under 23 genera, followed by Parmeliaceae with 207 species and 26 genera (Table 6.4). Among the 29 orders of lichens represented in Western Ghats, Lecanorales is the most speciose with 423 taxa under 67 genera and 13 families. It can also be noted that the Order Graphidales is another largest order in the region with 404 taxa under 60 genera and five families (Table 6.5). As mentioned earlier, the lichens of Western Ghats fall under seven classes, among them, Lecanoromycetes is the most dominant with 1288 taxa belonging to 17 orders, 52 families, and 233 genera.

The diversity of lichens in the Western Ghats is dominated by Graphidales and pyrenocarpous lichens with 404 and 224 taxa. These two groups of lichen are typical representative of lichen communities in tropical forest. According to the classification proposed by Jaklitsch et al. (2016) in "Syllabus of Plant Families" the genera, such as *Acanthothecis, Asteristion, Astrochapsa, Austrotrema, Carbacanthographis, Chapsa, Chroodiscus, Cruentotrema, Diploschistes, Dyplolabia, Fibrillithecis, Fissurina, Glaucotrema, Leucodecton, Melanotrema, Nadvornikia, Nitidochapsa, Ocellularia, Phaeographopsis, Pycnotrema, Reimnitzia, Rhabdodiscus, Sanguinotrema, Stegobolus, Thelotrema,* and *Wirthiotrema* belonged to Graphidaceae, hence making this family largest in the region. However, they are now segregated into families Diploschistaceae, Fissurinaceae, and Thelotremataceae (Wijayawardene et al., 2020). Among the pyrenocarpous lichens families, Pyrenulaceae, Trichotheliaceae, and Trypetheliaceae are well represented in the Western Ghats with 70, 54, and 33 taxa, respectively. It can also be noted that the region also has well representation of Arthoniales (Figure 6.3a) with 122 taxa under 28 genera and six families.

TABLE 6.1 Diversity Lichens and Their Hierarchical Arrangement Under Different Classes, Orders, and Families.

Class	Order	Family	Genus
Arthoniomycetes	Arthoniales	Arthoniaceae	*Arthonia* (9), *Arthothelium* (25), *Coniocarpon* (1), *Cryptothecia* (30), *Herpothallon* (6), *Stirtonia* (3), *Tylophoron* (3)
		Chrysotrichaceae	*Chrysothrix* (3)
		Genera incertae sedis	*Bactrospora* (3), *Synarthonia* (2)
		Lecanographaceae	*Alyxoria* (1), *Heterocyphelium* (1), *Lecanographa* (1), *Zwackhia* (4)
		Opegraphaceae	*Fouragea* (1), *Opegrapha* (8), *Sclerophyton* (1)
		Roccellaceae	*Chiodecton* (3), *Cresponea* (1), *Dichosporidium* (1), *Enterographa* (2), *Graphidastra* (1), *Mazosia* (5), *Pseudoschismatomma* (1), *Pulvinodecton* (1), *Roccella* (1), *Sigridea* (2), *Syncesia* (2)
Candelariomycetes	Candelariales	Candelariaceae	*Candelaria* (2)
		Pycnoraceae	*Pycnora* (1)
Dothideomycetes	Incertae sedis	Naetrocymbaceae	*Leptorhaphis* (1), *Naetrocymbe* (2)
	Monoblastiales	Monoblastiaceae	*Acrocordia* (1), *Anisomeridium* (16), *Megalotremis* (1), *Monoblastia* (1)
	Patellariales	Patellariaceae	*Patellaria* (1)
	Pleosporales	Arthopyreniaceae	*Arthopyrenia* (8)
		Mycoporaceae	*Mycoporum* (1)
	Strigulales	Strigulaceae	*Flavobathelium* (1), *Strigula* (15)
	Trypetheliales	Trypetheliaceae	*Astrothelium* (14), *Bathelium* (2), *Bogoriella* (5), *Constrictolumina* (2), *Dictyomeridium* (1), *Marcelaria* (1), *Nigrovothelium* (1), *Polymeridium* (3), *Trypethelium* (3), *Viridothelium* (1)
Eurotiomycetes	Mycocaliciales	Mycocaliciaceae	*Pyrgidium* (1), *Sphinctrina* (1)
	Pyrenulales	Pyrenulaceae	*Anthracothecium* (4), *Lithothelium* (4), *Pyrenula* (57), *Pyrgillus* (4), *Sulcopyrenula* (1)
Incertae sedis	Thelocarpales	Thelocarpaceae	*Thelocarpon* (1)
Lecanoromycetes	Acarosporales	Acarosporaceae	*Acarospora* (1)
		Arthrorhaphidaceae	*Arthrorhaphis* (2)
		Baeomycetaceae	*Baeomyces* (1)
		Hymeneliaceae	*Ionaspis* (1)
	Baeomycetales	Trapeliaceae	*Placynthiella* (1), *Trapelia* (2), *Trapeliopsis* (1)

TABLE 6.1 (Continued)

Class	Order	Family	Genus
	Caliciales	Xylographaceae	*Xylographa* (1)
		Caliciaceae	*Acolium* (1), *Amandinea* (6), *Baculifera* (1), *Buellia* (27), *Calicium* (3), *Cratiria* (5), *Diplotomma* (1), *Dirinaria* (10), *Hafellia** (2), *Pyxine* (27), *Sculptolumina* (1)
		Physciaceae	*Heterodermia* (26), *Hyperphyscia* (5), *Leucodermia* (2), *Phaeophyscia* (10), *Physcia* (15), *Physciella* (1), *Physconia* (1), *Polyblastidium* (9), *Rinodina* (3)
	Genera incertae sedis	Genera incertae sedis	*Piccolia* (1)
	Graphidales	Diploschistaceae	*Acanthothecis* (5), *Asteristion* (4), *Austrotrema* (2), *Carbacanthographis* (6), *Diploschistes* (8), *Fibrillithecis* (2), *Glaucotrema* (1), *Melanotrema* (1), *Myriotrema* (14), *Nadvornikia* (1), *Nitidochapsa* (1), *Ocellularia* (39), *Phaeographopsis* (1), *Reimnitzia* (1), *Rhabdodiscus* (5), *Sanguinotrema* (1), *Stegobolus* (2), *Topeliopsis* (2), *Wirthiotrema* (3)
		Fissurinaceae	*Cruentotrema* (2), *Dyplolabia* (1), *Fissurina* (33), *Pycnotrema* (1)
		Gomphillaceae	*Aderkomyces* (1), *Asterothyrium* (1), *Aulaxina* (3), *Bullatina** (1), *Calenia* (2), *Caleniopsis* (1), *Echinoplaca* (3), *Gyalectidium* (3), *Roluechia* (1), *Tricharia* (2)
		Graphidaceae	*Allographa* (32), *Anomomorpha* (1), *Creographa* (1), *Cryptoschizotrema* (1), *Diorygma* (27), *Glyphis* (2), *Graphina* *(1), *Graphis* (78), *Hemithecium* (11), *Kalbographa* (1), *Pallidogramme* (10), *Phaeographina**(2), *Phaeographis* (20), *Platygramme* (6), *Platythecium* (6), *Pseudochapsa* (2), *Pseudotopeliopsis* (1), *Sarcographa* (4), *Schistophoron* (1), *Thalloloma* (1), *Thecaria* (2)
		Thelotremataceae	*Astrochapsa* (2), *Chapsa* (6), *Chroodiscus* (2), *Leptotrema** (1), *Leucodecton* (7), *Thelotrema* (24)
	Gyalectales	Coenogoniaceae	*Coenogonium* (1)
		Gyalectaceae	*Cryptolechia** (1), *Gyalecta* (1)
		Phlyctidaceae	*Phlyctis* (5)
		Trichotheliaceae	*Clathroporina* (3), *Porina* (44), *Pseudosagedia* (3), *Segestria* (1), *Trichothelium* (3)
	Lecanorales	Catillariaceae	*Catillaria* (5)
		Cladoniaceae	*Cladonia* (27), *Lepraria* (16), *Stereocaulon* (1)
		Haematommataceae	*Haematomma* (1)
		Lecanoraceae	*Glaucomaria** (1), *Lecanora* (46), *Lecidella* (5), *Myriolecis* (1), *Omphalodina** (1), *Polyozosia** (1), *Straminella** (1)

TABLE 6.1 *(Continued)*

Class	Order	Family	Genus
		Malmideaceae	*Malmidea* (6)
		Megalariaceae	*Megalaria* (3)
		Parmeliaceae	*Austroparmelina* (1), *Bulbothrix* (7), *Canoparmelia* (6), *Cetraria* (1), *Cetrelia* (2), *Crespoa** (2), *Eumitria* (2), *Flavoparmelia* (1), *Flavopunctelia* (1), *Hypogymnia* (4), *Hypotrachyna* (30), *Melanelia* (1), *Melanohalea* (2), *Menegazzia* (1), *Myelochroa* (7), *Nephromopsis* (2), *Notoparmelia* (1), *Parmelia* (2), *Parmelinella* (6), *Parmotrema* (51), *Parmotremopsis* (1), *Punctelia* (4), *Relicina* (4), *Remototrachyna* (12), *Usnea* (51), *Xanthoparmelia* (5)
		Pilocarpaceae	*Bapalmuia* (1), *Brasilicia* (1), *Byssolecania* (1), *Byssoloma* (4), *Calopadia* (4), *Eugeniella* (1), *Fellhanera* (5), *Loflammia* (2), *Micarea* (1), *Sporopodium* (3), *Tapellaria* (2)
		Psoraceae	*Protoblastenia* (1)
		Ramalinaceae	*Bacidia* (23), *Bacidina* (3), *Badimia* (2), *Biatora* (1), *Bilimbia* (1), *Heppsora* (1), *Lopezaria** (1), *Phyllopsora* (17), *Psorella** (1), *Ramalina* (21), *Rolfidium* (1)
		Ramboldiaceae	*Ramboldia* (1)
		Sphaerophoraceae	*Bunodophoron* (1)
		Tephromelataceae	*Tephromela* (1)
	Lecideales	Lecideaceae	*Immersaria* (1), *Koerberiella* (1)
		Lopadiaceae	*Lopadium* (4)
	Leprocaulales	Leprocaulaceae	*Leprocaulon* (3)
	Ostropales	Stictidaceae	*Conotrema** (1), *Stictis* (1)
	Peltigerales	Coccocarpiaceae	*Coccocarpia* (3)
		Collemataceae	*Blennothallia* (2), *Collema* (17), *Enchylium* (2), *Lathagrium* (1), *Leptogium* (31), *Rostania* (2), *Scytinium* (4)
		Pannariaceae	*Erioderma* (2), *Fuscopannaria* (3), *Hispidopannaria** (1), *Kroswia** (1), *Leioderma* (1), *Lepidocollema* (4), *Pannaria* (5), *Parmeliella* (4), *Pectenia* (1), *Physma* (1), *Psoroma* (1)
		Peltigeraceae	*Dendriscocaulon** (1), *Dendriscosticta* (3), *Lobaria* (6), *Nephroma* (2), *Peltigera* (5), *Pseudocyphellaria* (7), *Ricasolia* (3), *Solorina* (1), *Sticta* (11)
		Vahliellaceae	*Vahliella* (3)
	Pertusariales	Megasporaceae	*Aspicilia* (2), *Circinaria* (1), *Megaspora* (1)

TABLE 6.1 *(Continued)*

Class	Order	Family	Genus
		Ochrolechiaceae	*Ochrolechia* (5)
		Pertusariaceae	*Pertusaria* (37)
		Varicellariaceae	*Varicellaria* (1)
		Variolariaceae	*Lepra* (7)
	Rhizocarpales	Rhizocarpaceae	*Rhizocarpon* (3)
	Teloschistales	Brigantiaeaceae	*Brigantiaea* (3), *Letrouitia* (6)
		Megalosporaceae	*Byssophragmia** (1), *Megaloblastenia* (1), *Megalospora* (5)
		Teloschistaceae	*Blastenia* (2), *Calogaya* (1), *Caloplaca* (13), *Flavoplaca* (1), *Gondwania* (1), *Huneckia* (1), *Ioplaca* (1), *Laundonia** (1), *Neobrownliella** (1), *Opeltia** (1), *Oxneriopsis* (1), *Polycauliona* (1), *Pyrenodesmia* (1), *Squamulea* (3), *Teloschistes* (1), *Upretia* (1), *Xanthoria* (1)
	Thelenellales	Thelenellaceae	*Thelenella* (1)
	Umbilicariales	Fuscideaceae	*Maronea* (1)
	Verrucariales	Verrucariaceae	*Catapyrenium* (1), *Clavascidium* (1), *Dermatocarpon* (1), *Endocarpon* (4), *Hydropunctaria* (1), *Normandina* (1), *Placidium* (1), *Staurothele* (4), *Verrucaria* (4)
Lichinomycetes	Lichinales	Lichinaceae	*Anema* (1), *Phylliscum* (3), *Thallinocarpon* (1)
		Peltulaceae	*Peltula* (5)

**Classification according to indexfungorum.org.*

Note: The numbers within the bracket indicates number species under that genus.

Source: Adapted from indexfungorum.org.

The Arthonioid lichens are considered a primitive group of lichen world and they do not form definite fruiting bodies (Smith, 1926). Another interesting feature of lichen biota of the Western Ghats is the occurrence of 117 taxa of cyanolichens represented by genera *Coccocarpia* (Figure 6.3b), *Collema, Enchylium, Lepidocollema, Leptogium, Pannaria, Parmeliella,* and *Sticta*, etc. The cyanolichens are the ecological indicators of old-growth forests and the availability of moisture in a habitat (Rikkinen, 2015).

TABLE 6.2 Diversity of Lichens and Their Hierarchical Arrangement of Lichenicolous Fungi Under Different Phyla, Classes, Orders, and Families.

Phylum	Class	Order	Family	Genus
Ascomycota	Arthoniomycetes	Arthoniales	Arthoniaceae	*Arthonia* (1)
			Opegraphaceae	*Opegrapha* (2)
	Dothideomycetes	Abrothallales	Lichenoconiaceae	*Abrothallus* (1), *Endococcus* (2), *Lichenoconium* (1), *Lichenothelia* (1)
		Asterinales	Stictographaceae	*Karschia* (1)
		Botryosphaeriales	Phyllostictaceae	*Phyllosticta* (1)
		Capnodiales	Mycosphaerellaceae	*Stigmidium* (2)
		Incertae sedis	Trichothyriaceae	*Lichenopeltella* (2)
			Pyrenidiaceae	*Pyrenidium* (1)
		Monoblastiales	Mycocaliciaceae	*Sphinctrina* (2)
		Pleosporales	Phaeosphaeriaceae	*Didymocyrtis* (1)
			Dacampiaceae	*Weddellomyces* (1)
	Eurotiomycetes	Sclerococcales	Dactylosporaceae	*Sclerococcum* (2)
	Genera incertae sedis	Genera incertae sedis	Genera Incertae Sedis	*Kalchbrenneriella* (1)
	Lecanoromycetes	Peltigerales	Massalongiaceae	*Polycoccum* (1)
	Sordariomycetes	Hypocreales	Bionectriaceae	*Cylindromonium* (1)
	Sordariomycetes	Hypocreales	Genera Incertae Sedis	*Roselliniella* (1)
Basidiomycota	Tremellomycetes	Filobasidiales	Filobasidiaceae	*Heterocephalacria* (1)
		Tremellales	Genera Incertae Sedis	*Biatoropsis* (1)
			Tremellaceae	*Tremella* (1)

Note: The numbers within the bracket indicates number species under that genus.

6.3.2 ENDEMISM AND RARITY

A total of 251 lichen taxa occurring in the Western Ghats are endemic to India, of which 129 are restricted in their distribution to this region only. In the case of lichenicolous fungi although four taxa are endemic, but it is too early to discuss about their endemism as they are newly described species. The Western Ghats also harbors some rare lichens. For example, *Arthothelium spectabile* A. Massal. is not collected since 1952, *Graphina canaliculata* (Fée) Müll. Arg., *Leptorhaphis epidermidis* (Ach.) Th. Fr. and *Bogoriella thelena* (Ach.) Aptroot & Lücking are not collected since 1832, while *Megaloblastenia marginiflexa* (Hook. f. Taylor) Sipman is not collected since 1892 (Singh and Sinha, 2010). These lichens may be extinct and suitable for inclusion in "Red Data Book."

TABLE 6.3 Most Speciose Genera of Lichens in the Western Ghats.

	Genus	Species		Genus	Species
1.	*Graphis* (Figure 6.3c)	78	6.	*Porina*	44
2.	*Pyrenula* (Figure 6.3g)	57	7.	*Ocellularia*	39
3.	*Parmotrema* (Figure 6.3f)	51	8.	*Pertusaria*	37
4.	*Usnea*	51	9.	*Fissurina*	33
5.	*Lecanora* (Figure 6.3e)	46	10.	*Allographa*	32

TABLE 6.4 Most Speciose and Dominant Families of Lichens in Western Ghats.

	Families	Species	Genus		Families	Species	Genus
1.	Graphidaceae	211	23	6.	Physciaceae	72	9
2.	Parmeliaceae	207	26	7.	Ramalinaceae	72	11
3.	Diploschistaceae	99	19	8.	Pyrenulaceae	70	5
4.	Caliciaceae	84	11	9.	Collemataceae	59	7
5.	Arthoniaceae	77	7	10.	Lecanoraceae	56	7

TABLE 6.5 Most Speciose Orders in Western Ghats and Number of Taxa Within.

	Order	Species	Genus	Families
1.	Lecanorales	423	67	13
2.	Graphidales	405	60	5
3.	Caliciales	156	20	2
4.	Peltigerales	128	29	5
5.	Arthoniales	122	28	6

6.3.3 SUBSTRATE PREFERENCE OF LICHENS

The bark (corticolous), rock (saxicolous), and soil (terricolous) are the main substrates for the growth of lichens in the Western Ghats. It is observed that the maximum number (1231) of lichens prefer to grow on bark followed by rock (90) while exclusive taxa growing on soil is less (22). Further, 76 taxa of lichens grow both on bark as well as on rock; 20 on bark and soil, while 50 taxa grow on all the three substrates (Figure 6.2). But lichens are also reported growing on leaves (foliicolous), dead wood (lignicolous) and on moss (muscicolous). Western Ghats has good diversity of foliicolous lichens (Figure 6.3h) represented by 88 taxa. Occurrence of good number of foliicolous lichens is another indication of healthy and undisturbed ecological conditions in parts of the Western Ghats. The foliicolous fungi prefer to grow in such undisturbed forests (Lücking, 1997). However, some foliicolous lichens also can grow on bark but not on rock or soil. Among the 22 taxa of muscicolous lichens, *Placynthiella icmalea* (Ach.) Coppins & P. James is exclusive to moss while others are found on rock, soil, or bark. A total of 12 species of lichens are reported from deadwood; however, they are also found on bark or soil.

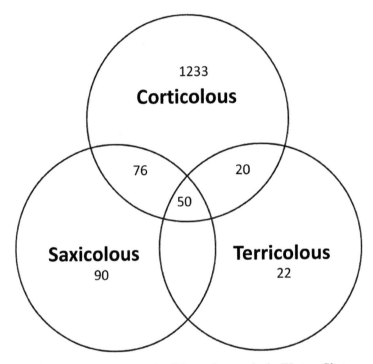

FIGURE 6.2   Species sharing among three major lichen substrates in the Western Ghats.

6.3.4 *DIVERSITY IN DIFFERENT STATES*

Among the six states through which the Western Ghats crosses Tamil Nadu represents the maximum number of lichens represented by 963 taxa, followed by Kerala with 783 and Karnataka with 658 taxa. Although the Western Ghats part of Maharashtra is one of the well-explored regions, it represents only 380 taxa. Goa is a smaller state with limited geographic area under the Western Ghats representingd 164 taxa. Goa is also an under-explored area for lichens. Similarly, the Western Ghats part of Gujarat also needs to be thoroughly explored for lichens. In Karnataka, Kerala, and Tamil Nadu, several protected areas and ecological interesting habitats are present. The lichens from the forests of these states are continuously explored for lichens since the British colonization period. There-fore, they represent a good number of lichens. In the case of lichenicolous fungi, it can be noted that there are no presentations from Goa, Gujarat, and Maharashtra states. The state Kerala represents the maximum number of lichenicolous fungi represented by 17 followed by Tamil Nadu with 14 species. Therefore, there is tremendous scope for the study of lichenicolous fungi in the Western Ghats.

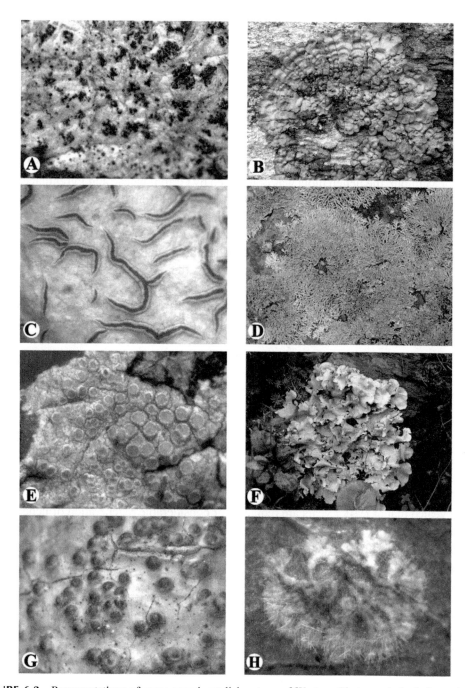

FIGURE 6.3 Representations of some prominent lichen taxa of Western Ghats. **a.** *Arthothelium abnorme* (Ach.) Müll. Arg., primitive lichens; **b.** *Coccocarpia palmicola* (Spreng.) Arv. & Galloway, cyanolichen; **c.** *Graphis cincta* (Pers.) Aptroot, member of most speciose genus; **d.** *Heterodermia diademata* (Taylor) D.D. Awasthi, common member of family Physciaceae; **e.** *Lecanora helva* Stizneb, member of a crustose and dominant genus; **f.** *Parmotrema nilgherrense* (Nyl.) Hale, member of most speciose family Parmeliaceae; **g.** *Pyrenula immissa* (Stirt.) Zahlbr., a dominant pyrenocarpous lichen; and **h.** *Strigula orbicularis* Fr., a foliicolous lichen. (Photos by authors)

6.3.5   AFFINITIES OF WESTERN GHATS LICHENS

The lichens of Western Ghats show much similarities to lichen biota of north-eastern India. Both regions show the dominance of graphidaceous and pyrenocarous lichens. For example, in a recent study conducted in Assam, out of 138 species, 59 belonged to Graphidaceae. Further, *Graphis* and *Pyrenula* were the prominent genera in Assam with 27 and 14 species, respectively (Behera et al., 2021). In Meghalaya, out of 337 species, 66 belonged to Graphidaceae, and *Graphis* and *Porina* were dominant genera with 30 and 18 species, respectively (Behera and Nayaka, 2020). Similarly, in a study conducted at the Senapati district of Manipur, the lichen biota was dominated by Graphidaceae and Pyrenulaceae with 16 and 15 species, respectively (Nayaka et al., 2020). Apart from the eastern Indian lichens, lichen biota of the Western Ghats shows affinities with Andaman Nicobar Islands. From these Islands, more than 160 taxa of lichens are reported so far and mostly they belong to crustose forms dominated by Graphidaceous and pyrenocarpous lichens (Niranjan and Sarma, 2018). The lichens of Western Ghats also show their affinities with tropical countries of the world, especially with Brazil, Colombia, Costa Rica, Indonesia, Mexico, Papua New Guinea, Paraguay, Thailand, and Vietnam. Several of the graphidaceous, pyrenocarpous, parmelioid, and physcioid lichens show their common occurrence in these countries as indicated by gbif.org. Needless to say, the lichens of Western Ghats are similar to that of Sri Lanka, and very few species have restricted distribution to these two regions.

6.4   CONSERVATION OF LICHENS IN WESTERN GHATS

The major utilization of lichens throughout the country is for spice. The lichens are slow-growing fungi and the growth rate of some of the lichens harvested for spice ranges from 5 mm to 2 cm per year. The lichens are also a lightweight organism and a large amount of raw material is required for various purposes (Upreti et al., 2005). A study conducted in the Western Ghats revealed that at least 7 years are required for considerable regeneration of lichens once harvested (Molleman et al., 2011). Moxham (1986) estimated that about 320 tons of lichens are collected from Nepal and adjoining areas. Shah (1997) recorded that approximately, 750 metric tons of lichens are harvested annually from Uttarakhand and another 800 metric tons from other regions of India including Himachal Pradesh, Sikkim, and Assam. However, such an estimate is not available for the Western Ghats, although, it is known that large quantities of lichens are harvested from this region. In Palni Hills of Tamil Nadu, the Paliyan tribes harvest a large amount of lichens and sell it to the local traders at the rate of 50 rupees per kg (Molleman et al., 2011). In Karnataka, lichens are harvested from forests of several districts including Chamarjpete, Chikmagalore, Mysore, and Shimoga by tribes and local people. The Bilirigi Rangana Hills of Mysore district is one of the hotspots for lichen harvest. The "Kuruba" tribes are known for lichen collection in parts of Karnataka. However, there is no serious effort for the conservation of lichens in the Western Ghats or in India. The lichens are treated as any other non-timber forest products and same guidelines apply for their harvest. Upreti and Nayaka (2008) discussed the lichen conservation issues in India and suggested creating lichen gardens and declaring lichen-rich sites as special sanctuaries. Although the Western Ghats has several lichen-rich sites, Upreti and Nayaka (2008) recommended Bhagwan Mahavir Wildlife Sanctuary, Goa; Jog Falls Forest area, Karnataka; Myladumpara Forest area, Kerala; and Nilgiri and Palni Hills of Tamil Nadu to be declared as lichen sanctuaries.

6.5 CONCLUSION

Undoubtedly, the Western Ghats has a rich diversity of lichens within India, which is equivalent to or more than that of north-east India. There are at least 10 research groups within the entire stretch of Western Ghats studying the lichens. Several new lichens are being added to the lichen biota of Western Ghats on regular basis. However, sometimes, there is question of authenticity due to their erroneous identification. Several ecologically interesting sites are being explored for lichens in the recent days. Nevertheless, there remains a number of interesting areas and special habitats, such as Shola forests of Kudremukh National Park in Karnataka, Sanctuaries in Goa, forests of Gujarat need to be explored. There is also need for revisiting some of the sites, such as Nilgiri and Palni Hills which are explored more than 25 years back. An addition of minimum 500 species can be expected if thorough exploration of lichens is carried out in the Western Ghats. There is also need for quantitative estimation of lichen diversity in Western Ghats so as to identify rare and threatened lichen taxa. In addition, the survey and quantification of lichen harvesting is also required for conservation. Some of the lichen species luxuriantly growing and available in bulk quantity may be utilized for bioprospecting. In conclusion, lichen diversity of Western Ghats needs to be explored intensively, their indicator values should be exploited, and they should be utilized in sustainable manner for the benefit of mankind.

ACKNOWLEDGMENTS

We are thankful to Directors, CSIR-NBRI, Lucknow, and KSCSTE-JNTBGRI for providing the laboratory facilities under in-house project OLP 0101; to Drs. G. P. Sinha, Siljo Joseph, and Komal K. Ingle for their clarification regarding some lichen taxa; to Mr. Roshnikumar Ngangom, Vishal Kumar, and Yogeshwar Sahu for helping in some data entry.

KEYWORDS

- **affinity**
- **biodiversity**
- **endemism**
- **lichenized fungi**
- **symbiosis**
- **taxonomy**

REFERENCES

Ahmadjian, V. Lichens Are More Important Than You Think. *BioScience* **1995**, *45* (3), 124.

Aptroot, A.; Sipman, H. J. M. Diversity of Lichenized Fungi in the Tropics. In *Biodiversity of Tropical Microfungi*; Hyde, K. D., Ed.; University Press, Hong Kong, **1997**; pp 93–106.

Awasthi, D. D. *Catalogue of Lichens from India, Nepal, Pakistan, and Ceylon.* Beihefte zur Nova Hedwigia, Heft 17. Verlag von J. Cramer, Weinheim, **1965**; p 137.

Balaji, P.; Hariharan, G. N. Checklist of Microlichens in Bolampatti II Forest Range (Siruvani Hills), Western Ghats, Tamil Nadu, India. *Czech Mycol.* **2013b**, *65* (2), 219–232.

Balaji, P.; Hariharan, G. N. Diversity of Macrolichens in Bolampatti II Forest Range (Siruvani Hills), Western Ghats, Tamil Nadu. *Int. Sch. Res. Notices.* **2013a**, *2013*, 1–7.

Behera, P. K.; Nayaka, S. Updated Checklist of Lichen Biota of Meghalaya, India with 93 New Distributional Records for the State. *J. Indian Bot. Soc.* **2020**, *100* (3&4), 134–147, DOI: 10.5958/2455-7218.2020.00033.9

Behera, P. K.; Nayaka, S.; Upreti, D. K.; Chauhan, R. S. New Distributional Records to Lichen Biota of Assam, India. *Indian For.* **2021**, *147* (4), 400–404, DOI: 10.36808/if/2021/v147i4/152523

Biju, H.; Bagool, R. G.; Nayaka, S. Additions to the Lichen Flora of Kerala State I: Parmelioid Macro-Lichens. *J. Econ. Taxon. Bot.* **2010**, *34* (4), 890–897.

Biju, H.; Bagool, R. G.; Nayaka, S. Additions to the Lichen Flora of Kerala State 2: Graphidaceae. *J. Econ. Taxon. Bot.* **2012**, *36* (4), 867–873.

Biju, H.; Bagool, R. G.; Nayaka, S. Diversity of Lichens in Idukki District with New Records to Flora of Kerala. *Indian J. For.* **2014a**, *37* (3), 333–340.

Biju, H.; Bagool, R. G.; Nayaka, S. New Records of Graphidaceous Lichens from Western Ghats, India. *Indian J. For.* **2014b**, *37* (4), 477–481.

Biju, H.; Sabeena, A.; Nayaka, S. New Records of Graphidaceae (Lichenized Fungi) from the Western Ghats of Kerala state, India. *Stud. Fungi.* **2021**, *6* (1), 213–223, DOI: 10.5943/sif/6/1/14

Chakarwarti, J.; Nayaka, S.; Srivastava, S. The Diversity of Endolichenic Fungi – A Review. *Asian J. Mycol.* **2020**, *3* (1), 490–511. DOI: 10.5943/ajom/3/1/18

Chitale, G.; Makhija, U. A New Species of the Lichen Genus *Phlyctis* from Maharashtra, India. *Mycotaxon* **2012**, *120* (1), 75–79.

Chitale, G.; Makhija, U.; Sharma, B. O. Additional Species of *Graphis* from Maharashtra, India. *Mycotaxon* **2011**, *115* (1), 469–480.

Chitale, G.; Makhija, U.; Sharma, B. O. New Combinations and New Species in the Lichen genera *Hemithecium* and *Pallidogramme*. *Mycotaxon* **2009**, *108* (1), 83–92.

Diederich, P.; Lawrey, J. D.; Ertz, D. The 2018 Classification and Checklist of Lichenicolous Fungi, with 2000 Nonlichenized, Obligately Lichenicolous Taxa. *Bryologist* **2018**, *121* (3), 340–425.

Easa, P. S. *Biodiversity Documentation of Kerala Part 3: Lichens.* Kerala Forest Research Institute: Peechi, Kerala, **2003**; p. 61.

Feuerer, T.; Hawksworth, D. L. Biodiversity of Lichens, Including a World-Wide Analysis of Checklist Data Based on Takhtajan's Floristic Regions. *Biodivers. Conserv.* **2007**, *16* (1), 85–98. DOI: 10.1007/s10531-006-9142-6

Hawksworth, D. L. Challenges in Mycology. *Mycol. Res.* **1995**, *99* (1), 127–128. DOI: 10.1016/s0953-7562(09)80326-2

Hawksworth, D. L.; Grube, M. Lichens Redefined as Complex Ecosystems. *New Phytol.* **2020**, *227* (5), 1362–1375. DOI: 10.1111/nph.16630

Ingle, K. K.; Nayaka, S.; Suresh, H. S. Lichens in 50 ha Permanent Plot of Mudumalai Wildlife Sanctuary, Tamil Nadu, India. *Trop. Plant Res.* **2016**, *3* (3), 694–700.

Jaklitsch, W. M.; Baral, H. O.; Lucking, R.; Lumbsch, H. T. Ascomycota. In *Syllabus of Plant Families – Adolf Engler's Syllabus der Pflanzenfamilien*; Frey, W., Ed.; Borntraeger: Stuttgart, **2016**; p 288.

Joseph, S.; Nayaka, S.; Ramachandran, V. S. Addition to the Lichen Flora of Nilgiris in Tamil Nadu. *ENVIS Newsl. (BSI)*, **2011**, *16 (1),* 6.

Joshi, Y. Documentation of Lichenicolous Fungi From India–Some Additional Reports. *Kavaka* **2018**, 51, 30–34.

Krishnamurthy, Y. L.; Subramanya, S. K. *Foliicolous Lichens of Central Western Ghats, India: Diversity, Distribution and Molecular Study.* Lambert Academic Publishing: Mauritius, **2017**.

Kumar, M. *Lichen (Macrolichen) Flora of Kerala Part of Western Ghats Kerala*. Forest Research Institute Research Report No. 194. KFRI: Peechi, **2000**.

Kumar, M.; Sequiera, S. An Enumeration of Macro Lichens from Palakkad District, Kerala State, India. *Indian J. For.* **2002**, *25* (3), 347–353.

Kumar, M.; Sequiera, S. Lichen of Western Ghats–An Overview. In *Biology of Lichens*; Mukerji, K. G., Chamola, B. P., Upreti, D. K., Upadhyay, R. K., Eds.; Aravali Books International: New Delhi, **1999**; pp. 297–331.

Lawrey, J. D.; Diederich, P. Lichenicolous Fungi–Worldwide Checklist, Including Isolated Cultures and Sequences 2018. http://www.lichenicolous.net (accessed Apr 11, 2020).

Lücking, R. The Use of Foliicolous Lichens as Bioindicators in the Tropics, with Special Reference to Microclimate. *Abstr. Bot.* **1997**, *21* (1), 99–116.

Lücking, R.; Hodkinson, B. P.; Leavitt, S. D. The 2016 Classification of Lichenized Fungi in the Ascomycota and Basidiomycota–Approaching One Thousand Genera. *Bryologist* **2017**, *119* (4), 361–416. DOI: 10.1639/0007-2745-119.4.361

Lücking, R.; Rivas Plata, E.; Chaves, J. L.; Umaña, L.; Sipman, H. J. M. How Many Tropical Lichens Are There ... Really? In *Diversity of Lichenology;* Thell, A., Seaward, M. R. D., Feuerer, T., Eds.; Bibliotheca Lichenologica. No. 100. J; Cramer in der Gebrüder Borntraeger Verlagsbuchhandlung: Berlin and Stuttgart, **2009**; pp 399–418.

Makhija, U.; Chitale, G.; Dube, A. *Lichens of Maharashtra*; Bishen Singh Mahendra Pal Singh: Dehra Dun, India, **2014**.

Mishra, G. K.; Dubey, N.; Bagla, H.; Bajpai, R.; Nayaka, S. An Assessment of Lichens Diversity from Bhimashankar Wildlife Sanctuary, Maharashtra, India. *Crypt. Biodivers. Assess.* **2017**, *2* (2), 11–17.

Molleman, L.; Boeve, S.; Wolf, J.; Oostermeijer, G.; Devy, S.; Ganesan, R. Commercial Harvesting and Regeneration of Epiphytic Macrolichen Communities in the Western Ghats, India. *Environ. Conserv.* **2011**, *38* (03), 334–341, DOI: 10.1017/s0376892911000142.

Money, N. P. Fungal Diversity. In *The Fungi,* 3rd ed.; Watkinson, S. C., Boddy, L., Money, N. P., Eds.; Academic Press: U.K., **2016**; pp 1–36.

Moxham, T. H. The Commercial Exploitation of Lichens for the Perfume Industry. In *Progress in Essential Oil Research*; Brunke, E. J., Ed.; Walter de Gruyter: Berlin, 1986; pp 491–503.

Nayaka, S.; Asthana, S. Diversity and Distribution of Lichens in India *vis a vis* Its Lichenogeographic Regions. In *Biodiversity Conservation, Status, Future and Way Forward;* Marimuthu, T., Ponmurugan, P., Subramanian, M., Mathivanan, N., Anita. S., Eds.; National Academy of Biological Science: Chennai, **2014**; pp 79–96.

Nayaka, S.; Ingle, K. K.; Bajpai, R.; Rawal, J. R.; Upreti, D. K.; Trivedi, S. Lichens of Gujarat State, India with Special Reference to Coastal Habitats. *Curr. Res. Environ. Appl. Mycol.* **2013**, *3* (2), 222–229. DOI: 10.5943/crea/3/2/4

Nayaka, S.; Joseph, S.; Ngangom, R.; Tilotama, K.; Arnold, P. K. Preliminary Studies on the Lichens Growing in FEEDS Campus and SB Garden in Manipur, India. *Stud. Fungi.* **2020**, *5* (1), 392–399. DOI: 10.5943/sif/5/1/20

Nayaka, S.; Raval, J.; Punjanai, B.; Ingle, K. K.; Upreti, D. K. In *Documenting Lichen Diversity in Marine National Park and Sanctuary, Gulf of Kutch, Gujarat*, Book of Abstracts, Fourth Indian Biodiversity Congress (IBC 2017), Mar 10–12, 2017; Pondicherry University: Puducherry, **2017**.

Nayaka, S.; Upreti, D. K. *Lichens Diversity in Western Ghats: Need for Quantitative Assessment and Conservation*; Commissioned Paper, Report of the Western Ghats Ecological Expert Panel, Ministry of Environment and Forests: New Delhi, 2011. http://varmaopinon.blogspot.com/2018/10/report-of-western-ghats-ecology-expert_14.html

Nayaka, S.; Upreti, D. K. *Status of Lichen Diversity in Western Ghats, India*; Sahyadri E-News, *Western Ghats Biodiversity Information System* – **2005** *Issue XVI*. http://wgbis.ces.iisc.ernet.in/biodiversity/newsletter/issue16/main_index.htm

Niranjan, M.; Sarma, V. V. A Check-List of Fungi from Andaman and Nicobar Islands, India. *Phytotaxa* **2018**, *347* (2), 101. DOI: 10.11646/phytotaxa.347.2.1

Patwardhan, P. G. Rare and Endemic Lichens in the Western Ghats, South Western India. In *An Assessment of Threatened Plants of India;* Jain, S. K., Rao, R. R., Eds.; Botanical Survey of India: Howrah, **1983**; pp 318–322.

Randive, P.; Joseph, S.; Gupta, P.; Nayaka, S.; Janarthanam, M. K. Additional Records of the Foliicolous Lichens to the State of Goa. *Indian For.* **2019,** *145* (7), 687–688.

Randive, P.; Joseph, S.; Nayaka, S.; Janarthanam, M. K. Notes on Foliicolous Lichens from Western Ghats Part of Goa, India. *Indian J. For.* **2017a,** *40* (3), 217–221.

Randive, P.; Nayaka, S.; Janarthanam, M. K. An Updated Checklist of Lichens from Goa with New Records from Cotigao Wildlife Sanctuary. *Crypt. Biodivers. Assess.* **2017b,** *2* (1), 26–36.

Randive, P.; Nayaka, S.; Janarthanam, M. K. Lichens of Cotigao Wildlife Sanctuary, Goa. *Int. J. Life Sci.* **2018,** (A9), 31–36.

Rashmi, S.; Rajkumar, H. G. First Report of Foliicolous Lichen Biota in South Karnataka-India. *Int. J. Curr. Microbiol. App. Sci.* **2015a,** *4* (6), 250–256.

Rashmi, S.; Rajkumar, H. G. Enumeration and New Records of Lichens in Kodagu district–A Micro Hotspot in Western Ghats of Karnataka, India. *Int. Res. J. Environ. Sci.* **2015b,** *4* (12), 17–25.

Rikkinen, J. Cyanolichens. *Biodivers. Conserv.* **2015,** *24,* 973–993.

Sequiera, S. Status of Lichen Diversity in Kerala, India. *Samagra* **2012,** *8,* 28–39.

Shah, N. C. Lichens of Economic Importance from the Hills of Uttar Pradesh. *J. Herbs Spices Med. Plants.* **1997,** *5,* 69–76.

Singh, A. *Lichenology in Indian Subcontinent 1966–1977.* Economic Botany Information Service, National Botanical Research Institute: Lucknow, **1980.**

Singh, K. P.; Sinha, G. P. *Indian Lichens: An Annotated Checklist.* Botanical Survey of India: Kolkata, 2010.

Singh, K. P.; Sinha, G. P. Lichens. In *Floristic Diversity and Conservation Strategies in India: (Cryptogams and Gymnosperms);* Mudugal, V., Hajra, P. K., Eds.; Botanical Survey of India: Howrah, **1997;** vol I; pp 195–234.

Singh, P.; Singh, K. P. A New Species of *Fissurina* and Additional Records of Graphidaceae (Ascomycota: Ostropales) from the State of Karnataka, India. *NeBIO* **2017,** *8* (1), 21–24.

Sinha, G. P.; Jagadeesh Ram, T. A. M. In *An Overview of the Current Status of Lichen Diversity in India and Identification of Gap Areas,* Book of Abstracts, International Symposium on Plant Taxonomy and Ethnobotany, Feb 13–14, 2020; Botanical Survey of India: Kolkata, **2020.**

Sinha, G. P.; Nayaka, S.; Joseph, S. Additions to the Checklist of Indian Lichens After 2010. *Crypt. Biodivers. Assess.* **2018,** *2008,* 197–206, DOI: 10.21756/cab.esp16

Sipman, H. J. M.; Aptroot, A. Where Are the Missing Lichens? *Mycol. Res.* **2001,** *105* (12), 1433–1439. DOI: 10.1017/s0953756201004932

Smith, A. L. Cryptotheciaceae. A Family of Primitive Lichens. *Transact. Brit. Mycol. Soc.* **1926,** *11,* 189–196.

Subramanya, S. K.; Vinayaka, K. S.; Udupa, K. E. S.; Shashirekha, B.; Praveena, V.; Krishnamurthy, Y. L. Diversity and Host Specificity of Lichens in Koppa Taluk of Central Western Ghats, Karnataka, India. *Indian J. For.* **2010,** *33* (3), 437–442.

Subramanya, S. K.; Krishnamurthy, Y. L. Diversity of Foliicolous Lichens in Shivamogga District, Karnataka. *Crypt. Biodivers. Assess.* **2017,** 176–181.

Subramanya, S. K.; Krishnamurthy, Y. L. New Distribution Records of *Strigula* (Strigulaceae, Ascomycota) From the Western Ghats in India. *Pol. Bot. J.* **2015a,** *60* (1), 99–103.

Subramanya, S. K.; Krishnamurthy, Y. L. Two New Lichen Species of the Genus *Coenogonium* (Ostropales: Coenogoniaceae) From the Western Ghats in India. *Kavaka* **2015b,** 44, 63–65.

Subramanya, S. K.; Krishnamurthy, Y. L. *Flavobathelium epiphyllum*: A New Addition to the Foliicolous Lichen Flora of India. *Indian For.* **2016a,** *142* (6), 615–616.

Subramanya, S. K.; Krishnamurthy, Y. L. Notes on Three New Records of Foliicolous Lichens from Karnataka Western Ghats, India. *J. Threat. Taxa.* **2016b,** *8* (6), 8950–8952. DOI: 10.11609/joft.2036.8.6.8950-8952

Upreti, D. K.; Divakar, P. K.; Nayaka, S. Commercial and Ethnic Use of Lichens in India. *Econ. Bot.* **2005,** *59* (3), 269–273. DOI: 10.1663/0013-0001(2005)059[0269:caeuol]2.0.co;2

Upreti, D. K.; Nayaka, S. Need for Creation of Lichen Gardens and Sanctuaries in India. *Curr. Sci.* **2008,** *94* (8), 976–978.

Vinayaka, K. S. Diversity and Distribution of Tropical Macrolichens in Shettihalli Wildlife Sanctuary, Western Ghats, Southern India. *Plant Sci. Today.* **2016,** *3* (2), 211–219.

Vinayaka, K. S.; Chetan, H. C.; Archana, R. M. Diversity and Distribution Pattern of Lichens in the Mid-Elevation Wet Evergreen Forest, Southern Western Ghats, India. *Intern. J. Res. Studies in Biosci.* **2016,** *4* (1), 15–20.

Vinayaka, K. S.; Krishnamurthy, Y. L.; Nayaka, S. Macrolichen Flora of Bhadra Wildlife Sanctuary, Karnataka, India. *Ann. For.* **2010,** 18, 81–90.

Vinayaka, K. S.; Krishnamurthy, Y. L. Ecology and Distribution of Lichens in Bhadra Wildlife Sanctuary, Central Western Ghats, Karnataka, India. *Biorem. Biodiver. Bioavail.* **2010a,** 5, 68–72.

Vinayaka, K. S.; Krishnamurthy, Y. L. In *A Preliminary Study on Macrolichens of Malnad Regions of Western Ghats, Karnataka, India,* Proceedings of International Conference on Biodiversity Conservation and Management, **2010b;** pp 118–121.

Vinayaka, K. S.; Nayaka, S.; Krishnamurthy, Y. L.; Upreti, D. K. A Report on Some Macrolichens New to Karnataka, India. *J. Threat. Taxa.* **2012,** *4* (1), 2318–2321.

Vinayaka, K. S.; Subramanya, S. K.; Udupa, S. K.; Krishnamurthy, Y. L. Diversity of Epiphytic Lichens and Evaluation of Important Host Species Exploited by Them in Tropical Semi-Evergreen and Deciduous Forests of Koppa, Central Western Ghats, India. *Asian Australas. J. Plant Sci. Biotechnol.* **2011,** *5* (1), 62–66.

Wijayawardene, N. N.; Hyde, K. D.; Al-Ani, L. K. T.; et al. Outline of Fungi and Fungus-Like Taxa. *Mycosphere* **2020,** *11* (1), 1060–1456. DOI: 10.5943/mycosphere/11/1/8

Zachariah, S.; Nayaka, S.; Gupta, P.; Varghese, S. K. The Lichen Genus *Pyxine* (Caliciaceae) in Kerala State with *P. dactyloschmidtii*as New to India. *Hattoria* **2019,** *10,* 109–117.

Zachariah, S. A.; Nayaka, S.; Joseph, S.; Gupta, P.; Thomas, S.; Varghese, S. K. New and Noteworthy Records of Lichens from Pathanamthitta District, Kerala, India. *Stud. Fungi.* **2018,** *3* (1), 349–356. DOI: 10.5943 / sif/3/1/35

Zachariah, S. A.; Nayaka, S.; Joseph, S.; Gupta, P.; Varghese, S. K. Eleven New Records of Lichens to the State of Kerala, India. *J. Threat. Taxa.* **2020,** *12* (10), 16402–16406. DOI: 10.11609/jot.5475.12.10.16402-16406

Zahlbruckner, A. *Catalogus Lichenum Universalis,* vol. (1–10), Borntraeger Leipzig: New York, **1922–1940.**

WEBSITES REFERRED

1. www.indexfungorum.org
2. www.gbif.org
3. www.mycobank.org
4. Singh, K.P. https://karunadu.karnataka.gov.in/kbb/english/Pages/Lichens-of-Karnataka.aspx
5. Hariharan, G.N.; Balaji, P. http://tnenvis.nic.in/tnenvis_old/Lichens/tamil-nadu.htm

CHAPTER 7

BRYOPHYTES OF THE WESTERN GHATS: A LOOK INTO THE DISTRIBUTION PATTERN

C. N. MANJU[1,2], K. P. RAJESH[2], K. M. MANJULA[2,3], B. MUFEED[1,2], and V. K. CHANDINI[2,4]

[1]*Bryology Laboratory, Department of Botany, University of Calicut, Calicut University P.O., Malappuram, Kerala, India*

[2]*PG and Research Department of Botany, The Zamorin's Guruvayurappan College (Affiliated to the University of Calicut), Kerala, India*

[3]*KSCSTE-Malabar Botanical Garden and Institute for Plant Science, Kozhikode-14, Kerala, India*

[4]*Department of Botany, St. Teresa's College (Autonomous), Ernakulam, Kerala, India*

ABSTRACT

The current status of the bryophytes of Western Ghats and their distribution pattern is examined. Like other plant groups, the distribution of Bryophytes also is on par with the moisture regime of the area. The diversity of this group varies in accordance with the altitudinal zones, and the nature of the macro vegetation types of the area. A total of 1224 bryophytes including 21 hornworts, 379 species liverworts, and 824 species mosses are known from the Western Ghats from the states of Maharashtra, Goa, Karnataka, Kerala, and Tamil Nadu along the Western Ghats. Among these 79 species are endemic to Western Ghats. The details on the distribution of Bryophytes were prepared based on the field studies conducted in the area and earlier works such as the checklists for Kerala (Manju et al., 2008), Tamil Nadu (Daniels, 2010), Karnataka (Frahm et al., 2013), and other published records. It also corresponds to the extent of works carried out in these areas.

7.1 INTRODUCTION

The bryophytes, being the major part of the *bryosphere*, play a prominent role in the ecosystem dynamics in most of the ecosystems of the world (Lindo and Gonzalez, 2010).

Biodiversity Hotspot of the Western Ghats and Sri Lanka. T. Pullaiah, PhD (Ed.)

During the last 20 years bryophytes received much attention in Western Ghats and the addition to the biodiversity increases day by day. The present chapter is an attempt to collate the distribution pattern on the bryophyte diversity of the Western Ghats. The first record of the bryophyte from the Western Ghats can be traced back to Van Rheede's monumental work *Hortus Indicus Malabaricus*. He described and illustrated one bryophyte, as *"poovan-peda"* (in volume 12 as tab. 37, p. 71.1693), earlier to the work by Linnaeus (1753), which later was assumed as *Bryum dichotomum* Dickson. But, after this collection there is a long gap in the recording of bryophytes from the Western Ghats. The earlier studies on the bryophytes of the Western Ghats are scanty and limited to random collection in the 19th century. The first attempt to explore the bryophytes of the Western Ghats started by some of the earlier settlers in the Nilgiri hills. Montagne (1842a,b) reported 66 species of mosses and 34 species of liverworts from the Nilgiri Hills (Tamil Nadu) in *Cryptogamae Nilgherienses*. Müller (1853) reported 56 species of mosses from the Nilgiris in *Musci Neilgherrensis*. Mitten (1859) reported over 700 species in his *Musci Indiae Orientalis*, of which 145 were from Tamil Nadu. Subsequently, Mitten (1861) reported 26 species of liverworts from Tamil Nadu in *Hepaticae Indiae Orientalis*. The 20th century also witnessed some bryophyte studies such as those of Dixon (1914) who reported 35 species of mosses from the Nilgiris. From the Western Ghats of Madurai (Tamil Nadu), Potier de la Varde (1922–1924) reported 48 species in his *Musci Madurensis*, 49 species in *Nouvelles herborisations dans le sud de l'Inde* (1925), and seven species in *Musci novi Indici* (1928). Dixon and Potier de la Varde's (1927, 1930) treatment on Indian bryophytes, *Contribution à la flore bryologique de l'Inde meridionale*, also includes many species from the Western Ghats of Tamil Nadu. They (Dixon and Potier de la Varde's 1930) also reported 56 species from Tamil Nadu in *Nouvelle contribution à la flore bryologique de l'Inde*. Another major contribution was made by Foreau (1930, 1961, 1964), who listed 368 species of mosses from the Palni hills (W. Ghats of Madurai) which included 95 new species and 15 new varieties.

The earlier studies on the bryophytes of Kerala are scanty and limited to random collections. The works of Dixon (1914), Bruhl (1931), Chopra (1938), Ellis (1989), etc. are major among them. Other bryologists such as Stephani (1898–1900, 1901–1906, 1907–1909, 1909–1912, 1912–1917, 1917–1924), Benedix (1953), Udar (1976), Udar and Srivastava (1975, 1977), Asthana and Srivastava (1991), Asthana et al. (1995), Nath and Asthana (1998), Singh (1994, 2002), Srivastava and Srivastava (2002), etc. also recorded some species of liverworts from Kerala, Tamil Nadu, and Karnataka. Gangulee's monumental work *Mosses of eastern India and adjacent regions* (1969–1980) provides information on the habitats and distribution of south Indian mosses. He occasionally mentions the occurrences of some species in Tamil Nadu, Kerala, Karnataka, but without exact localities. Dandotiya et al. (2011) presented a checklist of the Bryophytes of India, totaling 2489 taxa. However, it does not mention many species reported from South India.

Recent exhaustive studies on the bryophytes of Kerala by Nair et al. (2004, 2005a,b, 2006a,b, 2007a-c), Manju and Rajesh (2009a,b, 2011a,b, 2017), Manju et al. (2008, 2012a-d, 2015a,b, 2017a-c, 2019a,b, 2020, 2021), Daniels (2001, 2003, 2004, 2010, 2021), Daniels and Daniel (2002, 2003a-c, 2004, 2005, 2007a,b, 2008a,b, 2009a,b, 2013), Daniels et al.

(2010, 2011a,b, 2012, 2013, 2016, 2017), Daniels and Kariyappa (2007, 2013), Daniels and Felix (2009), Daniels and Mabel (2009, 2010), Daniels and Raja (2011), Daniels and Brijithlal (2012). Books by Daniels and Kariyappa (2019) and Daniels and Raja (2020) revealed the occurrence of a good number of species, which were added to the bryoflora of Kerala part of the Western Ghats. Nair et al. (2005a) published a book *"Bryophytes of Wayanad in Western Ghats,"* the first comprehensive taxonomic treatment on bryophytes from South India, mainly based on the Wayanad Wildlife Sanctuary. This includes a taxonomic account of 171 species and two varieties belonging to 105 genera and 47 families, including two new species *viz. Trichostomum wayanadense* Manju, K.P. Rajesh & Madhus. and *Amphidium gangulii* Manju, K.P. Rajesh and Madhus. Many of the species described were new distributional records of phytogeographical significance.

Daniels and Daniel (2013) published a book on the bryoflora of the Southernmost Western Ghats. It includes an account of 240 species of bryophytes with 145 mosses, 90 liverworts, and 5 hornworts. This work reports three new species and seven new records for India. Manju and Rajesh (2017) published a book as *Bryophytes of Kerala, Liverworts Part 1.* and reported 100 species of liverworts with photos. Daniels et al.'s (2018) book *"Bryoflora of Indira Gandhi National Park in Anamalai Hills of Western Ghats"* includes descriptions of 217 species with 135 mosses, 81 liverworts, and one hornwort. Among these seven species were reported as new for India and several species as new regional records and rediscoveries.

Daniels and Kariyappa (2019) published a book of the *Bryoflora of the Agasthyamalai Biosphere Reserve in the Southern Western Ghats* which stretches over the borders of Kollam and Thiruvananthapuram Districts in Kerala and Tirunelveli and Kanyakumari Districts in Tamil Nadu. They reported an account of 403 species and 15 intraspecific taxa belonging to 160 genera, 60 families, and 16 orders. Daniels and Raja (2020) published a book on the *Bryophytes of Silent Valley National Park in the Southern Western Ghats of Kerala* and dealt with 212 species and 3 infra-specific taxa belonging to 124 genera, 51 families, and 15 orders.

7.2   BRYOPHYTES OF WESTERN GHATS

A total of 1224 bryophytes including 21 species of hornworts, 379 species of liverworts, and 824 species of mosses are known from the Western Ghats. Among these 79 species are endemic to the Western Ghats. This is based on the updated data set from the earlier checklists such as Manju et al. (2008) for Kerala, Daniels (2010) for Tamil Nadu, and Frahm et al. (2013) for Karnataka and other published records. Details from the field studies conducted by us in the Western Ghats were also incorporated in collating the data on distribution of Bryophytes of the area. In recent years many additions were made to the Western Ghats and several new species were added from Kerala and Tamil Nadu. It also corresponds to the extent of works carried out in these areas. Some parts such as Kerala and Tamil Nadu are well explored for bryophytes in the recent past. However, efforts for documentation of this group are not extensive for areas such as Karnataka, Goa, and Maharashtra.

A total of 1224 Bryophyte taxa, including infra-specific categories, are now known from the states of the Western Ghats. The highest number known is 832 from the state of Tamil Nadu, followed by 713 from Kerala (Figure 7.1). Karnataka is known with 350 taxa. It is comparatively low, considering its vast area and mosaics of diverse habitats with a major part in the congenial, moist region of the Western Ghats. However, the low number, in any way, is not due to the poor quality of habitats or any other geographical or climatic factors, but it is due to the low amount of effort in documenting this plant group in Karnataka, compared to Kerala and Tamil Nadu. If explored in detail, chances are there in recording more than 800 taxa from Karnataka also. The low number of taxa from Maharashtra and Goa is as expected, due to the drier habitats. Out of the 1224, a total of 716 species are known from a single state only. Among these, Tamil Nadu has maximum nonshared ones with 374 taxa. Kerala and Karnataka are with 247 and 92 nonshared taxa respectively. Maharashtra is with two and Goa is with a single nonshared species. The cluster analysis based on the Sorensen Similarity Index also supports this trend (Figure 7.2). Maharashtra and Goa formed a group with 49.6% similarity. The other three states formed a group; among which Kerala and Tamil Nadu are with 59.3% similarity in the Bryophyte species composition.

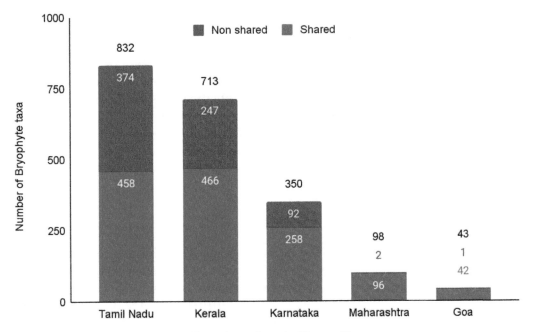

FIGURE 7.1   Pattern of distribution of bryophytes along the Western Ghats.

Among the total bryophyte taxa known from the Western Ghats, *ca.* 30 species are the most common ones, occurring in all these states. This includes 11 Liverworts (*Cyathodium cavernarum* Kunze, *Dumortiera hirsuta* (Sw.) Nees, *Heteroscyphus argutus* (Nees) Schiffn., *Lejeunea flava* (Sw.) Nees, *Lejeunea obfusca* Mitt., *Mastigolejeunea auriculata* (Wilson & Hook.) Schiffn., *Pallavicinia ambigua* (Mitt.) Steph., *Riccardia levieri* Schiffn., *Riccia billardieri* Mont. & Nees ex Gottsche, Lindenb. & Nees, *Riccia huebeneriana* Lindenb. and *Targionia hypophylla* L.) and 19 Mosses (*Barbula indica* (Hook.) Spreng., *Bryum cellulare* Hook., *Bryum coronatum* Schwägr., *Calymperes afzelii* Sw., *Calymperes erosum* Müll.Hal., *Fissidens bryoides* Hedw., *Fissidens ceylonensis* Dozy & Molk., *Fissidens crispulus* Brid. var. *crispulus*, *Fissidens flaccidus* Mitt., *Fissidens walkeri* Broth. var. *walkeri, Fissidens zollingeri* Mont., *Floribundaria walkeri* (Renauld & Cardot) Broth., *Garckea flexuosa* (Griff.) Margad. & Nork., *Hyophila involuta* (Hook.) A.Jaeger, *Isopterygium albescens* (Hook.) A.Jaeger, *Macromitrium sulcatum* (Hook.) Brid., *Octoblepharum albidum* Hedw., *Philonotis hastata* (Duby) Wijk & Margad. and *Racopilum orthocarpum* Wilson & Mitt.). Most of these taxa are with high light tolerance and adapted to the disturbed habitats. Except for a few, most of them are the common elements in the human habitats or human-modified habitats.

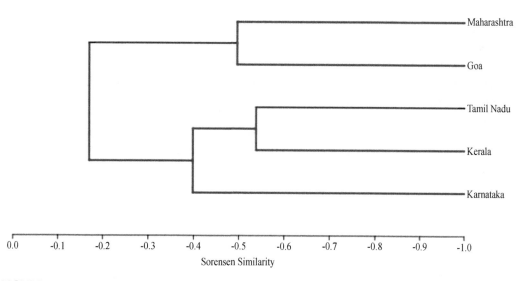

FIGURE 7.2   Dendrogram based on Sorensen similarity index showing the pattern of bryophyte distribution along the states of the Western Ghats.

In addition to the above-mentioned 30 taxa, 158 taxa are shared in the states of Karnataka, Tamil Nadu, and Kerala. This includes three Hornworts (*Anthoceros crispulus* (Mont.) Douin, *Notothylas dissecta* Steph. and *Phaeoceros laevis* (L.) Prosk. subsp. *laevis*), 63 Liverworts, and 92 Mosses. The Liverworts common to these states are: *Archilejeunea apiculifolia* Steph., *Asterella khasyana* (Griff.) (Griff.) Pandé et al., *Asterella leptophylla* (Mont.) Grolle, *Asterella wallichiana* (Lehm.) Grolle, *Caudolejeunea reniloba* (Gottsche) Steph., *Cheilolejeunea birmensis* (Steph.) Mizut., *Cheilolejeunea laeviuscula* (Mitt.) Steph., *Cheilolejeunea trapezia* (Nees) Kachroo & R.M.Schust., *Cololejeunea appressa* (A.Evans) Benedix, *Cololejeunea furcilobulata* (Berrie & E.W.Jones) R.M.Schust., *Cololejeunea hasskarliana* (Lehm. & Lindenb.) Steph., *Cololejeunea hyalina* G. Asthana & S.C. Srivast., *Cololejeunea lanciloba* Steph., *Cololejeunea latilobula* (Herzog) Tixier, *Cololejeunea minutissima* (Sm.) Schiffn., *Cololejeunea planissima* (Mitt.) Abeyw.,

Cololejeunea spinosa (Horik.) S.Hatt., *Cololejeunea udarii* G. Asthana & S.C. Srivast., *Cylindrocolea kiaeri* (Austin) Váňa, *Cylindrocolea tagawae* (N.Kitag.) R.M.Schust., *Fossombronia cristula* Austin, *Frullania tamarisci* (L.) Dumort. subsp. *obscura* (Verd.) S.Hatt., *Frullanoides tristis* (Steph.) van Slageren, *Herbertus dicranus* (Taylor ex Gottsche, Lindenb. & Nees) Trevis., *Jubula hutchinsiae* (Hook.) Dumort. subsp. *javanica* (Steph.) Verd., *Lejeunea lowriana* Steph., *Lejeunea neelgherriana* Gottsche, *Lejeunea wightii* Lindenb., *Leptolejeunea elliptica* (Lehm. & Lindenb.) Schiffn., *Leptolejeunea maculata* (Mitt.) Schiffn., *Lophocolea muricata* (Lehm.) Nees, *Lopholejeunea nigricans* (Lindenb.) Schiffn., *Lopholejeunea subfusca* (Nees) Steph., *Lunularia cruciata* (L.) Dumort. ex Lindb., *Marchantia emarginata* Reinw., Blume & Nees, *Marchantia linearis* Lehm. & Lindenb., *Marchantia polymorpha* L., *Mastigolejeunea humilis* (Gott.) Schiffn., *Mastigolejeunea repleta* (Taylor) A. Evans, *Metzgeria crassipilis* (Lindb.) A. Evans, *Metzgeria furcata* (L.) Corda, *Microlejeunea ulicina* (Taylor) Steph., *Notoscyphus lutescens* Schiffn., *Pallavicinia lyellii* (Hook.) Carruth., *Plagiochasma appendiculatum* Lehm. & Lindenb., *Plagiochasma rupestre* (J.R. Forst. & G. Forst.) Steph., *Plagiochila beddomei* Steph., *Plagiochila elegans* Mitt., *Plagiochila khasiana* Mitt., *Plagiochila nepalensis* Lindenb., *Plicanthus birmensis* (Steph.) R.M. Schust., *Porella acutifolia* (Lehm. & Lindenb.) Trevis., *Porella campylophylla* (Lehm. & Lindenb.) Trevis., *Ptychanthus striatus* (Lehm. & Lindenb.) Nees, *Radula javanica* Gottsche, *Reboulia hemisphaerica* (L.) Raddi, *Riccia discolor* Lehm. & Lindenb., *Riccia fluitans* L., *Riccia mangalorica* Ahmad ex Jovet-Ast, *Riccia stricta* (Gottsche, Lindenb. & Nees) Perold, *Schiffneriolejeunea pulopenangensis* (Gottsche) Gradst., *Schistochila aligera* (Nees & Blume) J. B. Jack & Steph. and *Spruceanthus semirepandus* (Nees) Verd.

The Mosses shared by the states of Karnataka, Kerala and Tamil Nadu are: *Aerobryidium aureonitens* (Hook. ex Schwägr.) Broth., *Aerobryidium filamentosum* (Hook.) M.Fleisch., *Anoectangium stracheyanum* Mitt., *Anomobryum auratum* (Mitt.) A.Jaeger, *Atrichum subserratum* (Hook.) Mitt., *Barbellopsis trichophora* (Mont.) W.R.Buck, *Barbula consanguinea* (Thwaites & Mitt.) A.Jaeger, *Brachymenium bryoides* Hook. ex Schwägr., *Brachymenium exile* (Dozy & Molk.) Bosch & Sande Lac., *Brachymenium nepalense* Hook., *Brachymenium systylium* (Müll.Hal.) A. Jaeger, *Brachythecium buchananii* (Hook.) A. Jaeger, *Bryum apalodictyoides* Müll.Hal., *Bryum apiculatum* Schwägr., *Bryum argenteum* Hedw. var. *argenteum*, *Callicostella papillata* (Mont.) Mitt., *Calymperes lonchophyllum* Schwägr., *Calyptothecium hookeri* (Mitt.) Broth., *Calyptothecium recurvulum* (Broth.) Broth., *Campylopus ericoides* (Griff.) A.Jaeger, *Campylopus fragilis* subsp. *goughii* (Mitt.) J.-P. Frahm, *Campylopus schmidii* (Müll.Hal.) A. Jaeger, *Chameleion peguense* (Besch.) L.T. Ellis & A. Eddy, *Claopodium prionophyllum* (Müll. Hal.) Broth., *Cryptopapillaria fuscescens* (Hook.) M. Menzel, *Cyathophorum adiantum* (Griff.) Mitt., *Diaphanodon blandus* (Harv.) Renauld & Cardot, *Dixonia orientalis* (Mitt.) H. Akiyama & Tsubota, *Ectropothecium cyperoides* (Hook. ex Harv.) A. Jaeger, *Entodon chloropus* Ren. & Card., *Entodon plicatus* Müll.Hal., *Entodontopsis anceps* (Bosch & Sande Lac.) W.R. Buck & Ireland, *Entodontopsis nitens* (Mitt.) W.R. Buck, *Entodontopsis wightii* (Mitt.) W.R. Buck & Ireland, *Entosthodon wichurae* M. Fleisch., *Erythrodontium julaceum* (Hook. ex Schwägr.) Paris, *Fissidens anomalus* Mont., *Fissidens crispus* Mont., *Fissidens hollianus* Dozy & Molk., *Floribundaria floribunda* (Dozy & Molk.) M. Fleisch.,

Foreauella orthothecia (Schwägr.) Dixon & P.de la Varde, *Funaria hygrometrica* Hedw., *Funaria hygrometrica* var. *calvescens* (Schwägr.) Mont., *Garovaglia plicata* (Brid.) Bosch. & Sande Lac., *Himantocladium cyclophyllum* (Müll.Hal.) M. Fleisch., *Himantocladium plumula* (Nees) M. Fleisch., *Homaliodendron exiguum* (Bosch. & Sande Lac.) M. Fleisch., *Homaliodendron flabellatum* (Sm.) M. Fleisch., *Leucobryum juniperoideum* (Brid.) Müll.Hal., *Leucoloma amoene-virens* Mitt., *Leucoloma mittenii* M. Fleisch., *Macgregorella indica* (Broth.) W.R. Buck, *Macromitrium moorcroftii* (Hook. & Grev.) Schwägr., *Macromitrium nepalense* (Hook. & Grev.) Schwägr., *Meteoriopsis reclinata* (Müll.Hal.) M. Fleisch., *Meteoriopsis squarrosa* (Hook. ex Harv.) M. Fleisch., *Microcampylopus khasianus* (Griffiths) Giese & J.-P. Frahm, *Neodicladiella pendula* (Sull.) W.R. Buck, *Oxyrrhynchium vagans* (A. Jaeger) Ignatov & Huttunen, *Philonotis mollis* (Dozy & Molk.) Mitt., *Philonotis thwaitesii* Mitt., *Pinnatella calcutensis* M. Fleisch., *Pogonatum aloides* (Hedw.) P. Beauv., *Pogonatum microstomum* (Schwägr.) Brid., *Pogonatum neesii* (Müll. Hal.) Dozy, *Pogonatum patulum* (Harv.) Mitt., *Pterobryopsis flexipes* (Mitt.) M. Fleisch., *Pterobryopsis orientalis* (Müll.Hal.) M. Fleisch., *Pterobryopsis schmidii* (Müll.Hal.) M. Fleisch., *Racopilum cuspidigerum* (Schwägr.) Ångstr., *Racopilum schmidii* (Müll.Hal.) Mitt., *Rhodobryum giganteum* (Schwägr.) Paris, *Rosulabryum billarderii* (Schwaegr.) J.R. Spence, *Rosulabryum wightii* (Mitt.) J.R. Spence, *Schoenobryum concavifolium* (Griff.) Gangulee, *Sematophyllum subhumile* (Müll.Hal.) M. Fleisch., *Sematophyllum subpinnatum* (Brid.) E.G. Britton, *Symphyodon perrottetii* Mont., *Symphysodontella involuta* (Thwaites & Mitt.) M. Fleisch., *Syntrichia fragilis* (Taylor) Ochyra, *Taxiphyllum taxirameum* (Mitt.) M. Fleisch., *Taxithelium nepalense* (Schwägr.) Broth., *Thamniopsis utacamundiana* (Mont.) W.R. Buck, *Thuidium cymbifolium* (Dozy & Molk.) Dozy & Molk., *Thuidium pristocalyx* (Müll.Hal.) A. Jaeger, *Trachyphyllum inflexum* (Harv.) A. Gepp, *Trachypodopsis serrulata* (P. Beauv.) M. Fleisch. var. *serrulata, Trachypodopsis serrulata* var. *crispatula* (Hook.) Zanten, *Trachypus bicolor* Reinw. & Hornsch., *Trematodon longicollis* Michx., *Vesicularia reticulata* (Dozy & Molk.) Broth., and *Weissia edentula* Mitt.

7.3 FUTURE PROSPECTS

The documentation of the Bryophyte diversity of the Western Ghats is still in its infancy. Compared to other groups of plants and animals, the Bryophytes had not received much attention in the past. The awareness on the diversity and the ecological role played by this group is meager or almost nil, as evidenced by the lack of policy documents or any ongoing programs for its conservation. It is high time to frame policy documents and also support the researchers in undertaking more efforts in documentation.

ACKNOWLEDGMENTS

The first author is grateful to the Department of Science & Technology (DST), New Delhi and Kerala State Council for Science Technology & Environment (KSCSTE) for financial support. The authors are indebted to Dr T. Pócs (Department of Botany, Eszterházy

College, EGER, Hungary), Dr L.T. Ellis and Dr Brian O'Shea (Natural History Museum, London), Dr Bill Buck (New York Botanical Garden), Dr Jésus Munoz and Dr Richard Zander (Missouri Botanical Garden), Dr M.L. So and Dr R.L. Zhu (Hong Kong Baptist University), Dr M. A. Bruggeman-Nannenga (Griffensteijnseplein 23, NL-3703 BE Zeist, Netherlands) Dr D. Christine Cargill (Curator of Cryptogam Herbarium, Australian National Herbarium, CANB, Australia), Dr Mohamed Bin Abdul Majid (University of Malaya), Dr Zen Iwatzuki (Hattori Botanical Laboratory, Japan), Dr Maria Ros (Universidad de Murcia, Epsinado), Dr Virendra Nath (Rtd.), Dr A.K. Asthana and Dr Ajith Pratap Singh (CSIR-National Botanical Research Institute, Lucknow, India) and Dr D.K. Singh (Rtd.), Dr Devendra Singh, and Dr. S.B. Singh (Botanical Survey of India), Dr Dulip Daniels (Scott Christian College, Nagercoil) for confirming the identity of some of our collections. We are also thankful to our friends who generously donated their bryophyte collections for our study. We are grateful to the staff members of the Kerala Forest Department for extending support during our field study.

KEYWORDS

- **Western Ghats**
- **bryophytes**
- **distribution**
- **current status**

REFERENCES

Asthana, A. K.; Srivastava, S. C. Indian Hornworts: A taxonomic study. *Bryophyt. Bibl.* **1991,** 42, 1–158.

Asthana, G.; Srivastava, S. C.; Asthana, A. K. The Genus *Cheilolejeunea* in India. *Lindbergia* **1995,** 20, 125–143.

Benedix, E. H. Indomalayische *Cololejeuneen. Feddes Repert. Beihefte.* **1953,** 134, 1–88.

Bruhl, P. A Census of Indian Mosses With Analytical Keys to the Genera. *Rec. Bot. Surv. India.* **1931,** *13* (1), 1–135; *13* (2), 1–152.

Chopra, R. S. Notes on Indian Hepatics. I. South India. *Proc. Indian Acad. Sci.* **1938,** *7* (5), 239–251.

Dandotiya, D.; Govindapyari, H.; Suman, S.; Uniyal, P. L. Checklist of the Bryophytes of India. *Arch. Bryol.* **2011,** 88, 1–126.

Daniels, A. E. D. *Acrolejeunea aulacophora* (Marchantiophyta), New to Asia From the Western Ghats. *Acta Bot. Hung.* **2021,** *63* (1–2), 152–155.

Daniels, A. E. D. Bryophytes. In *Tamil Nadu Biodiversity Strategy and Action plan. Wild Plant Diversity*; Annamalai, R., Ed.; Government of Tamil Nadu: Chennai, **2004,** pp 49–70.

Daniels, A. E. D. Checklist of the Bryophytes of Tamil Nadu, India. *Arch. Bryol.* **2010,** 65, 1–118.

Daniels, A. E. D. *Cololejeunea furcilobulata* (Berrie et Jones) Schuster and *Heteroscyphus argutus* (Reinw. et al.) Schiffn. From Mahendragiri Hills of Kanyakumari District of South India. In *Perspectives in Indian Bryology*; Nath, V., Asthana, A. K., Eds.; Bishen Singh Mahendrapal Singh: Dehra Dun, **2001,** pp 301–307.

Daniels, A. E. D. Studies on the Bryoflora of the Southern Western Ghats, India. Ph.D. Dissertation, Manonmaniam Sundaranar University: Tirunelveli, India, **2003.**

Daniels, A. E. D.; Brijithlal, N. D. Three Mosses Added to the Bryoflora of the Western Ghats From the Neyyar Wildlife Sanctuary, Kerala. *Indian J. For.* **2012,** *35* (4), 507–510.

Daniels, A. E. D.; Daniel, P. Addition to the Bryoflora of India. *Bull. Bot. Surv. India.* **2003a,** 45, 225–226.

Daniels, A. E. D.; Daniel, P. Additions to the Bryoflora of Peninsular India. *Indian J. For.* **2003c,** 26, 389–396.

Daniels, A. E. D.; Daniel, P. Additions to the Moss Flora of the Indian Mainland. *Bull. Bot. Surv. India.* **2005 [2006],** 47, 93–100.

Daniels, A. E. D.; Daniel, P. *Cololejeunea distalopapillata* and *C. vidaliana* (Lejeuneaceae) New to the Liverwort Flora of India. *Acta Bot. Hung.* **2009b,** *51* (1–2), 61–66.

Daniels, A. E. D.; Daniel, P. *Fissidens griffithii* Gangulee (Musci: Fissidentales) – An Addition to the Bryoflora of India. *Indian J. For.* **2003b,** 26, 193–194.

Daniels, A. E. D.; Daniel, P. *Frullania ceylanica* Nees (Frullaniaceae) – New to the Hepatic Flora of India. *Indian J. For.* **2008b,** 31, 637–639.

Daniels, A. E. D.; Daniel, P. *Leptolejeunea balansae* (Hepaticae: Jungermanniales) – A New Record of Bryoflora From the Indian Mainland. *J. Bombay Nat. Hist. Soc.* **2004,** 101, 333–334.

Daniels, A. E. D.; Daniel, P. Name Changes in Two Indian Liverworts. *Bull. Bot. Surv. India.* **2007a,** 49, 231–232.

Daniels, A. E. D.; Daniel, P. The Australian *Spruceanthus thozetianus* (Hepticae: Lejeuneaceae) Discovered in the Western Ghats of India. *Acta Bot. Hung.* **2009a,** *51* (3–4), 283–287.

Daniels, A. E. D.; Daniel, P. *The Bryoflora of the Southernmost Western Ghats, India.* Bishen Singh Mahendra Pal Singh: Dehra Dun, **2013.**

Daniels, A. E. D.; Daniel, P. The Liverwort *Schistochila aligera* (Nees & Blume) J.B. Jack & Steph. (Schistochilaceae) Rediscovered in India. *Cryptogam. Bryol.* **2008a,** 29, 307–310.

Daniels, A. E. D.; Daniel, P. The Mosses of the Southern Western Ghats. In *Current Trends in Bryology;* Nath, V.; Asthana, A. K., Eds.; Dehra Dun, **2007b,** pp 227–243.

Daniels, A. E. D.; Daniel, P. Two New Species of *Riccia* L. (Hepaticae: Marchantiales) From the Western Ghats of Tamil Nadu. *Bull. Bot. Surv. India.* **2002,** 44, 135–140.

Daniels, A. E. D.; Felix, R. The Endemic and Rare *Notothylas anaporata* Udar & DK Singh (Notothylaceae: Anthocerotae) Rediscovered. *Nelumbo* **2009,** 51, 217–218.

Daniels, A. E. D.; Kariyappa, K. C. *Bryoflora of the Agasthyamalai Biosphere Reserve, Western Ghats, India.* Bishen Singh Mahendra Pal Singh, **2019.**

Daniels, A. E. D.; Kariyappa, K. C. Bryophyte Diversity Along a Gradient of Human Disturbance in the Southern Western Ghats. *Curr. Sci.* **2007,** 93, 976–982.

Daniels, A. E. D.; Kariyappa, K. C. *Cheilolejeunea trapezia* (Nees) Kachroo & RM Schust. ex Mizut. var. *ceylanica* (Gottsche) AED Daniels & KC Kariyappa comb. et stat. nov.(Lejeuneaceae) from India. *Taiwania* **2013,** *58* (2), 140–145.

Daniels, A. E. D.; Kariyappa, K. C.; Daniel, P. The Moss Genus *Ochrobryum* Mitt. (Leucobryaceae) Added to the Bryoflora of the Western Ghats. *Nelumbo* **2011a,** 53, 155–160.

Daniels, A. E. D.; Kariyappa, K. C.; Daniels, P. The Liverwort *Thysananthus Spathulistipus* (Lejeuneaceae) Long-Lost in India Rediscovered. *Acta Bot. Hung.* **2011b,** *53* (3–4), 283–289.

Daniels, A. E. D.; Kariyappa, K. C.; Hyvönen, J.; Bell, N. The First Indian Record of *Pogonatum marginatum* Mitt. (Polytrichaceae) From the Western Ghats. *Bryoph. Divers. Evol.* **2016,** *38* (2), 41–46.

Daniels, A. E. D.; Kariyappa, K. C.; Mabel, J. L.; Raja, R. D. A.; Daniel, P. The Overlooked and Rare Liverwort *Frullania ramuligera* (Nees) Mont. (Frullaniaceae) Rediscovered in the Western Ghats of India. *Acta Bot. Hung.* **2010,** *52* (3–4), 297–303.

Daniels, A. E. D.; Kariyappa, K. C.; Sreebha, R. Rediscovery of *Trichostomum hyalinoblastum* (Bryophyta; Pottiaceae), an Elusive Endemic Moss of Western Ghats. *Nelumbo* **2013,** 55, 205–208.

Daniels, A. E. D.; Long, D. G.; Kariyappa, K. C.; Daniel, P. *Anastrophyllum aristatum* (Herzog ex N.Kitag.) A.E.D. Daniels et al., *comb. et stat. nov.* (Marchantiophyta: Anastrophyllaceae) from India and China. *J. Bryol.* **2012,** *34* (2), 146–149.

Daniels, A. E. D.; Mabel, J. L. Two Liverworts New to the Bryoflora. *Lindbergia* **2010,** 33, 77–80.

Daniels, A. E. D.; Mabel, J. L. Two Mosses New to the Bryoflora of the Indian Mainland. *Nelumbo* **2009,** 51, 179–182.

Daniels, A. E. D.; Mabel, J. L.; Daniels, P. The Erpodiaceae (Bryophyta: Isobryales) of India. *Taiwania* **2012**, *57* (2), 168–182.

Daniels, A. E. D.; Raja, R. D. A. *Bryoflora of the Silent Valley National Park Western Ghats, India.* Bishen Singh Mahendra Pal Singh: Dehra Dun, **2020**.

Daniels, A. E. D.; Raja, R. D. A. The New Guinean *Thysananthus appendiculatus* (Lejeuneaceae) Discovered in the Western Ghats of India. *Lindbergia* **2011**, 34, 40–43.

Daniels, A. E. D.; Sreebha, R.; Kariyappa, K. C. *Bryoflora of Indira Gandhi National Park in Anamalai Hills, India.* Bishen Singh Mahendra Pal Singh, **2018**.

Daniels, A. E. D.; Sreebha, R.; Kariyappa, K. C. *Syntrichia amphidiacea* (Pottiaceae) New to the Indian Mainland From the Western Ghats. *Lindbergia* **2016**, *39* (4), 35–38.

Daniels, A. E. D.; Sreebha, R.; Kariyappa, K. C. The Japanese *Fissidens neomagofukui* (Bryophyta: Fissidentaceae) – New to India from the Western Ghats. *Lindbergia* **2017**, *40*, 45–48.

Dixon, H. N. Report on the Mosses Collected by C.E.C. Fischer and Others from South India and Ceylon. *Records Bot. Surv. India.* **1914**, *6* (3), 75–91.

Dixon, H. N.; Potier de la Varde, R. Contribution à la Flore Bryologique de l'Inde méridionale. *Arch. Bot. Bull. Mens.* **1927**, *1* (8–9), 161–184.

Dixon, H. N.; Potier de la Varde, R. Nouvelle Contribution à la Flore Bryologique de l'Inde. *Annal. Crypt. Exot.* **1930**, 3, 168–193.

Ellis, L. T. A Taxonomic Revision of *Calymperes* in Southern India and Neighbouring Islands. *J. Bryol.* **1989**, 15, 697–732.

Foreau, G. Miscellaneous Notes: Some South Indian Mosses. *J. Bombay Nat. Hist. Soc.* **1964**, 61, 223–226.

Foreau, G. Notes on Bryological Geography for the Presidency of Madras. *J. Madras Univ.* **1930**, *2*, 238–250.

Foreau, G. The Moss Flora of the Palni Hills. *J. Bombay Nat. Hist. Soc.* **1961**, 58, 12–47.

Frahm, J.-P.; Schwarz, U.; Manju, C. N. A Checklist of the Mosses of Karnataka, India. *Arch. Bryol.* **2013**, 158, 1–15.

Gangulee, H. C. *Mosses of Eastern India and Adjacent Regions;* Fasc: Calcutta, 1–8, **1969–1980**, pp 1–2145.

Lindo, Z.; Gonzalez, A. The Bryosphere: An Integral and Influential Component of the Earth's Biosphere. *Ecosyst.* **2010**, 13, 612–627.

Linnaeus, C. *Species Plantarum, Vol. II.* Musci: Holmiae, **1753**.

Manju, C. N.; Aswathy, T.; Pournami, K. V.; Rajesh, K. P. *Amblystegium serpens* (Hedw.) Schimp. (Amblystegiaceae: Bryophyta) From Indian Peninsula. *Asian J. Biol. Life Sci.* **2017a**, 6, 412–413.

Manju, C. N.; Chandini, V. K.; Mufeed, B. *Physcomitrium immersum* (Funariales; Bryophyta) an Addition to the Bryoflora of Western Ghats. *Geophytology* **2017b**, *47* (1), 45–47.

Manju, C. N.; Chandini, V. K.; Rajesh, K. P. An Overview of the Family Calymperaceae (Bryophyta) in Western Ghats With Special Reference to Kerala and Its Status in India. In *Recent Advances in Botanical Science, Contemporary Research on Bryophytes*; Alam, A., Ed.; Bentham Science & Publishers Pte. Ltd: Singapore, **2020**, vol 1, pp 52–80.

Manju, C. N.; Chandini, V. K.; Rajesh, K. P. *Cololejeunea manilalia* (Lejeuneaceae: Marchantiophyta), A New Species from the Western Ghats of India. *Acta Bot. Hung.* **2017c**, *59* (1–2), 261–268.

Manju, C. N.; Chandini, V. K.; Rajesh, K. P. *Micromitrium vazhanicum* (Micromitriaceae; Bryophyta) A New Species from the Western Ghats of India. *Bryologist* **2019a**, *122* (2), 297–306.

Manju, C. N.; Chandini, V. K.; Rajesh, K. P.; Sivaprakasam, S. Rediscovery of *Oreoweisia brevidens* Herzog (Dicranaceae; Bryophyta) An Indian Endemic Species from Western Ghats After 75 years. *Lindbergia* **2019b**, *42*, 1–4.

Manju, C. N.; Leena, T. N.; Deepa K. M.; Rajesh, K. P.; Prakashkumar, R. *Lejeunea cocoes* (Lejeuneaceae; Marchantiophyta) in the Western Ghats of India. *Acta Bot. Hung.* **2012a**, 54, 341–343.

Manju, C. N.; Manjula, K. M.; Deepa K. M.; Rajesh, K. P.; Chandini V. K. *Aulacopilum beccarii* (Erpodiaceae: Bryophyta) from the Western Ghats of Kerala. *Geophytology* **2015a**, *45* (1), 62–65.

Manju, C. N.; Martin, K. P.; Sreekumar, V. B.; Rajesh, K. P. Morphological and Molecular Differentiation of *Aerobryopsis eravikulamensis* sp. nov. (Meteoriaceae: Bryophyta) and Closely Related Taxa of the Western Ghats of India. *Bryologist* **2012b**, *115* (1), 42–50.

Manju, C. N.; Nikesh, P. R.; Rajesh K. P. *Exormotheca tuberifera* (Marchantiophyta, Exormothecaceae) From the Western Ghats of Kerala, India. *Taiwania* **2011a**, *56* (4), 330–332.

Manju, C. N.; Pocs, T.; Rajesh, K. P.; Prakashkumar, R. Lejeuneaceae (Marchantiophyta) of the Western Ghats, India. *Acta biol. plant. Agriensis.* **2012c,** 2, 127–147.

Manju, C. N.; Prajitha, B.; Prakashkumar, R.; Ma, W.-Z. *Bryocrumia malabarica spec. nova.* (Bryophyta, Hypnaceae) a Second Species of the Genus From the Western Ghats of India. *Acta Bot. Hung.* **2021,** *63* (1–2), 165–170.

Manju, C. N.; Prajitha, B.; Rejilesh, V. K.; Anoop, K. P.; Prakashkumar, R. *Trichosteleum stigmosum* (Sematophyllaceae) from Silent Valley National Park, A New Record for India. *Taiwania* **2012d,** 57, 222–224.

Manju, C. N.; Rajesh, K. P. Bryophyte Diversity in the High Altitude Grasslands of the Western Ghats. *Acta Bot. Hung.* **2009b,** *51* (3–4), 329–335.

Manju, C. N.; Rajesh, K. P. *Bryophytes of Kerala, Liverworts vol. 1.* Malabar Natural History Society & Centre for Research in Indigenous Knowledge Science & Culture, **2017.**

Manju, C. N.; Rajesh, K. P. Contribution to the Bryophyte Flora of India: The Parambikulam Tiger Reserve in the Western Ghats. *Arch. Bryol.* **2011a,** 42, 1–10.

Manju, C. N.; Rajesh, K. P. *Grimmia funalis* (Schwägr.) Bruch & Schimp. (Grimmiaceae) From India. *Taiwania* **2011b,** *55* (2), 81–83.

Manju, C. N.; Rajesh, K. P. *Notoscyphus pandei* Udar & Kumar (Marchantiophyta, Hepaticae) from the Western Ghats of Kerala State, India. *J. Econ. Tax. Bot.* **2009a,** *33* (2), 342–344.

Manju, C. N.; Rajesh, K. P.; Madhusoodanan, P. V. *Checklist of the Bryophytes of Kerala, India.* Tropical Bryology Research Report No. 7, **2008,** pp 1–24.

Manju, C. N.; Rajilesh, V. K.; Deepa, K. M.; Prakashkumar, R. The Genus *Calycularia* (Marchantiophyta) in Kerala Part of Western Ghats. *Acta Bot. Hung.* **2015b,** *57* (3–4), 401–406.

Mitten, W. Hepaticae Indiae Orientalis. *J. Proc. Linn. Soc. Bot.* **1861 [1860],** 5, 89–128.

Mitten, W. Musci Indiae Orientalis, an Enumeration of the Mosses of the East Indies. *J. Proc. Linn. Soc. Bot.* **1859,** 1, 1–171.

Montagne, C. Cryptogamae Nilgherienses seu plantarum cellularium in Montibus peninsulae indicae Neel-Gherries dictis a Cl. Perrottet collectarum enumeratio. Musci. *Ann. Sci. Nat. Bot.,* ser. 2, **1842a,** *17,* 243–256.

Montagne, J. P. F. C. Cryptogamae Nilgherienses Plantarum cellularium in Montibus Peninsulae Indicae *Neel-Gherries* dictis a cl. Perrottet collectarum Enumeratio, Hepaticae. *Ann. Sci. Nat. Bot.,* ser. 2, **1842b,** *18,* 12–17.

Müller, C. Musci Neilgherrenses. *Bot. Ztg (Berlin).* **1853 [1854],** *11,* 17–24, 33–40, 57–62.

Nair, M. C.; Rajesh, K. P.; Madhusoodanan, P. V. *Bryophytes of Wayanad in Western Ghats,* Malabar Natural History Society: Kozhikode, **2005a.**

Nair, M. C.; Rajesh, K. P.; Madhusoodanan, P. V. *Bryum tuberosum* Mohamed & Damanhuri, a New Record for India. *Indian J. For.* **2004,** *27* (1), 39–40.

Nair, M. C.; Rajesh, K. P.; Madhusoodanan, P. V. *Duthiella* (Moss: Bryophyta) A New Record for Peninsular India. *Phytomorphology* **2007a,** 57, 13–14.

Nair, M. C.; Rajesh, K. P.; Madhusoodanan, P. V. *Lejeunea exilis* (Lejeuneaceae: Hepaticae): A New Record to India. *Acta Bot. Hung.* **2006a,** *48* (1–2), 85–87.

Nair, M. C.; Rajesh, K. P.; Madhusoodanan, P. V. Little Known *Plagiochila* (Dum.) Dum. (Plagiochilaceae: Hepaticae) From India. *Geophytology* **2005b,** 35, 39–44.

Nair, M. C.; Rajesh, K. P.; Madhusoodanan, P. V. Preliminary Exploration of Bryophytes of Chinnar Wildlife Sanctuary. *Geophytology* **2006b,** *36* (1&2), 7–15.

Nair, M. C.; Rajesh, K. P.; Madhusoodanan, P. V. The Genus *Philonotis* Brid. (Bartramiaceae: Bryopsida) in Peninsular India. *Folia Malays.* **2007b,** *18* (1), 21–32; (corrigendum) *18* (2), 137–140.

Nair, M. C.; Rajesh, K. P.; Madhusoodanan, P. V. Three New Bryophyte Records for Peninsular India. *Indian J. For.* **2007c,** 30, 349–352.

Nath, V.; Asthana, A. K. Diversity and Distribution of Genus *Frullania* Raddi in South India. *J. Hattori Bot. Lab.* **1998,** 85, 63–82.

Potier de la Varde, R. Musci Madurenses (suite et fin). *Rev. Bryol.* **1924,** 51, 10–14.

Potier de la Varde, R. Musci Madurenses (suite). *Rev. Bryol.* **1923,** 50, 17–27, 72–79.

Potier de la Varde, R. Musci Madurenses. *Rev. Bryol.* **1922,** 49, 33–44.

Potier de la Varde, R. Musci novi indici. *Ann. Cryptogam. Exot.* **1928,** 1, 37–47.

Potier de la Varde, R. Nouvelles herborisations dans le Sud de l'Inde (1er note). *Rev. Bryol.* **1925,** 52, 37–43.

Singh, D. K. Distribution of Family Notothylaceae in India and Its Phytogeographical Significance. *Adv. Plants Agric. Res.* **1994,** 2, 28–43.

Singh, D. K. *Notothylaceae of India and Nepal (A Morpho-Taxonomic Revision).* Dehra Dun, **2002.**

Srivastava, A.; Srivastava, S. C. *Indian Geocalycaceae (Hepaticae): A Taxonomic Study.* Bishen Singh Mahendra Pal Singh: Dehra Dun, India, **2002.**

Stephani, F. *Species Hepaticarum* 2: Genève *et* Bale. Georg et Cie: Lyon, **1901–1906.**

Stephani, F. *Species Hepaticarum* 3: Geneve *et* Bale. Georg et Cie: Lyon, **1907–1909.**

Stephani, F. *Species Hepaticarum* 4: Georg *et* Cie. meme Maison, Geneve & Bale: Lyon, **1909–1912.**

Stephani, F. *Species Hepaticarum* 5: Georg *et* Cie. meme Maison, Geneve & Bale: Lyon, **1912–1917.**

Stephani, F. *Species Hepaticarum* 6: Georg *et* Cie. meme Maison, Geneve & Bale: Lyon, **1917–1924.**

Stephani, F. *Species Hepaticarum* I. Georg *et* Cie. meme Maison, Geneve & Bale: Lyon, **1898–1900.**

Udar, R. *Bryology in India*; Primlane, P., Ed.; Chron. Bot: New Delhi, 1976.

Udar, R.; Srivastava, S. C. Notes on South Indian Hepaticae 1. *J. Bombay Nat. Hist. Soc.* **1975,** 72, 401–406.

Udar, R.; Srivastava, S. C. Notes on South Indian Hepaticae 2. The Genus *Herberta* Gray. *J. Bombay Nat. Hist. Soc.* **1977,** 74, 255–263.

CHAPTER 8

PTERIDOPHYTES OF THE WESTERN GHATS OF INDIA: A LOOK INTO THE PATTERNS

K. P. RAJESH

Department of Botany, the Zamorin's Guruvayurappan College (Affiliated to the University of Calicut), Kerala, India

ABSTRACT

The pattern of distribution of pteridophytes in the Western Ghats region of the Peninsular India is presented. The distribution of plants, especially the pteridophytes of the Western Ghats, is mainly based on the moisture regime, and the extent of rainless days. The northern parts of the Western Ghats are comparatively drier, with less vegetation cover, and pteridophyte diversity. The vegetation cover, and species diversity of pteridophytes, increases to its southern parts. At the extreme south, it again transforms to dry area with less forest cover, and a lesser number of pteridophytes. The pattern of pteridophyte distribution changes along the altitudinal gradients also. It shows a huped pattern with maximum in the altitudinal zone of above 1000 m. The extent of sharedness of the pteridophyte taxa, based on Sorensen Similarity Index, was calculated for the states of Gujarat, Maharashtra, Goa, Karnataka, and Tamil Nadu. The sharedness of pteridophytes was also verified for Sri Lanka.

8.1 INTRODUCTION

The Western Ghats, the 1600 km long chain of hills lying parallel to the western coast of Peninsular India, is the most influential structure that shapes the climate and culture. It was the search for the plant resources of this mountain that changed the course of history. Starting from the southern part of the state of Gujarat, it spreads through Maharashtra, Goa, Karnataka, Kerala, and Tamil Nadu. Its nature, however, varies greatly from the point of origin at the Tapti basin to the end at Kanyakumari of Tamil Nadu.

The assemblage of plants in the Western Ghats is directly related to the moisture regime of the area. Plant cover is comparatively lesser in the drier northern parts of the Western

Biodiversity Hotspot of the Western Ghats and Sri Lanka. T. Pullaiah, PhD (Ed.)
© 2024 Apple Academic Press, Inc. Co-published with CRC Press (Taylor & Francis)

Ghats, and increases toward the southern parts. It is maximum in the region, where the states of Karnataka, Kerala, and Tamil Nadu meet. The vast tracts of vegetation in this region provide ideal microclimates and microhabitats within the major habitats. At the extreme end, toward the Kanyakumari region, it again transforms into drier nature. It thus offers myriads of microclimatic zones, differing in the species composition, holding unique assemblages of plants and animals. Due to its long geological history, it is a centre of evolution, and holds a high degree of endemic elements in plants and animals.

The Western Ghats is the most influential structure in Peninsular India, which determines the pattern of climate, vegetation, and culture. The nature of plant assemblages or vegetation cover of the area is related to the soil and air moisture, and the aspect or slope of the area. Its western slopes are more moist and hold the best vegetation cover. The eastern slopes are comparatively less moist, leading to the formation of drier vegetation types. The number of rainless days is another factor influencing the vegetation cover, and associated local climate. The openings in the Western Ghats, such as the Palghat Gap, also cause changes in the moisture regime, leading to changes in the local climate and vegetation cover.

The distribution of Pteridophytes, being minor components in the vegetation types of the Western Ghats, is determined by the nature of the major vegetation of the area. As in other tropical parts of the world, the pteridophytes are a minor component in the vegetation types in the Western Ghats. Their distribution is thus highly dependent on the major vegetation type of the area. As the major vegetation of the Western Ghats changes along the gradients of altitude, so is the distribution of pteridophytes.

The Western Ghats is a complex system composed of the chains of mountains, which slopes on both sides, slowly toward the western part and abruptly toward the eastern parts, with innumerable streamlets and streams forming major rivers. On the western part, the crests slope to the coastal plains through an undulating midland. The natural vegetation of the coastal plains and midlands have largely been transformed for human habitation and others. The original vegetation cover, however, remains in some parts of the crests and folds of its mountains.

8.2 BRIEF HISTORY OF DOCUMENTATION OF PTERIDOPHYTE IN THE WESTERN GHATS

The earliest record on the pteridophytes of this region could be traced to the 17th century monumental work, *Hortus Indicus Malabaricus*, by Van Rheede. In its 12th volume (1693) illustrative account on the properties of 18 pteridophytes was given (Manilal, 2003). The pteridophytes were not valued much in the Indian socioeconomic and cultural scenario. However, perspectives of the Europeans toward them were different, but valuing them for their gracefulness. The colonial rulers and their associates, hence, started to document this plant group also with high significance. It resulted in publication of books on Ferns of this region. Richard Henry Beddome, a passionate natural historian associated with the forest service of British East India Company, brought treatises on ferns. His *Ferns of Southern India* (Beddome, 1863) is the most prominent among these. The popularity of his works reached great heights. It is evident from the fact that his books (Beddome, 1863,

1866, 1868, 1873, 1876, 1883, 1892) run into many editions and revisions. The *Ferns of Southern India* contains illustrative accounts of more than 270 taxa from this region. The book, *Ferns of Bombay* (Blatter and d'Almeida, 1922), was another major attempt to collate the data on this group from the northern part of the Western Ghat region. In the post independent period, the documentation on pteridophytes was continued by enthusiastic research groups of this region. It resulted in consolidative publications for states of Tamil Nadu and Kerala (Manickam and Irudayaraj, 1992; Nayar and Geevarghese, 1993; Nair et al., 1988, 1992a, b, 1994) and Karnataka (Rajagopal and Bhat, 1998, 2016), or for the Western Ghats (Benniamin and Sundari, 2020). Over these years, the taxonomy and nomenclature of the pteridophytes had undergone tremendous changes. It has also witnessed multifaceted approaches, upsurge of new taxa and sinking of many, etc. Many revisionary works (Nampy and Madhusoodanan, 1998; Hameed et al., 2003; Sreenivas et al., 2013; Patil et al., 2019, etc) and floristic documentations on this plant group as part of PhD programmes of universities in the region also greatly contributed toward improving the knowledge base on this plant group in the Western Ghat region. After passing many intermediary stages (Fraser-Jenkins, 1997, 2008a, b; Chandra, 2000; Chandra et al., 2008), the attempts for the preparation of a national pteridophyte flora also have progressed to a major stage by bringing out a detailed checklist in three parts (Fraser-Jenkins et al., 2017, 2018, 2021).

8.3 PTERIDOPHYTES OF THE WESTERN GHATS—THE PATTERNS

The present chapter tries to examine the patterns of distribution of pteridophytes of the Western Ghats. The pteridophytes are moisture-loving or moisture-adapted plants; their distribution also follows a similar trend. The area is drier toward its northern parts, in Gujarat and Maharashtra. The number of pteridophyte taxa and their assemblages are also comparatively lesser in these areas. About 396 taxa of pteridophytes in 106 genera of 29 families (Hassler, 2004–2021) are known to occur in the Western Ghats. The distribution of pteridophyte taxa varies along the Western Ghats. It shows a positive trend from the drier northern parts with Gujarat (22 taxa), Maharashtra (107), and Goa (51) to the moist southern part composed of Karnataka (229), Kerala (332), and Tamil Nadu (340) (Figure 8.1). The best vegetation cover is at the meeting points of three states, namely, Karnataka, Kerala, and Tamil Nadu. Except for a few, most of the pteridophytes of the region enjoy a wider distribution in the habitats of these three states. The 396 taxa of pteridophytes known from the Western Ghats include endemics, natural hybrids of intergeneric or interspecific nature (*Bolbitis* × *lancea* (Copel.) Ching, *Bolbitis* × *terminans* (Wall.) Gandhi & Fraser-Jenk, × *Chrinephrium insulare* (K.Iwats.) Nakaike, *Dryopteris* × *ghatakii* Fraser-Jenk, *Salvinia* x *molesta* Mitch, *Tectaria* × *pteropus-minor* (Bedd.) Fraser-Jenk.) and some naturalized exotics. It indicates the degree of variations in the cytological and genetic makeup, and the level of complexity of biology of this plant group in the area. Some of the naturalized exotics such as *Salvinia* x *molesta* and *Cyclosorus interruptus* had attained weedy status. Some others (such as *Blechnum occidentale, Adiantum latifolium*, etc) may appear like natural components in some habitats.

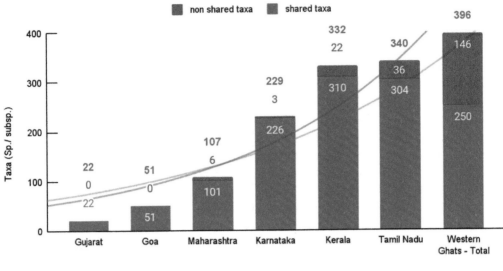

FIGURE 8.1   Distribution of Pteridophyte taxa (species/subspecies/varieties) in the Western Ghats.

8.3.1   DISTRIBUTION OF PTERIDOPHYTES ALONG VEGETATION TYPES

Moisture regimes and Pteridophytes: Number of rainless days is crucial in the distribution of pteridophytes, as it directly influences the amount of moisture in soil and air. In the drier parts, where the quality of the habitats is influenced by the amount of rain and the extent of rainless days, deciduous type of vegetation is prominent, which does not provide suitable microhabitats for pteridophytes. The pteridophytes of such habitats are of two major types. The first type are ephemerals that complete the major part of the life cycle during the wet season. The second type can withstand dryness by adaptations such as anatomical features and curling up. In the Western Ghats of Gujrat, both the types ephemerals (e.g., *Athyrium hoheneckerianum* (Kunze) T.Moore, *A. parasnathense* (C.B.Clarke) Ching ex Mehra & Bir, *Adiantum philippense* L.) and xerophytes (e.g., *Actiniopteris radiata* (Sw.) L.) are present.

Evergreen members, which thrive well throughout the season, are confined to the moist habitats such as evergreen and shola forests. The grasslands being open and exposed to dryness are less occupied by the pteridophytes. The number of species is comparatively less in grasslands. The *Pteridium*, [*P. aquilinum* (L.) Kuhn subsp. *wightianum* (Wall. ex J.Agardh) W.C.Shieh] which grows profusely in the disturbed and fire-affected grasslands, dries up during the summer period, but survives with the deeply buried long creeping rhizomes. Most of the other members may shed their spores and dry up as the moisture level decreases after the rainy days. Some other members such as *Oeosporangium* (syn. *Cheilanthes*) may curl up to protect the dormant rhizome part. Vegetative mode of repro-duction, such as bulbils, walking habit, etc are more prevalent among such members.

Evergreen and shola forests are at the apex of pteridophyte diversity. They provide innumerable microhabitats for the pteridophytes to thrive. The microhabitats in these major vegetation types range from variants of terrestrial (cuttings, floor), aquatic (stagnant, streams, and streamlets, etc), lithophytic (dry and wet), and epiphytic. In the finest scales, the terrestrial type may be with many variations occupied by different species, as per their preferences. The composition of the forest floor and cuttings of soil varies based on the moisture content, and accordingly the preference of species. The water sources of the forest area, both stagnant and flowing, are another factor determining the composition of the pteridophytes. Some members (e.g., *Leptochilus pteropus*) are confined to the boulders of flowing streams and streamlets. Some pteridophytes are adapted to the marshy areas formed by the stagnant waters. The epiphytic pteridophytes grow in abundance in the evergreen and shola forests.

8.3.2   PTERIDOPHYTES ALONG ALTITUDINAL GRADIENTS

The distribution of pteridophyte also varies along the altitudinal gradients. The lowlands and midlands are comparatively lesser in species diversity of this plant group. It increases to maximum in the higher altitude zones and attains a humped pattern. The maximum diversity is seen in the altitude zone of above 1500 m. Then it slightly slows down, forming the typical pattern of tropical areas. This may be due to the occurrence of the best vegetation cover with maximum microhabitats and congenial climate in the altitude zones above 1000 m.

8.4   PTERIDOPHYTES OF THE WESTERN GHATS—SHAREDNESS OF TAXA

The incidence data and list of pteridophytes in the states of the Western Ghat region (Gujarat, Maharashtra, Goa, Karnataka, Kerala, and Tamil Nadu) and Sri Lanka were prepared based on Hassler's database (Hassler, 2004–2021), omitting doubtful records. It is a comprehensive and up-to-date database incorporating details from a wide range of sources, including the checklists by Fraser-Jenkins et al. (2017, 2018, 2020). Details from recent publications and personal observations were also incorporated for some taxa to improve the present list. Analysis was performed for native taxa of the Western Ghats (378 of 396) and Sri Lanka (345 of 364), excluding cultivated, escape from garden, and invasive taxa. The Sorensen Similarity index and dendrogram were prepared in PAST software version 4.06b (Hammer et al., 2001).

Analysis of the distribution data of pteridophytes in the Western Ghats region using Sorensen Similarity Index provided three major clades (Table 8.1; Figure 8.2). The groups were formed based on the moisture regime. The Gujarat region lies in the driest zone, with the lowest number of taxa (19 native out of 22), and without much unique ones sorted out as the first group, with least similarity with other regions. The next group was formed by the states of Maharashtra (101 native taxa out of 107) and Goa (48 native taxa out of 51) with 53.16% similarity. The third group was formed by the states of Karnataka (219 native taxa out of 229), Kerala (317 native taxa out of 332), and Tamil Nadu (323 native taxa out of 340), which lies in the most moist zone, with high affinity with Sri Lanka (345 native taxa out of 364). Maximum similarity was shown between Kerala and Tamil Nadu (87.2%), and followed by Karnataka and Tamil Nadu (71.7%). This subgroup shows more than 50% similarity with Sri Lanka, *viz.*, Karnataka (51.94%), Kerala (64.37%), and Tamil Nadu (66.19%). Least sharedness was shown between Tamil Nadu and Gujarat (10.5%) and Kerala and Gujarat (11.30%).

TABLE 8.1   Sorensen Similarity Index Values of Sharedness of Pteridophyte Taxa in the Western Ghats Region.

	Karnataka	Kerala	Tamil Nadu	Maharashtra	Goa	Gujarat	Sri Lanka
Karnataka	1.000	**0.752**	**0.717**	0.548	0.357	0.143	**0.519**
Kerala		1.000	**0.872**	0.415	0.261	0.113	**0.644**
Tamil Nadu			1.000	0.407	0.246	0.105	**0.662**
Maharashtra				1.000	0.532	0.310	0.297
Goa					1.000	0.356	0.188
Gujarat						1.000	0.073
Sri Lanka							1.000

Note: Sorensen similarity index (SI) = 2c/(a+b), where c is the species common to the comparing localities, a is the total number of species in the first locality one, and b in the second locality.

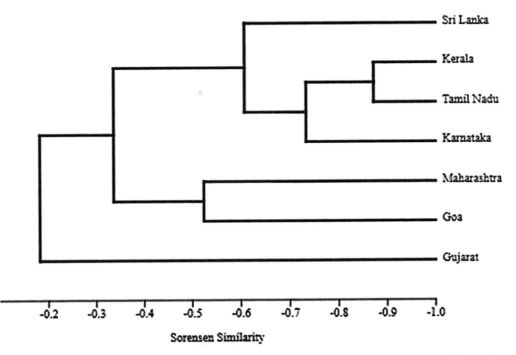

FIGURE 8.2   Dendrogram showing the pattern of sharedness of pteridophyte taxa in the Western Ghats region based on Sorensen Similarity Index. (Prepared in PAST ver. 4.06b with settings: Paired Group (UPGMA) and constraints set to none)

8.4.1 DISTRIBUTION IN THE REGION

Taxa such as *Aglaomorpha quercifolia* (L.) Hovenkamp & S.Linds., *Adiantum capillus-veneris* L., *Adiantum incisum* Forssk. subsp. *incisum, Adiantum philippense* L. subsp. *philippense, Aleuritopteris bicolor* (Roxb.) Fraser-Jenk., *Lygodium flexuosum* (L.) Sw., *Lycopodiella cernua* (L.) Pichi Serm., *Lepisorus nudus* (Hook.) Ching, *Actiniopteris radiata* (Sw.) L., *Pteris vittata* L. subsp. *vittata,* etc are the most frequent pteridophytes of the region. These are present in almost all the states of the Western Ghats. Invasive taxa such as *Salvinia x molesta* Mitch. and *Pityrogramma calomelanos* (L.) Link var. *calomelanos* had encroached into the habitats of this region, causing severe impact to the native taxa. Some taxa, which escaped from the garden such as *Adiantum latifolium* Lam., *A. concinnum* Humb. & Bonpl. ex Willd., *Blechnum occidentale* L., etc are also now well established in the mesic habitats of this region.

Benniamin and Sundari (2020) compared the degree of sharedness of the pteridophyte diversity of the Western Ghats with other regions of India such as North eastern India (56.61%), Eastern Himalaya (50.14%), North Western Himalaya (39.15%), Central India (21.69%), Western India (11.83%), Andaman and Nicobar Islands (6.76%), and Gangetic plains (6.19%). A general trend followed is that the degree of sharedness is proportional to the moisture regimes, and ever greenness of the vegetation types. The number of pteridophyte taxa is less in the dry zones, and accordingly shows less sharedness. This also indicates the significance of the Western Ghats, especially its moist region toward the southern part with the evergreen habitats, offering suitable microhabitats for supporting high pteridophyte diversity.

8.4.2 ENDEMISM

The Western Ghats is considered a center of evolution and endemism for plants and animals. However, the concept on the endemic pteridophytes from this region had been discussed at length (Fraser-Jenkins, 1997, 2008a, b; Chandra et al., 2008) and improved a lot in the recent past. Most of the novelties, described from this region, and considered as endemics, were now sunk into synonyms. It is now well accepted that the percentage of endemism among the pteridophytes is comparatively less in this region (*ca.* 16.41%). About 65 pteridophyte taxa are now considered endemics in this region, of which 29 are known from the Western Ghats only. *Athyrium hoheneckerianum* (Kunze) T.Moore is confined to the Indian region and Sri Lanka. Others (*ca.* 35) are known from Sri Lanka and Southern India.

List of endemic pteridophytes in the Western Ghats region (as adapted from the plants of the World Online http://www.plantsoftheworldonline.org/ and World Ferns by Hassler, https://www.worldplants.de/world-ferns/ferns-and-lycophytes-list)

Endemic to Sri Lanka and India:

1. *Athyrium hoheneckerianum* (Kunze) T.Moore: India (Andhra Pradesh, Chhattisgarh, Goa, Gujarat, Karnataka, Kerala, Madhya Pradesh, Maharashtra, Odisha, Rajasthan, Tamil Nadu) and Sri Lanka.

Endemic to Sri Lanka and South India:

1. *Aleuritopteris bullosa* (Kunze) Ching: India (Karnataka, Kerala, Tamil Nadu) and Sri Lanka.
2. *Aleuritopteris wollenwebri* Fraser-Jenk.: India (Kerala, Tamil Nadu) and Sri Lanka.
3. *Alsophila crinita* Hook.: India (Karnataka, Kerala, Tamil Nadu) and Sri Lanka. IUCN category: Endangered (EN).
4. *Arachniodes palmipes* (Kunze) Fraser-Jenk.: India (Tamil Nadu) and Sri Lanka
5. *Arachniodes sledgei* Fraser-Jenk.: India (Karnataka, Kerala, Tamil Nadu) and Sri Lanka.
6. *Asplenium decrescens* Kunze.: India (Karnataka, Kerala, Tamil Nadu) and Sri Lanka.
7. *Asplenium zenkerianum* Kunze: India (Kerala, Tamil Nadu) and Sri Lanka.
8. *Athyrium praetermissum* Sledge: India (Kerala, Tamil Nadu) and Sri Lanka
9. *Bolbitis asplenifolia* (Bory) K. Iwats.: India (Goa, Karnataka, Kerala, Maharashtra, Tamil Nadu) and Sri Lanka.
10. *Deparia petersenii* (Kunze) M.Kato subsp. *sledgei* Fraser-Jenk.: India (Andhra Pradesh, Karnataka, Kerala, Tamil Nadu) and Sri Lanka.
11. *Diplazium beddomei* C.Chr.: India (Kerala, Tamil Nadu) and Sri Lanka. IUCN category: Critically Endangered (CR) in India.
12. *Diplazium brachylobum* (Sledge) Manickam & Irud.: India (Kerala, Tamil Nadu) and Sri Lanka.
13. *Dryopteris approximata* Sledge: India (Kerala, Tamil Nadu) and Sri Lanka.
14. *Dryopteris deparioides* (T.Moore) Kuntze subsp. *deparioides*: India (Tamil Nadu) and Sri Lanka. IUCN category: Critically Endangered (CR) in India. Its other subspecies, *viz.*, subsp. *ambigua* (Sledge) Fraser-Jenk. and subap. *concinna* (Bedd.) C.Chr. are endemic to Sri Lanka.
15. *Dryopteris sledgei* Fraser-Jenk.: India (Tamil Nadu) and Sri Lanka. IUCN category: Endangered (EN) in India.
16. *Dryopteris wallichiana* (Spreng.) Hyl. subsp. *madrasensis* (Fraser-Jenk.) Fraser-Jenk.: India (Tamil Nadu) and Sri Lanka. Its other subspecies are widely distributed.
17. *Phlegmariurus niligaricus* (Spring) A.R. Field & Bostock: India (Karnataka, Kerala, Tamil Nadu) and Sri Lanka. IUCN category: Vulnerable (VU).
18. *Phlegmariurus vernicosus* (Hook. & Grev.) Á. & D. Löve.: (Kerala, Tamil Nadu) and Sri Lanka. IUCN category: Critically Endangered (CR).
19. *Hymenophyllum gardneri* Bosh: India (Kerala, Tamil Nadu) and Sri Lanka
20. *Lepisorus amaurolepidus* (Sledge) Bir ex Trikha: India (Karnataka, Kerala, Maharashtra, Tamil Nadu) and Sri Lanka
21. *Lindsaea venusta* Kaulf. ex Kuhn.: India (Karnataka, Kerala, Tamil Nadu) and Sri Lanka
22. *Oeosporangium thwaitesii* (Mett. ex Kunh) Fraser-Jenk.: India (Tamil Nadu) and Sri Lanka. IUCN category: Critically Endangered (CR) for India.
23. *Oreogrammitis attenuata* (Kunze) Parris: India (Tamil Nadu) and Sri Lanka. IUCN category: Endangered (EN) for India.
24. *Polystichum austropalaceum* Fraser-Jenk.: India (Tamil Nadu) and Sri Lanka.

25. *Polystichum harpophyllum* (Zenker ex Kunze) Sledge: India (Kerala, Tamil Nadu) and Sri Lanka.
26. *Pteris gongalensiis* T.G.Walker: India (Andhra Pradesh, Kerala, Tamil Nadu) and Sri Lanka.
27. *Pteris hookeriana* J. Agardh.: India (Kerala) and Sri Lanka. IUCN category: Critically Endangered (CR) for India.
28. *Pteris multiurita* J. Agardh.: India (Karnataka, Kerala, Tamil Nadu) and Sri Lanka.
29. *Pteris otaria* Bedd.: India (Karnataka, Kerala, Tamil Nadu) and Sri Lanka.
30. *Pteris praetermissa* T.G. Walker: India (Karnataka, Kerala, Maharashtra, Tamil Nadu) and Sri Lanka.
31. *Pteris reptans* T.G. Walker: India (Kerala) and Sri Lanka. IUCN category: Critically Endangered (CR) for India.
32. *Pyrrosia ceylanica* (Giesenh.) Sledge: India (Karnataka, Kerala, Tamil Nadu) and Sri Lanka.
33. *Selliguea hastata* (Thunb.) Fraser-Jenk. subsp. *montana* (Sledge) Fraser-Jenk. and Ranil: India (Karnataka, Kerala, Tamil Nadu) and Sri Lanka. Other subspecies, *viz.*, *hastata*, widely distributed.
34. *Tectaria × pteropus-minor* (Bedd.) Fraser-Jenk.: India (Kerala) and Sri Lanka
35. *Tectaria paradoxa* (Fee) Sledge: India (Andhra Pradesh, Karnataka, Kerala, Tamil Nadu) and Sri Lanka

Endemic to Western Ghats:

1. *Alsophila nilgirensis* (Holttum) R.M.Tryon: India (Karnataka, Kerala, Tamil Nadu). It also extends to Andhra Pradesh.
2. *Anemia schimperiana* subsp. *wightiana* (Gardner) Fraser-Jenk.: India (Kerala, Tamil Nadu). The other subspecies, *schimperiana*, is widely distributed in Africa.
3. *Asplenium andreisii* Fraser-Jenk.: India (Kerala, Tamil Nadu).
4. *Bolbitis beddomei* Fraser-Jenk. and Gandhi: India (Goa, Karnataka, Kerala, Maharashtra, Tamil Nadu). Also it extends to Madhya Pradesh.
5. *Bolbitis feeiana* (Copel.) Fraser-Jenk. and Gandhi: India (Goa, Karnataka, Kerala, Tamil Nadu).
6. *Dryopteris × ghatakii* Fraser-Jenk: India (Tamil Nadu).
7. *Dryopteris austroindica* Fraser-Jenk.: India (Karnataka, Kerala, Tamil Nadu). IUCN category: Endangered (EN).
8. *Dryopteris odontoloma* (T.Moore ex Bedd.) C.Chr.: India (Tamil Nadu). IUCN category: Near threatened (NT).
9. *Dryopteris peranemiformis* C.Chr.: India (Kerala, Tamil Nadu). IUCN category: Critically endangered (CR).
10. *Dryopteris scabrosa* (Kunze) Kuntze: India (Kerala, Tamil Nadu). IUCN category: Vulnerable (VU).
11. *Elaphoglossum beddomei* Sledge: India (Karnataka, Kerala, Tamil Nadu). IUCN category: Near threatened (NT).

12. *Elaphoglossum nilgiricum* Krajina ex Sledge: India (Karnataka, Kerala, Tamil Nadu). IUCN category: Endangered (EN).

13. *Elaphoglossum stigmatolepis* (Fee) T. Moore: India (Karnataka, Tamil Nadu). IUCN category: Endangered (EN).

14. *Hymenasplenium rivulare* (Fraser-Jenk.) Viane & S.Y. Dong: India (Karnataka, Kerala, Tamil Nadu). IUCN category: Near threatened (NT)

15. *Isoetes sahyadrii* Mahab. ex L.N. Rao: India (Karnataka, Maharashtra). Also it extends to Andhra Pradesh, Madhya Pradesh, and Rajasthan.

16. *Isoetes udupiensis* P.K. Shukla et al.: India (Karnataka).

17. *Oreogrammitis austroindica* (Parris) Parris: India (Tamil Nadu). IUCN category: Critically endangered (CR). Fraser-Jenkins (2012) assumes it as extinct. However, the chances of recollection of the so-called extinct taxa cannot be ruled out in India (Rajesh and Madhusoodanan, 1999).

18. *Oreogrammitis pilifera* (Ravi & Joseph) Parris: India (Karnataka, Kerala, Tamil Nadu). IUCN category: Vulnerable (VU).

19. *Polystichum anomalum (*Hook. & Arn.) J. Sm. subsp. *travancoricum* (Bedd.) Fraser-Jenk.: India (Kerala, Tamil Nadu). The other subspecies *anomalum* is endemic to Sri Lanka.

20. *Polystichum manickamianum* Benniamin et al.: India (Kerala, Tamil Nadu). IUCN category: Critically endangered (CR).

21. *Polystichum palniense* Fraser-Jenk.: India (Kerala, Tamil Nadu). IUCN category: Near threatened (NT).

22. *Polystichum subinerme* (Kunze) Fraser-Jenk.: India (Karnataka, Kerala, Tamil Nadu). IUCN category: Endangered (EN).

23. *Selaginella cataractarum* Alston: India (Kerala, Tamil Nadu). IUCN category: Critically endangered (CR).

24. *Selaginella coonooriana* R.D. Dixit: India (Kerala, Tamil Nadu).

25. *Selaginella ganguliana* R.D. Dixit: India (Kerala).

26. *Selaginella miniatospora* (Dalzell.) Baker: India (Goa, Karnataka, Maharashtra). IUCN category: Near threatened (NT).

27. *Selaginella proniflora* (Lam.) Baker: India (Goa, Karnataka, Kerala, Maharashtra, Tamil Nadu).

28. *Selaginella radicata* (Hook. & Grev.) Spring: India (Karnataka, Kerala, Tamil Nadu). Also extends to Andhra Pradesh.

29. *Tectaria wightii* (C.B.Clarke) Ching: India (Karnataka, Kerala, Tamil Nadu).

8.4.3 WESTERN GHATS AND SRI LANKA

Earlier the Western Ghats and Sri Lanka were considered separate entities for biogeo-graphical analyses (as in Myers, 1988, 1990). However, in later treatments these are usually considered a single unit of global biodiversity hotspots (Myers et al., 2000). This area shares a large number of taxa. Some plants and animals are confined to this region only. Pteridophytes also follow this trend. About 364 pteridophyte taxa including 79 endemics are known from Sri Lanka. More than 59% of the pteridophyte taxa (250 out of 396), including 36 endemic elements of the Western Ghats region, are shared with Sri Lanka.

8.4.4 THREATS TO THE PTERIDOPHYTES—PRESENT AND FUTURE

Some plants are at risks of extinction due to some inherent features. Pteridophytes with a long evolutionary history and with complex cytological and genetic makeup are more prone to the risk of extinction. The high level of polyploidy in members of Ophioglossaceae, especially in *Ophioglossum*, is an indication in this line. These are usually considered as at the fag end of their evolution. Their behavior is unpredictable in the changing climate. Another evidence of natural risk is due to the occurrence of chlorophyllous spores, which instantly germinate within the sporangium itself, without any dormancy period (viviparous condition) in some members. It has been reported in the *Grammitis attenuata* (Kunze) Parris of the Western Ghats (Irudayaraj et al., 2003). The narrow endemic, *Oreogrammitis austroindica* (Parris) Parris (*syn. Grammitis austroindica* Parris), was described based on a single collection by Beddome from the Nilgiris (Parris, 2001). It is now considered critically endangered or extinct (Fraser-Jenkins, 2012; Ebihara et al., 2012) as later workers could not locate it in the region. It is not sure whether it ended naturally.

Other natural causes that damage or deteriorate the habitats of pteridophytes include landslides and floods. Large areas are affected by such calamities, wiping out many populations of plants, including pteridophytes. The extent of damage to the pteridophytes due to flood and landslides is, however, not estimated in the Western Ghats. Other causes are man made such as degradation or conversion of the habitats for various purposes, fire, quarrying, etc. These are continuing in most parts of the Western Ghats in varying degrees. The vast network of protected areas under various categories such as National Parks, Tiger Reserves, and Wildlife Sanctuaries of the Western Ghats is a ray of hope in the conservation of the biodiversity of this region. The pteridophyte populations present in these protected areas are comparatively safer. However, the lack of charisma and less economic values, prompt many, including policy makers and biodiversity managers, to ignore this ecologically significant plant group.

8.4.5 DANGERS DUE TO CLIMATE CHANGE

The effect of climate change on plants, including pteridophytes, may not be promising and positive in tropics (Colwell et al., 2008). The predicted models usually reflect a declining trend, leading to the disappearance of many species (Sharpe, 2019; Yesuf et al., 2021). In the Western Ghats regions, however, such studies are yet to be started. The degree of risks is also yet to be assessed in finer scales. The distribution and survival of pteridophytes are highly dependent on the moisture content in the soil and air. Any changes in the quality of the habitat may lead to the disappearance of them. The extent of rainless days is crucial in determining the nature of the major habitats of the area, and thereby the distribution of pteridophytes. The detailed assessments are yet to be conducted. However, the predicted models of climate change indicate the reduction of rainy days in the Western Ghats region, which may lead to drying up, and losing the ever-green nature of the vegetation. The expansion of bird species, which prefer dry habitats, such as Pea fowl (*Pavo cristatus*), was predicted in the Western Ghats region (Jose and Nameer, 2020) in the near future itself. If it came true, a similar trend may be shown by some xerophytic pteridophytes, which may expand its distribution range in the Western Ghats. On the other hand, many moisture-dependent pteridophytes of the region may be wiped out. The notable feature is that more than 90% of the pteridophyte taxa of the Western Ghats region belong to the moisture-dependent category. They are confined to the evergreen and shola forests of the Western Ghats and may have largely been affected by such changes. More than 85% of the pteridophyte taxa (337 out of 396) of the Western Ghats are from 12 dominant families with more than 10 taxa (Figure 8.3). Among these, except for a few members of Pteridaceae, Polypodiaceae, and Thelypteridaceae all are moisture-loving taxa. Members of Dryopteridaceae (47), Aspleniaceae (35), Athyriaceae (25), and Hymenophyllaceae (22) are confined to the evergreen or semi-evergreen habitats, or with strategies of escaping the dry season by completing the major part of their life cycle during the wet season itself.

8.4.6 CONSERVATION

The pteridophytes, being noncharismatic elements, are least considered by the policy makers and other stakeholders of biodiversity conservation. It is evident from the fact that no major programmes are going on for the conservation of this plant group. Protocols for mass multiplication, either from spores or by other means, are yet to be developed for most of the pteridophytes of this region. The next step of reintroduction of rare and threatened pteridophytes into the habitats of the Western Ghats is yet to get momentum. The conservatories or gardens that maintain live collections of pteridophytes of the region are also very limited. The Jawaharlal Nehru Tropical Botanic Garden & Research Institute (JNTBGRI), Thiruvananthapuram, Kerala is one institution under the government sector, which maintains a good collection of live pteridophytes in its conservatory. The Kodaikanal Botanic Garden of St Xavier's College, Palayamkottai, Tamil Nadu maintains a good collection of pteridophytes (Benniamin and Sundari, 2020). The Gurukula Botanical Sanctuary, established by Wolfgang Theuerkauf (1948–2014) (see also Kumar, 2014; https://www.gbsanctuary.org/) at Periya in Wayanad district of Kerala, maintains a state-of-the-art conservatory of Pteridophytes. Some institutions such as University of Calicut (Kerala), Malabar Botanical Garden, and Institute for Plant Sciences (Kozhikode, Kerala), etc also hold some live collections of pteridophytes in their botanical gardens. But, it is high time to think whether these minimal efforts are enough to face the tasks of pteridophyte conservation in the coming ages of climate change (Sharpe, 2019).

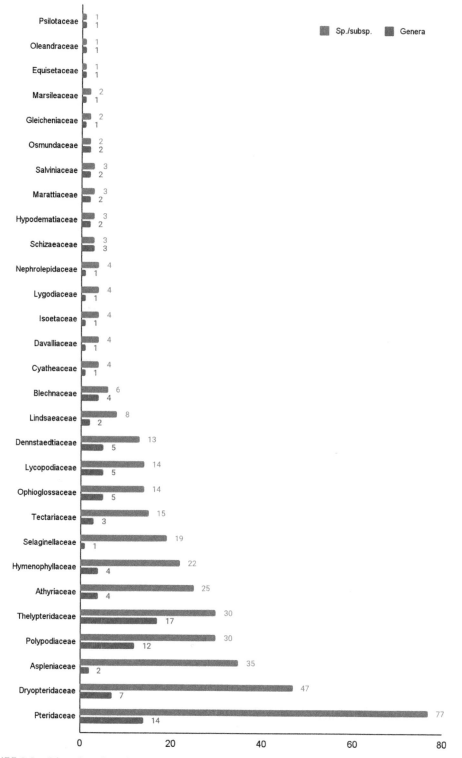

FIGURE 8.3 Diversity of Pteridophyte families in the Western Ghats.

8.4.7 TASKS AHEAD

Taxonomic ambiguity was solved to a great extent in areas where reassessments using advanced techniques, including molecular profiling, were adopted. In Japan, for instance, extensive reassessments performed utilizing high-quality data (based on *ca.* 216,687 specimens including 180,000 specimens collected by fern enthusiasts and deposited in the herbarium (TNS), cytotaxonomic information of 74% of taxa, *rbcL* sequence data of 97.9% taxa) helped in bringing more clarity on the knowledge base on pteridophyte flora (Ebihara and Nitta, 2019). Compared to such attempts, the situation in the Western Ghats region, or India at large, is far meagre and lagging behind to a great extent. The integrated approach of biological or evolutionary concepts in assigning the species is yet to be adopted in the Indian region. Morphology-based species concepts and methods are still prevalent in this region. A paradigm shift, from the morpho-anatomical approach to the morpho-molecular approach, is the need of the time in areas such as the Western Ghats region. It may be helpful in getting a better picture on the diversity of the pteridophyte taxa of a region of long evolutionary history, encompassing a high degree of complexities such as hybrids, apomictic, polyploids, etc. Effective approaches that provide deeper insights into the cytological and genetic make-up would be helpful in overcoming the limitations of the present-day morphology-based species concepts. However, the transition may not be easy as it involves heavy investments in terms of infrastructure, trained manpower, and other costs. Awareness on this plant group among the policy makers, biodiversity resource managers, and public at large is also to be improved, to get more attention in the age of habitat degradation and climate change.

KEYWORDS

- pteridophytes
- fern
- fern-allies
- lycophytes
- endemics
- Western Ghats
- Gujarat
- Maharashtra
- Goa
- Karnataka
- Kerala
- Tamil Nadu
- Sorensen similarity index
- Sri Lanka

REFERENCES

Beddome, R. H. *The Ferns of Southern India: Being Descriptions and Plates of the Madras Presidency*, 1st ed.; Gantz Brothers: Madras, **1863**.

Beddome, R. H. *The Ferns of British India: Being Figures and Descriptions of Ferns from All Parts of British India (Exclusive of Those Figured in "The Ferns of Southern India and Ceylon")*, Vol. I; Gantz Brothers: Madras, **1866**.

Beddome, R. H. *The Ferns of British India: Being Figures and Descriptions of Ferns from All Parts of British India (Exclusive of Those Figured in "The Ferns of Southern India and Ceylon")*, Vol. II; Gantz Brothers: Madras, **1868**.

Beddome, R. H. *The Ferns of Southern India: Being Descriptions and Plates of the Madras Presidency*; 2nd ed.; Higginbotham: Madras, **1873**.

Beddome, R. H. *Supplement to the Ferns of Southern India and British India: Containing a Revised List of All the Ferns of India, Ceylon, Burmah, and the Malay Peninsula and 45 Plates of Hitherto Unfigured Species*; Gantz Brothers: Madras, **1876**.

Beddome, R. H. *Handbook to the Ferns of British India, Ceylon and the Malay Peninsula*; Thacker, Spink and Co.: Calcutta, **1883**.

Beddome, R. H. *Handbook to the Ferns of British India, Ceylon and the Malay Peninsula; with Supplement*; Thacker, Spink and Co.: Calcutta, **1892**.

Benniamin, A.; Sundari, M. S. *Pteridophytes of Western Ghats—A Pictorial Guide*; Bishen Singh Mahendra Pal Singh: Dehra Dun, **2020**.

Blatter, E.; d'Almeida, J. F. *Ferns of Bombay*; D. B. Taraporevala Sons & Co.: Bombay, **1922**.

Chandra, S. *The Ferns of India (Enumeration, Synonyms and Distribution)*; International Book Distributors: New Delhi, **2000**.

Chandra, S.; Fraser-Jenkins, C. R.; Kumari, A.; Srivastava, A. A Summary of the Status of Threatened Pteridophytes of India. *Taiwania* **2008**, *53* (2), 170–209.

Colwell, R. K.; Brehm, G.; Cardelús, C.L; Gilman, A. C.; Longino, J. T. Global Warming, Elevational Range Shifts, and Lowland Biotic Attrition in the Wet Tropics. *Science* **2008**, *322* (5899), 258–261.

Ebihara, A.; Fraser-Jenkins, C. R.; Parris, B. S.; Zhang, X. C.; Yang, Y.H; Chiou, W. L.; Chang, H. M.; Lindsay, S.; Middleton, D.; Kato, M.; Praptosuwiryo, T. N. Rare and Threatened Pteridophytes of Asia 1: An Enumeration of Narrowly Distributed Taxa. *Bull. Natl. Museum Nat. Sci. Ser. B* **2012**, *38* (3), 93–119.

Ebihara, A.; Nitta, J. H. An Update and Reassessment of Fern and Lycophyte Diversity Data in the Japanese Archipelago. *J. Plant Res.* **2019**, *132* (6), 723–738.

Fraser-Jenkins, C. R. *New Species Syndrome in Indian Pteridology and the Ferns of Nepal*; International Book Distributors: New Delhi, **1997**.

Fraser-Jenkins, C. R. Endemics and Pseudo-Endemics in Relation to the Distribution Patterns of Indian pteridophytes. *Taiwania* **2008a**, *53* (3), 264–292.

Fraser-Jenkins, C. R. *Taxonomic Revision of Three Hundred Indian Subcontinental Pteridophytes: With a Revised Census List; a New Picture of Fern-Taxonomy and Nomenclature in the Indian Subcontinent*; Bishen Singh Mahendra Pal Singh, **2008b**.

Fraser-Jenkins, C. R. Rare and Threatened Pteridophytes of Asia 2: Endangered Species of India—The Higher IUCN Categories. *Bull. Natl. Museum Nat. Sci. Ser. B* **2012**, *38* (4), 153–181.

Fraser-Jenkins, C. R.; Gandhi, K. N.; Kholia, B. S. *An Annotated Checklist of Indian Pteridophytes, Part-2 (Woodsiaceae to Dryopteridaceae)*; Bishen Singh Mahendra Pal Singh: Dehra Dun, **2018**.

Fraser-Jenkins, C. R.; Gandhi, K. N.; Kholia, B. S.; Benniamin, A. *An Annotated Checklist of Indian Pteridophytes, Part-1 (Lycopodiaceae to Thelypteridaceae)*; Bishen Singh Mahendra Pal Singh: Dehra Dun, **2017**.

Fraser-Jenkins, C. R.; Gandhi, K. N.; Kholia, B. S.; Kandel, D. R. *An Annotated Checklist of Indian Pteridophytes, Part 3 (Lomariopsidaceae to Salviniaceae)*; Bishen Singh Mahendra Pal Singh: Dehra Dun, **2020**.

Hameed, C. A.; Rajesh, K. P.; Madhusoodanan, P. V. *Filmy Ferns of South India*; Penta Book Publishers & Distributors: Kozhikode, **2003**.

Hammer, Ø.; Harper, D. A. T.; Ryan, P. D. PAST: Paleontological Statistics Software Package for Education and Data Analysis. *Palaeontologia Electronica* **2001**, *4* (1), 9.

Hassler, M. World Ferns. Synonymic Checklist and Distribution of Ferns and Lycophytes of the World. Version 12.1; Last Update March 13th, **2021**, 2004–2021. www.worldplants.de/ferns/ (accessed 20 Apr 2021).

Irudayaraj, V.; Manickam, V. S.; Johnson, M. Vivipary, a Rare and Evolutionarily Important Phenomenon in a Rare Homosporous Fern *Grammitis medialis* from the Western Ghats, South India. *Curr. Sci.* **2003**, *85* (12), 1666–1667.

Jose, V.S; Nameer, P. O. The Expanding Distribution of the Indian Peafowl (*Pavo cristatus*) as an Indicator of Changing Climate in Kerala, Southern India: A Modelling Study Using MaxEnt. *Ecol. Indicators* **2020**, *110*, 105930.

Kumar, S. C. Obituary—Wolfgang Theuerkauf (1948–2014). *Rheedea* **2014**, 2492, 136–137.

Manickam, V. S.; Irudayaraj, V. *Pteridophyte flora of the Western Ghats, South India*; BI Publications Pvt. Ltd.: New Delhi, **1992**.

Manilal, K. S. *Van Rheede's Hortus Malabaricus: English Edition with Annotations and Modern Scientific Nomenclature*; University of Kerala: Thiruvananthapuram, **2003**.

Myers, N. Threatened Biotas: "Hot spots" in Tropical Forests. *Environmentalist* **1988**, *8*, 1–20.

Myers, N. The Biodiversity Challenge: Expanded Hot Spots Analysis. *Environmentalist* **1990**, *10*, 243–256.

Myers, N.; Mittermeier, R. A.; Mittermeier, C. G.; Da Fonseca, G. A.; Kent, J. Biodiversity Hotspots for Conservation Priorities. *Nature* **2000**, *403* (6772), 853–858.

Nair, N. C.; Ghosh, S. R.; Bhargavan, P. Fern-Allies and Ferns of Kerala, India- Part I. *J. Econ. Taxon. Bot.* **1988**, *12* (1), 171–209.

Nair, N. C.; Ghosh, S. R.; Bhargavan, P. Fern-Allies and Ferns of Kerala: Part II. *J. Econ. Taxon. Bot.* **1992a**, *16* (2), 251–282.

Nair, N. C.; Ghosh, S. R.; Bhargavan, P. Fern-Allies and Ferns of Kerala: Part III. *J. Econ. Taxon. Bot.* **1992b**, *16* (3), 501–550.

Nair, N. C.; Ghosh, S. R.; Bhargavan, P. Fern-Allies and Ferns of Kerala: Part IV. *J. Econ. Taxon. Bot.* **1994**, *18* (2), 449–476.

Nampy, S.; Madhusoodanan, P. V. *Fern Flora of South India: Taxonomic Revision of Polypodioid Ferns*; Daya Publishing House: New Delhi, **1998**.

Nayar, B. K.; Geevarghese, K. K. *Fern Flora of Malabar*; Indus Publishing Company: New Delhi, **1993**.

Parris, B. S. Two New Species and a New Combination in Indian and Sri Lankan Grammitidaceae. *Fern Gazette* **2001**, *16* (4), 201–202.

Patil, S. M.; Antony, R.; Nampy, S.; Rajput, K. S. Diversity, Distribution and Conservation Status of the Genus *Tectaria* Cav. from Deccan Peninsula and Western Ghats of India. *Notulae Scientia Biologicae* **2019**, *11* (3), 475–480.

Rajagopal, P. K.; Bhat, K. G. Pteridophytic Flora of Karnataka State, India. *Indian Fern J.* **1998**, *15*, 1–28.

Rajagopal, P. K.; Bhat, K. G. *Pteridophytes of Karnataka State, India*; Hemalatha Rajagopal: Udupi, **2016**.

Rajesh, K. P.; Madhusoodanan, P. V. Extinction Needn't Be Forever. *Nature* **1999**, *399* (6737), 631.

Sharpe, J. M. Fern Ecology and Climate Change. *Indian Fern J.* **2019**, *36*, 179–199.

Sreenivas, V. K.; Fraser-Jenkins, C. R.; Madhusoodanan, P. V. The Genus *Pteris* L. (Pteridaceae) in South India. *Indian Fern J.* **2013**, *30*, 268–308.

Yesuf, G. U.; Brown, K. A.; Walford, N. S.; Rakotoarisoa, S. E.; Rufino, M. C. Predicting Range Shifts for Critically Endangered Plants: Is Habitat Connectivity Irrelevant or Necessary? *Biol. Conserv.* **2021**, *256*, 109033.

CHAPTER 9

SPERMATOPHYTES OF THE WESTERN GHATS

S. KARUPPUSAMY

Department of Botany, The Madura College (Autonomous), Madurai, Tamil Nadu, India

ABSTRACT

The Western Ghats are a long chain of unbroken mountain ranges along the west coast of Indian Peninsula which covers rich vegetation with varied ecosystems. The recent account on the Flora of Western Ghats enumerated a total of 8080 taxa with 7402 species, 117 subspecies, and 476 varieties of seed plants in the area. Of these 376 are naturalized or introduced species, about 1438 species are cultivated or planted as ornamentals. Of a total of 5588 indigenous species, 2253 species are confined to India, from which 1273 species are exclusively restricted to the Western Ghats. A total account of Gymnosperms in the Western Ghats showed 9 families with 24 genera and 62 species. Among them *C. circinalis*, *C. annaikalensis*, *C. swamyi*, *Gnetum ula* (Gnetales), and *Nageia wallichiana* (Coniferales) are native Gymnosperms of Western Ghats range. This chapter reveals the floristic account, endemism, rarity, and conservation measures of spermatophytes of Western Ghats.

9.1 INTRODUCTION

The Western Ghats is being recognized as one of the global biodiversity hotspots of the world and with rich endemism, supported enormous biological wealth, which is undergoing great threatening due to anthropogenic pressure over the years. The entire stretch of Western Ghats covered six states such as Gujarat, Goa, Karnataka, Kerala, Maharashtra, and Tamil Nadu parallel to the Arabian coast of southern Peninsular India. The Western Ghats zone has enclaved a wide range of ecosystems and topographical features. The mountain ranges covered the hill area of around 1,29,037 km^2 lying between 8° N to 21° N–73° E to 78° E for about 1490-km-long stretch. The maximum height of 2800 m hill peaks before they merge to the east with the Deccan plateau at an altitude of 500–600 m. The mean width of the mountain range is about 100 km toward inland. The mountain range covered a considerable varied gradient of climatic conditions that resulted in the development of diverse nature of forest types ranging from the dry thorny scrub vegetation to the semi-evergreen and montane evergreen forests in high altitudes. The area includes

Biodiversity Hotspot of the Western Ghats and Sri Lanka. T. Pullaiah, PhD (Ed.)

a diversity of ecosystems containing an enormous number of medicinal plants and other important genetic bioresources such as wild relatives of cereal grains, fruits, and spices. The Ghats has been served the habitat for unique shola ecosystem that composed of montane grasslands interspersed with evergreen forest patches contains rich endemic gene pool.

9.2   FLORISTIC ACCOUNTS OF WESTERN GHATS

The Western Ghats is recognized as one among the "hottest hotspots" of world biodiversity with major tropical evergreen-forested area of Indian subcontinent (Rodgers and Panwar, 1988). The floristic richness and plant diversity of the region has been accounted out by Gamble and Fischer (1915–1936), Fyson (1932), Nair and Daniel (1986), Nayar (1996). Subsequently, several other State and District floras (Cooke, 1901–1908; Fyson, 1932; Ahuja and Singh, 1963; Manilal, 1988; Matthew, 1981–1984 and 1999; Mohanan and Henry, 1994; Nayar, 1996; Ramachandran and Nair, 1988; Rao, 1985–86; Sasidharan and Sivarajan, 1996; Rao and Razi, 1981; Saldanha and Nicolson, 1976; Saldanha, 1984, 1953; Keshava Murthy and Yoganarashiman, 1990; Yoga-narashiman *et al.,* 1981; Matthew, 1999–2003; Rao et al., 2004; Sasidharan, 2006; Sasidharan and Sivarajan, 1996; Nayar et al., 2008; Ganesan et al., 2019) highlighted the diversity and richness of the flowering plants in the region. Out of 5800 species of flowering plants, 2100 are endemic in this megaendemic area (Yoganarasimhan, 2000; Nair and Henry, 1983; Singh et al., 2015). It has covered approximately 27% of the total Indian flora. The endemic centers of the Ghats are Agasthyamalai (200 km^2) with 2000 species; The Nilgiris *ca* 2611 species, while Silent Valley (90 Km^2) supports 1300 species. Most of the District floras published in recent past reveal that most of them have more than 1200 species (Santapau, 1953, 1957; Rao and Razi, 1981; Keshava Murthy and Yoganarasimhan, 1990; Saldanha, 1984; Saldanha and Nicolson, 1976; Yoganarasimhan et al., 1981; Manilal, 1988; Ramachandran and Nair, 1988; Chandrabose et al.,1988; Mohanan and Henry, 1994; Mohanan and Sivadasan, 2002; Ramas-wamy et al., 2001; Sasidharan and Sivarajan, 1996; Sasidharan, 2004).

9.2.1   SOUTHERN WESTERN GHATS

The first authentic botanical account of southern Western Ghats can be found in the *Hortus Indicus Malabaricus* of Hendrik Andrian van Rheede published between 1678 and 1703. C. A. Barber, R. H. Beddome, T. F. Bourdillon, K. C. Jacob, M. A. Lawson, V. Narayanas-wamy, M. Rama Rao, K. Venkoba Rao, and R. Wight were the pioneer plant explorers of southern most Western Ghats. Wight's *Icones Plantarum Indiae Orientalis* (1840–1853) and *Illustrations to Indian Botany* (1840–1850) included accounts of many plants from this range. Bourdillon (1893) gave a detailed report on the Forests of Travancore, which formed the basis of his later monumental volume on the *Forest Trees of Travancore*. This work stimulated the taxonomic research and inspired many of the taxonomists who have worked out provincial floras like Dalzell and Gibson (1861), Cooke (1901–1908), Talbot (1909–11), Gamble and Fischer (1915–36), Parkinson (1923), and Fyson (1932). Rao (1914) enumerated the flowering plants of Travancore which contains a list of 3535 plant species. Fyson (1915) has contributed a major and important work on the hill top flora of Southern India and also his important contributions have been compiled in an illustrated book on *The Flora of the Nilgiri and Pulney Hill-Tops* (1915, 1921). He has published the revised volumes of the same in 1932. He has described the wild and introduced flowering plants around the hill stations of Kotagiri, Kodaikanal, and Ootacamund.

Botanical explorations carried out by different workers in the recent past in various states including different districts of southern parts of India mainly included the works of Saldanha (1984, 1996) *"Flora of Karnataka"* containing over 3400 species of flowering plants; Manilal (1988) *"Flora of Silent Valley"*; Sasidharan and Sivarajan (1996) *"Flowering Plants of Trissur Forest."* M. Mohanan published the Flora of Thiruvananthapuram, which formed an authentic report on the floristic diversity of southern part of Western Ghats. Mohanan and Sivadasan (2002) published an exclusive floristic study of Agasthyamalai, one of the three endemic hotspots recognized in Kerala. A study of Shenduruny flora by Sasidharan (1997) with emphasis on endemic species resulted in recognition of 951 taxa, of which 309 are Western Ghats endemics.

9.2.2 NORTHERN WESTERN GHATS

North of Kali River is the Deccan trap having relatively fragile rocks with hills not rising beyond 1650 m and having flat hill tops showing rocky outcrops of herbaceous vegetation. Monsoon herbs are abundant and show higher endemism as compared to the woody endemics of southern Western Ghats. Several workers have contributed the floristic account of northern Western Ghats including Cooke (1908), Blatter and McCann (1935), Caius (1936), Desai and Murthy (1950), Santapau (1953), Saldanha (1984, 1953, 1957), Saldanha and Santapau (1956), Auti et al. (2020), Bhushan et al. (2020), Potdar et al. (2012). Karthikeyan et al. (1989) enumerated the grass flora of northern Western Ghats that reported about 1254 species belonging to 260 genera. Some important floristic explorers of northern Western Ghats included Saldanha (1984 and 1996) *"Flora of Karnataka"* containing over 3400 species of flowering plants; Kulkarni (1988) *"Flora of Sindhudurg"*; Sahni (1990) *"Gymnosperms of India and Adjacent countries"*; Almeida (1996, 1998, 2001, 2003, 2009) *"Flora of Maharashtra"*; Naik (1998) *"Flora of Marathwada of Maharashtra."* Many researchers have reported about the floral diversity of plateaus in northern Western Ghats (Bor, 1960; Bharucha and Ansari, 1963; Patel, 1968; Sharma et al., 1984; Mahabale, 1987; Karthikeyan et al., 1981; Lakshminarasimhan and Sharma, 1991; Karthikeyan and Anandkumar, 1993; Kothari and Moorthy, 1993: Karthikeyan, 1996; Diwakar and Sharma, 1999; Singh and Karthikeyan, 2000, 2001; Yadav and Sardesai, 2002; Patil, 2003; Singh et al., 2001; Watve, 2013; Vinayaka and Krishnamurthy, 2016). Report on two basaltic plateaus of northern Western Ghats has been provided by Rahangdale and Rahangdale (2014, 2018). Their documentation chiefly focuses on flowering plant diversity from Durgawadi Plateau (600 taxa) and Naneghat Plateau (249 taxa). Specialized vegetation (Lekhak and Yadav, 2012), a new species (Malpure and Yadav 2009), and endemic flora (Joshi and Janarthanam 2004) were also reported from lateritic plateaus of northern Western Ghats.

9.3 DIVERSITY OF SPERMATOPHYTES

In Hooker's Flora of India (1904), he had drawn an attention to the distinctness of the Western Ghats, which he named the "Malabar" floristic province. The occurrence of Dipterocarpaceae, Clusiaceae, Myristicaceae, and Arecaceae has constituted to its characteristic vegetation of the region. Approximately 1215 species of woody taxa in the Western Ghats flora reflect the relict nature. Several arborescent taxa have more than five endemic species in each genus. The recent document on the Flora of Western Ghats accounted about 8080 taxa of higher plants with 7402 species, and 117 subspecies with 476 varieties of seeded plants in the region. There are 376 species of exotics or naturalized, 1438 species of cultivated or domesticated ornamentals. Out of 5588 indigenous flowering species, 2253 species are confined to India, from which 1273 species are exclusively endemic to the Western Ghats (Nayar et al., 2014). In recent decades, number of new species enumerated as additions to the region as new discoveries by various plant explorers.

The Nilgiri Biosphere Reserve covered approximately 265 km^2 area of altitudinal rages between 40 and 2600 m above msl. The hill ranges enclaved almost all the forest types of the Indian Peninsula with pristine relict vegetation. The floristic account of the area estimated a total of 1135 angiosperm taxa, belonging to 136 families and 644 genera. The major proportion of the plants constituted about 78% dicotyledons and 22% monocotyledons of India. A fascinating group of Orchids are very attractive among plant lovers for their long-lasting and beautiful flowers. They include different life forms such as epiphytic, saprophytic, and terrestrial species. Out of 1230 orchid species in India, about 300 species are expected to occur in the Western Ghats (Singh et al., 2019). Many of them are having high ornamental potential namely *Acanthephippium bicolor, Pecteilis gigantea, Rhynchostylis retusa, Vanda tessellata, Dendrobium* spp., *Aerides crispa, Eulophia* spp., and *Paphiopedilum druryi.* However, several new taxa and new reports of orchid species are being constantly described from the Western Ghats region indicating the further need of exploration in various regions.

Acanthaceae, Fabaceae, and Asteraceae are most dominant plant families and they are predominantly distributed in the Western Ghats. India has reported about 500 species of Acanthaceae, of which 38 genera and 130 species enumerated from Bombay and Khandala region alone (Santapau, 1951). The genus *Strobilanathes* of the family Acanthaceae is remarkable in southern Peninsular India and has about 54 species in the forested area of the Western Ghats (Nair and Daniel, 1986; Venu, 2006; Augustine, 2018). The family Fabaceae exhibits extraordinary varied life forms ranging from lofty rain forest trees to shrubs and herbs in the grasslands; from lianas to tender climbers and creepers along the fringes of forests. Several of them are economically important timber woods, edible seeds, medicinal drugs, and wild relatives of cultivated crops. In Asteraceae about 800 species in the country are contributed by nearly 50% of the total flowering plant in the Western Ghats. According to Rau and Narayana (1985), of about 50 species of tribe Vernonieae from south India alone, nearly 45 species occur in the Western Ghats.

The family Apocynaceae, with about 57 genera and 260 species in India is one among the dominant plant group with well represented in the Western Ghats with 16 endemic genera and about 72 species. The well-known endemic genera of the Western Ghats are *Ceropegia* (45 species), around 35 species are narrowly distributed in small ranges of the Western Ghats (Ansari, 1984). However, the genus *Strobilanthes* is highly diverse and with rich endemic in the Western Ghats, whereas the genus *Impatiens* shows richest endemism with 102 herbaceous species. Still, the more discoveries are coming from these largest genera of *Ceropegia, Impatiens,* and *Strobilanthes* continuously from the Western Ghats which showed the active diversification center of the area. In *Strobilanthes* genus alone about 11 new species have been discovered in the last decade itself such as *S. jomyi*, *S. agathyamalana*, *S. sainthomiana*, *S. malabarica*, *S. orbiculata*, *S. mullayanagiriensis*, *S. bislei*, *S. carmelensis*, *S. scopulicola*, *S. tricostata*, and *S. kannanii* (Augustine, 2018).

The monocot families are predominantly represented in the area namely Araceae, Commelinaceae, Arecaceae, and Zingiberaceae. In the family Araceae, about 29 genera and 126 species were reported in India of which more than 30% are from Western Ghats. Out of 42 species of *Arisaema* in India, around 18 species are distributed in the Western Ghats. The family Commelinaceae with 90 species reported in India, about 50 species in the Western Ghats, of which 20 species are endemic to the southern Western Ghats (Kammathy, 1983). A total of 285 climbing species comprising about 125 genera and 41 families were accounted from the southern Western Ghats. Out of 285 climbers, 33 species are enumerated as RET species, among them *Ceropegia mannarana* and *Gloriosa superba* are found to be endangered species; *Celastrus paniculatus*, *Aganosma cymosa*, *Smilax wightii*, *Corallocarpus gracilipes* are sporadic distribution in the range (Sarvalingam and Rajendran, 2016).

Similarly, the Western Ghats served as habitat for good diversity of biologically interesting plant groups of the insectivorous (Droseraceae, Lentibulariaceae), parasitic plants (Lauraceae, Cuscutaceae, Orobanchaceae, Scrophulariaceae, Viscaceae), and saprophytes (Burmanniaceae, some Orchids). A total of 173 woody species belonging to 58 families were accounted in the Kalakadu-Mundanthurai Tiger Reserve of the southern Western Ghats (Ganesh et al., 1996). The complete floristic diversity of the Western Ghats is incompletely known until now. Also, constantly added new distributional records and new species from the Western Ghats are still being made by taxonomists continuously. Therefore, taxonomic survey, documentation, and assessment of the floristic diversity of the Western Ghats is now an urgent task. Though, the threats and stresses are operating in this zone and also the significance associated with the biodiversity of the region under the National Biodiversity Action Plan. Gymnosperms diversity in Western Ghats showed comparatively lesser than angiosperms which represented three families with three genera and five species (Table 9.2). Many Gymnosperms are cultivated in gardens and introduced plants except three species of *Cycas* such as *C. circinalis*, *C. annikalensis*, *C. swamyi* and one Gnetophyte (*Gnetum ula*). The Bhimashankar wildlife sanctuary of the northern Western Ghats represented about 1142 angiospermic taxa, they belonging to 619 genera and 124 families. About 1094 taxa are wild growing, which represented about 118 families and 586 genera. Rest of the taxa are planted (34), and 14 introduced (Rahangdale and Raghangdale, 2017).

9.3.1 ENDEMISM

The nature of endemism in the Western Ghats has been debated by many botanists in different times since the past (Chatterjee, 1940, 1962; Nayar, 1980, 1982; Rao, 1972; Subramanyam and Nayar, 1974; Aahmedulla and Nayar, 1987; Nair and Daniel, 1986; Nayar and Ahmedulla, 1984; Bhushan et al., 2020). The Western Ghats are categorized only next to the Himalayan ranges in having a high number of endemic species. Although the Western Ghats are a part of the Gondwanaland, they are protected by the Arabian sea coast along the western side, the Vindhya and Satpura mountain ranges on the northern side, the semiarid Deccan plateau on the eastern side, and Indian Ocean on the southern end which act as a barrier for migration of plants. Hence, these kinds of oceanic island are supported with a large number of endemic species. The rich summits of the Western Ghats with their characteristic climates depend upon islands as regards the distribution of endemic plant species (Subramanyam and Nayar, 1974; Blasco, 1970, 1971). There are about 56 (now 60) endemic genera with around 2100 (38%) narrow endemic species in the Peninsular India (Nayar, 1982). Among these 49 genera are monotypic. Unlike the Himalayas, most of the endemic species in the Western Ghats are palaeoendemics. In the southern Western Ghats, Agastyamalai range are having richest in endemism followed by the Wyanad and Anamalai hill ranges. Of the 8080 higher plants species in the Western Ghats 1720 species are narrowly endemic. Nearly every third plant species are rare, threatened, and several are believed to be in the verge extinction for example *Dalbergia travancorica* and *Vanda wightii*. Of these 44 genera are "monotypic," they have represented only one species for example *Vanasushava pedata, Sivadasania josephiana*, etc. Further, the various endemic taxa revealed that Poaceae with 52 genera with 183 species is the largest among endemics. In the family Orchidaceae reported approximately 100 endemic species in the Western Ghats (Singh et al., 2019). Acanthaceae with 24 genera (*Andrographis, Gymnostachyum, Phlogacanthus* (Fig. 9.1d) and *Strobilanthes*) and Apocynaceae with 16 genera (*Decalepis, Kamettia, Seshagiria*) and 40 other families are considered as regards a large number of endemic plants concerned. From the 22 arborescent taxa having more than 5 endemic species in the Western Ghats, the evergreen tree species about 370 species (60% of the total evergreen species) are enumerated to be endemic to the Western Ghats (Pascal, 1991). The family Rubiaceae with 30 endemic genera with more than 180 species among them *Hedyotis* 23 species, *Psychotria* 18 species, *Ophiorrhiza* 16 species, *Pavetta* 15 species, *Ixora* and *Lasianthus* each 14 species and other genera have less than 10 species each in endemicity of the region.

The genus *Crotalaria* is a largest genus with about 30% of the species which are endemic to the region. Some other genera having high degree of endemic species in the Western Ghast are *Eriocaulon* 50 species; *Ceropegia* 33 species; *Iscahemum* 34 species; *Sonerila* 31 species; *Syzygium* 28 species; *Memecylon* 21 species; *Habenaria* 20 species; *Dimeria* 21 species; *Bulbophyllum* 19 species; *Oberonia* 16 species; *Isachne* 15 species; *Dichanthium* 11 species. Some of the woody endemic genera are *Blepharistemma, Erinocarpus, Meteoromyrtus, Otonephelium, Poeciloneuron*, and *Symplocos*. One among the unique features the Western Ghats is the representation of a rich number of arborescent endemic genera and species.

Northern Western Ghats accounted about 181 local endemics including 4 monospecific genera (Datar and Watve, 2018). They have found that a majority of the endemic species are therophytes, which complete their life cycle in a short period during monsoon. Western Ghats of Maharashtra revealed the distribution of 159 higher plant endemic species with 81 genera and 31 families (Mishra and Singh, 2001; Yadav and Kamble, 2008). The genus *Ceropegia* has the high number of species (17) with other endemic genera such as *Glypochloa, Dipacdi*, and *Eriocaulon* (Gaikwad et al., 2014). Most of these endemic taxa are restricted to a small range of geographical area and are rare in occurrence. Most of the narrow endemic taxa have been recorded only by their original type collection, which could not be recollected even after repeated botanical investigation by several botanists in their habitat of occurrence in last decades. Hence, they are assessed in the category of Data Deficient (DD) by IUCN.

According to Chatterjee (1940), only 34 endemic dicotyledonous genera have been listed from Peninsular India. Rao (1972) enumerated about 164 endemic genera to the Indian floristic region, from which nearly 60 genera are distributed to Peninsular India and Sri Lanka. Nayar (1977) accounted that a total of 2100 endemic flowering plant species are represented in Peninsular India, they contributed about 32% of its flora. However, Ahmedullah and Nayar (1986) have reported the exhaustive list on the endemic plants of Peninsular India. They have enlisted 1940 endemic species including infraspecific taxa from Peninsular India. The three volumes of Red Data Book of Indian Plants (Nayar and Sastry 1987, 1988, 1990) have listed 90 endemic taxa which are included from the northern Western Ghats. Nayar (1996) has revised the endemism in India and reported about 2150 endemic plants of Peninsular India. Tetali et al. (2000) recorded 439 endemic taxa of the northern Western Ghats.

Western Ghats harbor a number of endemic plants but highest number of endemic species were noted in Agasthyamalai Biosphere Reserve of southern most Western Ghats. The area of this region is highly diverse, the endemic species include *Andrographis rothii, Antidesma comptum, Biophytum longibracteatum, Cissampelos ansteadii, Dalbergia tinnevallaiensis, Drypetes porteri, Euonymus paniculatus, E. kanyakumariensis, Exacum wightianum* subsp. *uniflorum, Henckelia ovalifolia, H. repens* (Fig. 9.3b), *Impatiens disotis, I. rufescens, I. auriculata, I. tirunelveliensis, I. viridiflora* (Fig. 9.3e), *Indigofera tirunelvelica, Ophiorrhiza tirunelvelica, Rhynchosia jacobii, Sonerila clarkei, S. tinnevelica, S. travancorica, Syncolostemon comosus* (Fig. 9.1f), and *Taxocarpus beddomei*. Several woody endemic genera are also concentrated in this region which include *Aglaia, Apollinias, Cinnamomum, Cynometra, Eugenia, Garcinia, Litsea, Nothopegia, Polyalthia, Psychotria, Syzygium*, etc.

Sholas of Western Ghats are reservoirs of endemic species which possessed unique nature of climate and ecosystems. Several pockets of shola are isolated with patches of grasslands with concentrated number of trees and herbaceous species in small patches of shola vegetation. The best example is *Elaeocarpus blascoi* at the time of exploration of Flora of Palni hills noted only a single tree in the habitat (Matthew, 1997). These evergreen vegetation patches restore the number of endemic plants.

Monotypic genera are represented by only one species with its genus having no other closely allied species anywhere else in the world and hence they have high priority for conservation significance. There has been reported about 236 monotypic genera in India, of which 44 genera are distributed in the Western Ghats (Singh et al., 2015) (Table 9.1). Rana and Ranade (2009) have enumerated a detailed account of monotypic genera in India. According to them, Poaceae with 32 monotypic genera is the largest endemic family in India followed by Fabaceae (15 monotypic genera) and Asteraceae (with 12 monotypic genera). Important monotypic endemic genera are listed in the Western Ghats in Table 9.1.

9.3.2   RARE, ENDANGERED, AND THREATENED PLANTS

Unique features of Western Ghats are exceptionally high degree of endemism with rich biological diversity. A large number of plants are rare, endangered, and threatened due to heavy biological pressure and human interference. At least 350 globally threatened plant species occur in the Western Ghats (IUCN Red Data List). Out of 1500 narrow endemic plants of Western Ghats, 150 are classified as Vulnerable (VU), 173 are as Endangered (EN), and 36 as critically endangered (CR). About 15 plant species are possibly extinct since last decades.

Among the 5500 seed plants accounted from the southern Western Ghats, 2015 species are listed as endemics (Kiran Raj et al., 2003). The Southern Western Ghats constituting southern parts of Karnataka, Kerala, and southern Tamil Nadu are having richest biological diversity. There has been estimated over 4500 flowering plants, of which 1500 are the Western Ghats endemics that are distributed narrowly to this region. However, a precise account on the Rare, Endangered and Threatened (RET) flowering plants of the southern Western Ghats is being inadequate. A database launched by KFRI of Kerala estimated that 760 RET species are distributed over 332 genera in 109 families of the southern Western Ghats (http://redplants.kfri.res.in).

The largest endemic families of Western Ghats are Acanthaceae (24 genera with 127 species), Apocynaceae (16 genera with 70 species), Asteraceae (22 genera with 71 species), Euphorbiaceae (20 genera with 54 species), Fabaceae (36 genera with 97 species), Rubiaceae (30 genera with 150 species), Poaceae (52 genera with 193 species), Orchidaceae (27 genera with 94 species), and other 113 families contributed to more than 1000 species that showed the richest endemic nature of Western Ghats.

Endemic plants of Western Ghats include *Popowia beddomeana* (Fig. 9.1e), *Acranthera grandiflora* (Fig. 9.2a), *Ardisia blatteri* Gamble (Fig. 9.2b), *Arisaema agastyanum* (Fig. 9.2c), *Ceropegia omissa* (Fig. 9.2d), *Cinnamomum sulphuratum* (Fig. 9.2e), *Elaeocarpus gaussenii* (Fig. 9.2f), *Eugenia floccosa* Bedd. (Fig. 9.2g), *Eugenia singampattina* (Fig. 9.2h), *Garcinia travancorica* (Fig. 9.3a), *Henckelia repens* (Fig. 9.3b), *Impatiens levingei* (Fig. 9.3c), *I. travancorica* (Fig. 9.3d), *I. viridiflora* (Fig. 9.3e), *Ixora malabarica* (Fig. 9.3f), *I. ravikumariana* (Fig. 9.3g), *Medinilla malabarica* (Fig. 9.3h), *Neolitsea scrobiculata* (Fig. 9.4a), *Phyllanthus gageanus* (Fig. 9.4b), *P. singampattianus* (Fig. 9.4c), *Phyllocephalum sengaltherianum* (Fig. 9.4d), *Schefflera agathiyamalayana* (Fig. 9.4e), *Sonerila parameswaranii* (Fig. 9.4f), *Sonerila pulneyensis* (Fig. 9.4g), *S. travancorica* (Fig. 9.4h) and *Strobilanthes gracilis* (Fig. 9.4i).

Narrow endemic plants of Agasthyamalai are *Buchanania barberi*, *Dialium travanco-ricum*, *Humboldtia unijuga* var. *trijuga*, *Ixora agasthymalayana*, *Litsea beei*, *Memecylon agastyamalaianum*, *Memecylon sivadasanii*, *Pavetta bourdillonii*, *Symplocos macrophylla* ssp. *namboodirianus*, *Syzygium bourdillonii*, *Sanjappa cynometroides*, *Sageraea grandiflora*, *Vernonia beddomei* with characteristic woody habit substantiating the relic nature of the flora. The recent explorations are also continuously discovering many new species from that area such as *Memecylon sivadasanii*, *M. agastyamlaianum*, *M. manickamii*, *Ixora agasthyamalayana*, *Acrotrema agastyamalayanum*, *Litsea beei*, *Polyalthia shendurunii*, *Andrographis chendurunii*, and *Ophiorrhiza shendurunii*. There has been recorded 309 endemic species among them 100 are endangered (Warrier et al., 2020). Some notable endangered species are *Eugenia floccosa* (EN) (Fig. 9.2g), *E. discifera* (EN), *Hopea utilis* (EN), *Bentinckia condapanna* (VU), *Vateria indica* (CR), and *Syzygium travancricum* (CR). Genera with more than 15 endemic species are listed.

TABLE 9.1   The List of Monotypic Endemic Genera in the Western Ghats.

Adenoon indicum Dalz.	Asteraceae
Aenhenrya rotundifolia (Blatt.) C.S. Kumar & F.N. Rasmus. (Fig. 9.1a)	Orhidaceae
Agasthiyamalaia pauciflora (Bedd.) S. Rajkumar & Janarth. (Fig. 9.1b)	Clusiaceae
Bentinckia condapanna Berry	Arecaceae
Blepharistemma serratum (Dunnst.) Suresh	Rhizophoraceae
Calacanthus grandiflorus (Dalz.) Radlk.	Acanthaceae
Canscorinella bhatiana (K.S. Prasad & Raveendran) Shahina & Nampy	Gentianaceae
Chandrasekharania keralensis Nair et al.	Poaceae
Danthonidium gammiei (Bhide) C.E.Hub.	Poaceae
Deccania pubescens (Roth.) Triveng.	Rubiaceae
Erinocarpus nimmonii J. Graham	Tiliacae
Haplothismia exannulata Airy Shaw	Burmanniacae
Hardwickia binata Roxb.	Fabaceae
Indobanalia thyrsiflora (Moq.) Henry & B. Roy	Amaranthaceae
Indopoa paupercula (Stapf) Bor	Poaceae
Jerdonia indica Wight	Scrophulariaceae
Kamettia caryophyllata Nicolson & Suresh	Apocynaceae
Karnataka benthamii (C.B. Clarke) P.K. Mukh. & Constance	Apiaceae
Kingiodendron pinnatum (Roxb. ex DC.)	Fabaceae
Kunstleria keralensis Mohanan & Nair	Fabaceae
Lamprachaenium microcephalum (Dalz.) Benth.	Asteraceae
Limnopoa meeboldii (Fischer) Hubb.	Poaceae
Meteoromyrtus wynaadensis (Bedd.) Gamble	Myrtaceae
Munrochloa ritcheyi (Munro) M. Kumar & Remesh.	Poaceae
Moullava spicata (Dalz.) Nicolson	Fabaceae
Nanooravia santapaui (M.R. Almeida) Kiran Raj & Sivad.	Poaceae
Nanothamnus sericeus Thoms.	Asteraceae
Octotropis travancorica Bedd. (Fig. 9.1c)	Rubiaceae
Otonephelium stipulaceum (Bedd.) Radlk.	Sapindaceae
Paracaryopsis coelestina (Lindl.) R.R.Mill.	Boraginaceae
Pinda concanensis P.K. Mukh. & Constance	Apiaceae
Pogonachne racemosa Bor	Poaceae
Polyzygus tuberosus Dalz.	Apiaceae
Poeciloneuron indicum Bedd.	Clusiaceae
Sanjappa cynometroides (Bedd.) E.R. Souza & Kiranraj	Fabaceae
Seshagiria sahyadrica Ansari & Hemadri	Apocynaceea
Silentvaleya nairii Nair & Bhargavan	Poaceae
Sivadasania josephiana (Wadhwa & H.J. Chowdhery) N. Mohanan & Pimenov	Apiaceae
Solenocarpus indica Wight & Arn.	Anacardiaceae
*Trilobanche cookei (*Stapf) Sch. ex Henr.	Poaceae
*Triplopogon romasissimus (*Hack.) Bor	Poaceae
Vanasushava pedata (Wight) P.K. Mukh. & Constance (Fig. 9.1g)	Apiaceae
Willisia selaginoides (Bedd.) Warm. ex Willis	Podostemaceae

FIGURE 9.1 Endemic genera of Western Ghats. a. *Aenhenrya rotundifolia* (Blatt.) C. S. Kumar & F. N. Rasm. b. *Agasthiyamalaia pauciflora* (Bedd.) S. Rajkumar & Janarth. c. *Octotropis travancorica* Bedd. d. *Phlogacanthus albiflorus* Bedd. e. *Popowia beddomeana* Hook.f. & Thoms. f. *Syncolostemone comosus* (Wight & Benth.) D. F. Otieno. g. *Vanasushava pedata* (Wight) Mukh. & Constance. (Photos by author)

FIGURE 9.2   Endemic plants of Western Ghats. a. *Acranthera grandiflora* Bedd. b. *Ardisia blatteri* Gamble. c. *Arisaema agastyanum* C. S. Kumar & Sivadh. d. *Ceropegia omissa* Huber. e. *Cinnamomum sulphuratum* Nees, f. *Elaeocarpus gaussenii* Weibel. g. *Eugenia floccosa* Bedd. h. *Eugenia singampattiana* Bedd. (Photos by author)

FIGURE 9.3 Endemic plants of Western Ghats. a. *Garcinia travancorica* Bedd. b. *Henckelia repens* (Bedd.) Weber & Burtt. c. *Impatiens levingei* Gamble. d. *Impatiens travancorica* Bedd. e. *Impatiens viridiflora* Wight. f. *Ixora malabarica* (Dennst.) Mabb. g. *Ixora ravikumariana* (Gamble) Kottaim. h. *Medinilla malabarica* Bedd. (Photos by author)

FIGURE 9.4 Endemic plants of Western Ghats. a. *Neolitsea scrobiculata* (Meisn.) Gamble. b. *Phyllanthus gageanus* (Gamble) M.Mohanan. c. *Phyllanthus singampattianus* (Sebast. & A. N. Henry) Kumari & Chandrab. d. *Phyllocephalum sengaltherianum* (Narayana) Narayana. e. *Schefflera agasthiyamalayana* Manickam et al. f. *Sonerila parameswaranii* K. Ravik. & V. Lakshm. g. *Sonerila pulneyensis* Gamble. h. *Sonerila travancorica* Bedd. i. *Strobilanthes gracilis* Bedd. (Photos by author)

Nilgiris is an important geographic location known for floristic endemism in southern part of Peninsular India. There are three zones where a concentrated number of endemic species such as Nilgiris, Nilgiri-Wayanad, and Sigur plateau are observed. Wayanad microendemic center is represented by narrow endemic plant species: *Aglaia canarensis*, *Ariasema nilamburense*, *Cynometra beddomei*, *Desmodium wynaadense*, *Goniothalamus wynaadensis*, *Hedyotis leschenaultiana*, *Impatiens nataliae*, *Justicia wynaadenis*, *Meteoromyrtus wynaadensis*, *Osbeckia wynaadensis*, *Phyllanthus megacarpus*, *Sonerrila wynaadnesis*, *Strobilanthes virendrakumarana*, and *Tephrosia wynaadensis*. Some of the important endemic species concentrated in Nilgiri are *Impatiens laticornis, I. denisonii, I. levengii, I. orchioides, Actinodaphne lawsonii, A. salicina, Aglaia indica, Anapahalis beddomei, A. meeboldii, Andrographis lawsonii, Argyreia coonoorensis, A. nellygherrya, Atuna indica, Clematis theobromiana, Crotalaria formosa, Dalbergia congesta, Embelia gardneriana, Habenaria polydon, H. richardiana, Heracleum hookerianum,* and *Ilex gardneriana.*

Silent valley vegetation complex is rich with valuable gene pool of endemic species, where some notable endemic species are *Amomum nilgiricum, Antistrophe serratifolia, Bulbophyllum silentvalliensis, Calamus delessertianus, C. neelagiricus, C, renukae, Porpax microchilos, Cucumis silentvalleyi, Dalbergia beddomei, Gastrodia silentvalleyana, Impatiens sivarajanii, Liparis walakkadensis, Ranunculus subpinnatus, Robiquetia josephiana, Silentvalleya nairii, Valeriana hookeriana,* and *Willisia selaginoides.* Palni hill complex of Kodaikanal Wildlife Sanctuary covers patches of evergreen forests in high altitudes and have 49 endemic species solely endemic to this region. The significant endemic species of these hill ranges are *Actinodaphne bourneae, Alchemilla madurensis, Asparagus fysonii, Brachystelma bourneae, Christisonia saulierei, Elaeocarpus blascoi, Fimbristylis amplocarpa, Habenaria pallideviridis, Ixora saulierei, Liparis beddomei, Platostoma palniense, Plectranthus bishopianus,* and *Zehneria mysorensis.* Kanuvai hills of Coimbatore district revealed 51 endemic species with 39 genera, 28 families, and two subfamilies. Among them *Berberis nilghiriensis* one of the Critically Endangered species and others are *Crotalaria scabra, Murdannia lanuginosa, Smilax wightii, Elaeocarpus recurvatus, Litsea wightiana* var. *tomentosa,* and *Dalbergia congesta* (Prabhu Kumar et al., 2012; Warrier et al., 2020).

Anamalai is the highest peak of southern Western Ghats covering all ranges of forests in different altitudes where endemic taxa are represented. Some important endemic species are *Antistrophe serratifolia, Arundinella mesophylla, Begonia anamalaiensis, Commelina wightii, Impatiens konalarensis, I. tangache, Lasianthus strigilosus, Liparis platyphylla, Memecylon sisparense, Peucedanum anamalaiense, Smilax wightii, Sonerila nemakadensis,* and *Trichosanthes anamalaiensis.* About 352 species of endemic tree species have been reported from the Western Ghats, 50 species (14%) belonging to 19 families were listed as threatened categories and also included in the Red Data Book (Nayar and Sastry, 1987–1990). Five species with indeterminate status due to unavailability of information since the type collection were recorded prior to 1997 (Ramesh and Pascal 1997). These species are: *Memecylon sisparense, Miliusa nilagirica, Palaquium bourdillonii, Phaeanthus malabaricus,* and *Poeciloneuron pauciflorum* (Puyravaud et al., 2003).

Many endemic genera are confined to the northern Western Ghats of Maharashtra and these include *Calacanthus, Glypochloa, Haplanthodes, Hubbardia, Pinda, Seshagiria,* etc. *Ceropegia, Chlorophytum, Dipcadi,* and *Crinum* are showing higher representation of endemism in this region. The exclusively endemic species of the region are *Abutilon ranadei, Amorphophallus konkanensis, Ceropegia sahyadrica, C. rollae, C. vincaefolia, Chlorophytum indicum, Crinum brachynema, C. woodrowii, Curcuma caulina, Dicliptera nasikensis, Dipcadi minor,* and *Euphorbia katrajensis.* Southern zone of the northern Western Ghats comes into Karnataka, the vegetation dominated by thorny scrub species with stunted growth. The genera such as *Bhidea, Karnataka,* and *Polyzygos* are endemic to the region. Some species that are restricted to this range are *Calamus nagabettai, Callipedium magdalenii, Dyschoriste vagans, Hopea canarensis, Impatiens barberi, Isachne meeboldii, Ischameum tumidum, Justicia neesii,* and *Nervilia hispida.*

9.3.3 GYMNOSPERMS

In India, Gymnosperm diversity is rich in the alpine zone of Himalaya, but in the Western Ghats there is limited number of native Gymnosperms with many numbers of cultivated species for its ornamental and other purposes. A total of 43 species belonging to 20 genera and 10 families were reported in Nilgiri hills of southern Western Ghats (Jeevith et al., 2014). Most of the species are exotic and cultivated in gardens for ornamental purposes. *Gingko biloba* (maiden-hair tree) the sole survivor of living fossil among gymnosperms and other species such as *Araucaria bidwilli, Araucaria cookii, Zamia furfuracea,* and *Taxodium* are conserved in Botanical garden at Udhagamandalam (Jeevith et al., 2014).

A total account of native Gymnosperms in Western Ghats showed three families with three genera and five species (Table 9.2). Among them three species of *Cycas* such as *C. circinalis, C. annikalensis,* and *C. swamyi* are narrow endemic to the Western Ghats range. But *Nageia wallichiana* is only the conifer species reported in the native of southern Western Ghats. This species is reported very poor regeneration in the native range (Abilash et al., 2005; Abilash and Menon, 2009). All the native Gymnosperms along with exotic species are conserved in Tropical Botanical Garden and Research Institute, Kerala by *ex situ* conservation (Antony and Mohanan, 2010). The total gymnosperm diversity in India is 64 species of which Western Ghats has 5 native species and about 60 are introduced species (http://www.tnenvis.nic.in/tnenvis_old/biodiversity.htm).

TABLE 9.2 Wild Gymnosperms of Western Ghats.

Name of the species	Family	Native range	Habitat
Cycas annikalensis Singh & Radha	Cycadaceae	S. India	Vulnerable
Cycas circinalis L.	Cycadaceae	S.India	Endangered in wild
Cycas swamyi Singh & Radha	Cycadaceae	S. India	Vulnerable
Gnetum ula Brong.	Gnetaceae	S. India	Rare in wild
Nageia wallichiana (C.Presl) Kuntze	Podocarpaceae	S. India to Malaysia	Rare in wild

9.3.4 AFFINITY OF WESTERN GHATS FLORA

Representation of about 60 narrow endemic genera including 50 monotypic genera consti-tutes this region and is floristically unique and conservation significant. There have been listed more than 1286 endemic species in the southern Western Ghats alone (Nayar, 1996). In the present estimate pertaining to the political boundaries of the present-day India, it is accounted about 4300 (23.20%) species of endemics, from 18,532 flowering plant species estimated in the country. The flora of the Western Ghats revealed the region is closely associated with the floristic composition of E. Africa, Malaysia, and Sri Lankan flora. The Western Ghats flora evidenced the Gondwanaland origin which is close to the land-mass comprising S. America, Madagascar, India, Malaysia islands, Sri Lanka, Australia, and Antarctica. The oldest connection of Peninsular India with surrounding continents is evidenced from the distribution of some plant genera like *Hernandia, Lindenbergia, Pittos-porum, Acrotrema, Gomphandra, Nothapodytes, Sarcostigma, Hydnocarpus*, etc. in the Western Ghats, Africa, and some in S. America. The narrow endemic genus *Poeciloneuron* (Clusiaceae) in the Western Ghats has allied genera in S. America. Similarly, 10 other genera of Orchidaceae and 52 species of Andropogoneae (Poaceae) of the Western Ghats also extended their distribution in Africa. Grasses are very well represented in the Western Ghats and they are estimated that about 400 species are distributed in Kerala alone (Nair and Jayakumar, 2008). According to Mehrotra and Jain (1982), there are 329 species of Andropogoneae, about 250 species having narrow distribution in the Western Ghats. The genus *Isachne* is represented by 20 endemic species and *Garnotia* with 12 endemic species in the Western Ghats region (Prakash and Jain, 1979). Bamboos are also a prominent vegetation in the Western Ghats with about 25 species having common occurrence out of 100 species in India.

Phytogeographically, the floristic contribution of the Western Ghats is revealed with a very high endemism, and almost 21% of the total flowering plants of India were recorded in this region (Nayar, 1996). Among them, 12.5% of endemic species were restricted to southern part of the Western Ghats especially in the Agasthiyamalai region. Indo-Sri Lankan Peninsular elements have represented about 15% of the total flora, which showed the geographical affinity of the region with the adjacent Ocean Island. The pattern of distribution of various endemic species outside India: 66% of them were of Indo-Malayan and South and South-East Asian range about 30% of the total flora and they were constituted of pluri-regional species. Only 4% of the flora, which are either exotic weeds or naturalized from cultivation, showed the less disturbed status of the pristine vegetation (Nair and Jayakumar, 2008).

The floristic affinities of the Nilgiri Biosphere Reserve showed that it has strong Indo-Malayan (27.7%) and Indian (30.3%) relationships. Flora of moist vegetation types has strong affinities with Indomalayan (20.1%), Indian (38.6%), and Indo-Sri Lankan (9.7%) flora while the flora of dry evergreen vegetation types showed the Afrotropical and Pantropical affinities. Tropical montane forests have maximum endemic species (20%) suggesting that the unique vegetation types are highly specialized for conservation measures (Suresh and Sukumar, 1999). Another analysis on endemism revealed that 34 taxa are endemic to Western Ghats of Kerala and at the same time, broad scale phytogeographical affinities of the riverine flora extend to African, Australian, Holarctic, Indo-Pacific, and South American floristic kingdoms (Jisha and Nair, 2018).

9.3.5   CAUSES OF RARITY OF PLANTS

The Western Ghats is being much stressed on the threshold of development and with increased human pressure it has already lost much of its indigenous nature of vegetation and unique habitats. The entire area has already been listed under the world's "hottest hotspot" (Myers, 1988, Myers et al., 2000). Currently, there are several threats operating on the vegetation, which not only destroyed the unique habitats of flora and but also operating the spread of many invasives, alien species, and further deteriorating the plant wealth of the Western Ghats. Ever-increasing human population explosion, selective removal of valuable groups of plants, extensive practice of shifting agriculture in core areas in the forests, extension of urbanization, roads and dams construction on hills creating accessibility and availability of local resources, degradation and fragmentation of natural forests for various plantation crop cultivation namely coffee, fruits, vegetables, and spices are the major threats for the degradation of Western Ghats. Tourism inside the core zone and their greed for collection of selected groups of ornamental plants such as orchids, begonias, *Impatiens* spp., dependence of plant-based industries on wild plant resources, wrong policies of the government that allow unregulated export of timber, bamboos, and non-timber forest products lead to the impoverishing of biodiversity sink of the region, unplanned economic upliftment of the people, sudden spread of certain obnoxious alien invasive weeds such as *Eupatorium, Mikania, Parthenium,* and others are threatening the indigenous flora and these are all some of the noticeable causal threats in the Western Ghats biodiversity. Approximately 40% of natural forest vegetation in the Western Ghats has already been disappeared during the past 8–10 decades (Menon and Bawa, 1997). The *Dipterocarpus* spp. dominated low elevation evergreen forests have become the most vulnerable ecosystem in the region (Pascal, 1982, 1991), whereas some other low elevation forest species like *Buchanania barberi, Cynometra beddomei, Dialium travancoricum, Hopea jacobi, Inga cynometroides, Syzygium chavaran,* and *Buchanania lanceolata* are facing great threat with various degrees of extinction. However, some other high elevation species are rare, which include *Aglaia malabarica, Aporusa bourdillonii, Cynometra beddomei, Dialium travancoricum, Litsea travancorica, Polyalthia shendurunii, Phaeanthus malabaricus, Sageraea grandgora, Syzygium bourdillonii,* and *Vernonia beddomei.* Their rarity and possible extinction may be due to deforestation and development of agriculture or anthropogenic pressure.

Several endemic plant taxa are confined to the northern Western Ghats that are near threatened category because of highly isolated populations in small biogeographical area or in hillocks and hill peaks. Some of the Critically Endangered plant taxa are *Arisaema sivadasanii, Brachystelma malwanense, Brachystelama naorojii, Ceropegia anantii, C. fantastica, C. huberi, C. mahabalei, C. panchganiensis, C. santapaui, Dicliptera nasikensis, Scurrula stocksii,* and *Drimia razii.* These are the plant species known only from a single locality (type locality) with fewer number of individuals and occupied with less than 10 km^2 area (Gaikwad et al., 2014). The recent investigations are also not able to relocate their productive individuals despite critical field examinations undertaken by several taxonomists in Sahyadri ranges. Hence, these endemic populations seem to be on the verge of extinction.

In consequence of deforestation, several plant species have already become endangered or even presumed to be extinct, are not recollected after their type collection, and are also facing various levels of threat or extinction like ornamental plants, medicinal plants, biologically interesting plants, and aromatic species (Rao et al., 2004). Aromatic plant species namely *Pogostemon nilagiricus, P. travancoricus, P. wightii, Plectranthus nilgherricus, P. wightii, P. walkeri, Moonia heterophylla, Ocimum adscendens, Cinnamomum travancoricum,* and *C. wightii,* once abundant on the hill slopes of the Western Ghats, have now become very scarce due to their overexploitation for industrial purposes. Further, invasion of alien weeds like *Lantana camara, Parthenium hysterophorus, Mikania micrantha, Eupatorium odoratum, Mimosa instia,* etc., has affected several of our native and naturalized species in the Western Ghats. Invasive alien species are considered to be a second major threat to the native flora. The invasive alien plant species are severely competing with native flora for space, light, nutrients, and water. Again, clear-cut species extinction operating processes are not observed, and fragmentation of indigenous forests has pushed many endemic herbaceous species on road to extinction. It is a challenging but priority agenda such as assessment and identification of such fragmented species populations in different biogeographic zones of the Western Ghats for its conservation measurements before extinction.

9.3.6 WESTERN GHATS' SPECIAL HABITATS

The high elevation ranges of the Western Ghats are composed of plateaus and mesic forests amid rocky outcrops. These rocky outcrops are unique isolated rocky formations and known for their critical environment, rich with variable biodiversity as well as centers of species endemism. In the Western Ghats, rocky plateaus are considered special habitats, but very scarce information is available about their floristic composition and conservation importance. A few studies accounted their floristic richness and ecology. Still it has to be formulated the study design for sampling techniques, preservation models, and conservation strategies. A comprehensive botanical study of northern Western Ghats revealed a very high diversity of 443 flowering plant taxa in a small area of 2.87 km^2. The plant diversity of high altitudes is influenced by various factors such as soil composition, nature of rock, rainfall, velocity of wind, temperature, atmospheric pressure, etc.

Many researchers have reported the floristic richness of high-altitude plateaus of northern Western Ghats such as 600 taxa in Durgawadi plateau and 249 taxa in Naneghat plateau (Rahangdale and Rahangdale, 2014, 2018). But in Lateritic plateau (Joshi and Janarthanam, 2004; Bhattarai et al., 2012) about 300 plant species have been reported; similarly, Anjaneri outcrops was observed by Lekhak and Yadav (2012). Across the world, cliffs are considered distinct microhabitats but very little attention has been paid to the investigation of cliffs of the northern Western Ghats. A total of 102 species belonging to 35 families and 69 genera have been recorded for the cliff flora of the main range of northern

Western Ghats which pass through Sindhudurg, Ratnagiri, Kolhapur, Sangli, Satara, Pune, Nasik, Ahmednagar, Palghar, Thane districts, and Goa state (Yadav and Kamble, 2008; Watve, 2008, 2013). About 186 geophytes taxa were recorded in the high-altitude plateau of Maharashtra, of them 62 geophytes belonging to 22 families have adapted to a monsoon seasonality (Gaikwad et al., 2015).

High elevational ranges of the southern Western Ghats are a hottest hotspot of several endemic *Strobilanthes*. Possible reasons for a high number of species diversity and endemism could be geographically attributed to the isolation of highland valleys, sholas, and gorges coupled with altitudinal and eco-climatic variations within small geographical range. These high elevational hill ranges between 700 and 2690 m asl possess many hill peaks above 2000 m in this region. In *Strobilanthes*, a few species are only found to occur below 500 m namely *S. barbatus*, *S. warrensis*, *S. heyneanus*, and *S. rubicundus* and others may reach up to 1400 m asl. But, some other species like *S. homotropus*, *S. violaceous*, *S. zenkerianus*, *S. andersonii*, *S. gracilis*, *S. wightianus*, *S. urceolaris*, and *S. foliosus* are distributed narrowly above 2000 m asl. The altitudinal gradient of distribution of these 40 *Strobilanthes* species could be related to recent isolation of the various populations of the ancestral stock and speciation leads to these taxa as neoendemics (Agustine, 2012). However, grassland-specific species like *S. kunthianus* and *S. sessilis* have still more recent evolutionary affinity with grassland diversity speciation in southern Indian hill ranges. Micro-habitat specificity may also operate the phenomena toward a narrow distribution range and occurrence of a large number of *Strobilanthes* species in the southern Western Ghats. Several species like *S. tristis*, *S. warrensis*, *S. rubicundus*, *S. pulneyensis*, *S. micranthes*, *S. luridus*, *S. homotropus*, *S. heyneanus*, *S. decurrens*, *S. ciliatus*, *S. asperrimus*, *S. anceps*, *S. virendrakumarana*, *S. ixiocephalus*, *S. andersonii*, *S. barbatus*, and *S. dupenii* preferred the evergreen shola habitats. Similarly, the species namely *Strobilanthes foliosus*, *S. wightianus*, *S. urceolaris*, *S. caudatus*, *S. violacea*, *S. zenkerianus*, and *S. gracilis* preferred margins of the evergreen shola forests. The flowering interval and periodicity may vary from species to species in *Strobilanthes*; it ranged from annual to 16 years. Species namely *S. anamallaica*, *S. asperrimus*, *S. ciliatus*, *S. heyneanus*, *S. pulneyensis*, *S. sessilis*, *S. urceolaris*, and *S. wightianus* produce flowers every year, while *S. amabilis*, *S. gracilis*, *S. homotropus*, *S. kunthiana*, *S. micranthus*, *S. perrottetianus*, *S. violceus*, and *S. zenkarianus* are flowering with a gap of more than 10 years of vegetative growth (Venu, 2006). The montane forest of Western Ghats has rich endemic flora which revealed the occurrence of 286 species of plants belonging to 85 families. Among them, 88 rare, endangered, and threatened (RET) species belong to the families like Asteraceae (19 spp.), Euphorbiaceae (14 spp.), Fabaceae (15 spp.) Rubiaceae (nine spp.), Orchidaceae (eight spp.) Araceae (9 spp.) Acanthaceae (9 spp.), and Poaceae (14 spp.) (Brilliant et al., 2012).

9.3.7 CONSERVED AREAS

At present seven national parks in the Western Ghats region with a total area of about 2073 sq. km (1.3% of the region) and 39 wildlife sanctuaries covered an area of about 13,862 sq. km (8.1%). Also *in situ* conservation of several indigenous species is partly protected by the establishment of legally conserved areas like the Nilgiris Biosphere Reserve, Agasthya-malai Biosphere Reserve, Kalakad-Mundandurai Tiger Reserve, Indira Gandhi National Park, Silent Valley National Park, Bandipur National Park, Kudremukh National Park, Nagarahole National Park, Giant Squirrel Wildlife Sanctuary, Idukki Wildlife Sanctuary, Eravikulam Wildlife Sanctuary, Kodaikkanal Wildlife Sanctuary, Kanyakumari Wildlife Sanctuary, Wayanad Wildlife Sanctuary, Periyar Wildlife Sanctuary, Parimbikulam Wild-life Sanctuary, Shendurney Wildlife Sanctuary, Peppara Wildlife Sanctuary, etc. Though many numbers of endemic species are being outside the reserve, we do not find any *in situ* programmes for conservation. Even within the protected Biosphere Reserves, there is a severe threatening pressure by the invasive weeds. The Agasthiyamalai peak (1868 m) in southernmost highest range of the Western Ghats is recognized with one of the most popular hotspots; the significance of this area is that it contributed about 12% of the known plants narrowly endemic to the region. Ponmudi and Chemunji are another highest ridge in the Agasthiyamalai region which contributes a number of endemic biodiversity. In the same region two more other Wildlife Sanctuaries, viz. Peppara Wildlife Sanctuary and Neyyar Wildlife Sanctuary, are included, which are also main part of the Agasthiyamalai Biosphere landscape area and thus the major component of conservational vegetation complex in the southern Western Ghats with rich gene pool of endemic flora.

9.4 CONCLUSIONS

Conservation of the national biological resources is a challenging task and urgently in need of the attention of all biological scientists. The National Biodiversity strategy and Action Plan (Singh, 2002) rightly pointed out the course of action to be pursued for the conservation of the native flora of India. Biodiversity documentation is providing baseline data for conservation action plan and policy makers. Before exploiting the biodiversity, understanding the endemism and rarity of the biological resources would help to prepare sustainable utilization measures and conservation strategies. This chapter emphasized the endemic and endangered plant diversity of Western Ghats for preparing conservation action plan for rare, endangered, and threatened taxa of hotspot areas.

KEYWORDS

- gymnosperms
- angiosperms
- threatened plants
- endemism
- conserved areas
- special habitats

REFERENCES

Abilash, E. S.; Menon, A. R. R.; Balasubramanian, K. Regeneration Status in Natural Habitats of *Nageia wallichiana* (Presl.) O. Ktz, Goodrical Reserved Forest of Western Ghats of India. *Indian Forester* **2005**, *94*, 183–200.

Abilash, E. S.; Menon, A. R. R. Status Survey on *Nageia wallichiana* (Prsel) O. Kuntze in Natural Habit of Goodrical Reserve Forests, Western Ghats, India. *Indian Forester* **2009**, *135*, 281–286.

Ahmedullah, M.; Nayar, M. P. *Endemic Plants of the Indian Region*; Botanical Survey of India: Calcutta, **1986**.

Ahmedullah, M.; Nayar, M. P. *Endemic Plants of Indian Region*, Vol. 1, *Peninsular India*; Botanical Survey of India: Calcutta. **1987**.

Ahuja, B. S., Singh, K. P. 1963. Ecological Studies on the Humid Tropics of Western Ghats, India. *Proc. Natl. Acad. Sci.* **1963**, *32B*, 77–84.

Almeida, M. R. *Flora of Maharashtra, Vol. 1, 2, 3,4 and 5*; St. Xavier's College, Orient Press: Mumbai, **1996, 1998, 2001, 2003, 2009**.

Ansari, M. Y. Asclepiadaceae—Genus *Ceropegia. Fasc. Fl. India* **1984**, 16: 1–34.

Antony, R., Mohanan, N. N. Ex-Situ Conservation and Multiplication of *Podocarpus wallichianus* Presel.—a Threatened Conifer of Western Ghats. *Ind. J. Forestry* **2010**, *33*, 131–134.

Augustine, J. *Strobilanthes Bl. in the Western Ghats, India: The Magnificent Role of Nature in Speciation;* Malabar Natural History Society: Kerala, **2018**; 152pp.

Auti, S. G.; Kambale, S. S.; Gosavi, K. V. C.; Chandore, A. V. Floristic Diversity of Anjaneri Hills, Maharashtra, India. *J. Threat. Taxa* **2020**, *12* (10), 16295–16313. https://doi.org/10.11609/jott.3959.12.10.16295–16313

Bharucha, F. R.; Ansari, Y. Studies on the Plant Associations of Slopes and Screes of the Western Ghats, India. *Plant Ecol.* **1963**, *11*, 141–154.

Bhattarai, U.; Tetali, U.; Kelso, S. Contributions of Vulnerable Hydrogeomorphic Habitats to Endemic Plant Diversity on the Kas Plateau, Western Ghats. *Springer Plus* **2012**, *1*, 25. http:/dx.doi.org/10.1186/2193–1801–1-25

Bhushan K. S.; Kulkarni, A.; Vijayan, S.; Choudhary, R. K.; Datar, M. N. An assessment of the local endemism of flowering plants in the Northern Western Ghats and Konkan regions of India: checklist, habitat characteristics, distribution, and conservation. *Phytotaxa* **2020**, 440 (1), 25–54.

Blasco, F. Aspects of the Flora and Ecology of the Savannas of the South Indian Hills. *J. Bombay Nat. Hist. Soc.* **1970**, *67*, 522–534.

Blasco, F. Montagnes du sud deI'nde: forests, savanes, ecologie. *Inst, Fr. Pondicherry, Trv, Sec.Sci. Tech. Tome* **1971**, *10*, 436.

Blatter, E.; McCann, C. *The Bombay Grasses.* Sci. Monogr. No.5. Imp. Counc. Agric. Res. India. **1935**, *21*, 324.

Bor, N. L. *The Grasses of Burma, Ceylon, India and Pakistan*; Pergamon Press: London, **1960**.

Bourdillon, T. F. *Report on the Forest of Travancore*; Travancore Govt. Press: Trivandrum, **1893**.

Brilliant, R.; Varghese, V. M.; Paul, J.; Pradeepkumar, A. P. Vegetation Analysis of Montane Forests of Western Ghats with Special Emphasis on RET Species. *Int. J. Biodiver. Cons.* **2012**, *4*, 652–664.

Caius, J. F. The Medicinal and Poisonous Grasses of India. *J. Bombay Nat. Hist. Soc.* **1936**, *38*, 540–584.

Chandrabose, M.; Nair, N. C.; Chandrasekharan, C. *Flora of Coimbatore*; Bishen Singh Mahendra Pal Singh: Dehra Dun, **1988**.

Chatterjee, D. Studies on the Endemic Flora of India and Burma. *J. Asiat. Soc. Bengal* **1940**, *5*, 19–67.

Chatterjee, D. Floristic Pattern in Indian Vegetation. *Proc. Summer School Botany*, Darjeeling, **1962**, 32–42.

Cooke, T. *Flora of Presidency of Bombay*; Govt. of India, **1901–1908**.

Dalzell, N. A.; Gibson, A. *The Bombay Flora or Short Descriptions on All the Indigenous Plants Together with a Supplement of Introduced and Naturalized Species*; Education Society's Press: Byculla, **1861**.

Datar, M. N.; Watve, A. V. Vascular Plant Assemblage of Cliffs in Northern Western Ghats, India. *J. Threat. Taxa* **2018**, *10*, 11271–11284.

Desai, M. C.; Murthy, S. N. Some Grasses of Dharwar Agriculture College Estate, Dharwar. *Agric. Coll. Mag.* **1950**, *4*, 27.

Diwakar, P. G., Sharma, B. D. *Flora of Buldhana District, Maharashtra State*; Botanical Survey of India: Culcutta. **1999**.

Fyson, P. F. *The Flora of South Indian Hill Stations*, 2 vols.; Madras Govt. Press, **1932**.

Fyson, P. F. *The Flora of the Nilgiri and Pulney Hill Tops*, 3 vols; Govt. Press: Madras, **1915**.

Gaikwad, S.; Gore, R.; Garad, K.; Gaikwad, S. Endemic Flowering Plants of Northern Western Ghats (Sahyadri Ranges) of India: A Check List. *Check List* **2014**, *10*, 461–472.

Gaikwad, S.; Gore, R.; Garad, K.; Gaikwad, S.; Mulani, R. Geophytes of Northern Western Ghats (Sahyadri Range) of India: A Check List. *Check List* **2005**, *11* (1), DOI: 10.15560/11.1.1543

Gamble, J. S.; Fischer, C. E. C. *Flora of the Presidency of Madras*; Adlard & Son Ltd.: London, **1915–36**.

Ganesan, V.; Thangaraj, P. S.; Sivaprasad, P. S.; Balakrishnan, S. *Endemic Flora of Western Ghats—Anamalais*, Vol. 1. Anamalai Tiger Reserve Foundation: Tamil Nadu, 2019.

Ganesh, T.; Ganesan, R.; Soubadra Devy, M.; Davidar, P.; Bawa, K. S. Assessment of Plant Biodiversity at a Mid-Elevation Evergreen Forest of Kalakad–Mundanthurai Tiger Reserve, Western Ghats, India. *Curr. Sci.* **1996**, *71*, 379–392.

Jeevith, S.; Ramachandran, V. S.; Ramsundar, S. Gymnosperms of Nilgiris District, Tamil Nadu. *Res. Plant Biol.* **2014**, *4* (6), 10–16.

Jisha, K.; Nair, M. C. Diversity Analysis of Angiosperm in Riparian System Along Thuppanad River, Southern Western Ghats, Kerala, India. *Int. J. Adv. Res.* **2018**, 6, 531–539.

Joshi, V. C.; Janarthanam, M. K. The Diversity of Life-Form Type, Habitat Preference and Phenology of the Endemics in the Goa Region of the Western Ghats, India. *J. Biogeog.* **2004**, *31* (8), 1227–1237. https://doi.org/10.1111/j.1365–2699.2004.01067.x

Kammathy, R. V. Rare and Endemic Species of Indian Commelinaceae. In *An Assessment of Threatened Plants of India*; Jain, S. K., Rao, R. R., Eds.; Botanical Survey of India: Howrah, **1983**; pp 213–221.

Karthikeyan, S. Northern Western Ghats and Northern West Coast. In *Flora of India: Introductory Volume*; Hajra, P. K., Sharma, B. D., Sanjappa, M., Sastry, A. R. K., Eds.; Botanical Survey of India: Calcutta, **1996**; pp 375–390.

Karthikeyan, S.; Anandkumar, S. *Flora of Yavatmal District, Maharashtra State*; Botanical Survey of India, Fl. India ser. 3: Calcutta, **1993**.

Karthikeyan, S.; Nayar, M. P.; Sundara, R. A. Catalogue of Species Added to Cooke's Flora of the Presidency of Bombay During 1908–1978. *Rec. Bot. Surv. India* **1981**, *21*, 153–205.

Karthikeyan, S.; Jain, S. K.; Nayar, M. P.; Sanjappa, M. *Florae Indicae Enumeratio: Monocotyledonae*; Botanical Survey of India: Calcutta, **1989**.

Keshava Murthy, K. R.; Yoganarasimhan, S. N. *The Flora of Coorg (Kodagu), Karnataka, India*; Vimsat Publishers: Bangalore, **1990**.

Kiran Raj, M. S.; Sivadasan, M.; Ravi, N. Grass Diversity of Kerala—Endemism and Its Phytogeographical Significance. In *Plant Diversity, Human Welfare and Conservation*; Janarthanam, M. K., Narasiman, D., Eds.; Goa University: Goa, **2003**; pp 8–30.

Kothari, M. J.; Moorthy, S. *Flora of Raigad District, Maharashtra State*; Botanical Survey of India: Kolkatta, **1993**.

Kulkarni, B. G. *Flora of Sindhudurg*; Botanical Survey of India: Calcutta, **1988**.

Lakshminarasimhan, P.; Sharma, B. D. *Flora of Nasik District*. Fl. India ser. 3; Botanical Survey of India: Calcutta, **1991**.

Lekhak, M. M.; S. R. Yadav, S. R. Herbaceous Vegetation of Threatened High Altitude Lateritic Plateau Ecosystems of Western Ghats, Southwestern Maharashtra, India. *Rheedea* **2012**, *22*, 39–61.

Mahabale, T. S. Botany and Flora of Maharashtra. In *Maharashtra State Gazetteer*; Government of Maharashtra, **1987**.

Malpure, N. V.; Yadav, S. R. *Chlorophytum gothanense*, a New Species of Anthericaceae from the Western Ghats of India. *Kew Bull.* **2009**, *64*, 739–741. https://doi.org/10.1007/s12225–009–9165–8.

Manilal, K. S. Biodiversity of Silent Valley and Efforts for the Conservation of Tropical Rain Forests of India. In *Taxonomy and Biodiversity*; Pandey, A. K., Ed.; CBS Publishers & Distributors: New Delhi, **1995**; pp 322–338.

Manilal, K. S. *Flora of Silent Valley Tropical Rain Forest of India*; Department of Science & Technology: Calicut, **1988**.

Matthew, K. M. *The Flora of Tamil Nadu Carnatic*, 3 vols.; Rapinat Herbarium: Tiruchirapalli, **1981–84**.

Matthew, K. M. *The Flora of the Palni Hills, South India*, 3 Vols.; The Rapinat Herbarium: India, **1999–2003**.

Mehrotra, A.; Jain, S. K. Endemism in Indian Grasses—Tribe Andropogoneae. *Bull. Bot. Surv. India* **1982**, *22*, 51–58.

Menon, S.; Bawa, K. S. Application of Geographic Information Systems, Remote Sensing and a Landscape Ecology Approach to Biodiversity Conservation in Western Ghats. *Curr. Sci.* **1997**, *73* (2), 134–145.

Mishra, D. K., Singh, N. P. *Endemic and Threatened Flowering Plants of Maharashtra*; Botanical Survey of India: Calcutta, **2001**.

Mohanan, M.; Henry, A. N. *Flora of Thiruvananthapuram District*; BSI: Calcutta, **1994**.

Mohanan, M.; Sivadasan, M. *Flora of Agasthyamala*; BSI: Calcutta, **2002**.

Myers, N. Threatened Biotas: Hot Spots in Tropical Forests; *Environmentalist* **1988**, *8*, 1–20 and **1990**, *10*, 243 256.

Myers, N.; Mittermeier, R. A.; Mittermeier, C. G.; Fonesca, G. A. B.; Kents, J. Biodiversity Hotspots for Conservation Priorities. *Nature* **2000**, *403*, 853–858.

Naik, V. N. *Flora of Marathwada of Maharashtra*; Amurt Prakasahan: Aurangabad, 2 vols, **1998**.

Nair, K. K. N.; Jayakumar, R. Phytogeography, Endemism and Affinities of the Flora of New Amarambalam Reserve Forests in the Western Ghats of India. *Indian J. Forestry* **2008**, *31*, 85–94.

Nair, N. C., Daniel, P. Floristic Diversity of the Western Ghats and Its Conservation: A Review. *Proc. Indian Acad. Sci. Suppl.* **1986**, *3*, 127–163.

Nair, N. C., Henry, A. N. *Flora of Tamil Nadu, India, Series 1: Analysis*; Botanical Survey of India: Coimbatore, **1983**.

Nayar, M. P. Changing Patterns of the Indian Floras. *Bull. Bot. Surv. India* **1977**, *19*, 145–155.

Nayar, M. P. Endemic Flora of Peninsular India and Its Significance. *Bull. Bot. Surv. India* **1982**, *22*, 12–23.

Nayar, M. P. Endemism and Patterns of Distribution of Endemic Genera. *J. Econ. Taxon. Bot.* **1980**, *1*, 99–110.

Nayar, M. P. *Hotspots of Endemic Plants of India, Nepal and Bhutan*; Tropical Botanic Garden and Research Institute: Thiruvananthapuram, **1996**.

Nayar, M. P.; Sastry A. R. K. *Red Data Book of Indian Plants, Vol. 1, 2, 3*; BSI: Calcutta, **1987, 1988, 1990**.

Nayar, T. S.; Rasiya Begam, A.; Sibi, M. *Flowering Plants of the Western Ghats, India,* 2 Vols, Jawaharlal Nehru Tropical Botanic Garden and Research Institute: Thiruvananthapuram, Kerala, **2014**.

Nayar, T. S.; Sibi, M.; Rasiya Begam, A.; Mohanan, M.; Rajkumar, G. Flowering Plants of Kerala. Status and Statistics. *Rheedea* **2008**, *18* (2), 95–106.

Pascal, J. P. Floristic Composition and Distribution of Evergreen Forests in the Western Ghats, India. *Palaeobotanist* **1991**, *39*, 110–126.

Pascal, J. P. *Forest Map of South India—Sheet: Mercara-Mysore*; Karnataka & Kerala Forest Departments and the French Institute: Pondicherry. **1982**.

Patel, R. I. *Forest Flora of Melghat*; Bishen Singh Mahendra Pal Singh: Dehradun, **1968**.

Patil, D. A. *Flora of Dhule and Nandurbar Districts* (Maharashtra), Bishen Singh Mahendra Pal Singh: Dehradun, **2003**.

Potdar, G. G.; Salunkhe, C. B.; Yadav, S. R. *Grasses of Maharashtra*; Shivaji University: Kolhapur, **2012**.

Prabhu Kumar, K. M.; Sreeraj, V.; Binu, T.; Manudev, K. M.; Rajendran, A. Validation and Documentation of Rare Endemic and Threatened (RET) Plants from Nilgiri, Kanuvai and Madukkarai Forests of Southern Western Ghats, India. *J. Threat. Taxa* **2012**, *4* (15), 3436–3442.

Prakash, V.; Jain, S. K. Poaceae: Tribe –Garnotieae. *Fasc. Fl. India* **1979**, *14*, 1–42.

Puyravaud, J. P.; Davidar, P.; Pascal, J. P.; Ramesh, P. R. Analysis of Threatened Endemic Trees of the Western Ghats of India Sheds New Light on the Red Data Book on the Indian Plants. *Biodivers. Conserv.* **2003**, *12*, 2091–2106.

Rahangdale, S. S.; Rahangdale, S. R. Plant Species Composition on Two Rock Outcrops from the Northern Western Ghats, Maharashtra, India. *J. Threat. Taxa* **2014**, *6* (4), 5593–5612. http://doi.org/10.11609/JoTT. o3616.5593–612.

Rahangdale, S. S.; Rahangdale, S. R. Floristic Diversity of Bhimashankar Wildlife Sanctuary, Northern Western Ghats, Maharashtra, India. *J. Threat. Taxa* **2017**, *9* (8), 10493–10527. http://doi.org/10.11609/jott.3074.9.8. 10493–10527

Rahangdale, S. R.; Rahangdale, S. S. *Biodiversity of Durgawadi Plateaus*; Department of Forests, Government of Maharashtra, **2018**.

Ramachandran, V. S.; Nair, V. J. *Flora of Cannanore District*, Botanical Survey of India: Calcutta, **1988**.

Ramaswamy S. N.; Radhakrishana Rao, M.; Govindappa, D. A. *Flora of Shimoga District, Karnataka*; Prasaranga, University of Mysore, **2001**.

Ramesh, B. R.; Pascal, J. P. *Atlas of the Endemics of the Western Ghats (India): Distribution of Tree Species in the Evergreen and Semi Evergreen Forests*, Vol. 38; Institute Francais de Pondicherry, Publications du department d ecologie **1997**; p 403.

Rana, T. S.; Ranade, S. A. The Enigma of Monotypic Taxa and Their Taxonomic Implications. *Curr. Sci.* **2009**, *96* (2), 219–229.

Rao, C. K. Angiosperm Genera Endemic to Indian Floristic and Its Neighbouring Areas. *Indian For.* **1972**, *98*, 560–566.

Rao, R. R. *Studies on Flowering Plants of Mysore District*, 2 vols. Ph.D. Thesis submitted to Mysore University, **1973**.

Rao, R. R.; Razi, B. A. *A Synoptic Flora of Mysore District*; Today & Tomorrows Publishers: New Delhi, **1981**.

Rao, R. R.; Murugan, R.; Syamasundar, K. V.; Srinivasalu, B. Western Ghats,—A Major Emporium of Wild Aromatic Plants: Diversity, Conservation and Bioprospection. In *Concepts in Forestry Research*; Todaria, N. P., Chamola, B. P., Chauhan, D. S., Eds.; International Book Distributors: Dehra Dun, **2006**; pp 267–278.

Rao, R. S. *Flora of Goa, Diu, Daman, Dadra and Nagarhaveli*; Botanical Survey of India: Calcutta, **1985–1986**.

Rao, R. S. M. *Flowering Plants of Travancore*; Government Press: Travancore, **1914**.

Rau, M. A.; Narayana, B. M. A Review of the Tribe Vernonieae (Asteraceae) in South India. *Bull. Bot. Surv. India* **1985**, *25*, 19–25.

Rodgers, W. A.; Panwar, H. S. *Planning a Wildlife Protected Area Network in India*, 2 vols.; Project FO: IND/82/003. FAO, Dehra Dun. 339, **1988**.

Sahni, K. C. *Gymnosperms of India and Adjacent Countries*; Bishen Singh Mahendrapal Singh: Dehradun, **1990**.

Saldanha, C. J. *The Flora of Karnataka*, Vol. 1 and 2; Oxford & IBH Publishing Co: New Delhi, **1984, 1996**.

Saldanha, C. J., Nicolson, D. H. *Flora of Hassan District, Karnataka, India.* Amerind Publishers, New Delhi, **1976**.

Saldanha, C. J.; Santapau, H. New Plants Records for Bombay—III. *J. Bombay Nat. Hist. Soc.* **1956, 53**, 210–213.

Santapau, H. Acanthaceae of Bombay. *Bot. Mem. Univ. Bombay* **1951**, *2*, 1–104.

Santapau, H. *The Flora of Khandala on the Western Ghats of India*; Records of Botanical Survey of India, BSI: New Delhi, **1953**.

Santapau, C. J. *Flora of Purandhar*; Oxford Book Co.: New Delhi, **1957**.

Sarvalingam, A.; Rajendran, A. Rare, Endangered and Threatened (RET) Climbers of Southern Western Ghats, India. *Revista Chilena de Historia Natural* **2016**, *89*, 9. DOI: 10.1186/s40693–016–0058–6.

Sasidharan, N. *Flowering Plants of Kerala*; Kerala Forest Research Institute: Peechi, Thrissur, **2004**.

Sasidharan, N. *Illustrated Manual on the Tree Flora of Kerala Supplemented with Computer Aided Identification*; KFRI: Thrissur, 2006.

Sasidharan, N. *Studies on the Flora of Shendrruny Wildlife Sanctuary with Emphasis on Endemic Species*. KFRI Research Report: Peechi, Thrissur, **1997**, p 401.

Sasidharan, N.; Sivarajan, V. *Flowering Plants of Thrissur Forest*; Scientific Publishers: Jodhpur, **1996**.

Sharma, B. D.; Singh, N. P.; Raghavan, R. S.; Deshpande, U. R. *Flora of Karnataka: Analysis*; Botanical Survey of India: Howrah, **1984**.

Singh, J. S. (Co-Ordinator) *National Biodiversity Strategy and Action Plan (Natural Terrestrial Ecosystems)*—a Report Submitted to Ministry of Environment and Forests: New Delhi, **2002**.

Singh, N. P.; Karthikeyan, S. *Flora of Maharashtra State (Dicotyledones)* Vol. 1; Botanical Survey of India: Calcutta, **2000**.

Singh, N. P., Lakshminarasimhan, S., Karthikeyan, S. Prasanna, P. V., Eds. *Flora of Maharashtra State, Dicotyledons,* Vol. 2; Botanical Survey of India: Calcatta, **2001**.

Singh, P. *Floristic Diversity of India: An Overview*; Biodiversity of the Himalaya and Jammu and Kashmir State, **2020**; pp 41–79. DOI: 10.1007/978–981–32–9174–4_3.

Singh, P.; Karthigeyan, K.; Lakshminarasimhan, P.; Dash, S. S. *Endemic Vascular Plants of India*; Botanical Survey of India: Kolkata, **2015**.

Singh, S. K.; Agarwala, D. K.; Jalal, J. S.; Dash, S. S.; Mao, A. A.; Singh, P. *Orchids of India: A Pictorial Guide*; Botanical Survey of India: Kolkata, **2019**.

Subramanyam, K.; Nayar, M. P. Vegetation and Phytogeography of the Western Ghats. In *Ecology and Biogeography in India*; Mani, M. S., Ed.; The Hague, Dr. W. Junk Publishers, **1974**.

Suresh, H. S.; Sukumar, R. Phytographical Affinities of Flora of Nilgiri Biosphere Reserve. *Rheedea* **1999**, *9*, 1–21.

Tetali, P.; Tetali, S.; Kulkarni, B. G.; Prasanna, P. V.; Lakshminarasimhan, P.; Lale, M.; Kumbhojkar, M. S.; Kulkarni, D. K.; Jagtap, A. P. *Endemic Plants of India (A Status Report of Maharashtra State)*; Naoroji Godrej Centre for Research, Shindewadi: Satara India, **2000**.

Venu, P. *Strobilanthes* Blume (Acanthaceae) in Peninsular India; Botanical Survey of India: Kolkata, **2006**.

Vinayaka, K. S.; Krishnamurthy, Y. L. Floristic Composition and Regeneration Analysis of Halika Ghat Region, Central Western Ghats, Karnataka. *Trop. Plant Res.* **2016**, *3*, 654–661.

Warrier, R. R.; Geetha, S.: Sivakumar, V.; Gurudev Singh, B.; Anandalakshmi, R. Threatened Tree Species of the Western Ghats: Status, Diversity and Conservation. In *Conservation and Utilization of threatened medicinal plants*; Rajasekharan, P. E., Wani, S. H., Eds.; Springer International Publishing: New York, **2020**; pp 429–460.

Watve, A. Status Review of Rocky Plateaus in the northern Western Ghats and Konkan Region of Maharashtra, India with Recommendations for Conservation and Management. *J. Threat. Taxa* **2013**, *5* (5), 3935–3962; http://dx.doi.org/10.11609/JoTT.o3372.3935–62

Watve, A. Rocky Outcrops as Special Habitats in North Western Ghats, Maharashtra. In *Special Habitats and Threatened Plants of India*; Rawat, G. S., Ed.; ENVIS Bulletin: Wildlife and Protected Areas. Wildlife Institute of India: Dehradun, **2008**, *11* (1), 147–153.

Wight, R. *Icones Plantarum Indiae Orientalis;* J. B. Pharoah: Madras, **1840–1853**.

Wight, R. *Illustrations to Indian Botany*; Govt. Press: Madras, **1840–1850**.

Yadav, S. R.; Sardesai, M. M. *Flora of Kolhapur District*; Shivaji University: Kolhapur, **2002**.

Yadav, S. R.; Kamble, M. Y. Threatened Ceropegias of the Western Ghats and Strategies for Their Conservation. In *Special Habitats and Threatened Plants of India*; Rawat, G. S., Ed. *ENVIS Bull.* **2008**, *11* (1), 123–134.

Yoganarasimhan, S. N. *Medicinal Plants of India, Vol.2*, Tamil Nadu, R. R. I.: Bangalore, **2000**.

Yoganarasimhan, S. N.; Subramanyam, K., Razi, B. A. *Flora of Chickmagalur District, Karnataka*; India, International Book Distributors: Dehra Dun, **1981**.

CHAPTER 10

GENETIC DIVERSITY OF CROP PLANTS OF THE WESTERN GHATS

MAYA PERINGOTTILLAM and ALAGU MANICKAVELU

Department of Genomic Science, Central University of Kerala, Kasaragod, Kerala, India

ABSTRACT

Genetic diversity is the unlikeness in genetic constitution among individuals of a population, a species, or a community. It is an evolutionary adaptation that confers the very existence of organisms on the earth. Nevertheless, humans utilize this genetic diversity to create desirable changes in the characteristics of convenient organisms. Therefore, understanding crop genetic diversity is a prerequisite for improving the beneficial attributes of crop plants and fostering awareness of different ecologically significant concepts such as genetic drift, inbreeding, mutation, population fitness, extinction, stability of ecosystems, etc. The Western Ghats is an exceptionally diverse ecosystem in peninsular India, containing a large portion of the country's flora and fauna. This chapter focuses primarily on reporting the genetic diversity assessments conducted so far among crop plants across the Western Ghats. The chapter covers current trends and gaps in crop diversity assessments and suggests the most appropriate means for detecting high genetic differentiation.

10.1 INTRODUCTION

Genetic diversity is a broad term that encompasses all variations occurring within a single species or between different species in respect of their total genetic composition. It is the foundation of biodiversity that contributes to the survival of individuals and species under dynamic circumstances. The Western Ghats, the exceptionally diverse ecological region covering an area of 129,037 km^2, runs parallel to the west coast of peninsular India, traversing Tamil Nadu, Kerala, Karnataka, Goa, Maharashtra, and Gujarat. Although the Western Ghats constitute only 6% of India's total area, it contains more than 30% of significant flora and fauna (Molur et al., 2011), which is enough to showcase its enormous

Biodiversity Hotspot of the Western Ghats and Sri Lanka. T. Pullaiah, PhD (Ed.)
© 2024 Apple Academic Press, Inc. Co-published with CRC Press (Taylor & Francis)

heterogeneity. Moreover, the Western Ghats agroecosystem harbors high levels of ende-mism and is considered as one of the 36 biodiversity hotspots of the world (Ramachandra and Suja, 2006). Recently, Pradheep et al. (2021) have compiled the latest checklist of 306 angiospermic taxa cultivated in Kerala, highlighting the depth of genetic diversity in the Western Ghats region. The crop diversity of the Western Ghats is composed of diverse crop plants such as rice, sorghum, finger millet, small millet, black gram, green gram, cowpea, pigeon pea, *Dolichos* bean, horse gram, sword bean, okra, eggplant, cucumber, chilli/capsicum sp, taros, yams, elephant foot yam, jackfruit, banana, lemon, orange, *Syzygium* sp, sugarcane, black pepper, turmeric, ginger, coconut, areca nut, cotton, tea, coffee, rubber, and their wild relatives (Sivaraj et al., 2013). The propagation of selected excellent varieties of crop plants and ceaseless anthropogenic activities in these areas led to the depletion of wild genetic resources (Lathankumar et al., 2016), which specifies the need to evaluate the region's genetic diversity and devise appropriate strategies to conserve the depleting genetic resources.

A good apprehension of the molecular foundation of the critical biological phenomena in crop plants is pivotal for the effectual conservation, management, and utilization of crop plant genetic resources (PGR). Crop genetic diversity assessments have been conducted around for a long time and they are still being practiced, as they are essential in establishing the relationships and discrepancies among the genotypes of different crop plants. Here, an attempt is being made to summarize the genetic diversity characterizations performed so far in crop varieties and their wild relatives across the Western Ghats.

10.2 GENETIC DIVERSITY ASSESSMENTS IN RICE

Rice (*Oryza sativa*) is a major cereal crop and the most widely consumed staple food in south India. The more commonly followed DNA-based molecular markers to characterize the genetic diversity in different rice plants across the Western Ghats are random amplified polymorphic DNA (RAPD), simple sequence repeats (SSRs), insertion–deletions (InDels), and single nucleotide polymorphism (SNP). Vanaja et al. (2006) observed a declining genetic base of true breeding and high-yielding stocks from Kerala and suggested the need for future breeding programs focusing on genetically distant parents. Kumar et al. (2010) identified morphological, molecular, and biochemical markers for distinguishing Njavara ecotypes, a medicinal rice cultivar of Kerala, from other rice varieties. RAPD finger-printing was a practical approach for assessing genetic diversity among selected released rice varieties (Skaria et al., 2011). For assessing the genetic diversity of closely related rice plants, highly polymorphic SSR markers can be used (Thomas and Dominic, 2016). Comparatively, SSR markers are much more efficient than inter-simple sequence repeat (ISSR) and SNP markers in precise identification of genetic diversity because these are neutral, codominant, and multiallelic (Kumbhar et al., 2015). A recent and novel report of 96 rice accessions using SNP markers revealed the genetic diversity of these accessions in northern Kerala. Interestingly, InDels also proved to be the robust markers for uncovering genetic diversity in rice germplasm (Vasumathy et al., 2020).

The existence of the wild relatives of rice such as *Oryza rufipogon, Oryza nivara, Oryza officinalis, Oryza meyeriana* var. *granulata, Oryza officinalis* subsp. *malampuzhaensis,* and *Oryza sativa* f. *spontanea* engenders the Western Ghats region as a treasure trove of rice species (Semwal et al., 2019). IRRI has identified the forest reserves in the Western Ghats region of Kerala for the in situ conservation of these species (Vighneswaran and Nair, 2014). The RAPD analysis of the endemic tetraploid wild rice *Oryza malampuzhaensis* revealed their low genetic diversity highlighting the threat of extinction (Thomas et al., 2001). Also, the low gene flow and limited variation found in the evaluation of genetic diversity by RAPD of another endemic wild rice species, *Oryza rufipogon*, in the Western Ghats indicate the influence of its recent habitat destruction and fragmentation (Raj et al., 2010).

10.3 GENETIC DIVERSITY ASSESSMENTS IN SORGHUM

Sorghum (*Sorghum bicolor*) is the most important staple cereal crop for the food-insecure humankind in the world. RAPD markers are robust markers to delineate the vast genetic differentiation in sorghum collections (Sinha et al., 2014). Chakraborty et al. (2011) evaluated the concordance of RAPD and ISSR markers in analyzing the genetic diversity in sorghum collections and found that the combined application of these markers may be a useful tool to verify the genetic diversity. An integrated analysis using morphological methods and molecular markers like SSR can effectively differentiate individuals from collected germplasm (Rakshit et al., 2012). Ram et al. (2020) found a broad genetic variability in sorghum accessions from Maharashtra using SSR markers that could pave the path for further association analyses. Morphological and cytological studies of *Sorghum nitidum*, a least known species of the genus found in the Western Ghats, were also reported (Rao and Rao, 1990).

10.4 GENETIC DIVERSITY ASSESSMENT IN FINGER MILLET

Finger millet (*Eleusine coracana*) is an important food crop widely grown throughout south India as the grains are higher in protein, fat, and minerals than rice and sorghum. Minimal efforts have been made to assess the genetic diversity among the finger millet genotypes from India (Ramakrishnan et al., 2016). Genetic diversity was measured using RAPD primers in finger millet genotypes in peninsular India (Fakrudin et al., 2004). Analysis of population structure and genetic diversity of finger millet genotypes disclosed the genetic distinctiveness and higher genetic diversity in Indian genotypes than non-Indian genotypes (Ramakrishnan et al., 2016). A high similarity coefficient was reported among finger millet genotypes collected from south India (Das et al., 2007). Southern genotypes were also included in the population structure analysis of global collections of finger millet species using SSR markers (Babu et al., 2014). Recently, Pandian et al. (2018) analyzed the efficacy and consistency of three different marker systems, such as ISSR, RAPD, and DAMD, to show genome coverage and provide information on genetic similarity and variation. They also found that the amalgamation of one or more marker systems is advantageous in calculating phylogenetic relationships. Finger millet is known as "nutritious millet" containing proteins, mineral nutrients, and exceptionally high calcium than the grains of other cereals; only a few genetic and genomic studies have been performed to date (Caesar et al., 2018). So, this considerable scientific lacuna must be filled by employing advanced molecular techniques like whole genome sequencing to comprehend this wonder cereal crop completely.

10.5   GENETIC DIVERSITY ASSESSMENT IN SMALL MILLET/*PANICUM* SPP.

Small millet (*Panicum sumatrense*) is a domesticated species of the weedy plant *Panicum psilopodium*. The potential of little millet has not been exploited in India, indicating a greater scope for research relating to their yields and overall performance. A wide range of differences was observed for all the traits in 109 little millet accessions in the study conducted by Nirmalakumari et al. (2010). The environment has little effect on phenotypic expression, which indicates the possibility of using phenotypic data to improve the cultivars (Patel et al., 2018). Yadav et al. (2016) discovered low genetic diversity in a collection of 96 kodo millet accessions collected from different states of India.

10.6   GENETIC DIVERSITY ASSESSMENTS IN BLACK GRAM

Examining the diversity in black gram (*Vigna mungo*) using SSR markers found that the northern part of the Western Ghats was the pioneer center of domestication for cultivated black gram. Furthermore, the conservation of more than three-quarters of wild gene diversity in cultivated species suggests a little genetic bottleneck in black gram populations (Kaewwongwal et al., 2015). Gupta and Gopalakrishna (2009) dissected genetic diversity in black gram using amplified fragment length polymorphism (AFLP) and microsatellites from azuki bean and identified highly informative and polymorphic microsatellites that can be used for genomic study in black gram genotypes. Genetic diversity analysis and screening for yellow mosaic virus resistance of nine black gram genotypes using 42 SSR markers revealed 35.7% polymorphism, along with the identification of diverse resistant parents for breeding (Sathees et al., 2019).

10.7   GENETIC DIVERSITY ASSESSMENT IN GREEN GRAM

Formerly, Karuppanapandian et al. (2006a) investigated genetic diversity in a collection of green gram landraces and identified a narrow genetic base among the collections. Characterization of genetic diversity in elite (*Vigna radiata*) and wild species (*Vigna umbellata*) using a combination of morphological and molecular markers like SSR proved sufficient variability in green gram genotypes (Palaniappan and Murugaiah, 2012). More importantly, the combined use of these markers shows an increase in the efficiency of diversity analysis in green gram. Bisht et al. (2005) concisely portrayed the existence of diversity in wild *Vigna* species collected from four main phytogeographical zones including Western Ghats.

10.8 GENETIC DIVERSITY ASSESSMENTS IN HORSE GRAM

India is considered to be the focus of origin of horse gram (*Macrotyloma uniflorum*) (Sharma et al., 2015). Examination of genetic diversity and population structure in horse gram with its wild relatives, namely *M. axillare* and *M. gharwalensis* using RAPD and ISSR markers, revealed high genetic diversity in few accessions which can be exploited in crop breeding programs (Sharma et al., 2015). Recently, 30 polymorphic SSR markers were developed that could be able to pinpoint the rich genetic diversity in 360 horse gram accessions (Chahota et al., 2017). Significant genetic diversity has been assessed using agromorphological traits in horse gram species cultivated mainly in the southern states across India (Gomashe et al., 2018).

10.9 GENETIC DIVERSITY ASSESSMENTS IN COWPEA

Vigna unguiculata (cowpea) is a multipurpose crop, providing food for human and feed for livestock, and is extensively cultivated in south India. The feasibility of RAPD markers in analyzing the genetic diversity has been reported in cowpea landraces from diverse geographical locations of Tamil Nadu (Karuppanapandian et al., 2006b). The study revealed a unique genetic reservoir in cowpea landraces of Tamil Nadu, one of the states through which the Western Ghats pass. Recently, a more efficient and practical approach has been adopted in pure lines of local cowpea germplasm, which has shown that genetic diversity can be detected more accurately through the combined use of two types of molecular markers such as RAPD and SSR (Pradeepkumar et al., 2017). Ragul et al. (2018) promulgated a single SSR marker that can be used to differentiate a new cowpea variety VBN 3 from other genotypes. Devi and Jayamani (2020) reported the potentiality of SSR markers in evaluating the genetic diversity of cowpea genotypes.

10.10 GENETIC DIVERSITY ASSESSMENTS IN SUGARCANE

Phenotypic characterization of sugarcane (*Saccharum* sp) divulged variations among the germplasm in the southern states of India (Tawadare et al., 2019a). In contrast, Nair et al. (2002) assessed the genetic diversity in 28 major Indian sugarcane varieties and found that a large portion of the genome was similar among the types. Similarly, analysis of genetic diversity of *S. spontaneum* clones using 20 random, two ISSR, and two telomere primers disclosed less diversity between Tamil Nadu and Kerala (Mary et al., 2006). Therefore, there is an urgent necessity to diversify the genetic base by using new sources from the germplasm. Sequence-tagged microsatellite sites (STMS) primers have been used to analyze genetic diversity in sugarcane germplasm (Padmanabhan and Hemaprabha, 2018). High levels of polymorphism and heterozygosity were observed by analyzing genetic diversity using genomic and cDNA-derived microsatellite markers, and the results support the use of SSR markers as an excellent tool for diversity analysis and loci mapping in sugarcane germplasm (Singh et al., 2010). ISSR markers are also perfect in assessing genetic diversity (Devarumath et al., 2012). RAPD markers were also used to assess genetic diversity in sugarcane cultivars (Tawadare et al., 2019b).

10.11   GENETIC DIVERSITY ASSESSMENTS IN CUCUMBER

India is considered one of the primary and secondary centers of diversity for cucumber (*Cucumis sativus*), where the most extensive collection of cultivated cucumber and its wild species is maintained (Wang et al., 2018). Assessment of the genetic diversity of cucumber and its wild species using morphological and RAPD markers indicated free gene flow between the two, which signifies the importance of further population collection and analysis of this weedy species (Bisht et al., 2004). Pandey et al. (2013) reported low levels of polymorphism in a pool of cucumber accessions from six major agroclimatic zones using SSR markers, again reflecting the need for wild collections to broaden the narrow genetic base of these accessions. Similarly, a reduction in genetic diversity was noticed using the DIVA-GIS approach among the landraces of cucumber in India (Suma et al., 2019). In contrast, high variability was observed for different qualitative and quantitative traits in genotypes from Kerala and Rajasthan (Kumar et al., 2018).

10.12   GENETIC DIVERSITY ASSESSMENTS IN JACKFRUIT

Jackfruit (*Artocarpus heterophyllus*) is considered to be a native species of Western Ghats. The genetic diversity of 50 accessions was evaluated using AFLP markers, which discovered a moderate genetic diversity among the genotypes (Shyamalamma et al., 2008). Krishnan et al. (2015) studied genetic divergence and relationships among jackfruit genotypes from the Kuttanad region, Kerala using the RAPD technique. SSR markers were found to be effective in disclosing genetic diversity in jackfruit collections for pulp color (Kavya et al., 2019). Wide variability in morphological attributes of jackfruit germplasm has been reported earlier (Chandrashekar et al., 2018).

10.13   GENETIC DIVERSITY ASSESSMENTS IN GINGER

Ginger (*Zingiber officinale*) is a herbaceous perennial; the rhizomes of this crop are used as a spice. Molecular characterization of a global collection of ginger germplasm using RAPD and ISSR revealed a low genetic diversity in the accessions (Kizhakkayil and Sasikumar, 2010). Ashraf et al. (2014) assessed genetic diversity at an intraspecific level among 12 accessions of ginger using RAPD markers. The maximum similarity is observed among the accessions whose cultivation regions are very close compared to more distant genotypes. *Zingiber anamalayanum*, a new species from the southern Western Ghats, has recently been described (Sujanapal and Sasidharan, 2010).

10.14 GENETIC DIVERSITY ASSESSMENTS IN BLACK PEPPER

Black pepper (*Piper nigrum*) is a perennial vine raised for its berries, enormously used as a spice and has medicinal properties. The Western Ghats, especially Kerala, harbors maximum genetic diversity of black pepper and it is believed to be originated in the Western Ghats (Kumari et al., 2019). A total of 30 cultivars from Kerala were evaluated with AFLP markers showing high polymorphism (96.6%), indicating extensive diversity among the germplasm (Joy et al., 2007). A total of 44 prevalent black pepper genotypes from southern India were shown to have high morphologic and genetic diversity using SSR markers (Joy et al., 2011). Spike branching pepper type, an uncharacterized variant recently found in hilly regions of Western Ghats of Kerala, was examined using RAPD (Vimarsha et al., 2014). A combination of RAPD and SSR was used for the molecular characterization of interspecific hybrids in Kerala (Jagtap et al., 2016). Shivakumar and Saji (2019) characterized 82 black pepper genotypes based on morphological traits, indicated a wide range of genetic variation in black pepper germplasm and identified important traits for yield improvement. A recent study of 30 black pepper genotypes reported 69,126 NGS-based genomic SSRs from the assembled genomic sequence of *P. nigrum* and investigated genetic diversity using 50 SSR markers (Kumari et al., 2019).

10.15 ANALYSIS OF GENETIC VARIABILITY IN BANANA

Kerala has many different landraces and cultivars of *Musa*, domesticated over thousands of years. RAPD assays have been performed to detect genetic polymorphism in *Musa acuminata* wild accessions in Western Ghats (Mukunthakumar et al., 2013). This study reported ample genetic diversity in wild *Musa* accessions, further explaining the possibility of selecting a natural population for future studies. More molecular analyses have been recommended in wild *Musa* from Western Ghats (Mukunthakumar et al., 2013). Inter retrotransposons (IRAP) are transposable element-based markers, and they have been used to detect polymorphism in banana accessions from Tamil Nadu (Saraswathi et al., 2020). Padmesh et al. (2012) recommended the diverse set of *Musa acuminata* ssp. *burmannica* as a repository of useful resistant traits for their effective utilization in crop improvement. Das et al. (2018) analyzed genetic diversity in nematode-resistant and high-yielding banana cultivars and observed the capability of a few ISSR primers in providing the information needed for diversity analysis. Concordant results were obtained for genetic diversity assessments in banana cultivars collected from Karnataka (Babu et al., 2018).

10.16 GENETIC DIVERSITY ASSESSMENTS IN CITRUS

Citrus is a major food crop that is important for south India because of its diverse collection. Singh and Singh (2006) described the diversity of wild and cultivated forms of citrus in south India and the importance of conserving them. Diversity analysis of 38 accessions using SSR markers revealed low genetic differentiation among the collections (Rohini et al., 2020). Moreover, the collection from Karnataka represented a more diverse set of germplasm. Genetic relationships among three *Citrus* species such as *Citrus aurantifolia*, *Citrus maxima*, and *Citrus limon* collected from various parts of Kerala were analyzed using RFLP (Vazhacharickal et al., 2017).

10.17   GENETIC DIVERSITY ASSESSMENTS IN TURMERIC

Common turmeric (*Curcuma longa*) is the most economically valuable member of the genus *Curcuma*, widely cultivated in India. Sasikumar et al. (2005) reported that molecular characterization of *Curcuma* species is in a developing stage, despite biochemical, anatomical, and isozyme-based studies. An overview of the biological diversity in this genus has also been reported. Vijayalatha and Chezhiyan (2008) assessed genetic diversity in turmeric accessions using morphological and molecular techniques. They suggested a combined analysis using multivariate statistical techniques with a molecular marker like SSR to discriminate the genotypes. Siju et al. (2010) developed a set of polymorphic SSR markers that can be used to study genetic diversity in turmeric accessions. These markers were able to detect the duplicates in the collection that suggest a revisit of germplasm collection strategy based on local names. This fact was supported by Syamkumar and Sasikumar (2007), who analyzed genetic diversity in 15 *Curcuma* species using ISSR and RAPD.

10.18   GENETIC DIVERSITY ASSESSMENTS IN COCONUT

Coconut (*Cocos nucifera*) is an essential perennial plantation crop with no wild forms and constitutes Arecaceae. Thomas et al. (2013) analyzed genetic diversity in 90 local coconut germplasm selected by participatory approach from Kerala using SSR markers. "Mohachao narel" is a coconut variant in Maharashtra distinguished by sweet and soft endosperm. Genetic diversity analysis among this species revealed that palms of sweet kernel type had only 45% similarity (Samsudeen et al., 2013). This result specified a high level of genetic diversity between the palms, paving the way for inventing appropriate conservation and management measures for this particular coconut population and their use in future breeding programs. High levels of polymorphism have been reported for different attributes in coconut cultivars of south Travancore (Selvaraju and Jayalekshmy, 2011).

10.19   GENETIC DIVERSITY ASSESSMENTS IN ARECA NUT

Areca nut (*Areca catechu*) is one of the main cash crops maintained by ICAR-CPCRI, India. Rajesh and Ananda (2019) scrutinized the genetic diversity in 50 exotic and indigenous accessions, and suggested superior cultivars for areca nut improvement programs. RAPD is a rapid and sensitive technique for analyzing genetic variation within areca nut cultivars

(Bharath et al., 2015). Manimekalai et al. (2012) employed three marker systems such as ISSR, RAPD, and resistance gene-based markers to differentiate yellow leaf disease-resistant palms with susceptible ones. Moreover, the significance and precision of applying more than one molecular marker type in diversity assessments have been discussed.

10.20 GENETIC DIVERSITY ASSESSMENTS IN CASHEWNUT

Cashew (*Anacardium occidentale*) is an introduced tree species presently cultivated in Maharashtra, Goa, Karnataka, Kerala, Tamil Nadu, Andhra Pradesh, Orissa, and West Bengal. RAPD markers have been employed to assess the genetic diversity in 90 cashew germplasm and observed moderate to high genetic diversity in collected genotypes (Dhanaraj et al., 2002). Archak et al. (2003a) analyzed the genetic diversity in cashew varieties across coastal regions of south India using RAPD and ISSR markers. Their observations emphasized an urgent need to identify and include more parental lines for a successful hybridization program. Archak et al. (2003b) used a combination of AFLP, ISSR, and RAPD to discriminate 19 cashew accessions of India. AFLP had a high marker index, and hence was determined to be the marker of choice for cashew genetic analysis. Similarly, genetic divergence in 100 cashew accessions was analyzed using RAPD and ISSR markers that pointed out the efficiency of amalgamating two marker systems in assessing genetic diversity (Thimmappaiah et al., 2009). Archak et al. (2009) evaluated the cashew germplasm's introduction and diversification by investigating genetic diversity using AFLP markers and morphometric data.

10.21 GENETIC DIVERSITY ASSESSMENTS IN COCOA

Cacao (*Theobroma cacao*) is commonly known as "chocolate tree," being grown profitably as a mixed crop under areca nut and coconut germplasms of Karnataka, Kerala, Tamil Nadu, and Andhra Pradesh (Apshara, 2019). Although a high degree of genetic variability has been observed among cocoa trees in Tamil Nadu, it is surprising that these have not been scientifically documented. SSR analysis of cocoa plus trees in Tamil Nadu disclosed sufficient diversity that can be conveniently employed in cocoa breeding programs (Thondaiman et al., 2013). Given the immense economic importance of this crop, it is essential to study its response to environmental fluctuations using advanced molecular technologies.

10.22 CURRENT TRENDS, GAPS, AND SUGGESTIONS FOR BETTER KNOWLEDGE OF THE GENETIC DIVERSITY IN CROP CULTIVARS

The increasing population demands more production of foodstuffs bringing forth an urgent need to understand the distribution and diversity of plant genetic resources. Morphological, biochemical, cytological, and molecular characterizations are different methods to explore the variations in plant genetic resources and apply them to balance the food insecurity caused by the population explosion. Recently, Singh (2017) reexamined the biodiversity in the Western Ghats and reported 146 wild relatives of cultivated species. Nonetheless, studies on the genetic diversity of crop plants and their wild relatives in the Western Ghats are inadequate. Most of the published genetic diversity works are based on the analyses of morphological data such as Mahalanobis D² statistics, PCA, PCoA, cluster analysis, factor analysis, etc. (Jeevitha et al., 2018; Priya et al., 2018; Preethy et al., 2018; Vidya et al., 2018; Senthamizhselvi et al., 2019; Subramanian et al., 2019; Punithavathy et al., 2020; Wesly et al., 2020). There have been very few works on genetic diversity using molecular markers. RFLP, RAPD, and SSR were the markers of preference for genetic diversity assessments during the last two decades. Later, the attention has been changed to genetic diversity studies analyzed based on molecular markers that detect different alleles of genes in different loci. SNPs are the most abundant markers and are widely used in diversity studies because they can cause phenotypic diversity among individuals by making changes in exons, introns, promoter regions, etc. (Morgil et al., 2020). Therefore, identifying SNPs and their effect on phenotypic expression is very crucial.

Expeditious progress in sequence technologies, including SNP genotyping and genome sequencing, has provided new and powerful strategies for mapping complex phenotypes and identifying genes that cause this complexity. Genome-wide association mapping is a powerful and promising technique widely used in crop research that helps to detect the causal relation between SNPs and their complex phenotypes (Puranik et al., 2020; Tiwari et al., 2020). Lamentably, the existence of a significant research lacuna in this field of crop plants in the Western Ghats advertises its readiness to accept novel works. Therefore, this book chapter pointed out the need for research on genetic diversity among crop plants across the Western Ghats using advanced molecular techniques along with morphological characterization. It is also essential to understand the wild plant species for their conservation and subsequent use in crop improvement programs.

KEYWORDS

- **Western Ghats**
- **crop plants**
- **genetic diversity**
- **molecular tools**
- **assessments**

REFERENCES

Apshara, S. E. Cocoa Genetic Resources and Their Utilisation in Palm-Based Cropping Systems of India. In *Theobroma Cacao-Deploying Science for Sustainability of Global Cocoa Economy*; Intech Open, **2019**. DOI: 10.5772/intechopen.82077.

Archak, S.; Gaikwad, A. B.; Gautam, D.; Rao, E. V. V. B.; Swamy, K. R. M.; Karihaloo, J. L. DNA Fingerprinting of Indian Cashew (*Anacardium occidentale* L.) Varieties Using RAPD and ISSR Techniques. *Euphytica* **2003a,** *130* (3), 397–404.

Archak, S.; Gaikwad, A. B.; Gautam, D.; Rao, E. V. V. B.; Swamy, K. R. M.; Karihaloo, J. L. Comparative Assessment of DNA Fingerprinting Techniques (RAPD, ISSR and AFLP) for Genetic Analysis of Cashew (*Anacardium occidentale* L.) Accessions of India. *Genome* **2003b,** *46* (3), 362–369.

Archak, S.; Gaikwad, A. B.; Swamy, K. R.; Karihaloo, J. L. Genetic Analysis and Historical Perspective of Cashew (*Anacardium occidentale* L.) Introduction into India. *Genome* **2009,** *52* (3), 222–230.

Ashraf, K.; Ahmad, A.; Chaudhary, A.; Mujeeb, M.; Ahmad, S.; Amir, M.; Mallick, N. Genetic Diversity Analysis of *Zingiber officinale* Roscoe. by RAPD Collected from Subcontinent of India. *Saudi J. Biol. Sci.* **2014,** *21* (2), 159–165.

Babu, A. G.; Prabhuling, G.; Rashmi, S. K.; Satish, D.; Rohini, K. P.; Mulla, S. R.; Raghavendra, G.; Jagadeesha, R. C. Genetic Diversity Analysis Among Banana Cultivars Through ISSR Markers. *J. Pharmacogn. Phytochem.* **2018,** *7* (6), 1576–1580.

Babu, B. K.; Agrawal, P. K.; Pandey, D.; Kumar, A. Comparative Genomics and Association Mapping Approaches for *opaque2* Modifier Genes in Finger Millet Accessions Using Genic, Genomic and Candidate Gene-Based Simple Sequence Repeat Markers. *Mol. Breed.* **2014,** *34* (3), 1261–1279.

Bharath, B. G.; Ananda, K. S.; Rijith, J.; Nagaraja, N. R.; Chandran, K. P.; Karun, A.; Rajesh, M. K. Studies on Genetic Relationships and Diversity in Areca Nut (*Areca catechu* L.) Germplasm Utilizing RAPD Markers. *J. Plant. Crops.* **2015,** *43* (2), 117–125.

Bisht, I. S.; Bhat, K. V.; Lakhanpaul, S.; Latha, M.; Jayan, P. K.; Biswas, B. K.; Singh, A. K. Diversity and Genetic Resources of Wild *Vigna* Species in India. *Genet. Resour. Crop Evol.* **2005,** *52* (1), 53–68.

Bisht, I. S.; Bhat, K. V.; Tanwar, S. P. S.; Bhandari, D. C.; Joshi, K.; Sharma, A. K. Distribution and Genetic Diversity of *Cucumis sativus* var. *hardwickii* (Royle) Alef in India. *J. Hortic. Sci. Biotechnol.* **2004,** *79* (5), 783–791.

Ceasar, S. A.; Maharajan, T.; Krishna, T. A.; Ramakrishnan, M.; Roch, G. V.; Satish, L.; Ignacimuthu, S. Finger Millet [*Eleusine coracana* (L.) Gaertn.] Improvement: Current Status and Future Interventions of Whole Genome Sequence. *Front. Plant Sci.* **2018,** *9*, 1054.

Chahota, R. K.; Shikha, D.; Rana, M.; Sharma, V.; Nag, A.; Sharma, T. R.; Rana, J. C.; Hirakawa, H.; Isobe, S. Development and Characterisation of SSR Markers to Study Genetic Diversity and Population Structure of Horse Gram Germplasm (*Macrotyloma uniflorum*). *Plant Mol. Biol. Rep.* **2017,** *35* (5), 550–561.

Chakraborty, S.; Thakare, I.; Ravikiran, R.; Nikam, V.; Trivedi, R.; Sasidharan, N.; Jadeja, G. C. Assessment of Diversity Using RAPD and ISSR Markers in Sorghum Varieties Across Gujarat, India. *Electron. J. Plant Breed.* **2011,** *2* (4), 488–493.

Chandrashekar, K.; Vijayakumar, R. M.; Subramanian, S.; Kavino, M.; John Joel, A. Morphological Characterisation of Jackfruit (*Artocarpus heterophyllus* Lam.) Local Genotypes Under Coffee Ecosystem of Lower Pulney Hills. *Int. J. Curr. Microbiol.* **2018,** *7* (3), 2210–2224.

Das, S.; Mishra, R. C.; Rout, G. R.; Aparajita, S. Genetic Variability and Relationships Among Thirty Genotypes of Finger Millet [*Eleusine coracana* (L.) Gaertn.] Using RAPD Markers. *Z Naturforsch* **2007,** *62* (1–2), 116–122.

Das, S. C.; Balamohan, T. N.; Poornima, K.; Van Den Bergh, I. Evaluation of Genetic Diversity in Some Banana Hybrids Using ISSR Markers. *Int. J. Curr. Microbiol. Appl. Sci.* **2018,** *7* (1), 146–157.

Devarumath, R. M.; Kalwade, S. B.; Kawar, P. G.; Sushir, K. V. Assessment of Genetic Diversity in Sugarcane Germplasm Using ISSR and SSR Markers. *Sugar Tech.* **2012,** *14* (4), 334–344.

Devi, S. M.; Jayamani, P. Molecular genetic diversity in cowpea [*Vigna unguiculata* (L.) Walp.] genotypes. *J. Food Legumes.* **2020,** *33* (1), 6–9.

Dhanaraj, A. L.; Rao, E. V. V. B.; Swamy, K. R. M.; Bhat, M. G.; Prasad, D. T.; Sondur, S. N. Using RAPDs to Assess the Diversity in Indian Cashew (*Anacardium occidentale* L.) Germplasm. *J. Hort. Sci. Biotechnol.* **2002,** *77* (1), 41–47.

Fakrudin, B.; Shashidhar, H. E.; Kulkarni, R.S; Hittalmani, S. Genetic Diversity Assessment of Finger Millet, *Eleusine coracana* (Gaertn), Germplasm Through RAPD Analysis. *PGR Newslett.* **2004,** *138*, 50–54.

Gomashe, S. S.; Dikshit, N.; Chand, D.; Shingane, S. N. Assessment of Genetic Diversity Using Morpho-Agronomical Traits in Horse Gram. *Int. J. Curr. Microbiol. App. Sci.* **2018,** *7* (5), 2095–2103.

Gupta, S. K.; Gopalakrishna, T. Genetic Diversity analysis in Black Gram (*Vigna mungo* (L.) Hepper) Using AFLP and Transferable Microsatellite Markers from Azuki Bean (*Vigna angularis* (Willd.) OHWI & Ohashi). *Genome* **2009,** *52* (2), 120–129.

Jagtap, A. B.; Sujatha, R.; Nazeem, P. A.; Meena, O. P.; Pathania, S. Morpho-Molecular Characterisation of Putative Interspecific Crosses in Black Pepper (*Piper nigrum* L. and *Piper colubrinum*). *Plant Omics J.* **2016,** *9* (1), 73.

Jeevitha, S.; Karthikeyan, R.; Vignesh, M.; Malarkodi, A.; Tirumalai, R.; Nainu, A. J.; Anandan, R.; Prakash, M.; Murugan, S. Estimation of Morphological and Molecular Genetic Diversity in Black Gram [*Vigna mungo* (L.) Hepper] Under YMV Hotspot Regime. *Hortic. Biotechnol. Res.* **2018,** *4,* 6–9.

Joy, N.; Abraham, Z.; Soniya, E. V. A Preliminary Assessment of Genetic Relationships Among Agronomically Important Cultivars of Black Pepper. *BMC Genet.* **2007,** *8* (1), 1–7.

Joy, N.; Prasanth, V. P.; Soniya, E. V. Microsatellite Based Analysis of Genetic Diversity of Popular Black Pepper Genotypes in South India. *Genetica* **2011,** *139* (8), 1033–1043.

Kaewwongwal, A.; Kongjaimun, A.; Somta, P.; Chankaew, S.; Yimram, T.; Srinives, P. Genetic Diversity of the Black Gram [*Vigna mungo* (L.) Hepper] Gene Pool as Revealed by SSR Markers. *Breed Sci.* **2015,** *65* (2), 127–137.

Karuppanapandian, T.; Karuppudurai, T.; Sinha, P. B.; Kamarul, H. A.; Manoharan, K. Genetic Diversity in Green Gram [*Vigna radiata* (L.)] Landraces Analysed by Using Random Amplified Polymorphic DNA (RAPD). *Afr. J. Biotechnol.* **2006a,** *5* (13), 1214–1219.

Karuppanapandian, T.; Karuppudurai, T.; Sinha, P. B.; Haniya, A. M. K.; Manoharan, K. Phylogenetic Diversity and Relationships Among Cowpea (*Vigna unguiculata* L. Walp.) Landraces Using Random Amplified Polymorphic DNA Markers. *Gen. Appl. Plant Physiol.* **2006b,** *32* (3–4), 141–152.

Kavya, K.; Shyamalamma, S.; Gayatri, S. Morphological and Molecular Genetic Diversity Analysis Using SSR Markers in Jackfruit (*Artocarpus heterophyllus* Lam.) Genotypes for Pulp Colour. *Indian J. Agric. Res.* **2019,** *53* (1), 8–16.

Kizhakkayil, J.; Sasikumar, B. Genetic Diversity Analysis of Ginger (*Zingiber officinale* Rosc.) Germplasm Based on RAPD and ISSR Markers. *Sci. Hort.* **2010,** *125* (1), 73–76.

Krishnan, A. G.; Sabu, T. S.; Sible, G. V.; Xavier, L. Genetic Diversity Analysis in Jackfruit Selections of Kuttanad Region Using RAPD Technique. *Int. J. Sci. Res.* **2015,** *5* (4), 1–6.

Kumar, J. P.; Syed, S.; Reddy, P. S. S.; Lakshmi, L. M.; Reddy, D. S. Genetic Divergence Studies in Cucumber (*Cucumis sativus* L.) Genotypes for Yield and Quality. *Int. J. Curr. Microbiol. App. Sci.* **2018,** *7* (12), 2633–2643.

Kumar, P. S.; Elsy, C. R.; Nazeem, P. A.; Augustin, A. Use of Different Marker Systems to Estimate Genetic Diversity in the Traditional Medicinal Rice Cultivar of Kerala. *Intern. J. Plant Breed. Genet.* **2010,** *4* (2), 89–103.

Kumari, R.; Wankhede, D. P.; Bajpai, A.; Maurya, A.; Prasad, K.; Gautam, D.; Rangan, P.; Latha, M.; John K, J.; Bhat, K. V.; Gaikwad, A. B. Genome wide identification and characterisation of microsatellite markers in black pepper (*Piper nigrum*): A valuable resource for boosting genomics applications. *PLoS One.* **2019,** *14* (12), e0226002.

Kumbhar, S. D.; Kulwal, P. L.; Patil, J. V.; Sarawate, C. D.; Gaikwad, A. P.; Jadhav, A. S. Genetic Diversity and Population Structure in Landraces and Improved Rice Varieties from India. *Rice Sci.* **2015,** *22* (3), 99–107.

Lathankumar, K. J.; Sreekala, A. K.; Manikandan, K.; Preetha, T. S.; Padmesh, P. Intravarietal Diversity Analysis of a Western Ghat *Mangifera indica* L. Variety 'Kottoorkonam' Using ISSR Markers. *Intern. J. Biotech. Biochem.* **2016,** *12* (1), 85–94.

Manimekalai, R.; Deeshma, K. P.; Manju, K. P.; Soumya, V. P.; Sunaiba, M.; Nair, S.; Ananda, K. S. Molecular Marker-Based Genetic Variability Among Yellow Leaf Disease (YLD) Resistant and Susceptible Areca Nut (*Areca catechu.* L.) Genotypes. *Indian J. Hortic.* **2012,** *69* (4), 455–461.

Mary, S.; Nair, N. V.; Chaturvedi, P. K.; Selvi, A. Analysis of Genetic Diversity Among *Saccharum spontaneum* L. from Four Geographical Regions of India, Using Molecular Markers. *Genet. Resour. Crop Evol.* **2006,** *53* (6), 1221–1231.

Molur, S.; Smith, K. G.; Daniel, B. A.; Darwall, W. R. T. *The Status and Distribution of Freshwater Biodiversity in the Western Ghats, India*; Cambridge, UK and Gland, Switzerland: IUCN, and Zoo Outreach Organisation, Coimbatore, India, 2011.

Morgil, H.; Gercek, Y. C.; Tulum, I. Single Nucleotide Polymorphisms (SNPs) in Plant Genetics and Breeding. In *The Recent Topics in Genetic Polymorphisms.* IntechOpen, **2020.** DOI: 10.5772/intechopen.91886.

Mukunthakumar, S.; Padmesh, P.; Vineesh, P. S.; Skaria, R.; Kumar, K. H.; Krishnan, P. N. Genetic Diversity and Differentiation Analysis Among Wild Antecedents of Banana (*Musa acuminata* Colla) Using RAPD Markers. *Indian J. Biotechnol.* **2013**, *12*, 493–498.

Nair, N. V.; Selvi, A.; Sreenivasan, T. V.; Pushpalatha, K. N. Molecular Diversity in Indian Sugarcane Cultivars as Revealed by Randomly Amplified DNA Polymorphisms. *Euphytica* **2002**, *127* (2), 219–225.

Nirmalakumari, A.; Salini, K.; Veerabadhiran, P. Morphological Characterisation and Evaluation of Little Millet (*Panicum sumatrense* Roth. ex. Roem. and Schultz.) Germplasm. *Electron. J. Plant Breed.* **2010**, *1* (2), 148–155.

Padmanabhan, T. S.; Hemaprabha, G. Genetic Diversity and Population Structure Among *133* Elite Genotypes of Sugarcane (*Saccharum spp.*) for Use as Parents in Sugarcane Varietal Improvement. *3 Biotech.* **2018**, *8* (8), 339.

Padmesh, P.; Mukunthakumar, S.; Vineesh, P. S.; Skaria, R.; Kumar, K. H.; Krishnan, P. N. Exploring Wild Genetic Resources of *Musa acuminata* Colla Distributed in the Humid Forests of Southern Western Ghats of Peninsular India Using ISSR Markers. *Plant Cell Rep.* **2012**, *31* (9), 1591–1601.

Palaniappan, J.; Murugaiah, S. Genetic Diversity as Assessed by Morphological and Microsatellite Markers in Green Gram (*Vigna radiata* L.). *Afr. J. Biotechnol.* **2012**, *11* (84), 15091–15097.

Pandey, S.; Ansari, W. A.; Mishra, V. K.; Singh, A. K.; Singh, M. Genetic Diversity in Indian Cucumber Based on Microsatellite and Morphological Markers. *Biochem. Systemat. Ecol.* **2013**, *51*, 19–27.

Pandian, S.; Marichelvam, K.; Satish, L.; Ceasar, S. A.; Pandian, S. K.; Ramesh, M. SPAR Markers-Assisted Assessment of Genetic Diversity and Population Structure in Finger Millet (*Eleusine Coracana* (L.) Gaertn) Mini-Core Collection. *J. Crop Sci. Biotechnol.* **2018**, *21* (5), 469–481.

Patel, S. N.; Patil, H. E.; Modi, H. M.; Singh, Th. J. Genetic Variability Study in Little Millet (*Panicum miliare* L.) Genotypes in Relation to Yield and Quality Traits. *Int. J. Curr. Microbiol. App. Sci.* **2018**, *7* (6), 2712–2725.

Pradeepkumar, T.; Mathew, D.; Roch, C. V.; Veni, K.; Midhila, K. R. Genetic Interrelationship Among Cowpea Varieties Elucidated Through Morphometric, RAPD and SSR Analyses. *Legume Res.* **2017**, *40* (3), 409–415.

Pradheep, K.; John, K. J.; Latha, M.; Suma, A. Status of Crop Plants of Agricultural Importance in Kerala State, India: An Update. *Genet. Resour. Crop Evol.* **2021**, *68*, 1849–1873.

Preethy, T. T.; Aswathy, T. S.; Sathyan, T.; Dhanya, M. K.; Murugan, M. Performance, Diversity Analysis and Character Association of Black Pepper (*Piper nigrum* L.) Accessions in the High Altitude of Idukki District, Kerala. *J. Spices Aromatic Crops* **2018**, *27* (1), 17–21.

Priya, M.; Pillai, M. A.; Shoba, D.; Aananthi, N. Genetic divergence studies in black gram (*Vigna mungo* (L.) Hepper). *Legume Res.* **2018**, *5* (2), 1503.

Punithavathy, P.; Manivannan, N.; Subramanian, A.; Shanthi, P.; Prasad, V. Genetic divergence of black gram genotypes (*Vigna mungo* (L.) Hepper). *Electron. J. Plant Breed.* **2020**, *11* (1), 156–159.

Puranik, S.; Sahu, P. P.; Beynon, S.; Srivastava, R. K.; Sehgal, D.; Ojulong, H.; Yadav, R. Genome-Wide Association Mapping and Comparative Genomics Identifies Genomic Regions Governing Grain Nutritional Traits in Finger Millet (*Eleusine coracana* L. Gaertn.). *Plants People Planet* **2020**, *2* (6), 649–662.

Ragul, S.; Manivannan, N.; Mahalingam, A.; Prasad, V. B. R.; Narayanan, S. L. SSR Marker-Based DNA Fingerprinting for Cowpea Varieties of Tamil Nadu [*Vigna unguiculata* (L.) Walp.]. *Int. J. Curr. Microbiol. App. Sci.* **2018**, *7* (4), 641–647.

Raj, R. D.; Kiran, A. G.; Thomas, G. Population Genetic Structure and Conservation Priorities of *Oryza rufipogon* Griff. Populations in Kerala, India. *Curr. Sci.* **2010**, *98* (1), 65–68.

Rajani, J.; Deepu, V.; Nair, G. M.; Nair, A. J. Molecular Characterisation of Selected Cultivars of Rice, *Oryza sativa* L. Using Random Amplified Polymorphic DNA (RAPD) Markers. *Intern. Food Res. J.* **2013**, *20* (2), 919–923.

Rajesh, B.; Ananda, K. S. Study of Genetic Diversity and Identification of Promising Accessions of Areca Nut (*Areca catechu* L.). *Forest Res. Eng. Int. J.* **2019**, *3* (2), 39–44.

Rakshit, S.; Gomashe, S. S.; Ganapathy, K. N.; Elangovan, M.; Ratnavathi, C. V.; Seetharama, N.; Patil, J. V. Morphological and Molecular Diversity Reveal Wide Variability Among Sorghum Maldandi Landraces from India. *J. Plant Biochem. Biotechnol.* **2012**, *21* (2), 145–156.

Ram, K. C.; Sonam, S. K.; Narendra, R. C.; Ganesh, K. V. Analysis of Genetic Diversity in Sorghum [*Sorghum bicolor* (L.)] Accessions of Maharashtra as Estimated by Simple Sequence Repeats (SSR). *Int. J. Curr. Microbiol. App. Sci.* **2020**, *9* (04), 934–944.

Ramachandra, T. V.; Suja, A. Sahyadri: Western Ghats Biodiversity Information System. *Biodivers. Indian Scenario* **2006**, *1*, 1–22.

Ramakrishnan, M.; Ceasar, S. A.; Duraipandiyan, V.; Al-Dhabi, N. A.; Ignacimuthu, S. Using Molecular Markers to Assess the Genetic Diversity and Population Structure of Finger Millet (*Eleusine coracana* (L.) Gaertn.) from Various Geographical Regions. *Genet. Resour. Crop Evol.* **2016**, *63* (2), 361–376.

Rao, K. P.; Rao, N. K. *Sorghum nitidum* (Vah1) Pers., Occurrence, Morphology and Cytology. *Proc. Indian Acad. Sci. (Plant Sci.)* **1990**, *100* (5), 333–336.

Rohini, M. R.; Sankaran, M.; Rajkumar, S.; Prakash, K.; Gaikwad, A.; Chaudhury, R.; Malik, S. K. Morphological Characterisation and Analysis of Genetic Diversity and Population Structure in *Citrus* × *jambhiri* Lush. Using SSR Markers. *Genet. Resour. Crop Evol.* **2020**, *67*, 1259–1275.

Samsudeen, K.; Rajesh, M. K.; Nagwaker, D. D.; Reshmi, R.; Kumar, P. A.; Devadas, K.; Anitha, K. Diversity in Mohachao Narel, a Sweet Endosperm Coconut (*Cocos nucifera* L.) Population from Maharashtra, India. *Natl. Acad. Sci. Lett.* **2013**, *36* (3), 319–330.

Saraswathi, M. S.; Uma, S.; Ramaraj, S.; Durai, P.; Mustaffa, M. M.; Kalaiponmani, K.; Chandrasekar, A. Inter Retrotransposon Based Genetic Diversity and Phylogenetic Analysis Among the *Musa* Germplasm Accessions. *J. Plant Biochem. Biotechnol.* **2020**, *29* (1), 114–124.

Sasikumar, B. Genetic Resources of *Curcuma*: Diversity, Characterisation and Utilization. *Plant Gen. Resour.* **2005**, *3* (2), 230–251.

Sathees, N.; Shoba, D.; Saravanan, S.; Kumari, S. M. P.; Pillai, M. A. Assessment of the Genetic Diversity of Black Gram [*Vigna Mungo* (L.) Hepper] Collections for Yellow Mosaic Virus Resistance Using Simple Sequence Repeat Markers. *Legume Res.* **2019**. DOI: 10.18805/LR-4155.

Selvaraju, S.; Jayalekshmy, V. G. Morphometric Diversity of Popular Coconut Cultivars of South Travancore. *Madras Agric. J.* **2011**, *98* (1/3), 10–14.

Semwal, D. P.; Pradheep, K.; John, K. J.; Latha, M.; Ahlawat, S. P. Status of Rice (*Oryza sativa* L.) Gene Pool Collected from Western Ghats Region of India: Gap Analysis and Diversity Distribution Mapping Using GIS Tools. *Indian J. Plant Genet. Resour.* **2019**, *32* (2), 166–173.

Senthamizhselvi, S.; Muthuswamy, A.; Shunmugavalli, N. Genetic Divergence Studies in Black Gram [*Vigna mungo* (L.) Hepper]. *Electron. J. Plant Breed.* **2019**, *10* (4), 1606–1611.

Sharma, V.; Sharma, T. R.; Rana, J. C.; Chahota, R. K. Analysis of Genetic Diversity and Population Structure in Horse Gram (*Macrotyloma uniflorum*) Using RAPD and ISSR Markers. *Agric. Res.* **2015**, *4* (3), 221–230.

Shivakumar, M. S.; Saji, K. V. Association and Path Coefficient Analysis Among Yield Attributes and Berry Yield in Black Pepper (*Piper nigrum* L.). *J. Spices Aromatic Crops.* **2019**, *28* (2), 106–112.

Shivaramegowda, K. D.; Krishnan, A.; Jayaramu, Y. K.; Kumar, V.; Koh, H. J. Genotypic Variation Among Okra (*Abelmoschus esculentus* (L.) Moench) Germplasms in South India. *Plant Breed. Biotech.* **2016**, *4* (2), 234–241.

Shyamalamma, S. S. B. C.; Chandra, S. B. C.; Hegde, M.; Naryanswamy, P. Evaluation of Genetic Diversity in Jackfruit (*Artocarpus heterophyllus* Lam.) Based on Amplified Fragment Length Polymorphism Markers. *Genet. Mol. Res.* **2008**, *7* (3), 645–656.

Siju, S.; Dhanya, K.; Syamkumar, S.; Sheeja, T. E.; Sasikumar, B.; Bhat, A. I.; Parthasarathy, V. A. Development, Characterisation and Utilization of Genomic Microsatellite Markers in Turmeric (*Curcuma longa* L.). *Biochem. Systemat. Ecol.* **2010**, *38* (4), 641–646.

Singh, A. K. Revisiting the Status of Cultivated Plant Species Agrobiodiversity in India: An Overview. *Proc. Indian Acad. Sci. (Plant Sci.)* **2017**, *83* (1), 151–174.

Singh, I. P.; Singh, S. Exploration, Collection and Characterisation of Citrus Germplasm—A Review. *Agric. Rev.* **2006**, *27* (2), 79–90.

Sinha, S.; Kumaravadivel, N.; Eapen, S. RAPD Analysis in Sorghum [*Sorghum bicolor* (L.) Moench] Accessions. *Int. J. Bio-resour. Stress Manag.* **2014**, *5* (3), 381–385.

Sivaraj, N.; Pandravada, S. R.; Kamala, V.; Sunil, N.; Rameash, K.; Abraham, B.; Elangovan, M.; Chakrabarty, S. K. *Indian Crop Diversity. Managing Intellectual Property Under PVP and PGR.* Directorate of Sorghum Research (DSR), Hyderabad, **2013**.

Skaria, R.; Sen, S.; Muneer, P. M. A.; Analysis of Genetic Variability in Rice Varieties (*Oryza sativa* L.) of Kerala Using RAPD Markers. *Genet. Engg. Biotechnol. J.* **2011**, *24*, 1–9.

Subramanian, A.; Nirmal Raj, R.; Maheswarappa, H. P.; Shoba, N. Genetic Variability and Multivariate Analysis in Tall Coconut Germplasms. *J. Pharmacogn. Phytochem.* **2019**, *8* (3), 1949–1953.

Sujanapal, P.; Sasidharan, N. *Zingiber anamalayanum* sp. nov. (Zingiberaceae) from India. *Nord. J. Bot.* **2010,** *28* (3), 288–293.

Suma, A.; Elsy, C. R.; Sivaraj, N.; Padua, S.; Yadav, S. K.; John, K. J.; Krishnan, S. Genetic Diversity and Distribution of Cucumber (*Cucumis sativus* L.) Landraces in India: A Study Using DIVA-GIS Approach. *Electron. J. Plant Breed.* **2019,** *10* (4), 1532–1540.

Syamkumar, S.; Sasikumar, B. Molecular Marker Based Genetic Diversity Analysis of *Curcuma* species from India. *Sci. Hortic.* **2007,** *112* (2), 235–241.

Tawadare, R.; Thangadurai, D.; Khandagave, R. B.; Mundaragi, A.; Sangeetha, J. Phenotypic Characterisation and Genetic Diversity of Sugarcane Varieties Cultivated in Northern Karnataka of India Based on Principal Component and Cluster Analyses. *Braz. Arch. Biol. Technol.* **2019a,** *62,* e19180376.

Tawadare, R.; Thangadurai, D.; Khandagave, R.; Sangeetha, J.; Pandhari, R. RAPD Analysis of Sugarcane Cultivars for Early Maturation and Yield Improvement. *Plant Arch.* **2019b,** *19* (2), 2481–2486.

Thimmappaiah.; Santosh, W. G.; Shobha, D.; Melwyn, G. S. Assessment of Genetic Diversity in Cashew Germplasm Using RAPD and ISSR Markers. *Sci. Hortic.* **2009,** *120,* 411–417.

Thomas, G.; Joseph, L.; Kuriachan, P. Genetic Variation and Population Structure in *Oryza malampuzhaensis* Krish. et Chand. Endemic to Western Ghats, South India. *J. Genet.* **2001,** *80* (3), 141–148.

Thomas, J.; Dominic, V. Assessment of Genetic Diversity and Relationship of Coastal Salt Tolerant Rice Accessions of Kerala (South India) Using Microsatellite Markers. *J. Plant Mol. Breed.* **2016,** *4* (1), 35–42.

Thomas, R. J.; Rajesh, M. K.; Kalavati, S.; Krishnakumar, V.; George, D. J.; Jose, M.; Nair, R. V. Analysis of Genetic Diversity in Coconut and Its Conservation in Root (Wilt) Disease Affected Areas of Kerala: A Community Participatory Approach. *Indian J. Genet.* **2013,** *73* (3), 295–301.

Thondaiman, V.; Rajamani, K.; Senthil, N.; Shoba, N.; Joel, A. J. Genetic Diversity in Cocoa (*Theobroma cacao* L.) Plus Trees in Tamil Nadu by Simple Sequence Repeat (SSR) Markers. *Afr. J. Biotechnol.* **2013,** *12* (30), 4747–4753.

Tiwari, A.; Sharma, D.; Sood, S.; Jaiswal, J. P.; Pachauri, S. P.; Ramteke, P. W.; Kumar, A. Genome-Wide Association Mapping for Seed Protein Content in Finger Millet (*Eleusine coracana*) Global Collection Through Genotyping by Sequencing. *J. Cereal Sci.* **2020,** *91,* 102888.

Vanaja, T.; Randhawa, G. J.; Mammootty, K. P. Pedigree Evaluation and Molecular Diversity of Some True Breeding Rice (*Oryza sativa* L.) Genotypes of Kerala. *J. Trop. Agric.* **2006,** *44,* 42–47.

Vasumathy, S. K.; Peringottillam, M.; Sundaram, K. T.; Kumar, S. H. K.; Alagu, M. Genome-Wide Structural and Functional Variant Discovery of Rice Landraces Using Genotyping by Sequencing. *Mol. Biol. Rep.* **2020,** *47* (10), 7391–7402.

Vazhacharickal, P. J.; Sajeshkumar, N. K.; Mathew, J. J.; James, J. *Studies on Genetic Relationships Among Locally Cultivated Citrus Varieties in Kerala Employing matK and rbcL Gene Using PCR Technique and RFLP Markers. Scientific Study,* GRIN Publishers, Germany, **2017.**

Vidya, S. S.; Sabesan, T.; Saravanan, K. Genetic Divergence Studies in Blackgram (*Vigna mungo* L.) for Yield and Quantitative Traits. *J. Phytol.* **2018,** *10* (1), 24–26.

Vighneswaran, V.; Nair, A. S. Genetic Diversity in *Oryza* and Its Utilization. *J. Aquatic Biol. Fisheries* **2014,** *2,* 813–816.

Vijayalatha, K. R.; Chezhiyan, N. Multivariate Based Marker Analysis in Turmeric (*Curcuma longa* L.). *J. Hortic. Sci.* **2008,** *3* (2), 107–111.

Wang, X.; Bao, K.; Reddy, U. K.; Bai, Y.; Hammar, S. A.; Jiao, C.; Wehner, T. C.; Ramírez-Madera, A. O.; Weng, Y.; Grumet, R.; Fei, Z. The USDA Cucumber (*Cucumis sativus* L.) Collection: Genetic Diversity, Population Structure, Genome-Wide Association Studies, and Core Collection Development. *Hortic. Res.* **2018,** *5* (1), 1–13.

Wesly, K. C.; Nagaraju, M.; Lavanya, G. R. Estimation of Genetic Variability and Divergence in Green Gram *Vigna radiata* (L.) Germplasm. *J. Pharmacogn. Phytochem.* **2020,** *9* (2), 1890–1893.

Yadav, Y.; Lavanya, G. R.; Pandey, S.; Verma, M.; Ram, C.; Arya, L. Neutral and Functional Marker Based Genetic Diversity in Kodo Millet (*Paspalum scrobiculatum* L.). *Acta Physiol. Plant* **2016,** *38* (3), 75.

CHAPTER 11

PLANT BIODIVERSITY OF THE SILENT VALLEY, CORE ZONE OF NILGIRI BIOSPHERE RESERVE, WESTERN GHATS, KERALA

R. BALAJI, M.V. ANJU, R. PRASANNA, R. KURALARASAN, S. ARUN KUMAR, and C. KUNHIKANNAN

Institute of Forest Genetics and Tree Breeding, Forest Campus, R. S. Puram, Coimbatore, India

ABSTRACT

Silent Valley is the richest expression of life on earth and unique reserve of natural rain forests, and is declared as a National Park in 1984. The park harbors more than 1000 species of flowering plants, 3 gymnosperms, 100 pteridophytes, 210 species of bryophytes, 230 lichens, 110 fungi, and 140 algal species. Several endemic and threatened species are also reported from this National Park. Unique vegetation and climate prevailing in the Silent Valley provides well adaptable surrounding for the existing organisms of that area. Very few studies of algae and lichens are carried out, still some locations of the park are not yet explored in terms of biodiversity. Such study may lead to discovery of new species and add to the existing diversity of the Silent Valley National Park.

11.1 INTRODUCTION

Silent Valley represents pristine tropical rain forests and the core of the Nilgiri Biosphere Reserve (NBR), and an important heritage for all succeeding generations having academic and application values. In India, the rain forests are distributed only in a few locations like the Western Ghats and northeast regions. Among them, the Silent Valley and its environs in the Western Ghats are one of the remaining comparatively less damaged tracts of the rain forest and tropical moist evergreen forest in India today (Manilal, 1988).

The Silent Valley was declared as a National Park (Figure 11.1) on 15th November, 1984 and the park was formally inaugurated on 7th September 1985 with an area of 89 km^2 after cancelling the proposed hydroelectric project. Several people like nature lovers,

Biodiversity Hotspot of the Western Ghats and Sri Lanka. T. Pullaiah, PhD (Ed.)
© 2024 Apple Academic Press, Inc. Co-published with CRC Press (Taylor & Francis)

poets, environmentalists, artists, thousands of students, and media protested against the proposed hydroelectric dam across Kunthipuzha in the pristine forest of Silent Valley. The dam would have submserged about 830 ha of forestland with rich rain forest having many rare and endangered species. The Kerala government undertook the first survey for the hydroelectric project in 1951, and in 1973, the Planning Commission of India approved the project. In 1976, Kerala State Electricity Board announced the plan to initiate the construction of a 240 MW hydroelectric project over the Kunthipuzha originate in Silent Valley and flowing through the Palakkad and Malappuram districts after joining with Bharathapuzha. It triggered a wave of protest across the state. Rapidly, it became India's first most important environmental movement resulting in far-reaching changes (Manjusha, 2016). Silent Valley is the richest expression of life on earth and unique reserve of natural rain forests.

FIGURE 11.1   Entry gate of Silent Valley National Park in the core area. (Photo: R. Balaji, author)

11.1.1 LOCATION

Silent Valley having the status of a National Park is situated in Palakkad district of Kerala in the southern Western Ghats, located in Lat.11°05'N and Long.75°27'E. On September 1st, 1986, the Silent Valley National Park (SVNP) was included in the core area of the NBR. In 1988, on May 16th, SVNP was constituted as a separate division (area 89.52 km^2); on June 11th, 2007, a buffer zone (BZ) of 148 km^2 was added to SVNP to provide more protection and houses a rich mosaic of varied habitats (Figure 11.2). The area spreads toward east up to the Bhavani river in Attappadi reserve forest; toward south bordering the area of Mannarkkad forest division; toward west Nilambur south forest division; toward east both Nilambur south forest division and Mukurthi National Park of Tamil Nadu.

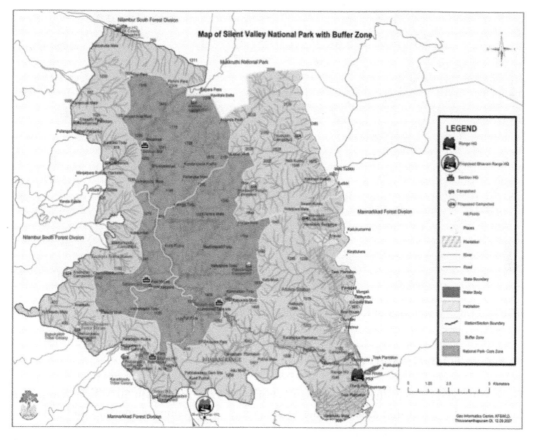

FIGURE 11.2 Map of Silent Valley; dark green area represents National Park, light green area represents buffer zone.

Source: Kerala Forest and Wildlife Department, Govt. of Kerala

A perennial river Kunthipuzha (Figure 11.3), a tributary of Bharathapuzha, passes through the park along the gorges originating at an altitude of 1861 m, near Kozhipara peak (2000 m) the river drains the entire valley. Kunthi river receives water from several streams like Kummattamthode, Valiyaparathode, Karingathode, Madirimaranthode, Kundamsalpuzha, and Alaisholaputtuthode, and drains down and merges with Thuthapuzha before draining to Bharathapuzha. Buffer zone has many plantations such as *Eucalyptus*, teak, cardamom, coffee, pepper, nutmeg, Malabar tamarind, and rubber. It is having four tribal groups called Kattunaiken, Kurumba, Irula, and Muduga settled in 10 different tribal hamlets namely Karadiyod, Ambalappara, Uppukulam, Anavai, Thudukki, Thadikkundu, Mukkali, Chindikki, and Karuvara.

FIGURE 11.3   General view of Kunthi river in Silent Valley. (Photo: R. Balaji, author)

11.1.2   GEOLOGY AND CLIMATE

The evolutionary age of the Silent Valley is believed to be more than 50 billion years. The formation of rock in this area is of the archean age comprising of gneisses and granites. Weathering of ancient crystalline and metamorphic rocks resulted in the formation of soil (Chand Basha, 1999). Soils of this area are blackish and slightly acidic with a good accumulation of organic matter (Nusrat et al., 2013). The forests get both southwest and northeast monsoon and former gives the highest precipitation. It receives 5000 mm rainfall yearly (Singh et al., 1984). The southwest monsoon sets during the first week of June. During June, July, and August the forest get the highest rainfall. Western side with high altitude hills of Silent Valley receives a minimum of 5045 mm rainfall, and near Walakkad the rainfall received goes up to 6500 mm (Adarsh et al., 2019), rarely reaches 10,000 mm

(Kunhikannan, 2010), and smaller quantity of rainfall also occurs during the drier months of January, February, and March (Champion and Seth, 1968; Von Lengerke, 1970, 1980; Kunhikannan, 2010). Average minimum temperature ranges from 8 to 14°C and maximum from 23 to 29°C (Manilal, 1988). The highest peaks in the park are Anginda (2383 m), Sispara (2206 m), Poochapara (1376 m), and Valiyamullumala (1237 m).

11.2 VEGETATION PATTERN OF SILENT VALLEY

The Silent Valley is one of the best protected forests in Palakkad region. The Indian taxonomist, Aiyar in 1932 studied the forest types and reported many types of vegetations in this area earlier (Sathish Kumar, 1999). The SVNP has vegetations like west coast tropical evergreen forest (at elevation of 600–1100 m above MSL), southern subtropical broad-leaved hill forest (1200–1800 m above MSL), southern montane wet temperate forest (above 1900 m), and grassland (Basha, 1999). The floristic diversity range from gigantic trees like *Ficus nervosa, Bischofia javanica, Acrocarpus fraxinifolius*, and *Elaeocarpus tuberculatus*, to lianas such as *Ancistrocladus heyneanus, Caesalpinia cucullata, Carissa inermis, Diploclisia glaucescens, Gnetum edule* (Figure 11.7(B)), *Oxyceros rugulosus, Smythea bombaiensis, Tetrastigma leucostaphylum*, and *Toddalia asiatica*. Some plants appear only for 1 or 2 days in a year, such as the saprophytic orchid *Epipogium roseum* and the tree root parasite *Balanophora fungosa*, some tiny lichens, bryophytes, and ferns (Kunhikannan et al., 2011). The area is very rich in plant diversity. Silent Valley supports almost all the types of life forms with great diversity (Sequiera and Kumar, 2008). The groupwise information is given under different headings.

11.2.1 ALGAE

Pushpangadan and Satish Kumar (1999) opined that the algal diversity of Silent Valley needs to be studied thoroughly, remaining untouched. From time immemorial, algae are being used as food, fodder, fertilizers as well as medicine. Because of their utility, algae are considered as an economically important group. These are the dominant group of the aquatic ecosystem. As a result of increasing interest, a study was carried out on the fresh water algae of Kerala (Rao and Gupta, 1997). Goyal (1982) published a report on algae of Silent Valley. Nasser and Sureshkumar (2012) reported 141 taxa of microalgae from Western Ghats including Silent Valley.

11.2.2 LICHEN

Lichens have a unique life form which acts as an indicator of biodiversity and climate change. Lichens are an interesting group, can grow anywhere with favorable climate, and found as an obligatory symbiotic organism, where fungi live with algae. They have ecological, economical as well as medicinal values. Vohra et al. (1982) discovered 11 lichen species from Silent Valley which are new addition to Indian lichen diversity. Silent Valley offers the suitable niche for the successful growth of lichens. A total of 40 different species of lichens were identified as macrolichens present in Silent Valley by Sequiera and Kumar (2008), and they have also studied certain host specificity among the lichen species and tree species. An intensive survey of the macrolichens of Kerala reported (Kumar, 2000) 254 species of macrolichens from Western Ghats, where 52 were from Silent Valley (Table 11.1).

TABLE 11.1 Lichen Species Recorded in Silent Valley.

1	*Bulbothrix isidiza*	26	*Parmelina wallichiana*
2	*Bulbothrix setschwanensis*	27	*Parmotrema eunetum*
3	*Cladonia cartilaginea*	28	*Parmotrema indicum*
4	*Coccocarpia erythroxyli*	29	*Parmotrema kamatii*
5	*Dirinaria confluens*	30	*Parmotrema sancti-angelii*
6	*Everniastrum cirrhatum*	31	*Parmotrema tinctorum*
7	*Everniastrum nepalense*	32	*Psorella psorina*
8	*Heterodermia coronata*	33	*Ramalina celastri*
9	*Heterodermia dactyliza*	34	*Ramalina inflata*
10	*Heterodermia diademata*	35	*Relicina abstrusa*
11	*Heterodermia dissecta*	36	*Stereocaulon sp.*
12	*Heterodermia flabellata*	37	*Sticta cyphellulata*
13	*Heterodermia hypocaesia*	38	*Sticta limbata*
14	*Heterodermia incana*	39	*Usnea albopunctata*
15	*Heterodermia isidiophora*	40	*Usnea complanata*
16	*Heterodermia leucomela*	41	*Usnea dentritica*
17	*Heterodermia leucomela* subsp.*boriyii*	42	*Usnea himalayana*
18	*Heterodermia obscurata*	43	*Usnea nepalensis*
19	*Heterodermia podocarpa*	44	*Usnea pictoides*
20	*Heterodermia pseudospeciosa*	45	*Usnea rigidula*
21	*Heterodermia togashii*	46	*Usnea sordida*
22	*Hypotrachyna awasthii*	47	*Usnea spinosula*
23	*Hypotrachyna crenata*	48	*Usnea splendens*
24	*Leptogium indicum*	49	*Usnea subflorida*
25	*Leptogium pichneum*	50	*Usnea vegae*

11.2.3 FUNGI

Fungi are eukaryotic organisms characterized by unique structural and physiological features, mode of nutrition (saprophytic or parasitic), etc. Generally, we consider fungi as harmful because of different diseases caused by them. But ecologically and economically they are very important group of organisms. Formerly, fungi were included in the plant kingdom, now they are treated separately. The group, fungi is the second largest (15,504) one among Indian plants other than flowering plants (21,849) on the basis of BSI (2021). Fungi in forest, mostly help in nutrient cycling as wood decaying fungi, several larger trees are found affected with wood rot fungus gradually deteriorating the strength of the tree and later fall off. After falling, wood decaying fungi (Figure 11.4) work on it along with insects that bore on the bark and help to disintegrate the wood into humus. Other groups like the foliicolous fungi, which can be seen on the leaves of the plants cause disease and death in plants (Hosagoudar and Riju, 2013). *Asteridiella oreocnidecola, Asterina chukrasiae, A. oreocnidegen, Meliola anodendricola, M. dolichi, M. silentvalleyensis,* and *Sarcinella oreocnidecola* are the seven new records and *Meliola daviesii* var. *longiseta* is a new variety recorded from Silent Valley area (Hosagoudar and Biju, 2006). *Asteridiella toddaliae, Diplococcinum atrovelutinum, Eupecte amicta, Irenopsis triumfetteae* var. *indica, Isthmospora* sp., *Leptospaerulina australis, Meliola aristolochigena, M. butleri, M. clausenigena M. pycnosporae, M. sairandhriana, M. strombosiigena, Teratosperma anacardii, T. theti, Phyllacora symploci, Prataprajella turpenicola, Spiropes armatellicola, and S. japonica* are new addition to the fungi group of Kerala from Silent Valley (Hosagoudar, 2006, 2007; Hosagoudar and Archana, 2009; Hosagoudar and Riju, 2011). *Asterina elaeocarpicola* belonging to genus *Asterina* and its anamorph on *Elaeocarpus* species was also recorded (Hosagoudar, 2009). *Sarcinella loranthacearum,* a schifffnerulaceous fungus was discovered on the leaves of *Loranthus* sps. (Hosagoudar et al., 2011). *Didymocyrtis ramalinae, Kalchbrenneriella cyanescens, Opegrapha brigantine* (Joshi et al., 2016), and *Phyllopsora albicans* (Mishra et al., 2011) are found present in Silent Valley National Park. *Meliola kakachiana* Hosag. var. *poochiparaensis* Hosag. & Sabeena, is a new variety and *Palawaniella jasmine* is a new record to India reported in this area (Hosagoudar and Sabeena, 2012). *Echidnodella myristicacearum* Hosag. & Divya, *Meliola neelikalluensis* Hosag. & Divya (Hosagoudar and Divya, 2013), *Amazonia symploci* Hosag., *Asteridiella fagraeae* Hosag. & Sabeena, and *Meliola glochidiifolia* Hosag. were recorded during 2013 (Hosagoudar, 2013). *Pyrenidium hypotrachynae* Y. Joshi, *Dactylospora protothallina, Didymocyrtis ramalinae, Endococcus apiciicola, Heterocephalacria physciacearum,* and *Stigmidium heterodermiae* are the recently added species to the fungi diversity of Silent Valley (Joshi et al., 2018). During 2019, Adarsh et al. conducted detailed study on the Aphyllophorales in Silent Valley and came up with the identification of 57 foliicolous fungi. The study on the Hypomycetes fungi revealed 255 species belonging to 152 genera in the Western Ghats and Silent Valley harbored the richest number of species (141), which is 55.4% belonging to 95 genera (Subramaniam, 1986).

11.2.4 BRYOPHYTES

Bryophytes are nonvascular plants where the plant body is a gametophyte. Taxonomically they are placed in between algae and pteridophytes, and are considered as amphibians of plant kingdom. They are classified in three classes namely, Musci the mosses, Hepaticae the liverworts, and Anthocerotae the hornworts. Medicinal and traditional uses of bryophytes were brought out and illustrated in "Hortus Malabaricus," the voluminous document on medicinal plants of Malabar region of Kerala (Remesh and Manju, 2009). Bryophytes play a vital ecological role by firmly attaching to the soil surface and preventing the soil from erosion (Verma et al., 2011). It also has a great role in water cycle of mountainous regions. The varied climate prevailing in Kerala favors the luxuriant growth of various bryophytes (Manju et al., 2008).

FIGURE 11.4   *Microporus* sp., a fungus. (Photo R. Balaji, author)

Cryptogamic flora of the Silent Valley was less explored until recently, and Srivastava and Sharma (2000) opined that detailed exploration can bring out many new additions to the location. Srivastava and Asthana (1986) reported a new species, *Folioceros udarii* from this area; also another species, *Handeliobryum setschwanicum* was collected for the first time from this area (Vohra et al., 1982). Species like *Pogonatum hexagonum* Mitt, *Asterella leptophylla* (Mont.) Pande, and *Fossombronia indica* Steph. are endemic species to India reported from Silent Valley (Vohra, 1981).

A great contribution to this line was provided by Manju and coworkers by publishing the bryophytic flora of Silent Valley. They documented 148 taxa having 109 mosses, 36 liverworts, and three hornworts. Out of these, nine species such as *Chrysocladium flammeum* (Mitt.) M. Fleisch., *Gymnostomum calcareum* Nees & Hornsch., *Glossadelphus bilobatus* (Dix.) Broth., *Hypnum flaccens* Besch., *Macromitrium turgidum* Dix., *Calyptothecium pinnatum* Nog., *Brotherella amblystegia* (Mitt.) Broth., *Notoscyphus paroicus*

Schiffn., and *Wijkia deflexifolia* (Ren. & Card.) Crum. were new reports to peninsular India, and species such as *Lejeunea cavifolia* (Ehrh.) Lindb., *Radula obscura* Mitt., *Radula meyeri* Steph., and *Barbella turgida* Nog. were new records for Kerala State. Two species such as *Trichostelium stigmosum* (Manju et al., 2012) and *Aerobryopsis wallichii* (Brid.) Fleisch. (Prajitha et al., 2013) have been reported as new records for India from Silent Valley. Later, an exhaustive account of bryoflora was given by Daniels and Raja (2020) which accounts for 212 species and three infraspecific taxa under 124 genera and 51 families. Out of these, 154 species and three infraspecific taxa under 88 genera and 31 families belong to the group mosses, three species under three genera in one family to hornworts, and 55 species under 33 genera and 19 families belong to liverworts. They further reported that there is one moss which is new to the bryoflora of India, six to peninsular India, 20 mosses and four liverworts to Kerala, and one liverwort rediscovered after more than 100 years.

As a result of studies carried by Manju and coworkers from 2001 to 2011, several new species were recorded and reported from different locations of Silent Valley. In addition, Daniels et al. (2010a) brought out a new species belonging to the tribe Hyophileae and named as *Indopottia zanderi*. It belongs to the family Pottiaceae. *Fossombronia indica, Spruceanthus semirepandus, Hypopterigium tenellum, Plagiochila fruticosa, Pogonatum microstomum, Campylopus ericoides, Leucobryum juniperoideum, Anomobryum auratum,* and *Racopilum orthocarpum* are the species recorded from Silent Valley (Manju and Prajitha, 2010; Manju et al., 2015). *Thysananthus appendiculatus* was discovered in Silent Valley, it was earlier considered to be endemic to New Guinea (Daniels and Raja, 2011). A new species, *Trichosteleum stigmosum* Mitt. belonging to the family Sematophyllaceae (Manju et al., 2012) and *Aerobryopsis wallichii* (Brid.) Fleischhas belonging to the family Meteoriaceae were reported as new records for India from Silent Valley (Prajitha et al., 2013). *Bryum apalodictyoides* was reported in Silent Valley during the study on the Genus *Bryum* (Bansal and Nath, 2014). *Schistochila aligera* was rediscovered after more than a century in Silent Valley and Agasthiyamalai Biosphere (Daniels and Daniels, 2008). *Frullania ramuligera,* a rare species, was rediscovered in Silent Valley which was also reported in Nilgiri hills and Agasthiyamalai Biosphre reserve (Daniels et al., 2010b).

11.2.5 PTERIDOPHYTES

Pteridophytes are the primitive vascular plant groups (Figures 11.5 and 11.6) which are cosmopolitan in distribution. They have a wide range of utility such as it serves as a source of food, medicine, fiber, construction materials, and its beautiful foliages can act as a decorative material. Some of the genera of pteridophytes possess secondary metabolites which are very unique and not present in higher plants. The forests of Silent Valley, with its suitable microhabitats, hold more than 100 species of pteridophytes. The diversity ranges from the minute filmy ferns to large tree ferns, including some rare elements such as *Elaphoglossum beddomei, E. nilgiricum, Blechnum colensoi,* etc.

FIGURE 11.5   *Bolbitis subcrenata* (Hook. & Grev.) Ching, a pteridophyte species. (Photo R. Balaji, author)

The first new fern taxon from Silent Valley may be *Pteris silentvalliensis* R.Ghosh & R.K.Ghosh (1982). It is now considered as *Pteris scabripes* Wall. A new fern genus *Nistarika* B.K. Nayar, Madhus. & Molly, with the species *N. bahupunctika* B.K.Nayar, Madhus. & Molly was published by Nayar et al. (1985). It was later transferred to *Leptochilus bahupunctika* (Nayar et al., 1985) Nampy & Madhus., and now is considered under *L. axillaris* (Cav.) Kaulf. Explorations in the area resulted in recording of some more unique fern taxa. Nayar and Geevarghese (1987) recorded *Pronephrium thwaitesii*, a little known fern from Silent Valley. Madhusoodanan and Jyothi (1993) described a new species as *Pellaea malabarica* Geev. ex Madhus. & Jyothi. It is now considered as *P. longipilosa* Bonap. Nayar & Geevarghese (1993) in their "Ferns of Malabar" mentioned many fern taxa from the Silent Valley, including novelties. The new species, named after the Kunthipuzha of Silent Valley, *Polystichum kunthianum* B.K.Nayar & Geev. is now considered under *P. subinerme* (Kunze) Fraser-Jenk. The new variety, as var. *kummatta* B.K. Nayar & Geev. is now considered under *Bolbitis appendiculata* (Willd.) K.Iwats. Nair et al. (1988, 1992a, 1992b, 1994) in their series of work on Fern-allies and Ferns of Kerala also mentioned many collections from the area. Kumar and Sequiera (1999) presented the details of epiphytic pteridophytes of the area. *Cyathea crinita* (Hook.) Cope has been recorded in Silent valley (Manickam, 1995). *Tectaria zeilanica, Drymoglossum heterophyllum,* and *Pteris confusa* are endemic to the study area (Sukumaran et al., 2009). *Polystichum subinerme* is endemic to south India and comes under very rare category. *Asplenium pellucidum* is considered as rare fern found in Silent valley (Chandra et al., 2008).

FIGURE 11.6   *Pteris pellucida* Presl. (Photo R. Balaji, author)

11.2.6   GYMNOSPERMS

Very few gymnosperms were identified in Silent Valley. The tree, *Cycas circinalis* L. (Cycadaceae), endemic to western peninsular India and a liana *Gnetum edule* (Willd.) Blume (Gnetaceae) (Figure 11.7(B)) are commonly present in the Silent Valley forests

(Vajravelu, 1990; Singh and Srivastava, 2013). Critically endangered species of *Cycas annaikalensis* Rita Singh & P.Radha endemic to Palakkad district is only present in the Annaikal hills with minimum population (Singh and Radha, 2006; Singh et al., 2007). It may occur in Silent Valley, so more studies for enumeration of gymnosperms species are essential. One ornamental species, *Cycas revoluta* Thunb. is commonly cultivated in the gardens all over the India and the study area too (Singh and Srivastava, 2013).

FIGURE 11.7   Some interesting plant species in Silent Valley. (A) *Eulophia emilianae* C.J.Saldanha, (B) *Gnetum edule* (Willd.) Blume, (C) *Rubia cordifolia* L., and (D) *Vanda thwaitesii* Hook.f. (Photo T. Balaji, author)

11.2.7   ANGIOSPERMS

The angiosperms, the flowering plant is the one group highly studied in Silent Valley. Before independence, only 21 species of plants were said to be identified within the Silent Valley, whereas 17 new plant species were reported only in a unique area, Sispara, in which seven new species were discovered exclusively by Robert Wight (Biju, 1999). After the declaration of Silent Valley as a National Park, taxonomic work by Dr. K. S. Manilal and team reported 966 angiosperms (Manilal, 1988). Dr. E. Vajravelu gave his contribution to Silent Valley and reported 950 plants species in core and buffer zone through the flora of Palghat district (Vajravelu, 1990). A total of 34 plant species were recorded from Medicinal Plants Conservation Area (MPCA), a piece of land on the way to Sairandri, solemnly maintained for the preservation of the medicinal plants, as new report to Silent Valley (Sasidharan and Anto, 1999).

The monographic work of the family Orchidaceae of Silent Valley reported 107 species and a subspecies belonging to 54 genera (Manilal, 1988; Sathish Kumar, 1999). The plants belonging to the family Araceae were also studied and reported 21 species covering nine species of *Arisaema*, four species of *Amorphophallus*, two species of *Lagenandra*, one species each of *Ariopsis, Remusatia, Theriophonum, Phothos, Anaphyllum*, and *Colocasia* (Sivadasan, 1999). The work on the genus *Ceropegia* revealed the presence of *Ceropegia candelabrum* var. *candelabrum, C. decaisneana, C. elegans* var. *elegans, C. ensifolia, C. fimbriifera*, and *C. hirsuta* from the area (Bruyns, 1997).

11.2.7.1 NEW DISCOVERIES AND REDISCOVERIES

Many new findings have been reported by many workers from Silent Valley totaling to 37 species and one variety. Sathish Kumar and team (Sathish Kumar, 1999; Sathish Kumar et al., 2008) discovered nine species of Orchidaceae, Sivadasan (1999) reported one member of Araceae, Shivamurthy and Sadanand (1997) gave one species of Podostemaceae. *Cucumella* sp., *Hydnocarpus* sp., *Liparis* sp., *Oberonia* sp., *Robiquetia* sp. (Manilal, 1988), *Silentvalleya* sp., *Hedyotis* sp., *Kanjaram* sp., and *Porpax* sp. (Vajravelu, 1990), one *Ophiorrhiza* sp., *Ophiorrhiza sahyadriensis* (Hareesh et al., 2015a), two *Impatiens* spp. (Hareesh et al., 2015b; Kumar and Sequiera, 1996) were identified. Two species of Orchidaceae such as *Ipsea malabarica* and *Vanda thwaitesii* (Figure 11.7 (D)) were believed to be extinct from the world, but it was rediscovered in the Silent valley forests (Sathish Kumar, 1999). Another new species of *Gentiana kurumbae* was recently found in the Cherukolmala, Attappady region (Anilkumar et al., 2015). There is a chance for the distribution of the same in Silent Valley since it is a nearest boundary to the study location. Table 11.2 summarizes new discoveries and rediscoveries.

11.2.7.2 NEW RECORDS FOR INDIA FROM SILENT VALLEY

Cyclea barbata, Dendrobium panduratum, Oberonia bicornis, Oberonia tenuis, Scutellaria oblonga, Syzygium makul, and *S. neesianum* were new records to the area (Manilal, 1988). In Orchidaceae, six species including *Aphyllorchis prainii* were doubtfully added to the "Survey of the flora of Anamalai hills," Coimbatore district by C.E.C. Fischer, *Oberonia brachyphylla* which was endemic to Karnataka and *Smithsonia straminea* were recorded to Mukkali forest of Silent Valley (Vajravelu, 1990); remaining *Eulophia emilianae* (Figure 11.7 (A)), *Liparis wrayi*, and *Oberonia bicornis* were recorded in Silent Valley forest and incorporated to flora of Kerala (Sathish Kumar, 1999).

TABLE 11.2 The List of New Discoveries of Plants from Silent Valley.

S. No.	Plant name	Distribution
1	*Amorphophallus nicolsonianus* Sivadasan	Silent Valley, endemic to south India
2	*Anisochilus suffruticosus* Wight (*Coleus suffruticosus* (Wight) A.J.Paton)	Silent Valley
3	*Bulbophyllum silentvalliensis* M.P.Sharma & S.K.Srivast.	Silent Valley
4	*Calamus neelagiricus* Renuka (*Calamus gamblei* Becc.)	Silent Valley
5	*Cassine kedarnathii* Sasidh. & Swarupan.	Silent Valley
6	*Cucumis silentvalleyi* (Manilal, T.Sabu, & P.Mathew) Ghebret. & Thulin (Syn.*Cucumella silentvalleyi* Manilal, T.Sabu, & P.Mathew)	Silent Valley
7	*Embelia gardneriana* Wight	Silent Valley
8	*Porpax microchilos* (Dalzell) Schuit., Y.P.Ng & H.A.Pedersen (Syn.Eria tiagii Manilal, C.S.Kumar, & J.J.Wood)	Silent Valley
9	*Garnotia courtallensis* (Arn. & Nees) Thwaites (Garnotia puchiparensis Bor)	Silent Valley
10	*Gastrodia silentvalleyana* C.S.Kumar, P.C.S.Kumar, Sibi, & S.Anil Kumar	Silent Valley
11	*Phyllanthus candolleanus* (Wight & Arn.) Chakrab. & N.P.Balakr. (Syn. Glochidion sisparense Gamble)	Silent Valley
12	*Hedyotis bourdillonii* (Gamble) R.S.Rao & Hemadri (Syn. *Hedyotis silentvalleyensis* Vajr., Rathakr. & Bhargavan)	Silent Valley
13	*Hedyotis leschenaultiana* DC. var. *leschenaultiana* (Syn. *Hedyotis sisaparensis* Gage)	Silent Valley
14	*Helichrysum wightii* C.B.Clarke ex Hook.f.	Silent Valley
15	*Hydnocarpus pendulus* Manilal, T.Sabu, & Sivar.	Silent Valley
16	*Ilex gardneriana* Wight	Silent Valley
17	*Impatiens denisonii* Bedd.	Silent Valley
18	*Impatiens munronii* Wight	Silent Valley
19	*Impatiens sahyadrica* V.B.Sreek., Hareesh, Dantas, & Sujanapal	Poochippara of Silent Valley
20	*Impatiens sivarajanii* M.Kumar & Sequiera (*Impatiens neobarnesii* C.E.C.Fisch.)	Sispara of Silent Valley
21	*Kanjarum palghatense* Ramam. (*Strobilanthes dupenii* Bedd. ex C.B.Clarke)	Silent Valley
22	*Kingidium niveum* C.S.Kumar (*Phalaenopsis mysorensis* C.J.Saldanha)	Silent Valley
23	*Lasianthus ciliatus* Wight	Silent Valley
24	*Liparis barbata* Lindl. (Syn.*Liparis indirae* Manilal & C.S.Kumar)	Silent Valley
25	*Memecylon sisparense* Gamble	Silent Valley
26	*Oberonia platycaulon* Wight (Syn.*Oberonia bisaccata* Manilal & C.S.Kumar)	Silent Valley

TABLE 11.2   *(Continued)*

S. No.	Plant name	Distribution
27	*Ophiorrhiza sahyadriensis* Hareesh, V.B.Sreek, & K.M.P.Kumar	Walakkad, Sispara, Anguinda of Silent Valley & Muthikulam, western slopes of Velliangiri hills
28	*Piper nigrum* L. (Syn.*Piper nigrum* var. *hirtellosum* Asokan Nair & Ravindran)	Silent Valley
29	*Piper silentvalleyense* Ravindran, M.K.Nair, & Asokan Nair	Silent Valley
30	*Porpax exilis* (Hook.f.) Schuit., Y.P.Ng & H.A.Pedersen (Syn. *Porpax chandrasekharanii* Bhargavan & C.N.Mohanan)	Silent Valley
31	*Rungia sisparensis* T. Anderson ex C.B.Clarke	Silent Valley
32	*Breynia saksenana* (Manilal, Prasann., & Sivar.) Chakrab. & N.P.Balakr. (Syn. *Sauropus saksenanus* Manilal, Prasann., & Sivar.)	Silent Valley
33	*Silentvalleya nairii* V.J.Nair, Sreek., Vajr. & Bhargavan	Silent Valley
34	*Sonerila elegans* Wight	Silent Valley
35	*Willisia arekaliana* Shivam. & Sadanand	Kunthipuzha dam site & endemic to Silent Valley
36	*Youngia nilgiriensis* Babc.	Silent Valley
37	*Robiquetia josephiana* Manilal & C.S.Kumar	Silent Valley, endemic to south India

11.2.7.3   THREATENED SPECIES

One species each from families such as Acanthaceae, Anacardiaceae, Arecaceae, Cappa-raceae, Elaeocarpaceae, Fabaceae, Icacinaceae, Lamiaceae, Liliaceae, Menispermaceae, Myristicaceae, Myrsinaceae, Myrtaceae, Poaceae, and Vitaceae; two species each from families like Celastraceae, Lauraceae, Meliaceae, and Zingiberaceae; three species each from families of Asclepiadaceae, Asteraceae, Balsaminaceae, Commelinaceae, and Melastomataceae; five species of Rubiaceae; seven species of Araceae; ten species of Orchidaceae were considered as threatened species (Venkatasubramanian et al., 1999). From Orchidaceae *Ipsea malabarica* was last collected in 1852, after that their population was not reported anywhere, but it was rediscovered in Silent Valley areas (Manilal, 1999). *Aerides crispa* and *Vanda thawitesii* (Figure 11.7(D)) were considered as endangered species (Sathish Kumar, 1999). The flora of Silent Valley reported 966 plant species, among them 20 tree species come under red list (Sasidharan, 2003). Sasidharan and Anto (1999) reported four species as threatened in MPCA. According to Conservation Assessment and Management Plan, few plants like *Aphanamixis polystachya, Cinnamomum sulphuratum, Embelia ribes, Garcinia gummi-gutta, Garcinia morella, Hydnocarpus alpina, Nothapo-dytes nimmoniana, Myristica dactyloides, Persea macrantha,* and *Symplocos racemosa* were considered as RET (Ravikumar and Ved, 2000), which were selected for the species recovery studies at MPCA of Silent Valley (Kunhikannan et al., 2011). Table 11.3 summa-rizes the total number of RET species enumerated from Silent Valley areas.

TABLE 11.3 The List of RET Species in Silent Valley.

S. No.	Plants names	Category
1	*Acanthephippium bicolor* Lindl.	Endangered
2	*Actinodaphne campanulata* Hook.f.	Rare and threatened
3	*Actinodaphne tadulingamii* Gamble	Rare
4	*Aerides crispa* Lindl.	Endangered
5	*Aglaia bourdillonii* Gamble	Threatened
6	*Aglaia elaeagnoidea* var. *bourdillonii* (Gamble) N.C.Nair	Rare
7	*Aglaia lawii* (Wight) C.J.Saldanha	Rare
8	*Amomum microstephanum* Baker	Rare and threatened
9	*Anaphalis beddomei* Hook.f.	Rare and threatened
10	*Anaphyllum wightii* Schott	Threatened
11	*Antistrophe serratifolia* (Bedd.) Hook.f.	Rare and threatened
12	*Arisaema attenuatum* E.Barnes & C.E.C.Fisch.	Threatened
13	*Arisaema barnesii* C.E.C.Fisch.	Rare and threatened
14	*Arisaema leschenaultii* Blume	Rare and threatened
15	*Arisaema murrayi* (J.Graham) Hook.	Rare and threatened
16	*Arisaema tuberculatum* C.E.C.Fisch.	Rare and threatened
17	*Arisaema tylophorum* C.E.C.Fisch.	Rare and threatened
18	*Asparagus fysonii* J.F.Macbr.	Rare and threatened
19	*Boesenbergia pulcherrima* (Wall.) Kuntze	Rare and threatened
20	*Bulbophyllum aureum* (Hook.f.) J.J.Sm.	Rare and threatened
21	*Calamus huegelianus* Mart. ex Walp.	Rare and threatened
22	*Canthium neilgherrense* Wight	Threatened
23	*Capparis rheedei* DC.	Rare
24	*Casearia wynadensis* Bedd.	Threatened
25	*Cayratia pedata glabra* Gamble	Rare
26	*Ceropegia decaisneana* Wight	Rare
27	*Ceropegia thwaitesii* Hook.	Rare
28	*Commelina indehiscens* E.Barnes	Rare
29	*Corymborkis veratrifolia* (Reinw.) Blume	Endangered
30	*Cyanotis burmanniana* Wight	Rare
31	*Cyclea fissicalyx* Dunn	Rare
32	*Derris thothathrii* Bennet	Rare
33	*Elaeocarpus munronii* (Wl.) Masters	Rare
34	*Elaeocarpus recurvatus* Corner	Threatened
35	*Embelia gardneriana* Wight	Threatened
36	*Eria albiflora* Rolfe	Rare and threatened
37	*Euonymus angulatus* Wight	Rare

TABLE 11.3 *(Continued)*

S. No.	Plants names	Category
38	*Hedyotis bourdillonii* (Gamble) R.S.Rao & Hemadri	Rare and threatened
39	*Hedyotis hirsutissima* Bedd.	Threatened
40	*Holigarna beddomei* Hook.f.	Threatened
41	*Holigarna grahamii* (Wight) Kurz	Rare
42	*Impatiens acaulis* Arn.	Threatened
43	*Impatiens lucida* B.Heyne ex Hook.f.	Threatened
44	*Impatiens viscosa* Bedd.	Rare and threatened
45	*Ipsea malabarica* (Rchb.f.) Hook.f.	Endangered
46	*Lasianthus ciliatus* Wight	Rare and threatened
47	*Memecylon flavescens* Gamble	Rare and endangered
48	*Memecylon lawsonii* Gamble	Rare and endangered
49	*Memecylon sisparense* Gamble	Critically endangered
50	*Miquelia dentata* Bedd.	Rare
51	*Murdannia lanceolata* (Wight) Kammathy	Threatened
52	*Murdannia lanuginosa* (Wall. ex C.B.Clarke) G.Brückn.	Rare
53	*Myristica malabarica* Lam.	Threatened
54	*Neanotis monosperma* (Wight & Arn.) W.H.Lewis	Rare and threatened
55	*Nostolachma crassifolia* (Gamble) Deb & Lahiri	Threatened
56	*Oberonia brachyphylla* Blatt. & McCann	Rare
57	*Ochlandra scriptoria* (Dennst.) C.E.C.Fisch.	Threatened
58	*Pogostemon gardneri* Hook.f.	Rare and threatened
59	*Porpax jerdoniana* (Wight) Rolfe	Rare
60	*Porpax reticulata* Lindl.	Rare
61	*Psychotria globicephala* Gamble	Rare and threatened
62	*Rungia linifolia* Nees	Rare and threatened
63	*Salacia beddomei* Gamble	Rare
64	*Saprosma fragrans* (Bedd.) Bedd.	Rare and threatened
65	*Schoenorchis jerdoniana* (Wight) Garay	Rare
66	*Sonerila elegans* Wight	Rare and threatened
67	*Syzygium benthamianum* (Duthie) Gamble	Rare and threatened
68	*Tainia bicornis* (Lindl.) Rchb.f.	Rare
69	*Toxocarpus palghatensis* Gamble	Rare and threatened
70	*Vanda thwaitesii* Hook.f.	Endangered
71	*Vateria indica* L.	Threatened
72	*Vernonia bourdillonii* Gamble	Rare and threatened
73	*Vernonia saligna* var. *nilghirensis* Hook.f.	Rare and threatened

11.2.7.4   MEDICINAL AND WILD CROP SPECIES

People belonging to tribes Irulas, Kurumbas, and Mudugars utilizes species like *Artemisia nilagirica, Buddleja asiatica, Cyclea barbata, Iphigenia indica,* and *Rubia cordifolia* (Figure 11.7(C)) for some treatments (Pushpangadan and Sathish Kumar, 1999). A total of 102 medicinal plants were studied by Yabesh et al. (2014), belonging to 53 families and 95 genera from area of Agali, Kottathara, Mannarkkad, Padavayal, and Sholayur of Palakkad district. In Silent Valley and adjacent area, totally 110 species, one subspecies, and six varieties of wild crop species were noted (Velayudhan et al., 1999). Species of genus *Alpinia, Amomum, Atylosia, Cinnamomum, Curcuma, Dioscorea, Elettaria, Garcinia, Mangifera, Musa, Myristica, Panicum, Piper, Rauvolfia, Zingiber,* etc., are medicinal and *Vateria indica, Canarium strictum, Pterocarpus marsupium, Sterculia guttata,* etc. yield resins and are commonly present in the study area; 26 plant species were lesser known edible such as *Baccaurea, Cucumella, Cucumis, Elaeocarpus, Garcinia, Mangifera, Rubus, Spondias,* and *Syzygium* (Pushpangadan and Sathish Kumar, 1999). The term MPCA denotes a forest area selected for conserving unique populations or diversity of medicinal plants in their natural ecosystem. The MPCA area of Silent Valley with west coast tropical evergreen type vegetation (Basha, 1999) covers an area of 200 ha all along the Kattivaramudi, down through Kummatanthodu to the Kunthi River (Kunhikannan et al., 2011).

11.2.7.5   WILD ORNAMENTAL SPECIES IN THE AREA

Every species have its own beauty in nature as its colorful foliage, flowers, growth pattern, etc., but a few species only are generally utilized for growing in our gardens for ornamental purpose. From Orchidaceae family, 30 species are having beautiful flowers and ornamental value (Sathish Kumar, 1999). Apart from Orchidaceae, some genera like *Aeschynanthus, Arisaema, Begonia, Didymocarpus, Exacum, Globba, Gymnostachyum, Hedychium, Hoya, Impatiens, Jasminum, Lillium, Medinilla, Rhyncoglossum, Sonerila, Strobilanthes,* and *Thunbergia* also can be used in gardens (Pushpangadan and Sathish Kumar, 1999).

11.3   CONCLUSIONS

In the Western Ghats of India, Silent Valley is one of the most diverse areas with different types of forests and a well-protected area located in the Palakkad district of Kerala. It is a nest for some important endemic species to survive. Many species of flora and fauna were recently discovered and some were rediscovered in Silent Valley National Park. Ethnobotanical knowledge of the tribes that live around Silent Valley National Park areas about several plants utilized by them is still maintained within them. Unique vegetation and climate prevailing in the Silent Valley provides well adaptable surrounding for the existing organisms of that area. Very few studies of algae and lichens were carried out; still some locations of Silent Valley National Park are not explored in terms of phyto as well as the faunal diversity. Many locations of this area are yet to be completely surveyed, which may lead to discovery of new species and add to the wealth of the existing diversity of the Silent Valley National Park. No doubt, Silent Valley is a treasure house of resources. We need to feel the essence and importance of such treasure and work for its conservation and make coming generations to feel it and love it.

KEYWORDS

- **Silent Valley National Park**
- **Nilgiri biosphere reserve**
- **plant biodiversity**
- **new discoveries**
- **rain forests**

REFERENCES

Adarsh, C. K.; Vidyasagaran, K.; Ganesh, P. N. The Diversity and Distribution of Polypores (Basidiomycota: Aphyllophorales) in Wet Evergreen and Shola Forests of Silent Valley National Park, Southern Western Ghats, India, with Three New Records. *J. Threat. Taxa* **2019**, *11* (7), 13886–13909.

Anilkumar, K. A.; Kumar, K. P.; Udayan, P. S. *Gentiana kurumbae*, a New Species of Gentianaceae from the Western Ghats of Kerala, India. *Taiwania* **2015**, *60* (2), 81–85.

Bansal, P.; Nath, V. Genus *Bryum* Hedw. in Peninsular India. *Frahmia* **2014**, *4*, 1–11.

Biju, S. D. Chronicle of Discoveries—The Pursuit of Plants. In *Silent Valley—Whispers of Reason*; Manoharan, T. M., Biju, S. D., Nayar, T. S., Easa, P. S., Eds.; Kerala Forest Department and Kerala Forest Research Institute. Thiruvananthapuram, Kerala, **1999**; pp 145–174.

Bruyns, P. V. A Note on *Ceropegia* L. (Asclepiadaceae) of Silent Valley, Kerala, India. *Rheedia* **1997**, *7*, 107–114.

BSI. *Plant Discovries-2020*; Botanical Survey of India: Kolkata, **2021**.

Champion, H. G.; Seth, S. K. *A Revised Survey of the Forest Types of India*; Govt.of India Publications: New Delhi, **1968**.

Basha, S. Forest Types of Silent Valley In: *Silent Valley—Whispers of Reason*; Manoharan, T. M., Biju, S. D., Nayar, T. S., Easa, P. S., Eds.; Kerala Forest Department and Kerala Forest Research Institute: Thiruvananthapuram, Kerala, **1999**; pp 109–116.

Chandra, S.; Fraser-Jenkins, C. R.; Kumari, A.; Srivastava, A. A Summary of the Status of Threatened Pteridophytes of India. *Taiwania*, **2008**, *53* (2), 170–209.

Daniels, A. E. D.; Raja, R. D. A. *Bryoflora of the Silent Valley National Park Western Ghats*; Bishen Singh Mahendrapal Singh: Dehradun, **2020**.

Daniels, A. E. D.; Raja, R. D. A.; Daniel, P. *Indopottia zanderi* (Bryophyta, Pottiaceae) gen. et sp. nov. from the Western Ghats of India. *J. Bryol.* **2010a**, *32* (3), 216–219.

Daniels, A.; Kariyappa, K.; Mabel, J.; Raja, R.; Daniel, P. The Overlooked and Rare Liverwort *Frullania ramuligera* (Nees) Mont. (Frullaniaceae) Rediscovered in the Western Ghats of India. *Acta Botanica Hungarica* **2010b**, *52* (3–4), 297–303.

Daniels, A. D.; Daniels, P. The liverwort *Schistochila aligera* (Nees & Blume) J.B. Jack & Steph. (Schistochilaceae) re-discovered in India. *Cryptogamie-Bryologie*, **2008**, *29* (3): 307–310.

Daniels, A. E. D.; Raja, R. D. A. The New Guinean *Thysananthus appendiculatus* (Lejeuneaceae) Discovered in the Western Ghats of India. *Lindbergia* **2011**, *34*, 40–43.

Ghosh, S. R.; Ghosh, R. K. A New Species of *Pteris* from Silent Valley, Kerala. *J. Bombay Nat. Hist. Soc.* **1982**, *79*, 385.

Goyal, S. K. Algal Flora of Silent Valley. *Phykos*, **1982**, *21*, 119–121

Hareesh, V. S.; Sreekumar, V. B.; Prabhukumar, K. M.; Nirmesh, T. K.; Sreejith, K. A., *Ophiorrhiza sahyadriensis* (Rubiaceae), a New Species from southern Western Ghats, Kerala, India. *Phytotaxa* **2015a,** *202* (3), 219–224.

Hareesh, V. S.; Sreekumar, V. B.; Dantas, K. J.; Sujanapal, P. *Impatiens sahyadrica* sp. nov (Balsaminaceae)—A New Species from Southern Western Ghats, India. *Phytotaxa* **2015b,** *207* (3), 291–296.

Hosagoudar, V.; Divya, B. Additions to Black Mildews. *Intern. J. Biol., Pharm. Allied Sci.* **2013,** *2* (2), 430–438.

Hosagoudar, V. B. The genus *Asterina* and Its Anamorph on *Elaeocarpus* Species in Southern Western Ghats of Peninsular India. *J. Appl. Nat. Sci.* **2009,** *1* (1), 27–30.

Hosagoudar, V. B. Additions to the Fungi of Kerala II. *Zoos' Print J* **2006,** *21* (11), 2475–2478.

Hosagoudar, V. B. Additions to the Fungi of Kerala–V. *Zoos' Print J.* **2007,** *22* (9), 2834–2836.

Hosagoudar, V. B. and Riju, M. C. Foliicolous Fungi of Silent Valley National Park, Kerala, India. *J. Threat. Taxa* **2013,** *5* (3), 3701–3788.

Hosagoudar, V. B. New and Noteworthy Black Mildews from the Western Ghats of Peninsular India. *Plant Pathol. Quarantine* **2013,** *3* (1), 1–10.

Hosagoudar, V. B.; Archana, G. R. Meliolaceae of Kerala, India-XXVIII. *J. Threat. Taxa* **2009,** *1* (6),348–350

Hosagoudar, V. B.; Biju, H. Studies on Foliicolous Fungi-XXII: Microfungi of Silent Valley National Park, Palghat District in Kerala State. *J. Mycopathological Res.* **2006,** *44* (1), 39–48.

Hosagoudar, V. B.; Riju, M. C. Three New Fungi from Silent Valley National Park, Kerala, India. *J. Threat. Taxa* **2011,** *3* (3), 1615–1619.

Hosagoudar, V. B.; Sabeena, A. New Fungi from Western Ghats, India. *Plant Pathol. Quarantine* **2012,** *2*, 10–14.

Hosagoudar, V. B.; Thomas, J.; Agarwal, D. K., Two New Schifffnerulaceous Fungi from Kerala, India. *J. Yeast Fungal Res.* **2011,** *2* (5), 85–87.

Joshi, Y.; Falswal, A.; Tripathi, M.; Upadhyay, S.; Bisht, A.; Chandra, K.; Bajpai, R.; Upreti, D. K. One Hundred and Five Species of Lichenicolous Biota from India: An Updated Checklist for the Country. *Mycosphere* **2016,** *7* (3), 268–294.

Joshi, Y.; Tripathi, M.; Bisht, K.; Upadhyay, S.; Kumar, V.; Pal, N.; Gaira, A.; Pant, S.; Rawat, K. S.; Bisht, S.; Bajpai, R. Further Contributions to the Documentation of Lichenicolous Fungi from India. *Kavaka,* **2018,** *50*, 26–33.

Kumar, M. Lichen (Macrolichen) Flora of Kerala Part of Western Ghats. *Kerala Forest Research Institute Research Report*, (194), **2000.**

Kumar, M.; Sequiera, S. *Impatiens sivarajanii*: A New Species of Balsaminaceae from Silent Valley National Park, Kerala, India. *Rheedea* **1996,** *6* (2), 51–54.

Kumar, M.; Sequiera, S. Observations on the Epiphytic Flora of Silent Valley. In *Silent Valley—Whispers of Reason*; Manoharan, T. M., Biju, S. D., Nayar, T. S., Easa, P. S., Eds., Kerala Forest Department and Kerala Forest Research Institute: Thiruvananthapuram, Kerala, **1999;** pp 251–256.

Kunhikannan, C. *Studies on Natural Regeneration of Important trees in Silent Valley National Park, Kerala*; Project Report, Institute of Forest Genetics and Tree Breeding: Coimbatore, **2010.**

Kunhikannan, C.; Venkatasubramanian, N.; Sivalingam, R.; Pramod Kumar, N.; Salvy Thomas; Sibin Thampan. Diversity of Woody Species in the Medicinal Plant Conservation Area (MPCA) of Silent Valley National Park, Kerala, Biodiversity, **2011,** *12* (2), 97–107.

Madhusoodanan, P. V.; Jyothi, P. V. *Pellaea malabarica* Geev. ex Madhus. & Jyothi. *Indian Fern J.* **1993,** *9* (1–2), 39.

Manickam, V. S. Rare and Endangered Ferns of the Western Ghats of South India. *Fern Gazette* **1995,** *15* (1), 1–10.

Manilal, K. S. Biodiversity and Ecological Status of Silent Valley as Revealed by Its Angiosperm Flora. In: *Silent Valley—Whispers of Reason*; Manoharan, T. M., Biju, S. D., Nayar, T. S., Easa, P. S., Eds.; Kerala Forest Department and Kerala Forest Research Institute: Thiruvananthapuram, Kerala, **1999;** pp 117–120.

Manilal, K. S. *Flora of Silent Valley, Tropical Rain Forest of India*; Department of Science and Technology, Government of India, **1988.**

Manju, C. N.; Rajesh, K. P.; Madhusoodanan, P. V. Checklist of the Bryophytes of Kerala, India. *Trop. Bryol. Res. Rep.* **2008,** *7*, 1–24.

Manju, C. N.; Rajilesh, V. K.; Prajitha, B.; Prakashkumar, R.; Rajesh, K. P. Contribution to the Bryophyte Flora of India : Silent Valley National Park in the Western Ghats, India. *Acta Biologica Plantarum Agriensis* **2015,** *3,* 73–98.

Remesh, M.; Manju, C. N. Ethnobryological Notes from Western Ghats, India. *Bryologist* **2009,** *112* (3), 532–537.

Manju, C. N.; Prajitha, B. *Systematic Studies on Bryophytes of Northern Western Ghats in Kerala,* **2010.** Final Report, Council Order No. (T) 155/WSC/2010/KSESTC DTD. 13.09.2010.

Manju, C. N.; Prakashkumar, R.; Prajitha, B.; Rajilesh, V. K.; Anoop, K. P. *Trichosteleum stigmosum* Mitt. (Sematophyllaceae) from Silent Valley National Park, a New Record for India. *Taiwania* **2012,** *57* (2), 222–224.

Manjusha, K. A. Silent Valley Movement in Kerala; A Study on the contributions of Kerala Sastra Sahitya Parishad. *Intern. J. Res. Soc. Sci.* **2016,** *6* (3), 2249–2496.

Mishra, G. K.; Upreti, D. K.; Nayaka, S.; Haridas, B., New Taxa and New Reports of *Phyllopsora* (lichenized Ascomycotina) from India. *Mycotaxon* **2011,** *115* (1), 29–44.

Nair, N. C.; Ghosh, S. R.; Bhargavan, P. Fern-Allies and Ferns of Kerala, India-Part 1. *J. Econ. Taxon. Bot.* **1988,** *12* (1), 171–209.

Nair, N. C.; Ghosh, S. R.; Bhargavan, P. Fern-Allies and Ferns of Kerala: Part 2. *J. Econ. Taxon. Bot.* **1992a,** *16* (2), 251–282.

Nair, N. C.; Ghosh, S. R.; Bhargavan, P., Fern-Allies and Ferns of Kerala: Part 3. *J. Econ. Taxon. Bot.* **1992b,** *16* (3), 501–550.

Nair, N. C.; Ghosh, S. R.; Bhargavan, P. Fern-Allies and Ferns of Kerala: Part *4 J. Econ. Taxon. Bot.* **1994,** *18* (2), 449–476.

Nasser, K. M. M.; Sureshkumar, S. Environmental Correlates of Microalgal Diversity in the High Ranges of the Biodiversity Hotspot, Western Ghats, India. *J. Aquatic Biol. Fisheries* **2012,** *1* (1 and 2), 174–183.

Nayar, B. K.; Geevarghese, K. K. *Fern Flora of Malabar*; Indus Publishing Company: New Delhi, **1993.**

Nayar, B. K.; Geeverghese, K. K. Rediscovery of *Pronephrium thwaitesii* (Thelypteridaceae), a Little Known and Long Lost Fern. *Blumea* **1987,** *32* (1), 213–220.

Nayar, B. K.; Madhusoodanan, P. V.; Molly, M. J. *Nistarika* a New Genus of Polypodiaceae from Silent Valley, South India. *Brit. Fern Gaz.* **1985,** *13,* 33–42.

Nusrat, T., Anjum, A., Ahamed, W. Mononechida (Nematoda) from Silent Valley National Park, India. *Zootaxa* **2013,** *3635* (3), 224–236.

Prajitha, B.; Manju, C. N.; Prakashkumar, R. *Aerobryopsis wallichii* (Bryophyta), a New Record for India. *Geophytology* **2013,** *42* (2), 147–149.

Pushpangadan P.; Sathish Kumar, C. Plant Wealth of Silent Valley. In: *Silent Valley—Whispers of Reason* Manoharan, T. M., Biju, S. D., Nayar, T. S., Easa, P. S., Eds.; Kerala Forest Department and Kerala Forest Research Institute: Thiruvananthapuram, Kerala, **1999;** pp 129–133.

Rao, P. S. N.; Gupta, S. L. Fresh water algae. In: *Floristic Diversity and Conservation Strategies in India*; Mudgal, V., Ajra, P. K., eds.; Botanical Survey of India: Calcutta, **1997.**

Ravikumar, K.; Ved, D. K. *Illustrated Field Guide-100 Red Listed Medicinal Plants of Conservation Concern in Southern India*; Foundation for Revitalization of Local Health Traditions, Bangalore, **2000;** pp 15–330.

Sasidharan, N. Red Listed Threatened Tree Species in Kerala: A Review. In *Conservation and Research Needs of the Rare, Endangered and Threatened (RET) Tree Species in Kerala Part of the Western Ghats. Proc. Workshop.* KFRI, Peechi, **2003;** pp 1–12.

Sasidharan, N.; Anto, P. V. Additions to the Flora of Silent Valley. In *Silent Valley—Whispers of Reason*; Manoharan, T. M., Biju, S. D., Nayar, T. S., Easa, P. S., Eds.; Kerala Forest Department and Kerala Forest Research Institute: Thiruvananthapuram, Kerala, **1999;** pp 135–144.

Sathish Kumar, C. Orchids of Silent Valley, In *Silent Valley—Whispers of Reason*; Manoharan, T. M., Biju, S. D., Nayar, T. S., Easa, P. S., Eds.; Kerala Forest Department and Kerala Forest Research Institute: Thiruvananthapuram, Kerala, **1999;** pp 191–216.

Sathish Kumar, C.; Suresh Kumar, P. C.; Sibi, M.; Anil Kumar, S. A New Species of *Gastrodia* R. Br. (Orchidaceae) from Silent Valley, Kerala, India. *Rheedea* **2008,** *18* (2), 107–110.

Sequiera, S.; Kumar, M. Epiphyte Host Relationship of Macrolichens in the Tropical Wet Evergreen Forests of Silent Valley National Park, Western Ghats, India. *Trop. Ecol.* **2008,** *49* (2), 211.

Shivamurthy, G. R.; Sadanand, K. B. A New Species of *Willisia* Warm. (Podostemaceae) from the Silent Valley, Kerala, India. *Kew Bull.* **1997**, *52*, 243–245.

Singh, J. S.; Singh, S. P.; Saxena, A. K.; Rawat, Y. S. India's Silent Valley and Its Threatened Rain Forest Ecosystem. *Environ. Conserv.* **1984**, *11* (3), 223–224.

Singh, N. P.; Srivastava, R. C. *Gymnosperms of India: Check List*; BSI: Kolkata, **2013**.

Singh, R.; Radha, P. A New Species of *Cycas* from the Malabar Coast, Western Ghats, India. *Brittonia* **2006**, *58* (2), 119–123.

Singh, R.; Radha, P.; Sharma, P. *Cycas annaikalensis* from the Western Ghats of India with a Discussion on the Specific Limits of *Cycas circinalis* L. *Memoirs of the New York Botanical Garden* **2007**, *97*, 542–556.

Sivadasan, M. Araceae of Silent Valley and neighbourhood, In *Silent Valley—Whispers of Reason*; Manoharan, T. M., Biju, S. D., Nayar, T. S., Easa, P. S., Eds.; Kerala Forest Department and Kerala Forest Research Institute: Thiruvananthapuram, Kerala, **1999**; pp. 225–250.

Srivastava, S. C.; Asthana, A. K. A new Folioceros from Silent Valley. Cryptogamie, Bryologie, Liche´nologie **1986**, *7*, 149–153.

Srivastava, S. C.; Sharma, D. A Preliminary Study on the Liverwort and Hornwort Flora of Silent Valley (Kerala). In *Prof D. D. Nautiyal Commemoration Volume Recent Trends in Botanical Researches*; Chauhan, Ed.; **2000**; pp 55–75.

Subramanian, C. V. The Progress and Status of Mycology in India. *Proc. Plant Sci.* **1986**, *96* (5), 379–392.

Sukumaran, S.; Jeeva, S.; Raj, A. D. S. Diversity of Pteridophytes in Miniature Sacred Forests of Kanyakumari District, Southern Western Ghats. *Indian J. Forestry* **2009**, *32* (3), 285–290.

Vajravelu, E. *Flora of Palghat District, Including Silent Valley National Park, Kerala*; Botanical Survey of India: Calcatta, **1990**; p 646.

Velayudhan, K. C.; Amalraj, V. A.; Abraham, Z.; John, K. J.; Nizar, M. A.; Asha, K. I. Wild Crop Genetic Resources of Silent Valley with Special Reference to In Situ Conservation of Piper species. In *Silent Valley—Whispers of Reason*; Manoharan, T. M., Biju, S. D., Nayar, T. S., Easa, P. S., Eds.; Kerala Forest Department and Kerala Forest Research Institute: Thiruvananthapuram, Kerala, **1999**; pp 217–224.

Venkatasubramanian, N.; Sasidharan, K. R.; Gurudev Singh, B.; Mahadevan, N. P. Rare and Threatened plants of Silent Valley, In *Silent Valley—Whispers of Reason*; Manoharan, T. M., Biju, S. D., Nayar, T. S., Easa, P. S., Eds.; Kerala Forest Department and Kerala Forest Research Institute: Thiruvananthapuram, Kerala, **1999**; pp 179–190.

Verma, P. K.; Alam, A.; Srivastava, S. C. Status of Mosses in Nilgiri Hills (Western Ghats), India. *Arch. Bryol.* **2011**, *102*, 1–16

Vohra, J. N. *Pogonatum hexagonum* Mitt., an Endemic Moss to India from Silent Valley, Kerala. *B. S. I. News Lett.* **1981**, *7* (1), 8.

Vohra, J. N.; Roychoudhary, K. N.; Ghosh, K.; Kar, R. K.; Singh, B. D.; Singh, R. K. Observations on the Cryptogamic Flora of Silent Valley. *Botanical Studies on Silent Valley (Cryptogams and Soil Study). Part I.* BSI, Howrah, **1982**; pp 19–30.

Von Lengerke, H. J. Heavy Rainfall Areas in Penisualar India. *Arch, met. Geoph. Biokl. Ser. B.* **1980**, *28*, 115–122.

Von Lengerke, H. J. The Nilgiris, Weather and Climate of a Mountain in South India. *Beitragezur sudasienforeschung.* 32. Weisbaden Steiner, **1970**.

Yabesh, J. E.; Prabhu, S.; Vijayakumar, S. An Ethnobotanical Study of Medicinal Plants Used by Traditional Healers in Silent Valley of Kerala, India. *J. Ethnopharmacol.* **2014**, *154* (3), 774–789.

CHAPTER 12

STATUS OF ICHTHYOFAUNA IN THE WESTERN GHATS

RAMAKRISHNA

Zoological Survey of India, India

ABSTRACT

UNESCO's World Heritage Convention benefits from worldwide recognition as a powerful legal instrument created to protect the world's unique natural heritage of outstanding universal value for humankind. One such area which holds an outstanding array of species is Western Ghats Hotspot. Fishes are the oldest and the largest group of vertebrates, having evolved some 450 million years ago. During the long course of their evolution fishes have kept pace with the development of a variety of aquatic environments both on the surface and in the subterranean waters. The present data analysis on the status of fishes in the WG reveals the presence of 495 species with 307 (62.0%) endemics to the WG and 16 species endemic to the Deccan Peninsula. Considering both the biogeographic zones 323 (65.2%) species are endemic to the south Indian states. Among the states of India, biogeoraphic zone, especially from WG, Karnataka is the richest in ichthyofaunal diversity and the state of Kerala has the most number of endemic species. Recent researches show that the migration of the Malayan stock of fishes to the Peninsula was mainly during Pliocene and Post-Pliocene. Hora (1944) drew attention to the possibility of this element having reached the Peninsula in a series of waves of migration (Hora's Satpura Hypothesis), besides Malayan elements. Recently the exotic species which has reached the aquatic ecosystem of WG has eliminated *Osteobrama belangari*, and the large indigenous *Hypselobarbus periyarensis* from the Periyar Lake, Kerala. Likewise, the introduction of larvicidal fishes *Gambusia affinis* and *Poecilia reticulata* found in many freshwater habitats replaced the indigenous *Aplocheilus parvus* and *Oryzias carnaticus*. Catfishes (family Loricariidae) are newly introduced species from South America, as ornamental fishes in aquarium trade. The north–eastern region and the WG are known for its rich repository of ornamental fish species. Out of the total 225 species reported, about 187 are known for their ornamental value and some of the important species are *Puntius conchonius, P. gelius, P. ticto, P. sophore, Brachydanio rerio, Botia almorhae, Carassius carassius, C. auratus, Badis badis, Barilius barna, B. vagra*. This chapter also includes capture fisheries and a culture fisheries.

Biodiversity Hotspot of the Western Ghats and Sri Lanka. T. Pullaiah, PhD (Ed.)
© 2024 Apple Academic Press, Inc. Co-published with CRC Press (Taylor & Francis)

12.1 INTRODUCTION

The study by Hamilton (1822) was the first on fish fauna in India followed by the most outstanding contribution was that of Dr. Sir Francis Day, a veterinary surgeon and naturalist who traveled extensively in India in the mid-nineteenth century and wrote several scientific papers and monographs such as the *Fishes of Malabar* in 1865. The most colossal work on Indian ichthyology is his work titled *The Fishes of India* in 1878 and *Fauna of British India* in 1889, describing 1418 species along with a separate volume on illustrations with 195 plates. Later studies on Fish fauna of India include Alcock and Jerdon (1848). Jayaram (1981) in his *Handbook of Freshwater Fishes of India* detailed 233 genera with 742 species falling under 64 families and 16 orders, subsequently in 2010, in his book of *The Freshwater Fishes of Indian origin* included 930 freshwater fishes. Menon (1999) listed 446 primary freshwater species from the Indian region alone.

12.2 DIVERSITY AND ENDEMISM

According to Raghavan (2011), Freshwater fishes of Western Ghats (WG) show high degree of diversity and endemism. Daniels (2001) listed 298 species from the WG of which 114 are endemic to WG. In contrast the checklist by Shaji et al. (2000), indicates the distribution of 287 freshwater teleosts from WG, of these, 192 were endemic and 17 exotic/transplanted to the area. Dahanukar *et al.* (2004) lists 288 freshwater fishes, of which 118 are endemic to WG, while Dayal et al. (2014) reports distribution of fish diversity under 48 families, and 143 genera, of which about 151 are endemic and 8 are exotics. Sreekantha (2013) (http:// wgbis.ces.iisc.ernet.in/ biodiversity/sahyadri_ enews/ newsletter/ issue 17) estimated 318 species of which 136 species are endemic to the region. Of this, about 27 species are critically endangered and 55 endangered while 128 are data deficient. Altogether 123 species of the freshwater fishes come under the category of critically endangered, endangered, and vulnerable. Of the 27 critically endangered species 24 are endemic to the region. Similarly, of the 55 endangered species, 37 are endemic. Yet 49 endemic species are data deficient. Molur et al. (2011) recorded 290 species of freshwater fishes in WG, belonging to 11 orders, 33 families, and 106 genera. The WG also has a rich endemic fish fauna of 189 species. Twelve genera, *Betadevario, Dayella, Horabagrus, Horalabiosa, Hypselobarbus, Indoreonectes, Lepidopygopsis, Longischistura, Mesonoemacheilus, Parapsilorhynchus, Rohtee,* and *Travancoria*, are endemic to the WG. According to Raghavan (2019), WG mountain ranges have unique freshwater fish fauna of more than 300 species, of which ~65% are endemic. Dahanukar and Raghavan (2013) recognized 320 species of freshwater fishes (including some secondary freshwater species, which can also live in brackish water and marine habitats. Jadhav (2020) documented an updated checklist with 42 fish families of 15 orders, encompassing 397 species belonging to 121 genera. This is about 1.1% of the total fish species (35,588), and about 2.3 % of total genera (5210).

The present data analysis on the status of fishes in the WG reveals the presence of 495 species with 307/495 (62.02%) endemic to WG and fishes Endemic to Deccan Peninsula are 16 and if these two taken together results in 65.25% (307 + 16 = 323) of the southern Indian states.

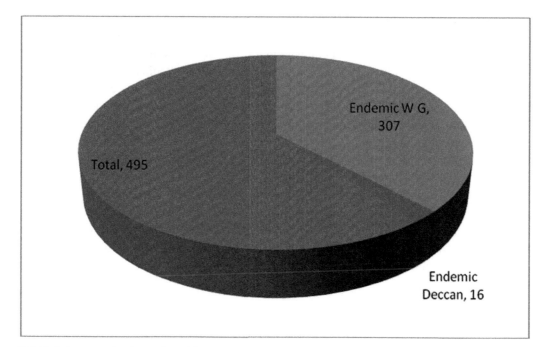

FIGURE 12.1   Endemic species of fishes in Western Ghats.

Among the States of India, Karnataka is the richest in ichthyofaunal diversity and the state of Kerala has the most number of endemic species. Among the various river systems in which the ichthyofaunal composition is known, it is observed that Ganges harbors the greatest ichthyofaunal diversity and River Cauvery the greatest number of endemic species. Of the different major ecoregions of India, WG is the richest in ichthyological diversity and endemicity.

12.3   DISTRIBUTION OF FISHES IN WESTERN GHATS

Dr. S. L. Hora, proposed his *Satpura hypothesis* for the distribution of the Malayan fauna and flora to Peninsular India (Hora, 1937). His hypothesis was based on the following fundamental conceptions, viz., (1) continuity of the Vindhya-Satpura ranges with the Assam Himalayas in the east and the WG in the west. (2) 5000 to 6000 ft elevation of the Vindhya Satpura ranges and the northern section of the WG. (3) continuity of an ecological belt of mountains with rainfall of about 254 cm or above and consequently tropical evergreen forests between Assam Himalayas and the mountains of Ceylon via the Vindhya-Satpura Trend and the WG. (4) dispersal of the fauna from east to west and the consequent changes in topography necessary thereof and the Garo-Rajmahal Gap is a very recent feature of the physiography of India. According to this hypothesis, India began to receive the Malayan forms only during the Pliocene Siwalik period (Menon, 1973). The highly specialized torrential fauna developed much later, probably in the Pleistocene.

Recent researches show that the migration of the Malayan stock of fishes to the Peninsula was mainly during Pliocene and Post-Pliocene. Hora (1944) drew attention to the possibility of this element having reached the Peninsula in a series of waves of migration. Bhimachar (1945) postulated about four waves of migration of fishes to the Peninsula and Ceylon from the region of the Assam Himalayas. On the basis of the distribution of certain freshwater fishes, they divided the WG into three divisions, namely, a northern division comprising the Deccan Trap area from the Tapti river down to 16°N. Latitude about the level of Goa; a central division, extending from 16°N. Latitude southwards and including the Malnad parts of Karnataka State, Coorg, Wynaad, and parts of South Canara District and the Nilgiris; and a southern division, comprising the Annamalai, Palni, and Cardomom Hills as shown in the figure. In the WG, freshwater fish distribution shows that the southern region is rich in species than the central and northern regions. The southern region is isolated from the remaining zones by the Palghat gap (11°N latitude). Bhimachar (1945) and Silas (1951) have discussed the importance of Palghat gap as a barrier for the distribution of hill stream fishes in the WG as it hampers their migration and thus the richness.

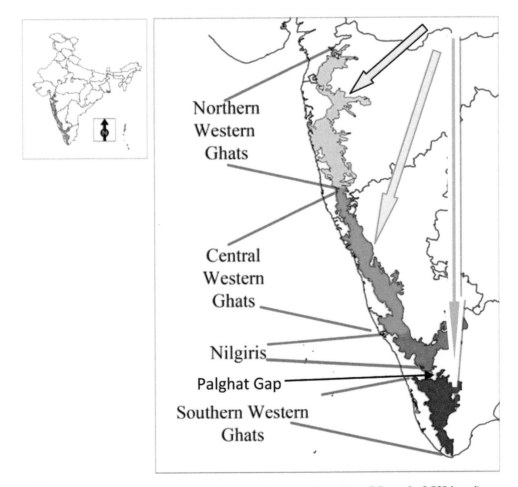

FIGURE 12.2   Possible route of migration of fishes to Western Ghats (Map: P Deepak, QGIS based)

The "*Himalayan Glaciation*" and trans migration of aquatic fauna, according to Hora (1949) to the Peninsular India was during the shorter interglacial period experienced in the southerly latitudes as a succession of cold "Pluvial Periods" of Pleistocene which were characterized by low temperature, increased humidity, and a fall in sea-level. As stated by Hora (1949), the physiographic changes facilitated the migration of even torrential fishes from the region of the Assam Himalayas to the Peninsula across the present day Garo-Rajmahal Gap. The succeeding phases of aridity and increased desiccation, after the glacial and interglacial period resulted in a rise in the sea-level, consequently not only isolated certain forms, but temporarily would have checked dispersal of freshwater forms from the Assam Himalayas to the Peninsula.

Menon (1951) investigating the migratory route of fishes to Peninsular India, namely, the possibility of the Eastern Ghats acting as an alternative route of migration, followed after a detailed study of the fish fauna of the Eastern Ghats and the Orissa Hills, that there were no typical representatives of the so-called Malayan element there and hence the migration of fishes through Eastern Ghats was ruled out. However, he adds by way of reconstructing the past drainage system of the northern part of the Peninsula, has shown that the Narmada-Tapti which flowed to the west as a single river, had Mahanadi and Godavari as its tributaries till comparatively recent times in geological history. The tilting of the Peninsula during the Pleistocene reversed this drainage system to its present-day pattern, thus affecting the fish dispersal.

Genera and species with the so-called Malayan affinities restricted to Peninsular India are: family Cyprinidae; *Osteochilus* Gunther *Osteochilus (Osteochilichthys) thomassi* (Day) O. *(Osteochilichthys) nashii* (Day) and *O. (Kantaka) brevidorsalia* (Day); *Schismatorhynchus* Bleeker *S. (Nukta) nukta* (Sykes); Rohtee Sykes, *R. (Bohtee) ogilbii* Sykes; *Puntius sarana & P. ticto*; *Tor mosal* Hamilton; *Acrossocheilus hexagonolepis* (McClelland); *Osteobrama cotio* Hamilton; *Labeo dero; Crossocheilus latius; Garra gotyla* (Gray); *Thymnichthys sandkhol* (Sykes). Other fish species with the Malayan affinities restricted to Peninsular India are *Psilorhynchus sucatio; Homatoptera montana; Bhavania australis; Travancoria jonesi; Lepidopygopsis typus; Silonia childreni; Silonia silondia; Pangasius pangasius; Eutroplichthys goongwaree; Neotropius khavalchor; Batasio travancoria; Pseudobagrus brachysoma; Gagata itchkeea; Gagata cenia; Gagata gagata; Glyptothorax horai; G. anandalei; Amblyceps mangois; Silurus goae; S. berdrnoeri* var. *wynaadensis; S. cochinchinensis.* Many of these species such as *Puntius, Tor, Garra, Nemachilus,* and *Clarias* are found in Ceylon and they extend as far west as Africa. It is believed that the temporary land connections with the mainland during the "Pluvial Periods," when the sea-level fell, would have enabled the migration of the Peninsular species to Ceylon as well as to Africa. *Homaloptera* has not so far been recorded from Ceylon, so with *Bhavania* and *Travancoria*, which evolved at a later date. This divergence would have taken place consequent to the tilting of the Peninsula which dated probably not later than the earlier part of the Middle Pleistocene. It is believed that the migration of the trans Himalayan genus *Silurus* in the southern Peninsula is due to the continuity of Narbada-Tapti River of the early Pleistocene. Similarly, the genera *Pseudobagrus, Osteochilus, Schismatorhynchus, Rohtee, Labeo (Morulius), Batasio, Silonia, Neotropius,* and *Tetraodon* from Southern China to the Peninsular India probably during the later part of the Lower Pleistocene. The "Pre-tilt forms" have generally diverged considerably and the earliest "Post-tilt forms" would have been species like, *Balitora brucei, Silurus berdmorei, Thynnichthys*, etc. (Silas, 1952).

It is clearly evident that the rich and diverse ichthyofauna in Peninsular India including WG is due to the geographical isolation as a factor of prime importance in the process of evolutionary divergence with climatic fluctuations during the Pleistocene, and Changes in the physiography of the Peninsula during the Pleistocene and consequent effects. Further, impetus to the rate of speciation was added when as a result of scrap faulting along the West Coast, the tilting of the Peninsula that took place. This tilting not only reversed the drainage system, but rejuvenated the streams of the WG. The disturbances that took place in the geologic history is substantiated by the best known waterfalls of India such as Gerusoppa falls (850 feet), Sivasamudram falls of the Cauvery in Mysore (300 feet), Gokak falls (180 feet); Yenna falls of the Mahableshwar hills descend 600 feet below in one leap and the Paikara in the Nilgiri Hills descend less steeply in a series of five cataracts over the gneissic precipice, besides many smaller ones.

12.4 INTRODUCTION OF EXOTICS

Exotics species such as *Oreochromis mossambicus, Cyprinus carpio* have no doubt contributed much to the country's fish production, but they have also had a negative impact on the indigenous fishes in the form of competition, predation, and environmental modification and so on. The unprecedented spread and colonization of the hardy *O. mossambica* has been posing severe threat to our indigenous fishes competing with the latter for space and food and quickly replacing them. The popular aquaculture and angling species *C. carpio* causes harm to other species by stirring up bottom sediments and creating turbid conditions in impoundments. In the Kashmir valley this species has replaced the indigenous Schizothoracine fishes and in the Loktak Lake of Manipur it has eliminated *Osteobrama belangari*, and the large indigenous *Hypselobarbus periyarensis* from the Periyar Lake, Kerala (Menon, 1989). Likewise, the introduction of larvicidal fishes *Gambusia affinis* and *Poecilia reticulata* found in many freshwater habitats replaced the indigenous *Aplocheilus parvus* and *Oryzias carnaticus*. Also the introduction of Gangetic carps into the peninsular rivers has affected the indigenous carps of the peninsula. These introductions generally cause zoogeographical pollution, loss of genetic identity of local populations, a high level of hybridization, and extinction or reduction of local communities of endemic species (Rema Devi, personal communication). Parasites and diseases have also been introduced with alien species. Introduction of species in the upper reaches will have more deleterious effect since the upper reaches harbor highly specialized and endemic species. A recent disturbing observation is the introduction and widespread culture of the African catfish *Clarias gariepinus*. This species has literally wiped out other species from temple tanks and ponds where they have been introduced. Species introduced for ornamental purposes are now well established in the wild as the cichlid the flowerhorn or *Cichlasoma* species and the scaled Loricarid catfish species.

Unlike the Grass carp and the Silver carp, which happens to be herbivores and plank-tivorous, the Common carp, an omnivore, utilizes different food niches of an aquatic ecosystem depriving the endemic fish species of their food. The species has a habit of browsing for food in the shallower areas of the impoundments which also are the sites of breeding and recruitment to the scores of indigenous fishes. On account of its omnivorous feeding habits, it devors the eggs, fry, etc., of the indigenous fishes reducing their population to a great extent. The fish also attains early maturity, in just about 3 months, and with a prolonged breeding season, increases its population with retarded growth on account of overpopulation and to compete for food and shelter with the fishes endemic to the system. This biological process of the species has directly affected the very existence of indigenous fish fauna in majority of the waters (e.g., Krishnaraja Sagar reservoir) in the State of Karnataka. Introduction of Tilapia (*Tilapia microlepis, Oreochromis mosambicus* as well as *Clarias garinpinneus* has created a sort of total destructive scene in all the tanks, rivers, and reservoirs in the State. The important point to be taken note of in this species is its prolific breeding habit resulting in over-crowding affecting the growth and population of fish indigenous to the biotope.

Catfishes (family Loricariidae) are now introduced, from South America, as ornamental fishes in aquarium trade. The armored catfishes such as *Pterygophlyctis multiradiatus, Pterygoplichthys disjunctivus, Pterygoplichthys pardalis* have been reported in India by several authors. In South India, *P. multiradiatus* has been reported from Kunnamkulam, Kerala (Ajithkumar et al., 1998) and Vylathur and the Chackai Canal, Kerala (Daniels, 2006; Krishnakumar et al., 2009).

12.5 ORNAMENTAL FISHES

Kerala is fast becoming popular for ornamental fish culture as many villagers in the districts of Thiruvanthapuram, Ernakulum, Thrissur, Allapuzha, and Kottayam have set up backyard ornamental fish production units. Similarly, Tamil Nadu (TN) is the second largest ornamental fish producer in the country after West Bengal. The village of Kolathur near Chennai is the epicenter of ornamental fish production of large varieties. The similar trend is being followed in Madurai, another major business city in TN.

The north–eastern region and the WG are known for its rich repository of ornamental fish species. Out of the total 225 species reported, about 187 are known for their ornamental value and some of the important species are *Puntius conchonius, P. gelius, P. ticto, P. sophore, Brachydanio rerio, Botia almorhae, Carassius carassius, C. auratus, Badis badis, Barilius barna, B. vagr*a. Ninety species of fishes were studied from the different river systems of the WG by Mercy et al. (2007). Based on the results, 85 of them were recommended as ornamental fishes. Most of them belong to the categories of barbs, loaches, danios, killifishes, and hill trouts. Of these, species belonging to *Puntius filamentosus, P. pookodensis, P. melanostigma, P. melanampyx, P. denisonii, P. fasciatus, P. arulius, P. narayani, P. sahyadriensis, P. punctatus, P. setnai, P. fraseri Garra mullya, Danio malabaricus, Chela fasciata, Nemacheilus triangularis, Nemacheilus semiarmatus, Pristolepis marginata, Horabagrus brachysoma,* all species of *Barilius, Danio malabaricus, D. neilgiriensis, Chela dadyburjori, Botia striata, B. macrolineata, Nangra ichthea, Tetraodon travancoricus, Pristolepis marginata, Scatophagus argus, Horaichthys setnai,* species of *Nemacheilus* and *Etroplus canarensis* were recommended for commercial production (Ponniah and Gopalakrishna, 2001). Similar exercises were made by Anandhi and Sharath (2013) from small streams located in Karnataka (Gundia river) and found 17 species ideal for ornamental purposes.

Freshwater ornamental fishes from the Western Ghats

Puntius fasciatus (Jerdon) or *P. melanampyx*. .... Sreams of Cauvery, Canara

Puntius filamentosus (Valenciennes) .... Rivers and streams in Karnataka

Puntius narayani (Hora) .... Coorg; Cauvery River (Karnataka)

Labeo potail (Sykes) .... Kabini river system

Barilius canarensis (Jerdon) .... Rivers and streams in Karnataka

Danio malabaricus (Jerdon) .... Rivers and streams in Karnataka

Esomus barbatus (Jerdon) .... Rivers and streams in Karnataka

Rasbora caverii Jerdon .... Cauvery, Karnataka

Garra mcclellandi (Jerdon) .... Cauvery river

Garra gotyla stenorhynchus (Jerdon) .... Cauvery and Krishna drainages

Garra bicornuta Rao .... Rivers and streams in Karnataka

Balitora shimogensis Menon *et al.*, ... River systems in Shimoga

Longischistura bhimachari (Hora) .... Thunga River, Shimoga, Karnataka

Schistura kodaguensis (Menon) .... Kodagu, Karnataka

Oreonectes evezardi (Day) .... Kodagu, Karnataka

Mesonoemacheilus petrubanarescui (Menon) .... Netravathi River, Dharmasthala, Karnataka

Mesonoemacheilus rueppelli (Sykes) .... Rivers and streams in Karnataka

Schistura semiarmatus (Day) .... Cauvery basin

Nemachilus poonaensis Menon .... Rivers and streams in Karnataka

Noemacheilus denisoni mukambbikaensis Menon .... Mookambika river

Macropodus cupanus dayi (Kohler) .... Rivers and streams in Karnataka

Parambassis (Chanda) thomassi(Day) .... Rivers and streams in Karnataka

Etroplus canarensis Day .... South Canara Karnataka

Etroplus maculatus (Bloch) .... South Canara Karnataka

Sicyopterus griseus (Day) .... South Canara Karnataka

Two indigenous ornamental fish resources of WG, namely, *Sahyadria denisonii* (Kerala queen) and *Dawkinsia filamentosa* (Filamentous barb) of Kerala origin are cultured among them *Sahyadria denisonii* is the most popular and highly priced freshwater ornamental fish, accounted for almost 60–65% of India's total ornamental fish exports. *Dawkinsia filamentosa* is yet another endemic fish to Kerala, Karnataka, and TN river streams and has a very good demand in the ornamental fish trade (Mercy et al., 2007).

12.6 OVER–EXPLOITATION AND DESTRUCTIVE COLLECTION PRACTICES

In in-land water bodies, the collection of fish in numbers more than what can be recruited will ultimately destroy the whole population. In order to boost fish production, exotic Grass carp such as *Ctenopharyngodon idella*, Silver carp–*Hypophthalmichthys molitrix*, and common carp such as *Cyprinus carpio* have been extensively stocked in almost all the in-land water areas of all the States in WG. Over-fishing has also resulted in the decline of several catfishes like *Bagarius bagarius, Silonia childreni, Pangasius pangasius, Sperata aor*, and *S. seenghala* in the lower reaches of the river Godavari. Indiscriminate use of poison to collect fish from pools and refugial pockets where fish take shelter when rivers dry up, and dynamiting to collect fish in large numbers, will result in complete elimination of the fish species, since both juveniles and breeding fish and other non-target species all fall prey to such destructive methods. In many lakes and tanks, plant derivatives as that of unriped fruits of *Randia dumetorum*, bark, seed, and root of *Barringtonia acutangula* and latex from *Euphoria tirucalli* are being extensively used to collect fish in such pools and congregated locations. Besides above, indiscriminate exploitation of certain ornamental fish species which have export potential and endemic to certain rivers; viz., Loaches in the river Tunga near Gajanur, Shimoga (WG, Karnataka) have been exploited, resulting in the depletion and even disappearance of indigenous fish species.

12.7 SPORTS FISHERIES IN THE WESTERN GHATS

British anglers have introduced first time brown trout (1860s) and rainbow trout (1909) in streams and rivers of the WG (Sehgal, 1999). Gupta et al. (2015) documented a number of species which is usually targeted by the anglers in India which include *Tor* sp., *T. putitora, T. khudree, Neolissochilus hexagonolepis*, and *Gibelion catla*. Occasionally caught other Indian native fish species include *Channa diplogramma, C. marulius, C. punctata, C. striata, Wallago attu, Hemibagrus maydelli, Bagarius bagarius, Labeo rohita, Cirrhinus cirrhosus, Hypophthalmichthys molitrix, Cyprinus carpio, Ctenopharyngodon idella*, and *Aristichthys nobilis*.

 Cyprinus carpio, Carrassius carassius, Tinca tinca, Orechromis mossambicus, and *Onchrhynchus mykiss* were introduced in the uplands of WG as sports and food fishes. Other native fishes of the major genera *Barilius, Puntius, Osteochilus, Labeo, Garra, Nemacheilus, Channa, Mystus*, and *Glyptothorax* dominate in the fisheries of this region (FAO, Technical Paper No. 385). *Tor* species also generally have high religious and cultural significance throughout S and SE Asia as well as some of these species attain over 50 kg, are considered as excellent sports fisheries. From among then *Tor* exhibits as many as 16 recognized species of which three species are now assessed as "Near Threatened," one "Vulnerable," three "Endangered," and one "Critically Endangered." However, eight species remain "Data Deficient" (Pinder et al., 2019).

12.8   FRESHWATER FISH CULTURE

Fish culture in the Indian subcontinent is perhaps as old as 321 and 300 B.C. with literature available on flourishing fish culture during this time. The end of the 19th century saw warm water fish culture in ponds involving collection and transport of carp spawn from rivers, and pond management, confined originally to Bengal, Bihar, and Orissa, subsequently, spreading to other states. The importance of fish culture as a source of food production was realized resulting in fish culture activities throughout the country. The beginning of the 20th century marked the introduction of several exotic species in Indian waters (Jhingran, 1983). The unplanned introduction today of several fast growing opportunistic species has now replaced our indigenous species. Further, the introduction of ornamental species has also been a cause in the decline of our own fauna.

Some important freshwater culture fishes	Exotic food fishes
Catla catla (Hamilton)	*Aristichthys nobilis* (Richardson)
Labeo rohita (Hamilton)	*Hypophthalmichthys molitrix* (Valenciennes)
Labeo calbasu (Hamilton)	*Cyprinus carpio* (Linnaeus)
Labeo fimbriatus (Bloch)	*Ctenopharyngodon idella* (Valenciennes)
Labeo bata (Hamilton)	*Gambusia affinis* (Baird & Girard)
Cirrhinus mrigala (Hamilton)	*Gambusia holbrooki* Girard
Cirrhinus reba (Hamilton)	*Tinca tinca* (Linnaeus)
Etroplus suratensis (Bloch)	*Carassius carassius* (Linnaeus)
Anabas testudineus (Bloch)	*Carassius auratus auratus* (Linnaeus)
Wallago attu (Bloch)	*Osphronemus goramy* Lacepede
Aorichthys seenghala (Sykes)	*Oreochromis mossambicus* (Peters)
Channa striatus (Bloch)	*Oreochromis niloticus niloticus*
Clarias batrachus (Linnaeus)	*Oncorhynchus mykiss* (Walbaum)
Heteropneustes fossilis (Bloch)	*Poecilia reticulata* Schneider
Larvicidal fishes	*Salvelinus fontinalis* (Mitchell)
Poecilia reticulata (Peters)	*Salmo trutta fario* (Linnaeus)
Gambusia affinis (Baird and Girard)	*Salmo trutta trutta* (Linnaeus)
	Xiphophorus hellerii (Heckel)

   The breeding and propagation of the mahseer was a success at the Tata Electric Company's Wulvhan Lake at Lonavala, Maharashtra, and the introduced rainbow trout in the cold waters of the Nilgiris. Freshwater Aquaculture proved successful with the spawn collection from natural spawning grounds by stripping, to induce breeding of Indian major and minor carps, catfishes, and other species including the minnow *Esomus danricus*. Induce-bred carps include *Labeo rohita, L. bata, Cirrhinus mrigala, C. reba,* and *Puntius sarana* (Silas, 1953). From carp pituitary extracts through human chorionic gonadotropin (HCG), isolation, characterization, and purification of fish gonadotropin (GtH) and fish gonodotropin releasing hormone (GnRH) have been used for fish breeding. Today, ovaprim prepared from salmon gonadotropin releasing hormone and domperidone are used for carp and catfish breeding. Ovaprim has been a boon for aquaculturists with standardized dosages for selected species increasing yield from 490 million fish fry in 1973–1974 to over 25,000 million fry today. Research is underway in the field of induced breeding using various other hormones and study of pheromones of fish species. Research projects on food and feeding habits of fishes, parasites and diseases and control measures, amenability to polyculture and composite species culture, sewage tolerant, and organic waste recycling species and integrated fish farming (crop-livestock-fish-prawns) in paddy field species and monosex culture have all yielded excellent results and boosted aquaculture practices in India (Silas, 2003). There are several colorful freshwater fishes which are used in ornamental trade.

12.9 THREAT

According to Jadhav (2020) of the 397 species of fishes recorded from WG, 288 species are found endemic, 96 (34%) are under threatened or near threatened categories, 124 (43%) are under data deficient or not evaluated categories, and 68 (23%) are under least concern categories. However, the present updated checklist of WG fishes clearly indicate 16 species are critically endangered, 65 species are endangered, and 35 are vulnerable thus taking the threatened species to 116 accounting 23.43%, while the near threatened 12 (2.42%); Least Concern 220 species (44.44%), and data deficient (28) and not evaluated 87, together accounting 115 species (23.23%) as detailed below.

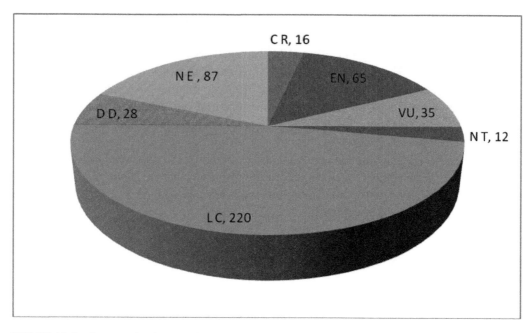

FIGURE 12.3   Status under threatened (redlist) category.

12.10   FISH SANCTUARIES IN WESTERN GHATS

1. Fish sanctuary at Sringeri maintained by Shri Sharada Peeta, Sringeri Mutt wherein the endangered Sahyadri Mahseer fish, *Tor khudree* is protected in a small river stretch of the river Tunga.

2. Chippalgudde Matsya Dhama, Tunga River, Teerthahalli, maintained by Siddivinayaka Trust on the banks of the river Tunga. The 4 km stretch protects more than 27 species of fish and significant species include the endangered Mahseer (*Tor kudree*) and *Puntius pulchellus*.

3. Shishisla Matsya Teertha, Kapila/Kumardhara River, Dakshina Kannada is probably the first protected area for conservation and protection in the country established in 1930. The conservation area supports nearly 18 species with a predominance of *Tor kudree*.

4. Bachananayakagundi and Dharmasthala at Dakshina Kannada.

5. Jammatagi Agrahara at Chikmagalore.

6. Thingale on Seethanadi, Thingale Temple Sanctuary set up by a Bhajani Mandal on the banks of Seethanadi, near Agumbe. Besides above, the wildlife sanctuaries and national parks (26 in total) are protected under Law in Karnataka which harbor many wetlands/water bodies of importance including the conservation of fishes.

7. Ramanathapura at Hassan District: Ramanathapura fish sanctuary situated in river Cauvery near Rameshwara Temple in Ramanathapura, Arakalgudu taluk, Hassan District also known as Vanhi Pushkarani. It was declared as sanctuary as per G.O. No. G. 1627/Ft. 296.35.2 of 11/17.06.1935 by the Government of Mysore under the provisions of Section 5 (1) of Mysore Game and Fish Preservation Regn. II of 1901 which states that fishing for one furlong on either side of the pond known as "Vanhi Pushkarani" in the Cauvery river by the side of the Rameshwara temple at Ramanathapura, Arakalgud taluk be prohibited." The sanctuary lies between $12°37'$ latitude and $76°5'$ longitude. Presently, the sanctuary has famous game-fish of India–Mahseer–*Tor khudree* and the Katli/Chocolate Mahseer-*Neolissoclilus hexagonolepis*. The latter species is observed to be in large numbers. Aghanashini River and Kabani River down to Panamaran are the potential sites for fish conservation.

8. Tekpowale fish sanctuarya near Mangaon village near Pune, to conserve the Mahseer species.

9. Walen Kondh fish sanctuary on Konkan side of the WG to conserve Deccan Mahseer.

10. Tilase fish sanctuarya sacred pool on the Vaitarna river in Thane district where a Mahseer fish is conserved. The sanctuary is associated with an ancient temple of Lord Shiva, Tilaseshwar temple.

12.11   GOVERNMENTAL AND NON-GOVERNMENTAL CONSERVATION INITIATIVES

India is committed to several international conventions, Code of conduct and guidelines toward conserving biodiversity resources. Indian Fisheries Act of 1897 is a landmark conservation strategy for those indulging in killing of fish by the use of poison and explosives. The delegated powers enabled the erstwhile Provinces (States) the responsibility of development, management, and conservation of fisheries in the inland and the territorial waters of the respective states. Under the 1901 Mysore Game and Fish Act (Established by Maharaja of Mysore to protect the game fish), the Department of Fisheries had limited power to enforce conservation measures. It is for this reason, the Karnataka Inland Fisheries Conservation Development Regulation Bill, 1996, revised as Karnataka Act No. 27 of 2003. This provides a uniform legislation to allow protection of Indigenous species in the recently integrated districts of Karnataka.

These regulations enabled the fisheries department to manage in the best possible ways by

- Declare closed-season.
- Prohibit harvesting of fish in any area for a period of 5 years.
- Prohibit poisoning or use of explosives in any stream, river, or lake.
- Regulate by notification, sale of fish within any specified local area.
- Make provisions to impose penalty to persons who contravenes conditions laid-down in licenses' granted.
- Empower Court of Magistrate to confiscate implements used for capture of fish and licensees granted.
- Empower officials delegated with powers to arrest without warrant any person who commits offence punishable under this regulation.

Though there are several Government Acts which impose restrictions on indiscriminate fish catching, scant respect is being given to the regulations and restrictions by the fishing community. Unless it dawns on each and every group involved in fishing that conservation is very essential for sustained catches, it is impossible to implement any legal measures. Environment awareness programmes could help in this regard. Under WG Development Scheme of Government of India to conserve the famous game fish of India–Mahseer–*Tor khudree*, a hatchery at Harangi near Kushalnagar in Coorg district at a total cost of Rs. 2.00 Crores was established. Efforts are being made to produce young ones of the species in captivity. Several non-governmental organisations, the industrial and corporate sectors have had both positive and negative impact on the ichthyofauna.

The allocable Special Central Assistance (SCA) for the Hill Area Development Programme (HADP) was being distributed between WG Development Programme and HADP in the proportion of 16% and 84%, respectively. With the declaration of WG as World Heritage site on July 1, 2012 with Kerala leads with 20 sites being inscribed in the heritage list followed by Karnataka with 10, TN five, and Maharashtra four thus 39 sites are found important in the Ghats. In order to identify eocosensitive areas in the WG, an Ecology Expert Panel was constituted by the Ministry (MOEF and CC) under the leader-ship of Prof. Madav Gadgil whose recommendation includes stoppage of hydroelectric and mining projects, which was opposed by several state governments.

12.12   RESOLVING TAXONOMIC AMBIGUITIES

The taxonomic status of several fish species is still unresolved. There remains to be done many revisionary studies for clearly defining a species/different stocks/subspecies and species complex. At the NBFGR, different stocks of *Catla catla* and *Labeo rohita* from different rivers of the Gangetic plains and northeast India have been identified using molecular and allozyme markers and in the endemic yellow catfish *Horabagrus brachysoma* from Chalakkudy and Meenachil rivers in the WG. Species which have been synonymized earlier are now known to be valid species as in the case of *Puntius, Tor, Aplocheilus, Oryzias,* and *Channa*. Taxonomic ambiguity across the *Tor* genus was related to the large number of chromosomes as *Tor* are tetraploid (Arai, 2011) and possess 100 diploid chromosomes (Mani et al., 2009). As a group, mahseer exhibits considerable phenotypic plasticity, including intra-specific morphological variation, trophic polymorphism, and sexual dimorphism, making precise, morphologically based identifications extremely difficult (Walton et al., 2017). Several such studies are going on and this is sure to change the concept of species, diversity of species, and identification of associated habitats for conservation measures. In the workshop conducted during 2001 (Ponnaiah and Gopalakrishnan, 2001), resolved that the *Tor khudree* and *T. mussullah* are two distinct species, based on morphometric counts and indicated that the latter exhibited a patchy distribution. Similarly, the validity of *Channa micropeltes, Channa leucopunctatus, Labeo nigriscens,* and *Macropodus cupanus dayi* as distinct species. The group also reported that *Puntius dobsoni* is a synonym of *P. jerdoni* while *P. pulchellus* is a valid species and *Osteochilithys* and *Kantaka* as separate genera.

One particular group of interest is the common freshwater genus *Puntius* of Hamilton (1822) which until recently included several species groups, have now been assigned to different genera viz. *Barbodes, Hypselobarbus, Gonoproktopterus, Tor, Neolissocheilus,* and *Puntius* based on biometric and osteological studies. Taxonomic ambiguity in the genus *Puntius* is yet to be resolved though several studies on systematic relationships have been done within the subfamily Barbinae (Gupta et al., 2018). Rainboth (1996) classified *Puntius* into another three genera, *Systomus, Barbodes,* and *Hypsibarbus*. Champasri et al. (2007) and Rajasekaran and Sivakumar (2014) stated that *Puntius* should not be split into three genera as Rainboth (1996) as it failed to provide distinct special characters to differentiate three new genera within the *Puntius* genus. Furthermore, Pethiyagoda et al. (2012) reported five well-supported clades as distinct genera within South Asian *Puntius* namely *Pethia, Dawkinsia, Dravidia, Systomus,* and *Puntius*.

During the last decade, several changes have been befallen in the order Siluriformes and Cypriniformes, particularly in the family's Siluridae and Sisoridae, and remarkably in families Cyprinidae with new generic taxa, like *Waikhomia (Waikhomia hira), Sahyadria (Sahyadria denisonii), Pethia (Pethia striata, Pethia sahit, Pethia sanjaymoluri, Pethia longicauda),* and creation of new family Nemacheilidae to accommodate Nemacheilian loaches, erection of new family Kryptoglanidae, and new genus *Aenigmachanna* under new family Aenigmachannidae (*Aenigmachanna gollum*).

12.13   CONSERVATION MEASURES

The conservation measures for protection of Ichthyofaunal diversity and sustainable use are summarized below

- Nature reserves
- Fish sanctuaries
  - Deep-pools
  - Temple tanks and fort moats
  - River stretches below dams as sanctuaries
  - Upper reaches of rivers as sanctuaries
- Captive breeding
- Habitat protection
- Abatement of pollution
- Ban of expansion of plantations in the hill ranges
- Monitoring stations
- Augmenting stocks in rivers
- Ecological and life history studies
- Wetlands as protected areas
- Ban on introduction of exotic species for ornamental or culture purposes

KEYWORDS

- **Western Ghats**
- **ichthyofauna**
- **fishes**
- **endemic**
- **teleosts**
- **aquarium**

REFERENCES

Ajithkumar, C. R.; Biju, C. R.; Thomas, R. *Plecostomus multiradiatus* - An Armoured Catfish from Freshwater Ponds Near Kunnamkulam, Kerala and its Possible Impact on Indigenous Fishes. *LAK* (*Limnological Association of Kerala) News*, **1998,** pp 1–2.

Anandhi, D. U.; Sharath, Y. G. Ornamental Fish Fauna of Adda Hole: Kabbinale Forest Range, Southern Western Ghats, Karnataka, India. *Intern. Res. J. Biol. Sci.* **2013,** *2* (11), 60–64.

Arai, R. In *Fish Karyotypes: A Check List*; Springer: Tokyo, **2011.**

Bhimachar, B. S. Zoogeographical Divisions of the Western Ghats as Evidenced by the Distribution of Hill Stream Fishes. *Curr. Sci.* **1945,** *14*, 12–16.

Champasri, T.; Rapley, R.; Duangjinda, M.; Suksri, A. Morphological Identification in Fish of the Genus *Puntius* Hamilton 1822 (Cypriniformes: Cyprinidae) of some Wetlands in Northeast Thailand. *Pak. J. Biol. Sci.* **2007,** *10*, 4383–4390.

Dahanukar, N.; Raghavan, R. R. *Freshwater Fishes of the Western Ghats: Checklist* v1.0 *Min | #01 |* August **2013** (file:///C:/Users/Downloads/Dahanukar Raghavan 2013. Min % 20(4).

Dahanukar, N.; Raut, R.; Bhat, A. Distribution, Endemism and Threat Status of Freshwater Fishes in the Western Ghats of India. *J. Biogeogr.* **2004,** *31* (1), 123–136.

Daniels, R. J. Endemic Fishes of the Western Ghats and the Satpura Hypothesis. *Curr. Sci.* **2001,** *81*, 240–244.

Daniels, R. J. R. Introduced Fishes: A Potential Threat to the Native Freshwater Fishes of Peninsular India. *J. Bombay Nat. Hist. Soc.* **2006,** *103*, 346–348.

Day, F. In *The Fishes of Malabar*; Bernard Quaritch: London, **1865**.

Day, F. In *The Fishes of India*; Being a Natural History of the Fishes Known to Inhabit the Seas and Fresh Waters of India, Burma, and Ceylon. B. Quaritch: London,**1878**.

Day, F. In *The Fauna of British India, Including Ceylon and Burma*; Publ. Taylor and Francis: London, 1889.

Dayal, R.; Singh, S. P.; Sarkar, U. K.; Pandey, A. K.; Pathak, A. K.; Chaturvedi, R. Fish Biodiversity of Western Ghats Region of India: A review. *J. Exp. Zool. India* **2014,** *17* (2), 377–399.

FishBase. Updated from Fishbase.org Accessed from Time to Time During Compilation, April–May **2021**.

Gupta, N.; Bower, S. D.; Raghavan, R.; Danylchuk, A. J.; Cooke, S. J. Status of Recreational Fisheries in India: Development, Issues, and Opportunities. *Rev. Fish. Sci. Aquac.* **2015,** *23* (3), 291–301.

Gupta, D.; Dwivedi, A. K.; Tripathi, M. Taxonomic Validation of Five Fish Species of Subfamily Barbinae from the Ganga River System of Northern India using Traditional and Truss Analyses. *PLoS ONE* **2018,** *13* (10), e0206031.

Hamilton, F. *An Account of the Fishes Found in the River Ganges and its Branches.* Hurst, Robinson, and Co.: Edinburgh, **1822**.

Hora, S. L. Distribution of Himalayan Fishes and its Bearing on Certain Palaeo- Geographical Problems. *Rec. Indian Mus.* **1937,** *39*, 251–259.

Hora, S. L. The Game Fishes of India III: The Mahseer or the Large Scaled Barbels of India. 1. The Putitor Mahseer *Barbus* (*Tor*) *putitora* Hamilton. *J. Bombay Nat. Hist. So*c. **1939,** *41*, 272–285.

Hora, S. L. On the Malayan Affinities of the Freshwater Fish Fauna of Peninsular India, and its Bearing on the Probable Age of the Garo-Rajmahal Gap. *Proc. Natl. Inst. Sci. India, Calcutta,* **1944,** *10 (*4), 423–439.

Hora, S. L. Satpura Hypothesis of the Distribution of Malayan Fauna and Flora of Peninsular India. *Proc. Natl. Inst. Sci. India* **1949,** *15*, 309–314.

Jadhav, S. S. Pisces. *Faunal Diversity of Biogeographic Zones of India: Western Ghats*; Zoological Survey of India: Kolkata, **2020**; pp 627–645.

Jayaram, K. C. In *Freshwater Fishes of India, Pakistan, Bangladesh, Burma and Sri Lanka*; Zoological Survey of India, **1981**.

Jerdon. On the Fresh-Water Fishes of Southern India. *Madras J. Lit. Sci.* **1848,** *15*, 329.

Jhingran, V. G. In *Fish and Fisheries of India.* (Revised second edition); Hindustan Publishing Corporation: New Delhi, **1983**.

Kottelat, M.; Whitten, T. *Freshwater Biodiversity in Asia: With Special Reference to Fish;* World Bank Technical Paper, 343; The World Bank: Washington, DC, USA, **1996;** p 59.

Krishnakumar, K.; Raghavan, R.; Prasad, G.; Bijukumar, A.; Sekharan, M.; Pereira, B.; Ali, A. When Pets become Pests-Exotic Aquarium Fishes and Biological Invasions in Kerala, India. *Curr. Sci.* **2009,** *97* (4), 474– 476.

Lévêque, C.; Oberdorff, T.; Paugy, D.; Stiassny, M. L. J.; Tedesco P. A. Global Diversity of Fish (Pisces) in Freshwater. In *Freshwater Animal Diversity Assessment*; Developments in Hydrobiology, vol 198; Balian, E. V., Lévêque, C., Segers, H., Martens, K., Eds.; Springer: Dordrecht, **2007**, pp 545–567. https://doi.org/ 10.1007/978-1-4020-8259-7_53

Mani, I.; Kumar, R.; Singh, M.; Kushwaha, B.; Nagpure, N. S.; Srivastava, P. K.; Lakra, W. S. Karyotypic Diversity and Evolution of Seven Mahseer Species (Cyprinidae) from India. *J. Fish Biol.* **2009,** *75*, 1079–1091.

Menon, A. G. K. Further Studies Regarding Hora', Satpura Hypothesis. 1. The Role of the Eastern Ghats in the Distribution of the Malayan Fauna and Flora to Peninsular India. *Proc. Nat. Inst. Sci. India* **1951,** *17* (6), 75–97.

Menon, A. G. K. Taxonomy and Speciation of Fishes. In *Fish Genetics in India*; Das and Jhingram; Eds.; Today & Tomorrow's Printers and Publishers: New Delhi, India, **1989**; pp 75–82.

Mercy, T. V.; Gopalakrishnan A.; Kapoor, D; Lakra, W. S. In *Ornamental Fishes of the Western Ghats of India*; National Bureau of Fish Genetic Resources: Lucknow, India, **2007**.

Molur, S.; Smith, K. G.; Daniel, B. A.; Darwall, W. R. T. (Compilers). The Status and Distribution of Freshwater Biodiversity in the Western Ghats, India; IUCN: Cambridge, UK and Gland, Switzerland and Zoo Outreach Organisation: Coimbatore, India, **2011**.

Pethiyagoda, R.; Meegaskumbura, M.; Maduwage, K. A Synopsis of the South Asian Fishes Referred to *Puntius* (Pisces: Cyprinidae). *Ichthyol. Explor. Freshwat.* **2012,** *23,* 69–95.

Pinder, A. C.; Britton, J. R.; Harrison, A. J.; Nautiyal, P.; Bower, S. D.; Cooke, S. J.; Lockett, S.; Everard, M.; Katwate, U.; Ranjeet, K.; Walton, S.; Danylchuk, A. J.; Dahanukar, N.; Raghavan, R. Mahseer (*Tor* spp.) Fishes of the World: Status, Challenges and Opportunities for Conservation. *Rev. Fish Biol. Fish.* **2019,** *29,* 417–452.

Ponniah, A. G.; Gopalakrishnan, A. *Endemic Fish Diversity of Western Ghats;* National Bureau of Fish Genetic Resources: Lucknow, India, **2000.**

Pramod, P. K.; Fang, F.; Devi, K. R.; Liao, T. Y.; Indra, T. J.; Beevi, K. S. J.; Kullander, S. O. *Betadevario ramachandrani,* A New Danionine Genus and Species from the Western Ghats of India (Teleostei: Cyprinidae: Danioninae). *Zootaxa* **2010,** *2519,* 31–47.

Raghavan, R. Need for Further Research on the Freshwater Fish Fauna of the Ashambu Hills Landscape: A Response to Abraham et al. *J. Threat. Taxa.* **2011,** *3* (5), 1788–1791.

Raghavan, R. *Conservation of Freshwater Fishes of the Western Ghats Hotspot, India.* Conference Abstract: XVI European Congress of Ichthyology; *Front. Mar. Sci.* **2019.** DOI: 10.3389/conf.fmars.2019.07.00151

Rajasekaran, N.; Sivakumar, R. Diversity of *Puntius* (Cyprinidae: Cypriniformes) from Lower Anicut, Tamil Nadu. *Int. J. Pure Appl. Biosci.* **2014,** *2* (6), 55–69.

Rainboth, W. J. *Taxonomy, Systematics and Zoogeography of Hypsibarbus, & New Genus of Large Barbs (Pisces, Cyprinidae) from the Rivers of Southeastern Asia;* University of California Press: Berkeley, USA, **1996.**

Sahyadri e-news: Issue xvii, Sahyadri : Western Ghats Biodiversity Information System ENVIS @CES, Indian Institute of Science, Bangalore. Downloaded on 16/06/2021

Sehgal, K. L. Coldwater Fish and Fisheries in the Indian Himalayas: Rivers and Streams. In *Fish and Fisheries at Higher Altitudes: Asia.* (Rome), FAO, Fisheries Technical Paper, no. 385, **1999.**

Shaji, C. P.; Easa, P. S.; Gopalakrishnan, A. Freshwater Fish Diversity of Western Ghats. In *Endemic Fish Diversity of Western Ghats;* Ponniah, A. G., Gopalakrishnan, A., Eds.; National Bureau of Fish Genetic Resources: Lucknow, India, **2000;** pp 33–35.

Silas, E. G. On a Collection of Fish from the Anamalai and Nelliampathi Hill Ranges (Western Ghats) with Notes on its Zoogeographical Significances. *J. Bombay Nat. Hist. Soc.* **1951,** *49,* 670–681

Silas, E. G. Further Studies Regarding Hora's Satpura Hypothesis. *Proc. Natl. Inst. Sci. India* **1952,** *18* (5), 423–447.

Silas, E. G. Classification, Zoogeography and Evolution of the Fishes of the Cyprinoid Families Homalopteridae and Gastromyzonidae. *Rec. Indian Museum* **1953,** *50* (2), 173–264.

Silas, E. G. History and Development of Fisheries Research in India. *J. Bombay Natl. History Soc.* **2003,** *100* (2 & 3), 502–520.

Singh, A. K. Emerging Alien Species in Indian Aquaculture: Prospects and Threats. *J. Aquatic Biol. Fish.* **2014,** *2* (1), 32–41.

Sreekantha. *Sahyadri Mathsya,* [Online] **2013.** http://wgbis.ces.iisc.ernet.in/biodiversity/sahyadri_enews/newsletter/issue

Walton, S. E.; Gan, H. M.; Raghavan, R.; Pinder, A. C.; Ahmad, A. Disentangling the Taxonomy of the Mahseers (*Tor* spp.) of Malaysia: An Integrated Approach using Morphology, Genetics and Historical Records. *Rev. Fish Sci. Aquac.* **2017,** *25,*171–183.

Website consulted [Online]

https://www.fishbase.in/summary/ accessed from time to time during compilation

CHAPTER 13

STATUS OF AMPHIBIAN DIVERSITY IN THE WESTERN GHATS

RAMAKRISHNA[1], P. DEEPAK[2], and K. P. DINESH[3]

[1]Zoological Survey of India, India

[2]Mount Carmel College, Bengaluru, Karnataka, India

[3]Zoological Survey of India (ZSI), Western Regional Centre (WRC), Pune, India

ABSTRACT

The Western Ghats (WG) is one of the biodiversity hotspots of the world known for its unique flora and fauna composition and their endemicity. In the Western Ghats, 253 species of amphibians are reported with 93.67% endemicity. These 253 species could be systematically arranged into 32 genera and 12 families. Among these 237 species, 17 genera, and four families are endemic to the escarpment. IUCN Red List conservation status suggests 24.4% as threatened species and for 66% of the species the data is either deficient or not assessed. The species accumulation curve for the escarpment shows the mixed trends since the inception of the studies, but the latest trends suggest the possibilities of many more new species descriptions in the near future.

13.1 INTRODUCTION

In India, the Western Ghats (WG) is an escarpment (Vijayakumar et al., 2019) running parallel to the west coastal Arabian sea which is about 1490 km long having one major break (Palghat Gap) at Palakkad. This undulating hill chain is discontinuous at Palghat Gap (which is around 10 km long) with a major break and a few minor passes in between. In the southern Western Ghats, "Shencottah pass" is the one and in the northern Western Ghats, "Goa pass" is the another phylogeographic break, wherein these major and minor breaks and passes might have acted as barrier in restricting the free movement of amphibian species on either side. These "gaps" and "passes" in addition to high altitudinal hill ranges would have contributed to high level of amphibian species endemicity (wide range, narrow range, and point endemics). In India, the Western Ghats massifs are the tallest following the Himalayas (Vijayakumar et al., 2014) with a very diverse forest vegetation pattern. The entire hill chains falls within the political boundaries of the states Gujarat, Maharashtra, Goa, Karnataka, Kerala, and Tamil Nadu (Table 13.1).

TABLE 13.1 Amphibian Species Diversity Pattern Across in Different Provinces and States of the Western Ghats.

Taxa	GJ	MH	GA	KA	KL	TN
	Gujarat	Maharashtra	Goa	Karnataka	Kerala	Tamil Nadu
Orders	2	2	2	2	2	2
Families (No.)	7	10	9	10	11	11
Genera (No.)	11	21	19	21	31	25
Species (No.)	20	43	38	101	194	90

The combination of huge massifs, mountain passes, the "Palghat Gap" and the climatic factors makes the flora and fauna of this escarpment unique. Due to availability of good number of species diversity and high degree of endemism, the WG escarpment is considered as one of the global biodiversity hotspots (Kunte et al., 1999).

Amphibians ("amphi" denoting "of both or double kinds" and "bios" referring "life") are the cold-blooded dual lifers which lead part of their early life in water (as tadpoles) and rest of the life on land (as adults) (AmphibiaWeb, 2018).

The Class Amphibia are the primitive tetrapod vertebrate group having legged and tailless frogs and toads (Order: Anura), tailed and legged salamanders (Order: Caudata), and the legless caecilians (Order: Gymnophiona). Being primitive vertebrates, having intermediate body plan between the vertebrate life in water and land, amphibians are considered as the living connecting link between the vertebrate life in water and land (Zardoya and Meyer, 2001). In the present-day context, they add immense value to the ecosystem services, wherein these creatures play a crucial role in the ecological pyramid both as prey (by eating insects and invertebrates) and predator (being eaten by birds and small mammals). Above all, amphibians are habitat specific and are sensitive to environmental variables, for this very reason amphibians are considered as "*ecological indicators*" (Simon et al., 2011).

Globally, 8324 species are documented of which 7344 species are frogs and toads, 766 species are salamanders, and 214 species are caecilians (Frost, 2021). In the Indian subcontinent, 454 species are documented including 413 species of frogs and toads, 2 species of salamanders, and 39 species of caecilians (Dinesh et al., 2021) (Table 13.2).

In the current account, all the amphibian species reported/described till April 2021 are considered following the taxonomy after Dinesh et al. (2021) and Frost (2021). Updated checklist of amphibians for India (Dinesh et al., 2021) is considered for all the analysis of the study (see Appendix 13.1).

TABLE 13.2 Amphibian Species Diversity in India and the Western.

Order	Species known from India	Species known from the Western Ghats	Species endemic to the Western Ghats
Anura	413	227	211
Caudata	2	Nil	Nil
Gymnophiona	39	26	26
	454	253	237

13.2 HISTORICAL RESUME

In the Western Ghats, the first amphibian species described was *Hydrophylax malabaricus* (as *Rana malabarica*) in 1838 by Tschudi from the "Malabar region" of the Western Ghats. Between the years 1799 and 1838, the species *Duttaphrynus melanostictus, Duttaphrynus scaber, Euphlyctis cyanophlyctis, Uperodon systoma, Hoplobatrachus tigerinus,* and *Polypedates maculatus* described from the rest of the Indian region were thought to be distributed in the Western Ghats.

Amphibian research and description of the species were initiated during the British rule in India. A total of 83 species of amphibians (Figure 13.1) were discovered during 1838–1947 by the British as well as Indian researchers. Most of these descriptions during the pre-independent Indian regime were done by the British and German researchers like Johann von Tschudi (1818–1889) (Tschudi, 1838); André Marie Constant Duméril (1774–1860) and Gabriel Bibron (1805–1848) (Dumeril and Bibron, 1841); Thomas C Jerdon (1811–1872) (Jerdon, 1853, 1870); Albert Günther (1830–1914) (Gunther, 1864, 1875); Richard Henry Beddome (1830–1911) (Beddome, 1870, 1878); Wilhelm Peters (1815–1883) (Peters, 1871, 1879); Ferdinand Stoliczka (1838–1874) (Stoliczka, 1872); George Albert Boulenger (1838–1937) (Boulenger, 1882, 1883, 1888, 1891, 1904); Nelson Annandale (1876–1924) (Annandale, 1909, 1913, 1919); Christoph Gustav Ernst Ahl (1898–1945) (Ahl, 1927), and George S. Myers (1905–1985) (Myers, 1942). Couple of Indian researchers at that time period who were involved in the amphibian research include C. R. Narayan Rao (1882–1960) (Rao, 1920, 1922, 1937); B. R. Seshachar (1910–1994) (Seshachar, 1939). Most of the species described during this period had an unresolved taxonomic status (till recently) due to lack of proper color description of the species as the specimens were not collected by the describer and usage of vague geographical locality details. As there were no laws prevailing during this period regarding the preservation of biological diversity, most of the type specimens were housed in the Museums of Universitat Humboldt, Zoologisches Museum (ZMB), Invalidenstrasse, Berlin, Germany; Museum National dHistoire Naturelle (MNHNP), Rue Cuvier, Paris, France; British Museum of Natural History (BMNH), London UK; Institut Royal des Sciences Naturelles de Belgique, Belgium; Stanford University collections (CAS-SU); Museum of Comparative Zoology (MCZ), Harvard University, Cambridge, Massachusetts, USA; Universitets Kobenhavn, Zoologisk Museum (ZMUC), Universitetsparken, Denmark including the Indian Central College Bangalore (CCB), Karnataka; Zoological Survey of India (ZSI), Kolkata.

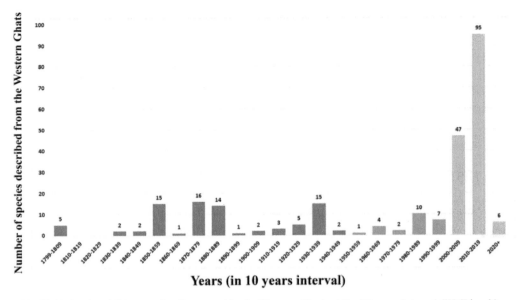

FIGURE 13.1   Amphibian species discovered in the Western Ghats at the 10 years interval (KP Dinesh)

After India attained independence from the British rule, a total of 170 species were described (between 1947 and 2021) in the Western Ghats. Only 23 species were described during 1947–1999, wherein 147 species were described in the new millennia (from 2000 to 2021). The last two decades witnessed the greatest number of species descriptions compared to any consecutive two decades of the Indian amphibian research. Interestingly, more than 75 species of amphibians were described by the S. D. Biju and his team members (Biju and Bossuyt, 2003, 2005a, 2005b, 2005c, 2006, 2009; Biju et al., 2007, 2008, 2009, 2010, 2011, 2014a, 2014b; Garg and Biju, 2016, 2017, 2019; Garg et al., 2017; Garg et al., 2019) and the rest of the species were reported and described by the consorted efforts of many other naturalists, researchers, and scientists of the country. Advent of molecular tools with the GPS technology in combination with the classical taxonomy helped the amphibian researchers in providing stable amphibian taxonomy compared to the earlier studies.

In the Western Ghats, addition to the relict family members of the family Nyctibatrachidae, Micrixalidae, and the Ranixalidae the noteworthy findings involve the discovery of the families Nasikabatrachidae by Biju and Bossuyt (2003) (with the species *Nasikabatrachus sahyadrensis*) and the family Astrobatrachidae by Vijayakumar et al. (2019) (with the species *Astrobatrachus kurichiyana*).

13.3   SPECIES DIVERSITY AND ENDEMISM

In the Western Ghats, the Class Amphibia is represented by only two orders, the order Anura (Frogs and Toads) and the order Gymnophiona (Caecilians). The frog and toad order Anura is represented by ten families with 28 genera and 227 species (Table 13.3).

The caecilian order Gymnophiona is represented by 26 species under four genera and two families (Table 13.3). The total species representation in the WG accounts to 253 species under 32 genera and 12 families.

TABLE 13.3 Amphibian Diversity in the Western Ghats.

Sl. no.	Family	No. of genera in the WG	No. of species in the WG	No. of species endemic to the WG
1	Bufonidae	4	12	10
2	Dicroglossidae	4	33	26
3	Micrixalidae	1	24	24
4	Microhylidae	4	15	10
5	Astrobatrachidae*	1	1	1
6	Nasikabatrachidae*	1	2	2
7	Nyctibatrachidae*	1	36	36
8	Ranidae	3	13	12
9	Ranixalidae*	2	18	18
10	Rhacophoridae	7	73	72
11	Grandisoniidae	2	13	13
12	Ichthyophiidae	2	13	13
	Total	**32**	**253**	**237 (93.67%)**

Among the 12 families reported for the Western Ghats, the family Micrixalidae, Astrobatrachidae, Nasikabatrachidae, Nyctibatrachidae, and Ranixalidae are endemic (Table 13.3). Of the 32 genera reported the following 17 genera *Beduka, Blaira, Pedostibes, Micrixalus, Melanobatrachus, Mysticellus, Nasikabatrachus, Astrobatrachus, Nyctibatrachus, Clinotarsus, Indirana, Walkerana, Beddomixalus, Ghatixalus, Mercurana, Indotyphlus,* and *Uraeotyphlus* are endemic. A total of 237 species are wide to narrow range endemics in the Western Ghats (93.67% endemism) (Table 13.3; Figure 13.2).

The tree frog/bush frog family Rhacophoridae is the specious group with 73 species showing 98% endemicity with varying degree of natural history pattern in breeding preference. Most of the tree frogs prefer terrestrial foam nest formation, single leaf foam nest formation, and multiple leaf foam nest formation with tadpole stage. The bush frogs show direct development without tadpole stage.

Second most specious group in the Western Ghats are the wrinkled frog family Nyctibatrachidae with 36 species, wherein the members are forested species preferring perennial source of water and are nocturnal.

Third most specious group in the Western Ghats are the cricket frog family Dicroglossidae with 33 species, wherein most of the species show aquatic habitat preference ranging from temporary mud puddles, swamps, streams, ponds, and agriculture habitats.

Fourth most specious group of the Western Ghats are the torrential stream frog family Micrixalidae with 24 species, wherein the members are forested species preferring perennial source of water and are diurnal.

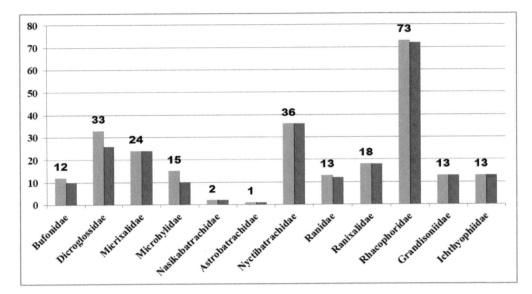

FIGURE 13.2 Amphibian species diversity (green column) and endemism (red column) family wise details for the amphibians of the Western Ghats (KP Dinesh)

In the legless amphibian groups, the Western Ghats is known to have 26 species under four genera and two families, wherein all the species and two genera *Indotyphlus* and *Uraeotyphlus* are endemic to the Ghats. These wormlike caecilians are poorly known due to their nocturnal and subterranean habitat.

13.4 THREATS AND CONSERVATION STRATEGIES

Taxonomically among the 12 families and 323 genera of the Ghats, periodic taxonomic revisions were being made for the species of the genera *Beduka, Blaira, Pedostibes, Micrixalus, Melanobatrachus, Microhyla, Mysticellus, Uperodon, Nasikabatrachus, Astrobatrachus, Nyctibatrachus, Clinotarsus, Hydrophylax, Indosylvirana, Indirana, Walkerana, Beddomixalus, Ghatixalus, Mercurana, Polypedates, Pseudophilautus, Raorchestes, Rhacophorus, Gegeneophis*, and *Indotyphlus*, addressing most part of the Linnaean shortfall. But field and taxonomic revisions are warranted for the members of the genera *Duttaphrynus, Euphlyctis, Hoplobatrachus, Minervarya, Sphaerotheca, Ichthyophis*, and *Uraeotyphlus*. Due to specific habitat preference by the species and availability of limited such habitats is a limiting factor for most of known species of amphibians in the Western Ghats.

One of the noteworthy aspects among the amphibians of the Western Ghats is that most of the known species are either known or reported from the protected areas, reserve forests and the forested areas. Fewer majorities of the species are in the agro biodiversity centers like plantations and homestead areas including urban landscapes. A holistic species-specific field survey studies are warranted to decipher the exact distribution range for each species to address the Wallacean shortfall.

Fewer studies were carried out in the tea plantations (Daniels, 2003); coffee plantations (Rathod and Rathod, 2013); and cardamom plantations (Kanagavel et al., 2017) to understand the amphibian community composition and their interactions. In the four dicroglossid frog species, abnormality studies were done by Gurushankara et al. (2007).

In the Schedule IV of the Wildlife Protection Act, 1972 (Arora, 2017), there is already a ban enforced on the frog leg trade (*Hoplobatrachus crassa, Hoplobatrachus tigerinus, Euphlyctis hexadactylus,* and *Euphlyctis karaavali*) and restrictions on the usage of amphibians in academic purposes like frog dissections (*Euphlyctis cyanophlyctis, Euphlyctis aloysii*) by the Indian government.

Although most of the species are described from the forested/protected areas of the Western Ghats "climate change" is considered as one of the major reasons for the decline of amphibian populations. Man-made developmental activities are resulting in the "habitat loss" and "habitat fragmentation" for these ecological indicators. Due to habitat loss, amphibian species are losing their breeding grounds, mostly fresh water bodies (including mud puddles, ponds, streams, and rivers) and habitat fragmentation is reducing the amphibian population interactions leading to genetic bottlenecks.

IUCN Red List conservation status for the Indian/Western Ghats amphibians was largely assessed during 2004–2012 (Table 13.4). As per the details available on the IUCN portal and the recent publications, 6.7% are considered under critically endangered, 11.4% endangered, 6.3% vulnerable, 1.6% near threatened, 12.9% least concerned, 22% data deficient, and 39.27% under not assessed category (Figure 13.3). From the earlier assessments, it is evident that more than 60% species in the Western Ghats needs assessment. Also, in the assessed categories cryptic species complexes require immediate attention to decipher the exact conservation status for the species of the Western Ghats. There is a need for field-based explorations and population level studies to address the problems of "Wallacean shortfall" for all the described species (Figure 13.3).

TABLE 13.4 IUCN Conservation Red List Status Family Wise Details for the Amphibians of the Western Ghats.

Sl. no.	Family	CE	EN	VU	NT	LC	DD	NA	
1	Bufonidae	1	4	2	1	2	2	Nil	12
2	Dicroglossidae	1	3	Nil	Nil	8	9	12	33
3	Micrixalidae	1	1	3	1	Nil	3	15	24
4	Microhylidae	Nil	2	1	1	7	1	3	15
5	Astrobatrachidae*	Nil	Nil	Nil	Nil	Nil	1	Nil	1
6	Nasikabatrachidae*	1	Nil	Nil	Nil	Nil	Nil	1	2
7	Nyctibatrachidae*	1	6	3	Nil	1	10	15	36
8	Ranidae	Nil	Nil	Nil	Nil	2	Nil	11	13
9	Ranixalidae*	2	3	1	Nil	2	1	9	18
10	Rhacophoridae	11	9	6	1	7	7	32	73
11	Grandisoniidae	Nil	Nil	Nil	Nil	1	12	Nil	13
12	Ichthyophiidae	Nil	Nil	Nil	Nil	3	10	Nil	13
	Total	**18**	**28**	**16**	**4**	**33**	**56**	**98**	**253**

CE, critically endangered; DD, data deficient; EN, endangered; LC, least concern; NA, not assessed; NT, near threatened; VU, vulnerable.

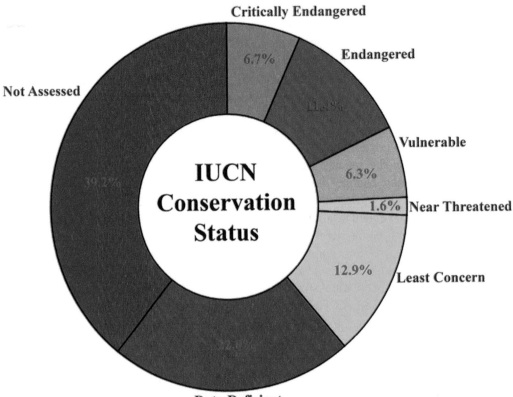

FIGURE 13.3   IUCN conservation Red List status for the amphibian species of the Western Ghats.

13.5   FUTURE TRENDS

In the history of amphibian species documentation in the Western Ghats, it took almost two centuries to document 253 species. Most of the species descriptions happened only in the new millennia which is almost 148 species compared to 105 species till 1999. For the analysis of species accumulation curve for 10 years interval, the curve was flat between the period 1889–1929 and 1940–1979 (Figure 13.4) reflecting a sort of saturation in amphibian studies. But, since 1980 the curve started showing an upward climbing trend suggesting the possibilities of few more new species reports (Figure 13.4). Since the beginning of the new millennia, the curve is showing a very steep climb suggesting the possibilities of many more new species descriptions in the near future (Figure 13.4).

As the access to the hilly terrain landscape of the Western Ghats is very tough for documentation and explorations, and amphibians being seasonal and nocturnal, it is very challenging to undertake the amphibian studies in the rainy season when most of them are very active for breeding. Accepting all the probabilities of occurrence of many more new species with the "integrative taxonomic approach," we are expecting near completion of addressing the issues of "Linnaean shortfall."

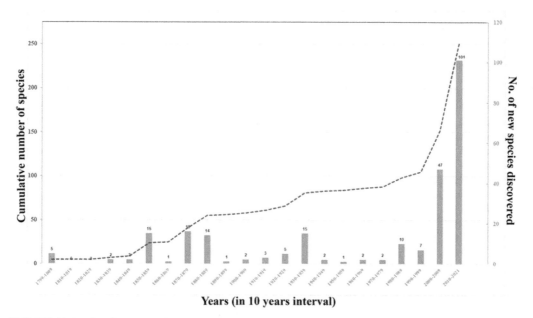

FIGURE 13.4 Species accumulation curve (green columns represent the number of new species discoveries and red-dotted line represent cumulative new species discovery pattern) for the amphibian species discovered from India between the year 1799 and 2021.

ACKNOWLEDGMENT

The authors are grateful to the Director, ZSI, Kolkata and the Officer-in-Charge, ZSI, WRC, Pune for their support. KPD acknowledges the field support of the staff of ZSI, WGRC, Calicut, ZSI, SRC, Chennai and ZSI, WRC, Pune; would like to place special thanks to the staff and officials of the Forest Departments of the State Gujarat, Maharashtra, Goa, Karnataka, Kerala, and Tamil Nadu for the support in field studies and is indebted to the DBT and SERB (SR/FR/LS-88/210/09.05.2012) for the fellowship and assistance to conduct part of this work. PD is grateful to the Principal and the Head Department of Zoology, Mount Carmel College, Autonomous, Bengaluru.

KEYWORDS

- **amphibia**
- **discovery pattern**
- **endemism**
- **hotspot**
- **relicts**
- **Western Ghats**

REFERENCES

Ahl, E. Zur Systematik der asiatischen Arten der Froschgattung *Rhacophorus*. *Sitzungsberichte der Gesellschaft Naturforschender Freunde zu Berlin* **1927,** *1927,* 35–47.

AmphibiaWeb. [Online] **2018**. <https://amphibiaweb.org> University of California, Berkeley, CA, USA. Accessed Aug 13, 2020.

Annandale, N. Miscellanea. Batrachia. Notes on Indian Batrachia. *Rec. Indian Museum* **1909,** *3,* 282–286.

Annandale, N. Some New and Interesting Batrachia and Lizards from India, Ceylon and Borneo. *Rec. Indian Museum* **1913,** *9,* 301–307.

Annandale, N. The Fauna of Certain Small Streams in the Bombay Presidency: Some Frogs from Streams in the Bombay Presidency. *Rec. Indian Museum* **1919,** *16,* 109–161.

Arora, V. The Wildlife (Protection) Act, 1972, Updated Edition 2014; Natraj Publishers: Dehra Dun, **2017;** pp 202.

Beddome, R. H. Descriptions of New Reptiles from the Madras Presidency. *Madras Mon. J. Med. Sci.* **1870,** *2,* 169–176.

Beddome, R. H. Description of a New Batrachian from Southern India, Belonging to the Family Phryniscidae. *Proc. Zool. Soc. London* **1878,** *1878,* 722–723.

Biju, S. D.; Bossuyt, F. New Frog Family from India Reveals an Ancient Biogeographical Link with the Seychelles. *Nature* **2003,** *425,* 711–714.

Biju, S. D.; Bossuyt, F. Two New *Philautus* (Anura: Ranidae: Rhacophorinae) from Ponmudi Hill in the Western Ghats of India. *Copeia* **2005a,** *2005,* 29–37.

Biju, S. D., Bossuyt, F. A New Species of Frog (Ranidae, Rhacophorinae, *Philautus*) from the Rainforest Canopy in the Western Ghats, India. *Curr. Sci.* **2005b,** *88,* 175–178.

Biju, S. D.; Bossuyt, F. New Species of *Philautus* (Anura: Ranidae, Rhacophorinae) from Ponmudi Hill in the Western Ghats of India. *J. Herpetol.* **2005c,** *39,* 349–353.

Biju, S. D.; Bossuyt, F. Two New Species of *Philautus* (Anura, Ranidae, Rhacophorinae) from the Western Ghats, India. *Amphib. Reptil.* **2006,** *27,* 1–10.

Biju, S. D.; Bossuyt, F. Systematics and Phylogeny of *Philautus* Gistel, 1848 (Anura, Rhacophoridae) in the Western Ghats of India, with Descriptions of 12 New Species. *Zool. J. Linn. Soc.* **2009,** *155,* 374–444.

Biju, S. D., Garg, S., Gururaja, K. V., Shouche, Y. S.; Walujkar, S. A. DNA Barcoding Reveals Unprecedented Diversity in Dancing Frogs of India (Micrixalidae, *Micrixalus*): A Taxonomic Revision with Description of 14 New Species. *Ceylon J. Sci.* **2014a,** *43,* 1–87.

Biju, S. D., Garg, S., Mahony, S., Wijayathilaka, N., Senevirathne, G.; Meegaskumbura, M. DNA Barcoding, Phylogeny and Systematics of Golden-Backed Frogs (*Hylarana*, Ranidae) of the Western Ghats-Sri Lanka Biodiversity Hotspot, with the Description of Seven New Species. *Contrib. Zool.* **2014b,** *83,* 269–335.

Biju, S. D., Roelants, K.; Bossuyt, F. Phylogenetic Position of the Montane Treefrog *Polypedates variabilis* Jerdon, 1853 (Anura: Rhacophoridae), and Description of a Related Species. *Org. Divers. Evol.* **2008,** *8,* 267–276.

Biju, S. D., Shouche, Y. S., Dubois, A., Dutta, S. K.; Bossuyt, F. A Ground-Dwelling Rhacophorid Frog from the Highest Mountain Peak of the Western Ghats of India. *Curr. Sci.* **2010,** *98,* 1119–1125.

Biju, S. D., Van Bocxlaer, I., Giri, V. B., Roelants, K., Nagaraju, J., Bossuyt, F. A New Night Frog, *Nyctibatrachus minimus* sp. nov. (Anura: Nyctibatrachidae): the Smallest Frog from India. *Curr. Sci.* **2007,** *93,* 854–858.

Biju, S. D., Van Bocxlaer, I., Giri, V. B., Loader, S. P.; Bossuyt, F. Two New Endemic Genera and a New Species of Toad (Anura: Bufonidae) from the Western Ghats of India. *BMC Res. Notes* **2009,** *2* (241), 1–10.

Biju, S. D., Van Bocxlaer, I., Mahony, S., Dinesh, K. P., Radhakrishnan, C., Zachariah, A., Giri, V. B.; Bossuyt, F. A Taxonomic Review of the Night Frog Genus *Nyctibatrachus* Boulenger, 1882 in the Western Ghats, India (Anura: Nyctibatrachidae) with Description of Twelve New Species. *Zootaxa* **2011,** *3029,* 1–96.

Boulenger, G. A. In *Catalogue of the Batrachia Salientia s. Ecaudata in the Collection of the British Museum,* Second Edition; Taylor and Francis: London, **1882**.

Boulenger, G. A. Description of New Species of Reptiles and Batrachians in the British Museum. *Ann. Mag. Nat. Hist.* (Series 5) **1883,** *12,* 161–167.

Boulenger, G. A. Descriptions of Two New Indian Species of *Rana. Ann. Mag. Nat. Hist.* (Series 6) **1888,** *2,* 506–508.

Boulenger, G. A. On New or Little-Known Indian and Malayan Reptiles and Batrachians. *Ann. Mag. Nat. Hist.* (Series 6) **1891**, *8*, 288–292.

Boulenger, G. A. Descriptions of Three New Frogs from Southern India and Ceylon. *J. Bombay Nat. Hist. Soc.* **1904**, *15*, 430–431.

Daniels, R. J. R. Impact of Tea Cultivation on Anurans in the Western Ghats. *Curr. Sci.* **2003**, *85* (10), 1415–1422.

Dinesh, K. P., Radhakrishnan, C., Deepak, P.; Kulkarni, N. U. A Checklist of Amphibians of India with IUCN Conservation Status; Version 3.0; Updated till April 2020. [Online] **2021**. https://zsi.gov.in/WriteReadData/userfiles/file/Checklist/Amphibians_2020.pdf (online only).

Duméril, A. M. C.; Bibron, G. Erpétologie Genérale ou Histoire Naturelle Complète des Reptiles, vol 8; Librarie Enclyclopedique de Roret: Paris, **1841**.

Frost, D. R. Amphibian Species of the World: an Online Reference, Version 6.1. [Online] **2021**. (accessed May 12, 2021). Electronic Database accessible at https://amphibiansoftheworld.amnh.org/ index.php. American Museum of Natural History, New York, USA.

Garg, S.; Biju, S. D. Molecular and Morphological Study of Leaping Frogs (Anura, Ranixalidae) with Description of Two New Species. *PLoS One* **2016**, *11* (11:e0166326), 1–36.

Garg, S.; Biju, S. D. Description of Four New Species of Burrowing Frogs in the *Fejervarya rufescens* Complex (Dicroglossidae) with Notes on Morphological Affinities of *Fejervarya* Species in the Western Ghats. *Zootaxa* **2017**, *4277*, 451–490.

Garg, S.; Biju, S. D. New Microhylid Frog Genus from Peninsular India with Southeast Asian Affnity Suggests Multiple Cenozoic Biotic Exchanges between India and Eurasia. *Sci. Rep.* **2019**, *9* (1906), 1–13.

Garg, S.; Suyesh, R.; Das, A.; Jiang, J. P.; Wijayathilaka, N.; Amarasinghe, A. A. T.; Alhadi, F., Vineeth, K. K., Aravind, N. A., Senevirathne, G., Meegaskumbura, M.; Biju. S. D. Systematic Revision of *Microhyla* (Microhylidae) Frogs of South Asia: A Molecular, Morphological, and Acoustic Assessment. *Vertebr. Zool.* **2019**, *69*, 1–71.

Garg, S.; Suyesh, R.; Sukesan, S.; Biju, S. D. Seven New Species of Night Frogs (Anura, Nyctibatrachidae) from the Western Ghats Biodiversity Hotspot of India, with Remarkably High Diversity of Diminutive Forms. *Peer J.* **2017**, *5* (e3007), 1–50.

Günther, A. C. L. G. In *The Reptiles of British India*; Ray Society by R. Hardwicke: London, **1864**.

Günther, A. C. L. G. Third Report on Collections of Indian Reptiles Obtained by the British Museum. *Proc. Zool. Soc. London* **1875**, *1875*, 567–577.

Gurushankara, H. P., Krishnamurthy, S. V.; Vasudev, V. Morphological Abnormalities in Natural Populations of Common Frogs Inhabiting Agroecosystems of Central Western Ghats. *Appl. Herpetol.* **2007**, *4*, 39–45.

Jerdon, T. C. Catalogue of Reptiles Inhabiting the Peninsula of India. *J. Asiatic Soc. Bengal* **1853**, *22*, 522–534.

Jerdon, T. C. Notes on Indian herpetology. *Proc. Asiatic Soc. Bengal* **1870**, *1870*, 66–85.

Kanagavel, A.; Parvathy, S.; Nirmal, N.; Divakar, N.; Raghavan, R. Do Frogs Really Eat Cardamom? Understanding the Myth of Crop Damage by Amphibians in the Western Ghats, India. *Ambio* **2017**, *46* (6), 695–705. DOI: 10.1007/s13280-017-0908-8.

Kunte, K. J.; Joglekar, A. P.; Utkarsh, G.; Pramod, P. Patterns of Butterfly, Bird and Tree Diversity in the Western Ghats. *Curr. Sci.* **1999**, *77*, 577–586.

Myers, G. S. A New Frog from the Anamallai Hills, with Notes on other Frogs and Some Snakes from South India. *Proc. Biol. Soc. Washington* **1942**, *55*, 49–56.

Peters, W. C. H. Über einige Arten der herpetologischen Sammlung des Berliner zoologischen Museums. Monatsberichte der Königlichen Preussische Akademie des Wissenschaften zu Berlin **1871**, *1871*, 644–652.

Peters, W. C. H. Über die Eintheilung der Caecilien und insbesondere über die Gattungen *Rhinatrema* und *Gymnopis*. Monatsberichte der Königlichen Preussische Akademie des Wissenschaften zu Berlin **1879**, *1879*, 924–945.

Rao, C. R. N. Some South Indian batrachians. *J. Bombay Nat. Hist. Soc.* **1920**, *27*, 119–127.

Rao, C. R. N. Notes on Batrachia. *J. Bombay Nat. Hist. Soc.* **1922**, *28*, 439–447.

Rao, C. R. N. On Some New Forms of Batrachia from S. India. *Proc. Indian Acad. Sci.* (Section B) **1937**, *6*, 387–427.

Rathod, S.; Rathod, P. Amphibian Communities in Three Different Coffee Plantation Regimes in the Western Ghats, India. *J. Threat. Taxa* **2013**, *5* (9), 4404–4413.

Seshachar, B. R. On a New Species of *Uraeotyphlus* from South India. *Proc. Indian Acad. Sci.* (Section B) **1939**, *9*, 224–228.

Simon, E.; Puky, M.; Braun, M.; Tóthmérész, B. Frogs and Toads as Indicators in Environmental Assessment. In *Frogs: Biology, Ecology and Uses*; Murray, J. L., Nova Science Publishers, Inc., **2011**.

Stoliczka, F. Observations on Indian Batrachia. *Proc. Asiatic Soc. Bengal* **1872**, *1872*, 101–113.

Tschudi, J. J. von. Classification der Batrachier mit Berücksichtigung der fossilen Thiere dieser Abtheilung der Reptilien; Petitpierre: Neuchâtel, **1838**.

Vijayakumar, K., Dinesh, K. P.; Prabhu, M. V.; Shanker, K. Lineage Delimitation and Description of 9 New Species of Bush Frogs (Anura: *Raorchestes*, Rhacophoridae) from the Western Ghats Escarpment. *Zootaxa* **2014**, *3893* (4), 451–488.

Vijayakumar, S. P., Pyron, R. A., Dinesh, K. P., Torsekar, V. R., Srikanthan, A. N., Swamy, P., Stanley, E. L., Blackburn, D. C.; Shanker, K. A New Ancient Lineage of Frog (Anura: Nyctibatrachidae: Astrobatrachinae subfam. nov.) Endemic to the Western Ghats of Peninsular India. *Peer J.* **2019**, *7* (e6457), 1–28.

Zardoya, R.; Meyer, A. On the Origin of and Phylogenetic Relationships Among Living Amphibians. *Proc. Natl. Acad. Sci.* **2001**, *98*, 7380–7383.

APPENDIX 13.1

An updated checklist of amphibians of the Western Ghats.

Sl. no.	Species	IUCN Red List status
1	*Beduka koynayensis* (Soman, 1963)	CE
2	*Beduka amboli* (Dubois et al., 2021)	EN
3	*Blaira ornata* (Günther, 1876)	EN
4	*Blaira rubigina* (Pillai and Pattabiraman, 1981)	VU
5	*Duttaphrynus beddomii* (Günther, 1876)	EN
6	*Duttaphrynus brevirostris* (Rao, 1937)	DD
7	*Duttaphrynus melanostictus* (Schneider, 1799)	LC
8	*Duttaphrynus microtympanum* (Boulenger, 1882)	VU
9	*Duttaphrynus parietalis* (Boulenger, 1882)	NT
10	*Duttaphrynus scaber* (Schneider, 1799)	LC
11	*Duttaphrynus silentvalleyensis* (Pillai, 1981)	DD
12	*Pedostibes tuberculosus* (Günther, 1876)	EN
13	*Euphlyctis aloysii* (Joshy et al., 2009)	NA
14	*Euphlyctis cyanophlyctis* (Schneider, 1799)	LC
15	*Euphlyctis karaavali* (Priti et al., 2016)	EN
16	*Euphlyctis mudigere* (Joshy et al., 2009)	NA
17	*Hoplobatrachus crassus* (Jerdon, 1853)	LC
18	*Hoplobatrachus tigerinus* (Daudin, 1802)	LC
19	*Minervarya agricola* (Jerdon, 1853)	NA
20	*Minervarya brevipalmata* (Peters, 1871)	DD
21	*Minervarya cepfi* (Garg and Biju, 2017)	NA
22	*Minervarya goemchi* (Dinesh et al., 2018 "2017")	DD
23	*Minervarya gomantaki* (Dinesh et al., 2015)	NA

APPENDIX 13.1 *(Continued)*

Sl. no.	Species	IUCN Red List status
24	*Minervarya kadar* (Garg and Biju, 2017)	NA
25	*Minervarya kalinga* (Raj et al., 2018)	DD
26	*Minervarya keralensis* (Dubois, 1981)	LC
27	*Minervarya krishnan* (Raj et al., 2018)	DD
28	*Minervarya kudremukhensis* (Kuramoto et al., 2008)	NA
29	*Minervarya manoharani* (Garg and Biju, 2017)	NA
30	*Minervarya marathi* (Phuge et al., 2019)	DD
31	*Minervarya modesta* (Rao, 1920)	NA
32	*Minervarya mudduraja* (Kuramoto et al., 2008)	NA
33	*Minervarya murthii* (Pillai, 1979)	CE
34	*Minervarya mysorensis* (Rao, 1922)	DD
35	*Minervarya neilcoxi* (Garg and Biju, 2017)	NA
36	*Minervarya nilagirica* (Jerdon, 1853)	EN
37	*Minervarya parambikulamana* (Rao, 1937)	DD
38	*Minervarya rufescens* (Jerdon, 1853)	LC
39	*Minervarya sahyadris* (Dubois et al., 2001)	EN
40	*Minervarya sauriceps* (Rao, 1937)	DD
41	*Minervarya syhadrensis* (Annandale, 1919)	LC
42	*Sphaerotheca dobsonii* (Boulenger, 1882)	LC
43	*Sphaerotheca leucorhynchus* (Rao, 1937)	DD
44	*Sphaerotheca maskeyi* (Schleich and Anders, 1998)	LC
45	*Sphaerotheca pluvialis* (Jerdon, 1853)	NA
46	*Micrixalus adonis* (Biju et al., 2014)	NA
47	*Micrixalus candidus* (Biju et al., 2014)	NA
48	*Micrixalus elegans* (Rao, 1937)	DD
49	*Micrixalus frigidus* (Biju et al., 2014)	NA
50	*Micrixalus fuscus* (Boulenger, 1882)	NT
51	*Micrixalus gadgili* (Pillai and Pattabiraman, 1990)	EN
52	*Micrixalus herrei* (Myers, 1942)	NA
53	*Micrixalus kodayari* (Biju et al., 2014)	NA
54	*Micrixalus kottigeharensis* (Rao, 1937)	CE
55	*Micrixalus kurichiyari* (Biju et al., 2014)	NA
56	*Micrixalus mallani* (Biju et al., 2014)	NA
57	*Micrixalus nelliyampathi* (Biju et al., 2014)	NA
58	*Micrixalus nigraventris* (Biju et al., 2014)	NA
59	*Micrixalus niluvasei* (Biju et al., 2014)	NA
60	*Micrixalus nudis* (Pillai, 1978)	VU
61	*Micrixalus phyllophilus* (Jerdon, 1853)	VU
62	*Micrixalus sairandhri* (Biju et al., 2014)	NA

APPENDIX 13.1   *(Continued)*

Sl. no.	Species	IUCN Red List status
63	*Micrixalus sali* (Biju et al., 2014)	NA
64	*Micrixalus saxicola* (Jerdon, 1853)	VU
65	*Micrixalus silvaticus* (Boulenger, 1882)	DD
66	*Micrixalus specca* (Biju et al., 2014)	NA
67	*Micrixalus spelunca* (Biju et al., 2014)	NA
68	*Micrixalus thampii* (Pillai, 1981)	DD
69	*Micrixalus uttaraghati* (Biju et al., 2014)	NA
70	*Melanobatrachus indicus* (Beddome, 1878)	LC
71	*Microhyla darreli* (Garg et al., 2019)	NA
72	*Microhyla nilphamariensis* (Howlader et al., 2015)	NA
73	*Microhyla ornata* (Duméril and Bibron, 1841)	LC
74	*Microhyla rubra* (Jerdon, 1853)	LC
75	*Microhyla sholigari* (Dutta and Ray, 2000)	EN
76	*Mysticellus franki* (Garg and Biju, 2019)	NA
77	*Uperodon anamalaiensis* (Rao, 1937)	DD
78	*Uperodon globulosus* (Günther, 1864)	LC
79	*Uperodon montanus* (Jerdon, 1853)	NT
80	*Uperodon mormoratus* (Rao, 1937)	EN
81	*Uperodon systoma* (Schneider, 1799)	LC
82	*Uperodon taprobanicus* (Parker, 1934)	LC
83	*Uperodon triangularis* (Günther, 1876)	VU
84	*Uperodon variegatus* (Stoliczka, 1872)	LC
85	*Nasikabatrachus bhupathi* (Janani et al., 2017)	NA
86	*Nasikabatrachus sahyadrensis* (Biju and Bossuyt, 2003)	EN
87	*Astrobatrachus kurichiyana* (Vijayakumar et al., 2019)	DD
88	*Nyctibatrachus acanthodermis* (Biju et al., 2011	NA
89	*Nyctibatrachus aliciae* (Inger et al., 1984)	EN
90	*Nyctibatrachus anamallaiensis* (Myers, 1942)	NA
91	*Nyctibatrachus athirappillyensis* (Garg et al., 2017)	DD
92	*Nyctibatrachus beddomii* (Boulenger, 1882)	EN
93	*Nyctibatrachus danieli* (Biju et al., 2011)	NA
94	*Nyctibatrachus dattatreyaensis* (Dinesh et al., 2008)	CE
95	*Nyctibatrachus deccanensis* (Dubois, 1984)	VU
96	*Nyctibatrachus deveni (*Biju et al., 2011)	NA
97	*Nyctibatrachus gavi* (Biju et al., 2011)	NA
98	*Nyctibatrachus grandis* (Biju et al., 2011)	NA
99	*Nyctibatrachus humayuni* (Bhaduri and Kripalani, 1955)	VU
100	*Nyctibatrachus indraneili* (Biju et al., 2011)	NA
101	*Nyctibatrachus jog* (Biju et al., 2011)	NA

APPENDIX 13.1 *(Continued)*

Sl. no.	Species	IUCN Red List status
102	*Nyctibatrachus karnatakaensis* (Dinesh et al., 2007)	EN
103	*Nyctibatrachus kempholeyensis* (Rao, 1937)	DD
104	*Nyctibatrachus kumbara* (Gururaja et al., 2014)	NA
105	*Nyctibatrachus major* (Boulenger, 1882)	VU
106	*Nyctibatrachus manalari* (Garg et al., 2017)	DD
107	*Nyctibatrachus mewasinghi* (Krutha et al., 2017)	NA
108	*Nyctibatrachus minimus* (Biju et al., 2007)	DD
109	*Nyctibatrachus minor* (Inger et al., 1984)	EN
110	*Nyctibatrachus periyar* (Biju et al., 2011)	NA
111	*Nyctibatrachus petraeus* (Das and Kunte, 2005)	LC
112	*Nyctibatrachus pillaii* (Biju et al., 2011)	NA
113	*Nyctibatrachus poocha* (Biju et al., 2011)	NA
114	*Nyctibatrachus pulivijayani* (Garg et al., 2017)	DD
115	*Nyctibatrachus radcliffei* (Garg et al., 2017)	DD
116	*Nyctibatrachus robinmoorei* (Garg et al., 2017)	DD
117	*Nyctibatrachus sabarimalai* (Garg et al., 2017)	DD
118	*Nyctibatrachus sanctipalustris* (Rao, 1920)	EN
119	*Nyctibatrachus shiradi* (Biju et al., 2011)	NA
120	*Nyctibatrachus sylvaticus* (Rao, 1937)	DD
121	*Nyctibatrachus vasanthi* (Ravichandran, 1997)	EN
122	*Nyctibatrachus vrijeuni* (Biju et al., 2011)	NA
123	*Nyctibatrachus webilla* (Garg et al., 2017)	DD
124	*Clinotarsus curtipes* (Jerdon, 1853)	LC
125	*Hydrophylax bahuvistara* (Padhye et al., 2015)	NA
126	*Hydrophylax malabaricus* (Tschudi, 1838)	LC
127	*Indosylvirana aurantiaca* (Boulenger, 1904)	NA
128	*Indosylvirana caesari* (Biju et al., 2014)	NA
129	*Indosylvirana doni* (Biju et al., 2014)	NA
130	*Indosylvirana flavescens* (Jerdon, 1853)	NA
131	*Indosylvirana indica* (Biju et al., 2014)	NA
132	*Indosylvirana intermedia* (Rao, 1937)	NA
133	*Indosylvirana magna* (Biju et al., 2014)	NA
134	*Indosylvirana montana* (Rao, 1922)	NA
135	*Indosylvirana sreeni* (Biju et al., 2014)	NA
136	*Indosylvirana urbis* (Biju et al., 2014)	NA
137	*Indirana beddomii* (Günther, 1876)	LC
138	*Indirana bhadrai* (Garg and Biju, 2016)	NA
139	*Indirana brachytarsus* (Günther, 1876)	EN
140	*Indirana chiravasi* (Padhye et al., 2014)	NA

APPENDIX 13.1   *(Continued)*

Sl. no.	Species	IUCN Red List status
141	*Indirana duboisi* (Dahanukar et al., 2016)	NA
142	*Indirana gundia* (Dubois, 1986)	CE
143	*Indirana leithii* (Boulenger, 1888)	VU
144	*Indirana longicrus* (Rao, 1937)	DD
145	*Indirana paramakri* (Garg and Biju, 2016)	NA
146	*Indirana salelkari* (Modak et al., 2015)	NA
147	*Indirana sarojamma* (Dahanukar et al., 2016)	NA
148	*Indirana semipalmata* (Boulenger, 1882)	LC
149	*Indirana tysoni* (Dahanukar et al., 2016)	NA
150	*Indirana yadera* (Dahanukar et al., 2016)	NA
151	*Walkerana diplosticta* (Günther, 1876)	EN
152	*Walkerana leptodactyla* (Boulenger, 1882)	EN
153	*Walkerana muduga* (Dinesh et al., 2020)	NA
154	*Walkerana phrynoderma* (Boulenger, 1882)	CE
155	*Beddomixalus bijui* (Zachariah et al., 2011)	NA
156	*Ghatixalus asterops* (Biju et al., 2008)	DD
157	*Ghatixalus magnus* Abraham, Mathew, Cyriac, Zachariah, Raju, and Zachariah, 2015	NA
158	*Ghatixalus variabilis* (Jerdon, 1853)	EN
159	*Mercurana myristicapalustris* (Abraham et al., 2013)	NA
160	*Polypedates maculatus* (Gray, 1830)	LC
161	*Polypedates occidentalis* (Das and Dutta, 2006)	DD
162	*Polypedates pseudocruciger* (Das and Ravichandran, 1998)	LC
163	*Pseudophilautus amboli* (Biju and Bossuyt, 2009)	CE
164	*Pseudophilautus kani* (Biju and Bossuyt, 2009)	LC
165	*Pseudophilautus wynaadensis* (Jerdon, 1853)	EN
166	*Raorchestes agasthyaensis* (Zachariah et al., 2011)	NA
167	*Raorchestes akroparallagi* (Biju and Bossuyt, 2009)	LC
168	*Raorchestes anili* (Biju and Bossuyt, 2006)	LC
169	*Raorchestes archeos* (Vijayakumar et al., 2014)	NA
170	*Raorchestes aureus* (Vijayakumar et al., 2014)	NA
171	*Raorchestes beddomii* (Günther, 1876)	NT
172	*Raorchestes blandus* (Vijayakumar et al., 2014)	NA
173	*Raorchestes bobingeri* (Biju and Bossuyt, 2005)	VU
174	*Raorchestes bombayensis* (Annandale, 1919)	VU
175	*Raorchestes chalazodes* (Günther, 1876)	CE
176	*Raorchestes charius* (Rao, 1937)	EN
177	*Raorchestes chlorosomma* (Biju and Bossuyt, 2009)	CE

APPENDIX 13.1 *(Continued)*

Sl. no.	Species	IUCN Red List status
178	*Raorchestes chotta* (Biju and Bossuyt, 2009)	DD
179	*Raorchestes chromasynchysi* (Biju and Bossuyt, 2009)	VU
180	*Raorchestes coonoorensis* (Biju and Bossuyt, 2009)	LC
181	*Raorchestes crustai* (Zachariah et al., 2011)	NA
182	*Raorchestes drutaahu* (Garg et al., 2021)	NA
183	*Raorchestes dubois* (Biju and Bossuyt, 2006)	VU
184	*Raorchestes echinatus* (Vijayakumar et al., 2014)	NA
185	*Raorchestes flaviocularis* (Vijayakumar et al., 2014)	NA
186	*Raorchestes flaviventris* (Boulenger, 1882)	DD
187	*Raorchestes ghatei* (Padhye et al., 2013)	NA
188	*Raorchestes glandulosus* (Jerdon, 1853)	VU
189	*Raorchestes graminirupes* (Biju and Bossuyt, 2005)	VU
190	*Raorchestes griet* (Bossuyt, 2002)	CE
191	*Raorchestes hassanensis* (Dutta, 1985)	NA
192	*Raorchestes honnametti* (Gururaja et al., 2016)	NA
193	*Raorchestes indigo* (Vijayakumar et al., 2014)	NA
194	*Raorchestes jayarami* (Biju and Bossuyt, 2009)	NA
195	*Raorchestes johnceei* (Zachariah et al., 2011	NA
196	*Raorchestes kadalarensis* (Zachariah et al., 2011)	NA
197	*Raorchestes kaikatti* (Biju and Bossuyt, 2009)	CE
198	*Raorchestes kakachi* (Seshadri et al., 2012)	NA
199	*Raorchestes kakkayamensis* (Garg et al., 2021)	NA
200	*Raorchestes keirasabinae* (Garg et al., 2021)	NA
201	*Raorchestes lechiya* (Zachariah et al., 2016)	NA
202	*Raorchestes leucolatus* (Vijayakumar et al., 2014)	NA
203	*Raorchestes luteolus* (Kuramoto and Joshy, 2003)	DD
204	*Raorchestes manohari* (Zachariah et al., 2011	NA
205	*Raorchestes marki* (Biju and Bossuyt, 2009)	CE
206	*Raorchestes munnarensis* (Biju and Bossuyt, 2009)	CE
207	*Raorchestes nerostagona* (Biju and Bossuyt, 2005)	EN
208	*Raorchestes ochlandrae* (Gururaja et al., 2007)	DD
209	*Raorchestes ponmudi* (Biju and Bossuyt, 2005)	CE
210	*Raorchestes primarrumpfi* (Vijayakumar et al., 2014)	NA
211	*Raorchestes ravii* (Zachariah et al., 2011)	NA
212	*Raorchestes resplendens* (Biju et al., 2010)	CE
213	*Raorchestes sanjappai* (Garg et al., 2021)	NA
214	*Raorchestes signatus* (Boulenger, 1882)	EN
215	*Raorchestes silentvalley* (Zachariah et al., 2016)	NA

APPENDIX 13.1 *(Continued)*

Sl. no.	Species	IUCN Red List status
216	*Raorchestes sushili* (Biju and Bossuyt, 2009)	CE
217	*Raorchestes theuerkaufi* (Zachariah et al., 2011)	NA
218	*Raorchestes thodai* (Zachariah et al., 2011)	NA
219	*Raorchestes tinniens* (Jerdon, 1853)	EN
220	*Raorchestes travancoricus* (Boulenger, 1891)	EN
221	*Raorchestes tuberohumerus* (Kuramoto and Joshy, 2003)	DD
222	*Raorchestes uthamani* (Zachariah et al., 2011)	NA
223	*Raorchestes vellikkannan* (Garg et al., 2021)	NA
224	*Rhacophorus calcadensis* (Ahl, 1927)	EN
225	*Rhacophorus lateralis* (Boulenger, 1883)	EN
226	*Rhacophorus malabaricus* (Jerdon, 1870)	LC
227	*Rhacophorus pseudomalabaricus* (Vasudevan and Dutta, 2000)	CE
228	*Gegeneophis carnosus* (Beddome, 1870)	DD
229	*Gegeneophis danieli* (Giri et al., 2003)	DD
230	*Gegeneophis goaensis* (Bhatta et al., 2007)	DD
231	*Gegeneophis krishni* (Pillai and Ravichandran, 1999)	DD
232	*Gegeneophis madhavai* (Bhatta and Srinivasa, 2004)	DD
233	*Gegeneophis mhadeiensis* (Bhatta et al., 2007)	DD
234	*Gegeneophis pareshi* (Giri et al., 2011)	NA
235	*Gegeneophis primus* (Kotharambath et al., 2012)	DD
236	*Gegeneophis ramaswamii* (Taylor, 1964)	LC
237	*Gegeneophis seshachari* (Ravichandran et al., 2003)	DD
238	*Gegeneophis tejaswini* (Kotharambath et al., 2015)	DD
239	*Indotyphlus battersbyi* (Taylor, 1960)	DD
240	*Indotyphlus maharashtraensis* (Giri et al., 2004)	DD
241	*Ichthyophis beddomei* (Peters, 1880)	LC
242	*Ichthyophis davidi* (Bhatta et al., 2011)	DD
243	*Ichthyophis kodaguensis* (Wilkinson et al., 2007)	DD
244	*Ichthyophis longicephalus* (Pillai, 1986)	DD
245	*Ichthyophis tricolor* (Annandale, 1909)	LC
246	*Uraeotyphlus bombayensis* (Taylor, 1960)	LC
247	*Uraeotyphlus gansi* (Gower et al., 2008)	DD
248	*Uraeotyphlus interruptus* (Pillai and Ravichandran, 1999)	DD
249	*Uraeotyphlus malabaricus* (Beddome, 1870)	DD
250	*Uraeotyphlus menoni* (Annandale, 1913)	DD
251	*Uraeotyphlus narayani* (Seshachar, 1939)	DD
252	*Uraeotyphlus oommeni* (Gower and Wilkinson, 2007)	DD
253	*Uraeotyphlus oxyurus* (Duméril and Bibron, 1841)	DD

FIGURE 13A.1 *Raorchestes glandulosus* from Coorg (KP Dinesh)

FIGURE 13A.2 *Raorchestes hassanesnis* from Coorg (KP Dinesh)

FIGURE 13A.3   *Raorchestes luteolus* from Chickkamagaluru (KP Dinesh)

FIGURE 13A.4   *Rhacophorus lateralis* from Chickkamagaluru (KP Dinesh)

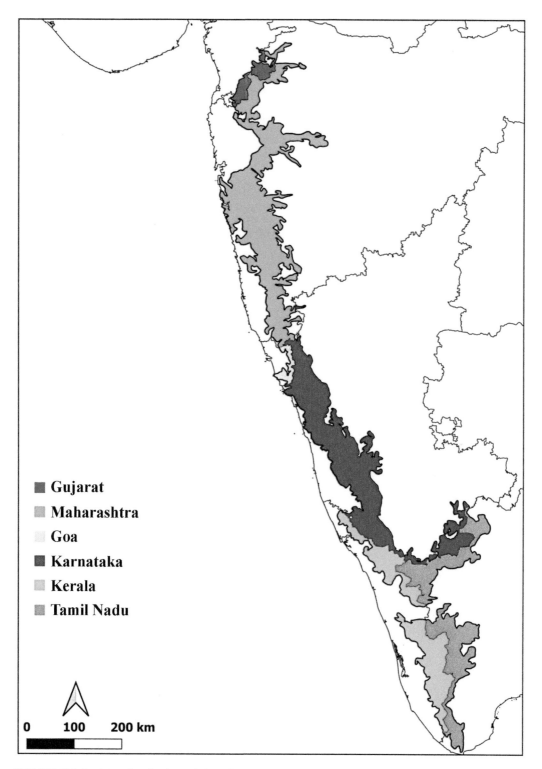

FIGURE 13A.5   Map showing boundaries of the Western Ghats (P Deepak using QGIS)

CHAPTER 14

REPTILE DISTRIBUTION IN THE WESTERN GHATS

RAMAKRISHNA

Zoological Survey of India, India

ABSTRACT

The status of all biogeographically distinct reptilian fauna in the Western Ghats (WG) of India has been assessed in the present chapter. A total of 305 species of reptiles occur in the WG biogeographic region of country, which include 9 species of chelonians, 152 species of lizards, and 143 species of snakes. Of these, nearly 50% (154) are endemic to the area and 48 species are endemic to Peninsular India. A considerable number of species are either included under least concern (156) or data deficient (33) thus requires immediate attention for their review based on the recent sightings. The lizards of the family Gekkonidae, Agamidae, and Scincidae (skinks) are very ancient and comprise of highly specialized lizards. The present-day distribution of this ancient genus of small geckos *Cnemapsis* from WG alone is 46 species of which 31 species are described new to science from WG between 2000 and 2020, signifying the enormous potentiality of cryptic species. Of the 46 species, 37 are endemic to WG (80%), 2 species are restricted to Deccan, and rest are Indian endemics. Recent introduction of invasive alien species includes *Trachemys scripta* (Thunberg in Schoepff, 1792), from South America needs further monitoring of its invasiveness. Endemic snake species from the family Uropeltidae are from the genera *Melanophidium, Platyplectrurs, Teretrurus, Plectrurus, Uropeltis, and Rhinophi,* a large scope exists for further discovery of this group.

14.1 INTRODUCTION

The Western Ghats (WG) is one of the rich and biodiverse tropical land mass, (mountain range) located along the western stretch of coastal area of Peninsular India. Several areas of WG are biotically significant in view of faunistic surveys as sporadic reptile collections were carried out. Historically, the earliest scientific references to the Indian snakes available must be credited to Patrick Russell. He has for the first time distinguished the venomous from

Biodiversity Hotspot of the Western Ghats and Sri Lanka. T. Pullaiah, PhD (Ed.)

the non-venomous snakes of India. His two volumes on Indian snakes (1796, 1801–1809) bear his stamp of authority on the subject at the time. William Theobald in 1876 was first to compile the *Descriptive Catalogue of the Reptiles of British India* wherein he mentions that "Catalogue is based on Gunther's Monograph, published by the Royal Society in 1864. This was followed by the publication "The *Reptiles of British India*" by Gunther (1864), "*Descriptive Catalogue of the Reptiles of British India*" by Blyth (1876), and *'The Fauna of British India: Reptilia and Batrachia'* by Boulenger (1890). The works of Malcolm Smith (1931, 1935a), though more than 85 years old, still remains the most important contribution. The foundation for the primary investigations on the chelonians (testudines) of India was firmly laid by Gray (1825–1875). The systematics and distribution of the saurian fauna of India has been worked out by earlier naturalists like Anderson (1871–1872), Gunther (1864–1875), Murray (1884–1887), Stoliczka (1870–1873), and Annandale (1904–1921).

Earlier studies on Indian reptiles include Hora (1924), Wall (1905–1919, 1928), Mahendra (1939), Gharpurey (1935), Inger et al. (1984), Das and Whitaker (1990), Malhotra and Davis (1991), Zachariah (1997), Ishwar et al. (2001), Hutton and David (2009), Chandramouli and Ganesh (2010), Bhupathy and Nixon (2011), Bhupathy et al. (2012), Daniel and Shull (1963), Nande and Deshmuk (2007), Mirza and Pal (2008), Giri (2008), and Giri and Bauer (2008).

14.2 DIVERSITY, DISTRIBUTION, AND ENDEMIC REPTILIA

14.2.1 TURTLE AND TORTOISE

Nine species of Chelonians are distributed in WG viz., *Vijayachelys silvatica* (Henderson, 1912), *Lessymis punctata* (Bonnaterre, 1789), *Pelochelys cantorii* (Gray, 1864), *Hardella thurjii* (Gray, 1831), *Melanochelys trijuga* (Schweigger, 1812), *Geochelone elegans* (Schoepff, 1795), *Indotestudo travancorica* (Boulenger, 1907), *Nilssonia leithii* (Gray, 1872), and *Trachemys scripta* (Thunberg in Schoepff, 1792). Among them *Nilssonia leithii* (Gray, 1872) a freshwater turtle endemic to peninsular India, distribution in the states of Karnataka (Cauvery, Krishna, and Godavari river system), Kerala (Chalakudy, Bharathapuzha, and Chaliyar), TN (Vaigai river system and Maharashtra (Pawna). Also found distributed in the Eastern Ghats of AP and Orissa. Because of the threat and trade, the species is placed under vulnerable and nationally protected. Henderson (1912) described the species *Vijayachelys silvatica* and placed them under the genus *Geoemyda* (Gray, 1834), found in WG. Praschag et al. (2006) based on phylogenetic distinctness proposed a new Genus *Vijayachelys* (named in honor of the late Indian herpetologist Jaganath Vijaya) which includes only *Vijayachelys silvatica* (a monotypic genus). The cane turtle is endemic to WG. Recently, they are sighted from many different localities: Chalakudy, Poyankutti, Kulathupuzha, and Nadukani reserve forests; Peechi-Vazhanai, Neyyar, Peppara, Idukki, and Aralam wildlife sanctuaries; Parambikulam Tiger Reserve in Kerala, Anamalai Tiger Reserve and Kodayar in TN, Mookambika Wildlife Sanctuary, Sharavathi, Kathlaekan, Agumbe, and Neria forest divisions in Karnataka (Deepak and Vasudevan, 2009). *Vijayachelys silvatica* is protected under Schedule I of Indian Wildlife (Protection) Act of 1972. IUCN 2008 Red List: Endangered. *Indotestudo travancorica* (Boulenger, 1907), originally described from the Travancore Hills, endemic to the WG, has been reported from Kerala, Tamil Nadu, and Karnataka states from 100–1000 m msl (Boulenger, 1907; Bhupathy and Choudhury, 1995). The species is hunted in most of its range, by the Tribes of the WG, yet another reason for their declining trend is due to the construction of hydroelectric dams and barrages for water storage, thus altering their habitat. *Indotestudo travancorica* is listed as "vulnerable."

Pelochelys cantorii (Gray, 1864) though cosmopolitan in its distribution, also found distributed in WG part, is endangered. *Hardella thurjii* (Gray, 1831), distributed in Indo Gangetic plain of North India, Nepal, Bhutan, and Bangladesh, a vulnerable with occasional reference of its distribution in WG. The Indian Black Turtle, *Melanochelys trijuga* a widely distributed in India, Bangladesh, Myanmar, Sri Lanka, Maldives, Chagos Islands, Nepal with six subspecies have been described of which four are found in India. The subspecies found in India are *Melanochelys trijuga trijuga* (Schweigger, 1812), *M. trijuga coronata* (Anderson, 1879), *M. trijuga indopeninsularis* (Annandale, 1913), *M. trijuga thermalis* (Lesson, 1830). The other two subspecies are *M. trijuga parkeri* (Deraniyagala, 1939), *M. trijuga edeniana* (Theobald, 1876) exhibit disjunct distribution in the hill ranges of the WG, south of Gujarat, and the southeast coast. *Geochelone elegans* is terrestrial in habit found in Indian subcontinent in western India and extreme southeastern Pakistan, in southeastern India (Karnataka, Kerala, and Tamil Nadu), and on the island of Sri Lanka. It is distinctive and attractive; growing up to around 30 cm across the carapace.

14.2.2 LIZARDS

One hundred and fifty two species are represented in WG of India, among the Family Gekkonidae is very ancient and comprises of highly specialized lizards exhibiting remarkable adaptive radiations, evolutionary history, distribution pattern, varied habitat selection, flexibility for change, and a tendency to evolve into new and more advance forms. The family included species belonging to *Hemidactylus* species, richness in *Hemidactylus* geckos is very high in places like Eastern Ghats and northern WG. The genus *Cyrtodactylus* comprise of six taxa in southern India, in fact need revisionary and population studies as they are distributed in low range hills and arid region of Peninsular India. *Cyrtodactyilis collegalensis,* a forest species which is becoming vulnerable. The other species distributed in WG are *Cyrtodactylus albofasciatus* (Boulenger, 1885) distributed in the southern WG of Karnataka (Castle Rock, Karwar, and Dakshina Kannada).

At the time of writing, the *Fauna of India* (*Sauria*) by (Tikader and Sharma, 1992), the genus *Cnemapsis* had only about 23 species. The present day distribution of this ancient genus of small Geckos *Cnemapsis* from WG alone is 46 species of which 31 species are described new to science from WG between 2000 and 2020, signifying the enormous potentiality of cryptic species. Of the 46 species, 37 are endemic to WG (80%), 2 species are restricted to Deccan, and rest are Indian Endemics. As the genus is restricted to hilly and forested terrain, crepuscular nature in rocks and caves, a still possibility of more species from the inaccessible, and underexplored area of WG.

The genus *Calaodactylus* has only one rare gecko *Calodactylus aureus* (Beddome, 1870) found in Eastern Ghats and Northern WG. This species is often reported to live and nest communally in large numbers in caves and rock formations, since the male changes its color to golden brown during the breeding season, popularly known as Golden Gecko. An ancient endemic genus *Dravidogecko* (Smith, 1933), known from the Indian peninsula since the late Cretaceous period, long thought to be a monotypic species, currently, seven species are known from the WG (Nilgiris and further south). *Dravidogecko anamallensis* (Gunther, 1875) a nocturnal species found in the hill tops of Anamalai, Palani and Tinnevelly Hills of South–western India. However, recently Chaitanya et al. (2019) described six (6) species viz., *D. douglasadamsi, D. janakiae, D. meghamalaiensis, D. septentrionalis, D. smithi,* and *D. tholpalli* from WG.

Species richness in *Hemidactylus* geckos is very high in places like Eastern Ghats and northern WG, mainly in the rocky hill tracts covered with dry, sparse forest country. As many as 30 species are distributed in WG; most of the *Hemidactylus* species are nocturnal and insectivorous. Of all, *Hemidactylus brooki* (Gray, 1845), a widely distributed species in India, *Hemidactylus prashadi* (Smith, 1935) distributed in deserted houses in Karwar, Jog (Karnataka), and Goa areas of WG, *Hemidactylus reticulatus* (Beddome, 1870), whose type locality is Kollegal in Karnataka part of WG as well as in hilly areas of Bababudin Hills, Gudikal Hills, and Shevaroy Hills. *Hemidactylus frenatus* (Schlegel, 1836) found in Peninsular India, West Bengal, Andaman and Nicobar Islands. *Hemidactyills flaviviridis* (Ruppell, 1835); *Hemidactylus garnoti* (Duméril and Bibron, 1836) distributed in whole of India, but widely distributed in North India above 20°N. *Hemidactylus giganteus* (Stoliczka, 1871) and *Hemidactylus bowringi* (Gray, 1845) distributed in Eastern and WG while the *Hemidactylus albofasciatlls* (Grandison and Soman, 1963) in the North WG.

Two species family Eublepharidae are known from WG viz., *Eublepharis fuscus* (Borner, 1981) and *Eublepharis hardwickii* (Gray, 1827).

Among the members of the family Agamidae, *Draco dussumieri* (Duméril and Bibron, 1837) distributed in WG (Malabar coast, Karwar, Goa, Mercara, Trivandrum); *Sitana ponticeriana* (Cuvier, 1844) distributed in Deccan plateau; *Otocryptis beddomii* (Boulenger, 1885) in Sivagiri Ghat, Cardamom Hills; *Salea horsfieldi* (Gray, 1845) in Nilgiri and Palni Hills, *Salea anamallayana* (Beddome, 1878) in Anaimalai, Palni, and other hills of South-western Ghats; *Calotes nemoricola* (Jerdon, 1853) distributed in Nilgiri Hills; *Calotes grandisquamis* (Gunther, 1875) in Anaimalai and Bramagherry Hills; *Calotes calotes* (Linnaeus, 1758) in Shevaroy Hills, Malabar coast; *Calotes rouxi* (Duméril and Bibron, 1837) in Matheran, Khandala, Kanara, Jog, Goa, and Malabar of south, central, and North WGs, while *Calotes elliotti* (Gunther, 1864) in Anaimalai, Tinnevelly, and Sivagiri Hills. *Psammophilus dosralis* (Gray, 1831) strictly terrestrial or rock dwelling species distributed in Eastern and WG of India south of 16°N (Malabar coast, Karnataka, Nilgiri and Nallamalai Hills, South Arcot, Bangalore); *Psammophilus blanfordanus* (Stoliczka, 1871) absolutely rock dwelling, lives in the crevices and holes in rocks up to 800 m of Eastern and WG. Work done by Macey et al. (2000) on agamids revealed three distinct clades, which largely correspond to the major Gondwanan landmasses. Many Southeast Asian agamids are nested within the Indian clade, suggesting an Indian origin.

The family Scincidae (skinks) is the largest group among lizards. *Kaestlea* is one of the endemic genera of skinks found in the WG. *Kaestlea* has five species all of which are restricted to the central and southern WG. *Ristella* is the only other endemic genus from the WG. It currently consists of four species, with a slightly wider distribution than that of *Kaestlea*. *Ristella beddomii* is found in parts of the northern WG (northernmost being Castle Rock in Karnataka) all the way till the southern WG. *Dasia* is distributed in South and Southeast Asia. In India, there are two species (*Dasia subcaerulea* and *Dasia johnsinghi)* found only in the WG. *Sphenomorphus* is one of the most specious genera of skinks and, in fact, for long it has been a catch-all genus with more than 130 species. *Sphenomorphus dussumieri* is endemic to Southern India and found quite frequently in the plains of Kerala. *Eutropis* is one of the most speciose genera in tropical Asia. *Eutropis clivicola*, is found only in mid to low elevations of the WG. *Eurylepis taeniolatus* is largely restricted to the arid northwestern part of India wherein its distribution extends to the Palaearctic. *E. poonaensis*, however, has a very restricted distribution in the northern WG and is known only from its type locality.

Mabuya is a genus of squamata in the family of long tailed skinks. *Mabuya bibroni* (Gray, 1838) generally inhabiting the sea coast of Kerala, Orissa, and Tamil Nadu; *Mabuya allapallensis* (Schmidt, 1926) distributed in Eastern and WG; *Mabuya carinata* (Schneider, 1801) known from Indian Peninsula, Assam, Bengal; *Mabuya beddomii* (Jerdon, 1870) is from the whole of Peninsular India, Anaimalai Hills, Sivagherry Hills, Tinnevelly Hills, Hills of Malabar coast, Salem and Karnataka and Southern portion of Madhya Pradesh, North-Eastern part of Maharashtra; *Mabuya trivittata* (Hardwicke and Gray) is from Eastern and WG extends up to Bengal belt. *Sphenomorphus dussumieri* (Duméril and Bibron, 1839), forest skink found in South-western portion from Kanara to Trivandrum, thus covering the complete WG and extends up to Sri Lanka.

The classification of the family Scincidae, and especially of the subfamily Lygosomine (skinks), has always presented many difficulties. Boulenger (1887) was the first who tried to arrange them into large groups. This arrangement is useful as a classification only. In India, the genus *Lygosoma* is represented by nine species, of which five are endemic (Datta-Roy et al., 2014) including Günther's Supple Skink *Lygosoma guentheri* (Peters, 1879) and the Lined Supple Skink *Lygosoma lineata* (Gray, 1839). *Lygosoma guentheri* distributed in Eastern and WG of Gujarat, Maharashtra, Goa, Karnataka, AP, and Telangana (Vyas, 2014). *Scincella travancoricum* (Beddome, 1870) extends its distribution to Anaimalai Hills, Palni Hills, and Malabar Coast; *Scincella beddomii* (Boulenger, 1887) extends southern WG; *Scincella laterimaculatum* (Boulenger, 1870) is in Tinnevelly Hills, Nilgiri Hills, Malabar coast; *Scincella bilineata* (Gray, 1846) known from Nilgiri Hills up to the summit at 2600 m. Species of *Riopa albopunctata* (=*Lygosoma albopunctatum*) (Gray, 1846) found almost whole of India and Sri Lanka; *Riopa guentheri* (Peters, 1879) (=*Lygosoma guentheri*) is in Maharashtra (Matheran, Sholapur, Kurduwadi), Karnataka (Belgaum, N. Kanara), AP, Maharashtra, South Gujarat, Karnataka, Telangana; *Lygosoma lineata* (Gray, 1839); extends its distribution to Bombay district between Poona and N Kanara, Gujarat, Madhya Pradesh, TN, Maharashtra, Karnataka, Jharkhand; *Lygosoma goaensis* (Sharma, 1976) has the restricted distribution to Goa and adjoining areas. *Eutropis carinata* is one of the largest species of this genus and is known to be widespread all over the Indian subcontinent. In general, *Eutropis* and *Lygosoma* suggest that, rather than being related to the African lineages, their ancestors came from Southeast Asia. The diversity of skinks in the WG appears to be relatively low compared to the rest of the Indian subcontinent. The zoogeography and evolutionary aspect of the group is interesting, the genus *Scincella* is supposed to have originated in Southeast Asia (Greer, 1974). Whether there already existed more than one species in southern Asia before India was joined to the mainland of Asia in the Miocene, is hard to tell, but not unlikely, considering the distance (in similarity) between the Chinese, the Thai and the South Indian species on the one hand and the Himalayan species on the other. Genus *Sphenomorphus* is included in the analysis because of Greer's (1974) statement that the recent genera *Sphenomorphus* and *Scincella* are developed from a Southeast Asian *Sphenomorphus*-like ancestor. The Southeast Asian species *Scincella reevesii, S. melanostica,* and S. *doriae* indeed very much resemble species of *Sphenomorphus* in characteristics of the scales and coloration (Ouboter, 1986).

Among the other species distributed in WG include *Cabrita leschenaulti* (Milne-Edwards, 1829), *Ophisops jerdoni* (Blyth, 1853), *Ophisops beddomii* (Jerdon, 1870), *Varanus bengalensis* (Linnaeus, 1758) and *Varanus salvator* (Laurenti, 1768).

14.2.3 SNAKES

Reptiles in WG are represented by 142 species of snakes. Family Uropeltidae comprises shield-tailed or rough-tailed snakes which are restricted to South Indian hilly areas of WG and Sri Lanka. Of the 55 known, extant species, 23 were described by Colonel Richard Henry Beddome (Wallach et al., 2014). Six genera of family Uropeltidae namely *Melanophidium, Platyplecturus, Teretrutus, Plecturus, Uropeltis, Rhinophi*s, and 3 genera of family Colubridae namely *Coronella, Pseudoxenodon,* and *Xylophis* are endemic. India harbors a combination of ancient and more recently dispersed lineages of typhlopoids. The genus *Gerrhopilus* is of Gondwanan origin that likely dispersed out of India into Southeast Asia. The other genera are intrusive elements that dispersed into India from Africa (*Grypotyphlops*) and Asia (*Indotyphlops* and possibly *Argyrophis*) post break-up of Gondwana (Sidharth and Karant, 2020). Moreover, even within a single hill range, different species of shield-tailed snakes allopatrically occur in varying elevational zones (Hutton, 1949; Wall, 1919). The diversity pattern of shieldtails in the WG is unlike any other Indian snake assemblage, but mirrors that of anuran amphibians (reviewed in Biju, 2001).

The distribution of the members of this family Uropeltidae is Peninsular India and Sri Lanka, primarily in the southern WG of India and southwestern and central Sri Lanka, whereas with a few species are distributed in the Eastern Ghats and northern WG of India, and northern Sri Lanka (Smith, 1943; Rajendran, 1985).

Teretrurus rhodogaster Wall, 1921 restricted to Anaimalai and Palni hills of the southern WG at elevations of c. 1280–2100 m; the Genus *Melanophidium* (Günther, 1864) includes four species in WG viz., *Melanophidium bilineatum* (Beddome, 1870) (Paralectotype: possibly Terrihioot peak, Wayanad District, Kerala state), *Melanophidium khairei* (Gower et al., 2016) (Amboli, Sindudurg district, Maharashtra, India. The Paratypes are identified from Patgaon, Kolhapur district, Maharashtra, Verle, and South Goa); "Jelewadi, Goa Frontier," likely Telewadi, Karnataka), *M. punctatum* (Beddome, 1871) (Kalakkad Mundanthurai Tiger Reserve in the Agasthyamalai hills, Kerala state, India). *M. wynaudense* (Beddome, 1863) distributed in Nilgiri hills to Agumbe, at elevations c. 600–2100 m (Pyron et al., 2016). *Platyplectrurus madurensis* (Beddome, 1877) distributed in Palni and Munnar hills at elevations > 1200 m; *Platyplectrurus trilineatus* (Beddome, 1867) in the Anaimalai and possibly Palni hills, at elevations > 1200 m. *Plectrurus aureus* (Beddome, 1880) restricted to Nilgiris at elevations c. 2050 m and Wyanad in Kerala; *Plectrurus guentheri* (Beddome, 1863) and *Plectrurus perrotetii* Duméril & Bibron in Duméril & Duméril, 1851 restricted to the Nilgiris (Walaghat and Coonoor) at elevations 1065–2250 m and recently recorded from Coorg (Kodagu). *Pseudoplectrurus canaricus* (Beddome, 1870) whose type locality is Kudremukh in the Kudremuk hills, Karnataka at elevations > 1800 m.

The distribution of *Rhinophis fergusonianus* (Boulenger, 1896) is restricted to type locality in the Cardamom hills of India, c. 1100 m; *Rhinophis goweri* (Aengals and Ganesh, 2013), whose type locality is Bodamalai hills, and the nearby Kolli hills, in the southern Eastern Ghats of India at elevations c. 900 m; *Rhinophis sanguineus* (Beddome, 1863) is known from the Nilgiris northwards to the Agumbe hills, c. 750–1065 m; *Rhinophis travancoricus* (Boulenger, 1893) is from southern WG from Thattekad to Kanyakumari, typically found mainly on of western slopes < 800 m, but ranging from c. 0 to 1335 m. The species of the genus *Teretrurus* (Beddome, 1886) viz., *Teretrurus sanguineus* (Beddome, 1867) exhibits disjunct distribution in India, including the Anaimalai-Munnar hills at elevations > 1000 m, and the Travancore–Agasthyamalai Hill complex south of the Sencotta Gap.

The distributional pattern of Uropeltids is interesting, as species such as *Melanophidium punctatum, Teretrurus sanguineus, Rhinophis travancoricus, Uropeltis ellioti, U. ceylanica, U. arcticeps, U. rubrolineata, U. myhendrae, U. liura, U. pulneyensis* distributed in Agasthyamalai range (8°N 77°E; 1868 m asl); *Melanophidium punctataum, Teretrurus sanguineus, Rhinophis travancoricus, R. fergusonianus, U. ceylanica, U. madurensis, U. rubrolineata, U. liura, U. pulneyensis* distributed in Cardamom hills (9°N 77°E; 1978 m asl); *Melanophidium punctatum, Platyplectrurus madurensis, P. trilineatus, Brachyophidium rhodogaster, Teretrurus sanguineus, Rhinophis travancoricus, Uropeltis ellioti, U. nitida, U. ocellata, U. beddomei, U. macrorhyncha, U. woodmasoni, U. ceylanica, U. rubromaculata, U. rubrolineata, U. broughami U. maculata, U. petersi, U. pulneyensis* distributed in Anaimalai hills (10°N 76–77°E 2695 m asl); *Melanophidium punctatum, M. bilineatum, M. wynaudense Plectrurus perroteti, P. aureus, P. guentheri, Rhinophis sanguineus, Uropeltis ellioti, U. ocellata, U. ceylanica* distributed In Nilgiris hills (11°N 76°E 2600 m asl); *Melanophidium punctatum, M. wynaudense, Plectrurus perroteti, Pseudoplectrurus canaricus, Rhinophis sanguineus, Uropeltis ellioti, U. ceylanica* in Central WG and *Rhinophis goweri, Uropeltis ellioti, U. dindigalensis, U. shorttii* in Eastern Ghats. Very interesting pattern of distribution has been observed that those with compressed and tapering tail like *Platyplectrurus, Plectrurus*, and *Pseudoplectrurus* are endemic to high elevation Shola landscape only. On the other hand, taxa with a rounded, convex, and truncate tail like *U. ceyanica* group, *U. ellioti* group, and *Rhniophis* are more widespread in South India occurring in both the Eastern and the WG (Ganesh, 2015). These features make shieldtail fauna to fairly fit into an adaptive radiation scheme, that is reported (Schulter, 2000) to be characterized by common ancestry, rapid speciation, ability to live in variety of environments, phenotype-environment ecomorphological correlations.

The Family Colubridae: *Coronella brachyura, Cercaspis carinatus, Balanophis ceylonensis, Macropisthodon plumbicolor, Aspidura brachyorrhus, Aspidura copii, Aspidura trachyprocta, Aspidura drummondhayi, Aspidura guentheri, Haplocercus ceylonensis,*

Xylophis perroteti, and *Xylophis stenorhynchus.* Snake species which are common to the oriental and palaearctic regions: Family Typhlopidae which includes *Indotyphlops exiguus* (Daudin, 1803), *Indotyphlops exiguus* (Jan, 1864), *Indotyphlus fletcheri* (Wall, 1919); the Family Boidae includes *Python molurus, Eryx conicus, Eryxjohni.* Family Colubridae: *Elaphe helena, Elaphe hodgsoni, Ptyas mucosus, Argyrogena fasciolata* (Shah, 1802), *Sibynophis subpunctatus* (Dumeril and Bibron, 1854), *Oligodon arnensis* (Shaw, 1802), *Oligodon brevicauda* (Gunther, 1862), *Oligodon taeniolatus* (Jerdon, 1853), *Oligodon travancoricus* (Beddome, 1877), *Oligodon venustus* (Jerdon, 1853), the Genus *Lycodon* which includes 7 species, *Xenochrophis piscater, Amphiesma stolata, Boiga trigonata, Psammophis condanarus* (Merrem, 1820); the Family Elapidae which includes *Bungarus caeruleus* and *Naja naja.* Family Viperidae: *Vipera russelli, Vipera lebetina, Echis carinatus* and the snake species of oriental origin (Indo-Chinese and Indian subregion), intruded into palaearctic region are *Elaphe helena, Dendrelaphis pictus,* and *Macropisthodon plumbicolor.*

Among the members of the family Boidae, *Eryx whitakeri* (Das, 1991) distributed in SW India from sea level up to 2050 ft in WG, the type locality "Mangalore, Karnataka state, India." *Coelognathus helena monticollaris* (Schulz, 1992; Family Colubridae) the Montane Trinket Snake is endemic to WG; *Coluber gracilis* (Gunther, 1862), *Coronella brachyura* (Gunther, 1866) distributed in Deccan Plateau in the region of Gujarat, Maharashtra, and Madhya Pradesh; *Oligodon venustus* (Jerdon, 1853) in WG south of Goa; *Oligodon travancoricus* (Beddome, 1877) in WG south of Palghat gap, *Oligodon affinis* (Gunther, 1862), *Oligodon brevicaudus* (Gunther, 1862) in WG, south of the Goa gap, *Oligodon nikhili* (Whitaker and Dattatri, 1982) restricted to Palni Hills, Tamil Nadu. *Dendrelaphis ashoki* (Rooijen and Vogel, 2011), *Dendrelaphis girii* (Vogel and Van Rooijen, 2011), *Dendrelaphis grandoculis* (Boulenger, 1890) in Southern WG in Anamali hills, Hills of Karnataka (Castle Rock, Belgaum district, Karnataka, India).

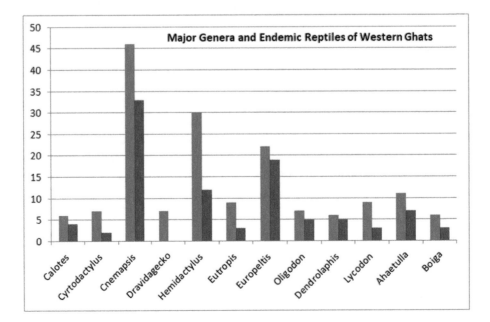

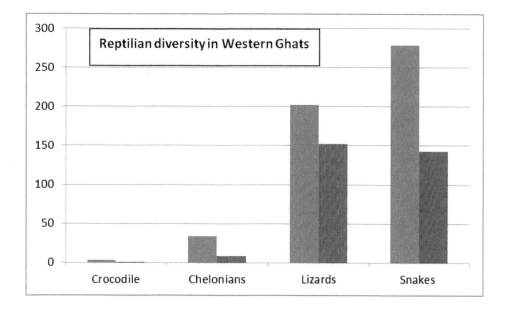

14.3 AFFINITIES OF THE REPTILES

Reptiles were the dominant group of vertebrates during the Mesozoic period and most orders of reptiles were established by the end of Triassic and some became extinct at that time. Of the 19 orders of Reptiles only four survive today (Crocodelia, Testudines, Squamata, and Rhynchocephalia). Ancestry of modern lizards can be traced back to the middle of Jurassic period of Mesozoic era in the Diapsid reptilian stock, about 155 mya. Eosuchia is the order of fossil reptile which inhabited the earth between Upper Triassic and Upper Permian period (160–190 mya). All Eusuchia perished by the beginning of the Cretaceous period or early Mesozoic era, it gave a chance to the lizards and snakes. An abrupt climatic change such as temperature, humidity, rainfall, and the elevation of the mountain ranges contributed to the change in the ecology of that period resulting in the dispersal of the lizards to different ecological niches. Furthermore, the habitat changes brought about the variations in their forms and adaptation to arboreal, scansorial, sartorial, cursorial, terrestrial, cave dwelling, rock dwelling, volant mode of life. It's a well-known fact that at the end of Cretaceous period South America, Africa, India, and Antarctica were completely separated, reptiles in India, particularly, the WG evolved tremendously and became extremely diversified. According to Ramesh (2005), India as a whole in general was the original home of the common ancestral stock of reptiles, including the lizards, much prior to the changes which took place in the northern WG. The dispersal to Madagascar and South Africa took place in the early Cenozoic time, since the peninsula had land connections. It is also believed that most of the species flourished in the WG and then the inter transmission took place between this region and other parts like Tropical South Africa, Madagascar, Western Asia, Assam, Myanmar, Malayan Peninsula, Java, Sumatra, and Borneo.

It seems that in the beginning, the main concentration of the reptiles in India was at lower portion of WG and from there the migration took place to Assam hills through Satpura-Vindhya ranges, many migrated species which could remain unaltered in both the regions till present times, reached up to Java, Sumatra, and Borneo (Ramesh, 1998). Species exclusively to the Indian subregion: *Hardella thurgi* is endemic to this subregion among Chelonians; Species which are common to the oriental and palaearctic regions, present day distribution of Chelonins in WG are *Vijayachelys silvatica* (Henderson, 1912) *Lessymis punctata* (Bonnaterre, 1789), *Pelochelys cantorii* (Gray, 1864), *Geochelone elegans* (Schoepff, 1795), *Indotestudo travancorica* (Boulenger, 1907) and *Nilssonia leithii* (Gray, 1872) and the species common to Indian, Indo-Chinese, and Malayan subregion. Testudines: *Melanochelys trijuga* (Schweigger, 1812). Similarly, it seems that in the beginning, the main concentration of the lizards in India was at lower portion of WG and from there the migration took place to Assam hills through Satpura-Vindhya ranges.

Many migrated species which could remain unaltered in both the regions till present times reached up to Java, Sumatra, and Borneo and are as follows: *Cnemaspis kandiana, Hemidactylus frenatus, Hemidactylus bowringi, Cosymbotus platyurus* (=*Hemidactylus platyurus*), *Gehyra mutilata, Hemiphyllodactylus typus, Lepidodactylus lugubris, Mabuya macularia,* and *Riopa lineata* (=*Lygosoma lineata*). The trans migration can be easily noticed in the following species, now inhabiting the Peninsular India and the Malayan region: *Draco dussumieri* and *Draco norvilli; Draco subcaerulea* and *Draco olivacea; Cophotis ceylanica* and *Cophotis sumatrana* (=*Pseudocophotis sumatrana); Sphenomorphus dussumieri,* and *Sphenomorphus maculatum*: and *Riopa albopunctata* (=*Lygosoma albopunctata*) and *Riopa bowringi* (*Lygosoma bowringi*); *Dasia olivacea* and *Dasia subcaerulea.* Members of the Family: genera *Cyrtodactylus, Cnemaspis, Calodactylodes, Dravidogecko, Hemidactylus, Lophopholis, Eublepharis hardwickii* are endemic to the Indian sub-Region including WG. Similarly, Agamid members *Draco, Sitana, Otocryptis, Japalura, Salea, Calotes* species distributed in WG are of Indian origin. Members of Family Scincidae viz., *Mabuya, Dasia, Sphenomorphus, Scincella, Lygosoma,* and *Risterlla* are of Indian origin. However, some of this oriental origin (Indo-Chinese and Indian subregions), intruded in to palaearctic region as well as Indo-Malayan region.

14.4 CONCLUSION

An updated checklist of 518 species of reptiles includes 3 species of crocodiles, 34 species of turtles and tortoises, 202 species of lizards, and 279 species of snakes belonging to 28 families recorded from India. Recent publication of Deuti et al. (2020) reveals the distribution record of reptilian fauna of the WG consisting of 270 species as of now with 173 endemic species (including the Indian peninsular endemics). This includes one species of crocodile, 7 species of turtles (with 2 endemics), 133 species of lizards (with 91 endemics), and 129 species of snakes (with 78 endemics). The present listing of reptiles of WG revealed that 152 species of lizards (Reptilia: Sauria) and 143 species of snakes, 9 species of chelonians.

Diversity of reptiles of Western Ghats		Conservation status	
Crocodile	1	Endangered species	9
Chelonia	9	Vulnerable species	15
Lizards	152	Near threatened species	5
Snakes	143	Least concerned	87
Endemic to Western Ghats	154 (50.49%)	Data deficient	33
Endemic to Peninsular India	48	Not evaluated	156
Invasive species	1		

KEYWORDS

- affinity
- diversity
- distribution
- endemic
- status

REFERENCE

Agarwal, I.; Giri, V. B.; Bauer, A. M. A Cryptic Rock Dwelling *Hemidactylus* (Squamata: Geckkonidae) from South India. *Zootaxa* **2011,** *2765,* 21–37.

Bhupathy, S.; Choudhury, B. C. Status, Distribution and Conservation of the Travancore Tortoise, Indotestudo Forstenii in Western Ghats. *J. Bombay Natl. History Soc.* **1995,** *92,* 16–21.

Biju, S. D. *A Synopsis to the Frog Fauna of the Western Ghats, India*; Occasional Publication 1, Indian Society for Conservation Biology (ISCB), **2001;** pp 1–24.

Boulenger, G. A. Report on a Collection of Reptiles and Batrachians from the Timor Laut Islands, formed by Mr. H. O. Forbes. *Proc. Zool. Soc. London* **1883,** *1883,* 386–388.

Boulenger, G. A. *Catalogue of Lizards in the British Museum (Natural History)*; London, **1887;** vol 3, pp 477–512.

Boulenger, G. A. A New Tortoise from Travancore. *J. Bombay Nat. Hist. Soc.* **1907,** *17,* 560–565.

Chaitanya, R.; Giri, V.; Deepak, V.; Datta Roy, A.; Murthy, B. H. C. K.; Karanth, P. Diversification in the Mountains: A Generic Reappraisal of the Western Ghats Endemic Gecko Genus *Dravidogecko* Smith, 1933 (Squamata: Gekkonidae) with Descriptions of Six New Species. *Zootaxa* **2019,** *4688,* 1–56. https://doi.org/10.11646/zootaxa.4688.1

Das, I. Growth of Knowledge on the Reptiles of India, with an Introduction to Systematics, Taxonomy and Nomenclature. *J. Bombay Nat. Hist. Soc.* **2003,** *100* (2-3), 446–502.

Datta-Roy, A. On the Trail of Skinks of the Western Ghats. *Resonance* **2014,** *19* (8), 753–763.

Datta-Roy, A.; Singh M.; Karanth, K. P. Phylogeny of Endemic Skinks of the Genus *Lygosoma* (Squamata: Scincidae) from India Suggests an In Situ Radiation. *J. Genet.* **2014,** *93,* 163–167.

Deepak, V.; Vasudevan, K. Endemic Turtles of India. In *ENVIS Bulletin: Wildlife and Protected Areas*; Vasudevan, K., Ed.; chapter 3. Freshwater Turtles and Tortoises of India, Wildlife Institute of India: Dehradun, India. **2009,** *12* (1), 25–42.

Deuti, K.; Palot, M. J.; Sethy, P. G. S.; Sarkar, S. Reptilia. In *Faunal Diversity of Biogeographic Zones of India: Western Ghats*; Zoological Survey of India: Kolkata, **2020;** pp 663–697.

Ganesh, S. R. Shieldtail Snakes (Reptilia: Utopeltidae)–The Darwin Finches of South Indian Snake Fauna? *Workshop on Snakes–*2015; August 26–27, **2015.**

Greer, A. E. The Generic Relationships of the Scincid Lizard Genus *Leiolopisma* and its Relatives. *Austr. J. Zool. (Suppl.)* **1974,** *31,* 1–67.

Hutton A. F. Notes on the Snakes and Mammals of the High Wavy Mountains, Madura District, South India. Part I–Snakes. *J. Bombay Nat. Hist. Soc.* **1949,** *48,* 454–460.

Ouboter, P. E. A Revision of the Genus *Scincella* (Reptilia Scincidae) of Asia with Some Notes on its Evolution. *Zoologische verhandellingen* **1986,** *229,* 4–63.

Prasad, B. Annual Address of the President to the National Institute of Sciences of India. *Proc. Nat. Inst. Sci. India Calcutta* **1942,** *8* (1), 27–43.

Praschag, P.; Schmidt, C.; Fritzsch, G.; Muller, A.; Gemel, R.; Fritz, U. *Geoemyda silvatica,* an Enigmatic Turtle of the Geoemydidae (Reptilia: Testudines), Represents a Distinct Genus. *Org. Divers. Evol.* **2006,** *6,* 151–162.

Pyron, R. A.; Ganesh, S. R.; Sayyed, A.; Sharma, V.; Wallach, V.; Somaweera R. A Catalogue and Systematic Overview of the Shield-Tailed Snakes (Serpentes: Uropeltidae). *Zoosystema* **2016,** *38* (4), 453–506. https://doi.org/10.5252/z2016n4a2

Rajendran, M. V. A Survey of Uropeltid Snakes. *J. Madurai Univ.* **1977,** *6,* 68–73.

Sharma, R. C. *Fauna of India and the Adjacent Countries–Reptilia (Testudines and Crocodilia);* Zoological Survey of India: Kolkata, **1998;** vol 1.

Sharma, R. C. *The Fauna of India and the Adjacent Countries-Reptilia (Sallria)*; Zoology Survey of India: Kolkata, **2005;** vol 2, pp: 1–430.

Schluter, D. In *The Ecology of Adaptive Radiation*; Oxford University Press: Oxford, **2000.**

Sidharthan, C.; Karanth, K. P. India's Biogeographic History Through the Eyes of Blindsnakes-Filling the Gaps in the Global Typhlopoid Phylogeny. *Mol. Phylogenet. Evol.* **2021,** *157,* 107064. DOI: 10.1016/j.ympev.2020.107064.

Smith, M. A. *Fauna of British India including Ceylon and Burma*; vol–III Serpentes; Taylor & Francis: London, UK, **1943.**

Tikader, B. K.; Sharma, R. C. *Handbook of India Lizards;* Zoological Survey of India, **1992.**

Vyas, R. Notes and Comments on the Distribution of Two Endemic *Lygosoma* Skinks (Squamata: Scincidae: Lygosominae) from India. *J. Threat. Taxa* **2014,** *6* (14), 6726–6732.

Wall, F. Notes on a Collection of Snakes made in the Nilgiri Hills and Adjacent Wynaad. *J. Bombay Nat. Hist. Soc.* **1919,** *26,* 552–584.

Wall, F. Notes on a Collection of Snbakes from Shenbaganur, Palni Hills. *J. Bombay Nat. Hist. Soc.* **1928,** *29,* 388–398.

Wallach, V.; Williams, K. L.; Boundy J. *Snakes of the World: A Catalogue of Living and Extinct Species*; CRC Press: Boca Raton, **2014.**

CHAPTER 15

BUTTERFLIES AS ECOSYSTEM ENGINEERS IN THE WESTERN GHATS: A BRIEF REVIEW WITH SYSTEMS PERSPECTIVE

PAULRAJ SELVA SINGH RICHARD[1], HOPELAND P.[2], UMESH PAVUKANDY[3], and NITHYA BASITH[1]

[1]Department of Botany, Madras Christian College (Autonomous), Chennai, Tamil Nadu, India

[2]Arihant Heirlooms, Thalambur, Tamil Nadu, India

[3]Pavukandy House, Kozhikode, Kerala, India

ABSTRACT

Butterflies are among the most studied invertebrate groups in India. The Western Ghats alone hosts 336 species of butterflies of which about 12% are endemic. The Western Ghats is considered significant and marked among the hottest biodiversity hotspots with very high levels of endemism. Studies on various aspects of the Western Ghats including its diversity are a few centuries old. Though several efforts have been undertaken across the Western Ghats on various aspects, considerable critical information is still lacking. In an effort to bridge the gap in the knowledge of the butterflies of the Western Ghats, we provide a brief review of studies undertaken in the Western Ghats, along with notes on endemic species of butterflies and plants, their interactions, associations as larval host plants, nectar sources, and pollinators. This chapter aims to highlight the knowledge of the complex interactions of species in a geographically significant landscape such as the Western Ghats and specific interactions between endemic species of butterflies and endemic plants. We highlight these three levels of interaction in the hope to draw attention to the multilayered complex interactions and possible species-specific co-evolution in this larger landscape and thus presenting their role as ecosystem engineers. The loss or disruption of interaction between butterflies and their environment, which if lost can have a domino effect across the ecosystems of Western Ghats. Understanding such complex interactions across taxa

Biodiversity Hotspot of the Western Ghats and Sri Lanka. T. Pullaiah, PhD (Ed.)
© 2024 Apple Academic Press, Inc. Co-published with CRC Press (Taylor & Francis)

groups and a systems approach in this topographically significant landscape is imperative and has strong potential to help in the long-term conservation of not just butterflies but also the larger sensitive ecosystems and biodiversity hotspots within the Western Ghats and beyond.

15.1 INTRODUCTION

Butterflies are among the most explored invertebrate groups in India and particularly in the Western Ghats. However, many aspects remain unexplored like the status or decline in butterfly populations, if any. Globally, studies currently indicate a global decline in insects especially butterflies (Warren et al., 2021; Wepprich et al., 2019). In India, however, empirical data on the decline is lacking. Further, global concerns regarding species' ability to adapt in the context of global change and climate change further raise conservation concerns of ecosystems and species. In this context, butterflies present various opportunities to understand the ecosystems including being ecological indicators as they are very sensitive to the changes in the environment and specifically the scenario in the Western Ghats (Larsen, 1988; Pollard, 1991; Kunte, 1997). It therefore remains important to establish wholistic baseline information, diversify study approaches, and gain the understanding of butterflies and the Western Ghats to be able to successfully keep up with conservation challenges the landscape faces. Information, such as diversity at the local level, population dynamics, annual fluctuations, and regulators among many other important questions are still lacking and have not been studied or understood (Padhye et al., 2012; Anto et al., 2021). However, butterflies being charismatic species, have gained popularity among the general public in the recent past. Understanding their distribution is currently gathering momentum through Citizen Science programs and networks, particularly in the last 10 years. Although butterflies and the Western Ghats have drawn attention and make excellent model systems given their character and sensitivity, the role of butterflies beyond being charismatic species and model organisms for conservation as significant ecosystem engineers is something frequently overlooked (Ehrlich and Raven, 1964; Rao and Girish, 2007; Bonebrake et al., 2010).

Their role in ecosystems is multitiered. Through pollination, they engage and maintain reproductive success, genetic diversity, and promote disease resistance among other benefits, thus playing a vital role in the population biology of plants through this interaction. As herbivores (at the larval stage), they influence the population dynamics of plants and diversity at the local level (Agrawal et al., 2006; Bagchi et al., 2014). They form a critical prey base themselves in both larval and adult stages for many species including migratory species in India and the Western Ghats (Katti and Price, 2003; Bonebrake et al., 2010). They also have critical links with the members of Hymenoptera (wasps) and Tachinidae (flies) and possibly other taxa groups as parasite and parasitoid hosts where some of these links were realized only recently from the Western Ghats (Bonebrake et al., 2010; Gupta et al., 2011, 2015). Over 75% of the world's Lycaenidae butterflies are also known to have close associations with ants in their larval stage. These interactions can be mutualistic, parasitic, or predatory (Pierce et al., 2002). Besides such interactions and associations, competition at the intra and interspecific level of butterflies themselves is suggested to be a structuring process impacting the community dynamics at the local and eventually larger landscape (Bonebrake et al., 2010). This multilevel association of butterflies with various taxa groups suggests their significance as an important link- as ecosystem engineers in a diverse landscape with high endemism such as the Western Ghats.

The Western Ghats along with Sri Lanka is one of the 36 global biodiversity hotspots and runs parallel along the West Coast of India from southern Gujarat in Central India to the southernmost point of India and extends to Sri Lanka (Myers et al., 2000). It is one of the 10 hottest biodiversity hotspots covering an area of 129,037 sq. km. area. This 6% of the land area alone holds nearly 30% of Indian biodiversity and is known for its rich biodiversity and high endemism (CEPF, 2016; Groombridge, 1992). Gunawardene et al. (2007) suggested that at least 28% of mosses, 43% of liverworts, 56% of evergreen trees, and ~40% of lianas; among fauna, at least ~20% of ants, 40% of odonates, 11% of butterflies, 76% of molluscs, 41% of freshwater fishes, 78% of amphibians, 62% of reptiles, 4% of birds, and 11% of mammals are considered endemic to the Western Ghats. Since then many more endemic species across all taxa groups have been described.

Traversing six states of India, the Western Ghats is not only known for its biological resources and levels of endemism but also for culture and is recognized as one of the world heritage sites by UNESCO. It hosts many Tiger reserves with very high levels of protection along with wildlife sanctuaries, national parks, and also holds transitional zone in the form of reserve forests with lower levels of protection. It also hosts several indigenous communities who have historically lived there. However, today, the landscape grapples with landscape transformation tangled with colonial histories among others. It also has geographic elements like the Palghat gap—a prominent pass and other minor passes like the Ariankavu pass which have been noted as geographic elements shaping species distribution and endemism. This presents the Western Ghats as significant locations by sociopolitical construct for the scientific inquiry and high conservation value with various opportunities to understand and work with butterflies and the Western Ghats as model systems to further conservation goals.

To highlight these important taxa (butterflies) in this conservationally significant landscape (Western Ghats), we briefly review the body of work on butterflies in the Western Ghats and attempt to make a case for them as ecosystem engineers with a systems approach and understanding. We provide a list of endemic butterflies of the Western Ghats, their known larval host plants, nectar sources, and distribution of these endemic butterflies across the Western Ghats based on literature and original work (Appendix 15.1). Further, we also provide a list of endemic larval host plants and their known associated butterflies adapted from Nitin et al. (2018) along with their known distribution, occurrence likelihood in various forest/habitat types and altitudes (Appendix 15.2). We also provide information about the interaction of endemic species based on the original work of over 50 lifecycle studies in Agasthiyamalai Biosphere Reserve with species, such as *Pratapa deva* (White Tufted Royal) and *Curetis siva* (Shiva Sunbeam) undertaken for the first time in the southern Western Ghats with additions from literature in the Agasthiyamalai Biosphere Reserve (Pavukandy, Pers. Obs.) (Appendix 15.1). We also draw notes from Citizen Science portals like the "Butterflies of India" of the "Biodiversity Atlas—India" and provide collated data based on the observation of photographs uploaded of nectaring butterflies and egg-laying butterflies. Through these data, we highlight the significance of butterflies and their interactions with their environment specifically endemic species interactions, and more importantly, interactions between endemic butterflies and endemic plants. These complex interactions of butterflies have been highlighted to show their potential co-evolution with plants as host plants in their larval stage and as nectar sources in their adult stage including their role as pollinators and thus their role as ecosystem engineers. Understanding such complex species-specific interactions is imperative to understand the function of complex ecosystem. Such complex, sensitive interconnectedness of physical and biological components, co-evolution of taxa groups including endemic species and their drivers in the region present the Western Ghats and butterflies as a potential model system for study and conservation.

15.2   EARLY EXPLORATION OF BUTTERFLIES

The knowledge of butterflies particularly grew in the early stages of formal science in the 18th and 19th century because of hobby specimen collectors in Europe (Hancock, 2006). It gathered interest and momentum worldwide alongside the age of exploration and the rise of colonialism in the rest of the world (Longstaff, 1912; Goankar, 1996). In India, Civil officers of the British East India Company and others with colonial links actively engaged in collecting specimens. Many of these were collected and sent to British and European museums. In India, the establishment of the Asiatic Society in 1784 and later the Indian Museum or Imperial Museum in 1814, the Madras Museum in 1851, and eventually the establishment of the Bombay Natural History Society in 1883 and their respective journals were pivotal in not just gathering of specimens but also the amalgamation of information and baseline knowledge that is available today.

In the Western Ghats, the studies on butterflies date back to the 18th century. "*Systema Natræ—10th Edition*" was the first publication on butterflies in India and Western Ghats by Carl Linnaeus—the father of taxonomy who had then just formulated a modern binomial taxonomic system of classification. Crimson Rose (*Pachliopta hector*) was one of the butterflies' first species described by Linnaeus (Gaonkar, 1996). Linnaeus' system of classification was also timely and it helped formalize the discoveries (Kunte et al., 2019). Several species of butterflies were described from then by Johan Christian Fabricius, a student of Linnaeus and Pieter Cramer, a Dutch merchant and an amateur entomologist primarily based on collections or illustrations by others who made it available to them from locations where the Dutch had colonial or trading links as they had never traveled to India. Fabricius also described 50 butterflies based on collections made between 1767 and 1785 by Johann Gerhard König, who was also a student of Linnaeus, who served initially as a physician in the Tranquebar Mission in Tamil Nadu before working as a naturalist widely in India (Gaonkar, 1996; Kunte et al., 2019). Later, several studies on butterfly diversity covered various parts of the Western Ghats, such as North Canara (Watson, 1890; Davidson and Aitken, 1890; Davidson et al., 1896, 1897, 1898; Bell, 1909–1927); Coorg (Hannyngton, 1916; Home, 1934; Yates 1929–1933); Nilgiris (Larsen, 1987a,b,c;

Larsen, 1988) among others took place. Overall, over 400 publications on butterflies in the Western Ghats have been made to date (Figure 15.1). They have primarily been of four forms—species descriptions, checklists of butterflies, their choice of habitats, and larval host plants (Padhye et al., 2012). Some studies also explored other aspects like migrations (Larsen, 1987a,b,c; Kunte, 2005). Between 1750 and 1880, there were just under 10 publications. From the 1880s to 2000, as natural history knowledge and interests grew, they were rightly facilitated by the setting up of journals and museums in India, such as the Journal of Bombay Natural History Society, the popularity of the Journal of Bombay Natural History publications, and works gathered pace and an average of 13 publications per decade were made. In the last two decades, however, studies on butterflies gathered momentum and have grown exponentially with over 100 publications. Studies in the last two decades have also moved from traditional frameworks of checklists and distribution to topics, such as molecular biology, physiology, and evolutionary biology among others.

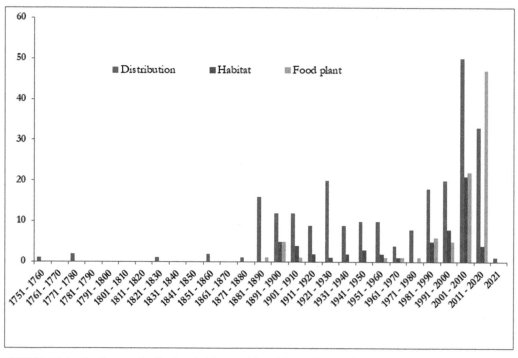

FIGURE 15.1   Studies on distribution, habitat, and larval host plants of butterflies of Western Ghats, India. (Figure prepared by authors)

Among the aspects studied, which induced curiosity and observation historically and today is migration. The migration patterns of butterflies in India have been recorded since the early 1900s. Large number of butterflies migrate regularly in the southern parts of the Western Ghats (Evershed, 1910; Willams, 1927, 1938; French, 1943; Fisher, 1945), which is in contrast with the northern parts where small-level migration occurs (Williams, 1938; Chaturvedi, 1993; Bharosa, 2000). Studies also record the short-distance north–south migrations from Anamalais–Palni and Nilgiris (Williams, 1938; Briscoe, 1952; Larsen,

1987a,b,c) and long-distance east–west migrations from the Western Ghats (Kunte, 2005). In the Western Ghats, about 36 species of butterflies have been reported to show migratory behavior (Evershed, 1910; Williams, 1927; Larsen, 1977). Many more are likely to show their migratory habits. Butterflies like the Painted lady (*Cynthia cardui*) migrate in small numbers from northwestward of the Himalayas during spring and this is in contrast to the large migration of the same species recorded from the Western Ghats (Williams, 1938; Smetacek, 2002). Two primary reasons have been proposed to be the cause of migrations; the availability of food resources and the other reason is to escape torrential weather (Smetacek, 2002; Kunte, 2005; Bonebrake et al., 2010). Kunte (2005) suggested that the Danaine butterflies migrate to scrub or dry deciduous forests in the plains and east coast of India from evergreen or semi-evergreen forests of southern Western Ghats during April or May and mid-October to escape severe southwest monsoon. However, a deeper understanding of its aspects, scale, significance, and implications is required. Additionally, drivers behind other subfamily and families of butterflies showing migratory habits in the Western Ghats needs further research. More recent work on migration has focused on the evolutionary biology of butterflies (Bhaumik and Kunte, 2018, 2020). Much is still unknown regarding butterfly migration in India. Many aspects still are open to be researched including basic information, such as species that migrate, the distance they migrate, locations they commence migration from, locations they disperse to, and triggers of migration among several others (Kunte, 2005; Bonebrake et al., 2010; Bhaumik and Kunte, 2018, 2020). More widely spread, systematic long-term studies and a network of Citizen Science observations show promise and can help understand other poorly known migratory butterfly species and their time of arrival and regions of dispersal as preliminary observations in the coastal plains like Chennai (Hopeland pers. obs).

Concerning the species descriptions, a review by Aravind et al. (2007) suggests that larger and colorful groups, such as papilionids and Nymphalids were discovered much earlier when compared with smaller Lycanids and Hesperids. They also noted that endemic species were discovered significantly later than non-endemics. The latest descriptions from the Western Ghats also agree with the findings. Recent descriptions include the endemic Nilgiri Grass Yellow (*Eurema nilgiriensis*) in 1990; a subspecies of One-spot Grass Yellow (*Eurema andersonii*), the Sahyadri One-spot Grass Yellow (*Eurema andersonii shimai*) in 1999 and more recently, Ramaswami's six-lineblue (*Nacaduba sinhala ramaswamii*) which is considered a subspecies of Ceylon six-lineblue (*Nacaduba sinhala*) known originally from Sri Lanka in 2021 (Yata, 1990; Yata and Gaonkar, 1999; Sadasivan et al., 2021). Although studies may have grown many folds in the last two decades on butterflies, species discovery curves for butterflies in the Western Ghats have been noted to be largely saturated (Aravind et al., 2007).

Other aspects affecting the trend on knowledge and conservation of butterflies include Citizen Science platforms. Such programs at the international and national level include the "Butterflies of India" the "Biodiversity Atlas—India," India Biodiversity Portal and iNaturalist, and Project Noah among others. It also has several regional and local efforts by natural history groups and state forest departments. This collaborative measure among scientists, forest managers, and the public show much promise and is scalable to gather knowledge at both local and landscape level including achieving larger conservation goals in the Western Ghats (Aves, 2021; Kunte et al., 2021; Sekhsaria, 2019; Vattakaven et al., 2016).

15.3 DIVERSITY AND DISTRIBUTION OF BUTTERFLIES IN THE WESTERN GHATS

Over 17,280 butterfly species are known to exist in the world while 1677 species occur in India (Shields, 1989; Gasse, 2013 unpub). The Western Ghats holds 336 species of butterflies belonging to 6 families and 164 genera (Kunte et al., 2021). Among these, 58 species of butterflies are widely distributed throughout the Western Ghats (Padhye et al., 2012). When compared with northern Western Ghats, butterfly diversity and endemism are rich in the Southern Western Ghats (Kunte, 2008; Padhye et al., 2012). Rich floral diversity and varied habitats in the Southern Western Ghats could be significant factors for the higher butterfly diversity in this region (Ramesh and Pascal, 1997; Ramesh, 2001; Gimaret-Carpentier et al., 2003; Davidar et al., 2005; Padhye et al., 2012). This process of biogeographical understanding has initiated the zoning of Western Ghats—the Northern Western Ghats that is, Gujarat to Goa (16°–23°N), the Central Western Ghats that is, Goa to Palghat Gap (11°–16°N), and the Southern Western Ghats that is, South of Palghat Gap to Nagerkoil (8°–11°N) (Gaonkar, 1996; Padhye et al., 2012). Diversity across these three biogeographic regions based on previous studies suggests an overall diversity of about 200 species in the Northern Western Ghats (Gaonkar, 1996; Padhye et al., 2012), about 316 species in the Central Western Ghats (Gaonkar, 1996; Larsen, 1987; Naik and Mustak, 2016; Padhye et al., 2012), while the Southern Western Ghats holds the highest number of about 317 species (Gaonkar, 1996; Ghorpadé and Kunte, 2010; present study). The authors of this chapter through a combination of literature survey and surveys over the last decade in the Agasthyamalai landscape, south of the Shencottah alone note the occurrence of over 284 species. Ghorpadé and Kunte (2010) reported 310 species from Palni hills alone. This combined landscape of Munnar–Anamalais–Palni hills–Periyar Tiger Reserve–Megamalai–Srivilliputhur–Agasthyamalai which forms the Southern Western Ghats likely accounts for more number of species.

15.4 ENDEMIC BUTTERFLIES OF WESTERN GHATS

A minimum of 33 endemic species are known from the Western Ghats. Some works indicate the presence of at least 54 species that are endemic based on species currently noted as subspecies with the likelihood of elevation to species in the future (Padhye et al., 2012; Bhakare and Ogale, 2018) (Appendix 15.1). Many species are also noted to be confined regionally. For example, butterfly species, such as *Appias wardii* (Sahyadri Albatross), *Baracus hampsoni* (Spotted Hedge Hopper), *Celaenorrhinus fusca* (Dusky/Tamil Spotted Flat), *Cheritra freja butleri* (Sahyadri Common Imperial), *Elymnias caudata* (Tailed Palmfly), *Eurema nilgiriensis* (Nilgiri Grass Yellow), *Graphium teredon* (Narrow-banded Bluebottle), *Kallima horsfieldii* (Sahyadri Blue Oakleaf), *Lethe drypetis* (Tamil Treebrown), *Mycalesis junonia*

(Malabar Glad-eye Bushbrown), *M. orcha* (Pale-brand Bushbrown), *Spindasis abnormis* (Abnormal Silverline), *Thoressa sitala* (Nilgiri Plain/Sitala Ace), *Ypthima tabella* (Sahyadri Baby Five-ring), *Zipaetis saitis* (Tamil/Banded Catseye), and *Troides minos* (Sahyadri Birdwing; Figure 15.2k) are known to occur throughout the Western Ghats. Species such as *Arhopala alea* (Sahyadri Rosy/Kanara Oakblue), *Caltoris canaraica* (Kanara Swift), *Curetis siva* (Shiva Sunbeam), *Halpe hindu* (Sahyadri Banded/Indian Ace), *Halpemorpha hyrtacus* (White-branded Ace), *Oriens concinna* (Sahyadri/Tamil Dartlet; Figure 15.2b), *Pachliopta pandiyana* (Malabar Rose), *Papilio budda* (Malabar Banded Peacock), *Papilio dravidarum* (Malabar Raven), *Papilio liomedon* (Malabar Banded Swallowtail), *Thoressa astigmata* (Unbranded Ace), and *T. honorei* (Sahyadri Orange/Madras Ace) are noted to be occurring in the Central and Southern Western Ghats. Species such as *Aeromachus dubius* (Dingy Scrub Hopper), *Arnetta mercara* (Kodagu/Coorg Forest Hopper), *Baracus subditus* (Striped Hedge Hopper), *Celatoxia albidisca* (White-disc Hedge Blue), *Colias nilgiriensis* (Nilgiri Clouded Yellow), *Hyarotis coorga* (Kodagu Brush Flitter), *Mycalesis igilia* (Sahyadri/Small Long-branded Bushbrown), *Neptis palnica* (Palni Sailer), *Parantica nilgiriensis* (Nilgiri Tiger), *Parantirrhoea marshalli* (Travancore Evening Brown), *Quedara basiflava* (Golden/Yellow Base Tree Flitter; Figure 15.2c), *Thoressa evershedi* (Travancore Tawny Ace), *Telinga adolphei* (Red-eye Bushbrown), *T. davisoni* (Palni Bushbrown), *T. oculus* (Red-disc Bushbrown), *Ypthima chenu* (Nilgiri Four-ring), and *Y. ypthimoides* (Palni Four-ring) are noted to be confined only to the Southern Western Ghats. Butterflies such as *Cethosia mahratta* (Sahyadri/Tamil Lacewing; Figure 15.2g), *Cirrochroa thais* (Tamil Yeoman), *Discophora lepida* (Southern Duffer), *Hypolycaena nilgirica* (Nilgiri Tit), *Prioneris sita* (Painted Sawtooth), and *Rapala lankana* (Malabar Flash) are found in both the Western Ghats and Sri Lanka. Some are considered highly narrow endemic species and occur in a very small geographical range. The ecological roles of such narrow endemics likely remain critical to the long-term stability of the ecosystems they are part of. Their species-specific interaction between plants and butterflies is important for the long-term sustenance of not just the species themselves but also the ecosystems as one loss can have a domino effect of species extinctions and ecosystem degradation. Roles can include nectar sources and plant's role as larval host plants and ecological roles in the co-reproductive success of each of the species (Appendixes 15.1 and 15.2 for notes on interactions).

15.5 BUTTERFLY–PLANT INTERACTIONS

Butterflies apart from being charismatic, play key and underrecognized roles in ecosystem function. They do so by shaping plant communities through their interaction with plants as herbivores (at the larval stage), pollinators among other functions. These processes are believed to directly drive species richness and species persistence providing resilience against diseases and other elements, biotic and abiotic, and hence through such processes and functions play the role of ecosystem engineers (Ehrlich and Raven, 1964; Bonebrake et al., 2010; Becerra, 2015).

FIGURE 15.2 Butterflies of the Western Ghats: HESPERIIDAE: **a.** *Erionota torus* (rounded palm-redeye), **b.** *Oriens concinna* (sahyadri dartlet), **c.** *Quedara basiflava* (golden flitter); LYCAENIDAE: **d.** *Cheritra freja* (common imperial), **e.** *Loxura atymnus* (yamfly); NYMPHALIDAE: **f.** *Elymnias caudata* (tailed palmfly), **g.** *Cethosia mahratta* (Tamil lacewing), **h.** *Cyrestis thyodamas* (common map), **i.** *Vanessa indica* (Indian red admiral), **j.** *Argynnis hybrida* (Nilgiri fritillary); PAPILIONIDAE: **k.** *Troides minos* (Southern birdwing); PIERIDAE: **l.** *Prioneris sita* (painted sawtooth). (Photos: Authors)

Larval host plants—The nature of phytophagous insects and their host plants are closely associated and highly specialized (Mitter et al., 1988; Ehrlich and Raven, 1964). The butterfly larvae depend on their host plants for their nutrients along with potential chemicals for their display and defense (Boppré, 1984). Some male butterflies of the Danaine family like Crows and Tigers popularly called milkweed butterflies swarm and forage damaged leaves or stem and exposed roots to acquire pyrrolizidine alkaloids which are essential for courtship as females tend to avoid males that lack these chemicals (Kunte, 2005; Honda et al., 2018). The butterfly–plant interactions specifically the role of larval host plants in the Western Ghats have been fairly well documented by pioneers, such as Davidson, Bell, and Aitken (Davidson and Aitken, 1890; Davidson et al. 1896–1898). More recently, Kunte (2000, 2006), Sadasivan (2007–2021) among several others have made additions to this list. Regions largely represented through these works include the Northern Western Ghats around Satara and Sindhudurga (Bhakare and Ogale, 2018); the Central Western Ghats around Karwar (Bell, 1909–1927); Nilgiris (Wynter-Blyth, 1957), parts of Kerala side of the Western Ghats (Sadasivan, 2007, 2015); parts of Southern Western Ghats around Topslip and Parambikulam Tiger Reserve in the Anamalais (Kunte, 2000, 2006), and more recently from parts of Agasthiyamalai Bioreserve from Shendurney Wildlife Sanctuary (Sujitha et al., 2019) and Kalakad–Mundanthurai Tiger Reserve (Pavukandy, Pers. Obs.). The advancement in this knowledge has resulted in a comprehensive compilation and report of the larval host plants of the Western Ghats revealing an association of 320 species of butterflies and 833 species of plants (Nitin et al., 2018). It also notes that larval host plants are not known for 16 butterfly species. A more recent and interesting work by Sujitha et al. (2019) highlighted the larval host plant associations from the *Myristica* swamps of Shendurney Wildlife Sanctuary. It is notable as it highlighted the interactions in an ecotope, the *Myristica* swamps. It also recorded 18 butterfly endemics associated with the swamp. Understanding these complex interactions between taxa groups remains important from a holistic conservation viewpoint (Tiple et al., 2011). A notable interaction includes the larvae of the endemic butterfly species, *Pachliopta pandiyana* which depends only on *Thottea siliquosa* (Aristolochiaceae) plant that occurs only in the Western Ghats and Sri Lanka (Goanker, 1996). The number of such monophagous species is known to be greater than polyphagous species (Veenakumari et al., 1997; Tiple et al., 2011). This kind of exclusive interaction between plants and butterflies is very significant in the Western Ghats. Observations in various regions reveal that the endemic monophagous butterflies also have a second larval host plant (Susanth, 2005; Kunte, 2006). This underlines the need for understanding such interactions at the local level and this knowledge is still lacking. Based on the existing data, we note 72 endemic plants in the Western Ghats that support 116 butterfly species as larval hosts (Appendix 15.2). The understanding of the nature of species-specific butterfly–plant interaction has greater implications for conservation as the loss of one species can also result in cascading extinctions of other associated species. Such multilayered interactions and associations of species, ecosystems, and habitats have much scope and are going to be crucial in understanding and providing insights into a systems approach to conserving sensitive landscapes like the Western Ghats.

Nectar and pollen as dietary sources for adult butterflies—Butterflies are holometabolous insects; their mutualistic association with food plants is highly complex and obligate. Adult butterflies depend on various sources, such as floral nectar, extra-floral nectar, ripened fruits, tree saps, carcasses, urine, fecal matters, and mud puddles for their dietary requirements (Beck et al., 1999; Krenn et al., 2001; Molleman et al., 2005). Each source attracts a specific type of butterfly that is, both sexes prefer nectar, ripened fruits, as well as tree sap, whereas carcasses and fecal matters mostly attract male butterflies. Likewise, butterflies that visit flowers are rarely found on feeding tree sap and avoid carcasses and fecal matters (Smetacek, 2002). In general, butterflies mostly rely on nectar or other fluid resources that are usually composed of 15–30% simple sugars (Kingsolver, 1985). Nectar also fulfills the nitrogen requirements of butterflies (Baker, 1978). According to Baker and Baker (1973; 1975), the nectar from butterfly-visited flowers is usually rich in sucrose, amino acids, and less viscous. The frequent floral visits of butterflies for nectar during the time of inadequate precipitation indicate that nectar serves as a water source for butterflies (Percival, 1965). The mouth parts of the butterflies are highly evolved by having long, thin proboscis that is suitable for foraging nectar from flowers with narrow corolla tubes (Reddi and Bai, 1984; Malleshappa and Richard, 2011). Wynter-Blyth (1957) recorded that all the members of Danainae, Nymphalinae of Nymphalidae, and only the males of Lycaenidae, Papilionidae, Pieridae, and Hesperidae visit flowers. In the Western Ghats, studies focusing on the nectar sources of butterflies have been undertaken at a very preliminary level. Shahabuddin (1997) recorded that 20 plant species of agriculture and forest habitats have been foraged by 58 butterfly species in Palni Hills. Jothimani et al. (2014) recorded 26 species of butterflies to be depending on 36 nectar plants from Maruthamalai hills, the Southern Western Ghats. Jeevith and Samydurai (2015) documented 84 nectar plants foraged by 65 butterflies from the Glenmorgan region, Nilgiris; these include endemics, such as *Anaphalis brevifolia*, *Leucas lamifolia*; other notable taxa, such as *Desmodium repandum, Nothopodytes nimmoniana, Jasminum mesyni, Rhododendron arboreum*, and *Rhodomyrtus tomentosa*; invasive species, such as *Ageratina adenophora* (Crofton weed), *Cestrum aurantiacum* (Orange Cestrum), *Lantana camara* (Common Lantana), *Solanum mauritianum* (Wild Tobacco Tree), *Parthenium hysterophorus* (Parthenium weed). *Ariadne merione, Junonia hierta, Papilio polytes*, and *Graphium agamemnon* butterflies were found to be relying on 86 plant species from Chamarajnagar district, Karnataka (Basavarajappa, 2016). D Sousa et al. (2018) studied the nectaring behavior of 45 butterflies from 19 common plant species from Permude village in Dakshina Kannada, Karnataka. Recently, Hari (2020) observed 21 nectar source plants for 138 butterflies from Coimbatore, Tamil Nadu; this includes plant species, such as *Santalum album, Wrightia tinctoria, Plumbago zeylanica, Heliotropium indicum, Oxalis corniculata*, etc. Endemic butterflies in the disturbed habitats of wet evergreen forests are attracted to several taxa of invasive plants, such as *Bidens pilosa, Chromolena odorata, Lantana camara, Stachytarpheta jamaicansis, Tridax procumbens* (Hopeland and Richard, pers. obs.). Endemic plants, such as *Canthium travancoricum, Hedyotis purpurescens, Olea paniculata, Vernonia ramaswamii, V. travancorica, Leucas lanceifolia, Psychotria nilgiriensis* var. *astephana, Symplocos cochinchinensis, Syzygium densiflorum, S. mundagam, Clerodendrum infortunatum, Maesa indica*, and *Leea indica* are known to act as important nectar sources for endemic butterflies, such as *Parantica nilgiriensis* (Nilgiri Tiger), *Appias wardii* (Sahyadri albatross), *Celaenorrhinus fusca* (Tamil spotted flat), *Notocrypta paralysos mangla* (Sahyadri common banded demon), *Ypthima ypthimoides* (Palni four-ring), *Troides minos* (Sahyadri birdwing), *Thoressa astigmata* (Southern spotted ace), *Cethosia mahratta* (Sahyadri lacewing), and *Cirrochroa thais* (Tamil yeoman) (Richard, pers. obs.). Such notes on species-specific interactions of endemic plants and endemic butterflies need more attention to understand the key links and networks of ecosystem function.

Pollination—Butterflies are usually diurnal and often visit tubular flowers with flat rim or open blossoms that bloom during the day (Swihart, 1971; Faegri and Pijl, 1979; Reddi and Bai, 1984; Richard et al., 2011). Studies indicate that the butterflies show preferences for different floral colors. The members of Nymphalidae and Pieridae are attracted to yellow/orange and red flowers; Hesperiid members prefer only yellow/orange flowers (Ilse, 1928; Ilse and Vaidya, 1956). Though butterflies visit flowers mainly for foraging nectar, they also sometimes help pollinate plants (Reddi and Bai, 1984; Richard et al., 2011). In the Southern Western Ghats, narrow endemic taxa with tubular blossoms and copious nectar are pollinated by endemic butterfly species. Malleshappa and Richard (2011) recorded 11 butterfly species foraging nectar from the tubular blossoms of *Pychotria nilgiriensis* var. *astephana*, a Southern Western Ghats endemic taxon. Blossoms of other narrow endemic species, such as *Vernonia ramaswamii* and *V. travancorica* are mostly visited by small Hesperiid and Papilionid members (Richard, pers. obs.). Studies also have shown that butterflies visit other blossom types for nectar and aid in pollinating them (Reddi and Bai, 1984; Devy and Davidar, 2003; Richard et al., 2011). The Glassy Tiger butterfly is one of the pollinators of brush-shaped blossomed *Syzygium mundagam* (Richard et al., 2011). Endemic butterflies *Appias wardii* (Sahyadri albatross) and *Ypthima ypthimoides* (Palni four-ring) are significant pollinators of an endemic plant species, *Psychotria nilgiriensis* var. *astephana* as they have high rates of visiting frequency and compelling pollinating behavior on the blossom (Malleshappa and Richard, 2011). Any disruption or loss of one species or link in such ecosystem can have cascading species loss, and this suggests that such interactions are key factors to be studied. The difference between nectar stealers and true pollinators is another aspect that studies would need to differentiate while undertaking such efforts. Further interests in canopy ecology and canopies as nectar sources and the role of butterflies as pollinators is an aspect that has not been explored and is a great research opportunity for future research. Such studies can highlight community dynamics of floral communities such as trees which are seldom looked at with reference to butterflies. A deep understanding of this association of butterfly–plant interactions will not just add another layer of knowledge into various aspects of co-evolution, evolutionary ecology, and biology of the plants and butterflies of the Western Ghats but also suggest their role as ecological engineers of this landscape and consequently knowledge towards their conservation.

15.6 THREATS TO BUTTERFLIES OF WESTERN GHATS

Butterflies in the Western Ghats face threats in various forms. They range from historical anthropogenic-driven events that disrupted ecosystem process to modern day challenges to ecosystems such as global change and climate change. Though threats to butterflies have been largely indicated, studies in the tropics suggest varied responses by butterflies and therefore are poorly understood (Bonebrake et al., 2010). Threats include habitat loss to economic activities, such as mining, monocultures, agriculture, invasive species, fragmentation, fire, grazing and linear intrusions, such as roads, railways, and electric lines among others (Kunte, 1997; Bonebrake et al., 2010). The rate of deforestation in the Western Ghats of Karnataka, Kerala, and Tamil Nadu alone showed 25.6% forest cover loss between 1973 and 1995. It also noted a loss of dense forest and open forests at 19.5 and 33.2% while degraded forests increased at 26.64% (Jha et al., 2000). The loss of different forest types will affect butterflies too (see Section 15.4). The gradual disappearance of tropical rain forest in the recent past is suggested to be the reason for low butterfly diversity in the northern parts of Western Ghats (Gaonkar, 1996). Most of the butterfly species including 16 endemics in the Western Ghats occur in evergreen forests (Padhye et al., 2012). Among the six families of butterflies in the Western Ghats, most of the members of Papilionidae including its endemics occur in evergreen forests; the endemic Pieridae species occur in both evergreen forests and grasslands (Padhye et al., 2012). Species of Nymphalidae, Lycaenidae, and Hesperiidae occur in both evergreen and deciduous forests (Padhye et al., 2012). Therefore, any loss of forests and transition of landscapes from one form to another is likely to impact butterflies. Studies have also noted the species richness to be relatively low in highly disturbed habitats and altered landscapes due to human activities (Kunte, 2008; Padhye et al., 2012). Devy and Davidar (2001) reported high species diversity of common butterflies in the logged forest sites than in the unlogged forests in Kakachi, Kalakad Mudanthurai Tiger Reserve, and the Southern Western Ghats. Disturbed forests with sparse canopy cover allowed partial illumination which was noted to be favorable for some butterflies over others. However, interior forest species such as *Idea malabarica* (Malabar Tree-Nymph) were very rarely noticed in the logged sites due to the distance from the intact forest. The frequency and abundance of inner forest butterfly species declined critically in forests where the canopy was partially exposed due to frequent logging (Devy and Davidar, 2001). The availability of rich nectar sources in the exposed secondary forest was also reported to attract diverse ubiquitous butterfly taxa than pristine habitats (Davidar et al., 1993). This observation by Devy and Davidar (2001) suggests the overall rise in diversity at the local level by the arrival of generalist butterfly species but potential long-term deterioration of overall diversity, including plant associations and larger ecosystem deterioration. Lovejoy et al. (1986) suggested a similar response by butterflies where forest fragmentation and edge effect-driven diversity of butterflies competed with several forest interior species.

Another result of logging history is the invasive species problem driven by anthropogenic landscape change. The effect of logging history has roles in amplifying the edge effect and stronger invasive species dispersal, which in turn disrupts the native butterfly and plant interactions. Over 100 species of invasive species are known across the Western Ghats (Aravindhan and Rajendran, 2014; Adhikari et al., 2015). Many are known to occur in the Agasthyamalai Biosphere Reserve too. Although, their role as a disruptor to butterfly ecology remains to be assessed. Levels of disruption include invasive species of plants attracting butterflies as larval host plants and nectar sources and therefore disrupting the herbivory and pollination-induced population dynamics and community ecology at the local level.

Another problem is fragmentation by increasing the linear networks like roads which magnify the edge effect but also have additional consequences like roadkills. Such aspects have not been explored at the landscape level but can be far reaching. This can be manifold specifically for migratory butterfly species and possibly more so for the butterflies of Danainae. Many indicative observations have been made on the eastern plains of India specifically between Bengaluru and Chennai that are even known to the common public where migrating butterflies impede vehicular traffic annually during the migratory period. However, the scale and extent of the threat have not been assessed. More than 50 species have been reported to be roadkills in the Western Ghats (Rao and Girish, 2007; Roshnath and Cyriac, 2013; Sony and Arun, 2015). Migrating butterflies, particularly Danaine and emigrants seem to be highly vulnerable to becoming roadkills (Roshnath and Cyriac, 2013; Sony and Arun, 2015). This aspect requires further attention to know the scale and extent of roadkills and needs potential solutions where necessary.

Bonebrake et al. (2010) suggested agriculture as a potential and major threat to butterflies in the form of both habitat loss and anthropogenic toxins, such as pesticides, herbicides, pollution, and genetically modified organisms in the tropics. This is an area that requires further exploration in the Western Ghats. Genetically modified plants, for example, are already in cultivation in populated high-elevations in the Western Ghats and unregulated pesticide use is already known to cause mass deaths of larger mammals annually in the Western Ghats. Their effect on butterflies is likely severe; however, it needs attention and examination to understand the scale and extent of the severe effect.

Much of the Western Ghats hold tangled histories with colonial period when landscape level transformation occurred resulting in large-scale teak, tea, coffee, and plantations across the Western Ghats. A net loss of over 35% of the forest cover in the Western Ghats has been suggested to have occurred between 1920 and 2013 (Reddy et al., 2016). The scale and extent of logging inspite of such studies may still largely be undermined where seemingly green and protected forests exists but lack in ecological function. This has largely remained a blindspot among researchers, policy makers, and the public. This is compounded by the threats of today, particularly to low-elevation evergreen forests by various anthropogenic activities, such as alteration of habitats to plantations, monoculture, pesticide use, forest fire, pollution, grazing and global challenges like climate change (Ambrose and Raj, 2005). The combination of threats past and present hold a new set of challenges and scenarios. Therefore, the true ecosystem health and status of butterflies remain unknown in the Western Ghats due to the absence of systematic work on diversity, population, and threats. An attempt to understand their true status requires further immediate attention.

15.7 CONSERVATION MEASURES AND CONCLUSION

Butterflies are sensitive to changes in the environment and habitats. In this era of global change and climate change, they remain susceptible to population crashes and or extinctions. In the recent decades, studies on butterfly diversity, host plant interaction, and distribution have increased several fold. Yet, the distribution and population data on butterflies in the Western Ghats is still lacking at the local level. Studies of their ecology and complex interactions will remain crucial to determine the actions toward conservation (Tiple et al., 2011).

Butterflies probably had never been viewed as ecosystem engineers. Through their role in interactions across various taxa groups and thus playing a vital role in shaping diversity and ecosystems at various levels, we suggest their role as ecosystem engineers. Goankar (1996) suggested that butterflies interact with over 1000 species of plants. Through this chapter, we confirm that butterflies interact with at least 1000 species of plants in the Western Ghats alone. Over 833 species play the role of larval host plants, of which at least 72 are endemic and over 200 species play the role as nectar source, of which many are endemic. We highlight this role of butterflies in the Western Ghats as a segue for further investigation as ecosystem engineers, indicators of ecosystem status specifically ecosystem deterioration, and a call to identify similar taxa groups and their role in Western Ghats to positively identify critical links and hold a nuanced way to think of conservation in the Western Ghats. Understanding such complex interactions and integrating the multilevel knowledge systems across taxa groups, communities and ecosystems and geographies both within and outside the Western Ghats with a systems approach is crucial not just for the conservation of butterflies but also for sensitive ecosystems of the Western Ghats on the whole.

Jha et al. (2000) suggested over 25% loss of forest cover in the Western Ghats and over 25% increase in forest degradation. A study in Agasthiyamalai Biosphere Reserve suggested similar impacts (Dutta et al., 2016). However, there are limitations to GIS studies where the status of ecosystem functions cannot be detected. The situation is likely far worse with implication for the status of butterflies at the landscape level across the Western Ghats. Behind this problem lies a hidden potential with ecological restoration of forests as a conservation action beneficial to butterflies. Butterfly focused measures, such as butterfly parks and similar measures may remain band-aid measures on a bullet wound, but for studies that more actively seek answers with systems and community ecology approaches to solve problems that butterflies face. This would benefit both butterflies and the larger Western Ghats. This chapter is but a layer to highlight the significant multitiered role of butterflies in the Western Ghats. The future of butterflies and their conservation depends on drawing larger public attention and conservation toward larger landscape where they range which is very wide and larger than the Western Ghats for many species like Danaine butterflies. It also hangs on a fine balance on how we as species tackle the larger problems of humanity, such as global change and climate change which can make a deep cut on these highly sensitive taxa group as studies in Europe have shown. But with evidence-based approach to conservation, drawing strong public attention and applying the evidence on the ground with systems approaches at the larger landscape level and tapping habitat restoration potential remains much hope.

In India, butterflies have gained more popularity in the recent past across various sections of the society. Citizen Science has helped further garner support among the public. Five states have currently declared their state butterflies. They are *Papilio polymnestor* (Blue Mormon) for Maharashtra in 2015; *Troides minos* (Southern birdwing) for Karnataka in 2016 (Kasambe, 2018); *Papilio buddha* (Malabar banded peacock) for Kerala, and

Papilio bianor (Common peacock) for Uttarakhand in 2018. Recently, *Cirrochroa thais* (Tamil yeoman) was declared as state butterfly of Tamil Nadu. With the support of citizen scientists, Forest department in the states of Tamil Nadu and Kerala currently co-organize butterfly surveys in various parts of the states especially the Western Ghats. Through collaborative efforts of natural history groups and butterfly enthusiasts, local forest departments are now conducting annual surveys in various wildlife sanctuaries, particularly in the states of Tamil Nadu and Kerala over the last few years. This collaborative effort in terms of sharing of logistics support and knowledge base among such groups is covering more ground at the local level. This has tremendous potential to fill in the knowledge gaps at the local level, such as diversity, annual fluctuations including systems approaches among others. This also has implications positively to further science and conservation through open science, data sharing, public literacy, and participation including informing policy makers and biodiversity data representation in judicial court proceedings as recent cases in India have shown (Aves, 2021; Kunte et al., 2021; Sekhsaria, 2019; Vattakaven et al., 2016). This collaborative measure among scientists, public, and forest managers bring positivity and hope to lead larger conservation measures in this sensitive landscape.

ACKNOWLEDGMENTS

The authors thank Dr. T. Pullaiah, Emeritus Professor of Botany, Sri Krishnadevaraya University, Anatapur, Andhra Pradesh, and Dr. S. Karuppusamy, Professor of Botany, The Madura College, Madurai, Tamil Nadu for providing constant support throughout this work; Ms. Bhanumathy, a puppeteer and wildlife photographer for her encouragement and technical advice. The authors would also like to thank V. K. Chandrashekhar for identity verifications of butterflies. The authors would also like to thank Dr. Pratheesh C. Mammen, State Program Coordinator, Kerala Disaster Management Authority, Jason Christie Paul, an independent researcher for reviews and comments on the manuscript and Rishiddh Jhaveri, a PhD Student at the Center for Molecular Biology for discussions on butterflies.

KEYWORDS

- **biodiversity hotspot**
- **endemic butterflies**
- **endemic plants**
- **migration**
- **nectar plants**
- **pollination**
- **larval host plants**
- **ecosystem restoration**

REFERENCES

Adhikari, D.; Tiwari, R.; Barik, S. K. Modelling Hotspots for Invasive Alien Plants in India. *PLoS One* **2015,** *10* (7), e0134665.

Agrawal, A. A.; Lau, J. A.; Hambäck, P. A. Community Heterogeneity and the Evolution of Interactions Between Plants and Insect Herbivores. *Q. Rev. Biol.* **2006,** *81* (4), 349–376.

Ambrose, D. P.; Raj, D. S. Butterflies of Kalakad Mundanthurai Tiger Reserve, Tamil Nadu. *Zoo's Print J.* **2005,** *20* (12), 2100–2107.

Anto, M.; Binoy, C. F.; Anto, I. Endemism-Based Butterfly Conservation: Insights from a Study in Southern Western Ghats, India. *J. Basic Appl. Zool.* **2021,** *82* (1), 1–19.

Aravind, N. A.; Tambat, B.; Ravikanth, G.; Ganeshaiah, K. N.; Shaanker, R. U. Patterns of Species Discovery in the Western Ghats, A megadiversity Hot Spot in India. *J. Biosci.* **2007,** *32,* 781–790.

Aravindhan, V.; Rajendran, A. Diversity of Invasive Plant Species in Boluvampatti Forest Range, the Southern Western Ghats, India. *Am.-Eurasian J. Agric. Environ. Sci.* **2014,** *14* (8), 724–731.

Aves, Y. The New Scientific Revolution. In *Perspectives. Alternatives India*; Vikalp Sangam, **2021.** https://vikalpsangam.org/article/the-new-scientific-revolution/.

Bagchi, R.; Gallery, R. E.; Gripenberg, S.; Gurr, S. J.; Narayan, L.; Addis, C. E.; Freckleton, R. P.; Lewis, O. T. Pathogens and Insect Herbivores Drive Rainforest Plant Diversity and Composition. *Nature* **2014,** *506,* 85–88.

Baker, H. G. Chemical Aspects of the Pollination Biology of Woody Plants in the Tropics. In *Tropical Trees as Living Systems*; Tomlinson, P. B., Zimmermann, M. H., Eds.; Cambridge University Press: Cambridge, 1978.

Baker, H. G.; Baker, I. Aminoacids in Nectar and Their Evolutionary Significance. *Nature* **1973,** *241,* 543–545.

Baker, H. G.; Baker, I. Studies of Nectar Constitution and Plant-Pollinator Coevolution. In *Coevolution of Animals and Plants*; Gilbert, L. E., Raven, P. H., Eds.; University Texas Press: Austin and London, **1975;** pp 100–140.

Basavarajappa, S. S. Study on Nectar Plants of Few Butterfly Species at Agriculture Ecosystems of Chamarajnagar District, Karnataka, India. *Int. J. Entomol. Res.* **2016,** *1* (5), 40–48.

Bean, A. E. Occurrence of *Spindasis abnormis* (Moore), (Lepidoptera—Lycaenidae) on the Western Ghats. A Revised Description, Including Male Genitalia and Notes on Early Development. *J. Bombay Nat. Hist. Soc.* **1968,** *65* (3), 618–632.

Becerra, X. J. On the Factors That Promote the Diversity of Herbivorous Insects and Plants in Tropical Forests. *Proc. Natl. Acad. Sci. (USA)* **2015,** *112* (19), 6098–6103.

Beck, J.; Mühlenberg, E.; Fiedler, K. Mud-Puddling Behaviour in Tropical Butterflies: In Search of Proteins or Minerals. *Oecologia* **1999,** *119,* 140–148.

Bell, T. R. The Common Butterflies of the Plains of India (Including Those Met with the Hill Stations of the Bombay Presidency). *J. Bombay Nat. His. Soc.* **1909–1927,** *20,* 279–330, 1115–1136; *21,* 517–544, 740–766, 1131–1157; *22,* 92–100, 320–344, 517–531; *23,* 73–103, 481–497; *24,* 656–672; *25,* 430–453, 636–664; *26,* 98–140, 438–487, 750–769, 941–954; *27,* 26–32, 211–227, 431–447, 778–793; *29,* 429–455, 703–717, 921–946; *30,* 132–150, 285–305, 561–586, 822–837; *31,* 323–351, 655–686, 951–974.

Bhakare, M.; Ogale, H. *A Guide to the Butterflies of Western Ghats (India) Includes Butterflies of Kerala, Tamil Nadu, Karnataka, Goa, Maharashtra and Gujarat States.* Bhakare and Ogale: Satara and Sindhudurga, Maharashtra, India, **2018.**

Bharosa, K. M. Large Scale Emergence and Migration of the Common Emigrant Butterflies, *Catopsilia pomona* (Family: Pieridae). *J. Bombay Nat. Hist. Soc.* **2000,** *97,* 301.

Bhaumik, V.; Kunte, K. Female Butterflies Modulate Investment in Reproduction and Flight in Response to Monsoon-Driven Migrations. *Oikos* **2018,** *127,* 285–296.

Bhaumik, V.; Kunte, K. Dispersal and Migration Have Contrasting Effects on Butterfly Flight Morphology and Reproduction. *Biol. Lett.* **2020,** *16,* 20200393.

Bonebrake, T. C.; Ponisio, L. C.; Boggs, C. L.; Ehrlich, P. R. More Than Just Indicators: A Review of Tropical Butterfly Ecology and Conservation. *Biol. Conserv.* **2010,** *143* (8), 1831–1841.

Boppré, M. Chemically Mediated Interactions Between Butterflies. In *The Biology of Butterflies*, Symposium of the Royal Entomological Society; Vane-Wright, R. I., Ackery, P. R., Eds.; Academic Press, **1984;** vol 11, pp 259–275.

Briscoe, M. V. Butterfly Migration in the Nilgiris. *J. Bombay Nat. Hist. Soc.* **1952,** *50,* 417–418.

CEPF. *Final Assessment of CEPF Investment in the Western Ghats Region of the Western Ghats and Sri Lanka Biodiversity Hotspot – A Special Report.* ATREE and Critical Ecosytem Partnership Fund: Bangalore, **2016**; p 73.

Chandrasekharan, V. K. *Neptis palnica* Eliot–Palni Sailer. In *Butterflies of India, v. 3.11*; Kunte, K., Sondhi, S., Roy, P., Eds.; Indian Foundation for Butterflies, 1969. http://www.ifoundbutterflies.org/sp/3004/Neptis-palnica (accessed May 09, 2021).

Chandrasekharan, V. K.; Haneesh, K. M. *Rapala lankana* (Moore, 1879)–Malabar Flash. In *Butterflies of India, v. 3.11*; Kunte, K., Sondhi, S., Roy, P., Eds.; Indian Foundation for Butterflies, 2021. http://www.ifoundbutterflies.org/sp/1021/Rapala-lankana (accessed May 20, 2021).

Chaturvedi, N. Northward Migration of the Common Indian Crow Butterfly *Euploea core* (Cramer) in and Around Bombay. *J. Bombay Nat. Hist. Soc.* **1993**, *90*, 115–116.

D Sousa, J. M.; Mayikho, B.; D Silva, P. In *Butterfly Diversity and Their Host Nectar Plants of Permude Village in Dakshina Kannada.* Conf. Proc. Lake Conservation and Sustainable Management of Ecologically Sensitive Regions in Western Ghats, Sahyadri Conservation Series 65, Karnataka, India, Dec 28–30, 2016; Ramachandran, T. V., Chandran, M. D. S., Alva, M., Eds.; CES, Indian Institute of Science: Bangalore, **2018**; pp 392–397.

Davidar, P.; Devy, M. S.; Ganesh, T.; Krishnan, R. M. In *Proceedings of International Symposium on Pollination*; Veeresh, G. K., Uma Shaankar, R., Ganeshaiah, K. N., Eds.; IUSSI Publication, **1993**; pp 325–334.

Davidar, P.; Puyravaud, J. P.; Leigh Jr. E. G. Changes in Rain Forest Tree Diversity, Dominance and Rarity Across a Seasonality Gradient in the Western Ghats, India. *J. Biogeogr.* **2005**, *32*, 493–501.

Davidson, J.; Aitken, E. H. Notes on the Larvae and Pupae of Some of the Butterflies of the Bombay Presidency. J. Bombay Nat. Hist. Soc. **1890**, 5, 349–375.

Davidson, J.; Bell, T. R.; Aitken, E. H. The Butterflies of the North Canara District of the Bombay Presidency. Part I. J. Bombay Nat. Hist. Soc. **1896**, 10, 237–259.

Davidson, J.; Bell, T. R.; Aitken, E. H. The Butterflies of the North Canara District of the Bombay Presidency. Part II. J. Bombay Nat. Hist. Soc. **1897**, 10, 372–393.

Davidson, J.; Bell, T. R.; Aitken, E. H. The Butterflies of the North Canara District of the Bombay Presidency. Part IV. J. Bombay Nat. Hist. Soc. **1898**, 11, 22–63.

Davis, S. D.; Heywood, V. H.; Hamilton, A. C., Eds. *Centres of Plant Diversity: A Guide and Strategy for Conservation.* Vol II. Asia, Australia and the Pacific. The World Wildlife Fund (WWF) and IUCN. The World Conservation Union, IUCN Publication Unit: Cambridge, UK, **1995**.

Devy, M. S.; Davidar, P. Response of Wet Forest Butterflies to Selective Logging in Kalakad-Mundanthurai Tiger Reserve: Implications for Conservation. *Curr. Sci.* **2001**, *80* (3), 400–405.

Devy, M. S.; Davidar, P. Pollination Systems of Trees in Kakachi: A Mid-Elevation Wet Evergreen Forest of Western Ghats, India. *Am. J. Bot.* **2003**, *90* (4), 650–657.

Dutta, K.; Reddy, C. S.; Sharma, S.; Jha, C. S. Quantification and Monitoring of Forest Cover Changes in Agasthyamalai Biosphere Reserve, Western Ghats, India (1920–2012). *Curr. Sci.* **2016**, *25* (2), 508–520.

Ehrlich, P. R.; Raven, P. H. Butterflies and Plants: A Study in Coevolution. *Evolution* **1964**, *18* (4), 586–608.

Evershed, J. Note on Migrations. *J. Bomb. Nat. Hist. Soc.* **1910**, *20* (3), 390–391.

Faegri, K.; van der Pijl, L. *The Principles of Pollination Ecology*; Pergamon Press: Oxford, 1979.

Ferguson, H. S. A List of the Butterflies of Travancore. *J. Bombay Nat. Hist. Soc.* **1891**, *6*, 432–448.

Fisher, D. J. Four Butterfly Migrations in India and Ceylon. *Proc. Royal. Entomol. Soc. London* **1945**, *20* (A), 110–116.

French, W. L. Butterfly Migration (*Danais melissa dravidarum* and *Euploea core core*). *J. Bombay Nat. Hist. Soc.* **1943**, *44*, 310.

Gaonkar, H. *The Butterflies of Western Ghats, India, Including Sri Lanka: A Biodiversity Assessment of Threatened Mountain System*; Centre for Ecological Science, Indian Institute of Science: Bangalore, **1996**; p 93.

Gasse, P. V. *Butterflies of India-Annotated Checklist.* Unpublished, 2013

Ghorpadé, K.; Kunte, K. Butterflies (Lepidoptera—Rhopalocera) of the Palni Hills, Southern Western Ghats in peninsular India: An Updated Review, with an Appreciation of Brigadier W.H. Evans. *Colemania* **2010**, *23*, 1–19.

Gimaret-Carpentier, C.; Dray, S.; Pascal, J.-P. Broad-Scale Biodiversity Patterns of the Endemic Tree Flora of the Western Ghats (India) Using Canonical Correlation Analysis of Herbarium Records. *Ecography* **2003**, *26*, 429–444.

Groombridge, B., Ed. *Global Biodiversity–Status of the Earth's Living Resources: A World Conservation Monitoring Centre Report*; Chapman and Hall: London, **1992**.

Gunawardene, N. R.; Daniels, A. E.; Gunatilleke, I. A.; Gunatilleke, C. V.; Karunakaran, P. V.; Nayak, K. G.; Prasad, S.; Puyravaud, P.; Ramesh, B. R.; Subramanian, K. A.; Vasanthy, G. A Brief Overview of the Western Ghats-Sri Lanka Biodiversity Hotspot. *Curr. Sci.* **2007**, *93* (11), 1567–1572.

Gupta, A.; Gawas, S. M.; Bhambure, R. On the Parasitoid Complex of Butterflies with Descriptions of Two New Species of Parasitic Wasps (Hymenoptera: Eulophidae) from Goa, India. *Syst. Parasitol.* **2015**, *92* (3), 223–240.

Gupta, A.; Pereira, B.; Churi, P. V. Illustrated Notes on Some Reared Parasitic Wasps (Braconidae: Microgastrinae) with New Host and Distribution Records from India Along with Reassignment of Glyptapanteles Aristolochiae (Wilkinson) as a New Combination. *Entomol. News* **2011**, *122* (5), 451–468.

Hancock, M. W. Boffin's Books and Darwin's Finches: Victorian Cultures of Collecting. Ph.D. Thesis, Department of English, Kansas State University, **2006**; p 327.

Hannyngton, F. 1916. Notes on Coorg Butterflies with Detailed List of the Hesperidae. *J. Bombay Nat. Hist. Soc.* **1916**, *24* (3), 578–581.

Hari, T. Butterfly Diversity of Amrita Vishwa Vidyapeetham, Coimbatore, Tamil Nadu, India. *J. Bombay Nat. Hist. Soc.* **2020**, *117*, 66–78.

Home, W. M. C. Notes on Coorg Butterflies. *J. Bombay Nat. Hist. Soc.* **1934**, *37*, 669–674.

Honda, K.; Matsumoto, J.; Sasaki, K.; Tsuruta, Y.; Honda, Y. Uptake of Plant-Derived Specific Alkaloids Allows Males of a Butterfly to Copulate. *Sci Rep.* **2018**, *8* (5516), 1–10.

Ilse, D. Uber den Farbensinn der Tagfalter. *Z. Vgl. Physiol.* **1928**, *8*, 658–692.

Ilse, D.; Vaidya, V. G. Spontaneous Feeding Response to Colours in *Papiliodemoleus* L.; *Proc. Indian Acad. Sci.* **1956**, *43* (B), 23–31.

Jeevith, S.; Samydurai, P. Butterflies Nectar Food Plants from Glenmorgan, The Nilgiris, Tamil Nadu, India. *Int. J. Plant Animal Env. Sci.* **2015**, *5* (4), 150–157.

Jha, C. S.; Dutt, C. B. S.; Bawa, K. S. Deforestation and Land Use Changes in Western Ghats, India. *Curr. Sci.* **2000**, *79* (2), 231–238.

Jothimani, K.; Ramachandran, V. S.; Rajendran, A. Role of Butterflies in Maruthamalai Hills of Southern Western Ghats. *Acad. J. Entomol.* **2014**, *7* (1), 7–16.

Kalesh, S.; Prakash, S. K. Additions to the Larval Host Plants of Butterflies of the Western Ghats, Kerala, South India (Rhopalocera, Lepidoptera): Part 1. J. Bombay Nat. Hist. Soc. **2007**, 104, 235–238.

Kalesh, S.; Prakash, S. K. Early Stages of the Travancore Evening Brown Parantirrhoea marshalli Wood-Mason (Satyrinae, Nymphalidae, Lepidoptera), An Endemic Butterfly from the Southern Western Ghats, India. J. Bombay Nat. Hist. Soc. **2009**, 106, 142–148.

Kalesh, S.; Prakash, S. K. Additions to the Larval Host Plants of Butterflies of the Western Ghats, Kerala, South India (Rhopalocera, Lepidoptera): Part 2. J. Bombay Nat. Hist. Soc. **2015**, 112, 111–114.

Kasambe, R. *Butterflies of Western Ghats*, 2nd ed.; Author, **2018**. p 372.

Katti, M.; Price, T. D. Latitudinal Trends in Body Size Among Over-Wintering Leaf Warblers (Genus *Phylloscopus*). *Ecography* **2003**, *26* (1), 69–79.

Kingsolver, J. G. Butterfly Engineering. *Sci. Am.* (New York) **1985**, *25*, 106–113.

Krenn, H. W.; Zulka, K. P.; Gatschnegg, T. Proboscis Morphology and Food Preferences in Nymphalid Butterflies (Lepidoptera: Nymphalidae). *J. Zool.* **2001**, *254*, 17–26.

Kunte, K. Seasonal Patterns in Butterfly Abundance and Species Diversity in Four Tropical Habitats in Northern Western Ghats. *J. Biosci.* **1997**, *22*, 593–603.

Kunte, K. *Butterflies of Peninsular India (India: A Lifescape, Fascicle 1. Series Editor, Madhav Gadgil)*; Universities Press, Hyderabad and Indian Academy of Sciences: Bangalore, **2000**; p 254.

Kunte, K. Species Composition, Sex-Ratios and Movement Patterns in Danaine Butterfly Migrations in Southern India. *J. Bombay Nat. Hist. Soc.* **2005**, *102* (3), 280–286.

Kunte, K. Additions to Known Larval Host Plants of Indian Butterflies. *J. Bombay Nat. His. Soc.* **2006**, *103* (1), 119–122.

Kunte, K. The Wildlife (Protection) Act and Conservation Prioritization of Butterflies of the Western Ghats, Southern Western India. *Curr. Sci.* **2008**, *94*, 729–735.

Kunte K. Biogeographic Origins and Habitat Use of the Butterflies of the Western Ghats, South-Western India. In *Invertebrates in the Western Ghats-Diversity and Conservation*; Priyadarshan, D. R., Subramanian, K. A., Devy, M. S., Aravind, N. A., Eds.; Ashoka Trust for Research in Ecology and the Environment: Bengaluru, **2016**.

Kunte, K.; Basu, D. N.; Kumar, G. S. G. Taxonomy, Systematics, and Biology of Indian Butterflies in the 21st Century. In *Indian Insects: Diversity and Science*; Ramani, S., Mohanraj, P., Yeshwanth, H. M., Eds.; CRC Press, **2019**.

Kunte, K., Sondhi, S.; Roy, P., Eds. *Butterflies of India, v. 3.11*; Indian Foundation for Butterflies, **2021**. http://www.ifoundbutterflies.org (accessed May 20, 2021).

Larsen, T. B. Butterfly Migrations in the Nilgiri Hills of South India. *J. Bombay Nat. Hist. Soc.* **1977**, *74*, 546–549.

Larsen, T. B. The Butterflies of the Nilgiri Mountains of the Southern India (Lepidoptera: Rhopalocera). *J. Bombay Nat. Hist. Soc.* **1987a**, *84* (1), 26–54.

Larsen, T. B. The Butterflies of the Nilgiri Mountains of the Southern India (Lepidoptera: Rhopalocera). *J. Bombay Nat. Hist. Soc.* **1987b**, *84* (2), 291–316.

Larsen, T. B. The Butterflies of the Nilgiri Mountains of the Southern India (Lepidoptera: Rhopalocera). *J. Bombay Nat. Hist. Soc.* **1987c**, *84* (3), 560–584.

Larsen, T. B. The Butterflies of the Nilgiri Mountains of the Southern India (Lepidoptera: Rhopalocera). *J. Bombay Nat. Hist. Soc.* **1988**, *85* (1), 26–43.

Longstaff, G. B. *Butterfly-Hunting in Many Lands*; Longmans, Green and Co.: London, **1912**; p 728.

Lovalekar, R.; Agavekar, G.; Kunte, K. Spindasis abnormis, the Endemic Abnormal Silverline Butterfly of the Western Ghats, South Western India (Lepidoptera, Lycaenidae). In Parthenos Newsletter Diversity India, **2011**; pp.13–16.

Lovejoy, T. E.; Bierregaard, R. O. Jr.; Rylands, A. B.; Malcolm, J.; Quintela, C. E.; Harper, L. H.; Brown, K. S. Jr.; Powell, A. H.; Powell, G. V. N.; Schubert, H. O.; Hays, M. B. Edge and Other Effects of Isolation on Amazon Forest Fragments. In *Conservation Biology: The Science of Scarcity and Diversity*; Soule, M. E., Ed.; Sinauer Associates: Sunderland, Mass, **1986**: pp 287–292.

Malleshappa, H.; Richard, P. S. S. Floral Biology and Floral Visitors of *Psychotria nilgiriensis* var. *astephana* (Hook.f.) Deb. et Gang. from Kalakad Mundanthurai Tiger Reserve, Southern Western Ghats, India. *Indian Forest* **2011**, *137* (8), 1049–1055.

Mitter, C.; Farrel, B.; Wiegmann, B. The Phylogenetic Study of Adaptive Zones: Has Phytophagy Promoted Insect Diversification? *Am. Nature* **1988**, *132*, 107–128.

Molleman, F.; Krenn, H. W.; Van Alphen, M. E.; Brakefield, P. M.; De Vries, P. J.; Zwaan, B. J. Food Intake of Fruit-Feeding Butterflies: Evidence for Adaptive Variation in Proboscis Morphology. *Biol. J. Linn. Soc.* **2005**, *86*, 333–343.

Myers, N.; Mittermeier, R. A.; Mittermeier, C. G.; da Fonseca, G. A.; Kent, J. Biodiversity Hotspots for Conservation Priorities. *Nature* **2000**, *403*, 853–858.

Naik, D.; Mustak, M. S. A Checklist of Butterflies of Dakshina Kannada District, Karnataka, India. *J. Threat. Taxa* **2016**, *8* (12), 9491–9504.

Narayanan, S. Checklist of Butterflies in Tiger Land–Kalakad Mundanthurai Tiger Reserve, Tamil Nadu, India. *J. Biosci. Res.* **2015**, *5* (1), 138–143.

Nayar, M. P. *Hotspots of Endemic Plants of India, Nepal and Bhutan*; Tropical Botanic Garden Research Institute: Trivandrum, India, **1996**.

Nitin, R.; Balakrishnan, V. C.; Churi, P. V.; Kalesh, S.; Prakash, S.; Kunte, K. Larval Host Plants of the Butterflies of the Western Ghats, India. *J. Threat. Taxa* **2018**, *10* (4), 11495–11550.

Padhye, A.; Shelke, S.; Dahanukar, N. Distribution and Composition of Butterfly Species Along the Latitudinal and Habitat Gradients of the Western Ghats of India. *Check List* **2012**, *8* (6), 1196–1215.

Percival, M. *Floral Biology*; Pergamon Press: London, **1965**.

Pierce, N. E.; Braby, M. F.; Heath, A.; Lohman, D. J.; Mathew, J.; Rand, D. B.; Travassos, M. A. The Ecology and Evolution of Ant Association in the Lycaenidae (Lepidoptera). *Ann. Rev. Entomol.* **2002**, *47* (1), 733–771.

Pollard, E. Monitoring Butterfly Numbers. In *Monitoring for Conservation and Ecology*; Goldsmith, F. B., Ed.; Chapman and Hall: London, **1991**; p. 87–111.

Ramesh, B. R. Patterns of Richness and Endemism of Arborescent Species in the Evergreen Forests of the Western Ghats, India. In *Tropical Ecosystems: Structure, Diversity and Human Welfare*, Proceedings of the

International Conference on Tropical Ecosystems; Ganeshayer, K. N., Umashankar, R., Bawa, K. S. Eds.; IBH: Oxford, India, **2001**; pp 539–544.

Ramesh, B. R.; Pascal, J.-P. *Atlas of the Endemics of the Western Ghats (India): Distribution of Tree Species in the Evergreen and Semi-Evergreen Forests*; Institut français de Pondichéry, Publications du department d'écologie, **1997**; p 38.

Rao, R. S.; Girish, M. S. Road Kills: Assessing Insect Casualties Using Flagship Taxon. *Curr. Sci.* **2007,** *25* (3), 830–837.

Reddi, C. S.; Bai, G. M. Butterflies and Pollination Biology. *Proc. Ind. Acad. Sci. (Anim. Sci.)* **1984,** *93* (4), 391–396.

Reddy, C. S.; Jha, C. S.; Dadhwal, V. K. Assessment and Monitoring of Long-Term Forest Cover Changes (1920–2013) in Western Ghats Biodiversity Hotspot. *J. Earth Syst. Sci.* **2016,** *125* (1), 103–114.

Richard, P. S. S.; Muthukumar, S. A.; Malleshappa, H. Relationship Between Floral Characters and Floral Visitors of Selected Angiospermic Taxa from Kalakad Mundanthurai Tiger Reserve, Southern Western Ghats, India. *Indian For.* **2011,** *137* (8), 962–975.

Roshnath, R.; Cyriac, V. P. Way Back Home-Butterfly Roadkills. *Zoo's Print* **2013,** *28,* 18–20.

Sadasivan, K.; Kochunarayanan, B.; Khot, R.; Naicker, S. R. K. A New Taxon of Nacaduba Moore, 1881 (Lepidoptera: Lycaenidae: Polyommatini) from Agasthyamalais of the Western Ghats, India. *J. Threat. Taxa* **2021,** *13* (3), 17939–17949.

Sekhsaria, P.; Thayyil, N. *Citizen Science in Ecology in India: An Initial Mapping and Analysis*; DST-Centre for Policy Research: IIT Delhi, **2019**.

Shahabuddin, G. Habitat and Nectar Resource Utilisation by Butterflies Found in Siruvattukadu Kombei, Palni Hills, Western Ghats. *J. Bombay Nat. Hist. Soc.* **1997,** *94,* 423–428.

Shamsudeen, R. S. M.; Mathew, G. Diversity of Butterflies in Shendurney Wildlife Sanctuary, Kerala (India). *World J. Zool.* **2010,** *5* (4), 324–329.

Shields, O. World Numbers of Butterflies. *J. Lepid. Soc.* **1989,** *43* (3), 178–183.

Smetacek, P. The Study of Butterflies: Congregations, Courtship and Migration. *Resonance* **2002,** *7* (5), 6–14.

Sony, R. K.; Arun, P. R. A Case Study of Butterfly Road Kills from Anaikatty Hills, Western Ghats, Tamil Nadu, India. *J. Threat. Taxa* **2015,** *7* (14), 8154–8158.

Sujitha, P. C.; Prasad, G.; Sadasivan, K. Butterflies of the Myristica Swamp Forests of the Shendurney Wildlife Sanctuary in the Southern Western Ghats, Kerala, India. *J. Threat. Taxa* **2019,** *11* (3), 13320–13333.

Susanth, C. *Parsonsia spiralis*, New Larval Host Plant of Endemic Butterfly Malabar Tree Nymph, Idea Malabarica Moore (Danainae, Nymphalidae). *J. Bombay Nat. Hist. Soc.* **2005,** *102* (3), 354–355.

Swihart, C. A. Colour Discrimination by the Butterfly *Heliconius charitonius* Linn. *Anim. Behav.* **1971,** 19, 156–164.

Tiple, A. D.; Khurad, A. M.; Dennis, R. L. H. Butterfly Larval Host Plant Use in a Tropical Urban Context: Life History Associations, Herbivory, and Landscape Factors. *J. Insect Sci.* **2011,** *11,* 65.

Vattakaven, T.; George, R.; Balasubramanian, D.; Réjou-Méchain, M.; Muthusankar, G.; Ramesh, B.; Prabhakar, R. India Biodiversity Portal: An Integrated, Interactive and Participatory Biodiversity Informatics Platform. *Biodivers. Data J.* **2016,** *4,* e10279.

Veenakumari, K.; Mohanraj, P.; Sreekumar, P. V. Host Plant Utilization by Butterfly Larvae in the Andaman and Nicobar Islands (Indian Ocean). *J. Insect Conserv.* **1997,** *1,* 235–246.

Warren, M. S.; Maes, D.; Swaay, C. A. M. V.; Goffart, P.; Dyck, H. V.; Bourn, A. N. D.; Wyckoff, I.; Hoare, D.; Ellis, S. The Decline of Butterflies in Europe: Problems, Significance, and Possible Solutions. *Proc. Nat. Acad. Sci.* **2021,** *118* (2), e2002551117.

Watson, E. Y. A Preliminary List of Butterflies of Mysore. *J. Bombay Nat. Hist. Soc.* **1890,** *5* (1), 28–37.

Wepprich, T.; Adrion, J. R.; Ries. L.; Wiedmann, J.; Haddad, N. M. Butterfly Abundance Declines Over 20 Years of Systematic Monitoring in Ohio, USA. *PLoS One* **2019,** *14* (7), e0216270.

Williams, C. B. A Study of Butterfly Migration in South India and Ceylon, Based Largely on records by Messrs. a Evershed, E. E. Green, J. C. F. Fryer and W. Ormiston. *Trans. Entomol. Soc. London* **1927,** *75,* 1–33.

Williams, C. B. The Migration of Butterflies in India. *J. Bombay Nat. Hist. Soc.* **1938,** *40,* 439–457.

Wynter-Blyth, M. A. *Butterflies of the Indian Region*; Oxford-Bombay Natural History Society: Bombay, 1957.

Yata, O. A New *Eurema* Species from South India (Lepidoptera: Pieridae). *Esakia* **1990,** (1), 161–165

Yata, O.; Gaonkar, H. A New Subspecies of *Eurema andersoni* (Lepidoptera, Pieridae) from South India. *Entomol. Sci. Entomological Society of Japan,* **1999,** *2* (2), 281–285.

Yates, J. A. Some Notes on the Travancore Evening Brown Butterfly (*Parantirrhoea marshalli*) in Coorg. *J. Bombay Nat. Hist. Soc.* **1929,** *33,* 455–457.

Yates, J. A. Notes on *Parantirrhoea marshalli* and *Prioneris sita. J. Bombay Nat. Hist. Soc.* **1930,** *34,* 832–833.

Yates, J. A. Butterflies of Coorg–Part I. *J. Bombay Nat. Hist. Soc.* **1931a,** *34,* 1003–1014.

Yates, J. A. Butterflies of Coorg–Part II. *J. Bombay Nat. Hist. Soc.* **1931b,** *35,* 103–114.

Yates, J. A. *Appias libythea libythea* and *A. albina albina,* the Dry and Wet Season Forms of *Appias sita. J. Bombay Nat. Hist. Soc.* **1932,** *35,* 698–701.

Yates, J. A. The Butterflies of Bangalore and Neighbourhood. *J. Bombay Nat. Hist. Soc.* **1933,** *36* (2), 450–459.

APPENDIX 15.1

List of Endemic Butterflies of the Western Ghats Including Sri Lanka, Their Known Larval Host Plants, Nectar Source Plants, and Their Distributions Across the Western Ghats.

S. no.	Family and scientific name	Common name	Larval host plants	Nectar source plants	Distribution
HESPERIIDAE					
1.	*Aeromachus dubius* (Elwes and Edwards, 1897)	Dingy Scrub Hopper	Poaceae *Cyrtococcum trigonum* (Kalesh and Prakash, 2015)	*Bidens pilosa*	SWG
2.	*Arnetta mercara* (Evans, 1932)	Kodagu/Coorg Forest Hopper			SWG
3.	*Baracus hampsoni* Elwes and (Edwards, 1897)	Spotted/Hampson's Hedge Hopper	Poaceae (Bell, 1926) *Imperata cylindrica* (Kalesh and Prakash, 2015)		WG
4.	*Baracus subditus* (Moore, 1884)	Striped Hedge Hopper	Poaceae (Bell, 1926) *Imperata cylindrica* (Kalesh and Prakash, 2015)		SWG
5.	*Caltoris canaraica* (Moore, 1884)	Kanara Swift	*Bambusa* sp. (Wynter-Blyth, 1957)		CWG, SWG
6.	*Celaenorrhinus fusca* (Hampson, 1889)	Dusky/Tamil Spotted Flat	*Strobilanthes callosus* (Wynter-Blyth, 1957) *S. asperimus* (Kalesh and Prakash, 2015)	*Vernonia* sp. *Olea* sp. *Stachytarpheta jamaicansis Leea indica*	WG
7.	*Halpe hindu* (Evans, 1937)	Sahyadri Banded/ Indian Ace	*Bambusa* sp. (Wynter-Blyth, 1957) *Ochlandra scriptoria* (Kalesh and Prakash, 2015)		CWG, SWG
8.	*Halpemorpha hyrtacus* (de Nicéville, 1897)	White-branded/ Bicolour Ace	*Ochlandra talbotii* (Bell, 1926) *O. travancorica* (Kalesh and Prakash, 2007)		CWG, SWG
9.	*Hyarotis coorga* Evans, 1949	Kodagu Brush Flitter	Unknown		SWG
10.	*Oriens concinna* (Elwes and Edwards, 1897)	Sahyadri/Tamil Dartlet	Poaceae		CWG, SWG
11.	*Quedara basiflava* (de Nicéville, 1889)	Golden/Yellow Base Tree Flitter	*Calamus rotang* (Pavukandy, Pers. Obs.) *C. thwaitesii* (Kunte, 2006) *C. hookerianus* (Kalesh and Prakash, 2015)		SWG
12.	*Thoressa astigmata* (Swinhoe, 1890)	Unbranded Ace	*Ochlandra talbotii* (Bell, 1926) *O. travancorica* (Nithin et al., 2018)	*Maesa indica Justicia* sp.	CWG, SWG

APPENDIX 15.1 *(Continued)*

S. no.	Family and scientific name	Common name	Larval host plants	Nectar source plants	Distribution
13.	*Thoressa evershedi* (Evans, 1910)	Travancore Tawny/Evershed's Ace	*Ochlandra scriptoria O. travancorica* (Nithin et al., 2018)		SWG
14.	*Thoressa honorei* (de Nicéville, 1897)	Sahyadri Orange/Madras Ace	Poaceae		CWG, SWG
15.	*Thoressa sitala* (de Nicéville, 1885)	Nilgiri Plain/Sitala Ace	Poaceae		WG
LYCAENIDAE					
16.	*Arhopala alea* (Hewitson, 1862)	Sahyadri Rosy/Kanara Oakblue	*Syzygium salicifolium* (Pavukandy, Pers. Obs.) *Terminalia paniculata Hopea* sp. (Davidson et al., 1896)		CWG, SWG
17.	*Ancema sudica* (Evans, 1926)	Sahyadri Silver Royal	*Viscum angulatum* (Wynter-Blyth, 1957) *V. capitellatum* (Wynter-Blyth, 1957)		CWG, SWG
18.	*Curetis siva* (Evans, 1954)	Shiva Sunbeam	*Aganope thyrsiflora* (Pavukandy, Pers. Obs.) *Desmodium oojeinense* (Wynter-Blyth, 1957)		CWG, SWG
19.	*Celatoxia albidisca* (Moore, 1884)	White-disc Hedge Blue	*Aganope thyrsiflora* (Pavukandy, Pers. Obs.)		SWG
20.	*Hypolycaena nilgirica* (Moore, 1884)	Nilgiri Tit			SWG. SL
21.	*Rapala lankana* (Moore, 1879)	Malabar Flash			SWG, SL
22.	*Spindasis abnormis* (Moore, 1884)	Abnormal Silverline	*Acacia caesia Ziziphus rugosa* (Pavukandy, Pers. Obs.) (Chandrasekharan and Haneesh, 2021) *Entada* sp. *Terminalia* sp. (Bean, 1968) *Cassia fistula* (Lovalekar et al., 2011)	*Lantana camara*	WG
NYMPHALIDAE					
23.	*Argynnis hybrida* (Evans, 1912)	Nilgiri Fritillary	*Smythea bambaiensis*	*Chromolaena odorata*	SWG
24.	*Cethosia mahratta* (Moore, 1872)	Sahyadri Lacewing	*Adenia hondala* (Pavukandy, Pers. Obs.) *Passiflora edulis P. subpeltata* (Wynter-Blyth, 1957)	*Stachytarpheta jamaicansis Lantana camara Chromolaena odorata Bidens pilosa Leea indica*	WG
25.	*Cethosia nietneri* (C. & R. Felder, 1867)	Tamil Lacewing			WG, SL

APPENDIX 15.1 *(Continued)*

S. no.	Family and scientific name	Common name	Larval host plants	Nectar source plants	Distribution
26.	*Cirrochroa thais* (Fabricius, 1787)	Tamil Yeoman	*Hydnocarpus wightianus* (Wynter-Blyth, 1957)	*Leea indica Mimosa pudica Lantana camara*	WG, SL
27.	*Discophora lepida* (Moore, 1858)	Southern Duffer	*Oclandra scriptoria O. travancorica* (Kalesh and Prakash, 2015; Pavukandy, Pers. Obs.) *Dendrocalamus strictus* (Wynter-Blyth, 1957)		CWG, SWG, SL
28.	*Elymnias caudata* (Butler, 1871)	Tailed Palmfly	*Arenga wightii* (Wynter-Blyth, 1957) *Cocos nucifera* (Wynter-Blyth, 1957) *Calamus rotang* (Kunte, 2006) *C. thwaitesii* (Kunte, 2006)		WG
29.	*Idea malabarica* (Moore, 1877)	Malabar Tree-nymph	*Aganosma cymosa* (Wynter-Blyth, 1957) *Parsonia alboflavescens* (Pavukandy, Pers. Obs.)		CWG, SWG
30.	*Kallima horsfieldi* (Kollar, 1844)	Sahyadri/South Indian Blue Oakleaf	*Pseuderanthemum malabaricum Lepidagathis cuspidata Strobilanthes callosus* (Wynter-Blyth, 1957)		WG
31.	*Lethe drypetis* (Hewitson, 1863)	Two-eyed/Tamil Treebrown	*Bambusa bambos* (Wynter-Blyth, 1957)		WG
32.	*Mycalesis igilia* (Fruhstorfer, 1911)	Sahyadri/Small Long-branded Bushbrown			SWG
33.	*Mycalesis junonia* (Butler, 1868)	Malabar Glad-eye Bushbrown	*Oplismenus compositus* (Kalesh and Prakash, 2015)		WG
34.	*Mycalesis orcha* (Evans, 1912)	Pale-brand Bushbrown	Poaceae		WG
35.	*Neptis palnica* (Eliot, 1969)	Palni Sailer	*Trema orientalis* (Chandrasekharan, 2021)		SWG
36.	*Parantica nilgiriensis* (Moore, 1877)	Nilgiri Tiger	*Tylophora indica* (Kalesh and Prakash, 2015) *T. subramanii* (Richard, Pers. Obs.)	*Syzygium mundagam Vernonia travancorica Hedyotis purpurascens Chromolaena odorata*	SWG
37.	*Parantirrhoea marshalli* (Wood-Mason, 1881)	Travancore Evening Brown	*Ochlandra scriptoria* (Wynter-Blyth, 1957) *O. travancorica* (Kalesh and Prakash, 2009; Pavukandy, Pers. Obs.)		SWG

APPENDIX 15.1   *(Continued)*

S. no.	Family and scientific name	Common name	Larval host plants	Nectar source plants	Distribution
38.	*Telinga adolphei* (Guérin-Méneville, 1843)	Red-eye Bushbrown	Poaceae		SWG
39.	*Telinga davisoni* (Moore, 1891)	Palni Bushbrown	Poaceae		SWG
40.	*Telinga oculus* (Marshall, 1881)	Red-disc Bushbrown	*Oplismenus compositus* (Kalesh and Prakash, 2015)		SWG
41.	*Ypthima chenu* (Guérin-Méneville, 1843)	Nilgiri Four-ring	Poaceae (Wynter-Blyth, 1957)		SWG
42.	*Zipaetis saitis* (Hewitson, 1863)	Banded/Tamil Catseye	*Ochlandra scriptoria O. travancorica* (Kalesh and Prakash, 2015; Pavukandy, Pers. Obs.)		WG
43.	*Ypthima tabella* (Marshall, 1883)	Sahyadri Baby Five-ring	Poaceae	*Justicia* sp. *Spermacoce* sp. *Tridax procumbens*	WG
44.	*Ypthima ypthimoides* (Moore, 1881)	Palni Four-ring	Poaceae	*Justicia* sp. *Leucas* sp.	SWG
	PAPILIONIDAE				
45.	*Graphium teredon* (C. & R. Felder, 1865)	Narrow-banded Bluebottle	*Alseodaphne semecarpifolia Cinnamomum camphora C. macrocarpa C. verum Litsea glutinosa Persea macarantha* (Wynter-Blyth, 1957)	*Chromolaena odorata Lantana camara Stachytarpheta jamaicensis*	WG, EG
46.	*Pachliopta pandiyana* (Moore, 1881)	Malabar Rose	*Thottea siliquosa* (Wynter-Blyth, 1957; Pavukandy, Pers. Obs.)	*Stachytarpheta jamaicansis Lantana Lantana camara Chromolena odorata Clerodendrum viscosum Justicia beddomeii*	CWG, SWG
47.	*Papilio buddha* (Westwood, 1872)	Malabar Banded Peacock	*Zanthoxylum rhetsa* (Wynter-Blyth, 1957)	*Helecteres isora Lantana camara Stachytarpheta jamaicensis Clerodendrum infortunatum Tabernaemontana heyneana*	CWG, SWG

APPENDIX 15.1 *(Continued)*

S. no.	Family and scientific name	Common name	Larval host plants	Nectar source plants	Distribution
48.	*Papilio dravidarum* (Wood-mason, 1880)	Malabar Raven	*Glycosmis pentaphylla* (Wynter-Blyth, 1957)	*Glycosmis pentaphylla* (Wynter-Blyth, 1957)	CWG, SWG
49.	*Papilio liomedon* (Moore, 1875)	Malabar Banded Swallowtail	*Acronychia pedunculata, Melicope lunu-ankenda* (Wynter-Blyth, 1957)	*Mussaenda* sp. *Chromolaena odorata Clerodendrum infortunatum*	CWG, SWG
50.	*Troides minos* (Cramer, 1779)	Southern Birdwing	*Aristolochia griffithii* (Wynter-Blyth, 1957) *A. tagala* (Wynter-Blyth, 1957, Kunte, 2000; Pavukandy, Pers. Obs.) *Thottea siliquosa* (Wynter-Blyth, 1957)	*Lantana camara Tabernaemontana heyneana Clerodendrum infortunatum Syzygium mundagam Stachytarpheta jamaicansis*	WG
	PIERIDAE				
51.	*Appias wardii* (Moore, 1884)	Sahyadri/Lesser Albatross	*Capparis baducca Drypetes venusta* (Wynter-Blyth, 1957)	*Tridax procumbens*	WG
52.	*Colias nilagiriensis* (C. & R. Felder, 1859)	Nilgiri Clouded Yellow	*Parochetus communis* (Wynter-Blyth, 1957)	*Justicia* sp. *Leucas* sp. *Bidens pilosa*	SWG
53.	*Eurema nilgiriensis* (Yata, 1990)	Nilgiri Grass Yellow		*Justicia* sp. *Tridax procumbens Leucas* sp.	WG
54.	*Prioneris sita* (C. & R. Felder, 1865)	Painted Sawtooth	*Capparis zeylanica* (Wynter-Blyth, 1957)	*Chromolaena odorata Lantana camara*	SWG, SL

Butterfly list adapted from Padhye et al. (2012); Bhakare and Ogale (2018); Nitin et al. (2018); Kunte et al. (2021).
Distribution adapted from Gaonkar (1996); Padhye et al. (2012); ifoundbutterflies.org (Kunte et al., 2021).

APPENDIX 15.2

Endemic Larval Host Plant and Their Dependent Butterfly Species.

S. No.	Larval host plants	Distribution and IUCN	Forest/habitat	Altitude	Butterfly species
	ACANTHACEAE				
1.	Asystasia mysorensis (A. lawiana)	WG	E	700–1000	Hypolimnas misippus
2.	Barleria courtallica	WG	SE, E	700–900	Junonia atlites, J. hierta, J. lemonias
3.	Cynarospermum asperrimum (Blepharis asperrima)	WG	MD	500–800	Sarangesa dasahara, S. purendra
4.	Justicia neesii	WG	MD	600–800	Junonia iphita, J. lemonias, J. orithya
5.	Lepidagathis cuspidata	WG	E	800–1100	Kallima horsfieldii
6.	Lepidagathis keralensis	SWG	Lateritic rocks	900–1000	Junonia lemonias, J. orithya
7.	Lepidagathis prostrata	SWG	MD	400–600	Junonia orithya
8.	Strobilanthes barbatus (Nilgirianthus barbatus)	SWG	SE, E	300–1000	Celaenorrhinus leucocera
9.	Strobilanthes heyneana (N. heyneana)	SWG	SG	700–2100	Celaenorrhinus leucocera
10.	Strobilanthes asperrinus	SWG	SE	500–800	Celaenorrhinus fusca
11.	Strobilanthes callosus	WG	SE, E	1300–2400	Celaenorrhinus leucocera, C. fusca, C. ambareesa, C. putra, Kallima horsfieldii
12.	Strobilanthes ciliata	WG	SE, E	400–1000	Junonia iphita, Celaenorrhinus leucocera, C. ambareesa, C. putra, Kallima horsfieldii
13.	Strobilanthes ixiocephalus	SWG	E	800–1300	Celaenorrhinus leucocera
	ACHARIACEAE				
14.	Hydnocarpus pentandrus (H. wightianus)	SWG	SE	800–1400	Cirrochroa thais
	ANNONACEAE				
15.	Meiogyne pannosa	WG	E	700–1400	Rathinda amor
16.	Polyalthia fragrans	WG	SE, E	500–1000	Graphium agamemnon, G. doson, G. nomius
	APOCYNACEAE				
17.	Aganosma cymosa	WG, SL	E, SE	800–1200	Idea malabarica
18.	Ceropegia fantastica	WG (Cr)	SE	800–900	Danaus genutia genutia

APPENDIX 15.2 (Continued)

S. No.	Larval host plants	Distribution and IUCN	Forest/habitat	Altitude	Butterfly species
19.	*Ceropegia lawii*	WG (En)	SE, E	1000–1400	*Danaus genutia genutia, Parantica aglea*
20.	*Ceropegia vincaefolia*	WG	E	1000–1400	*Danaus genutia genutia*
21.	*Hoya pauciflora*	SWG	E, SG	900–1400	*Euploea sylvestre, Tirumala limniace*
22.	*Hunteria zeylanica*	WG	E	1100–1600	*Graphium doson eleius*
23.	*Tylophora subramanii*	SWG	SG	1200–1600	*Parantica nilgiriensis*
	ARALIACEAE				
24.	*Schefflera wallichiana*	WG	E, SG	700–2000	*Burara gomata kanara*
	ARECACEAE				
25.	*Arenga wightii*	WG	SE, E	600–1100	*Elymnias hypermnestra undularis*
26.	*Calamus brandisii*	SWG	SE, E	500–1500	*Suastus minuta bipunctatus*
27.	*Calamus hookerianus*	WG	SE	700–900	*Suastus gremius gremius, Quedara basiflava, Hyarotis adrastus praba*
28.	*Calamus vattayila (C. pseudofeanus)*	SWG	E	1000–1400	*Quedara basiflava*
29.	*Calamus thwaitesii*	WG	SE, MD	700–900	*Quedara basiflava*
30.	*Calamus travancoricus*	WG	SE, E	500–1400	*Suastus minuta bipunctatus*
31.	*Pinanga dicksonii*	WG	E	400–1000	*Gangara thyrsis, Elymnias caudata, Suastus gremius*
	ARISTOLOCHIACEAE				
32.	*Thottea siliquosa*	WG, SL	SE, E	600–1100	*Pachliopta pandiyana, Troides minos*
	CAPPARACEAE				
33.	*Capparis cleghornii*	WG	SE	500–900	*Appias libythea, Appias lyncida latifasciata, Cepora nadina remba, Cepora nerissa phryne, Hebomoia glaucippe australis*
34.	*Capparis moonii*	WG			*Cepora nadina remba, Hebomoia glaucippe australis*
35.	*Capparis roxburghii*	WG	DD	400–700	*Appias libythea, A. lyncida latifasciata, Cepora nadina remba*
	CELASTRACEAE				
36.	*Salacia fruticosa*	WG	SE, E	600–1400	*Bindahara moorei*
37.	*Salacia macrosperma*	WG	SE, E	700–1500	*Bindahara moorei*

APPENDIX 15.2 *(Continued)*

S. No.	Larval host plants	Distribution and IUCN	Forest/habitat	Altitude	Butterfly species
	CONNARACEAE				
38.	*Connarus wightii*	SWG	E	800–1300	*Deudorix epijarbas epijarbas, Nacaduba beroe gythion*
	DIPTEROCARPACEAE				
39.	*Hopea parviflora*	SWG	SE	450–700	*Arhopala bazaloides, A. centaurus, A. amantes, Rathinda amor*
40.	*Hopea ponga*	WG	SE	400–800	*Arhopala bazaloides bazaloides, A. centaurus pirama, A. amantes amantes*
41.	*Vateria indica*	WG	SE, E	700–1100	*Nacaduba kurava canaraica*
	FABACEAE				
42.	*Acrocarpus fraxinifolius*	WG	SE, E	700–1100	*Eurema blanda silhetana*
43.	*Bauhinia phoenicea*	WG	SE, E	700–1700	*Acytolepis puspa, Cheritra freja butleri*
44.	*Dalbergia benthamii (D. rubiginosa)*	SWG	E	900–1400	*Tapena thwaitesi*
45.	*Derris canarensis*	WG (En)	SE, E	600–1100	*Curetis thetis, Hasora badra*
46.	*Desmodium heterophyllum*	WG	MD	500–700	*Zizina otis indica*
47.	*Humboldtia decurrans*	SWG	E	900–1400	*Jamides celeno*
48.	*Moullava spicata*	WG	MD	500–800	*Acytolepis puspa gisca, Neptis hylas varmona*
	LAMIACEAE				
49.	*Clerodendrum infortunatum*	WG	E, SG	800–1700	*Rapala manea schistacea*
	LAURACEAE				
50.	*Alseodaphne semecarpifolia*	WG	SE, E	800–1000	*Graphium teredon, Papilio clytia clytia*
51.	*Cinnamomum macrocarpum*	SWG (Vu)	E, SG	900–1800	*Graphium doson eleius, G. teredon, Papilio clytia clytia, Cheritra freja butleri*
52.	*Cinnamomum malabatrum*	WG	E	900–1400	*Graphium doson eleius, G. teredon, Papilio clytia clytia clytia*
53.	*Cinnamomum verum*	WG, SL	E, SG	900–1400	*Graphium teredon*
54.	*Litsea travancorica*	SWG	E	600–1100	*Graphium teredon, Papilio clytia*
55.	*Persea macarantha*	WG, SL	SE, E	800–1400	*Graphium teredon*
	LORANTHACEAE				
56.	*Helixanthera wallichiana*	WG	E, SG	900–1400	*Tajuria cippus cippus*

APPENDIX 15.2 *(Continued)*

S. No.	Larval host plants	Distribution and IUCN	Forest/habitat	Altitude	Butterfly species
57.	Taxillus tomentosus (Loranthus tomentosus) **MELIACEAE**	WG, SL	MD, E, SG	700–1500	Tajuria melastigma, Pratapa deva deva
58.	Aglaia lawii **MYRTACEAE**	WG	E	1100–1400	Charaxes psaphon imna, C. schreiber wardii
59.	Syzygium salicifolium **OLEACEAE**	WG	E	900–1200	Arhopala alea
60.	Olea dioica **ORCHIDACEAE**	WG	E, SG	1000–1600	Athyma ranga karwara
61.	Cottonia peduncularis **PHYLLANTHACEAE**	WG	E, SG	900–1400	Chliaria othona othona
62.	Glochidion heyneanum **POACEAE**	SWG	E, SG	900–1400	Athyma inara inara, A. perius perius
63.	Ochlandra scriptoria	SWG	E, SG	800–1400	Baoris farri, Caltoris kumara kumara, Thoressa astigmata, T. evershedi, Parantirrhoea marshalli, Halpe hindu, H. porus, Discophora lepida, Sovia hyrtacus, Zipaetis saitis
64.	Ochlandra talbotii	WG	SG	900–1600	Thoressa astigmata, Parantirrhoea marshalli, Baoris farri, Sovia hyrtacus, Matapa aria, Halpemorpha hyrtacus
65.	Ochlandra travancorica	SWG	Ochlandra reed brakes	1100–1800	Thoressa astigmata, T. evershedi, T. honorei, Parantirrhoea marshalli, Baoris farri, Sovia hyrtacus, Discophora lepida, Matapa aria, Halpemorpha hyrtacus, Zipaetis saitis
	PUTRANJIVACEAE				
66.	Drypetes venusta **RHAMNACEAE**	WG	SE, E	1200–1500	Appias albina swinhoei, A. wardii
67.	Smythea bombaiensis (Ventilago bombaiensis)	WG	SE, E	200–1000	Eurema nilgiriensis

APPENDIX 15.2 *(Continued)*

S. No.	Larval host plants	Distribution and IUCN	Forest/habitat	Altitude	Butterfly species
	RUBIACEAE				
68.	*Hymenodictyon orixense*	WG	MD, SE, E	500–900	*Moduza procris procris*
69.	*Ochreinauclea missionis*	SWG (Vu)	E, SG	1300–1900	*Moduza procris procris*
	RUTACEAE				
70.	*Atalantia wightii*	WG	SE, E	600–1100	*Papilio polymnestor, Chilades lajus lajus*
	ZINGIBERACEAE				
71.	*Curcuma pseudomontana*	WG	SG	1100–1600	*Udaspes folus*
72.	*Elettaria cardamomum*	WG	Cultivated	800–1100	*Jamides celeno, J. alecto*

Distribution: WG, Western Ghats; SWG, southern Western Ghats; CWG, Central Western Ghats; SL, Sri Lanka.

IUCN Category: Cr, Critically Endangered; En, Endangered; Vu, Vulnerable.

Forest/Habitat type: DD, Dry Deciduous; E, Evergreen; MD, Moist Deciduous; SE, Semi-evergreen; SG, Shola-grassland.

Source: Adapted from Nitin et al. (2018); Sujitha et al. (2019).

CHAPTER 16

STATUS OF AVIAN DIVERSITY IN THE WESTERN GHATS

RAMAKRISHNA

Zoological Survey of India, India

ABSTRACT

The decline in many bird species of the Western Ghats (WG) is due to the destruction, degradation, landscape alteration, and fragmentation of habitat and wide expanding conflicts of interest. A thorough and comprehensive refined checklist of birds of the WG together with the endemic status is considered an accurate document of that region's bird life. This human dominated world biological heritage site has been analyzed for the rich avifauna (683 species) for the first time together with the distribution range in different states of the WG.

16.1 INTRODUCTION

According to the checklist of birds of India, the subcontinent has 1232 species (which includes earlier 2123 species and subspecies) under 78 families, and 20 orders (1998) now updated to 1317 species (zsi.gov.in, 2019). India with a geographical land area of 3,287,263 sq. km ranks among the top 10 countries in the world (Leepage, 2016); it is covering 2.2% of the world's terrestrial landmass; India is known to harbor about 12.5% of its avifauna. This spectacular diversity is believed to have arisen from multiple factors that include its unique biogeographical and ecological history, its heterogeneity of physical features, and a high degree of eco-climatic variations ranging from tropical to temperate (Praveen et al., 2016).

16.2 DIVERSITY OF AVIFAUNA

Of the 1317 bird species recorded in India, 72 are endemic to the country, over 80% of migratory birds (307 species) visit India through CAF, and among these, 87 species are of high conservation concern, including two Critically Endangered, five Endangered and 13

Biodiversity Hotspot of the Western Ghats and Sri Lanka. T. Pullaiah, PhD (Ed.)

Vulnerable species (Personal communication from SACON). India is home to 1317 species of birds of which about 30% are migratory.

The Western Ghats (WG) is comparatively poorer in bird life compared with Northeast India or Southeast Asia. For instance, 12 species of Pittas are found in Southeast Asia, but a single species is seen in the WG (Daniels, 1997). Only five species of sunbirds are reported from the WG. Of the 10 species of hornbills in the Indian subcontinent, only four species are in the WG. The prominent among them is the great Indian hornbill and the others are the Great Pied Hornbill *Buceros bicornis*, Malabar Pied Hornbill *Anthracoceros coronatus*, and Malabar Gray Hornbill *Ocyceros griseus* are restricted to the wet zone of the WG. Similarly, Blue-eared Kingfisher *Alcedo meninting* and the Malabar Trogon *Harpactes fasciatus*, both uncommon species are absent in the Area.

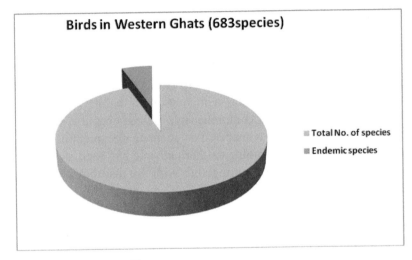

FIGURE 16.1   Birds in the Western Ghats.

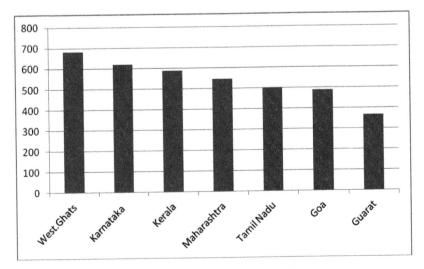

FIGURE 16.2   Diversity of avifauna in different states of the Western Ghats.

16.3 ENDEMIC BIRDS

16.3.1 NILGIRI WOOD PIGEON

Columba elphinstonii (Sykes, 1832), an endemic bird found throughout the WG. It is found mainly from the foothills to 2000 m and occasionally in the hills of Eastern Ghats. The Uncommon smooth gray pigeon inhabits wet forests in hilly and montane areas; occasionally it descends down to feed on the ground at the hills of Nandi in Chikkaballapur, Karnataka which is a part of the Eastern Ghats. The Nilgiri Wood Pigeon is largely a frugivorous bird, with its animal component (invertebrates) in the diet small enough to be neglected. In their analyses, Somasundaram and Vijayan (2010) further show that the Nilgiri Wood Pigeon feeds on fruits of different sizes, leaf buds, flowers, snails, coleopteran grubs, and soil. About 85 samples showed that the Nilgiri Wood Pigeon acted as a seed predator for many plants, as the seeds were crushed, as well as a seed dispersal agent with uncrushed seeds.

16.3.2 TRERON AFFINIS

Treron affinis (Jerdon, 1840), gray-fronted green pigeon species is endemic and non-migrant resident to southern India and is distributed in peninsular India in the WG and Eastern Ghats. These gray-fronted green pigeon species have high forest dependence. *Treron affinis* is closely related to Sri Lankan green pigeon (*Treron pompadora*) and was previously considered conspecific with *T. pompadora* (Birds of India, 2021). According to Birdlife International (2021), several subspecies are recognized earlier, such as *Treron pompadora, T. affinis, T. chloropterus, T. phayrei, T. axillaris,* and *T. aromaticus* (del Hoyo et al., 2014) were previously lumped as *T. pompadora.*

16.3.3 MALABAR GRAY HORNBILL

Ocyceros griseus (Latham, 1790) is found distributed all over the WG, south from the West Central Maharashtra (Bombay) plains to 1600 m. According to Birdlife International (2020), Malabar gray hornbill is placed under Vulnerable category of IUCN, but the species is generally a common and widely distributed one along the WG (Mudappa and Raman, 2009).

16.3.4 MALABAR BARBET

Megalaima malabarica = Psilopogon malabaricus (Blyth, 1847), a monotypic species of small barbet is endemic to SW India. This species is found in the WG from south Goa to southern Kerala. It is locally common in evergreen hill forest and shade-coffee plantations.

16.3.5   MALABAR PARAKEET

Psittacula columboides (Vigors, 1830) is endemic to the WG, locally, it is common in evergreen broadleaf forest and occasionally seen in moist deciduous forest. It is found mainly in the WG, in the plains to 1600 m; it is limited to the WG from Thana district (Maharashtra) south to Kerala.

16.3.6   MALABAR WOODSHRIKE

Tephrodornis virgatus sylvicola (Jerdon, 1830) is found in the Surat Dangs in Southeast Gujarat and southwest of the WG from Goa south, western Mysore, and Kerala. It is common in edges and glades of evergreen forests. Malabar Woodshrike typically inhabits tropical and subtropical lowland forests, both humid and dry. It is endemic to the WG region. Current nomenclature: *Tephrodornis sylvicola* (Jerdon, 1839) (Bird Life International, 2021).

16.3.7   DENDROCITTA LEUCOGASTRA

Dendrocitta leucogastra is a long-tailed Rufous Treepie inhabitant of the WG, typically found well away from human habitation in humid hill forests, where it inhabits both primary and secondary patches. It is mainly found in the WG from Goa; plains, North Karnataka – 1500 m. It is also reported from the Surat Dangs.

16.3.8   INDIAN BROAD-TAILED GRASS-WARBLER

Schoenicola platyurus (Jerdon, 1844) is an Old World warbler belongs to the family Locustellidae (Bush warblers), mainly found in the WG from south Goa. It is found within 900–2000 m altitude. It is a resident species in the WG of India and is suspected to have local movement (Rasmussen and Anderton, 2012). Though there are isolated records from Point Calimere and Vishakapatanam Ghats, but the distribution is highly debated. The previous records from Maharashtra are from Bopdeo Ghat, Lonavla, and Rajgurunagar in Pune District; Ramshej Ghat in Nashik District, Amba Ghat in Kolhapur District, and Dhule District. *Schoenicola platyurus* is classified as Vulnerable and is endemic to India (Birdlife International, 2016).

16.3.9   FLAME-THROATED BULBUL

Pycnonotus gularis (Gould, 1836) is distributed in the WG from the south of Goa to the south of Travancore up to 900 m.

16.3.10   GRAY-HEADED BULBUL

Pycnonotus priocephalus (= *Brachypodius priocephalus* Jerdon, 1839) a resident bird and distributed in the WG, South of Goa, in the plains to 1200 m. It has a very limited distribution in the heavy rainfall areas along the Southwestern region of India, from Belgaum and south Goa through Kerala and east to Nilgiris and Palnis, western Mysore and Coorg; from the plains to .1000 m, rarely to 1800 m; optimum zone between 600 and 900 m (Ali and Ripley, 1987). This species is listed as a Near Threatened species (Birdlife International, 2014) owing to habitat loss and complex life. The family *Pycnonotidae* (bulbuls) comprises about 140 species and 355 taxa, widespread in southern Asia, Africa, Madagascar and islands of the western Indian Ocean (Fishpool and Tobias, 2005; Woxvold et al., 2009). It has a very limited distribution in the heavy rainfall areas along the southwestern side of India from Belgaum and Goa south through Kerala including the Nilgiris, Palnis, western Mysore, and Coorg from plains to 1400 m (Ali and Ripley 1987; Balakrishnan 2009). Although Gray-headed Bulbuls are reported from the moist deciduous, scrub, and evergreen forests in the rain shadow areas, the breeding of the species is restricted to the mid-elevation (c. 50 km^2) between 700 and 1400 m in Silent Valley National Park (Balakrishnan, 2007).

16.3.11   INDIAN RUFOUS BABBLER

Turdoides subrufus (Jerdon, 1839) (= *Argya subrufa* (Jerdon, 1839) is endemic to the WG of southern India. It is a member of the family Leiothrichidae with the type locality Manantoddy, Wynaad, Kerala, India. It is mainly found in the WG and associated ranges (e.g., Palnis and Shevaroy hills), from about Belgaum south, 825–1220 m. Two subspecies have been recognized *Argya subrufa subrufa*: SW India (WG of Karnataka, Kerala, TN (Nilgiri Hills). *Argya subrufa hyperythra*: SW India in Kerala, south of Palghat Gap of WG.

16.3.12   WYNAAD LAUGHINGTHRUSH

Dryonastes delesserti (= *Garrulax delesserti* Jerdon, 1839) is an endemic species found in the WG from south Goa, 155–1220 m and is fairly common but local to Karnataka, Kerala, and TN. These laughingthrushes move in groups in dense forests.

16.3.13   KERALA LAUGHINGTHRUSH

Trochalepteron fairbanki (= *Strophocincla fairbanki* Blandford, 1869) (= *Montecincla fairbanki* Blandford, 1869) is a resident bird and endemic to the WG, distributed in the south of the Palghat Gap. *Montecincla fairbanki* (Blandford, 1869) and *Montecincla meridionalis* (Blandford, 1880), commonly known as Ashambu Laughingthrush are found distributed in the hills and forests of SW India (southern Kerala). According to Birdlife International,

Trochalopteron fairbanki and *T. meridionale* were previously lumped as *Strophocincla fairbanki* following Rasmussen and Anderton (2005), Conservation status: Near Threatened (IUCN); although specifically not mentioned, its family Muscicapidae is included under Schedule IV of the Indian Wildlife (Protection) Act. Several ornithologists consider the Kerala Laughingthrush as the nominate subspecies and is found in the Palni Hills and *meridionale* with a shorter white brow is found in the high hills south of the Achankovil Gap.

16.3.14   MONTECINCLA JERDONI

Montecincla jerdoni (Blyth, 1851, *Trochalopteron jerdoni*) (WG in Coorg region) is commonly known as the Banasura laughingthrush; the species is categorized as Endangered according to IUCN, as this laughingthrush has a very small and severely fragmented range. According to Birdlife International, the population is estimated to fall in the range of 250–2500 mature individuals. Each subpopulation likely has > 250 mature individuals based on the same assumptions and so the species' population size is placed in the range of 500–2500 mature individuals. It is endemic to southern India and is restricted to high elevations in the districts of Wayanad (Kerala) and Coorg (Karnataka). It is found in several localities, but is severely fragmented and has likely gone extinct at a few locations (Praveen and Nameer, 2012).

16.3.15   MONTECINCLA CACHINENSIS

Montecincla cachinensis (= *Garrulax cachinnans* (Jerdon, 1839)) is the Nilgiri laughingthrush, is endemic to SW Peninsular India (in the high-elevation areas of Nilgiris and adjoining hill ranges in Peninsular India). The Nilgiri Laughingthrush *Trochalopteron cachinnan* is found in the WG between Southern Karnataka and the Palghat gap, mainly in the Nilgiri Hills between 1200 and 2285 m. There are unconfirmed sight records from Goa and Munnar. The Nilgiri laughingthrush has a very small and severely fragmented range.

16.3.16   MALABAR WHITE-HEADED STARLING

Sturnia blythii (Jerdon, 1845) is endemic, mainly found in the WG from south Goa, Karnataka (Dandeli, Uttara Kannada) with a few records from as far north as Bombay.

16.3.17   CHESTNET–TAILED STARLING

Sturnus malabaricus (Gmelin, 1789) with the type locality of Malabar Coast is found to migrate within the continent. This is mainly an arboreal species in the plantations, gardens, and cultivated lands.

16.3.18 NILGIRI THRUSH

Zoothera neilgherriensis (Blyth, 1847) is found mainly in the WG, south from northwest Karnataka; at 600–2100 m to Kerala and TN. It is uncommon in dense evergreen forest and sholas.

16.3.19 NILGIRI BLUE ROBIN

Myiomela major (Jerdon, 1844) (= *Sholicola major* (Jerdon, 1844) having the type locality of Nilgherris (TN) is found distributed in the hills of southern Karnataka, Nilgiris, Kerala and western TN. Also known as Nilgiri Shortwing, *Myiomela major* is endemic to the Shola forests of the higher hills of southern India, mainly north of the Palghat Gap.

16.3.20 WHITE-BELLIED BLUE ROBIN

Myiomela albiventris (= *Sholicola albiventris* (Blanford, 1868)) is a resident bird distributed within a tiny range in the Palni to Ashambu hills, endemic to the Shola forests of the higher hills of Kerala and TN (south of the Palghat Gap), mostly above 1600 m.

16.3.21 NILGIRI FLYCATCHER

Eumyias albicaudatus (Jerdon, 1840) is distributed in the WG from south Karnataka up to the summits of many ranges in Kerala and TN with a very restricted range in the hills of southern India.

16.3.22 WHITE-BELLIED BLUE FLYCATCHER

Cyornis pallipes (Jerdon, 1840) (= *Muscicappa pallipes*), a small passerine bird, endemic and found in the WG from Central south, Maharashtra from 150 to 1700 m, especially found distributed from Bhimashankar to Kerala and adjacent hills of TN Coffee, Cardamom, Reeds, and Myristica swamps.

16.3.23 BLACK-AND-ORANGE FLYCATCHER

Ficedula nigrorufa (Jerdon, 1839) is distributed from the south of the WG from Nilgiris to Travancore at 750–1800 m, endemic to the central and southern WG, the Nilgiris and Palni hill ranges. Inhabits Montane Shola forests from 1000 m upwards.

16.3.24 NILGIRI FLOWERPECKER

Dicaeum concolor (Jerdon, 1840), a tiny bird in the flower pecker family is endemic to the WG, found from the south of south Maharashtra, western and southern Karnataka, Kerala and Western TN in the foothills to 1300 m.

16.3.25 SMALL SUNBIRD

Leptocoma minima (= *Nectarenia minima* Sykes, 1832), a tiny resident, found in the WG; chiefly in the foothills in the southern part of northwest Maharashtra, 300–2100 m to the southernmost hills of Karnataka, Kerala, and western TN. It is locally common in evergreen and moist evergreen forests, Shola forests, garden shade plantations, scrubby secondary growth.

16.3.26 VIGORS'S SUNBIRD

Aethopyga vigorsii (Sykes, 1832) is a resident and altitudinal migrant found in the mono-typic species; it is endemic to the northern WG from Gujarat to Goa, and in the western Satpuras (Khandesh), foothills to c.1000 m, probably farther south till the Nilgiris.

16.3.27 NILGIRI PIPIT

Anthus nilghiriensis (Sharpe, 1885) is locally very common and found in the WG of Kerala, TN and found breeding in Nilgiris and Palni Hills; it is restricted to the southern end of the WG, where it inhabits open grassy areas at 1000–2300 m. This species is listed as Vulnerable. Confirmation that the area of suitable habitat totals less than 400 km^2 would likely make the species eligible for uplisting. It is locally fairly common within its small range, particularly above 2000 m (Vinod, 2007).

16.3.28 WHITE-CHEEKED BARBET

Megalaima viridis (= *Psilopogon viridis* Boddaert, 1783) is found from the Surat Dangs, from the south of Gujarat to the western base of Satpuras, the entire WG strip. It is endemic to the forest areas of the WG and adjoining hills.

16.3.29 MALABAR LARK

Galerida malabarica (Scopoli, 1786) is mainly found southeast of Gujarat from Kathiawar and throughout the (Endemic) WG; up to 2000 m in peninsular India: south to Maharashtra,

Karnataka (Dakshina Kannada) Kerala, and Western TN. *Galerida deva* (Sykes, 1832) is found distributed in Rajasthan, Katchchh, and TN (Type locality: Hospet (Karnataka).

WG region is the one of the most important endemic bird areas of the world (Satterfield et al., 1998). Based on the new taxonomic changes (Rasmussen and Anderton 2005, 2012; del Hoyo et al., 2014), 27 endemic species are known from the WG. The present analysis on the WG holds 29 endemic birds. However, some birds, such as Red Spurfowl *Galloperdix spadicea* (Gmelin, 1919), Gray Junglefowl *Gallus sonneratii*, Yellow-throated Bulbul *Pycnonotus xantholaemus*, Malabar Whistling thrush *Myiophonus horsfieldii*, Indian Scimitar Babler *Pomatorhinus horsfieldii* have a distribution range beyond the region, mostly in the Deccan Peninsula, and some of them extend the Eastern Ghats up to Orissa, but most of them are seen predominantly in the WG as detailed below.

16.3.30 PAINTED BUSH-QUAIL

Perdicula erythrorhyncha is mainly found in the WG and associated hills to the east of Bombay and south, hills of Central India and northeastern India from southwest Bengal to Northern AP.

16.3.31 RED SPURFOWL

Galloperdix spadicea (Gmelin, 1919), endemic to India has three subspecies

- *Galloperdix spadicea spadicea*: N India (Uttar Pradesh) and terai of w Nepal to s India
- *Galloperdix spadicea caurina*: western India (Aravalli Hills of southern Rajasthan)
- *Galloperdix spadicea stewarti*: S India (Kerala coast excluding Wynaad)

Painted Spurfowl *Galloperdix lunulata* (Valenciennes, 1825) found in the eastern base of the WG, low hills up to 1000 m.

16.3.32 GRAY JUNGLEFOWL

Gallus sonneratii is distributed throughout southern Peninsula.

16.3.33 YELLOW-THROATED BULBUL

Pycnonotus xantholaemus (Jerdon, 1845), distributed at 600–1300 m, is endemic to very specialized habitat in South India. Only found in hills with dry rocky scrub and boulders, co-occurring locally with the more widespread White-browed Bulbul throughout its range. The Yellow-throated Bulbul *Pycnonotus xantholaemus* (YTB) is endemic to peninsular India and is known to occur in southern Andhra Pradesh, Karnataka, and TN with a few records from Kerala, thus patchily distributed. They are found on scrub habitats on steep, rocky hills. From Karnataka, the species has been recorded from Nandi Hills, Kanganahalli Betta, Kendatti State Forest, Adichunchunagiri range, Ragihalli State Forest, and Biligiri-rangan hills, which is the meeting point of Eastern and WG in the Nilgiri range.

16.3.34 MALABAR WHISTLING THRUSH

Myiophonus horsfieldii (Vigors, 1831) is a resident to WG, from the foothills to 2200 m, down to the plains (Grimmett et al., 2011; Narayanan and Manchi, 2007) in rain-related movements. There are some isolated records from the Chamundi and Nandi Hills of Karnataka (Praveen 2006), Karikilli Bird Sanctuary, coastal region of TN near Chennai.

16.3.35 INDIAN SCIMITAR BABLER OR OLD WORLD BABBLER

Pomatorhinus horsfieldii (Sykes, 1832), having the type locality of Mahabaleshwar in WG, is mainly distributed in peninsular India. It has been treated in the past as subspecies of the White-browed Scimitar Babbler which is found along the Himalayas, but now it is separated into two species, the peninsular Indian species and the Sri Lanka Scimitar Babbler (*Pomatorhinus melanurus*). Distribution of Subspecies are

- *Pomatorhinus horsfieldii horsfieldii*: W India (WG from Satpura Range to Goa)
- *P. horsfieldii travancoreensis*: SW India (WG from North Kanara (Karnataka) to Kerala)

16.3.36 WHITE-NAPED TIT

Parus nuchalis (= *Machlolophus nuchalis* (Jerdon, 1845)) is found in Southern peninsula (northwest Karnataka to northwest TN). It is very local and scarce in the northwest of its range and very rare in the south. It is Globally Vulnerable and endemic to India.

16.3.37 ROCK BUSH-QUAIL

Perdicula argoondah (Sykes, 1832) Central and western India, Represented by two subspecies in WG (1) *P. a. argoondah* (Sykes)—Penninsular India from Berar south to TN and (2) *P. a. salimalii* Whistler-East Central Karnataka, Kerala (Wynaad district), subspecies *P. a. salimalii* is restricted to the WG and endemic to India.

16.4 THREATS

Disturbance in ecosystems plays a critical role in structuring habitat conditions for wild species because it alters vegetation heterogeneity and resource availability (Adeney et al., 2006). Certain events are fine-grained and affect individuals in a population, whereas other disturbances are larger grained and affect the assemblages of species in a community. Still larger disturbance events affect the entire landscapes and ecosystems (Ross et al., 2002). One disturbance that is strongly related to human activity is fire. Fire is the major disturbance agent in many ecosystems because it represents an unpredictable threat that can spread across large areas over a short period (Barlow and Peres, 2004). Birds are especially sensitive to habitat disturbances, such as fire, because community assemblage is strongly related to vertical and horizontal vegetation structure. The nature, amount, and spatial distribution of ignitable fuel largely govern the character of the fire in any forest location (Goldammer, 1990). In a similar situation at Madagascar, where recurrent fires turned a rich specialized forest avifauna, with some restricted-range species, into one of opportunistic and open habitat birds, characteristic of savannas (Pons and Wendenburg, 2005). Similarly, nesting guilds show diverse responses to fire regimes, closed and open nesting birds that use trees and shrubs to nest were notably underrepresented in the site due to the absence of canopy and understory nesting substrate for these species. It is for this reason, one need to understand the species–habitat associations.

16.4.1 FRAGMENTATION

Across the WG, 40% of the natural vegetation was lost between 1920 and 1990. This has resulted in increased fragmentation of the remaining habitats, with the number of patches growing fourfold and a concomitant 83% reduction in mean patch size during this period (Menon and Bawa, 1998). This fragmentation of the forests makes them more vulnerable to escaped agricultural fires along their extensive edges and that the reduced patch size makes it more likely that entire fragments will burn during each fire event (Kodandapani et al., 2004). Plantations are an important cause of habitat fragmentation in the WG.

Within the WG, mountain valleys have always been thought to be most important biogeographic barriers. The widest valley, the Palghat Gap, has been identified as a major biogeographic barrier in the WG for birds (Robin et al., 2010). Two subspecies are recognized: (1) *Brachypteryx major major* (Jerdon, 1844) (= *Myiomela major* (Jerdon, 1844)) occurs in the hills of southern Karnataka (Babubadan, Bramhagiri and the Nilgiri hills) from c. 1300 to 1900 m alt., and in Kerala (Palni hills) replaced by the *B. m. albiventris* (Blandford) about c. 900 m alt. The impact of biogeographical barriers on the distribution of White-bellied Shortwing *Brachypteryx major* is restricted only to the high-elevation *Shola* forests (above 1400-m). The study using cytochrome *b* (cyt *b*), cytochrome *c* oxidase subunit I (COI) and Control Region (D-loop) around the separated area of Palghat (40 km wide) and the Shencottah (7 km wide) Gaps as also by great geographical distances (>80 km) provides insights into processes that may have impacted the speciation and evolution of the endemic fauna of this region. The validity of two subspecies— *Brachypteryx major major* and *B. m. albiventris* revealed the northern species be called the Rufous-bellied Shortwing *Brachypteryx major* and the southern species the White-bellied Shortwing *Brachypteryx albiventris* (Robin et al., 2010). Their study also concludes that ancient geographical gaps clearly form barriers for dispersal and gene flow and have caused deep divergences in a montane species, while habitat and climate refugia have influenced the recent population history. Further investigation on this aspect is the need of the hour.

In India, over 370 bird species are reportedly traded in more than 900 markets making the country the third highest in bird trade globally (Source: TRAFFIC India). This unlawful trade is one of the major threats to Indian birds, with 100 species classified as threatened by IUCN in 2018. On an average, only 1% of the cases of wildlife crime/trade are convicted by the Court of Law in India. This low conviction rate is largely attributed to lack of legally admissible evidences with respect to authenticity of species identification by enforcement agencies.

TABLE 16.1   Threatened Birds and their Distribution in Western Ghats.

Sl. no	IUCN category of threatened Birds	Common name	Distribution in states
	Critically endangered birds		
1	*Gyps bengalensis*	White-rumped Vulture	GA, KA, KE, MH, TN
2	*Gyps indicus*	Long-billed (Indian) Vulture	GA, KA, KE, MH, TN
3	*Aegypius calvus*	Red-headed Vulture	KA, KE, MH, TN
4	*Eurynorhynchus pygmeus*	Spoon-billed Sandpiper (old record)	MH, TN
5	*Ardeotis nigriceps*	Great Indian Bustard	KA
	Endangered birds		
6	*Neophron percnopterus*	Egyptian Vulture	KA, KE, MH, TN
7	*Tringa guttifer*	Spotted Greenshank	MH, TN
8	*Strophocincla cachinnans*	Nilgiri (Black-chinned) Laughingthrush	KE, MH, TN
9	*Sterna acuticauda*	Black-bellied Tern	KA, KE, MH, TN
10	*Myiomela albiventris*	White-bellied Blue Robin	KE, MH, TN
11	*Myiomela major*	Nilgiri Blue Robin	KE, MH, TN
12	*Sypheotides indicus*	Lesser Florican	KA
	Vulnerable		
13	*Leptoptilos javanicus*	Lesser Adjutant	GA, KA, KE, MH, TN
14	*Ciconia episcopus*	Asian Woollyneck	GA, KA, KE
15	*Clanga hastata*	Indian Spotted Eagle	GA, KE
16	*Clanga clanga*	Greater Spotted Eagle	KE, MH, TN
17	*Aquila heliaca*	Eastern Imperial Eagle	GA, KA, KE
18	*Chlamydotis macqueenii*	Macqueen's Bustard	KE
19	*Gallinago nemoricola*	Wood Snipe	KA, KE
20	*Columba elphinstonii*	Nilgiri Woodpigeon	GA, KA, KE, MH, TN
21	*Pycnonotus xantholaemus*	Yellow-throated Bulbul	KA, KE, MH, TN
22	*Ficedula subrubra*	Kashmir Flycatcher	KE, MH, TN
23	*Chaetornis striata*	Bristled Grassbird	KE
24	*Schoenicola platyurus*	Indian Broad-tailed Grass-warbler	KA, KE, TN

TABLE 16.1 *(Continued)*

Sl. no	IUCN category of threatened Birds	Common name	Distribution in states
25	*Parus nuchalis*	White-naped Tit	KA, MH, TN
26	*Anthus nilghiriensis*	Nilgiri Pipit	KE, MH, TN
27	*Calidris tenuirostris*	Great Knot	GA
28	*Rynchops albicollis*	Indian Skimmer	GA
	Near threatened birds		
29	*Bulweria fallax*	Jouanin's Petrel	KE, TN,
30	*Pelecanus philippensis*	Spot-billed Pelican	KA, KE, MH, TN
31	*Anhinga melanogaster*	Oriental Darter	GA, KA, KE, MH, TN
32	*Mycteria leucocephala*	Painted Stork	GA, KA, KE, MH, TN
33	*Ephippiorhynchus asiaticus*	Black-necked Stork	MH, TN
34	*Threskiornis melanocephalus*	Black-headed Ibis	GA, KA, KE, MH, TN
35	*Phoeniconaias minor*	Lesser Flamingo	MH, TN
36	*Aythya nyroca*	Ferruginous Duck	KE
37	*Ichthyophaga ichthyaetus*	Gray-headed Fish-eagle	KA, KE, MH, TN
38	*Ichthyophaga humilis*	Lesser Fish-eagle	KA, KE
39	*Gyps himalayensis*	Himalayan Griffon	KA
40	*Aegypius monachus*	Cinereous Vulture	KE
41	*Sterna aurantia*	River Tern	KE
42	*Circus macrourus*	Pallid Harrier	GA, KA, KE, MH, TN
43	*Esacus recurvirostris*	Great Thick-knee	KA, MH, TN
44	*Vanellus duvaucelii*	River Lapwing	TN
45	*Falco chicquera*	Red-headed Falcon	KA, MH, TN
46	*Numenius arquata*	Eurasian Curlew	KA, KE, MH, TN
47	*Limosa limosa*	Black-tailed Godwit	GA, KA, KE, MH, TN
48	*Sterna aurantia*	River Tern	GA, KA, MH, TN
49	*Psittacula eupatria*	Alexandrine Parakeet	GA, KA
50	*Coracias garrulus*	European Roller	KA, KE
51	*Anthracoceros coronatus*	Malabar Pied Hornbill	GA, KA, KE, MH, TN
52	*Buceros bicornis*	Great Pied Hornbill	KA, KE, MH, TN
53	*Pycnonotus priocephalus*	Gray-headed Bulbul	GA, KA, KE, MH, TN
54	*Strophocincla (Trochalopteron) fairbanki*	Palni Laughingthrush	KA, KE, MH, TN
55	*Phylloscopus tytleri*	Tytler's Leaf Warble	KE
56	*Ficedula nigrorufa*	Black-and-Orange Flycatcher	KA, KE, MH, TN
57	*Eumyias Albicaudata*	Nilgiri Flycatcher	GA, KA, KE, MH, TN
58	*Phylloscopus tytleri*	Tytler's Leaf-warbler	KA, MH, TN
59	*Tryngites subruficollis*	Buff-breasted Sandpiper	GA

GA, Goa; KA, Karnataka; KE, Kerala; MH, Maharashtra; TN, Tamil Nadu.

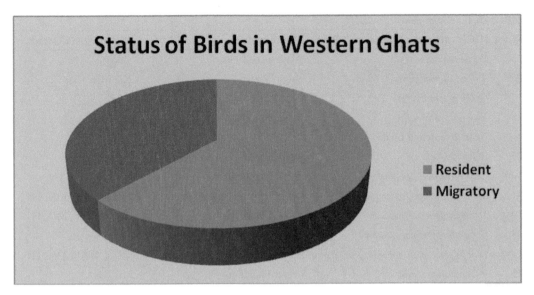

FIGURE 16.3   Status of birds in Western Ghats.

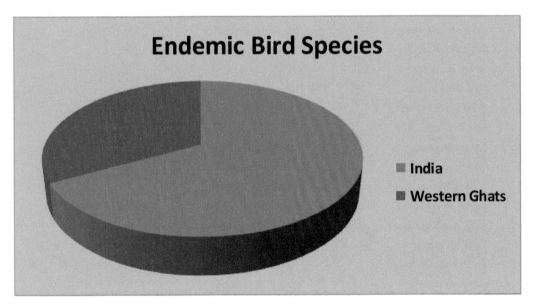

FIGURE 16.4   Endemic birds of Western Ghats vis-a-vis with India.

16.5   IMPORTANT BIRD AREA (IBA)

BirdLife International, to date, has identified and documented more than 13,000 sites in over 200 countries and territories worldwide, as well as in the marine environment. The Endemic Bird Area (EBA) extends along the WG from just north of Bombay, south to the tip of the peninsula, in the states of Maharashtra, Goa, Karnataka, Kerala, and TN;

although, a few of the restricted-range species present are also recorded from disjunct localities. Based on four important criteria viz., A1 = Threatened species; A2 = Restricted-Range species; A3 = Biome species; A4 = Congregatory species, all national parks besides area, including Wildlife Sanctuaries are declared as Important Bird Areas (IBA) in the state of Goa, Karnataka, Maharashtra, TN, and Kerala.

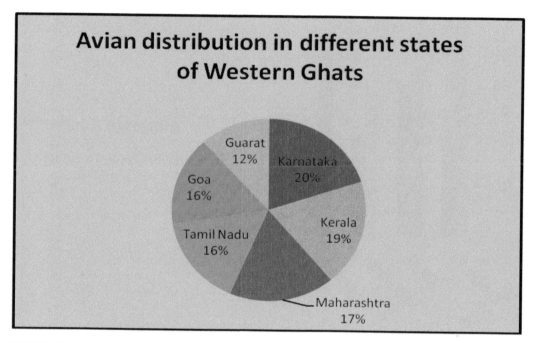

FIGURE 16.5 Aviation distribution in different states of Western Ghats.

There is a network of c.40 protected areas in the WG, many of which include extensive areas of the EBA's characteristic habitats (Birdlife International, 2021). They are located along the entire length of the Ghats, and support populations of all of the restricted-range birds. Among the 467 Important Bird Areas (IBAs) reported from India, many of them are located in the WG biogeographic region. TN has a total of 39 IBAs, of which 18 are within the biogeographic zone of WG. All the five IBAs reported from Goa are within the zone. Of the 28 IBAs recognized from Maharashtra state, 20 are within the biogeographic zone. Although the state of Gujarat is having 17 IBAs, none reported from this biogeographic zone. Karnataka has the maximum IBAs of 44 reported, of which 18 are within the region. While, all the 33 IBA sites located in the states of Kerala, all are in the zone, of which 30 are within the WG mountain ecosystem.

16.6 THREATENED BIRDS

Of the 683 species recorded from the WG, 35 are categorized in the Globally Threatened species (28 in WG, BirdLife International, 2018), of which five species are listed in the

Critically Endangered category, seven are Endangered and 16 are in Vulnerable category and 31 species are in Near Threatened of the IUCN Red List.

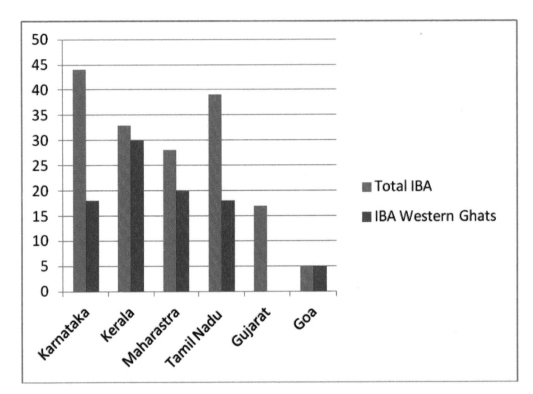

FIGURE 16.6   Total IBA in each state vis-à-vis with Western Ghats.

16.7   WAY FOREWORD IN AVIFAUNA

Subspecies are generally designated as geographical variants of a wide-ranging form, which merge into one another where their ranges join, that is, incipient species, produced by peculiar environments, but which are not yet entirely isolated from one another. In practice, a form is judged to be a species or subspecies by the degree of difference exhibited between it and its nearest geographic relative. In earlier studies, the concept of trinomials also related to the study of territoriality and behavior of birds, especially the migrating behavior that is, predictable timing of these migrations, the great distances traversed in some instances, and the uncanny accuracy in navigation displayed even by young migrating birds, combine to add to the fascination of the subject. Physiology, again, has an important relationship to ecology, since the environmental relations of birds, and consequently their distributions, are governed in part by physiological tolerances. The segregation of geographic races and the tracing of evolutionary development constitute one of the most valuable and instructive phases of modern systematic work, but we should realize that all the facts so discovered cannot be embodied in our nomenclature and that if we give up the effort to so embody them, we in no sense mean to belittle them.

More recently phylogenetic studies of birds have formed the vanguard of evolutionary biology, besides molecular approaches to phylogenetics. Recently, the egg white approach to molecular phylogenetics has gained importance. Molecular biology of birds remains at the forefront of both academic and applied research, with the extraction of DNA from blood, feathers, bone or tissue of live and dead birds now commonplace (Dickinson, 2016). The other suggestion made by the author is to have continued bibliographic research and data sharing. Lastly, it is the nomenclature that is the foundation of all further research and therefore should have a basic and sound description of the species is the need of the hour.

KEYWORDS

- important bird area
- endemism
- threat
- diversity
- distribution
- status

REFERENCES

Adeney, J. M.; Ginsberg, J. R.; Russell, G. J.; Kinnaird. M. F. Effect of an ENSO-Related Fire on Birds of a Lowland Tropical Forest in Sumatra. *Anim. Conserv.* **2006,** *9,* 292–301.

Ali, S.; Ripley, S. D. *Handbook of the Birds of India and Pakistan (Compact edition);* Oxford University Press: Delhi, **1987.**

Balakrishnan, P. Status, Distribution and Ecology of the Grey-Headed Bulbul *Pycnonotus priocephalus* in the Western Ghats, India. PhD Thesis, Bharathiar University, Coimbatore, India, **2007.**

Balakrishnan, P. Breeding Biology and Nest Site Selection of Yellow-Browed Bulbul *Iole Indica* in Western Ghats, India. *J. Bombay Nat. Hist. Soc.* **2009,** *106* (2), 176–183.

Barlow, J.; Peres, C. A. Avifaunal Response to Single and Recurrent Wildfires in Amazonian Forests. *Ecol. Appl.* **2004,** *14,* 1358–1373.

BirdLife International **2009, 2014, 2016, 2020,** downloaded from time to time during compilation.

Bor, N. L. The Vegetation of the Nilgiris. *Indian For.* **1938,** *64,* 600–609.

Daniels, R. J. R. *A Field Guide to the Birds of Southwestern India,* Oxford University Press, New Delhi, **1997.**

del Hoyo, J.; Collar, N. J.; Christie, D. A.; Elliott, A.; Fishpool, L. D. C. *HBW and BirdLife International Illustrated Checklist of the Birds of the World. Volume 1: Non-Passerines.* Lynx Edicions BirdLife International: Barcelona, Spain, and Cambridge, UK, **2014.**

Dickinson, E. C. Reinforcing the Foundations of Ornithological Nomenclature: Filling the Gaps in Sherborn's and Richmond's Historical Legacy of Bibliographic Exploration, *ZooKeys* **2016,** *550,* 107–134.

Fishpool, L. D. C.; Tobias, J. A. Family Pycnonotidae (Bulbuls). *Handbook of the Birds of the World;* del Hoyo, J., Eliott, A., Christie, D. A., Eds.; Lynx Edicions: Barcelona, **2005;** vol 10, pp 124–253.

Goldammer, J. G. *Fire in Tropical Biota: Ecosystem Processes and Global Challenges,* Springer-Verlag: Berlin, **1990.**

Grimmett, R.; Inskipp, C.; Inskipp, T. *Birds of the Indian Subcontinent,* Oxford University Press: New Delhi, **2011.**

IUCN, https://www.iucnredlist.org/ consulted from time to time during the compilation, **2018**.

Kodandapani, N.; Cochrane, M. A.; Sukumar, R. Conservation Threat of Increasing Fire Frequencies in the Western Ghats, India. *Biol. Conserv.* **2004**, *18*, 1553–1561.

Leepage, D. Avibase, the world Database website. https://avibase.bsc-eoc.org/species (accessed from time to time, **2016**).

Menon, S.; Bawa, K. S. Tropical Deforestation: Reconciling Disparities in Estimates for India. *Ambio* **1998**, *27*, 576–577.

Mudappa, D.; Raman, T. R. S. A Conservation Status Survey of Hornbills (Bucerotidae) in the Western Ghats, India. *Indian Birds* **2009**, *5* (4), 90–102.

Mudappa, D.; Raman, T. R. S. Malabar Gray Hornbill (*Ocyceros griseus*), version 2.0. In *Birds of the World*, Billerman, S. M.; Keeney, B. K., Eds.; Cornell Lab of Ornithology: Ithaca, NY, **2020**.

Narayanan, S. P.; Manchi, S. Sighting of Malabar Whistling Thrush from the Deccan. *Malabar Trogon* **2007**, *5* (3), 13.

Noble, W. A. The Shifting Balance of Grasslands, Shola Forests, and Planted Trees in the Upper Nilgiris, Southern India. *Indian For.* **1967**, *93*, 691–693.

Pons, P.; Wendenburg, C. The Impact of Fire and Forest into Savanna on the Bird Communities of West Madagascan Dry Forests. *Anim. Conserv.* **2005**, *8*, 183–193.

Praveen, J. Post-Monsoon Dispersal of Malabar Whistling Thrush *Myiophonus horsfieldii* (Vigors) to Chamundi Hill and Nandi Hills, Karnataka, Southern India. *Zoos' Print J.* **2006**, *21* (9), 2411.

Praveen, J.; Jayapal, R.; Pittie, A. A Checklist of the Birds of India. *Indian Birds* **2016**, *11* (5&6), 113–172.

Praveen J.; Nameer, P. O. Strophocincla Laughing Thrushes of South India: A Case for Allopatric Speciation and Impact on Their Conservation. *J. Bombay Nat. Hist. Soc.* **2012**, *109* (1&2), 46–52.

Rasmussen, P. C.; Anderton, J. C. *Birds of South Asia; The Ripley Guide*, 2nd ed.; Smithsonian Institutions and Lynx Edicions, **2012;** vol 2.

Rasmussen, P. C.; Anderton, J. C. *Birds of South Asia—The Ripley Guide*, Smithsonian Institution and Lynx Edicions: Washington, DC and Barcelona, **2005;** pp 414–415.

Ripley, S. D. *A Synopsis of the Birds of India and Pakistan—Together with Those of Nepal, Bhutan, Bangladesh and Sri Lanka;* Bombay Natural History Society, Oxford University Press: Bombay, **1961**.

Robin, V. V.; Sinha, A.; Ramakrishnan, U. Ancient Geographical Gaps and Paleo-Climate Shape the Phylogeography of an Endemic Bird in the Sky Islands of Southern India. *PLoS One* **2010**, *5*, e13321.

Ross, K. A.; Fox, B. J.; Fox, M. D. Changes to Plant Species Richness in Forest Fragments: Fragment Age, Disturbance and Fire History May be as Important as Area. *J. Biogeogr.* **2002**, *29*, 749–765.

Satterfield, A. J.; Cropsby, M. J.; Wege, D. C. *Endemic Bird Areas of the World; Priorities for Biodiversity Conservation*; Cambridge, U.K, **1998**.

Somasundaram, S.; Vijayan, L. Foraging Ecology of the Globally Threatened Nilgiri Wood Pigeon (*Columba elphinstonii*) in the Western Ghats, India. *Chin. Birds.* **2010**, *1* (1), 9–21.

Vinod, U. J. *Status and Ecology of the Nilgiri Pipit in the Western Ghats*, PhD Thesis, Bharathiar University: Coimbatore, India, **2007**.

Woxvold, I. A.; Duckworth, J. W.; Timmins, R. J. An Unusual New Bulbul (Passeriformes: Pycnonotidae) from the Limestone Karst of Lao PDR. *Forktail* **2009**, *25*, 1–12.

Website consulted

https://avibase.bsc-eoc.org/species
https://avibase.bsc-eoc.org/checklist.jsp?region=hol
http://www.bnhsenvis.nic.in/Database/Endemic-Birds-of-Western-Ghats
http://datazone.birdlife.org/home
https://www.iucnredlist.org/
https://www.wwfindia.org/

CHAPTER 17

STATUS OF MAMMALS IN THE WESTERN GHATS

RAMAKRISHNA

Zoological Survey of India, India

ABSTRACT

The mega biodiverse India includes 429 (excluding 26 species of marine mammals) species of mammals with a distribution of as many as 165 species (182 including subspecies 45.16%) in the Western Ghats (WG) hotspot. A number of factors are operating for their distribution; threat and survival, besides the taxonomic ambiguity of few and crepuscular nature making difficulty, are suggested for further studies.

17.1 INTRODUCTION

In the geological history, mammals originated from reptilian ancestors from "*mammal-like reptiles*," and more formally known as the "pre-mammals." The Permo-Triassic mammal-like reptile faunas of southern Africa, and the Jurassic mammalian faunas of Europe and North America, both discovered in the last century, have been central to the development of theories on the reptile–mammal transition. Though several workers attempted to study the Indian mammals, the real picture emerged for the first time from the work of Jerdon (1874). Later a consolidated account of the Indian mammals emerged from the work of Sterndale (1884), Blanford (1888), Anderson (1881), and Sclater (1891). Ryley (1913) published the Coorg and Dharwar (Report No. 11, Western Ghats part) based predominantly on the collections and field notes by G.C. Shortridge, who worked extensively for 2 months, December–January of 1912–1913 in the small districts of Western Ghats (WG).

17.2 DIVERSITY OF MAMMALS

The mammalian fauna of the world is represented by 5487 species belonging to 1148 genera, 154 families, and 29 orders (Wilson and Reeder, 2005) (5863 species, zsi.gov.in, 2019). India, the largest among the South Asian countries, has the maximum number of species recorded with as many as 429 species and subspecies. The other countries, among the Indian continent with species richness in descending order, are Nepal (197 species), Pakistan (190 species), Bangladesh (134 species), Afghanistan (124 species), Sri Lanka (122 species), Bhutan (112 species), and Maldives (21 species). Out of the total number of 429 species of mammals recorded in India, which is about 7–32% of the global mammalian species represented by 48 families and 14 orders (zsi.gov.in/checklist, 2019). According to Nameer (2000), WG is known to harbor 137 species with 13 endemics. The recent publication by ZSI on WG (2020) also revealed the distribution of 137 species under 77 genera belonging to 32 families from the region. Out of 137 species, 17(12%) are endemic to the zone, but with a difference in the number of endemic mammals. The present compilation of 2022 includes 165 species of which 146 species are endemic to WG.

17.3   DISTRIBUTION PATTERN IN THE WESTERN GHATS

17.3.1   PRIMATES

India is home to at least 24 species of nonhuman primates that include two species of Loris, 10 species of langurs, 10 species of macaques, and two species of small apes (Molur et al., 2003). Subsequently, one more species was added Arunachal macaque (*Macaca munzala*) in 2005, thus 25 species are distributed in India.

The first complete account of Indian nonhuman Primates was provided by Blanford (1888). Later, a detail review of world's Primates was made by Elliot (1913). Pocock (1939), Roonwal and Mohnot (1977), Seth (1983), Roonwal et al. (1984) gave a detailed taxonomic description along with keys to the identification of Indian Primates. In terms of population abundance, the rhesus macaque and the Hanuman langur have also been considered among the largest populations of any nonhuman primate, sharing this distinction with the baboons of Africa. The nationwide primate survey conducted by the ZSI from 1978 to 1983 showed the common langur to be the most abundant monkey species in India, but still, its population was estimated to be only 233,800 individuals (Tiwari, 1983).

The slender lories of India (*Loris lydekkerianus)* have two subspecies, *Loris lydekkerianus lydekkerianus,* and *L. l. malabaricus*, with different morphological traits occur in the eastern drier region and the western wet region of the country, respectively (Kumara et al., 2006). They are accorded the status of Near Threatened on the IUCN Red List of threatened species, and at the national level, they have been assigned the highest level of protection under Schedule-I, Part I, of the Indian Wildlife (Protection) Act, 1972, and are found distributed in WG. The potential distribution of *L. l. malabaricus* extends over a relatively smaller geographic range and shows no overlap with *L. l. lydekkerianus*. It tends to be confined to the western side of the WG (Karnataka), a region dominated by wetter climate, receiving summer (June to September) rainfall through southwest monsoons (Kumara et al., 2009).

Macaca radiata (É. Geoffroy Saint-Hilaire, 1812), type locality: India. Two subspecies are recognized; Dark-bellied Bonnet Macaque *M. r. radiata* (É. Geoffroy Saint-Hilaire, 1812) and the Pale-bellied Bonnet Macaque *M. r. diluta* Pocock, 1931. *M. r. radiata* distributed in south and western India, the northern limit is the Tapti River, south to the Palni Hills and southeast as far as Timbala, inland of Puducherry; introduced to the Mascarene Islands, including Mauritius, probably in the 16th Century; *M. r. diluta*: SEt India (states of Kerala and TN), from the southern tip and the southeast coast, north to Kambam at the SW foot of the Palni Hills, and Puducherry on the E coast. Conservation status: least concern (both subspecies).

Malabar Sacred Langur *Semnopithecus hypoleucos* Blyth, 1841 Type locality: India, Travancore. Three subspecies are provisionally recognized: Southern Malabar Sacred Langur *S. h. hypoleucos* Blyth, 1841; Northern Malabar Sacred Langur *S. h. achates* (Pocock, 1928); Black-legged Malabar Sacred Langur *S. h. iulus* (Pocock, 1928). *S. h. hypoleucos* distributed in southwestern India (Southern WG), from around Brahmagiri Hills in Karnataka state to south to the north of Silent valley in Kerala state; *S. h. achates* distributed in western India (Western Deccan Plateau and Eastern slopes of WG) up to the Roonwal line that follows the Tapti and Godavari rivers in the north; limited by the distribution of *S. entellus* in the NE and *S. priam* priam in the east; *S. h. iulus*: Southwest India (WG), from Jog Falls in Karnataka state at 440 m, and south along the hilly wet zones to the Brahmagiri Hills. Conservation status: vulnerable; *S. h. hypoleucos* (as *S. hypoleucos*) Vulnerable; *S. h. achates*, and *S. h. iulus* Least Concern.

Tufted Sacred Langur *Semnopithecus priam* Blyth, 1844 Type locality: India, Coromandel Coast. Three subspecies are recognized: Madras Tufted Sacred Langur *S. p. priam* Blyth, 1844; Central Indian Tufted Sacred Langur *S. p. anchises* Blyth, 1844; Sri Lankan Tufted Sacred Langur *S. p. thersites* (Blyth, 1847) (described as *Presbytis*). Distribution of *S. p. priam*: South and southeast India (Andhra Pradesh, Karnataka, Kerala, and TN states), a highly fragmented distribution ranging from the Krishna River in Andhra Pradesh south to Tirunelveli in TN; *S. p. anchises*: South central India (Southern Deccan Plateau) found in the districts of Kurnool, Andhra Pradesh, and in Pavagada in the district of Tumkur, Karnataka; *S. p. thersites* found in dry Zone of Sri Lanka, ranging from Jaffna in the north to the south coast. Conservation status: Near Threatened; *S. p. priam* Near Threatened; *S. p. anchises* (as synonym of *S. dussumieri*) Least Concern; *S. p. thersites* Endangered. Nilgiri Langur *Semnopithecus johnii* (Fischer, 1829) Type locality: India, Tellicherry with no subspecies/color variants. This species occurs only in the WG in southwestern India, rather unevenly distributed in the hill country of the WG from the Aramboli Pass north to Srimangala (Groves, 2001; Kumara and Singh, 2004). Molur et al. (2003) estimate a total population of less than 20,000. The population since then was relatively stable, but in the last decade there are some declines suspected (The IUCN Red List of Threatened Species: *Semnopithecus johnii*—published in 2020, iucn.org). This species is known to occur in numerous protected areas, including: Aaralam Sanctuary, Brahmagiri (Kodagu, Karnataka), Sanctuary, Chimmony Sanctuary, Chinnar Sanctuary, Eravikulam National Park, Grizzled Giant Squirrel Sanctuary, Idukki Sanctuary, Indira Gandhi Sanctuary, Kalakkad Sanctuary, Mudumalai Sanctuary, Mukurthi National Park, Mundanthurai Sanctuary, Neyyar Sanctuary, Parambikulam Sanctuary, Peechi Sanctuary, Peppara Sanctuary, Periyar National Park, Periyar Sanctuary, Shendurney Sanctuary, Silent Valley National Park, Thattekadu Sanctuary, and Wayanad Sanctuary (Molur et al., 2003).

17.3.2   ORDER CARNIVORA

Detailed taxonomic and distributional studies on Indian Carnivora were made by Blanford (1888), Pocock (1939, 1940, 1941), Ellerman and Morrison-Scott (1951), Mivart (1890), and Corbet and Hill (1992). Wozencraft (Wilson and Reeder, 2005) provided a detailed list of Indian species in the world's checklist along with distribution and synonyms.

The Genus *Canis* Linnaeus, 1758 includes two species *Canis aureus* and *Canis lupus,* both distributed in WG. Genus *Vulpes* Frisch, 1775 includes four species *Vulpes bengalensis, Vulpes cana, Vulpes ferrilala* and *Vulpes vulpes,* of which only *V. bengalensis* distributed in WG. Genus *Felis* Linnaeus, 1758 includes two Indian species *Felis chaus* and *Felis silvestris*, both distributed in WG. Genus *Prionailurus* Severtzov, 1858 includes three Indian species *Prionailurus bengalensis; Prionailurus rubiginosus,* and *Prionailurus viverrinus* all distributed in WG. Genus *Panthera* Oken, 1816 contains three Indian species (1) *Panthera leo* (2) *Panthera pardus* (3) *Panthera tigris,* the details of the *P. pardus* and *P tigris* is given below.

17.3.2.1   TIGERS AND LEOPARDS

WG region holds the world's single-largest tiger population (981 in 2018), according to a government report (moefcc.gov.in). There are 18 tiger reserves in WG. India is home to 70% of tigers in the world. The Mudumalai-Bandipur-Nagarhole-Wayanad complex holds the world's single-largest tiger population, currently estimated at over 981 tigers (2019 tiger assessment report). The report said that in the WG landscape complex, Karnataka has 524 tigers while Kerala (190) and TN (264) (New Delhi, Jan 20, 2019 (PTI)). The latest tiger census figures also show that Karnataka has the highest number of tigers in the age group of 1.5 years and more in its five Tiger Reserves namely, Bandipur, Bhadra, Nagarahole, Dandeli-Anshi, and BRT Tiger Reserves. In addition to the above, NTCA has accorded final approval to Kudremukh National Park in Karnataka as the new tiger reserve during 2015.

According to the Indian government's status of Leopard in India, 2018 report Central India and the Eastern Ghats range has 8071 leopards and WG range has 3386. WG harbors nearly 2487 leopards, of which the state of Karnataka numbers is 1129, highest among the WG states.

17.3.2.2   MONGOOSE

In the family of Herpestidae, seven species are known from India and four from the WG (Mudappa, 2013; Nameer et al., 2015). Four are known from southern India, Indian (Common) Gray Mongoose *H. edwardsii,* Brown Mongoose *H. fuscus,* Ruddy Mongoose *H. smithii,* and Stripe-necked Mongoose *H. vitticollis.* Out of these four, Indian Gray Mongoose is the most widespread, seen near human habitation and along forest edges, as well as in the forest interior. While *H. smithii* Stripe-necked Mongoose is seen in most forested areas of the WG (Mudappa, 2013), Brown Mongoose *H. fuscus*, and Ruddy Mongoose *H. smithii* reported to have more restricted distribution. Ruddy Mongoose *Herpestes smithii and the* Gray Mongoose *Herpestes edwardsii* are the two species commonly occurring in many parts of TN (Kalakkad Mundunthurai), Kerala (Chinar WLS), and scrub jungles of Karnataka part. The mongoose is killed and traded for its hair. The hair is used for making paint brushes, which are pliant and soft. Each animal yields about 10 g of hair.

TABLE 17.1   Number of Large Mammals of Western Ghats from Recent Census.

Name of the tiger reserve	Area (sq. km)	Based on tiger census of 2018; elephant census of 2017; leopard census of 2015
Periayar Tiger Reserve, Kerala	925.00	The state of Kerala is home to 3054 of the elephants, 190 of the tigers and 472 of the leopards in the state of Karnataka alone is home to 6049 (22%) of the elephants, 524 (18%) of the tigers and 1131 (14%) of the leopards in India
Parambikulam Tiger Reserve, Kerala	643.66	
Bandipur, Tiger Reserve Karnataka	872.24	
Bhadra Tiger Reserve Karnataka	500.16	
Nagarahole Tiger Reserve Karnataka	643.39	
Dandeli-Anshi Tiger Reserve Karnataka	475.00	
Biligir Ranga Temple Tiger Reserve Karnataka	539.52	
Kudremukh Tiger Reserve Karnataka	600.32	
Melghat Tiger Reserve Maharashtra	2758.52	The state of Maharashtra is home to 6 of the elephants, 317 of the tigers and 908 of the leopards in India
Tadoba Andheri Tiger Reserve Maharashtra	1727.59	
Pench Tiger Reserve Maharashtra	483.96	
Nawegaon - Nagzira Tiger Reserve Maharashtra	653.67	
Sahyadri Tiger Reserve Maharashtra	1165.57	
Bor Tiger Reserve Maharashtra	138.12	
Anamalai Tiger Reserve Tamil Nadu	1479.87	The state of Tamil Nadu is home to 2791 of the elephants, 264 of the tigers, and 815 of the leopards in India
Kalakkad Mundanthurai Tiger Reserve Tamil Nadu	1601.54	
Madumalai Tiger Reserve Tamil Nadu	688.59	
Satyamangalam Tiger Reserve Tamil Nadu	1408.40	
Total Number from Tiger Reserves, Wildlife Sanctuaries, National Parks of Western Ghats		**11,900/27,312 Elephants (43.57%) 1295/2967 Tigers (6.33%) 3326 /9265 Leopards (35.89%)**

The brown mongoose *Herpestes fuscus* is a species with restricted distribution and is seen in the southern WG and Sri Lanka. Four subspecies of the brown Mongoose have been recognized by Corbet and Hill (1992), of which the one seen in the WG is *H. f. fuscus*, while the remaining three subspecies are confined to Sri Lanka (Gilchrist et al., 2009). The primary habitat of *H.f. fuscus* is the evergreen forests, high altitude grasslands-shola forests, and adjoining tea and coffee plantations (Mudappa, 2002; Mudappa et al., 2008).

Family Mustelidae Fischer, 1817 includes Otters. Eurasian otter, *Lutra lutra* (Linnaeus, 1758), has been historically reported from the WG of Coorg in Karnataka state, and the Anamalai hills, Nilgiris, and Palni hills of TN state (Pocock, 1941; Prater, 1971). Two subspecies *Lutra* are reported by Pocock (1941), (1) *L. p. perspicillata*—in northeast and southern India, Myanmar, and Sumatra (2) *L. p. sindica* in north and northwestern India and Pakistan (Sindh province). The smooth coated otter occurs along the large rivers and lakes, in mangrove forests along the coast and estuaries, and in WG, it even uses the rice fields for foraging (Foster-Turley, 1992; Sivasothi and Nor, 1994). Along the larger perennial water bodies in India, it shows preference for rocky and sandy stretches in all the seasons, since these stretches provide sites for denning and grooming (Hussain, 1999; Nawab, 2009). *L. p. perspicillata*—inhabits plain areas near swamps, irrigation canals, rivers, dams, and lakes. They are at home in mangroves and shallow, flooded rice paddies, inhabits plain areas near swamps, irrigation canals, rivers, dams, and lakes. They are known to prefer sloping banks with vegetation. WG records are from River Cauveri from its source in the hills of Coorg, down 330 km to where it enters the plains of TN. Two species exhibits sympatry occupying the same or overlapping geographic areas (Smooth-coated Otters (*Lutrogale perspicillata*) and Asian Small-clawed Otters (*Aonyx cinereus*), found in linear riverine habitats however, both otter species have different micro and macro habitat preferences. In the Eravikulam region of WG in Kerala, *A. cinereus* found to use pools more extensively than cascades and riffles in the stream types, and 2nd-order streams with substrate and grass cover were preferred over 1st and 3rd or higher-order streams (Perinchery et al., 2011). *L. perspicillata* found in Periyar Lake in Kerala, WG were confined to the shallower and narrower regions of the lake (Anoop and Hussain, 2004).

Subfamily Mellivorinae Gray, 1865, includes *Mellivora capensis* (Schreber, 1776) Honey Badger, a widely distributed species placed under Least Concern of IUCN, Appendix-III of CITES and Schedule-I of Wildlife Protection Act. The Subfamily Mustellinae Fischer, 1817 includes Genus *Martes* Pinel, 1792 with three Indian species *Martes flavigula, Martes foina* and *Martes gwatkinsi. Martes flavigula* (Boddaert, 1785) commonly known by Yellow-throated Marten is Least Concern (IUCN), Appendix-III of CITES and placed under Schedule-II, most common in the forests all over India. *Martes gwatkinsii* Horsfield, commonly known Nilgiri Marten distribution extends up to the Charmadi Reserved Forest (13°00′–07′N, 75°23–28′E) (Kudremukh; Wildlife Foundation India) is a globally threatened (IUCN: Vulnerable) mustelid endemic to the WG, southern India in the WG of Kerala, Karnataka, and TN. This area corresponds to the reported northernmost extent of the species at 13°N (Schreiber et al., 1989).

Family Ursidae Fischer, 1817, subfamily Ursinae Fischer, 1817, Genus *Melursus* Meyer, 1793 includes Sloth Bear *Melursus ursinus* (Shaw, 1791) is Vulnerable species (IUCN) placed under Appendix-I of CITES and Schedule-I of Wildlife (Protection), Act 1972, Endemic to the Indian subcontinent, found distributed in many drier and thorny forests of WG. Sloth bear (*Melursus ursinus*) is one of the four bear species found in India and is entirely tropical in distribution. Wildlife Conservation Society analyzed the occupancy pattern in WG, sloth bears occupied nearly 61% (nearly 15,000 km²) of the landscape's forests. Sloth bears are conserved by way of Bear Sanctuaries in the state of Gujarat and Karnataka.

Family Viverridae Gray, 1821; Subfamily Paradourinae Gray, 1865 includes Genus *Paradoxurus* Cuvier, 1821 with two Indian species *Paradoxurus hermaphroditus* and *Paradoxurus jerdoni. Paradoxurus hermaphrodites:* The Common Palm Civet found distributed in WG from Anamalai hill range (TN) to Shimoga in Karnataka and extends further up to Chinar Wildlife Sanctuary, Kerala. *Paradoxurus jerdoni* an endemic carnivore species, was once fairly plentiful in Coorg, Karnataka. Karanth (1986) opined that Malabar Civet still maintains a large distributional range. Brown Palm Civet *P. jerdoni* is endemic to the WG, where it occurs in wet evergreen forests and adjacent coffee estates at altitudes of 500–2000 msl (Rajamani et al., 2002). The Brown Palm Civet replaces the Common Palm Civet (*P. hermaphroditus*) in tropical rainforests of the WG. It may be sympatric with the Common Palm Civet only in the transition zones between the rainforests and drier habitats (Nandini and Muddappa 2002).

Subfamily Viverinae Gray, 1821 includes Genus *Viverra* Linnaeus, 1758 with *Viverra civettina* and *Viverra zibetha* as two Indian species. *Viverra civettina* Blyth, 1862 commonly known by the name Malabar Civet is Endemic and Critically Endangered species distributed in south India (Karnataka, Kerala, and TN) recorded in WG from Honnavar in Karnataka to Kanyakumari in TN. It is also placed in CITES Appendix-III and Schedule-I of Indian Wildlife (Protection) Act, within its reported geographic range of southern India, Malabar Civet remains the least common of the four sympatric civets; the two scansorial palm civets—Common Palm Civet *Paradoxurus hermaphroditus* and Brown Palm Civet *P. jerdoni* (subfamily Paradoxurinae)—and the ground-dwelling Small Indian Civet *Viverricula indica* (subfamily Viverrinae) being relatively common and widely distributed (Mudappa, 1999). Genus *Viverricula* Hodgson, 1838 Monotypic genus contains one species *Viverricula indica,* IUCN Least Concern, placed under Appendix-III and Schedule-II distributed throughout the country in suitable habitats.

The family Muridae includes five subfamilies. Subfamily Platacanthomyinae includes two very distinct genera, namely, *Platacanthomys* from southern India and *Typhlomys* from southern China. *Platacanthomys lasiurus* Blyth, 1859 distributed in southwest Peninsular India, north up to 14^0 N latitude. Reported from Kerala (Trivandrum) and Karnataka (Shimoga, Mysore-Kanara border, and S. Coorg). Recently reported from Peppara WLS, Trivandrum district, Kerala. Species occurs in moist evergreen forest, lives in holes in the large trees or in the clefts among rocks. It is a pest of pepper (*Piper nigrum*), cashew nut (*Allacardium occidentale*), cardamom, and jackfruit; also likes fermented palm juice. The Ethiopian subfamily Gerbillinae consists of gerbils, which are found in arid and sub-arid tracts. In India, one species has extended its distribution to non-arid areas too. *Tatera indica cuvieri* (Waterhouse, 1838), whose distribution in WG Maharashtra (Satara, Kolaba, Ratnagiri), Goa, Karnataka (Dharwar, N. Kanara, Bellary, Shimoga, Mysore, Coorg), TN (Arcot, Salem, Madurai, Coimbatore, Rameshwaram), and Kerala (Trivandrum). Also found distributed in Sri Lanka, Eastern Ghats of India (Andhra Pradesh; Cuddapah, Kurnool, Palakonda Hills).

Subfamily Murinae includes one species, which is endemic to India. *Vandeleuria nilgiriaca* Jerdon, 1867 [Endangered B1ab (iii)+2ab (iii) ver 3.1] is endemic to the northern WG, especially the Nilgiris (Musser and Carleton, 2005) and the WG of Brahmagiri and Pushpagiri areas of Coorg, Karnataka. It ranges from 900 to 2100 msl and is restricted in its extent of occurrence to less than 5000 sq km. and its area of occupancy to less than 500 sq km. Specimens of *Vandeleuria* in central and northern Coorg (Kodagu) match the descriptions of *V. nilagiriaca*, while those in the south and southeast in the low-lying areas are *V. oleracea*. Ellerman (1961) maintained six subspecies of *Vandeleuria oleracea*, namely, *oleracea, dunleticola, modesta, ruhida, spadicea,* and *nilagirica* from India on the basis of body-color and lengths of tail and occipito-nasal. Subsequently based on further studies, only two subspecies are recognized *V. oleracea oleracea* from south India and Gujarat, and *V. oleracea dumeticola* from north India, on the basis of color of dorsum and the chromosome number.

Subfamily Sciuridae includes *Funambulus tristriatus* an endemic to forested tracts of WG. It occurs at elevations of 700–2100 msl, and is widely distributed in the mountain ranges of WG, in Kerala, Karnataka, and TN.

Family Erinacidae Fischer von Waldheim, 1817 Subfamily Erinacinae and the Genus *Hemiechinus* Fitzinger, 1866 contains three Indian species. (1) *Hemiechinus collaris,* (2) *Hemiechnus micropus,* and (3) *Hemiechinus nudiventris.* Only *Hemiechinus nudiventris* (Horsfield, 1851 = *Paraechinus nudiventris* (Horsfield, 1851)) found its distribution in WG especially in TN and Kerala. A small mammal endemic to India. The study reports of Kumar et al. (2018) indicate the distribution data of the south Indian hedgehogs habitat includes pasture lands, edges of agriculture fields, scrublands, grasslands, its distribution is restricted to few districts within southern Peninsular India (Nameer et al., 2015; Chakraborty et al., 2017).

17.3.3 PROBASCIDIA

Elephant (*Elephas maximus*) was found to be around in 23,543 km^2 of the WG landscape. Three major populations (one to the north of the Palakkad Gap and the two to the south of it include: (1) the northern population extends through the Protected Areas of Pushpagiri-Talakaveri-Nagarahole-Bandipur-Mudumalai-Wayanad-Biligiri Ranga TR and Cauvery Wildlife Sanctuary, and their intervening forests, extending up to the Eastern Ghats. (2) The two southern populations south of the Palakkad Gap were those extending across Protected Areas of (1) Peechi-Vazhani-Parambikulum-Indira Gandhi Wildlife Sanctuary-Idukki, and (2) Periyar-Srivilliputhur-Shendurney-Peppara to KMTR. Elephant signs were also recorded from parts of Anshi and Bhadra Protected Areas and from the forests of Mudigere tehsil of Chikmagalur district of Karnataka. The comparative figures for the states show that the estimated population of wild elephants in the country has increased from 29,391 to 30,711 as compared to 27,657–27,682 in 2007 (MoEFCC, 2015) Karnataka known to harbor 4035 elephants during 2007 and 5648–6488 numbers during 2012 and 6049, in the recent 2017 census.

TABLE 17.2 Brahmagiri-Nilgiri Landscape (Karnataka-Kerala-Tamil Nadu).

Karnataka		6049
Kerala		3054
Tamil Nadu		2761
Total Number in India:	26,786; Western Ghats:	11, 864 (44.29%)

17.3.4   ORDER PERISSODACTYLA

Genus *Equus* Linnaeus, 1758 and the Genus *Rhinoceros* Linnaeus, 1758 are not represented in WG.

17.3.5   ORDER ARTIODACTYLA

Detailed account Systematic list of the species and taxonomy and distribution of Indian Artiodactyls are available in Blanford (1891), Lydekker (1914), Walker et al. (1968), Prater (1980), Ellerman and Morrison-Scott (1951), Corbet and Hill (1992), and Grubb (in Wilson and Reeder, 2005). The order Artiodactyla is represented in India by 31 species, 20 genera, and five families, out of which one species is endemic. Family Suidae Gray, 1821 includes genus *Sus,* which contains two Indian species. (1) *Sus salvanius,* (2) *Sus scrofa,* of which the later species distributed in WG. *Sus scrofa* a widely distributed species across the world.

Family Tragulidae Milne-Edwards, 1864, Genus *Moschiola* Hodgson, 1843 includes *Moschiola indica* Gray, 1852 distributed from TN to Madhya Pradesh placed in Schedule-I and Least Concern (IUCN). Indian chevrotains are among the most frequent hunted animals by indigenous and local.

Family Cervidae Goldfuss, 1820 Subfamily Cervinae Goldfuss, 1820 and the Genus *Axis* H. Smith, 1827 includes two Indian species viz., *Axis axis* and *Axis porcinus. Axis axis* (Chital or spotted deer) is distributed in Peninsular India, northwards to Kumaon and Sikkim including West Bengal, introduced in Andaman Islands, and also found in Nepal, Sri Lanka as well as introduced in many parts of the world. The species occupancy was recorded from within 20,760 km^2 in the WG landscape. The species occurred as three large populations north of the Palakkad Gap: (1) Anshi-Dandeli complex, (2) Bhadra-Kudremukh complex, and (3) Nagarahole to Cauvery complex. Chital occurrence was low in the south of the Palakkad Gap and restricted to open canopy forests, although the species was recorded from within Periyar and Kalakad-Mundanthurai Tiger Reserves. Sambar (*Rusa unicolor* Kerr, 1792) occupancy of this species is known from 37,899 km^2 in the WG. A contiguous distribution was recorded from Anshi–Dandeli up to the Palakkad Gap in the south and eastwards into Cauvery Wildlife Sanctuary and beyond. The Genus *Cervus* Linnaeus, 1758 includes (1) *Cervus duvauceli,* (2) *Cervus elaphus,* (3) *Cervus eldii,* and (4) *Cervus unicolor* (Change in nomenclature from *Cervus* to *Rusa). Cervus unicolor* Kerr, 1792 (=*Rusa unicolor* (Kerr,1792 *(Rusa unicolor unicolor* Kerr, 1792 in WG))) commonly known as Sambar distributed in TN, northwards to Uttar Pradesh, east to Northeastern States according to IUCN, vulnerable, and placed under Schedule-III. Subfamily Munti-acinae Pocock, 1923 includes Genus *Muntiacus* Rafinesque, 1815 contain one Indian species *Muntiacus muntjak* (Zimmermann, 1780) found and distributed almost throughout the country except Jammu and Kashmir and desert region as well as in Bangladesh, Bhutan, China, Indo-china, Indonesia, Malaysia, Nepal, Pakistan, and Sri Lanka.

Family Bovidae Gray, 1821 Subfamily Antilopinae Gray, 1821 a Monotypic genus contain one species *Antilope cervicapra* (Linnaeus, 1758), which has a wide distribution

and found in open plains, Indian antelope, and is generally found in herds; these are sometimes extremely numerous and comprise occasionally several thousand animals. *Gazella
bennettii* (Sykes, 1831) commonly known as Chinkara whose type locality is Deccan
Plateau, India.

Subfamily Bovinae Gray, 1821 Genus *Bos* Linnaeus, 1758 contains Indian species
of *Bos frontalis* (Mithun), *Bos grunniens* (Yak), and *Bos gaurus* (Gaur). There is a
taxonomic ambiguity of the distribution of the subspecies, according to Groves (2003)
Groves and Grubb (2011), proposed two sub species; *B. gaurus gaurus*, which inhabits in
India and Nepal. Genus *Bubalus* Smith, 1827 includes *Bubalus bubalis* (Linnaeus, 1758)
[Water Buffalo or Indian Buffalo], large bovid found distributed in many parts of WG.
Tetracerus quadricornis (de Blainville, 1816), commonly known as four-horned antelope
or chousingha. It is monotypic and endemic to Peninsular India and small parts of lowland
Nepal. Considered Vulnerable by the IUCN and placed in Schedule-I at the national level.

The WG constitute one of the most extensive extant strongholds of Gaur (*Bos gaurus)*,
with good numbers in Wynaad-Nagarahole-Mudumalai-Bandipur complex (Ranjitsinh,
1997). Ranjitsinh (1997) estimated 12,000–22,000 in India and Gaur (*Bos gaurus*)
occupancy was recorded in 23,225 km². Major populations of about 2000 individuals have
been reported in both Nagarahole and Bandipur National Parks, over 1000 individuals in
Tadoba Andhari Tiger Project, 500–1000 individuals in both Periyar Tiger Reserve and
Silent Valley and adjoining forest complexes, and over 800 individuals in Bhadra Wildlife
Sanctuary, and found distributed in northern parts of WG landscape (mostly within
Protected Areas) and as contiguous populations in:

a.   Nagarahole-Mudumalai-Wayanad-BRT Wildlife Sanctuary,
b.   Peechi-Vazhani-Parambikulum-Indira Gandhi Wildlife Sanctuary,
c.   Periyar-Srivilliputhur-Shendurney-Peppara to Kalakad Mundanthurai.

Major populations have been reported in Nagarahole National Park (probably over
2000), Bhadra Wildlife Sanctuary (over 800), Bandipur National Park (over 2000),
Biligirirangangswamy Wildlife Sanctuary and Malemahadeswara Hill range (over 1000),
Anshi–Dandeli Tiger Reserve and adjoining forest mosaic (perhaps about 400), Kudremukh
National Park and Someshwara Wildlife Sanctuary (about 200–400), Brahmagiri–Pushpa-
giri–Talakaveri Sanctuary and Mukurti National Park (Choudhury, 2002).

Genus *Hemitragus* Hodgson, 1841 includes two Indian species: (1) *Hemitragus
hylocrius*; (2) *Hemitragus jemlahicus,* the former distributed in the Himalayan range and
the latter in the high ranges of WG. While the forests of the region underwent massive
alterations due to altered land-use, the wildlife too suffered immensely. Apart from habitat
loss, extensive uncontrolled hunting became a major problem. Areas around Wayanad and
in parts of Mudumalai comprising of Karguli, Theppakkadu, and Masinagudi were known
for their big game shooting comprising of tiger, elephant, and gaur while the Nilgiris were
known for the tahr. In fact, around 1890s wildlife numbers dwindled to such an extent that
a group of sports hunters formed the Nilgiri Game Association and enforced strict quotas
for shooting animals (Rangarajan, 2001). By 1879, the Nilgiri Game Act was initiated. It
was due to this Act that the Nilgiri tahr population recovered in this area and continues to
survive even today. In the Anamalais of WG, the High Range Wildlife and Environment

Preservation Association has been providing similar protection to the tahr since 1895 (Johnsingh, 2006). *Hemitragus hylocrius* (Ogilby, 1838) [=*Niligiritragus hylocris*] once ranged over the greater part of the WG, the present range is restricted to TN and Kerala, from Nilgiri Hills in the north to Ashambu hills in the South (11°30′N to 8°20′N). It has been reported from at least at 20 different locations, but its number has depleted to an alarming stage. At the beginning of this century, the range of tahr probably extended northward at least to the Brahmagiri hills of southern Karnataka (Shackleton, 1997). The animals are more or less confined to altitudes of 1200–2600 m (Nilgiri Tahr Trust); populations as low as 900 may or may not represent prehuman extent of occurrence in elevation (Rice, 1988). The number appear to have been declining as cited by Hopeland et al. (2016).

Family Manidae Gray, 1821 Genus *Manis* Linnaeus, 1758 includes two species *Manis crassicaudata* and *Manis pentadactyla*. *Manis crassicaudata* Gray, 1827 whose type locality is India (Locality not specified) with distribution throughout India, elsewhere in Bangladesh, China, Pakistan, Sri Lanka. The species is placed under Near Threatened (IUCN), Appendix-II of CITES and Schedule-I of Wildlife Protection Act.

Family Soricidae includes *Feroculus feroculus* (Kelaart, 1850) Kelaart's Long-clawed Shrew, endemic to Sri Lanka and southern India and Endangered species found distributed in WG of Karnataka, Kerala, and TN. Among the Genus *Suncus* Ehrenberg, 1832 includes five Indian species (1) *Suncus dayi* (2) *Suncus etruscus* (3) *Suncus montanus* (4) *Suncus Inurinus* (5) *Suncus stoliczkanus* all distributed in WG, two species are endemic to the Ghats.

Tupaiidae Gray, 1825 includes *Anathana ellioti* Endemic to India found distributed in Peninsular India, north to Bihar and West Bengal in the east and the Satpura Hills, Madhya Pradesh in the west. The species is placed in Least Concern, Appendix-II and Schedule-V.

Chiroptera (Bats): Of the 119 species of bats known from the Indian subcontinent, 65 species of bats are recorded in WG which represents >55% of bat species. Out of 65 species recorded from WG, 48% roosts in caves, 34% roosts in crevices, 6% roosts in logs of dry trees, and 12% roosts in foliage and trees, only one species found in bamboo thickets. Of the 52 species, 84% are insectivorous, 12% are frugivorous, and 2% are carnivorous (Korad et al., 2007). Chiroptera along with the order Insectivora and Rodentia constitute more than 60% of the mammalian fauna.

Kerivoula picta Pallas, 1767 (Painted Bat, Painted Woolly Bat) is a widespread species that has been recorded from southern and northeastern South Asia, southern China, most of mainland Southeast Asia, and some major islands in insular Southeast Asia. In South Asia, the species is known from Bangladesh (Sarkar and Sarkar, 2005; Srinivasulu and Srinivasulu, 2005), India. *Otomops wroughtoni* (Thomas) is known from a single location (Type locality) in Barapede cave near Talewadi, Uttara Kannada, Karnataka (Ramakrishna et al., 2004). In 2013, Girish et al. visited Mullayangiri peak in Chikkamagalur district of Karnataka and found Hodgson's Bat (*Myotis formosus*) for the first time.

17.4 THREATS

Ellerman and Morrison-Scott (1951) considered *Viverra civettina* as a subspecies of *V. megaspila* Blyth. However, Lindsay (1928), Pocock (1939), and Wozencraft (in Wilson and Reeder, 2005) treated *V. civettina* as a distinct species. Distributional range of certain species is still a mystery for example *Viverra civettina*, it is for this reason Nandini and Muddappa (2010) suggests the study of the morphometrics and molecular phylogenetics of the genus, particularly of all *Viverra civettina, Viverra megaspila,* and *Civettictis civetta* skins labeled as being from India before further field surveys are commissioned.

Killing, habitat loss, development, fragmentation, and restricted distribution, pose a considerable threat to mammal fauna of WG (Nameer et al., 2001). Yet another threat for the survival of smaller and medium mammals is the conversion of grasslands, which are often tagged as "degraded" or "low quality forest."

TABLE 17.3   Status of number species documented under Wildlife Protection Act, 1972 and IUCN RED List Category for the Mammals of the WG

Mammals under Wildlife Protection Act, 1972		Mammals under Red List Category of IUCN	
Schedule – I	30	Critically endangered	1
Schedule – II	22	Endangered	11
Schedule – III	6	Vulnerable	17
Schedule – IV	7	Near threatened	9
Schedule – V	67	Data deficient	5
Mammals in CITES Appendix		Least concern	97
Appendix – I 15		Not evaluated	25
Appendix – II 19			
Appendix – III 8			

KEYWORDS

- **mammalia**
- **distribution**
- **threat**
- **inventory/checklist**

REFERENCES

Anderson, J. *Catalogue of Mammalia in the Indian Museum, Calcutta, Part I*, Indian Museum: Calcutta, **1881**.
Anoop, K. R.; Hussain, S. A. Factors Affecting Habitat Selection by Smooth-Coated Otters (*L. perspicillata*) in Kerala, India. *J. Zool.* **2004,** *263*, 417–423.
Blanford, W. T. *The Fauna of British India, Including Ceylon and Burma. Mammalia. Part I & II*, Taylor and Francis: Red Lion Court Fleet Street, London, **1888–1891**.
Chakraborty, S.; Srinivasulu, C.; Molur, S. *Paraechinus nudiventris*, The IUCN Red List of Threatened Species, **2017**; e.T39594A22326706. DOI: 10.2305/IUCN.
Choudhury, A. Distribution and Conservation of the Gaur Bos Gaurus in the Indian Subcontinent, *Mamm Rev.* **2002**, *32*, 199–226.

CITES (Convention on International Trade in Endangered Species of Wild Fauna and Flora) Appendices I, II & III (valid from Apr 04, **2017**), p 21.

Corbet, G. B.; Hill, J. E. *Mammals of the Indo-Malayan Region: A Systematic Review;* Oxford University Press: Oxford, UK, **1992**.

Ellerman, J. R. *The Fauna of Indian Including Pakistan, Burma and Ceylon. Mammalia*, 3 [Rodentia], Govt. of India: Delhi, **1961**.

Ellerman, J. R.; Morrison-Scott, T. C. S. *Checklist of Palaearctic and Indian Mammals 1758 to 1946*, British Museum Natural History: London, UK, **1951**.

Elliot, D. G. *A Review of the Primates*; Amerrican Museum of Natural History, New York, NY, vol 3, **1913**.

Foster-Turley, P. A. Conservation Aspects of the Ecology of Asian Small-Clawed and Smooth Otters on the Malay Peninsulas. *IUCN Otter Spec. Group Bull.* **1992**, *7*, 26–29.

Gilchrist, J. S., Jennings, A. P.; Veron, G.; Carvallini, P. Family Herpestidae, In *Handbook of the Mammals of the World-vol. 1. Carnivores*; Wilson, D. E., Mittermeier, R. A., Eds.; Lynx Edicions: Barcelona, **2009**; pp 262–329.

Groves, C. P. *Primate Taxonomy*; Smithsonian Institution Press: Washington, DC, **2001**.

Groves, C. P. Taxonomy of Ungulates of the Indian Subcontinent. *J. Bombay Nat. Hist. Soc.* **2003**, *100*, 341–362.

Groves, C. P. The Taxonomy, Distribution and Adaptations of Recent Equids. In *Equids in the Ancient World*; Meadow, R. H., Uerpmann, H. P., Eds.; Ludwig Reichert Verlag: Weisbaden, **1986**; pp 11–65.

Groves, C. P.; Mazak, V. On Some Taxonomic Problems of the Asiatic Wild Asses; with Description of a New Subspecies (Perissodactyla, Equidae). *Z. Siiugetierk.* **1967**, *32*, 321–355.

Groves, C.; Grubb, P. *Ungulate Taxonomy*; John Hopkins University Press: Baltimore, MD, **2011**.

Grubb, P. Order Perissodactyla. In *Mammal Species of the World A Taxonomic and Geographic Reference*; Wilson, D. E., Reeder, D. M., Eds.; Smithsonian Inst. Press: Washington and London, **1993**; pp 369–372.

Hopeland, P.; Puyravaud, J.-P.; Davidar, P. Nilgiri Tahr in the Agasthamalai Range, Western Ghats. Population, Status and Threats. *J. Threat. Taxa* **2016**, *8* (6), 8849–8952.

Hussain, S. A. Otter Conservation in India. *ENVIS Bull.–Wildlife and Protected Areas* **1999**, *2* (2), 92–97.

IUCN, *IUCN Red List Categories and Criteria: Version 3.1*, 2nd ed.; Gland: Switzerland and Cambridge, UK, **2012**; pp iv + 32.

Jerdon, T. C. *The Mammals of India; A Natural History of all the Animals Known to Inhabit Continental India*, John Weldon: London, UK, **1874**.

Johnsingh, A. J. T. *Field Days: A Naturalists Journey Through South and Southeast Asia*, Universities Press (India) Pvt. Ltd: Chennai, **2006**.

Karanth, K. U. Status of Wildlife and Habitat Conservation in Karnataka. *J. Bombay Nat. Hist. Soc.* **1986**, *83*, 166–179.

Korad, V. S.; Yardi, K. D.; Raut, R. N. Diversity and Distribution of Bats in Western Ghats. *Zoo's Print J.* **2007**, *22*, 2752–2758.

Kumar, A.; Chellam, R.; Choudhury, B. C.; Mudappa, D.; Vasudevan, K.; Ishwar, N. M.; Noon, B. R. *Impact of Rainforest Fragmentation on Small Mammals and Herpetofauna in the Western Ghats, South India*; WII-USFWS Collaborative Project Final Report Wildlife Institute of India: Dehradun, **2002**.

Kumar, B.; Togo, J.; Singh, R. The South Indian Hedgehog *Paraechinus nudiventris* (Horsfield, 1851): Review of Distribution Data, Additional Localities and Comments on Habitat and Conservation. *Mammalia* **2018**, **2018**, 1–11.

Kumara, H. N.; Irfan-Ullah, M.; Kumar, S. Mapping Potential Distributions of Slender Loris Subspecies in Peninsular India. *Endanger. Species Res.* **2009**, *7*, 29–38.

Kumara, H. N.; Singh, M. The Influence of Differing Hunting Practices on the Relative Abundance of Mammals in Two Rainforest Areas of the Western Ghats, India. *Oryx* **2004**, *38*, 321–327.

Kumara, H. N.; Singh, M.; Kumar, S. Distribution, Habitat Correlates, and Conservation of Loris lydekkerianus in Karnataka, India. *Intern. J. Primatol.* **2006**, *27* (4), 941–969.

Lindsay, H. M. Note on Vive"a Civettina. *J. Bombay Nat. Hist. Soc.* **1928**, *33*, 146–148.

Lydekker, R. *Catalogue of the Ungulate Mammals in the British Museum (Natural History)*; Trustees of the British Museum: London, United Kingdom, **1914**; vol iii.

Mivart, St. G. *Dogs, Jackals, Wolves, and Foxes A Monograph of the Canidae*; R. H. Porter: London, **1890**.

Molur, S.; Brandon-Jones, D.; Dittus, W.; Eudey, A.; Kumar, A.; Singh, M.; Feeroz, M. M.; Chalise, M.; Priya, P.; Walker, S., Eds. *Status of South Asian Primates: Conservation Assessment and Management Plan (C.A.M.P.) Workshop Report*; Zoo Outreach Organization/CBSG-South Asia: Coimbatore, **2003**.

Mudappa, D. Herpestids, Viverrids, and Mustelids. In *Mammals of South Asia, 1*; Johnsingh, A. J. T., Manjrekar, N., Eds.; Universities Press: Hyderabad, India, **2013**, pp 471–498.

Mudappa, D. Lesser-Known Carnivores of the Western Ghats. *ENVIS Bul.: Wildlife and Protected Areas, Mustelids, Viverrids and Herpestids of India* **1999**, *2* (2), 65–70.

Mudappa, D. Observations of Small Carnivores in the Kalakad-Mundanthurai Tiger Reserve, Western Ghats, India. *Small Carniv. Conserv.* **2002**, *27*, 4–5.

Mudappa, D.; Prakash, N.; Pawar, P.; Srinivasan, K.; Ram, M. S.; Kittur, S.; Umapathy, G. First Record of Eurasian Otter *Lutra lutra* in the Anamalai Hills, Southern Western Ghats, India. *IUCN Otter Speci. Group Bull.* **2018**, *35* (1), 47–56.

Muddapa, D.; Choudhury, A.; Wozencraft, C.; Yonzon, P. *Herpestes fuscus*. In *IUCN 2014*; 2014 IUCN Red List of Threatened Species, **2008**. http://www.iucnredlist.org/details/41612/0 (accessed Mar 09, 2015).

Musser, G. G.; Carleton, M. D. Rodentia: Myomorpha: Muroidea: Murinae. In *Mammal Species of the World: A Taxonomic and Geographic Reference, 3rd ed.; vol. 1 & 2*; Wilson, D. E., Reeder, D. M., Eds.; The Johns Hopkins University Press: Baltimore, MD, **2005**; pp 1189–1531.

Nameer, P. O. *Checklist of Indian Mammals*; Kerala State Forest Department and Kerala Agricultural University, **2000**.

Nameer, P. O.; Molur, S.; Walker, S. Mammals of Western Ghats: A Simplistic Over View. *Zoos' Print J.* **2001**, *16* (11), 629–639.

Nameer, P. O.; Praveen, J.; Bijukumar, A.; Palot, M. J.; Das, S.; Raghavan, R. A Checklist of the Vertebrates of Kerala State, India. *J. Threat. Taxa* **2015**, *7*, 7961–7970.

Nandini, R.; Muddappa, D. Mystery or Myth: A Review of History and Conservation Status of the Malabar Civet *Viverra civettina* Blyth, 1862. *Small Carniv. Conserv.* **2010**, *43*, 47–59.

Nandini, R.; Muddappa, D.; Van Rompaey, H. Distribution and Status of the Brown Palm Civet in the Western Ghats, South India. *J. Small Carniv. Conserv.* **2002**, *27* (2), 6–10.

Perinchery, A.; Jathanna, D.; Kumar, A. Factors Determining Occupancy and Habitat Use by Asian Small-Clawed Otters in the Western Ghats, India. *J. Mammal.* **2011**, *92* (4), 796–802.

Pocock, R. I. Description of a New Race of Puma (*Puma concolor*), with a Note on an Abnormal Tooth Growth in the Genus. *The Annals and Magazine of Natural History.* **1940**, *11* (6), 307–313.

Pocock, R. I. *The Fauna of British India, Including Ceylon and Burma. Mammalia.* vol. II, Taylor and Francis: London, **1941**.

Pocock, R. I. *The Fauna of British India including Ceylon and Burma—vol. 1*; Taylor and Francis: London, **1939**.

Prater, S. H. *The Book of Indian Animals*; Bombay Natural History Society: Bombay, India, 1980.

Rajamani, N.; Muddappa, D.; Van Rompaey, H. Distribution and Status of the Brown Palm Civet in the Western Ghats, South India. *J. Small Carniv. Conserv.* **2002**, *27* (2), 6–10.

Rangarajan, M. *India's Wildlife History*; Permanent Black: New Delhi, **2001**.

Ranjitsinh, M. K. *Beyond the Tiger: Portraits of Asian Wildlife*; Birajbasi Printers: New Delhi, **1997**.

Rice, C. G. Habitat, Population Dynamics and Conservation of Nilgiri Tahr Hemitragus Hyalocris. *Biol Conserv.* **1988**, *44*, 137–156.

Roonwal, M. L.; Mohnot, S. M. *Primates of South Asia: Ecology, Sociobiology, and Behavior*; Harvard University Press: Cambridge, **1977**.

Roonwal, M. L.; Mohnot, S. M.; Rathore, N. S., Eds. *Current Primate Researches*; University of Jodhpur Press: Jodhpur, **1984**.

Ryley, K. V. Bombay Natural History Society's Mammal Survey of India. No. 9–14. *J. Bombay nat. Hist. Soc.* **1913**, *22*, 283–295; 464–513; **1914**, *22*, 684–725.

Sarkar, S. U.; Sarkar, N. J. Bats of Bangladesh with Notes of Their Status, Distribution and Habitat. *Bat Net-CCINSA Newsl.* **2005**, *6* (1), 19–20.

Schreiber, A.; Wirth, R.; Riffel, M.; Van Rompaey, H. *Weasels, Civets, Mongooses and Their Relatives: An Action Plan for the Conservation of Mustelids and Viverrids*; IUCN, Gland: Switzerland, **1989**; pp 99.

Sclater, W. L. *Catalogue of Mammalia in the Indian Museum, Calcutta, Part II*; Indian Museum: Calcutta, India, **1891**.

Seth, P. K., Ed. *Perspectives in Primate Biology*; Today and Tomorrow: New Delhi, **1983**.

Shackleton, D. M. *Wild Sheep and Goats and Their Relatives: Status Survey and Conservation Action Plan for Caprinae*; IUCN, Gland: Switzerland and Cambridge, UK, **1997**.

Sivasothi, N.; Nor, B. H. M. A Review of Otters (Carnivora: Mustelidae: Lutrinae) in Malaysia and Singapore. *Hydrobiologia* **1994**, *285*, 151–170.

Srinivasulu, C.; Srinivasu, B. A Review of Chiropteran Diversity of Bangladesh. *Bat Net-CCINSA Newsl.* **2005**, *6* (2), 6–11.

Sterndale, R. *Natural History of the Mammalia of India and Ceylon*, Thacker, Spink & Co.: Calcutta, India, **1884**.

Tiwari, K. K. *Report on Census of Rhesus Macaque and Hanuman Langur of India*; Mimeo Report, National Primate Survey, Zool Soc of India: Calcutta, **1983**.

Walker, E. P.; Warnick, F.; Hatnet, S. E.; Lange, K. L.; Davis, M. A.; Vible, H. E.; Wright, P. F. *Mammals of the World*; John.Hopkins Press: Baltimore, MD, vol 2, **1968**.

Wilson, D. E.; Reeder, D. M., Eds. *Mammal Species of the World: A Taxonomic and Geographic Reference*; Third Edition Johns Hopkins University Press: Baltimore, MD, **2005**; vol 2, pp 1–2141.

CHAPTER 18

PHYTOPLANKTON DIVERSITY OF THE WESTERN GHATS OF INDIA

SEETHARAMAIAH THIPPESWAMY

Department of Environmental Science, Mangalore University, Mangalore, Karnataka, India

ABSTRACT

All the major groups of phytoplankton such as Bacillariophyceae, Chlorophyceae, Cyanophyceae, Dinophyceae, Euglenophyceae, and Xanthophyceae inhabit lotic and lentic water bodies in the Western Ghats. The western part of India from Tapti river of Gujarat is under the influence of west flowing drainage system, and the abundance of phytoplankton in Tapti river is in the decreasing order as Diatoms > Cyanophyceae > Chlorophyceae > Desmids. The phytoplankton composition in the west flowing rivers of Karnataka such as Sharavathi, Sita, and Nethravathi is documented. Phytoplankton diversity from Irrity river, tributary of river Valapattanan; river Tirur, tributary of river Bharathapuzha; rivers Meenachil and Chalakudy is available. Phytoplankton diversity in the Parambikulam–Aliyar irrigational canals in Tamil Nadu is also available. A total of 35 species of phytoplankton belonging to 14 genera of Bacillariophyceae, 15 genera of Chlorophyceae, and six genera of Myxophyceae from river Godavari near Nashik in the Western Ghats region of Maharashtra were reported. Phytoplankton diversity from river Mula, represented by Chlorophyceae, was the dominant and most diversified group and is represented by the orders Volvocales, Tetrasporales, Oedogonials, Ulothrichales, Desmidiales, Chaetophorales, Cladophorales, Chlorococcales, and Zygnematales. A total of 88 species (25 families) and 97 species (29 families) of phytoplankton from rivers Tunga and Bhadra, respectively, have been reported in the Western Ghats of Karnataka. A total of 53 species of phytoplankton belonging to Bacillariophyceae, Chlorophyceae, Cyanophyceae, Desmidaceae, Euglenophyceae, and Ulvophyceae have been reported at Cauvery Nisargadhama, Abbey Falls, Bhagamandala, and Talakaveri regions of river Cauvery in Coorg district, Karnataka. The microalgal species assemblages in the catchment region of river Noyyal, tributary of river Cauvery, in the Western Ghats region of Tamil Nadu revealed a total of 142 species belonging to 10 different families. Chlorophyceae was predominant in Hemavathy reservoir. Next to

Biodiversity Hotspot of the Western Ghats and Sri Lanka. T. Pullaiah, PhD (Ed.)
© 2024 Apple Academic Press, Inc. Co-published with CRC Press (Taylor & Francis)

Chlorophyceae, Dinophyceae (*Ceratium* sp.) occurred in significant density in Nugu and Harangi reservoirs. Myxophyceae, which is generally abundant in tropical reservoirs, was not very significant in reservoirs of Karnataka. The plankton community of Bhatghar reservoir in Maharashtra; Linganamakki, Anjanapura, Hemavathy, Bhadra, Nugu, Harangi, and Kabini reservoirs in Karnataka; Pazhassi, Mullaperiyar, Peringalkuthu, Parambikulam, and Idukki reservoirs in Kerala; Upper Aliyar reservoir in Tamil Nadu is presented. Further, phytoplankton diversity in Panvel, Venna, and Dhamapur lakes of Maharashtra; Syngenta, Lotus, Curtorim, and Khandola lakes of Goa; Vattakkayal, Mullaperiya, and Periyar Lakes in Kerala regions of the Western Ghats is documented. Phytoplankton found in various ponds including temple ponds, and also in freshwater wetlands including Kola lands of Kerala are presented.

18.1 INTRODUCTION

Plankton inhabit in all types of aquatic environments such as ponds, lakes, rivers, estuaries, lagoons, backwaters, sea, etc. Some species flourish in highly eutrophic waters while others are very sensitive to organic or chemical substances. The plankton species assemblage will be useful in assessing water quality of a given water body, and plankton respond quickly to environmental changes due to their short life cycles, standing crops, and species compositions. Based on the nutrition point of view, the aquatic algae are included under phytoplankton which are unicellular, colonial or filamentous, photosynthetic, and are grazed upon by zooplankton (free floating heterotrophic animals) and other aquatic higher organisms (nekton). The Western Ghats or Sahyadris is one among the global biodiversity hotspots. The Western Ghats runs parallel to the western coast of the Indian Peninsula from north to south along the western edge of the Deccan Plateau, and separates the plateau from a narrow coastal plain, called Konkan in the northern part and Malabar in the southern part, along the Arabian sea, and forms the catchment area for complex riverine drainage systems in the states of Gujarat, Maharashtra, Goa, Karnataka, Kerala, and Tamil Nadu. The freshwater algal diversity in the Western Ghats region of Gujarat (Solanki and Manoj, 2014, 2016; Sarang and Manoj, 2017), Maharashtra (Kamat, 1964, 1968; Balakrishnan and Chougule, 2002; Ghadage and Karande, 2008; Thakur and Behere, 2008; Priyadarshani et al., 2014; Kamble and Karande, 2014), Goa (Kerkar and Madkaiker, 2003; Gaunker and Kerkar, 2004; Geeta and Kerkar, 2009; Kerkar and Lobo, 2009; Kerkar, 2009), Karnataka (Hegde, 1986; Hegde and Issacs, 1988a, 1988b; Rajeshwari and Rajashekhar, 2012; Sharathchandra and Rajashekhar, 2015), Kerala (Suxena et al., 1973; Jose and Patel, 1989; Easa, 2004; Jose and Francis, 2007, 2010; Tessy and Sreekuma, 2008, 2009, 2011, 2017; John and Francis, 2013; Nasser and Sureshkumar, 2013; Jisha and Tessy, 2014; Ray and Krishnan, 2016; Tessy and Anu, 2016; Preshanthkumar et al., 2017; Tessy and Sreenisha, 2020), and Tamil Nadu (Venkataraman, 1939; Suresh et al., 2012; Sureshkumar and Thomas, 2019) is fairly documented in the literature covering different aquatic environs. In the present chapter, phytoplankton inhabiting the freshwater habitats such as rivers, reservoirs, lakes, tanks, ponds, and wetlands of the Western Ghats of India are presented based on the information available in the literature.

18.2 BACKGROUND

The water bodies have various types of macro- and microorganisms such as nektonic, benthic, and planktonic forms inhabiting in the water column and bottom sediments. The nektonic forms are able to migrate freely and swim against the water current. On the other hand, benthic forms inhabit the bottom of water from the littoral zone to the abyssal zone in the marine environment, and many invertebrates inhabit the bottom of freshwater bodies such as ponds, lakes, reservoirs, rivers, and streams. While on the contrary, planktonic organisms are generally microscopic, aggregate, passively floating, drifting, or somewhat motile and move with the mercy of water currents, tides, and waves in a body of water. The term plankton does not include pleuston, organisms on the surface of the water, and nekton, organisms swimming actively against the water current. The name plankton is derived from the Greek word and was coined by Victor Hensen in 1887 (Hensen, 1887). Plankton are grouped into various categories using different criteria (Hardy and Milne, 1938; Mann and Carr, 1992; Omori and Ikeda, 1992; Wommack and Colwell, 2000; Robert, 2001; Morris et al., 2002; Wang et al., 2012, 2014; Smith, 2013; Stubner et al., 2016; Leles et al., 2018). They are often divided based on their size into megaplankton (> 20 cm), macroplankton (2–20 cm), mesoplankton (0.2–20 mm), microplankton (20–200 µm), nanoplankton (2–20 µm), picoplankton (0.2–2 µm), and femtoplankton (< 0.2 µm). Plankton are also divided into various types based on their life stages: Holoplankton are those organisms that are in planktonic stage for their entire life cycle, meroplankton are those organisms which have both planktonic and benthic stages in their life cycles, pseudoplankton are those organisms that attach themselves to planktonic organisms or any floating objects, tychoplankton are those organisms which are free-living or attached benthic organisms, and other nonplanktonic organisms that are carried into the plankton through a disturbance of their benthic habitat. Further, plankton are grouped again into four types based on their habitat namely, marine plankton, aeroplankton, geoplankton, and freshwater plankton. Furthermore, plankton are classified into five types based on trophic levels such as phytoplankton, zooplankton, mycoplankton, bacterioplankton, and virioplankton.

A scattered information is available on different aspects of phytoplankton inhabiting various types of freshwater bodies of the Western Ghats region from Gujarat (Solanki and Manoj, 2014, 2016; Sarang and Manoj, 2017), Maharashtra (Prajapathi et al., 2014), Goa (Sawaiker and Rodrigues, 2016, 2017), Karnataka (Rao and Madhyastha, 1990; Krishnamurthy and Reddy, 1996; Malathi and Thippeswamy, 2008, 2012; Thippeswamy and Malathi 2008a, 2008b; Thippeswamy et al., 2008; Desai et al., 2008; Ramesh and Thippeswamy, 2009; Basavaraja et al., 2013; Divya et al., 2013), Kerala (Krishanan, 2012; Tessy and Sreekumar, 2013; Divya et al., 2013; Nasser and Sureshkumar, 2014; Thomas and Tessy, 2015; Ray and Krishnan, 2016; Sebastain, 2016; Tessy, 2019; Ray et al., 2020; Sherly et al., 2020; Tessy and Sreenisha, 2020), and Tamil Nadu (Manikam et al., 2012; Palaniswamy et al., 2015). A semienclosed water body found in the coastal region with one or more rivers or streams connections and with a free connection to the open sea forming a transition zone between riverine and marine environments is called an estuary, and this water body from river mouth extending toward the low lands of river course is generally referred as estuarine system where salinity variation is very common depending on the tides and freshwater influx. The planktonic organisms inhabiting this complex dynamic environment from Indian region have been reviewed by Thippeswamy and Malathi (2009). In this chapter, the diversity of phytoplankton inhabiting the freshwater bodies in the Western Ghats, excluding the estuarine and mangrove swamp ecosystems, is presented with the available information in the literature published by various authors.

18.3   PHYTOPLANKTON IN LOTIC WATER BODIES

Freshwater is one of the limited resources on our planet. The freshwater in the Western Ghats in both small and large-sized lotic and lentic water bodies is slowly drying up due to various anthropogenic activities in the region and thereby the biota in these water bodies is also vanishing at a faster rate. A river is a stream of water that generally originates in the mountains and flows downward in a definite course on earth's surface until it reaches a lake or another river or may dry up on its course of flow or drains into the sea. The Western Ghats has many both east and west-flowing perennial and seasonal rivers. These are mostly rainfed water bodies flooded during the southwest monsoon season, and receive most of the water during rainy season from June to September. The water resource of many of these rivers is exploited for production of hydel power and also for irrigation purposes by creating very large, medium, and small reservoirs by constructing dams across these rivers. Many of these small rivers and also many tributaries dry up during summer season due to lack of natural flow, rainwater, and also due to depletion of groundwater sources. The east-flowing river system of the Western Ghats has three major river systems, namely, the river Godavari in Maharashtra state, the river Krishna in Maharashtra and Karnataka states, and the river Cauvery in Karnataka, Kerala, and Tamil Nadu states originating in the Western Ghats and drains into the Bay of Bengal. In addition to these river systems, there are small rivers, namely Tambraparni and Vaigai rivers also originating in the Western Ghats and drain into the Bay of Bengal. The west-flowing river system consists of small and medium-sized rivers, compared to east-flowing rivers, from Gujarat to Kerala and, originates in the Western Ghats and drains into the Arabian sea. Heavy rainfall during monsoon season brings great volume of water in these rivers due to the prevailing high gradient in the hilly terrain of the Western Ghats region, and these rivers are prone to tidal effects to considerable lengths in the inland area.

18.3.1   PHYTOPLANKTON IN WEST-FLOWING RIVER SYSTEM

Among the west-flowing rivers of the Western Ghats, some of the rivers namely Purna, Auranga, and Par in the state of Gujarat; Vaitarna, Ulhas, Pathalganga, Savitri, Vashishti, Gad, Kajali, Kundalika, and Arjuna in the state of Maharashtra; Mandovi (Mahadayi) and Zuari riverine complex, Tiracol, Chapora, and Talpone in the state of Goa; Kali, Gangavali (Bedthi), Aganashini, Sharavathi, and Kollur-Chakra-Kubja-Halade riverine complex, Sita-Madisala-Swarna riverine complex, Mulki (Shambavi river)-Pavanji (Nandini river) riverine complex and Gurpura/Phalguni-Nethravati riverine complex in the state of Karnataka and Chandragiri, Thejaswini, Perumba, Valapattanam, Chaliyar, Meenachi, Bharatapuzha, Chalakudi, Periyar, and Pamba in the state of Kerala are important and somewhat larger in size. Many of these west-flowing rivers are influenced by tidal waters from the Arabian sea and form estuarine ecosystems near the mouth of these rivers.

18.3.1.1   PHYTOPLANKTON IN WEST-FLOWING RIVERS OF GUJARAT

Pollution-tolerant phytoplankton species such as *Amphora* sp., *Anabaena* sp., *Ankistrodesmus* sp., *Cocconeis* sp., *Caloneis* sp., *Cosmarium* sp., *Chlorella* sp., *Cyclotella* sp., *Fragilaria* sp., *Gomphonema* sp., *Gyrosigma* sp., *Melosira* sp., *Microcystis* sp., *Merismopodia* sp., *Navicula* sp., *Nitzschia* sp., *Oscillatoria* sp., *Skeletonema* sp., *Scenedesmus* sp., *Spirogyra* sp., *Synedra* sp., *Nostoc* sp., *Pediastrum* sp., *Pleurosigma* sp., *Pinnularia* sp., *Pandorina* sp., and *Euglena* sp. were reported from Tapi river in Gujarat by Solanki and Manoj (2014). The diversity of diatoms, blue-green algae, green algae, and Desmids of Tapi river in relation to water quality was studied by Solanki and Manoj (2016) and observed that the abundance of these groups of algae was in the decreasing order as Diatoms > Cyanophyceae > Chlorophyceae > Desmids. They have further reported that *Chlorella* sp., *Anabaena* sp, *Navicula* sp., *Nitzschia* sp., *Oscillatoria* sp., *Fragilaria* sp., *Pediastrum* sp., *Melosira* sp., *Pleurosigma* sp., and *Mastagloea* sp. were present during the entire study period and suggested, furthermore, that the species such as *Spirogyra* sp., *Oscillatoria* sp., *Scenedesmus* sp., *Pinnularia* sp., *Gomphonema* sp., and *Euglena* sp. may be used as a bioindicator of water pollution. A total of five groups of phytoplankton such as Bacillariophyceae, Chlorophyceae, Cyanophyceae, Euglenophyceae, and Dinophyceae were recorded in Tapti River in Gujarat by Sarang and Manoj (2017).

18.3.1.2   PHYTOPLANKTON IN WEST-FLOWING RIVERS OF KARNATAKA

Ramesh and Thippeswamy (2009) have recorded a total of 88 species of phytoplankton in the west-flowing Seeta river at Seethanadi village near Hebri town, foothills of the Western Ghats in Karnataka and reported that among 22 families of phytoplankton recorded, Naviculaceae (13 spp.), Desmidaceae (13 spp.), Zygnemataceae (seven spp.), and Fragilaraceae (seven spp.) had the highest species representation, followed by Cymbellaceae (four spp.), Ulotrichaceae (three spp.), and Hydrodictyaceae (three spp.) during the study period from April 2005 to May 2006. The most commonly encountered species in river Seeta were *Navicula minuta*, *Navicula* sp., *Fragilaria capucina*, *Tabellaria* sp., *Ulothix* sp., and *Amphora ovalis* during the study period. Desai et al. (2008) reported a total of 216 species belonging to 59 genera of phytoplankton under Bacillariophyceae, Desmidials, Chlorococcales, Cyanophyceae, Dinophyceae, Euglenophyceae, and Chrysophyceae groups in the Sharavati river basin in the central Western Ghats of Karnataka. Among Bacillariophyceae, *Anomoeoneis sphaerophora*, *Gyrosigma attenuaium*, *G. gracila*, *Gomphonema lanceolatum*, *G. longiceps*, *Navicula viridula*, *Nitzschia abtusa*, *N. palea*, *Pinnularia lundii*, *P. maharashtrensis*, *Surirella ovata*, *Synedra acus*, and *S. ulna* were common during all collections from riverine environs. The representatives of Desmidiales were *Arthrodesmus psilosporus*, *Closterium ehrenbergii*, *Cosmarium decoratum*, *Desmidium baileyi*, *Staurastrum limneticum*, *S. freemanii*, *S. multispinceps*, *S. peristephes*, *S. tohopekaligense*, and *Iriploceros gracile*. The representatives of Chlorococcalean (*Eudorina elegans*, *Muogeotia punctate*, *Pediastrum simplex*, and *Spirogyra rhizobrancchialis*), Dinophycean (*Ceratium hirundnella*), and Cyanophycean (*Microcystis aeruginosa*) were also presented. The monthly variation of phytoplankton was investigated from Netravathi river of the Western Ghats region by Rao and Madhyastha (1990), and revealed that the diatoms were dominant during winter (November and December) and blue-green algae during summer (March and April), and Myxophyceae, Dinophyceae, and Xanthophyceae were sparsely represented in this river basin. Ramesh and Thippeswamy (2009) have recorded phytoplankton species belonging to Chlorophyta, Chrysophyta, Rhodophyta, and Charophyta from Kempuhole stream of river Kumaradhara, tributary of river Nethravathi, in Western Ghats region of Karnataka, and reported that Chrysophyta was dominant throughout the study period with 67.58% followed by Chlorophyta with 31.68%. Rhodophyta and Charophyta ranked third and fourth positions with 0.48 and 0.32%, respectively. Of the 27 families of phytoplankton, Naviculaceae was represented by 14 species followed by Desmidaceae (13 spp.), Zygnemataceae (11 spp.), Fragilaraceae (eight spp.), Ulotrichaceae (four spp.), Cymbellaceae (four spp.), and Hydrodictyaceae (three spp.). The most commonly encountered species in Kempuhole stream were *Navicula minuta*, *Navicula* sp., *Fragilaria capucina*, and *Tabellaria* sp.

18.3.1.3 *PHYTOPLANKTON IN WEST-FLOWING RIVERS OF KERALA*

Phytoplankton diversity from Irrity river, tributary of river Valapattanan, in the Western Ghats region of Kerala was studied by Divya et al. (2013) and found that Bacillariophyceae was the dominant group followed by Desmidaceae, Chlorophyceae, Euglenophyceae, Cyanophyceae, and Ulvophyceae. A total of 16 species of algae were identified from the river Meenachil, one of the key rivers in Kottayam district, emerges from the Western Ghats and drains into Vembanad lake in Kerala (Sebastian, 2016). He further revealed that genus *Cosmarium* (*C. granulopolaris*, *C. bikopa*, *C. spinoreniformis*, and *C. bumeranga*) was more diverse followed by *Oscillatoria* (*O. rubescens*, *O. splendida*, and *O. laetevirens*), *Anabaena* (*A. circinalis*), and *Lyngbya* (*L. ceylanica* and *L. kuetzingii*). A total of 57 species of phytoplankton belonging to 30 genera comprising four classes, namely, Bacillariophyceae with 39 species belonging to 21 genera, Cyanophyceae with 14 species belonging to 7 genera, and Chlorophyceae and Euglenophyceae with two species each belonging to one genus were observed by Tessy and Sreenisha (2020) in river Tirur, tributary of river Bharathapuzha, from the Western Ghats region of Kerala. The phytoplankton diversity in relation to water quality at three sites, namely Ichipara, Kanjirappilly, and Vettukadavu from Chalakudy river in the Western Ghats region of Kerala was studied by Thomas and Tessy (2015) and revealed that Bacillariophyceae was the dominant group of algae found at all the study sites during the study period from March to August 2007.

18.3.1.4 PHYTOPLANKTON IN WEST-FLOWING RIVERS IN TAMIL NADU

The Aliyar stream is the tributary of the river Kannadipuzha, one of the main tributaries of the river Bharathapuzha, and originates from Aliyar dam at Aliyar near Pollachi in the Western Ghats region of Tamil Nadu. The Parambikulam river is one of the four tributaries of the Chalakudy river and originates in the Western Ghats region of the Coimbatore district in Tamil Nadu. The Parambikulam dam has been constructed across the river at Anamalai, located in the Western Ghats of Kerala. Parambikulam–Aliyar (10°15' and 10°30' N; 76°50' and 77°10' E) irrigational canal project was undertaken in 1961 by both Kerala and Tamil Nadu governments to provide water facilities to local population of both the states. Phytoplankton diversity in the Western Ghats region of Parambikulam–Aliyar irrigational canals at Kulanaickenpatti canal, Seelakkampatti canal, Poosaripatti canal, Kongalnagaram canal, and Pethappampatti canal in Tamil Nadu was studied by Manickam et al. (2012) and revealed a total of 22 species of phytoplankton belonging to Cyanophyceae (nine spp.), Chlorophyceae (seven spp.), and Bacillariophyceae (six spp.). Cyanophyceae member was found to be dominant followed by Chlorophyceae and Bacillariophyceae. Among Cyanophyceae members, *Microcystis aeruginosa* was recorded at all five stations studied, followed by *Chroococcus minutes*, *Oscillatoria subbrevis*, *Phormidium fragile*, and *Synechococcus aeruginosus* in four stations, *Oscillatoria cortiana* and *Spirulina labyrinthiformis* in three stations, *Aphanocapsa pulchra* in two stations, and *Synechocystis aqutilisi* in one station only. Similarly, Chlorophycean member *Scenedesmus quadricauda* was recorded in all five stations studied, followed by *Spirogyra hyalina* in four stations, *Pediastrum duplex* and *Rhizoclonium crassipellitum* in three stations, and *Chlorella vulgaris*, *Ulothrix aequalis*, and *Ulothrix zonata* in two stations. Among Bacillariophycean members, *Cyclotella meneghinianai* was recorded at four stations studied, followed by *Amphora coffeaformis*, *Fragilaria brevistriata*, *Gomphonema lanceolatum*, *Melosira granulate*, and *Navicula radiosa* in three stations.

18.3.2 PHYTOPLANKTON IN RIVER GODAVARI

River Godavari raises at Brahmagiri hills in Sahyadri ranges of the Western Ghats at Trymbakeshwar near Nashik city in the state of Maharashtra about 50 miles from Arabian sea, flows eastward in the Deccan Plateau in south India, and falls into the Bay of Bengal in Andhra Pradesh. Among the seven major tributaries of Godavari, river Pravara is the smallest and the only tributary originating in the Western Ghats akin to Godavari. River Pravara has a subtributary which also rises in the Western Ghats. Important dams built in the Western Ghats on Godavari, Darna, and Mula rivers are Kashypi, Gangapur, Gautami Godavari, Darna, Mukana, Upper Vaitarna, Ghatghar, Nilawande, Mula, and Kadwa. A total of 35 species of phytoplankton belonging to 14 genera of Bacillariophyceae, 15 genera of Chlorophyceae, and six genera of Myxophyceae from river Godavari near Nashik in the Western Ghats region of Maharashtra were reported by Nalawade and Bagul (2018). They further revealed that the family Bacillariophyceae was dominant and represented by *Ceratoneis*, *Amphora*, *Caloneis*, *Fragilaria*, *Navicula*, *Synedra*, *Diatoms*, *Gomphonema*,

Pinnularia, Melosira, Tabellaria, Denticula, Cymbella, and *Cyclotella* followed by family Chlorophyceae with *Chlorella, Chlaymydomonas, Spirogyra, Ulothrix, Hydrodictyon, Cladophora, Cosmarium, Chlorococcum, Oedogonium, Microspora, Desmidium, Chara, Zygnema, Syndesmus,* and *Volvox,* and Myxophyceae with *Nostoc, Anabaena, Oscillatoria, Rivularia, Coccochlori,* and *Phormidium.*

18.3.3   PHYTOPLANKTON IN RIVER KRISHNA

The river Krishna originates in the Western Ghats near Mahabaleshwar in the state of Maharashtra, flows through the states of Karnataka, Telangana, and Andhra Pradesh, and drains into the Bay of Bengal at Vijayawada. The largest tributary of the Krishna river is Tungabhadra and the longest is Bhima; other important tributaries are Panchganga, Warna, Yerla, Malaprabha, Ghataprabha, and Bhavanasi. The Varada, Vedavathi, and Handri are the main tributaries of the river Tungabhadra. The rivers Bhadra and Tunga rise at "Gangamula" on the Varaha Parvata mountain in the Aroli-Gangamula region at an elevation of about 1198 m above MSL in the Sahyadri range of the Western Ghats of India. The river Bhadra flows eastward initially and then turns northeast in the Western Ghats, whereas river Tunga initially flows northward in the scenic Tunga valley, then turns northeast in the Western Ghats, and joins the Bhadra river at Koodli, north of Shivamogga town in the plain terrain. Thereafter, the combined flow of rivers Tunga and Bhadra is called river Tungabhadra. The river Vedavathi rises from the Baba Budangiri mountains in the Western Ghats of Chikkamagalur district and joins the Tungabhadra river in Siruguppa taluk of Bellary district. Ultimately, the river Tungabhadra joins the river Krishna near Kurnool in the state of Andhra Pradesh.

18.3.3.1   PHYTOPLANKTON IN RIVER MULA

River Mula originates at Deoghar in the Mulshi region of the Western Ghats of Maharashtra and meets Pavana at Dapodi, which also has its source in the Western Ghats. The Mula then joins the Mutha at Sangam, and together they flow as the Mula–Mutha as a single river for a distance of 56 km to meet river Bhima, which later merges into river Krishna and falls into the Bay of Bengal. Gunale and Balakrishnan (1981) carried out study on the freshwater ecology of algae related to water pollution in Pavana, Mula, and Mutha rivers, and indicated that certain algal groups were indicative of level of organic enrichment. Kshirsagar et al. (2012) made observations on phytoplankton diversity from river Mula at Puna city and reported that the Chlorophyceae was the dominant and most diversified group at all stations, and is represented by the orders Volvocales, Tetrasporales, Oedogonials, Ulothrichales, Desmidiales, Chaetophorales, Cladophorales, Chlorococcales, and Zygnematales. Genera such as *Scenedesmus, Pandorina, Pediastrum, Coelastrum, Chlorella, Ankistrodesmus, Micractinium, Spirogyra, Closterium, Staurastrum, Cosmarium, Closterium,* and *Actinastrum* were common. *Scenedesmus quadricauda, Chlorella vulgaris,* and *Pediastrum duplex* were abundant at stations Aundh and Dapodi. Bacillariophyceae was the second largest group with *Gomphonema, Melosira, Cymbella,* and *Fragilaria* found frequently and abundantly at all stations, whereas species like *Melosira granulata, Synedra ulna, Cymbella affinis, Nitzschia palea,* and *Cyclotella meneghiniana* were frequently found at stations Aundh and Dapodi, which are organically polluted, and *Diatomella, Cocconeis,* and *Caloneis* were found at station Wakad. They have further reported that Cyanophyceae members such as *Merismopedia glauca, Merismopedia elegans, Merismopedia minima, Arthrospira massartii, Oscillatoria limosa, Oscillatoria tenuis, Oscillatoria princeps,* and *Microcystis aeruginosa* were observed during the study period, and the Euglenophyceae consisted of *Euglena oxyuris, E. acus, Phacus acuminatus, Phacus,* sp., *Trachelomonas, Lepocinclis* sp., *Petalomonas* sp., *Trachelomonas volvocina, T. horrid,* and *T. dubia.* Polluted water showed the occurrence and dominance of *Scenedesmus quadricauda, Chlorella vulgaris, Oscillatoria limosa,* and *Melosira granulata* throughout the year. Polluted water of the Mula river near Poona city revealed that these species could be used as indicators of organic pollution (Kshirsagar, 2013).

18.3.3.2 *PHYTOPLANKTON IN RIVER TUNGA*

A total of 63 species of phytoplankton belonging to Cynophyceae (six spp.), Bacillariophyceae (24 spp.), Chlorophyceae (32 spp.), and Rhodophyceae (one sp.) have been reported by Krishnamurthy and Reddy (1996) from Tunga river near Sringeri town in Karnataka from the Western Ghats region. Malathi and Thippeswamy (2008) have reported that the phytoplankton diversity was composed of Chlorophyceae (39 spp.), Chrysophyceae (34 sp.), Rhodophyceae (two spp.), Cyanophyceae (six spp.), and unidentified algae (two spp.) in the Sita river, a tributary of river Tunga, at B. G. Katte near Koppa in the Western Ghats of Karnataka. The most commonly encountered species were *Spirogyra* sp., *Zygnema* sp., *Desmidium swartzi, Pediastum* sp. *Chetophora incrassata, Uronema africana, Microsterias americana,* and *Closterium* sp. in river Sita. The most abundant group in Sita river was Chlorophyta. Of the 39 taxa of Chlorophyceae, a total of eight species (*Spirogyra* sp., *Zygnema pectinatum, Closterium* sp., *Desmidium swartzi, Microsterias americana, Uronema africana, Cheatiphora incrassata,* and *Pediastrum dupex*) followed by Chrysophyta with six species (*Navicula* sp., *Pinnularia* sp., *Fragilaria capucina, Synedra ulna, Tabellaria fenestrata,* and *Surirella tenera*) of Chrysophyceae were found. Malathi and Thippeswamy (2012) have reported that the phytoplankton diversity in Malathi river, a tributary of river Tunga, at Kalmane in the Western Ghats of Karnataka consisted of Chlorophyceae (43 spp.), Bacillariophyceae (26 spp.), Cyanophyceae (11 spp.), Chrysophyceae (two spp.), and unidentified algae (one sp.). Among Chlorophyceae, *Spirogyra* sp., *Clostridium lingissimus, Cosmarium cucurita, C. premorsum, Desmidium swartium, Pleurotaenium trabecular, Pediastrum duplex,* and *Sphaerocystis schroeteri* were abundant followed by Bacillariophyceae (*Navicula* sp., *Cymbella cistula, Fragilaria capucina, Synedra ulna, Tabellaria fenestrata,* and *Surirella elegans*), Crysophyceae (*Hydrudus foetidus* and *Dinobryon divergens*), and Cyanophyceae (*Aphanocapsa* sp.).

18.3.3.3   PHYTOPLANKTON IN BHADRA RIVER

Thippeswamy and Malathi (2008a) have revealed that the Bilegal (Kochche) stream, tributary of river Bhadra, is composed of 17 species of phytoplankton belonging to four families, and the algal diversity is composed of Bacillariophyceae (*Diatoma vulgare*, *Diatoma hiemale*, *Gomphonema truncatum*, *Synedra ulna*, *S. acus*, *Fragilaria capucina*, *Navicula minuscula*, *N. radiosa*, *N. mutica*, *Navicula* sp., *Amphora ovalis*, and *Stauroneis* sp.), Chlorophyceae (*Ulothrix* sp., *Spirogyra* sp., *Zygnema* sp., and *Spondylosium planum*), and Myxophyceae (*Microcoleus lyngbyus*). Family Fragilariaceae was dominant with 12 species followed by Ulotrichaceae (four spp.), Desmidiaceae (one sp.), and Chrococcaceae (one sp.). Among phytoplankton, family Naviculaceae was dominant with 18 species followed by Desmidiaceae (10 spp.), Fragilariaceae (eight spp.), and Chroocoecaceae (seven spp.). Chlonococcaceae (*Desmatractum induration*), Scenedesmaceae (*Actinansteem* sp.), Endospharecae (*Chlorochytrium* sp.), Schizogoniaceae (*Schizogonium murale*), Coccomysoaceae (*Tetraspora cylindrica*), Characiaceae (*Dictyosphaeruim* sp.), Hydruraceae (*Hydrurus foetidus*), Tribonemataceae (*Tribonema* sp.), Nitzschiaceae (*Nitzschia* sp.), Ephthemiaceae (*Rhoplodia gibba*), and Thoreaceae (*Thorea ramosissima*) were represented by a single species each. Thippeswamy and Malathi (2008b) have reported a total of 36 species of phytoplankton belonging to 23 families comprising 31 genera in Bakri stream, tributary of river Bhadra, near N. R. Pura in the Western Ghats of Karnataka, and reported that the phytoplankton diversity was composed of Chlorophyceae (10 spp.), Chysophyceae (one sp.), Bacillariophyceae (nine spp.), Myxophyceae (two spp.), and unidentified algae (two spp.). Among phytoplankton, Chlorophyta composed of 10 species (*Monostroma* sp., *Nitrium digitus*, *Spirogyra inflata*, *S. nitida*, *Spirogyra* sp., *Closterium leibleinii*, *C. longissima*, *Spondylosium planum*, *Hydrodictyon reticulum*, and *Pediastrum duplex*) followed by Bacillariophyta with nine species (*Diatoma hiemale*, *Fragilaria capucina*, *Synedra acus*, *S. ulna*, *Tabellaria fenestrata*, *Navicula* sp., *Surirella elegans*, *S. tenera*, and *Nitzchia* sp.), Myxophyta with two species (*Microcoleus lyngbyaceus* and *Spirolina* sp.), Chrysophyta with a single species (*Hydrurus foetidus*), and two species of unidentified algae during the study period.

18.3.4   PHYTOPLANKTON IN CAUVERY RIVER

The river Cauvery rises at Talakaveri in the Brahmagiri hills range in the Western Ghats in Kodagu district of the state of Karnataka. Many tributaries of Cauvery also originate in the Western Ghats, namely, Harangi, Hemavati, Lakshmana Thirtha, and Kabini flow in the state of Karnataka; Kabini, Noyyal, Bhavani, and Amaravathi originate in the Western Ghats of the states of Kerala and Tamil Nadu. A total of 53 species of phytoplankton belonging to Bacillariophyceae, Chlorophyceae, Cyanophyceae, Desmidaceae, Euglenophyceae, and Ulvophyceae have been reported at Cauvery Nisargadhama, Abbey Falls, Bhagamandala, and Talakaveri regions of river Cauvery in Coorg district, Karnataka (Divya et al., 2013). The microalgal species assemblages and their associated physico-chemical parameter dynamics in the catchment region of river Noyyal, tributary of river

Cauvery, in the Western Ghats region of Tamil Nadu were studied by Sureshkumar and Thomas (2019), and revealed a total of 142 microalgal species belonging to 10 different families from five different sites. They have further reported that the relative percentage distribution of algae was dominated by Scenedesmaceae (36.6%), and concluded that the physicochemical parameters influenced the dominant taxa of microalgae belonging to Chlorellaceae, Scenedesmaceae, and Chlorococcaceae in river Noyyal.

18.4 PHYTOPLANKTON IN RESERVOIRS

A reservoir is a man-made lake that is created when a dam is built across a river and the river water is stored toward the tail end of river behind the dam creating a reservoir. The major reservoirs in the Western Ghats region are Gangapur, Lonavala, Walwahn, Darna, Khadakawasla, Bhatghar, Nira, Balkawadi, and Koyna in the state of Maharashtra; Supa, Bommanahalli, Tattihalli, Kodasalli, Kadra, Lingamakki, Gersoppa, Bhadra, Tunga, Hemavathi, Harangi, and Kabini reservoirs in the state of Karnataka; Upper Bhavani, Bhavanisagar, Aliyar, Amaravathi, Palar, and Upper Sholayar reservoirs in the state of Tamil Nadu; Parambilkulam, Sholayar, Poringalkuttu, Pariyar, Idukki, Idamalayar, Nayyar, and Pepper reservoirs in the state of Kerala. The plankton community of Bhatghar reservoir on river Nira in Maharashtra is composed of Chlorophyceae (*Gonatozygon, Hormidium, Eudorina, Closteridium,* and *Kirchneriella*), and is the major group. Diatoms (*Navicula, Nitzchia, Syndera,* and *Surirella*), Dinophyceae (*Ceratium*), and Desmids (*Staurastrum*) are the other forms recorded. The presence of blue-green algae, *Microcystis aeruginosa* indicates the organic enrichment of water body. Phytoplankton diversity of Dhom reservoir on Krishna river in the state of Maharashtra was represented by *Chlorella, Navicula, Nitzschia, Synedra,* and *Phormidium*. However, the most important genus in the reservoir is *Raphidiopsis*, blue-green algae, usually recorded from unpolluted water bodies (CICFRI, 1997). Ramakrishniah et al. (2000) have explored the ecological aspects and fish yield potential of selected reservoirs of Karnataka and revealed the phytoplankton comprising major groups such as Myxophyceae, Chlorophyceae, Dinophyceae, and Bacillariophyceae. Further, phytoplankton were significant in Hemavathy and Bhadra reservoirs. The initial phytoplankton dominance in Hemavathy reservoir persisted even after 13 years of its formation. Among the four groups, Chlorophyceae was predominant, especially in Hemavathy reservoir. Next to Chlorophyceae, Dinophyceae (*Ceratium* sp.) occurred in significant density in Nugu and Harangi reservoir. Myxophyceae, which is generally abundant in tropical reservoirs, was not very significant in reservoirs of Karnataka. Chlorophyceae was represented by 21 forms of which *Ulothrix* and *Pediastrum* were important. Diatoms were represented by 10 forms and they occurred rarely. *Ceratium hirundinella* is the only species representing Dinophyceae. Myxophyceae was represented by *Microcystis* sp. (Ramakrishniah et al., 2000).

Basavaraja et al. (2013) reported a total of 152 phytoplankton species belonging to 59 genera in the Anjanapura reservoir in Karnataka, and the dominant genera were *Crucigenia, Pediastrum, Scenedesmus, Tetraedron, Cyclotella, Gyrosigma, Melosira, Navicula, Fragilaria, Pinnularia, Synedra, Closterium, Cosmarium, Euastrum, Staurastrum,*

Gleocapsa, Merismopedia, Microcystis, Oscillatoria, Euglena, and *Phacus.* The phyto-plankton diversity of Linganamakki reservoir consisted of Myxophyceae (*Microcystis* sp.), Chlorophyceae (*Pediastrum* sp., *Ulothrix* sp., *Desmidium* sp., and *Staurastrum* sp.), Bacillariophyceae (*Gonatozygon* sp.), and Dinophyceae (*Ceratium* sp.) (Ramakrishniah et al., 2000). Desai et al. (2008) reported a total of 216 species belonging to 59 genera of phytoplankton comprising Bacillariophyceae, Desmidials, Chlorococcales, Cyanophy-ceae, Dinophyceae, Euglenophyceae, and Chrysophyceae groups in the Sharavati (Linga-namakki) reservoir. Among Bacillariophyceae, *Anomoeoneis sphaerophora, Gyrosigma attenuaium, G. gracila, Gomphonema lanceolatum, G. longiceps, Navicula viridula, Nitzschia obtusa, N. palea, Pinnularia lundii, P. maharashtrensis, Surirella ovata, Synedra acus,* and *S. ulna* were common. The representatives of Desmidials were *Arthrodesmus psilosporus, Closterium ehrenbergii, Cosmarium decoratum, Desmidium baileyi, Staura-strum limneticum, S. freemanii, S. multispinceps, S. peristephes, S. tohopekaligense,* and *Iriploceros gracile.* The representatives of Chlorococcalean (*Eudorina elegans, Muogeotia punctata, Pediastrum simplex,* and *Spirogyra rhizobrachialis*), Dinophycean (*Ceratium hirundnella*), and Cyanophycean (*Microcystis aeruginosa*) were also presented.

The phytoplankton diversity of the Bhadra reservoir consisted of Myxophyceae (*Microcystis* sp.), Chlorophyceae (*Pediastrum* sp., *Zygnema* sp., *Pachycladon* sp., *Arthrodesmus* sp., *Sphaerocystis* sp., *Micrasterias* sp., *Ulothrix* sp., *Tetraedron* sp., *Cosmarium* sp., *Coelastrum* sp., and *Ankistrodesmus* sp.), Bacillariophyceae (*Synedra* sp.), and Dinophyceae (*Ceratium* sp.) (Ramakrishniah et al., 2000). The phytoplankton diversity in Hemavathy reservoir was reported by Ramakrishniah et al. (2000) and revealed that the phytoplankton consisted of Myxophyceae (*Microcystis* sp.), Chlorophyceae (*Pediastrum* sp., *Pachycladon* sp., *Ulothrix* sp., *Nephrocystium* sp., and *Desmidium* sp.), Bacillariophyceae (*Pinnularia* sp. and *Gonatozygon* sp.), and Dinophyceae (*Ceratium* sp.). They have further made the observation on phytoplankton in the Harangi reservoir and revealed that the phytoplankton composed of Myxophyceae (*Microcystis* sp.), Chlorophyceae (*Closterium* sp., *Ulothrix* sp., *Nephrocystium* sp., *Staurastrum* sp., *Coelastrum* sp., and *Ankistrodesmus* sp.), Bacillariophyceae (*Navicula* sp. and *Pinnularia* sp.), and Dinophyceae (*Ceratium* sp.). The phytoplankton diversity of Kabini reservoir consisted of Myxophyceae (*Microcystis* sp.), Chlorophyceae (*Pediastrum* sp., *Ulothrix* sp., *Pachycladon* sp., *Arthrodesmus* sp., *Oedogonium* sp., *Micrasterias* sp., and *Cosmarium* sp.,), Bacillariophyceae (*Pinnularia* sp.), and Dinophyceae (*Ceratium* sp.), whereas Nugu reservoir consisted of Myxophyceae (*Microcystis* sp.), Chlorophyceae (*Ulothrix* sp.), Bacillariophyceae (*Synedra* sp.), and Dinophyceae (*Ceratium* sp.) (Ramakrishniah et al., 2000).

Phytoplankton diversity of Pazhassi reservoir in Wynad district from the Western Ghats region of Kerala was reported by Divya et al. (2013). They have further reported that Bacillariophyceae was the dominant group followed by Desmidaceae, Chlorophyceae, Euglenophyceae, Cyanophyceae, and Ulvophyceae. Krishnan (2012) made observations on phytoplankton of Mullaperiyar reservoir in the Western Ghats of Kerala and observed that the density of phytoplankton was positively correlated with the nutrients in almost all seasons, and the diversity was dependent on the seasonal fluctuations in the environment and also on the increased anthropogenic activities in and around the reservoir. A total of 94 species of microalgae belonging to 42 genera comprising Bacillariophyceae (14 spp.),

Chlorophyceae (10 spp.), Cyanophyceae (five spp.), Desmidiaceae (11 spp.), and Euglenophyceae (two spp.) were recorded from the Peringalkuthu reservoir, Western Ghats, Kerala (Nasser and Sureshkumar, 2013). A total of 89 taxa of phytoplankton comprising Chlorophyceae, Desmidiaceae, Bacillariophyceae, Cyanophyceae, and Euglenophyceae were recorded during the study period in Parambikulam reservoir, built across the Parambikulam river in the Western Ghats in the Palghat district of Kerala, by Nasser and Sureshkumar (2014), and found that the Bacillariophyceae group was dominant with 42 taxa followed by Desmidiaceae with 26 taxa. They have further noted that the dominant genera were *Pinnularia* and *Navicula* of Bacillariophyceae, and *Closterium* and *Cosmarium* of Desmidiaceae during the study period. A total of 95 algal species belonging to Cyanophyceae, Chlorophyceae, Bacillariophyceae, Dinophyceae (Dinoflagellates), and Desmids from Idukki reservoir in Anjunadu valley of Kerala were recorded, and majority of the species were belonging to the classes Phaeophyceae, Conjugatophyceae, Florideophyceae, Ulvophyceae, Zygnemataceae, and Cyanophyceae (Joshi, 2019).

18.5 PHYTOPLANKTON IN LAKES

A lake is an area filled with water, localized in a basin, surrounded by land, apart from any river or other outlet that serves to feed or drain the lake. Most lakes are fed and drained by rivers and streams. Natural lakes are generally found in mountainous areas, and other lakes are found along the courses of rivers. Many lakes are artificial and are constructed for industrial or agricultural use, for hydroelectric power generation, or water supply for domestic or agricultural use, or recreational and other activities. They are generally larger and deeper than ponds and have a larger surface area depending on local topography. In lakes, sunlight can't reach the bottom of all areas within the lakes, and usually, lakes have aphotic zones, which are deep areas of water that receive no sunlight. Generally, lakes range from 2 ha (5 acres) to 8 ha (20 acres) in size. However, Elton and Miller (1954) reported that lakes may have water bodies of 40 ha (99 acres) or more. The Western Ghats has several man-made and natural lentic water bodies. The important lakes present in the Western Ghats are the Ooty (2500 m altitude, 34.0 ha) in Nilgiris Hills; the Kodaikanal (2285 m, 26 ha) and the Berijam in the Palani Hills; the Devikulam (6.0 ha) and Letchmi (2.0 ha) lakes in the Munnar range in Tamil Nadu; the Pookode and the Karlad lakes of Wayanad in Kerala; the Vagamon, the Devikulam (6 ha), and the Letchmi (2 ha) lakes in Idukki in Kerala state.

A total of 16 genera of phytoplankton belonging to Chlorophyceae (seven), Bacillariophyceae (four), and Myxophyceae (five) in Panvel lakes (Vishrale, Krishnale, and Dewale lakes) in the Western Ghats of Maharashtra were recorded (Prajapati et al., 2014). Algal species belonging to Chlorophyceae, Cyanophyceae, Bacillariophyceae, and Euglenophyceae were recorded in Venna lake from Western Ghats in Maharashtra by Patil et al. (2015) and found that the members of Chlorophyceae like *Pediastrum*, *Scenedesmus*, and *Staurastrum* showed constant occurrence throughout the year, while some of the bloom-forming algae like *Microcystis* observed only during the summer season, and Euglenoids showed presence during winter season only. Korgaonkar and Bharamal (2016) have reported a total of 10 species of phytoplankton belonging to Chlorophyceae (seven spp.), Myxophyceae (two spp.), and Euglenophyceae (one sp.) from Dhamapur lake from Malvan taluka in Sindhudurg district of Maharashtra. The phytoplankton diversity in Syngenta, Lotus, Curtorim, and Khandola lakes of Goa was observed by Sawaiker and Rodrigues (2016) and revealed a total of 71 algal species belonging to Cyanophyceae (15 spp.), Dinophyceae (one sp.), Bacillariophyceae (17 spp.), Euglenophyceae (five spp.), and Chlorophyceae (33 spp.). Sawaiker and Rodrigues (2017) have identified the indicator species of diatoms, namely, *Gomphonema parabolum*, *Navicula halophila*, *Navicula microcephala*, and *Navicula mutica* for organic pollution, and *Amphora ovalis*, *Stauroneis phoenicenteron*, and *Synedra ulna* for anthropogenic pollution in Syngenta, Lotus, and Curtorim lakes of Goa. Further, they have also recorded the indicator species of organic pollution (*Navicula mutica*) and anthropogenic pollution (*Navicula microcephala*) in Khandola pond in Goa. The diversity and abundance of plankton community of Vattakkayal lake in the Maruthady area of Kollam in Kerala were studied by Sherly et al. (2000) and reported that among the phytoplankton species identified, Bacillariophyceae (35.41%) formed the dominant group, followed by Chlorophyceae (27.52%), Cyanophyceae (23.10%), and Euglenophyceae (13.97%). They have further reported that the high dominance of the pollution indicator plankton species such as *Closterium* sp., *Nitzschia* sp., *Navicula* sp., *Oscillatoria* sp., *Microcystis* sp., *Brachionus* sp., etc. indicated the eutrophication of water due to organic pollution. The Periyar wildlife sanctuary (PWS), a major international tourist center in Kerala, in which Mullaperiya and Periyar lakes were studied in order to explore the nutrient status and associated phytoplankton, and observed that the nutrient and phytoplankton of the lakes are dependent on the seasonal fluctuation in the environment and also influenced by the increased anthropogenic activities in and around the Mullaperiyar (Krishnan, 2012) and Periyar lakes (Ray and Krishnan, 2016).

18.6   PHYTOPLANKTON IN TANKS AND PONDS

A natural or man-made tank (village or temple) is part of an ancient traditional way of water harvesting and preserving of local rainfall or surface runoff and also water from streams and rivers for subsequent use, primarily for agricultural, drinking, sacred bathing, and ritual activities; these tanks are found over a quite large area, usually in India. Generally, tanks are constructed across a slope in such a way that water will be collected and stored for future activities by taking advantage of local topography. A pond is a man-made or natural water body with size varied from 1 m^2 to 2 ha (5 acres). However, the international Ramsar wetland convention sets the upper limit for pond size as 8 ha (20 acres) (Karki, 2008). The water in a pond is usually shallow and completely a photic zone. The sunlight is penetrated to reach the bottom of pond through water column.

The biodiversity of filamentous blue-green algae from various localities of Satara district was observed by Kamble and Karande (2014) and reported a total of 23 species belonging to 10 genera (*Spirulina*, *Oscillatoria*, *Microcoleus*, *Nostoc*, *Anabaena*, *Plectonema*, *Scytonema*, *Tolypothrix*, *Calothrix*, and *Westiellopsis*) under the orders of Nostocales and

Stigonematales. A total of 96 species representing 42 genera belonging to 18 families of 10 orders of Chlorophyta were collected from Pernem taluka of Goa from various habitats such as paddy fields, temple pond, and temporarily formed puddles and streams by Geeta and Kerkar (2009) and reported that the genera, namely, *Chlorogonium*, *Coleochaete*, *Chlamydomonas*, *Volvulina*, *Eudorina*, *Chaetophora*, *Pandorina*, *Pediastrum*, *Hormidium*, *Gloeocystis*, *Apiocystis*, *Phacotus*, *Chlorosarcinopsis*, and *Desmidium* were recorded exclusively in temple pond, and *Draparnaldiopsis* and *Spirotaenia* were found in stream. Gaunker and Kerkar (2004) have reported algal species from temple ponds from north Goa. A total of 10 genera with 38 species of Desmids were recorded from north Goa by Kerkar and Lobo (2009) and the representatives of the taxa were *Cosmarium* (10 spp.), *Closterium* (nine spp.), *Micrasterias* (five spp.), *Staurastrum* (five spp.), *Euastrum* (four spp.), *Netrium* (one sp.), *Desmidium* (one sp.), *Pleurotaenium* (one sp.), *Arthodesmus* (one sp.), and *Triploceros* (one sp.). A total of 46 species belonging to 23 genera representing seven families, namely, Chroococcaceae, Stigonemataceae, Oscillatoriaceae, Nostocaceae, Scytonemataceae, Microchaetaceae, and Rivulariaceae were recorded from various habitats in Goa (Kerkar and Madkaiker, 2003). Kerkar (2009) studied the algal diversity in temple ponds from Ponda taluka, permanent water bodies (lakes) from Salcete taluka of Goa, and reported that algal flora from ponds and water bodies mainly consisted of *Scenedesmus*, *Pediastrum*, *Euastrum*, and various other Desmids. Species such as *Cladophora glomerata* and *Chaetophora* sp. were also collected from these water bodies.

Thippeswamy et al. (2008b) reported a total of 12 phytoplankton belonging to eight families comprising eight species of Chlorophyceae (*Rhizoclonium* sp., *Pleurodiscus* sp., *Sprogyra* sp., *Scenedesmus* sp., *Monostroma quaternarium*, *Arthodesmus octocornis*, *Closterium striolatum*, and *C. setaceum*), followed by three species of Mysophyceae (*Aphanocapsa* sp., *Gloeocapsa minuta*, and *Dactylothece* sp.), and one species of Chrysophyceae (*Hydrurus foetidus*) from Kodagi tank, a seasonal tank, near N. R. Pura in the Western Ghats of Karnataka. Family Desmidiaceae was dominant with three spp. (*Arthrodesmus octocornis*, *Closterium striolatum*, and *C. setaceum*), followed by Zygnemaceae (*Pleurodiscus* sp. and *Spirogyra* sp.), Chroococaceae (*Aphanocapsa* sp. and *Gleocapsa minuta*), Cladophoraceae (*Rhizoclonium* sp.), Scenedesmaceae (*Scenedesmus* sp.), Ulvaceae (*Monostroma quaternarium*), and Hydruraceae (*Hydrurus foetidus*). The phytoplankton diversity in Koramagudda tank of Lakkavalli range of Bhadra wildlife sanctuary was studied by Gowda and Vijayakumar (2017) and reported a total of 40 species belonging to 26 genera comprising Chlorophyceae, Bacillariophyceae, Cyanophyceae, and Euglenophyceae.

Seasonal dynamics of the phytoplankton community in seven different kinds of eutrophic waters from ponds (rural, urban, and temple), lakes, reservoirs, rivers, and rock pools in four districts of Kerala, south India, were studied by Ray et al. (2021) and reported a total of 297 algal species belonging to eight phyla, 11 classes, and 26 orders. Chlorophyta, which comprised 138 species, was the dominant group, followed by Charophyta (95 spp.), Cyanophyta (30 spp.), Bacillariophyta (20 spp.), Euglenophyta (10 spp.), and Ochrophyta (two spp.). The microalga *Cosmarium*, which comprised 29 species, was the most diverse algal genus, followed by *Scenedesmus* (22 spp.). They have further reported that the most common algal species found in these eutrophic water bodies of Kerala were *Ankistrodesmus falcatus, Ankistrodesmus spiralis, Chlorococcum humicola, Coelastrum microporum, C. reticulatum, Cosmarium contractum, C. retusiforme, Crucigeniella irregularis, Gleocystis gigas, Kirchneriella lunaris, K. obese, Melosira granulata, Microcystis aeruginosa, Monoraphidium arcuatum, M. contortum, M. griffithii, Oocystis lacustris, Pediastrum duplex, P. tetras, Radiococcus nimbatus, Scenedesmus dimorphus, S. perforatus, S. quadricauda, Selenastrum bibarianum, S. gracile, Tetraedron gracile,* and *T. trigonum,* and the high abundance of the green microalgal species *Kirchneriella lunaris, Ankistrodesmus falcatus, Radiococcus nimbatus, Scenedesmus dimorphus, Coelastrum microporum,* and the diatom *Melosira granulata* was observed in eutrophic waters throughout the study area. Algal diversity in ponds of Thrissur district, Kerala was represented by six divisions namely Chlorophyta, Chrysophyta, Bacillariophyta, Euglenophyta, Pyrrhophyta, and Cyanophyta comprising 114 genera with 602 species taxa belonging to eight classes, namely, Chlorophyceae, Xanthophyceae, Chrysophyceae, Cryptophyceae, Bacillariophyceae, Euglenophyceae, Dinophyceae, and Cyanophyceae (Tessy, 2019). She has further reported that the Chlorophyceae was the major group comprising 284 taxa belonging to 51 genera, of which 175 spp. belonging to 16 genera were Desmids followed by Cyanophyceae (113 spp. with 31 genera), Bacillariophyceae (112 spp. with 21 genera), and Euglenophyceae (85 spp. with five genera. Dinophyceae (three spp. with two genera), Cryptophyceae (two spp. with one genus), Xanthophyceae (one sp. belonging to one genus), and Chrysophyceae (two spp. belonging to two genera) were also reported in the ponds.

Arulmurugan et al. (2010) reported a total of 61 algal taxa from 37 temple tanks of Palakkad and Thrissur districts of Kerala, of which 12 taxa were belonging to Cyanophyceae. Maya et al. (2000), Prameela et al. (2001), Jose and Sreekumar (2005), Jose et al. (2008), Ajayan et al. (2013), and Jisha and Tessy (2014) studied the algae from the temple ponds of Kerala. A total of 54 species of blue-green algae belonging to 17 genera were recorded in temple ponds from Koodalmanikyam and Kodungallur temple ponds in Thrissur district, Kerala by Jisha and Tessy (2014), and of these, a total of 40 species were belonging to 14 genera of Koodalmanikyam temple pond and 15 species to six genera of Kodungallur temple pond. A total of 41 algal species belonging to 20 genera were recorded from the Guruvayur Sree Krishna temple pond by Tessy and Anu (2016), and of these, 20 were belonging to Chlorophyceae (48.8%) followed by eight in Euglenophyceae (19.5%), six each in Bacillariophyceae (14.6%), Cyanophyceae (14.6%), and one belonging to Dinophyceae (2.5%). Further, they have reported a total of 15 pollution-tolerant algal species (*Cyclotella meneghiniana, Euglena gracilis, Euglena proxima, Lepocinclis ovum, Melosira, Nitzschia palea, Oscillatoria princeps, Phacus pleuronectes, Phacus pyrum, Scenedesmus acuminatus, Scenedesmus dimorphus, Scenedesmus quadricauda, Synedra ulna, Tetraedron muticum,* and *Trachelomonas volvocina*). A total 93 species of algae belonging to the phylum Cyanophyta, Chlorophyta, Charophyta, Bacillariophyta, and Glaucophyta were recorded from 10 temple ponds in the Western Ghats of Kerala (Prasanthkumar et al., 2017). In the summer season, the highest relative abundance was shown by *Kirchneriella lunaris* (50%) followed by *Pediastrum tetras, Botryococcus braunii,* and *Coelastrum reticulatum* (30% each), whereas in the monsoon season, the highest relative abundance of 40% was shown by *Ankistrodesmus spiralis, Kirchneriella obese, Pediastrum tetras,* and *Scenedesmus dimorphus.*

18.7 PHYTOPLANKTON IN FRESHWATER WETLANDS

A wetland is a distinct ecosystem that is flooded by water, either permanently or seasonally, where oxygen-free processes prevail (Keddy, 2010). Freshwater wetlands are generally grouped into those water bodies in which there is an accumulation of organic matter (peat) to form bogs and fens, where the decomposition of organic matter is low or those water bodies in which the rooting of plants is mainly through the medium of mineral-rich soil to form marshes and swamps, where organic matter accumulation is very negligible. It is known that approximately 6% of the world's area is covered by wetlands (Bateman et al., 1992); according to the world conservation monitoring center (WCMC, 1992), there are approximately 5.3–5.7 million km² of freshwater wetlands in the world. The Ramsar international convention defines wetlands as areas of marsh, fen, peat land or water, whether natural or artificial, permanent or temporary, with water that is static or flowing, fresh, brackish or salt, including areas of marine water, the depth of which at low tide does not exceed 6 m (Barbier et al., 1997). However, Ramsar convention had not defined wetlands in the context of a river basin as is widely followed now (Srinivasan, 2010). Gowda and Vijayakumara (2009) have made observations on phytoplankton from Lakkavali range of Bhadra wildlife sanctuary in the Western Ghats region of Karnataka, and recorded Chlorophyceae (26 spp.), Bacillariophyceae (17 spp.), Cyanophyceae (11 spp.), and Euglenophyceae (four spp.).

18.7.1 PHYTOPLANKTON IN KOLE WETLANDS OF KERALA

The Kole lands which form one of the rice granaries of Kerala are part of unique Vembanad–Kole wetland ecosystem of humid tropical wetland ecosystem in the southwest coast of India (Srinivasan, 2010). The Kole lands remain submerged under flood water for about 6 months in a year and this seasonal alteration gives it both terrestrial and water-related properties which determine the ecosystem structure and process (Easa, 2004). A network of main and cross canals provides external drainage and connects the different regions of the Kole to the rivers. The Vembanad–Kole system is fed by 10 rivers and all these rivers originate from the Western Ghats, flow westward and drain into the Arabian sea through the Vambanad–Kole water body. A total of 188 species of algae belonging to 64 genera comprising six taxonomic groups in the Thrissur Kole wetlands (part of Vembanad Kole, Ramsar site) of Kerala have been identified by Tessy and Sreekumar (2008), and reported that Desmidiaceae was the dominant with 83 species belonging to 19 genera followed by Chlorophyceae with 41 species of 20 genera. They have further reported that the pollution-tolerant algae such as *Melosira, Oscillatoria, Pandorina, Pediastrum, Closterium, Navicula, Microcystis*, and *Scenedesmus* were considered as indicator of enriched waters. However, Bacillariophyceae was less dominant throughout the study in the study area with only 13 species belonging to 11 genera. The freshwater algal taxa belonging to Xanthophyceae, Chrysophyceae, and Dinophyceae in the Kole regions of Thrissur district of Kerala consisted of *Centritractus belanlophorus, Ceratium hirundinella, Ophiocytium capitatum*, *Peridinium cinctum*, *Peridinium cinctum f. tuberosum*, and *Uroglenopsis americana* (Tessy and Sreekumar, 2011). Tessy and Sreekumar (2017) have studied the freshwater diatoms from the Kole lands of Thrissur which is a part of the Vembanad–Kole, a declared Ramsar site of Kerala, represented by a total of 74 species of diatoms belonging to 20 genera, namely, *Achnanthes* (two spp.), *Amphora* (two spp.), *Anomoeneis* (one sp.), *Caloneis* (two spp.), *Cocconies* (one sp.), *Cymbella* (three spp.), *Eunotia* (six spp.), *Fragilaria* (three spp.), *Gomphonema* (four spp.), *Gyrosigma* (one sp.), *Melosira* (three spp.), *Navicula* (15 spp.), *Nitzschia* (seven spp.), *Pinnularia* (nine spp.), *Pleurosigma* (one sp.), *Rhopalodia* (one sp.), *Stauronies* (four spp.), *Surirella* (five spp.), *Synedra* (three spp.), and *Tabellaria* (one sp.).

18.8   CONCLUSIONS

The Western Ghats or Sahyadris of India in their tropical humid forests harbors a large number of endemic species of macroscopic and microscopic organisms in both aquatic and terrestrial environs. The Western Ghats has a large number of natural and man-made water bodies which swell to the maximum extent of their capacity during monsoon season due to assured rainfall every year. These water bodies shrink to the maximum extent during the summer season due to overexploitation of water resources by mankind, and also due to natural processes like evaporation, etc. The organisms inhabiting these water bodies must be able to cope up with such environmental variations and complete their lifecycle, if they are short-lived. One such group of organisms is the phytoplankton inhabiting various types of freshwater aquatic environs in the Western Ghats. The commercial production of Spirulina tablet, Spirulina energy, Spirulina iced tea, etc. as supplementary food for human beings from phytoplankton is noteworthy. Even the application of omics is also not available on this group of taxa from this region. Many phytoplankton species are being used as bioindicators of aquatic pollution. The documentation of phytoplankton diversity in the Western Ghats has been established based on traditional taxonomical techniques. The application of modern taxonomical techniques using molecular and biotechnological methods to identify the phytoplankton is not yet available from this region. Further, even the clarification of taxonomical ambiguity of some species or even identification of cryptic species of phytoplankton using modern technique like DNA barcoding is missing even to date from this region. Furthermore, identification of plankton using traditional approach over a large riverine ecosystem or any other water body covering various microhabitats in different months, seasons, and years is also not available from this biodiversity hotspot. However, the identification of algal diversity or plankton diversity from different freshwater bodies in the Western Ghats region has been accumulated. The data on identification or description, using traditional taxonomy, of algal and/or phytoplankton species from the Western Ghats region are available in gray literature and also published in predatory journals.

KEYWORDS

- freshwater algae
- phytoplankton
- lotic water body
- reservoirs and lakes
- tanks and ponds
- wetlands
- Sahyadris
- India

REFERENCES

Ajayan, K. V.; Selvaraju, M.; Thirugnanamoorthy, K. Phytoplankton Population of Ananthapura Temple Lake of Kasaragod, Kerala. *Insight Bot.* **2013,** *3* (1), 6–14. DOI: 10.5567/BOTany-IK-2013.6.14.

Arulmurugan, P.; Nagaraj, S.; Anand, N. Biodiversity of Freshwater Algae from Temple Tanks of Kerala. *Recent Res. Sci. Tech.* **2010,** *2* (6), 58–71.

Balakrishnan, M. S.; Chaugule, B. B. Checklist of Algae. In *Biodiversity of the Western Ghats of Maharashtra.* Jagtap, A. P.; Singh, N. P.; (Eds.); M/S Bishen Singh Mahendrapal Singh. Dehra Dun. **2002,** pp. 113–122.

Barbier, E. B.; Acreman, M. C.; Knowler, D. *Economic Valuation of Wetlands: A Guide for Policy Makers and Planners*; Ramsar Convention Bureau: Gland, Switzerland, **1997.**

Basavaraja, D.; Narayana, J.; Puttaiah, E. T.; Prakash, K. Phytoplankton Species Diversity Indices in Anjanapura Reservoir, Western Ghats Region, India. *J. Environ. Biol.* **2013,** *34* (4), 805–810.

Bateman, I. J.; Folke, C.; Gren, I-M.; Turner, R. K. Wetland Ecosystems: Primary and Secondary Values for Sustainable Management. *Second Conference on the Ecology and Economics of Biodiversity Programme*; Stockholm, Beijer international institute for ecological economics, Royal Swedish Academy of Sciences, 1992.

Central Inland Capture Fisheries Research Institute. *Ecology and Fisheries of Bhatgar Reservoir.* Bull. No. *73,* (CICFRI, Barrackpore, West Bengal), **1997,** p 28.

Desai, S. R.; Subashchandran, M. D.; Ramachandra, T. V. Phytoplankton Diversity in Sharavati River Basin, Central Western Ghats. *ICFAI Univ. J. Soil Water Sci.* **2008,** *1* (1), 7–28.

Divya, K. S.; Murthy, S. M.; Puttaiah, E. T. A Comparative Study of the Growth of Phytoplanktons in Surface Water Samples and in the Formation of Algal Blooms. *Int. J. Innov. Res. Sci. Eng. Technol.* **2013,** *2* (7), 2736–2747.

Easa, P. S. *Biodiversity Documentation for Kerala. Part 1: Algae.* KFRI handbook no. 17. Kerala Forest Research Institute, Peechi, Kerala, India, **2004.**

Elton, C. S.; Miller, R. S. The Ecological Survey of Animal Communities: With a Practical System of Classifying Habitats by Structural Characters. *J. Ecol.* **1954,** *42* (2), 460–496. https://doi.org/10.2307/2256872.

Gaunker, T.; Kerkar, V. Studies on Algal Diversity in Temple Ponds from North Goa. *Indian Hydrobiol.* **2004,** *7* (1 and 2), 67–71.

Geeta, K.; Kerkar, V. Freshwater Green Algal Flora from Parsem (Pernem) Goa, India, *Indian Hydrobiol.* **2009,** *12,* 114–119.

Ghadage, S. J.; Karande, C. T. **2008,** Chroococcales from Satara district (M. S.), *Bioinfolet* **2008,** *5* (4), 336–340.

Gowda, H. T. G.; Vijayakumara. Overview of Some Wetlands in the Lakkavali Range of Bhadra Wildlife Sanctuary, mid-Western Ghats Region, Karnataka: Threats, Management and Conservation Issues. *Funct. Plant Sci. Biotechnol.* **2009,** *3* (1), 60–69.

Gowda H. T. G.; Vijayakumara. Phytoplankton Diversity Indices and Seasonal Variations of Koramagudda Kere in Lakkavalli Range of Bhadra Wildlife Sanctuary, Karnataka. *Global J. Curr. Res.*, **2017**, *5* (1), 22–26.

Gunale, V. R.; Balakrishnan, M. S. Biomonitoring of Eutrophication in the Pavana, Mula & Mutha Rivers Flowing Through Poona. *Indian J. Environ. Health.* **1981**, *23*, 316–322.

Hardy, A. C.; Milne, P. S. Studies in the Distribution of Insects by Aerial Currents. *J. Anim. Ecol.* **1938**, *7* (2), 199–229. doi.org/10.2307/1156.

Hegde, G. R.; Issacs, S. W. Certain Interesting Desmids Taxa from Uttara Kannada District of Karnataka State. *Phykos*, **1988a**, *27*, 8–12.

Hegde, G. R.; Issacs, S. W. Freshwater Algae of Karnataka state I, *Phykos*, **1988b**, *27*, 96–103.

Hegde, U. K. Contribution of the Knowledge of Fresh Water Algae of Karnataka- Some New Desmid Taxa from Shimoga District, *Phykos* **1986**, *25*, 108–112.

Hensen, V. Uber die Bestimmung des Planktons oder des im Meere treibenden Materials an Pflanzen und Thieren. *V. Bericht der Commission zur Wissenschaftlichen Untersuchung der Deutschen Meere*, Jahrgang 12–16, **1887**, pp 1–108.

Jisha, C. R.; Tessy, P. P. Blue Green Algae of Koodalmanikyam and Kodungallur Temple Ponds, Thrissur district, Kerala. *Proceedings of National Seminar Cum Workshop on Plant Systematics and Herbarium Techniques*, September 24–25, **2014**, OPGS, (Department of Botany, KKTM Government College, Pullut, Kodungallur, Kerala), **2014**; pp 66–71.

John, J.; Francis, M. S. *An Illustrated Algal Flora of Kerala*, Vol. I, Idukki District; Green Carmel Scientific Books: Kochi. **2013**; 281 pp.

Jose, J.; Francis, M. S. Investigation in the Algal Flora of Thodupuzha Thaluk, Kerala. *Indian Hydrobiol.* **2007**, *10* (1), 79–86.

Jose, J.; Francis, M. S. Wetland Algal Resources of Western Ghats (Idukki District Region), Kerala, India. *J. Basic Appl. Biol.* **2010**, *4* (3), 34–41.

Jose, L.; Mathew, S. C.; Sreekumar, S. M., Studies on Organic Pollution Based on Physicochemical and Phycological Characteristics of Some Temple Ponds of Ernakulum, Kerala, India. *Nature Environ. Pollut. Technol.* **2008**, *7* (1), 97–100.

Jose, L.; Patel, R. J. Contribution to the Freshwater Diatom Flora of Kerala. *J. Phytol. Res.***1989**, *2* (1), 45–51.

Jose, L.; Sreekumar, S. M. A Study on Phytoplankton Constitution And Organic Pollution in Some Rural and Temple Ponds of Ernakulum. *STARS Intern. J.* **2005**, *6* (2), 36–39.

Joshi, S. C. *Munnar Landscape Project, Kerala, First Year Progress Report.* Submitted to UNDP, India, Kerala State Biodiversity Board, Thiruvananthapuram, **2019**.

Kamat, N. D. The Algae of Mahabaleshwar. *J. Univ. Bombay* **1964**, *31*, 28–41.

Kamat, N. D. The Algae of Alibag, Maharashtra. *J. Bombay Nat. Hist. Soc.* **1968**, *65*, 88–104.

Kamble, P.; Karande, V. C. Observations on Filamentous Blue Green Algae from Satara District, Maharashtra, India. *J. Indian Bot. Soc.* **2014**, *93* (1 and 2), 120–125.

Karki, J. Koshi Tappu Ramsar Site: Updates on Ramsar Information Sheet on Wetlands. *The Initiation* **2008**, *2* (1), 10–16, https://doi.org/10.3126/init.v2i1.2513.

Keddy, P. A. *Wetland Ecology: Principles and Conservation*, 2nd ed.; Cambridge University Press: New York, **2010**.

Kerkar, V. Preliminary Survey on Diversity of Green Algae from the Various Habitats of Goa, In *Biology and Biodiversity of Microalgae*; Ananda, N., Ed.; Centre for Advanced Studies in Botany, University of Madras: Chennai, **2009**; pp 190–195.

Kerkar, V.; Lobo, A. Desmid Diversity for Northern Goa, India. *Proceedings of the International Conference on Algal Biomass, Resources and Utilization*, **2009**; pp 132–141.

Kerkar, V.; Madkaiker, S. Freshwater Blue Green Algae from Goa. *Indian Hydrobiol.*, **2003**, *6* (1 and 2), 45–48.

Korgaonkar, D. S.; Bharamal, D. L. Seasonal Variation in Plankton Diversity of Dhamapur Lake (Malvan) of Sindhudurg District (MS), India. *Int. J. Curr. Micrbiol. Appl. Sci.* **2016**, *5* (3), 884–889. http://dx.doi.org/10.20546/ijcmas.2016.503.102

Krishnamurthy, S. V.; Reddy, S. R. Phytoplankton Diversity in the Drift of a Tropical River Tunga, Western Ghats (India). *Proc. Indian Natl. Sci. Acad. B* **1996**, *62* (2), 105–110.

Krishnan, R. J. Nutrient (N&P) Enrichment Coupled with Phytoplankton Dynamics of Mullaperiyar Reservoir in the Western Ghats of Kerala. *Biosci. Biotech. Res. Asia* **2012**, *9* (1), 379–385.

Kshirsagar, A. D. Use of Algae as a Bioindicator to Determine Water Quality of River Mula from Pune City, Maharashtra (India). *Univers. J. Environ. Res. Technol.* **2013**, *3* (1), 79–85.

Kshirsagar, A. D.; Ahire, M. L.; Gunale, V. R. Phytoplankton Diversity Related to Pollution from Mula River at Pune City. *Terrestrial Aquatic Environ. Toxicol.* **2012**, *6* (2), 136–142.

Leles, S. G.; Polimene, L.; Bruggeman, J.; Blackford, J.; Ciavatta, S.; Mitra, A.; Flynn, K. J. Modelling Mixotrophic Functional Diversity and Implications for Ecosystem Function. *J. Plankton Res.* **2018**, *40* (6), 627–642, https://doi.org/10.1093/plankt/fby044.

Malathi, S.; Thippeswamy, S. Plankton Diversity of River Sita, a Tributary of River Tunga in the Western Ghats, India. *Presented at the Lake 2008*; Indian Institute of Science: Bangalore, **2008**; pp 63–64.

Malathi, S.; Thippeswamy, S. Plankton Diversity of River Malathi, a Tributary of River Tunga, in the Western Ghats, India. *Int. J. Environ. Sci.* **2012**, *3* (2), 211–223.

Manickam, N.; Bhavan, P. S.; Vijayan, P.; Sumathi, G. Phytoplankton Species Diversity in the Parambikulam-Aliyar Irrigational Canals (Tamil Nadu, India). *Int. J. Pharma Bio. Sci.* **2012**, *3* (3), (B) 289–300.

Mann, N. H.; Carr, N. G. *Photosynthetic Prokaryotes*; Springer: Boston, **1992**. DOI: 10.1007/978-1-4757-1332-9.

Maya, S.; Prameela, S. K.; Sarojini, M. V. A Preliminary Study on the Algal Flora of Temple Tanks of southern Kerala. *Phykos* **2000**, *39* (1 and 2), 77–83.

Morris, R. M.; Rappe, M. S.; Connon, S. A.; Vergin, K. L.; Siebold, W. A.; Carlson, C. A.; Giovannoni, S. J. SAR11 Clade Dominates Ocean Surface Bacterioplankton Communities. *Nature* **2002**, *420* (6917), 806–810. DOI: 10.1038/nature01240.

Nalawade, P. M.; Bagul, A. B. Physico-Chemical Conditions and Plankton Diversity of Godavari River in Nashik City area of Maharashtra: A Comparative Assessment. *Intern. J. Ecol. Environ. Sci.* **2020**, *2* (4), 373–379.

Nasser, K. M. M.; Sureshkumar, S. Interaction Between Microalgal Species Richness and Environmental Variables in Peringalkuthu Reservoir, Western Ghats, Kerala. *J. Environ. Biol.* **2013**, *34*, 1001–1005.

Nasser, K. M. M.; Sureshkumar, S. Seasonal Variation and Biodiversity of Phytoplankton in Parambikulam Reservoir, Western Ghats, Kerala. *Int. J. Pure Appl. Biosci.* **2014**, *2* (3), 272–280.

Omori, M.; Ikeda, T. *Methods in Marine Zooplankton Ecology*; Krieger Publishing Company: Malabar, **1992**.

Palaniswamy, R.; Manoharan, S.; Mohan, A. Characterisation of Tropical Reservoirs in Tamil Nadu, India in Terms of Plankton Assemblage Using Multivariate Analysis. *Indian J. Fisheries* **2015**, *62* (3), 1–13.

Patil, S. V.; Karande, V. C.; Karande, C. T. Limnological Study of Venna Lake, Mahabaleshwar, Maharashtra, India. *Int. Res. J. Environ. Sci.* **2015**, *4* (8), 45–49.

Prajapati, S. R. S.; Jadhav, A. S.; Anilkumar, U. Study of Phytoplankton Biodiversity in Panvel Lakes (Vishrale, Krishnale and Dewale Lake) at Dist.–Raigad (Maharashtra) India. *Int. J. Res. Nat. Soc. Sci.* **2014**, *2* (7), 113–120.

Prameela, S. K.; Maya, S.; Menon, S. V. Phytoplankton Diversity of Temple Tanks of Four Coastal Districts of Kerala. *Proc. of XIII Science Congress*, Thrissur, **2001**; pp 203–204.

Prasanthkumar, S.; Santhoshkumar, K.; Ray, J. G. Diversity of Fast Growing Algae of Bloomed Temple Ponds in Kerala, In *Mainstreaming Biodiversity for Sustainable Development*; Cheruvat, D., Nilayangode, P., Oommen, O. V., Eds.; Kerala State Biodiversity Board: Thiruvananthapuram, India, **2017**; pp 240–247.

Priyadarshani, K.; Ghadge, S.; Karande, C. T.; Karande, V. C. Biodiversity of Blue Green Algae from Satara District (M. S.). *Int. J. Appl. Biol. Pharma. Technol.* **2014**, *5* (3), 239–246.

Rajeshwari, K. R.; Rajashekhar, M. Diversity of Cyanobacteria in Some Polluted Aquatic Habitats and a Sulfur Spring in the Western Ghats of India. *Algol. Stud.* **2012**, *138*, 37–56. DOI: 10.1127/1864–1318/2012/0005.

Ramakrishniah, M.; Rao, D. S. K.; Sukumaran, P. K.; Das, A. K. *Ecology and Fish Yield Potential of Selected Reservoirs of Karnataka*; Central Inland Capture Fisheries Research Institute: Barrackpore, West Bengal, India, Bull. No. *94*, **2002**; p 41.

Ramesha, M. M.; Thippeswamy, S. Plankton Diversity in the West Flowing Rivers of Karnataka, Western Ghats, India. *International Symposium on Environmental Pollution, Ecology and Human Health*, Held at S. V. University, Tirupati, 25–27 July, **2009**, Department of Zoology, S. V. University, Tirupati, BD17; p 135.

Rao, J.; Madhyastha, M. N. Seasonal Succession of Phytoplankton in a River of Western Ghats, India. *Acta Hydroch Hydrob.* **1990**, *18* (4), 433–442. doi.org/10.1002/aheh.19900180406.

Ray, J. G.; Krishnan, R. J. Seasonal Trophic Status of Mullaperiyar Lake in the Western Ghats of Kerala, India. *Plant Arch.* **2016**, *16* (1), 459–463.

Ray, J. G.; Santhakumaran, P.; Kookal, S. Phytoplankton Communities of Eutrophic Freshwater Bodies (Kerala, India) in Relation to the Physicochemical Water Quality Parameters. *Environ. Dev Sustain.* **2021,** *23,* 259–290. https://doi.org/10.1007/s10668–019–00579-y.

Robert, G. W. *Limnology: Lake and River Ecosystems,* 3rd ed.; Academic Press: San Diego, **2001.**

Sarang, T.; Manoj, K. Phytoplankton Population in Relation to Physico-Chemical Properties of River Tapi, Surat, Gujarat, India. *Res. J. Recent Sci.* **2017,** *6* (3), 35–37.

Sawaiker, R. U.; Rodrigues, B. F. Physico-Chemical Characteristics and Phytoplankton Diversity in Some Freshwater Bodies of Goa, India. *J. Environ. Res. Dev.* **2016,** *10* (4), 706–711.

Sawaiker, R. U.; Rodrigues, B. F. Biomonitoring of Selected Freshwater Bodies Using Diatoms as Ecological Indicators. *J. Ecosyst. Echography,* **2017,** *7* (2), 234. DOI: 10.4172/2157–7625.1000234

Sebastain, S. Algal Diversity of River Meenachil in Kerala, India. *Indian J. Appl. Res.* **2016,** *6* (3), 203–204. doi. org/10.36106/ifor.

Sharathchandra, K.; Rajashekhar, M. Distribution Pattern of Freshwater Cyanobacteria in Kaiga Region of Western Ghats of Karnataka. *J. Phytol.* **2015,** *7,* 10–18. DOI: 10.5455/jp.2015–09–17.

Sherly, W. E.; Nisha, T. P.; Vineetha, S. A First Report on the Plankton Status of Vattakkayal Lake, Kerala. *Int. J. Sci. Technol. Res.* **2000,** *9* (3), 260–264.

Smith, D. J. 2013. Aeroplankton and the Need for a Global Monitoring Network. *BioScience 63* (7), 515–516. doi.org/10.1525/bio.2013.63.7.3

Solanki, M.; Manoj, K. Algal Biodiversity with Reference to Heavy Metals in Tapi River from Surat District, Gujarat. *J. Environ. Bio-Sci.* **2014,** *28* (2), 267–270.

Solanki, M.; Manoj, K. Phytoplankton Biodiversity with Reference to Physicochemical Characteristics of Tapi in Surat District, Gujarat. *Int. J. Appl. Pure Sci. Agric.* **2016,** *2* (2), 9–14.

Srinivasan, J. T. *Understanding the Kole Lands in Kerala as a Multiple use Wetland Ecosystem.* Working Paper No. *89,* Centre for Economic and Social Studies, Hyderabad, **2010.**

Stubner, E. I.; Soreide, J. E.; Reigstad, M.; Marquardt, M.; Blachowiak-Samolyk, K. Year-Round Meroplankton Dynamics in High-Arctic Svalbard. *J. Plankton Res.* **2016,** *38* (3), 522–536. https://doi.org/10.1093/plankt/fbv124.

Suresh, A.; Kumar, R. P., Dhanasekaran, D.; Thajuddin, N.; Biodiversity of Microalgae in the Western and Eastern Ghats, India. *Pak. J. Biol. Sci.* **2012;** *15* (19), 919–928.

Sureshkumar, P.; Thomas, J. Seasonal Distribution and Population Dynamics of Limnic Microalgae and Their Association with Physico-Chemical Parameters of River Noyyal Through Multivariate Statistical Analysis. *Sci. Rep.* **2019,** *9,* 15021. https://doi.org/10.1038/s41598–019–51542-w.

Suxena, M. R.; Venkateswarlu, V.; Raju, N. S.; Rao, V. S. The Algae and Testacea of Cranganore, Kerala State, India. *J. Indian Bot. Soc.* **1973,** *52,* 316–341.

Tessy, P. P. Systematic Assessment of Freshwater Phytoplankton in the Perennial Ponds of Thrissur District, Kerala, India. *Albertian J. Multidiscip. Res.* **2019,** *1* (1), 116–118.

Tessy, P. P.; Anu, P. K. Algal Diversity of Guruvayur Temple Pond, Thrissur District, Kerala. *Int. J. Adv. Life Sci.* **2016,** *9* (3), 302–306.

Tessy, P. P.; Sreekumar, R. A Report on the Pollution Algae from the Thrissur Kol Wetlands (Part of Vembanad Kol, Ramsar Site), Kerala. *Nat. Environ. Pollut. Technol.* **2008,** *7* (2), 311–314.

Tessy, P. P.; Sreekumar, R. Assessment of the Biodiversity and Seasonal Variation of Freshwater Algae in the Thrissur Kol Wetlands (Part of Vembanad-Kol, Ramsar Site), Kerala. *J. Econ. Taxon. Bot.* **2009,** *33* (3), 721–732.

Tessy, P. P.; Sreekumar, R. Diversity and Distribution of Freshwater Algae Belonging to Xanthophyceae, Chrysophyceae and Dinophyceae from the Kole lands of Thrissur, Kerala. In: *Biodiversity and Bioprospecting with Reference to Plants and Microbes.* BIOPROS – 11, 2011; pp 140–146.

Tessy, P. P.; Sreekumar, R. Assessment on Hydrographic Parameters and Phytoplankton Abundance of Thrissur Kole Lands (Part of Vembanad—Kol, Ramsar Site), Kerala. *Int. J. Adv. Life Sci.* **2013,** *6* (5), 583–593.

Tessy, P. P.; Sreekumar, R. Seasonal and Spatial Distribution of Freshwater Diatoms from Thrissur Kole Lands (Part of Vembanad—Kol, Ramsar Site), Kerala. *Phykos* **2017,** *47* (2), 129–134.

Tessy, P. P.; Sreenisha, K. S. Phytoplankton Diversity of Tirur River, Malappuram District, Kerala. *Indian Hydrobiol.* **2020,** *19* (1 and 2), 9–16.

Thakur, H. A.; Behere, K. H. Study of Filamentous Algal Biodiversity at Gangapur Dam, Nashik Dist. (M. S.), India. *Proceedings of Taal (2007), the 12th Lake Conference* **2008;** pp 456–461.

Thippeswamy, S.; Malathi, S. Plankton Diversity of Bilagal (Kochche) Stream, a Tributary of River Bhadra, in Kudremukh National Park. *Presented at the 24th National Conference of the ISCAP on Biotechnol. Strategies for Biodiversity Conservation,* Held at Kuvempu University, 13–15 March, 2008, Department of Applied Zoology, Kuvempu University, **2008a;** pp 75–76.

Thippeswamy, S.; Malathi, S. Plankton of Bakri Stream, Tributary of River Bhadra in the Western Ghats of India. In *International Conference on Biodiversity Conservation and Management;* Jayachandran, K. V., Augustine, A., Eds.; 22–26 August, 2008, CUSAT, Cochin, **2008b;** pp 86–92.

Thippeswamy, S.; Malathi, S. Plankton Diversity in Indian Estuaries—A Review. In *Plankton Dynamics of Indian Waters;* Hosetti, Ed.; Daya Publishing House: Jaipur, India, **2009;** pp 306–333.

Thippeswamy, S.; Sequira, M. G.; Suresh, G. C. Hydrobiology of Kodige Tank in the Western Ghats of India. In *Proceedings of TAAL 2007: The World Lake Conference,* **2008;** pp 12–18.

Thomas, M. L.; Tessy, P. P. An Assessment of Phytoplankton and Physico-Chemical Characteristics of Chalakudy River, Kerala. *Int. J. Adv. Life Sci.* **2015,** *8* (2), 197–203.

Venkataraman, G. A Systematic Account of Some South Indian Diatoms. *Proc. Indian Acad. Sci.* **1939,** *10* (6), 293–368.

Wang, G.; Wang, X.; Liu, X.; Li, Q. Diversity and Biogeochemical Function of Planktonic Fungi in the Ocean. In *Biology of Marine Fungi;* Raghukumar, C., Ed.; Springer-Verlag: Berlin, Heidelberg, **2012;** pp 71–88. DOI: 10.1007/978-3-642-23342-5.

Wang, X.; Singh, P.; Gao, Z.; Zhang, X.; Johnson, Z. I.; Wang, G. Distribution and Diversity of Planktonic Fungi in the West Pacific Warm Pool. *PLoS One,* **2014,** *9* (7), e101523. DOI: 10.1371/journal.pone.0101523.

Wommack, K. E.; Colwell, R. R. Virioplankton: Viruses in Aquatic Ecosystems. *Microbiol. Mol. Biol. Rev.* **2000,** *64* (1), 69–114. DOI: 10.1128/MMBR.64.1.69–114.2000.

World Conservation Monitoring Centre. *Global Biodiversity: Status of the Earth's Living Resources;* Chapman and Hall: London, **1992.**

CHAPTER 19

ZOOPLANKTON DIVERSITY IN THE WESTERN GHATS

RAMAKRISHNA

Zoological Survey of India, Bangalore, India

ABSTRACT

Zooplankton are the heterogeneous assemblage of often phenotypically close but varied plankter groups; they are crucial elements of freshwater ecosystems as they occupy the center of the aquatic food web in the ecological pyramid being as an important food for almost all freshwater fishes and other aquatic species. The present analysis, first of its kind for biogeographic region of the country, includes the groupwise species composition (diversity) of zooplankton. In terms of diversity of zooplankton, the dominant group is Rotifera with 238 species followed by Copepoda (231) > Protozoa (161 species) > Cladocera (129 species) > Ostracoda with 73 species among the states of Western Ghats (WG) of India, a biological hotspot of the world. Among the zooplankton groups, a degree of overlapping of species in freshwater and estuarine ecosystems has been noticed in the present chapter, which needs to be addressed by subsequent workers.

19.1   INTRODUCTION

Lakes, ponds, and rivers are freshwater habitats for aquatic plants, animals, and microorganisms which play a key role in maintaining regional biodiversity. The important component of the ecological pyramid of the freshwater ecosystem is plankton. Plankton refer to those nonmotile or insufficiently motile aquatic organisms to overcome the transport of currents, either they will be phytoplankton or zooplankton. Planktonic communities in natural aquatic ecosystems serve as a key group for energy production (Alikunhi et al., 1955). Phytoplankton fix carbon by photosynthesis and are consumed by a variety of different types of protozoa (including ciliates and flagellates), by microzooplankton such as rotifers and immature stages (nauplii) of copepods, and by macrozooplankton (cladocerans and copepods). A second route of energy flow occurs from protozoa to microzooplankton and macrozooplankton (Havens, 2014). Zooplankton themselves are fed upon by fish and are

Biodiversity Hotspot of the Western Ghats and Sri Lanka. T. Pullaiah, PhD (Ed.)
© 2024 Apple Academic Press, Inc. Co-published with CRC Press (Taylor & Francis)

thus the vital transition between primary production (phytoplankton) and fish. Without these primary consumers, herbivores and other levels of food chain would collapse (Wetzel, 2001). They occupy a higher trophic position either directly or indirectly via protozoo-plankton, so are subject to either bottom–up or top–down control in the wetlands, as energy transfer depends heavily on secondary productivity in aquatic ecosystem (Montagnes et al., 2010). Zooplankton communities are also sensitive to anthropogenic impacts. They play vital roles in biogeochemical cycling of carbon and nitrogen, and aid the stability of aquatic food webs. The zooplankton association, richness, abundance, seasonal variation, and diversity can be used for the assessment of water quality and for pisciculture manage-ment practices.

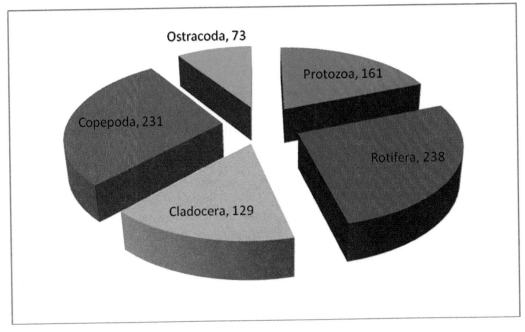

FIGURE 19.1   Zooplankton diversity in the Western Ghats.

19.2 PROTOZOA

The group protozoa has now been raised to the level of subkingdom under the kingdom Protista (Cavalier-Smith, 2004). Protozoa comprise a highly heterogeneous assemblage of several phyla, namely, Sarcomastigophora, Labyrinthomorpha, Apicomplexa, Microspora, Ascetospora, Myxozoa, and Ciliophora, according to the revised classification developed in 1980 by the society of Protozoologists (http://protozoa.uga.edu/). Complexity in protozoans evolved through specialization of different parts of the cell organelles and cytoskeleton in particular. The term "free-living heterotrophic flagellates" refers to a group of free-living protists moving and/or feeding through the use of flagella exclusively by heterotrophic means or, if with plastids, also capable of ingesting particles (Patterson and Zölffel, 1991).

Heterotrophic flagellates form an important and diverse component of the communities living in aquatic ecosystems (Lee and Patterson, 2000). Heterotrophic flagellates hold the key to the evolution of protozoa and higher organisms, and also this group plays an important role in planktonic processes, such as biochemical cycling of nutrients and carbon and the flow of energy (Laybourn-Parry and Parry, 2009). Dinoflagellates among protozoa are usually marine, ephemeral, and often form bloom in those waters, a few are estuarine species, for example, *Ceratium hirudinella* (Muller, 1773), *Ceratium furcoides* (Levander) Langhans 1925, *Ceratium cornutum* (Ehrenberg) Claparède and J. Lachmann 1859, and *Peridinium willei* Huetfeld-Kaas 1900 reported along the west coast and tributaries flowing along it in Western Ghats (WG).

19.3 DIVERSITY OF FRESHWATER PROTOZOA

Naidu (1966a, 1966b) worked extensively on flagellate species of south India. Free-living protozoan diversity in Indian wetlands studied by Bindu (2010) mentions about the present estimate of 1247 species from marine, estuarine, and freshwater systems. The inventory of free-living protozoans from various parts of WG of India is representing altogether 76 species including flagellates, rhizopods (testate amoebae), and ciliates. As per the present record, the greatest number was represented by testate amoebae (43 species) followed by ciliates (31 species) and flagellates (two species). Maharashtra recorded the highest number of species (44 species) followed by Kerala (29 species), Tamil Nadu (25 species), and Karnataka (11 species), and there are no reports from Goa. It is definite that many species are still to be explored and can be well revised in the light of future thorough research. No consolidated work on the distribution of free-living protozoa from the WG area is available; the stray analysis from the Meghamalai wildlife sanctuary of Tamil Nadu, part of the WG, indicates the presence of 25 species of testate amoebae (Bindu, 2018). Ravindra and Savita (2019) cataloged 91 protozoan taxa. A total of 20 were flagellates, 32 rhizopods, and 39 ciliates from WG of Maharashtra; 41 species of freshwater protozoa from a water body in Satara district of Maharashtra (Ravindra and Savita, 2018). The present analysis indicates the distribution of 161 species from WG as indicated with Karnataka—78, Kerala—41, Goa—7, Maharashtra—96, and Tamil Nadu with 59 species.

The ciliates embrace heterotrophic species with a heterokaryotic nuclear apparatus. The macronucleus, which is usually highly polyploid (except of the curious *Karyorelictea*, where it is diploid and does not divide) and divides amitotically during asexual reproduction, controls mainly somatic functions, such as RNA synthesis and ontogenesis. The diploid micronucleus is active mainly during sexual reproduction, called conjugation (Raikov, 1972; Corliss, 1979).

Importance of protozoa as bioindicators for pollution and environmental biomonitoring has been recognized since long (Kelkwitz and Marson, 1908). Several field and experimental studies have been carried out in this regard and results obtained therefrom support that protozoa may be conveniently used for environmental biomonitoring, particularly for ecological monitoring of water quality. Occurrence of the ciliate, *Metopus* sp. in any water body indicates the presence of hydrogen sulfide (Bick, 1972). Presence of this species and its associated ciliates belonging to the genera *Caenomorpha, Epalxella, Pelodinium,* and

Sprodinium in putrefying sludge is the indicator of the self-purification process which has been stopped due to lack of oxygen and presence of high concentration of H_2S.

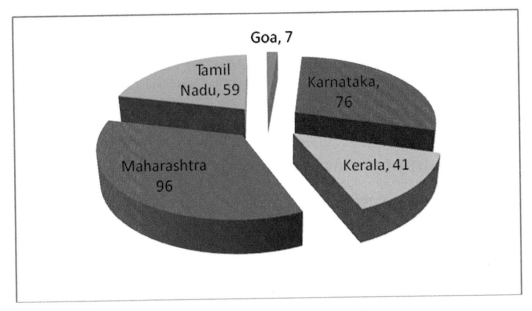

FIGURE 19.2   Species diversity and distribution of protozoa in Western Ghats.

19.4   ROTIFERA

Rotifers are extensively used as a model organism to study the aquatic toxicology (Snell and Janssen, 2018), especially the genus *Brachionus* (Grosell et al., 2006; Preston et al., 2000) and *Lecane* (Pérez-Legaspi and Rico-Martínez, 2001). They are playing a vital role in the trophic tiers of freshwater impoundments and serve as living capsules of nutrition.

19.4.1   ROTIFERA DIVERSITY

According to Segers (2008), 2031 Rotifer species are reported so far, the Oriental region comprises of 78 Monogononta genera with 486 species and nine Bdelloida genera with 58 species. As per the compiled valid list of Indian rotifers, 419 species belonging to 65 genera under 25 families are reported from India (Sharma and Sharma, 2017), and also species richness of Indian Rotifera comprises 81 and 24% of species of the taxon reported from the Oriental region and worldwide (Chitra, 2018). The present analysis of its distribution in WG is given graphically. Now the present checklist is by far the most comprehensive one recording the occurrence of 492 species belonging to 78 genera and 26 families.

Among the rotifers, the biodiversity importance of two genera *Lecane* and *Brachionus* is distinctly observed in the rotifer taxocoenosis of Tamil Nadu. *Lecane* (40 species, 22.6%) is the most speciose genus; its richness represents about 50% of total species so far known from India. The Oriental *Keratella edmondsoni*, described from Tamil Nadu (Ahlstrom, 1943) as *Keraella quadrata* var. *edmondsoni*, was raised to the status of a distinct species by Nayar (1965). *Brachionus donneri*, another Oriental species described by Brehm (1951), was erroneously listed as pantropical species (Sharma and Sharma, 2001, 2005). A total of 14 species, namely *Anuraeopsis fissa, Brachionus angularis, B. bidentatus, B. budapesti-nensis, B. caudatus, B. calyciflorus, B. falcatus, B. forficula, B. plicatilis, B. quadridentata, Keratella cochlearis, K. tropica, Plationus patulus,* and *Platyias quadricornis* are widely or nearly widely distributed in India (Sharma, 2014; Sharma and Sharma, 2021), and are found distributed in WG too.

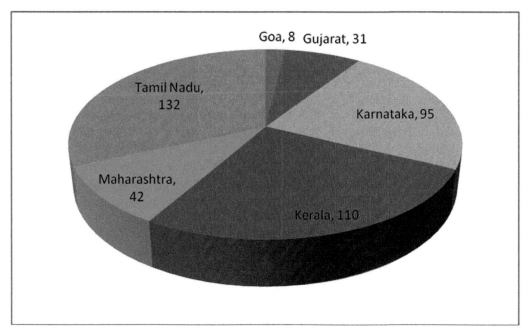

FIGURE 19.3   Species composition of Rotifera in Western Ghats.

It was understood that the various anthropogenic activities such as entry of agricultural runoffs (insecticides and pesticides) from surrounding agricultural field seem to be the major cause of eutrophication, which ultimately determines the diversity and abundance. *B. angularis, B. calyciflorus,* and *B. rubens* represent eutrophic indicators from this country. The last two species are known for their sporadic blooms in the rural areas of peninsular India. The epizoic nature of some species has been commented upon by a few workers from this country. *B. rubens* has been reported (Nayar and Nair, 1969) on the fairy shrimp.

19.5   OSTRACODA

The subclass Ostracoda belongs to class Crustacea and Phylum Arthropoda. Indian tempo-rary freshwater habitats in particular have high ostracod diversity (Deb, 1983; Victor and

Michael, 1975; Victor and Fernando, 1979; George et al., 1993; Shinde, 2012). Studies on Ostracods of the Indian subcontinent were made and their status was critically evaluated by Victor and Fernando (1979).

19.5.1 DIVERSITY AND ENDEMISM

The Oriental region is the hotspot of freshwater Ostracoda diversity with 271 species, of which nearly 58% of the species are endemic to India within the oriental region (Karuthapandi and Rao, 2014).

Despite many limnological investigations in India, the freshwater ostracods have remained, to quote Kesling (1956), "neglected little crustaceans." The study by Victor and Fernando (1981) attempts to provide an initial picture of freshwater ostracod distribution in India, however restricted to peninsular India. Studies on Indian Ostracoda commenced in 1859, and since then till now 208 species of nonmarine Ostracoda, both extinct and extant, have been reported. A total of 152 valid species belonging to 39 genera, five families, and two superfamilies are listed by Karuthapandi et al. (2014). Latest checklist indicates the distribution of 208 species in 43 genera from India. Tilak (2020) in her checklist on the basis of available records found only 37 species of Ostracods belonging to one order, 15 genera under three families and 11 subfamilies with current valid nomenclature records from WG of India. Stray instances of records of Ostracods from WG are known from the work of Patil and Talmalae (2005), with 38 species of Ostracoda belonging to 16 genera under four families from Maharashtra; 41 species belonging to 14 genera under three families from a seasonal pond in the Savitribai Phule Pune University Campus, Pune, Maharashtra (Kulkarni et al., 2015); 17 Ostracoda species belonging to 11 genera from Pune by Shinde et al. (2014); six species of Ostracods were reported by Patil (2001) from the Nilgiri biosphere reserve; records from Dharwar, Karnataka with 24 Ostracod species by Vaidya (1996). Of the 17 species of Ostracods recorded from north WG by Shinde et al. (2014), 13 were found in the pools on the basaltic mesa. *Plesiocypridopsis dispar* (Hartmann, 1964) was found almost on every plateau made up of basaltic mesa. *Chrissia biswasi* (Deb, 1972) is the largest (giant Ostracod) species found on basaltic mesa. Some of the species which were found commonly on basaltic mesa were not found on ferricretes. *Sclerocypris* sp. is a large taxon found only in the pools formed on ferricretes rock, while *P. dispar* is only known from basaltic mesa. The distributional data on freshwater ostracods are meagre, even for peninsular India, since a vast area remains to be explored.

The Indian Eucypridines are represented by two genera, *Eucypris* and *Strandesia*. The genus *Oncocypris* was recorded only in the extreme southwest coast of the Peninsula. A certain faunal overlap of freshwater Ostracods in peninsular India and the Indo-Malaysian subregion is noted in the present study. Examples are *Candonopsis putealis*, *Pseudocypretta maculata*, *Chrissia humilis*, *Strandesia flavescens*, *Hemicypris pyxidata*, and *Cypridopsis dubia*. These species were originally described from Indonesia, Celebes, and Sumatra (Sars, 1903). So also, *Cypretta fontinalis* and *Strandesia bicornuta* were described from the mountainous terrains of the WG in the Peninsula, but never recorded in the plains.

Endemism in Indian freshwater Ostracods is worthy of consideration. The occurrence of "giant" forms with a length of 3 mm or more is considered to be indicative of endemism in freshwater Ostracods (McKenzie, 1971). If this view is accepted, then *Chrissia krishnakantai, Stenocypris biswasi, Stcnocypris distincta*, and *Candollocypris dentatus* can be considered as "giant" endemic species for India. However, no endemic genera of freshwater Ostracods are definitely known for India, unlike Australia (*Mytilocypris*, McKenzie), southern Africa (*Megalocypris*; *Acocypris*; *Pseudocypris*; *Liocypris*; *Sclerocypris*, *Afrocypris*, Sars; *Aptelocypris*, Rome; *Hypselecypris*, Rome), and south America (*Amphicypris*). Some species are very restricted in their distribution and the species, namely *Cypretta fontinalis* Hartmann, 1965 and *Strandesia bicornuta* Hartmann, 1964 are described from the mountainous terrains of the WG, but never from the plains (Victor and Fernando, 1979). Three species, namely *Ilyocypris nagamalainsis* Victor and Michael, 1975; *Cypretta alagarkoilensis* Victor and Michael, 1975; *Eucypris bispinosa* Victor and Michael, 1975 are recorded only from the hill regions of Madurai district (type locality), WG of Tamil Nadu (Victor and Fernando, 1979). On this basis, five species, namely *Cypretta fontinalis* Hartmann, 1965; *Cypretta alagarkoilensis* Victor and Michael, 1975; *Strandesia bicornuta* Hartmann, 1964; *Eucypris bispinosa* Victor and Michael, 1975; *Ilyocypris nagamalainsis* Victor and Michael, 1975 are found to be endemic species to WG (Tilak, 2020).

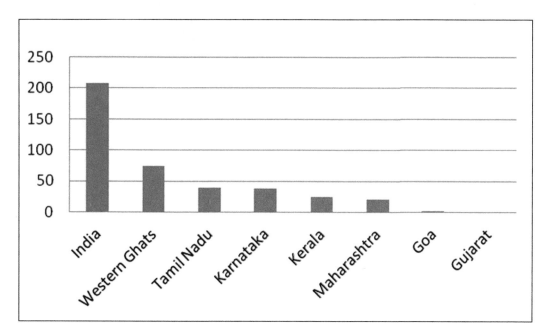

FIGURE 19.4 Ostracod diversity in Western Ghats.

19.6 CLADOCERA

The order name Cladocera belongs to the order Cladocera of the subclass Branchiopoda under the subphylum Crustacea, and is grouped into 11 families. The Cladocerans are a predominately aquatic group, swim using their antennae, with a series of jerks similar to the hops of a flea. Some have adapted to terrestrial habitats with free water, such as bromeliad basins. Others are able to use the film of water from the capillary spaces and leaf surfaces of bryophytes. Not only are the antennae important for swimming, but they are also powerful chemical sensory organs (Ecomare, 2014).

19.6.1 DIVERSITY

Water fleas have been reported from all over the world. While it is true that several genera and some species are cosmopolitan, certain species are endemic. In the Indian context, the studies on Cladocera usually form a part of general limnological investigations in lakes, ponds, and reservoirs, wherefrom several genera and species are reported from time to time. Michael and Sharma (1988) recorded 93 species in India. An account of 190 species belonging to 49 genera and 10 families is listed by Raghunathan and Suresh Kumar (2003). Recent updated checklist by Sharma and Sharma (2017) includes 131 species belonging to 47 genera, 11 families, and four orders.

Species diversity and distribution of Cladocera in the freshwater rock pools of the northern WG have been reported. A total of 59 samples collected from 12 different localities contained 22 species belonging to five families (Padhye and Victor, 2015). A total of 22 species of Cladocera from Dharwad lakes of WG of Karnataka were recorded by Patil and Gouder (1988), of which 17 species are first record for south India and *Guernella raphaelis* Richard is new to the subcontinent. With ca. 130 species reported from India (an estimate yet again) (Chatterjee et al., 2013), 51 species from NWG represent roughly 40% of total species richness known (Padhye and Dumont, 2015). The number of species found can thus be said to be low for Asia, and certainly for a biodiversity-rich area like the WG. The reasons for this lesser number might be due to incomplete or biased sampling and lack of systematic and/or repeated sampling of large water bodies, each of which should harbor about 50 Cladoceran species at a time (Dumont and Segers, 1996).

The diversity and distribution of Cladocera are related to the physicochemical factors. Dhembare (2011) in his study on the diversity and density of zooplankton with water factors in Mula dam in Maharashtra reported that the density and diversity of zooplankton depend upon water temperature and conductivity. Following is list of species distributed in WG that needs special mention in their geographic distribution.

Indian endemic: *Indialona ganapati* Petkovski

Asian endemics: *Daphnia (Ctenodaphnia) similoides* Hudec, 1991

Australasian species: *Daphnia (Ctenodaphnia) cephalata* King, 1853; *Simocephalus (Echinocaudus) acutirostratus* (King, 1853)

Indo-Malaysian species: *Oxyurella singalensis* (Daday, 1862)

Indo-Ethiopian species: *Leberis punctatus* (Daday, 1898)

Vietnam, Hainan, Malaysia, and India: *Camptocercus vietnamensis* Sinev, 2012

Africa and tropical Asia: *Leydigia (Neoleydigia) ciliata* Gauthier, 1939; *Notoalona globulosa* (Daday, 1898)

Palearctic: *Kurzia (Kurzia) latissima* (Kurz, 1875)

Species with "dubious reports" or "reports inquirendae": *Leydigia (Neoleydigia) australis* Sars, 1885

Misidentifications and Lapsi: *Alona quadrangularis* (O. F. Muller, 1776) = *Alona kotovi* Sinev, 2012. All Indian reports to be rechecked (Sharma and Sharma, 2014). *Pseudosida bidentata* Herrick, 1884 = *Pseudosida sazalyi* Daday, 1898. All Indian reports to be rechecked (Chatterjee et al., 2013; Sharma and Sharma, 2013)

Distribution that needs confirmation: *Pseudocida bidentata; Simocephalus (Aquipiculus) latirostris* Stingelin, 1906; *Alona costata* Sars, 1862; *Alona vericosa* Sars, 1901; *Anthalona verrucosa* (Sars, 1901); *Camptocercus uncinatus* Smirnov, 1971.

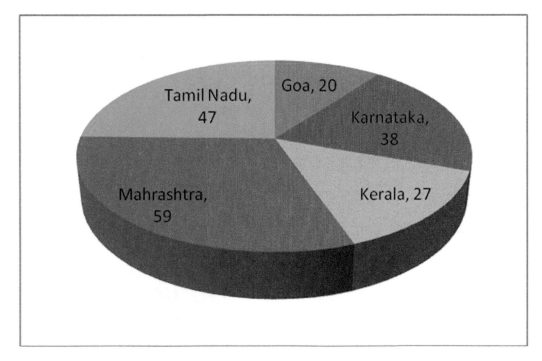

FIGURE 19.5   Species distribution of Cladocera in Western Ghats.

19.7   COPEPODA

Copepods are small aquatic crustaceans often dominant to other fauna and one of the most successful groups comparable with insects in terms of diversity and adaptive radiation.

19.7.1   DIVERSITY

A total of 181 species and subspecies of Cyclopoid copepods belonging to 38 genera and nine families, and 403 species and subspecies of harpacticoids under 163 genera and 35 families have been reported for India (Roy and Venkataraman, 2018). The catalog Calanoid copepod of India includes 287 species and subspecies belonging to 82 genera and 25 families. The Indian copepods are represented by 871 species under 283 genera and 69 families. In India, the diversity of freshwater inland copepods is represented by 200 species distributed in 60 genera (Karuthapandi et al., 2020).

A total of 82 species in 29 genera and five families of copepods were documented from WG, of which 27 species belong to order Calanoida, 39 species to Cyclopoida, and 16 species to Harpacticoida.

In Indian context, strictly speaking endemism cannot be reliably determined in free-living inland copepods due to lack of extensive and intensive biogeographic investigations. But based on the above review, presently two genera *Keraladiaptomus* and *Megadiaptomus* are found to be strictly endemic to India. *Allodiaptomus mirabilipes*, *Keraladiaptomus rangareddyi*, and *Phyllodiaptomus (Ctenodiaptomus) sasikumari* are endemic to southern WG of Kerala, while *Megadiaptomus montanus* is endemic to northern WG (Karuthapandi et al., 2020). The present analysis includes 231 species of Copepoda from WG. Details of distribution among the WG states are as follows:

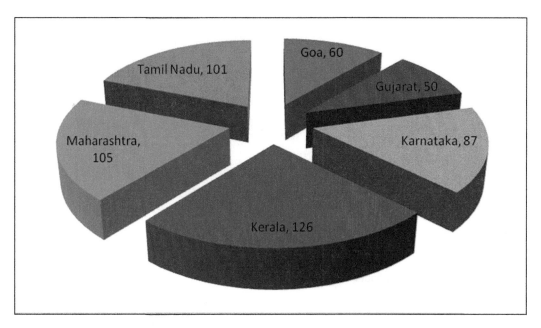

FIGURE 19.6   Copepod distribution in the states of Western Ghats.

Most of the copepods inhabit in temporary pools, ponds, and lakes, which are considered as highly endangered inland water ecosystem of the world (Reid et al., 2002). The invisibility of these ecosystems to environmental protection organizations (Marrone et al., 2006), and further poor public understanding of inland water ecosystems' importance and services, especially in the context of India has further exacerbated conservation challenges/issues. Four genera showing a typical Oriental distribution (*Heliodiaptomus*, *Neodiaptomus*, *Eodiaptomus*, and *Allodiaptomus*), two genera showing a Gondwanan distribution (*Paradiaptomus* and *Tropodiaptomus*), one genus distributed both in the Palaearctic and Oriental regions (*Phyllodiaptomus*), and one genus restricted to peninsular India and Sri Lanka (*Megadiaptomus*) were observed in the NWG (Kulkarni and Pai, 2016). The dominance of Oriental elements in the present day fauna of peninsular India is not surprising, and it has been noted earlier for various taxa such as insects, fish, reptiles (Mani, 1974), and diaptomid copepods as well (Ranga Reddy, 2013).

Megadiaptomus montanus n. sp. was described (Kulkarni et al., 2018) based on plankton samples collected from the WG of Maharashtra state of India. This is the third species of *Megadiaptomus* Kiefer, 1936, the other two being *M. hebes* Kiefer, 1936 from Sri Lanka and southern India (Karnataka) and *M. pseudohebes* Ranga Reddy, 1988 from Andhra Pradesh.

19.8 CONCLUSIONS

The word "limnology" is a Greek word meaning "lake knowledge." History of limnology begins with the lake studies basically involved in the beginning by studying the biological characters; however, later the study was extended to other aspects of the lake ecosystem. According to Ramsar convention, all inland water bodies (lakes, rivers, ponds, and other aquatic bodies) were considered as "wetlands" that gained worldwide recognition. At the same time, clean water Act in 1972 introduced amendments to Federal water pollution control Act in United States of America primarily focusing on reducing water pollution from point sources. As it advanced to 1980s, many lakes and wetlands were investigated for management aspects and many were recorded as polluted due to the expansion of human settlements, increased population, and increased water use for various purposes. This led to degradation of water bodies, conditions extending beyond the pollution sources (primarily direct wastewater discharges into lakes and wetlands).

Zooplankton in a freshwater ecosystem temporary or permanent, lentic or lotic, and its dynamics of quantity and biomass, change in the complexity of its species diversity; trophic nature is driven by the physicochemical milieu (abiotic factors) of the freshwater body. The population densities are also driven by climatic and seasonal factors, as normally high and low population densities were recorded in summer and early monsoon season, respectively in tropics, especially in south Indian states.

KEYWORDS

- **Cladocera**
- **Copepoda**
- **Ostracoda**
- **Protozoa**
- **Rotifera**
- **Western Ghats**

REFERENCES

Ahlstrom, E. H. A revision of the Rotatorian Genus *Keratella* with Description of Three New Species and Five New Varieties. *Bull. Am. Mus. Nat. Hist.* **1943**, *80*, 411–457.

Alikunhi, K. H.; Chaudhuri, H.; Ramchandran V. On the Mortality of Carp Fry in Nursery Pond and the Role of Plankton in Their Survival and Growth. *Indian J. Fisheries* **1955**, *2*, 257–313.

Bick, H. *Ciliated Protozoa, An illustrated Guide to the Species Used as Biological Indicators in Freshwater Biology;* Publ. World Health Organisation, **1972**.

Bindu, L. On Some Testacids (Protozoa) of Melghat Wildlife Sanctuary, Maharashtra, India. *J. Threat. Taxa* **2010**, *2* (4), 827–830.

Bindu, L. On Some New Records of Testate Amoebae (Protozoa: Rhizopoda) from Meghamalai Wildlife Sanctuary, Western Ghats, Tamil Nadu, India. *Uttar Pradesh J. Zool.* **2018**, *39* (1), 8–11.

Brehm, V. Einer neuer *Brachionus* aus Indien (*Brachionus donneri*). *Zoologischer Anzeiger* 1951, *146*, 54–55.

Cavalier-Smith, T. Only Six Kingdoms of Life. *Proc. Royal Soc. B Biol. Sci.* **2004**, *271*, 1251–1262.

Chatterjee, T.; Kotov, A. A.; Van Damme, K.; Chandrasekhar, S. V. A.; Padhye, S. An Annotated Checklist of the Cladocera (Crustacea: Branchiopoda) from India. *Zootaxa* **2013**, 3667, 1–89.

Chitra J. Rotifera. In: *Faunal Diversity of Biogeographic Zones: Islands of India*; Chandra, K., Raghunathan, C., Eds.; Zoological Survey of India: Kolkata; **2018**; pp 125–131.

Corliss, J. O. *The Ciliated Protozoa*, 2nd ed.; Pergamon Press: Oxford, **1979**.

Deb, M. Brief Descriptions of New Species of Ostracoda: Crustacea from Maharashtra State (India). *J. Zool. Soc. India* **1983**, *81*, 135–166.

Dhembare, J. A. Statistical Approaches for Computing Diversity and Density of Zooplankton with Water Factors in Mula Dam, Rahuri, MS, India. *Eur. J. Exp. Biol.* **2011**, *1* (2), 68–76.

Dumont, H. J.; Segers, H. Estimating Lacustrine Zooplankton Species Richness and Complementarity. *Hydrobiologia* **1996**, *314*, 125–132.

Ecomare. *Encyclopedia: Cladocera*, **2014** http:/www.ecomare.nlindex.php?id=3531 (accessed 11 March 2014).

Foissner, W. Two New "flagship" Ciliates (Protozoa, Ciliophora) from Venezuela: *Sleighophrys pustulata* and *Luporinophrys micelae*. *Eur. J. Protistol.* **2005**, *41*, 99–117.

Forró, L.; Korovchinsky, N. M.; Kotov, A. A.; Petrusek, A. Global Diversity of Cladocerans (Cladocera; Crustacea) in Freshwater. *Hydrobiologia* **2008**, *595*, 177–184.

George, S.; Martens, K.; Nayar, C. K. G. Two New Species of Freshwater Ostracoda of the Genus *Parastenocypris* Hartmann, 1964 from Kerala, India. *Hydrobiologia* **1993**, *23*, 183–193.

Grosell, M.; Gerdes, R. M.; Brix, K. V. Chronic Toxicity of Lead to Three Freshwater Invertebrates—*Brachionus calyciflorus*, *Chironomus tentans*, and *Lymnaea stagnalis*. *Environ. Toxicol. Chem.* **2006**, *25*, 97–104.

Havens, K. E. *Lake Eutrophication and Plankton Food Webs Eutrophication: Causes, Consequences and Control*; Springer Science+Business Media: Dordrecht, **2014**; pp 73–79. DOI 10.1007/978–94–007–7814–6_6

Karuthapandi, M.; Rao, D. V. Innocent, B. X. Freshwater Ostracoda (Crustacea) of India—A Checklist. *J. Threat.Taxa* **2014**, *6* (12), 6576–6581.

Karuthapandi, M.; Shabuddin, S.; Deepa, J. Crustacea: Copepoda. In *Faunal Diversity of Biogeographic Zones of India: Western Ghats*; Zool. Surv. India: Kolkata, **2020**, pp 167–177.

Kelkwitz, R.; Marsson, M. Okologie der pflanzlichemsaprobein. *Ber. Deut. Bot. Ges.* **1908**, *26*, 505–519.

Kesling, R. V. The Ostracod, a Neglected Little Crustacean. *Turtox News.* **1956**, *34*, 4–6.

Kulkarni, M. R.; Padhye, S.; Vanjare, A. I.; Jakhalekar, S. S.; Shinde, Y. S.; Paripatyadar, V. S.; Sheth, D. S.; Kulkarni, S.; Phuge, K. S.; Bhakare, K.; Kulkarni, A. S.; Pai, K.; Ghate, V. H. 2015. Documenting the Fauna of a Small Temporary Pond from Pune, Maharashtra, India. *J. Threat. Taxa* **2015**, *7* (6), 7196–7210.

Kulkarni, M. R.; Pai, K. The Freshwater Diaptomid Copepod Fauna (Crustacea: Copepoda: Diaptomidae) of the Western Ghats of Maharashtra with Notes on Distribution, Species Richness and Ecology. *J. Limnol.* **2016**, *75* (1), 135–143.

Kulkarni, M. R.; Shaik, S.; Ranga Reddy, Y.; Pai, K. A New Species of *Megadiaptomus* Kiefer, 1936 (Copepoda: Calanoida: Diaptomidae) from the Western Ghats of India, with Notes on Biogeography and Conservation Status of the Species of the Genus. *J. Crustacean Biol.* **2018**, *38* (1), 66–78.

Laybourn-Parry, J.; Parry, J. Flagellates and the Microbial Loop. In *The Flagellates*; Leadbeater, B. S. C., Green, J. C., Eds.; Taylor & Francis: London, **2009**.

Lee, W. J.; Patterson, D. J. Heterotrophic Flagellates (Protista) from Marine Sediments of Botany Bay. *Australia J. Nat. Hist.* **2000,** *34,* 483.

Lee, W. J.; Patterson, D. J. Abundance and Biomass of Heterotrophic Flagellates, and Factors Controlling Their Abundance and Distribution in Sediments of Botany Bay. *Microbial Ecology* **2002,** *43,* 467.

Mani, M. S. *Ecology and Biogeography in India*; W. Junk Publ, **1974**.

Marrone, F.; Barone, R.; Flores, L. N. Ecological Characterization and Cladocerans, Calanoid Copepods and Large Branchiopods of Temporary Ponds in a Mediterranean Island (Sicily, Southern Italy). *Chem. Ecol.* **2006,** *22* (Supplement 1), 5181–5190.

Martens, K.; Schon, I.; Meisch, C.; Horne, D. J. Global Diversity of Ostracods (Ostracoda, Crustacea) in Freshwater. *Hydrobiologia* **2008,** *595,* 185–193.

McDonald, R. I.; Green, P.; Balk, D.; Feket, B. M.; Revenga, C.; Todd, M.; Montgomery, M. Urban Growth, Climate Change, and Freshwater Availability. *Proc. Natl. Acad. Sci. USA* **2011,** *108* (15), 6312–6317.

Mckenzie, K. G. Palaeozoogeography of Freshwater Ostracoda. *Bull. Centre Rech. Pau.* **1971,** *5,* 207–237.

Michael, R. G., Sharma, B. K. *Fauna of India and Adjacent Countries: Indian Cladocera (Crustacea: Branchiopoda: Cladocera)*; Zoological Survey of India: Calcutta, **1988**.

Montagnes, D. J. S.; Dower, J. F.; Figueiredo, G. M. The Protozooplankton-Ichthyoplankton Trophic Link: An Overlooked Aspect of Aquatic Food Webs. *J. Eukaryot. Microbio.* **2010,** *57* (3), 223–228.

Naidu, K. V. Some Thecamoebae (Rhizopoda: Protozoa) from India. *Hydrobiolia* **1966a,** *27,* 465–478.

Naidu, K. V. Studies on the Freshwater Protozoa of South India III: Euglenoidina 2. *Hydrobiologia,* **1966b,** *27* (1–2), 23–32.

Nayar, C. K. G. Taxonomic Notes on Indian Species of *Keratella* (Rotifera). *Hydrobiologia* **1965,** *26,* 457–462.

Nayar, C. K. G.; Nair, K. K. N. A Collection of Brachionid Rotifers from Kerala. *Proc. Indian Acad. Sci.* **1969,** *69,* 223–233.

Padhye, S. M.; Dumont, H. J. Species richness of Cladocera (Crustacea: Branchiopoda) in the Western Ghats of Maharashtra and Goa (India), with Biogeographical Comments. *J. Limnol.* **2015,** *74* (1), 182–191.

Padhye, S. M.; Victor, R. Diversity and Distribution of Cladocera (Crustacea: Branchiopoda) in the Rock Pools of Western Ghats, Maharashtra, India. *Ann. Limnol.* **2015,** *51* (4), 315–322.

Patil C. S.; Gouder, B. Y. M. Cladocera of Dharwad, Karnataka. *J. Bombay Nat. Hist. Soc.* **1988,** *85, 112* – 117.

Patil, S. G. 2001.Ostracoda. In *Fauna of Nilgiri Biosphere Reserve: Fauna of Conservation Area Series.* No.11. Zoological Survey of India, Kolkata, 2001; pp 29–30.

Patil, S. G.; Talmalae, S. S. A Checklist of Freshwater Ostracods (Ostracoda: Crustacea) of Maharashtra, India. *Zoos' Print J.* **2005,** *20* (5), 1872–1873.

Patterson, D. J.; Zölffel, M. Heterotrophic Flagellates of Uncertain Taxonomic Position. In *The Biology of Freeliving Heterotrophic Flagellates*; Patterson, D. J.; Larsen, J., Eds.; Systematic Association: Oxford, 1991; p 427.

Pérez-Legaspi, I. A.; Rico-Martínez, R. Acute Toxicity Tests on Three Species of the Genus *Lecane* (Rotifera: Monogononta). *Hydrobiologia* **2001,** *446,* 375–381.

Preston, B. L.; Snell, T. W.; Robertson, T. L.; Dingmann, B. J. Use of Freshwater Rotifer *Brachionus calyciflorus* in Screening Assay for Potential Endocrine Disruptors. *Environ. Toxicol. Chem.* **2000,** *19,* 2923–2928.

Raghunathan M. B.; Suresh Kumar, R. Checklist of Indian Cladocera. *Zoos' Print J.* **2003,** *18* (8), 1180–1182.

Ranga Reddy, Y. *Neodiaptomus prateek* n. sp., a New Freshwater Copepod from Assam, India, with Critical Review of Generic Assignment of *Neodiaptomus* spp. and a Note on Diaptomid Species Richness (Calanoida: Diaptomidae). *J. Crust. Biol.* **2013,** *33* (6), 849–865, figs. 1–10.

Raikov, I. B. Nuclear Phenomena During Conjugation and Autogamy in Ciliates. In *Research in Protozoology*; Chen T-T., Ed., Vol 4; Pergamon Press: New York, **1972;** pp 146–289.

Ravindra, B.; Savitha, N. Protozoan Fauna of Freshwater Habitat in Water Body on Ajinkyatara Fort, Satara, Maharashtra in December, 2017. *Intern. J. Recent Sci. Res.* **2018,** *9* (3), 25395–25398.

Ravindra, B.; Savitha, N. Freeliving Protozoan Diversity in Selected Stations of Krishna River in Satara District, Maharastra. *J. Emerg. Technol. Innov. Res.* **2019,** *6* (4), 91–96.

Reid, J. W.; Bayly, I.; Pesce, G. L.; Rayner, N. A.; Reddy, Y. R.; Rocha, C. E. F.; Morales, E. S.; Ueda, H. *Conservation of Continental Copepods*; Kluwer Academic/Plenum Publishers, **2002;** pp 1–9.

Roy, M. K. D.; Venkataraman, K. Catalogue on Copepod fauna of India, Part 2. Cyclopoidea and Harpacticoida (Arthropoda; Crustacea). *J. Environ. Sociobiol.* **2018,** *15* (2), 109–194.

Sars, G. O. Freshwater Entomostraca from China and Sumatra. *Arch. Mathematik Og Naturvidenskab, Kristiania* **1903,** *25,* 3–44.

Segers, H. The Biogeography of Littoral Lecane Rotifera. *Hydrobiologia* **1996,** *323,* 169–197.

Segers, H. Zoogeography of the Southeast Asian Rotifera. *Hydrobiologia* **2001,** *446/447,* 233–246.

Segers H. Global Diversity of Rotifers (Rotifera) in Freshwater. *Hydrobiologia* **2008,** *595,* 49–59.

Sharma, B. K. Rotifers (Rotifera: Eurotatoria) from Wetlands of Majuli—The Largest River Island, the Brahmaputra River Basin of Upper Assam, Northeast India. *Check List* **2014,** *10* (2), 292–298.

Sharma, B K.; Sharma, S. Biodiversity of Rotifera in Some Tropical Floodplain Lakes of the Brahmaputra River Basin, Assam (N. E. India). *Hydrobiologia* **2001,** *446–447,* 305–313.

Sharma, B K.; Sharma, S. Biodiversity of Freshwater Rotifers (Rotifera: Eurotatoria) from North-Eastern India. Mitteilungen aus dem Museum für Naturkunde Berlin, *Zoologische Reihe* **2005,** *81,* 81–88.

Sharma, S.; Sharma, B. K. Faunal Diversity of Aquatic Invertebrates of Deepor Beel (a Ramsar Site), Assam, Northeast India. *Zool. Surv. India, Wetland Ecosyst. Ser.* **2013,** *17,* 1–226.

Sharma, B. K.; Sharma, S. Faunal Diversity of Cladocera (Crustacea: Branchiopoda) in Wetlands of Majuli (the Largest River Island), Assam, Northeast India. *Opusc. Zool. Budapest* **2014,** *45,* 1–9.

Sharma, B. K.; Sharma, S. Rotifera: Eurotatoria. In *Current Status of Freshwater Faunal Diversity in India*; Kailash Chandra, K., Gopi, C., Rao, D. V., Valarmathi, K., Alfred, J. R. B., Eds.; Zoological Survey of India: Kolkata, **2017;** pp 93–113.

Sharma, B. K.; Sharma, S. Biodiversity of Indian Rotifers (Rotifera) with Remarks on Biogeography and Richness in Diverse Ecosystems, *Opusc. Zool. Budapest* **2021,** *52* (1), 69–97.

Shinde, Y. S. *Studies on Freshwater Ostracoda (Crustacea) of Pune District Maharashtra*, PhD Thesis; Department of Zoology, University of Pune, **2012.**

Shinde, Y. S.; Victor, R.; Pai, K. Freshwater Ostracods (Crustacea: Ostracoda) of the Plateaus of the Northern Western Ghats, India. *J. Threat. Taxa* **2014,** *6* (4), 5667–5670.

Snell, T. W.; Janssen, C. R. Microscale Toxicity Testing with Rotifers. *Microscale Testing in Aquat. Toxicol.* **2018,** 1st ed.; CRC Press: New York, pp. 409–422.

Thilak, J. Crustacea: Branchiopoda: Cladocera. In: *Faunal Diversity of Biogeographic Zones of India: Western Ghats.*; Zool. Surv. India: Kolkata, **2020;** pp 179–186.

Vaidya, A. S. A Note on Zoogeographic Distribution of Recent Freshwater Ostracoda from the Lakes of Dharwad, Karnataka. Contrs.XV. *Indian Colloq. Micropol. Strat. Dehradun,* **1996;** pp 439–445.

Victor, R.; Michael, R. G. Nine New Species of Freshwater Ostracods in Madurai Area in Southern India. *J. Nat. Hist.* **1975,** *9,* 361–376.

Victor, R.; Fernando, C. H. The freshwater Ostracods. (Crustacea: Ostracoda). *Rec. Zool. Surv. India* 1979, *74* (2) 147–242.

Victor, R.; Fernando, C. H. *An Illustrated Generic Key to the Freshwater Ostracoda of the Oriental Region*; University of Waterloo Biology Series 23, **1981;** 92 pp.

Wetzel, R. G. *Limnology: Lake and River Ecosystems*; 3rd ed.; Academic: San Diego, **2001.**

CHAPTER 20

PRUDENT MANAGEMENT OF PROTECTED AREAS IN INDIA THROUGH VIRTUAL SPATIAL DECISION SUPPORT SYSTEM

RAMACHANDRA T. V.[1,2,3] and BHARATH SETTURU[1]

[1]Energy and Wetlands Research Group [CES TE 15], Centre for Ecological Sciences, Indian Institute of Science, Bangalore, India

[2]Centre for infrastructure, Sustainable Transportation and Urban Planning [CiSTUP], Bangalore, India

[3]Centre for Sustainable Technologies (astra), Indian Institute of Science, Bangalore, India

ABSTRACT

The forests in protected areas (PAs) of Karnataka are undergoing tumultuous changes due to unplanned activities with tourism, mining, and other activities. The current research analyzes the spatiotemporal changes in the select PAs such as Someshwara Wildlife Sanctuary (SWLS) and Bhadra Wildlife Sanctuary (BWLS). Land use and land cover (LULC) analyses highlight as the evergreen cover has reduced from 38.75 to 31.43% by 2016 in SWLS and loss of deciduous forests from 50.09 to 37.43% in BWLS (1973–2016) with an increase of plantations, agriculture activities, which are threatening the ecological integrity of the region. The Markov cellular automata (MCA)-based visualization reveals loss of deciduous forest cover from 37.43 to 34.71% in BWLS by 2026 with an increase in agricultural activities along the edges of forests. SWLS shows loss of evergreen cover (31.43–28.84%) due to penetrations of horticulture crops (constitute 18.5%) and settlements.

Spatial decision support system (SDSS) is designed taking advantage of the recent advances in information and open-source web technologies, through the integration of spatial with attribute information. This enhances governance transparency while meeting societal needs, which helps in the prudent planning of the protected areas (PA). Visualization of landscape dynamics through synthesis and integration of information would enable

Biodiversity Hotspot of the Western Ghats and Sri Lanka. T. Pullaiah, PhD (Ed.)
© 2024 Apple Academic Press, Inc. Co-published with CRC Press (Taylor & Francis)

understanding the current status and predict likely changes, which are essential for effective decision-making toward sustainable management of natural resources. Web-based spatial decision support system (WSDSS) would help in accomplishing effective dissemination of the ecological, socio, economic, biodiversity, and environmental information. Integration of temporal, spatial information with the simulation techniques facilitates visualization to analyze the viability of decisions through exploratory analysis based on multiple criteria, which enable a detailed comparative analysis of alternate scenarios.

20.1 INTRODUCTION

India with 2.4% of the global land area accounts for 7–8% of the total recorded species and ranks seventh among the 17 mega biodiversity countries of the world. A network of protected areas such as national parks, sanctuaries, conservation reserves, and community reserves has been created to purpose in-situ conservation under sections 18, 35, 36A, and 36C of the Wildlife Protection Act (1972). There are 662 Protected Areas (PAs) extending over 1,58,508 km^2 (4.83% of the total geographic area), comprising 103 National Parks, 514 Wildlife Sanctuaries, 44 Conservation Reserves, and 4 Community Reserves. 39 Tiger Reserves and 28 Elephant Reserves have been designated for species-specific management of tiger and elephant habitats. However, globalization and the opening up of Indian markets during the nineties have led to large-scale land cover changes. Conservation and management endeavor is currently facing numerous challenges such as habitat loss due to forest fragmentations, conversion of forests to monoculture plantations, overuse of biomass resources due to the biotic pressures, increasing human–wildlife conflicts, poaching, and illegal trade, etc. The accelerated anthropogenic pressure with the unplanned developmental activities has triggered changes in the ecosystem function evident from barren hilltops, conversion of perennial streams to seasonal streams, lowered pollination services, etc., because of the changes in ecosystem integrity and impacting the ecology, biodiversity, nutrient cycling, socioeconomic systems (Brooks et al., 2002; Haddad et al., 2015; Ramachandra et al., 2016, 2017, 2018a).

Decision-making pertaining to natural resources has been a challenging task due to the requirement of complex data and the need for more transparency of decision processes to participate in the process. To enhance the individual or group of person's ability to make decisions and support decision-making activities necessitates the design of Decision Support Systems (DSS) to support decision-making activities toward prudent management of natural resources. Spatial decision support systems (SDSS) have evolved over the last few decades, taking advantage of advances in spatial and communication technologies to integrate location-specific data (spatial data). SDSS combines analytical tools with functions available in GIS (Geographic Information System) as well as models for geo-visualizing likely scenarios associated with the decisions.

SDSS is an integrated spatial database management system with spatial and attribute database, analytical and operational research models, visualization options, reporting capabilities, and the expert knowledge of decision-makers to assist in prudent management of ecosystem resources and solving specific problems in a simple and effective manner (Crossland et al., 1995; Sugumaran and Degroote, 2010; Ferretti and Montibeller, 2016). SDSS offers insights into the structure of spatial decision problems by helping decision makers to frame alternatives and strategies in a problem-solving process (Matthies et al., 2007). Recent advances in geo-informatics, the internet, information technology, and the availability of spatial data have led to effective knowledge dissemination. Free and Open Source Software (FOSS) with the availability of multi-resolution spatial data has helped in capturing the landscape dynamics and lending a platform to assess climate change, water resources, urban development, natural disaster mitigation, etc. (Ramachandra and Kumar, 2008; Steiniger and Hay, 2009; Bharath et al., 2012).

Globally, PA networks or conservation areas have proved to be the most effective means of conservation to arrest the detrimental effect of anthropogenic impacts (Gray et al., 2016; Ramachandra et al., 2018b). Conservation through appropriate conservation framework has been regarded globally as a key strategy to address the growing extinction crisis (Jenkins and Joppa, 2009). Change analyses through temporal remote sensing data have confirmed that the PAs have been successful in controlling the rapid land use changes (Nelson and Chomitz, 2011), successfully maintaining the integrity of forest ecosystem, regulating the loss of carbon sequestering capabilities (Nolte et al., 2013), and mitigating instances of forest fires (Joppa and Pfaff, 2011). PAs networks are final refuges for threatened species, natural ecosystem processes, and a positive effect on species' richness and abundance, evident from the review of 86 case studies (Coetzee et al., 2014) and are aiding as a cornerstone to conserve tropical biodiversity despite burgeoning anthropogenic stresses. These measures, solely regulatory with the organizational and political difficulties, are associated with imposing and enforcing restrictions over a broad range of detrimental human activities (Kiesecker et al., 2015). Unfortunately, still the fragmentation of forests, human encroachment, mismanagement, and other environmental stresses within PAs are imperiling biodiversity (Laurance et al., 2012). This necessitates assessing current status, landscape dynamics within PAs, ecology, management, and enrichment, which remains a critical knowledge gap due to the lack of comprehensive data and appropriate conservation approaches (Green et al., 2015). The present endeavor develops an interactive Spatial Decision Support System for sustainable management of protected areas in Karnataka (KPASDSS) by taking advantage of the latest advancements in geo-informatics with the integration of spatial and attributes information from disparate sources. Visualizing the scenario associated with the proposed decisions helps the decision-makers to acquire a holistic view for policy synthesis. Geo-visualization through the interactive decision maps with the assigned target criteria levels and information about potentially feasible combinations of attribute data. This allows decision-makers to gain insights of the respective region's ecological, economic, and social interactions, which helps in generating the information required for developing ecologically sound policies. This helps in (i) the prudent governance of PAs while enhancing the transparency and meeting the societal needs, and (ii) visualization of landscape dynamics with the integration of attribute information enables understanding the current status and predicting likely changes, which are quintessential for sustainable management of natural resources.

20.2   MATERIALS AND METHOD

20.2.1   STUDY AREA

Karnataka is located between 11 and 18° N latitudes and 74–78° E longitudes with an area of 1, 91,790 km². It accounts 5.8% of the total geographic area of India, with 300 km coastline. Karnataka has 38,284 km² forest area, supporting diverse biodiversity because of its great variability in climate, topography, and soil. It has around 10% of tiger and 25% of the elephant population of the country. The State has diverse species of flora and fauna evident from the occurrence of 4500+ species of flowering plants, 600 bird species, 160 species each of mammal and reptiles, 800 species of fish, 70 species of frogs, etc. The region is endowed with the major rivers such as Krishna, Cauvery (east-flowing), Sharavathi, Netravathi, Kali, Aghnashini, Chakra, and Varahi (West flowing). The Western Ghats alone covers around 60% of total forests of the state. Figure 20.1 illustrates the location of various PAs in Karnataka accounting 16% of forest area under protection. The State has 6 National Parks (NP), 23 Wild Life Sanctuaries (WLS), one biosphere reserve, and one community reserve. These regions are under various levels of ecological stress due to the escalation of unplanned and unrealistic developmental activities with the mismanagement of resources during the postindependence period.

FIGURE 20.1   Protected areas in the Karnataka state.

20.2.2 METHOD

Figures 20.2 and 20.3 outline the framework and method involved in designing a spatial decision support system for sustainable management of protected areas in Karnataka (SDSS-KPA). SDSS was developed in the Cento Linux environment (http://www.centos.org) using Free and Open Source Software for Geoinformatics (FOSS4G) such as GeoServer (http://geoserver.org/download), PostgreSQL (http://postgresql.org), Post GIS (http://postgis.net), and Apache Tomcat web server (http://tomcat.apache.org). Database (spatial and attribute data) management is achieved through FOSS PostgreSQL and Post GIS (Brovelli and Magni, 2003). GeoServer is an open-source software server written in Java that allows users to share, edit, publish geospatial data and runs in an integrated Apache Tomcat Web server environment. This includes implementation of the Open Geospatial Consortium (OGC) standards (http://www.opengeospatial.org/standards; http://www.opengeospatial.org/blog/2034) such as Web Feature Service (WFS) and Web Coverage Service (WCS) standards, Web Map Service (WMS) that are integrated in client-server architecture with the request and response process through a simple HTTP interface OpenLayers (http://openlayers.org), which is open source JavaScript API integrated as a default map environment that renders dynamic maps over the web for visualization of spatial data in a web browser. The integration of numerous datasets has been possible through the default store options of GeoServer.

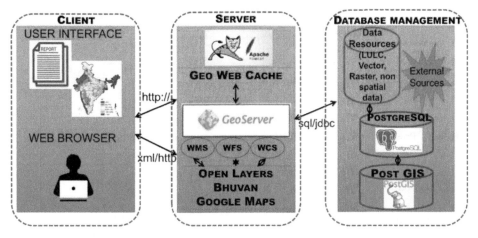

FIGURE 20.2 Spatial decision support system framework for prudent management of PAs in Karnataka.

Land use dynamics of PAs were assessed by classifying temporal remote sensing data with training data (collected from the field and online portal—Google, Bhuvan) using a supervised classifier based on GMLC. Vector data include the administrative boundaries with the associated attribute information. These vector and raster datasets were imported to GeoServer for visualization with OGC web services. Geo-visualization with the modeling capability helps explore the complex behavior of agents (Bharath et al., 2014), which helps understand the causes and consequences of changes. Modeling and prediction have been useful in evolving appropriate mitigation measures with an accurate assessment of likely LULC changes, the drivers of changes, and likely impacts on the ecosystem, etc. (Haase and Schwarz, 2009). The Markov-based Cellular Automata (CA-Markov) modeling technique has been implemented by computing transition probability and transition areas matrices based on earlier land uses. Accurate assessment and validation of the prediction is done by comparing the simulated land use (2016) with the actual land uses (2016) and computation of kappa statistics. Based on the better accuracy, the projection with the same technique is made for predicting likely changes in 2026.

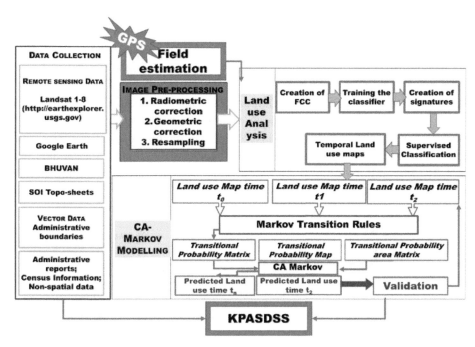

FIGURE 20.3   Protocol for the design of KPASDSS for visualization of natural resources in PAs.

20.3   RESULTS

SDSS is designed to focus on the conservation of protected areas in Karnataka (KPASDSS), which allows data management (updating and manipulation), data visualization with options of OGC-based web protocols—WMS and WFS. The servers request-response fetches vector and raster data through OGC web mapping standards including GML, and SLD (Styled Layer Descriptor is an XML-based markup language). KPASDSS main window is shown in Figure 20.4 with default Open Layers backdrop. The user has options to select layers based on themes. Data for creating these layers were compiled through ecological field investigations and compilation from disparate sources (published literatures). The user will be able to visualize the spatial information and also analyze multi-layers simultaneously. This consists of the options to change backdrop with Bhuvan (ISRO's Geo-Portal available in the public domain for providing visualization services), Google maps, Bing maps, etc. Figure 20.5 shows the PAs with land use data available at Bhuvan. In addition, the end users have access to numerous HTML pages such as photographs, help, glossary, existing publications and management plans of each PA, external links, references list, and brief textual information for viewing and querying spatial layers.

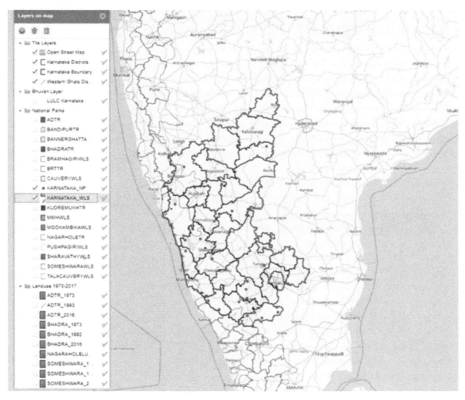

FIGURE 20.4   KPASDSS and layers available.

SDSS manages the information of the PAs in Karnataka with options for analyses and visualization. Figure 20.6 shows land use details of various PAs selected and land uses of the different time periods may also be viewed. The report option provides land use details across each category of selected PAs. For example, Figure 20.7 depicts forest cover from 1973 to 2016 and likely land use of 2026 in Someshwara WLS (SWLS) and Bhadra WLS (BWLS). The land use change analysis using temporal remote sensing data reveals that the evergreen forest cover has declined from 38.75 (1973) to 31.43% (2016) in the SWLS. Similarly, deciduous forest cover has declined from 50.09 to 37.43% in BWLS during the past four decades (1973–2016) with an increase of plantations, agriculture activities, which are threatening the ecological integrity of the PA. CA Markov modeling highlights the likely shift from agriculture to horticulture across PAs. SWLS would witness further erosion of evergreen cover from 31.43 to 28.84% (2016–2026), increasing the area under horticulture from 21.46 to 28.08% (2026). BWLS also will witness an increase in monoculture plantations followed by horticulture. Deciduous forest would disappear from 37.43 to 34.71%

(2016–2026) with an increase of horticulture from 15 to 19% in BWLS. The changes will impact wild fauna and the region currently sustains tiger population. The earlier coffee estates in buffer region with native vegetation supporting diverse herbivorous populations are being transformed into commercial rubber plantations, causing a serious impact on wildlife. The conversion from tea estate to rubber plantations is another major threat to the fragile ecosystem in recent times. Query-based visualization will help decision-makers to frame appropriate policies to arrest further degradations of PAs.

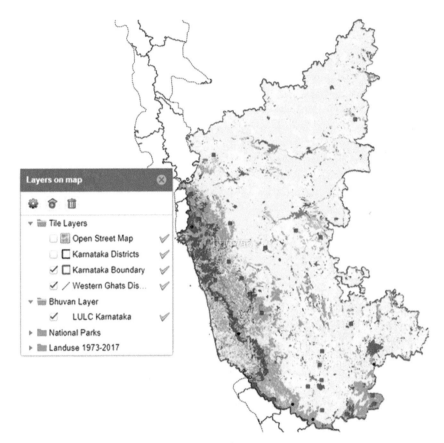

FIGURE 20.5   Visualization of PAs on Bhuvan WMS layer.

20.4   DISCUSSION

Geovisualization offers a visual interface with computational analysis capabilities and modeling based on human–computer interaction, geographic information science, operations research, data mining and machine learning, decision science, cognitive science, and other disciplines in solving complex decision problems (Andrienko et al., 2007). SDSS provides an opportunity to integrate multi-criterion evaluation (MCE) approaches of decision-making with a set of alternatives, which can be evaluated by the pros and cons of each option's properties for effective resource assessment and planning (Ruda, 2016; Gebetsroither-Geringer et al., 2018; Beard et al., 2018). SDSS has emerged as a powerful and flexible tool developed for different environmental applications and supports in framing policy measures for the global biodiversity hotspots (Karnatak and Roy, 2019).

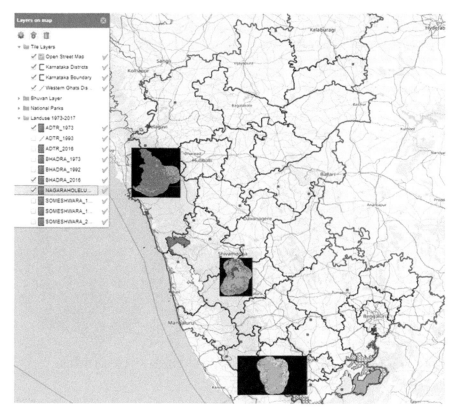

FIGURE 20.6   Visualization of various PAs with land uses.

PAs have been playing vital role in ecosystem conservation as a refuge for sensitive wild flora and fauna, while maintaining ecosystem balance. Long-term conservation and management of PAs depend on how well they are protected from the threats such as encroachment, fire, hunting/poaching, and other unsustainable resource harvesting. The human-induced land use land cover (LULC) changes adversely impact these regions and result in loss of productivity and biodiversity. LULC changes in the forested areas can abruptly influence the underlying ecological processes, carbon sequestration potential, increase emissions, induce climate change, and affect a broad range of ecosystem service supply (Ramachandra and Bharath, 2019). Understanding landscape dynamics and visualization of likely LULC transitions would help take location-specific ecosystem approaches of conservation measures and manage natural resources through stakeholder's participation (Ramachandra et al. 2016) and maintain ecological integrity. Merging of scientific knowledge through spatial technologies provides an important step toward using science to inform stakeholders and assist decision-makers in the effective policy decisions.

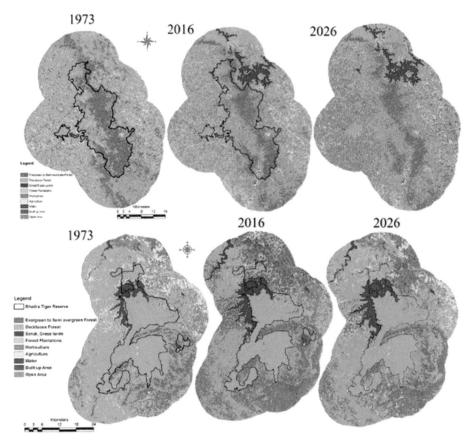

FIGURE 20.7   Land use dynamics with likely LU changes by 2026 in SWLS and BWLS.

20.5   CONCLUSIONS

A spatial Decision Support System designed for the prudent governance of protected areas (PA) will enhance the transparency in natural resources management while meeting societal needs. Visualization of landscape dynamics through synthesis and integration of information would enable understanding the current status and predict likely changes, which are essential for effective decision-making toward sustainable management of natural resources. Visualization of spatiotemporal changes indicates the decline in evergreen forest cover from 38.75 (1973) to 31.43% (2016) in SWLS. Similarly, there is a loss of deciduous forests from 50.09 to 37.43% in BWLS during 1973–2016 with an increase of monoculture plantations and agriculture activities. Modeling and predicting likely LU change through CA-Markov highlights that the SWLS would witness further erosion of evergreen cover from 31.43 to 28.84% (2016–2026) with increases in the area horticulture from 21.46 to 28.08% (2026). BWLS also will witness an increase in monoculture plantations followed by horticulture. Deciduous forest would disappear from 37.43 to 34.71% (2016–2026) with an increase of horticulture from 15 to 19% in BWLS. Advance visualization of likely LU changes would help to evolve appropriate and prudent management strategies to conserve ecologically sensitive PAs to sustain biodiversity.

ACKNOWLEDGMENTS

This research was supported by the grant from (i) The Karnataka State Audit Accountant office, [Pr.AG(E&RSA), Karnataka, Bengaluru, http://www.agkar.cag.gov.in/] and ENVIS division, The Ministry of Environment, Forests and Climate Change, GoI. We are grateful to the official languages section at IISc for the assistance in language editing. We acknowledge the sustained infrastructure support from (i) NRDMS division, The Ministry of Science and Technology (DST), Government of India, (ii) Indian Institute of Science, and (iii) ENVIS Division, the Ministry of Environment Forests and Climate Change, Government of India.

KEYWORDS

- **protected areas**
- **SDSS**
- **land use land cover**
- **CA-Markov**
- **sustainability**
- **conservation**

REFERENCES

Andrienko, G.; Andrienko, N.; Jankowski, P.; Keim, D.; Kraak, M. J.; MacEachren, A.; Wrobel, S. Geovisual Analytics for Spatial Decision Support: Setting the Research Agenda. *Int. J. Geogr. Inf. Sci.* **2007**, *21*, 839–857. DOI: 10.1080/13658810701349011.

Beard, R.; Wentz, E.; Scotch, M. A Systematic Review of Spatial Decision Support Systems in Public Health Informatics Supporting the Identification of High Risk Areas for Zoonotic Disease Outbreaks. *Int. J. Health. Geogr.* **2018**, *17*, 17–38. DOI: 10.1186/s12942-018-0157-5.

Bharath, S.; Bharath, A. H.; Rajan, K. S.; Ramachandra, TV. Cost Effective Mapping, Monitoring and Visualisation of Spatial Patterns of Urbanisation Using FOSS. In *Proceeding of FOSS 4G, First National Conference on Open Source Geospatial Resources to Spearhead Development and Growth, India* **2012**, *1*, 12–20.

Bharath, S.; Rajan, K. S.; Ramchandra, T. V. Visualization of Forest Transition in Uttar Kannada. In *Proceedings of: Lake* **2014**, *Conference on Conservation and Sustainable Management of Wetland Ecosystem in Western Ghats,* **2014**, *14*, 187–190.

Brooks, T. M.; Mittermeier, R. A.; Mittermeier, C. G.; Da Fonseca, G. A.; Rylands, A. B.; Konstant, W. R.; Flick, P.; Pilgrim, J.; Oldfield, S.; Magin, G.; Hilton-Taylor, C. Habitat Loss and Extinction in the Hotspots of Biodiversity. *Conserv. Biol.* **2002**, *16*, 909–923. DOI: 10.1046/j.1523-1739.2002.00530.x.

Brovelli, M. A.; Magni, D. An Archaeological Web GIS Application Based on Mapserver and PostGIS. *Int. Arch. Photogramm. Remote Sens. Spat. Inf. Sci.* **2003,** *34,* 89–94.

Coetzee, B. W.; Gaston, K. J.; Chown, S. L. Local Scale Comparisons of Biodiversity as a Test for Global Protected Area Ecological Performance: A Meta-Analysis. *PLoS One.* **2014,** *9,* 2–12, e105824. DOI: 10.1371/journal.pone.0105824.

Crossland, M. D.; Wynne, B. E.; Perkins, W. C. Spatial Decision Support Systems: An Overview of Technology and a Test of Efficacy. *Decis. Support Syst.* **1995,** *14,* 219–235. DOI: 10.1016/0167-9236(94)00018-N

Ferretti, V.; Montibeller, G. Key Challenges and Meta-Choices in Designing and Applying Multi-Criteria Spatial Decision Support Systems. *Decis. Support. Syst.* **2016,** *84,* 41–52. DOI: 10.1016/j.dss.2016.01.005.

Gebetsroither-Geringer, E.; Stollnberger, R.; Peters-Anders, J. Interactive Spatial Web-Applications as New Means of Support for Urban Decision-Making Processes. *Int. Ann. Photogramm. Remote Sens. Spat. Inf. Sci.* **2018,** *4,* 59–66.

Gray, C. L.; Hill, S. L.; Newbold, T.; Hudson, L. N.; Börger, L.; Contu, S.; Hoskins, A. J.; Ferrier, S.; Purvis, A.; Scharlemann, J. P. Local Biodiversity Is Higher Inside Than Outside Terrestrial Protected Areas Worldwide. *Nat. Commun.* **2016,** *7,* 12–20, 12306. DOI: 10.1038/ncomms12306.

Haddad, N. M.; Brudvig, L. A.; Clobert, J.; Davies, K. F.; Gonzalez, A.; Holt, R. D.; Lovejoy, T. E.; Sexton, J. O.; Austin, M. P.; Collins, C. D.; Cook, W. M. Habitat Fragmentation and Its Lasting Impact on Earth's Ecosystems. *Sci. Adv.* **2015,** *1,* 23–30, e1500052. DOI: 10.1126/sciadv.1500052.

Haase, D.; Schwarz, N. Simulation Models on Human-Nature Interactions in Urban Landscapes: A Review Including Spatial Economics, System Dynamics, Cellular Automata and Agent-Based Approaches. *Living Rev. Landsc. Res.* **2009,** *3,* 1–45. DOI: 10.12942/lrlr-2009-2.

Jenkins, C. N.; Joppa, L. Expansion of the Global Terrestrial Protected Area System. *Biol. Cons.* **2009,** *142,* 2166–2174. DOI: 10.1016/j.biocon.2009.04.016.

Joppa, L. N.; Pfaff, A. Global Protected Area Impacts. In: *Proceedings of the Royal Society of London B: Biological Sciences.* **2011,** *278,* 1633–1638. DOI: 10.1098/rspb.2010.1713.

Karnatak, H.; Roy, A. Himalayan Spatial Biodiversity Information System. In *Remote Sensing of Northwest Himalayan Ecosystems*; Springer: Singapore, **2019;** pp 237–249.

Kiesecker, J. M.; McKenney, B.; Kareiva, P. Offsets: Factor Failure into Protected Areas. *Nature.* **2015,** *525* (7567), 33–37. DOI: 10.1038/525033c.

Laurance, W. F.; Useche, D. C.; Rendeiro, J.; Kalka, M.; Bradshaw, C. J.; Sloan, S. P.; Laurance, S. G.; Campbell, M.; Abernethy, K.; Alvarez, P.; Arroyo-Rodriguez, V. Averting Biodiversity Collapse in Tropical Forest Protected Areas. *Nature.* **2012,** *489* (7415), 290–294.

Matthies, M.; Giupponi, C.; Ostendorf, B. Environmental Decision Support Systems: Current Issues, Methods and Tools. *Environ Model Softw.* **2007,** *22,* 123–127. DOI: 10.1016/j.envsoft.2005.09.005.

Nelson, A.; Chomitz, K. M. Effectiveness of Strict vs. Multiple Use Protected Areas in Reducing Tropical Forest Fires: A Global Analysis Using Matching Methods. *PLoS One.* **2011,** *6* (8), 1–14, e22722. DOI: 10.1371/journal.pone.0022722.

Nolte, C.; Agrawal, A.; Silvius, K. M.; Soares-Filho, B. S. Governance Regime and Location Influence Avoided Deforestation Success of Protected Areas in the Brazilian Amazon. *Proc. Natl. Acad. Sci. USA* **2013,** *110,* 4956–4961. DOI: 10.1073/pnas.1214786110.

Ramachandra, T. V.; Bharath, S. Global Warming Mitigation Through Carbon Sequestrations in the Central Western Ghats. *Remote Sens. Earth Syst. Sci.* **2019,** 1–25. DOI: 10.1007/s41976-019-0010-z.

Ramachandra, T. V.; Bharath, S.; Chandran, M. D. S. Geospatial Analysis of Forest Fragmentation in Uttara Kannada District, India. *For. Ecosyst.* **2016,** *3* (10), 1–15. DOI: 10.1186/s40663-016-0069-4.

Ramachandra, T. V.; Bharath, S.; Rajan, K. S.; Chandran, M. D. S. Modelling the Forest Transition in Central Western Ghats, India. *Spat. Inf. Res.* **2017,** *25,* 117–130. DOI: 10.1007/s41324-017-0084-8.

Ramachandra, T. V.; Bharath, S. Geoinformatics Based Valuation of Forest Landscape Dynamics in Central Western Ghats, India. *J. Remote Sens. GIS.* **2018a,** *7* (1), 227–235. DOI: 10.4172/2469-4134.1000227.

Ramachandra, T. V.; Bharath, S.; Gupta, N. Modelling Landscape Dynamics with LST in Protected Areas of Western Ghats, Karnataka. *J. Environ. Manage.* **2018b,** *206,* 1253–1262. DOI: 10.1016/j.jenvman.2017.08.001.

Ramachandra, T. V.; Kumar, U. Spatial Decision Support System for Land Use Planning. *ICFAI Univ. J. Environ. Sci.* **2008,** *2,* 7–19.

Ruda, A. Spatial Decision Support Using Data Geo-Visualization: The Example of the Conflict Between Landscape Protection and Tourism Development. *J. Maps.* **2016,** *12*, 1262–1267. DOI: 10.1080/17445647. 2016.1152915.

Steiniger, S.; Hay, G. J. Free and Open Source Geographic Information Tools for Landscape Ecology. *Ecol. Info.* **2009,** *4*, 183–195. DOI: 10.1016/j.ecoinf.2009.07.004.

Su, S.; Yang, C.; Hu, Y.; Luo, F.; Wang, Y. Progressive Landscape Fragmentation in Relation to Cash Crop Cultivation. *Appl Geogr.* **2014,** *53*, 20–31. DOI: 10.1016/j.apgeog.2014.06.002.

Sugumaran, R.; Degroote, J. *Spatial Decision Support Systems: Principles and Practices*; CRC Press, **2010;** p 508. DOI: 10.1201/b10322.

CHAPTER 21

FACTORS AFFECTING DIVERSITY AND DISTRIBUTION OF ANIMALS IN THE WESTERN GHATS BIODIVERSITY HOTSPOT

RAMAKRISHNA[1], K. P. DINESH[1], and P. DEEPAK[2]

[1]*Zoological Survey of India, India*

[2]*St. Carmels College, Bangalore, India*

ABSTRACT

The Indian Peninsula is a compact natural unit of geomorphological and biogeographical evolution, and this phenomenon is exhibited with the distribution of fauna from Madagascar and tropical East Africa as well as a considerable number of Indo–Chinese and a small number of Malayan forms in Peninsular India, in addition to its own assemblage of many groups of animals. Indian landmass after collision with the Eurasian landmass, a significant biotic interchange of faunal groups took place, with India being a both biotic "ferry" and biotic "sink." This hypothesis of "Out of India" and "Into India" is often used to explain patterns of distribution among Southeast Asian taxa. The present-day disjunct distribution of a large assemblage of genera and species of Malayan affinities has long attracted the attention of naturalists and efforts have been made to explain the causes of similarities. One well-known explanation for this is the Satpura Hypothesis of Sunderlal Hora (1949). The Western Ghats (WG) is the home for many of the original Gondwana relicts, the autochthonous fauna of Peninsular India, the transmigrants from the Palearctic and later Indo–Chinese and Malayan species and some Himalayan reflects which reached Peninsular India during the glacial periods have found refugium in the WG.

21.1 INTRODUCTION

Tropical countries of the world contain a high percentage of the world's species and many of these areas are recognized as "Mega diversity Countries," India is one among the 18 mega-diversity countries of the world. Norman Myers (1988) who introduced the concept

Biodiversity Hotspot of the Western Ghats and Sri Lanka. T. Pullaiah, PhD (Ed.)

of International Conservation Priority Areas, namely, "hotspots" based on species richness, endemism, level of threat to an area, and natural habitats from human activities, especially considering that of vertebrates (mammals, birds, reptiles, and amphibians) among animals and angiosperms in the plant kingdom. In all 36 such biodiversity-rich areas (hotspots) are identified throughout the world. Four of the hotspots are partly represented in India, namely, (1) Western Ghats (WG), (2) Himalaya falling within India's geographical region, (3) Parts of Northeast India and the Andaman Islands representing the Indo–Burma Hotspot, and (4) Nicobar Islands which forms a part of the Sundaland Hotspot.

21.2   THE WESTERN GHATS

The WG is a UNESCO World Heritage Site (declared on July 2, 2012) and is one among of the eight hottest "hotspots" of biodiversity, 39 cluster sites are identified, spread over 7953.15 km² in India's WG to its World Heritage List. The cluster of sites is in Agasthyamalai, Periyar, Anamalai, Nilgiris, Upper Cauvery in Kodagu, Kudremukh, and Sahyadri region. Sites are selected for their outstanding universal value, based on four important criteria viz., Critical habitat for several globally threatened flagship species (Asian elephant, Gaur, and Tiger); endangered species such as the Lion-tailed Macaque, Nilgiri Tahr, and Nilgiri Langur; exceptionally high levels of plant and animal diversity and endemism (17,000 animal species and 7200 plants [angiosperms] and endemism varying from 20–70% among different groups of animal kingdom and plants); exceptionally high levels of speciation and evolutionary radiation and large scale biological and ecological processes (Myers et al., 2000).

WG is a long chain of hills and mountains aligned along the edge of the Deccan Plateau, separating the plateau from the Malabar and Konkan coasts, adjoining the Arabian Sea. The mountain range begins at about 21° North latitude near the river Tapti in the state of Gujarat and runs down about 1490 km towards the coast of Kanyakumari on the southern tip of India at 8 North latitude, interrupted only by the 30 km wide Palghat Gap at around 11°N from Tapti Valley of Gujarat in the north to Kanyakumari, Tamil Nadu in the south. The argument that the WG, despite their appearance, are not true mountains but rather faulted edge of the Deccan plateau that may have formed during the break-up of the supercontinent of Gondwana some 150 million years ago (mya), is widely accepted by geologists, Krishnan (1974) explains the geomorphological, geological, and geophysical evidence supporting this interpretation.

The WG ranges form a barrier to the monsoon winds originating in the Indian Ocean and moving northeast, which results in heavy rainfall (8000 mm) in the region during the South West Monsoon, which is between June and October. This mountain range, popularly known as *Sahyadri mountains/Sahyadri range* and one of the richest hotspots of biological diversity in the world, harbors 38 east-flowing and 27 west-flowing rivers to drain the entire watershed region and other wetlands which is a natural abode of numerous aquatic animals including fishes and the main watershed of Peninsular India.

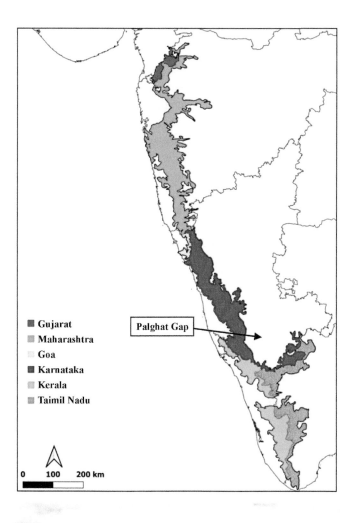

FIGURE 21.1 Sholas in Bababudengiri Hills of WG, a critical habitat for plants and animals (Photo: Ramakrishna).

21.3 FAUNAL UNIQUENESS IN THE WESTERN GHATS

The zoogeographical affinities of the Peninsula are with the Gondwana faunas of Madagascar and tropical east Africa (Cretaceous fragment of the India–Madagascar landmass). There are considerable Indo–Chinese, and small numbers of Malayan forms also, formerly continuously distributed from Myanmar, Assam through Eastern Ghats (EG) to the South Peninsula and Sri Lanka, but now with the destruction of the habitats confined as relicts in small, isolated pockets of the south Indian hills. Thus, the WG is the home for many of the original Gondwana relicts, the autochthonous fauna of Peninsular India, the trans-migrants from the Palearctic and later Indo–Chinese and Malayan species together with the Himalayan relicts that reached Peninsular India during the glacial periods have found refugium in the WG. Based on the geologic history, the rainforests of WG are unique having an admixture of the following biotic elements, the resultant of the biogeographic evolution of the country, namely, the following:

- Gondwana derivatives
- Peninsular endemics
- Madagascan elements
- Indo-Chinese derivatives
- Malayan elements
- Palearctic elements

The microcenters of endemism in WG are as follows:

- Agasthyamalai Hills
- Anaimalai Hill ranges
- Palani Hills
- Nilgiris Silent Valley–Wyanad–Kodagu
- Mahabaleshwar–Khandala ranges
- Konkan Raghad
- Southern Deccan

21.3.1 IMPACT OF GEOGRAPHIC BARRIERS IN THE WESTERN GHATS

In the WG there are a series of geologically heterogeneous forest gaps which are the valleys that break the continuity of the mountain ranges. These three gaps are the Shencottah pass, the Palghat gap, and the Goa pass divide the range into southern, central, and northern WG (Robin et al., 2010). The Palghat gap, Moyar pass, and the Shencottah pass have resulted in preventing the spread of certain species and have hence facilitated local speciation and endemism leading to micro-evolutionary processes. Nilgiri tahr is distributed in this landscape and is endemic. The genetic characteristics, population structure, and impact of these gaps on the species were analyzed (Joshi et al., 2018) and reveal the presence of two different populations of Nilgiri tahr from the north and south of Palghat gap in the WG, India. Similarly, the impact of Biogeographical barriers on the distribution of white-bellied shortwing *Brachypteryx major* restricted only to the high-elevation *Shola* forests (above

1400 m) studied using cytochrome *b*, cytochrome *c* oxidase subunit I, and D-loop control region around the separated area of Palghat (40-km wide) and the Shencottah (7-km wide) gaps as also by great geographical distances (>80 km) provide insights into processes that may have impacted the speciation and evolution of the endemic fauna of this region. The validity of two subspecies, *B. major* and *B. m. albiventris*, revealed that the northern species be called the rufous-bellied shortwing *B. major* and the southern species white-bellied shortwing *Brachypteryx albiverntris* (Robin et al., 2010). Their study concludes that ancient geographical gaps clearly form barriers for dispersal and gene flow and have caused deep divergences in a montane species, while habitat and climate refugia have influenced recent population history. However, a study on a species of a group of caecilians, showed that the gap did not function as a barrier for the species (Gower et al., 2007).

21.3.2 FACTORS AFFECTING DISTRIBUTION

The present-day distribution of biota in WG is a resultant of series of volcanic eruptions which occurred around 65 mya, giving rise to Deccan traps which led to mass extinctions (Samant and Mohabey, 2009). The fishes, dinosaurs, and micromammals perished out of the peninsula while several invertebrate groups escaped from the hostile environment and retreated to the areas south of the Deccan trap (Menon, 1992). Thus, South WG could have served as refugia for most dominant wet evergreen plants and some aquatic species of animals which could have once been widely distributed during the prevolcanic period, later confined predominantly to the western escarpment mountain range (Pascal, 1988).

21.3.2.1 DISJUNCT DISTRIBUTION

The phenomenon where closely related organisms are geographically separated, and their distribution is discontinuous is often referred to as disjunct distribution. Such occurrences were observed in plants and animals (Mani, 1974; Wanntrop et al., 2006; Ali, 1935; Gaston and Zacharias, 1996; Jayaram, 1974; Kurup, 1974; Joshi and Kunte, 2014). The disjunct distribution of animal and plant species in WG and northeastern India is explained by way of proposed two models, namely, dispersal via corridors followed by extinction in the corridors have caused disjunctions (Puri et al., 2013). Alternatively, second model (Ripley and Beehler, 1990), suggests that the current disjunctions reflect relicts of a once continuous distribution. The fossil record suggests that some taxa in northeastern India may have become extinct because of such changes in climate due to the upliftment of the Himalayas. For example, a fossil leaf like structure that of *Poeciloneuron indicum* Beddome, an endemic species to the WG, is found in the Oligocene sediments of north-eastern India (Tinsukia, Assam). India was once covered continuously by tropical humid forests that in recent times have been fragmented into isolated patches due to climatic changes (Mani, 1974). The Continuous Range Hypothesis or Brij hypothesis explains the tropical fauna similarities between the WG and Eastern Himalayas (Dilger, 1952). Apparently, this hypothesis is not widely accepted because the WG populations display very little differentiation compared to their eastern counterparts (Puri et al., 2013).

The Himalayan Glaciation theory explains the disjunct distribution and the occurrence of Himalayan elements in WG. During the Pleistocene period, massive environmental changes took place, which were responsible for the dispersal, fragmentation, and isolation of all kinds of biological populations, leading to speciation and evolution. Pleistocene Glaciations on the Himalaya caused rapid dispersion of European and Boreal Asiatic species deep South into the peninsula. These animals occupied certain hilltops in the WG. Bovids, elephants, horses, and deer are found at most post-Siwalik fossil-yielding sites in India. Nilgiri tahr found in the southern WG is closely related to the Himalayan tahr, despite the large gap.

The disjunct distribution is further elaborated by Randhawa (1945), who based his conclusions on a variety of evidence. Numerous excavations in the Brij districts of central India, including Mathura, have unearthed ancient sculptures which clearly show several species of trees characteristic of the wet tropics. Four of the most often depicted trees are the Asoka, *Saraca indica*; the Kadamba, *Anthocephalus indicus*; the Champak, *Michelia champaca* and the Nagkesar, *Mesua ferrea*. These trees are found at present in the wet tropics of the eastern Himalayas, Assam, and Burma as well as in the WG; thus, they have a distribution like that of the animal anomalies. "The 'Brij" country which was once covered with lush evergreen tropical forests about 2000 years ago now has completely changed. The jungles which were the abode of the rhinoceros and the wild elephant have disappeared and in their place, we find sandy wastes haunted by flocks of black buck. In view of the above evidence showing the existence of a former wet tropical belt across India between the eastern Himalayas and the west coast north of the Vindhya–Satpura trend which served as the connecting link between the now anomalous faunas. The prevailing hypothesis for birds is that Himalayan taxa dispersed to the WG during the time when peninsular India was covered in moist forest and subsequently these populations were isolated due to progressive aridification of the subcontinent (Reddy et al., 2017). Since this connection existed during historical times the small amount of differentiation now found between many of the Eastern and Western forms is perfectly understandable.

21.3.2.2 *PRIMARY GATEWAY OF INDIA (ASSAM GATEWAY)*

Assam has always been treated as the "gateway" to India for the Indochinese/Malayan flora and fauna, often called the Primary gateway. After entering through this gateway, the species has to negotiate secondary routes to make further entry into the peninsular region. The identified four secondary entry points or the "secondary gateway" are

1. Chittagong Hills–Sundarbans
2. Garo Hills–Rajmahal Hills
3. Darjiling Hills–Rajmahal Hills
4. Himalyan–Aravali Range.

Hora (1949a) proposed the *Satpura Hypothesis*, based on his work on the torrential fishes of the area in question to which he attempted to explain the existing similarities between the two widely separated faunas by presuming certain past modifications of the present Vindhya–Satpura trend of hills which loosely connect the eastern Himalayas with the west coast of India. He believed that this range of hills was the connecting link that existed in the past between the east and the west.

The earliest stock of the present-day freshwater fishes of India and other aquatic organisms seems to have invaded Southeast Asia during the Eocene period but India began to receive Malayan forms only during the Pliocene Siwalik period (Menon, 1973). The highly specialized torrential fauna developed much later, probably in the Pleistocene. The origin and distribution of this fauna is associated with the later Himalayan orogenic movements (Hora, 1949a). The list of fishes mentioned by Hora (1944, 1949a) is reproduced below.

List of fishes common to India and the Malayan Region (Hora, 1949)	
Notopterus Lacepede	*Bagarius* Bleeker
Bariliua Hamilton	*Calichrous* Hamilton
Rasbora Bleeker	*Chaca* Cuv. and Val.
Barbus Cuv. and Val.	*Glytosternum* McCleeland
Labeo Cuvier	*Leiocassis* Bleeker
Clarias Gronov	*Macronus* Dumeril
Haplochilus McClell	*Pangasius* Cuv. and Val.
Eleotris Gronov	*Pseudeutropius* Bleeker
Gobius Artedi	*Silurus* Artidi
Periopthalmus Bl. and Sch.	*Wallago* Bleeker
Ambasis Cuv. and Val.	*Monopterus* Lacepede
Mastacembulus Cuv. and Val.	*Symbranchus* Bloch
Chela Ham.	*Namdus* Cuv. and Val
Cirrihina Cuvier	*Pristolepis* Jerdon
Dangila Cuv. and Val.	*Ophiocephalus* Bloch
Osteochilus Gunther	*Anabas* Cuvier
Thimnichthys Bleeker	*Polyacanthus* Cuv. and Val.
Acanthopsis v. Hasselt	*Osphronemus* Lacepede
Botia Gray	*Mugil* Linn.
Lepidocephalichthys Bleeker	*Sciaena* (Artedi) Cuvier
Nemachilus v. Hass	*Sicydium* Cuv. and Val.
Homoloptera v. Hass	*Rhynchobdella* Bloch and Schn.

Source: Adapted from Nitin et al. (2018); Sujitha et al. (2019).

Hora suggested that soon after India joined mainland Eurasia in the Eocene (56–35 mya) the westward migration of Malayan freshwater fauna began. Dispersal of hill-stream and torrent fishes was enabled by (a) river capture, (b) longitudinal river valleys, and (c) tilting of mountain blocks. Hora has also drawn attention to the Garo–Rajmahal gap, through which the rivers Ganges and Brahmaputra drain. He has shown that the gap was nonexistent till the late Miocene and hence had directly served as a bridge in the movement of Malayan hill–stream fishes through Assam, over the Satpuras, and finally down the WG (Hora, 1944). Hora proposed the large Malayan stream fishes, including *Wallago*, *Silonia*,

Cirrhinus, etc., have thus entered India during the Eocene. However, Hora completely ignored fishes that are not strictly inhabitants of torrents, and yet have migrated from the Malayan Archipelago and largely evolved in the WG. Forty-eight genera of freshwater fishes have got at least one species endemic in the WG. Fishes in the genera *Labeo, Puntius, Mystus, Pristolepis, Monopterus, Pseudosphromenus*, etc., do show Malayan affinity. These are, however, not typically torrential stream fishes. They could have reached the WG through "normal" means (Daniels, 2001). There are five invasions of fishes to peninsular India corresponding to the five glacial periods of the Pleistocene (Silas, 1952) and it is quite possible that the route of migration of torrential and other aquatic fauna lay along the Narmada–Tapti drainage of the Vindya–Satpura mountains (Menon, 1951). Kuttapetty et al. (2014) supports the Horas hypothesis with the distribution pattern of *Rhododendron arboreum* spp. *arboreum* and as *R. arboreum* spp. *nilagiricum* in SWG and further down in Sri Lanka as *R. arboreum* spp. *zeylanicum* testifies to the migration of taxa during the early Pleistocene or Pliocene periods. There are many Indian plant taxa with disjunct distributions in northeastern India and WG, raising several systematic and biogeographic questions—nine genera (*Arisaema* Mart., *Begonia* L., *Ceropegia* L., *Hoya* R. Br., *Impatiens* L., *Indigofera* L., *Rubus* L., *Strobilanthes* Blume, and *Vitis* L.). These genera were chosen as they contain species which are endemic and occur in either northeastern India or WG, with at least one species being included in global phylogenetic analyses (Puri et al., 2016).

The distribution of pattern of an endemic fish *Schismatorynchos nukta* of WG is highly intriguing (Daniels, 2001; Jadhav et al., 2011; Raghavan et al., 2013), as this fish found its distribution listed in the states of Assam, Arunachal Pradesh, and Tripura of Northeast India (Goswami et al., 2012), as well as in Bangladesh and Nepal. It may be one of the links explaining the Malayan affinity of Indian peninsular fish fauna; dispersal of Malayan fish fauna along the Satpura mountain range; example of convergent evolution; and basis of Satpura Hypothesis (Hora, 1942, 1944, 1949; Silas, 1952; Menon 1980; Daniels, 2001; Karanth, 2003).

21.3.2.3   BIOTIC FERRY HYPOTHESIS

As India collided with Asia, there was significant biotic interchange between both landmasses, with India being both biotic "ferry" and biotic sink. The *"Out of India"* and *"into India"* hypothesis often used to explain patterns of distribution among Southeast Asian taxa. According to this hypothesis, Southeast Asian taxa originated in Gondwana and diverged from their Gondwanan relatives when the Indian subcontinent drifted from Gondwana in the Late Jurassic period. They later colonized Southeast Asia when it collided with Eurasia in the early Cenozoic era. According to Mani (1974), these relatively young intrusive elements are largely derived from the Indo-Chinese and Malayan subregion of tropical Asia. Numerous researchers have reported on the overall faunal similarity between peninsular India and SE Asia (Blanford, 1901; Hora, 1949b; Hora and Jayaram, 1949; Mani, 1974). This out of Asia and into India scenario has been discussed in vertebrates based on paleontological evidence (Clyde et al., 2003). Such taxa include, for example, snakes of the genus *Eryx* and killifishes *Apocheilus* and *Pachypanchax* (Murphy and

Collier, 1997); the unusual frog *Nasikabatrachus* from the WG, which is closely related to the endemic Seychellean family Sooglossidae (Biju and Bossuyt, 2003); and two sister genera of cichlid fishes, *Etroplus* in southern India and *Paretroplus* in Madagascar (Sparks, 2004). The discovery of a new subfamily (Astrobatrachinae), new genus (*Astrobatrachus*) and a species *Astrobatrachus kurichiyana* (Vijayakumar et al., 2019) from Wayanad is another good example of living fossil frog existence in the central WG. The newly discovered microhylid from Wayanad, Kerala part of WG; *Mysticellus franki* by Garg and Biju (2019), the new microhylid frog genus from Peninsular India with Southeast Asian affinity suggests phylogenetic evidence of Microhylinae of multiple dispersal events between India and Asia through postulated land links right from Eocene up to Miocene through final Indo-Asia mass, also termed as the "Eocene exchange hypothesis" which also suggests prolonged isolation of Indian Peninsula by the Deccan traps along with intermitted periods of isolation during Eocene–Oligocene, and limited Miocene exchange between northern Indian regions (presently comprising of Himalaya–Tibetan plateau and Northeast India) and Asia due to complex geological and paleoclimatological events associated with Indo–Asia collision—explains early colonization of certain faunal groups from India to Southeast Asia, as well as the observed patterns of regional endemism and widespread distributions (Garg and Biju, 2019).

Menon (1951) based on the suggestion of Earnst Mayr carried out extensive surveys in EG and Orissa hills and found no Malayan fishes, he concluded that the fishes in EG derived fishes from Vindhya–Satpura mountains and northern part of WG. However, this theory does not apply to the birds distributed in WG. The Satpura Hypothesis (Hora, 1949; Ali, 1949) envisages the Vindhya–Satpura range in central India as a "corridor" for dispersal of taxa from the eastern Himalayas to the northern end of the WG. The route along the EG (Abdulali, 1949; Mani, 1974) proposes that some taxa dispersed from the eastern Himalayas along the EG to the southern tip of the WG (Eastern Ghats Hypothesis). The "southern route across the Indian Ocean" (Croizat, 1949) hypothesizes that source taxa from the eastern Himalayas dispersed across what is today the Bay of Bengal to reach peninsular India. However, other taxa may have adopted more than one route simultaneously to disperse to the WG. Based on the distribution of Hornbills in Biligiri Ranga Hills, meeting point of EG and WG in the Nilgiri range of hills Srinivasan and Prashanth (2006) concludes that the bird dispersal in WG has taken two routes, namely, along the EGs and then northward along the WG, or both. The vicariance model (Karanth, 2003) holds that present species distributions are "relict populations" of a formerly continuous range, with species having suffered extinction in the intervening areas. It may be noted that no hypothesis dealing with faunal dispersal is contradictory to the other—while one may hold true for the dispersal of particular taxa, another may satisfactorily explain the dispersal of others.

ACKNOWLEDGMENTS

Authors wish to convey thanks to Director, Zoological Survey of India for the facilities and support and Deepak to the Authorities of St. Carmel's College, Bangalore for their help and support.

KEYWORDS

- Satpura hypothesis
- Gondwanaland
- geographic barriers
- biotic ferry
- glaciation

REFERENCES

Ali, S. In *Birds of Kerala*; Oxford University Press: Madras, **1935**.

Biju, S. D.; Bossuyt, F. New Frog Family from India Reveals an Ancient Biogeographical Link with the Seychelles. *Nature* (London) **2003**, *425*, 711–714.

Biswas, S. K. A Review on the Evolution of the Rift Basins in India during Gondwana with Special Reference to Western Indian Basins and their Hydrocarbon Prospects. *Proc. Indian Natl. Sci. Aacad.* **1999**, *65* (3), 261–283.

Blanford, W. T. The Distribution of Vertebrate Animals in India, Ceylon, and Burma. *Phil. Trans. Roy. Soc. London* **1901**, *194*, 335.

Chatterjee, S. P. Fluctuations of Sea Level Around the Coasts of India During the Quaternary Period. *Tropical Geomorphol.* **1961**, *3*, 48–56.

Chauhan, M. S. Holocene Vegetation and Climatic Changes in Southeastern Madhya Pradesh, India. *Curr. Sci.* **2002**, *83*, 1444–1445.

Clyde, W. C.; Khan, I. H.; Gingerich, P. D. Stratigraphic Response and Mammalian Dispersal During Initial India-Asia Collision: Evidence from the Ghazij Formation, Balochistan, Pakistan. *Geology* **2033**, *31*, 1097–1100.

Conti, E.; Eriksson, T.; Schönenberger, J.; Sytsma, K. J.; Baum, D. A. Early Tertiary Out-of-India Dispersal of Crypteroniaceae: Evidence from Phylogeny and Molecular Dating. *Evolution* **2002**, *56*, 1931–1942.

Daniels, R. R. Endemic Fishes of the Western Ghats and the Satpura Hypothesis. *Curr. Sci.* **2001**, *81* (3), 240–244.

Dilger, W. C. The Brij Hypothesis as an Explanation for the Tropical Faunal Similarities Between the Western Ghats and the Eastern Himalayas, Assam, Burma and Malaya. *Evolution* **1952**, *6*, 125–127.

Ferguson, H. S.; Bourdillon, T. F. The Birds of Travancore with Notes on their Nidification. *J. Bombay Nat. Hist. Soc.* **1903**, *15*, 249.

Garg, S.; Biju, S. D. New Microhylid Frog Genus from Peninsular India with Southeast Asian Affinity Suggests Multiple Cenozoic Biotic Exchanges between India and Eurasia. *Sci. Rep. (Nature, London)* **2019**, *9* (1906), 1–13.

Gaston, A. J.; Zacharias, V. J. The Recent Distribution of Endemic and Disjunct Birds in Kerala State: Preliminary Results of an Ongoing Survey. *J. Bombay Nat. Hist. Soc.* **1996**, *93*, 389–400.

Goswami, U. C.; Basistha, S. K.; Bora, D.; Shyamkumar, K.; Saikia, B.; Changsan, K. Fish Diversity of North East India, Inclusive of the Himalayan and Indo Burma Biodiversity Hotspots Zones: A Checklist on their Taxonomic Status, Economic Importance, Geographical Distribution, Present Status and Prevailing Threats. *Inter. J. Biodivers. Conserv.* **2012**, *4* (15), 592–613.

Gower, D. J.; Dharne, M.; Bhatta, G.; Giri, V.; Vyas, R.; Govindappa, V.; et al. Remarkable Genetic Homogeneity in Unstriped, Long-Tailed *Ichthyophis* Along 1500 km of the Western Ghats, India. *J. Zool.* **2007**, *272*, 266–275.

Hora S. L. On the Malayan Affinities of the Fresh-Water FishFauna of Peninsular India, and its Bearing on the Probable Age of the Garo-Rajmahal Gap. *Proc. Nat. Inst. Sci. India* **1944**, *15*, 362–364.

Hora, S. L. Satpura Hypothesis of the Distribution of the Malayan Fauna and Flora to Peninsular India. *Proc. Nat. Inst. Sci. India* **1949a**, *15* (8), 309–314.

Hora, S. L. Dating the Period of Migration of the So-Called Malayan Element in the Fauna of Peninsular India. *Proc. Nat. Inst. Sci. India* **1949b**, *15* (8), 345–351.

Hora, S. L.; Jayaram, K. C. Remarks on the Distribution of Snakes of Peninsular India with Malayan Affinities. *Proc. Nat. Inst. Sci. India* **1949c**, *15* (8), 399–402.

Jadhav, B. V. Kharat, S. S.; Raut, R.; Paingankar, M.; Dahanukar, N. Fresh Water Fish Fauna of Koyna River, Northern Western Ghats, India. *J. Threat. Taxa.* **2011**, *3* (1), 1449–1455.

Jayaram, K. C. Ecology and Distribution of Freshwater Fishes, Amphibians, and Reptiles. In *Ecology and biogeography in India*; Mani, M. S., Ed.; Dr. W. Junk b.v. Publishers: The Hague, **1974**; pp 517–580.

Joshi, S.; Kunte, K. Dragonflies and Damselflies (Insecta: Odonata) of Nagaland, with an Addition to the Indian Odonate Fauna. *J. Threat. Taxa* **2014**, *6*, 6458–6472.

Joshi, B. M.; Matura, R.; Predit M. A.; De, R.; Pandav, B.; Sharma, V. Palghat Gap Reveals Presence of Two Diverged Populations of Nilgiri Tahr (*Nilgiritragus hylocrius*) in Western Ghats, India. *Mitochondrial DNA Part B* **2018**, *3* (1), 245–249.

Karanth, K. P. Evolution of Disjunct Distributions Among Wet-Zone Species of the Indian Subcontinent: Testing Various Hypotheses using a Phylogenetic Approach. *Curr. Sci.* **2003**, *85*, 1276–1283.

Krishnan, M. Geology. In *Ecology and Biogeography in India*; Mani, M. S., Ed.; Dr W Junk: The Hague, **1974**; pp 60–98.

Kurup, G. U. Mammals of Assam and the Mammal-Geography of India. In *Ecology and biogeography in India*; Mani, M. S., Ed.; Dr. W. Junk b.v. Publishers: The Hague, **1974**; pp 585–613.

Kuttapetty, M.; Pillai, P. P.; Varghese, R. J.; Seeni, S. Genetic Diversity Analysis in Disjunct Populations of *Rhododendron arboreum* from the Temperate and Tropical Forests of Indian Subcontinent Corroborate Satpura Hypothesis of Species Migration. *Biol. Sect. Botany* **2014**, *69* (3), 311–322.

Lele, K. M. In *The Problem of Middle Gondwana in India*, Proceedings of the 22nd International Geological Congress, New Delhi, **1964**; Section 9, pp 181–202.

Li, Z. X.; Powell, C. Mc. A. Late Proterozoic to Early Palaeozoic palaeomagnetism and the Formation of Gondwanaland. In *Gondwana Eight: Assembly, Evolution and Dispersal*; Findlay, R. H., Unrug, R., Banks, M. R., Veevers, J. J., Eds.; A.A. Balkema: Rotterdam, **1993**; pp 9–21.

Maheshwari, H. K. Provincialism in Gondwana Floras. *Palaeobotanist* **1992**, *40*, 101–127.

Mani, M. S. Biogeographical Evolution in India. In *Ecology and biogeography in India*; Mani, M. S., Ed.; Dr. W. Junk b.v. Publishers: The Hague, 1974; pp 698–724.

Meher-Homji, V. M. On the Indo-Malaysian and Indo-African Elements in India. *Feddes Repert.* **1983**, *94*, 407–424.

Meher-Homji, V. M. History of Vegetation of Peninsular India. *Man Environ.* **1989**, *13*, 1–10.

Menon, A. G. K. Further Studies Regarding the Hora's Satpura Hypothesis; 1. The role of Eastern Ghats in the Distribution of Malayan Fauna and Flora in Peninsular India. *Proc. Nat. Inst. Sci. India* **1951**, *17* (6), 75–97.

Menon, A. G. K. The Satpura Hypothesis. *Proc. Indian Nat. Sci. Acad. B.* **1980**, *46* (1), 27–32.

Menon, A. G. K. *Zoogeography of India in Taxonomy in Environment & Biology*; Special Volume Published by Director; Zoological Survey of India, **1992**; pp 59–75.

Metcalfe, I. Late Palaeozoic and Mesozoic Palaeogeography of Southeast Asia. *Palaeogeogr. Palaeoclimatol. Palaeoecol.* **1991**, *87*, 211–221.

Murphy, W. J.; Collier, G. E. A Molecular Phylogeny for Aplocheiloid Fishes (Atherinomorpha, Cyprinodonti-formes): The Role of Vicariance and the Origins of Annualism; *Mol. Biol. Evol.* **1997**, *14*, 790–799.

Myers, N. Threatened Biotas: Hot-Spots in Tropical Forests. *Environmentalist* **1988**, *8*, 187–208.

Myers, N.; Mittermeier, R. A.; Mittermeier, C. G.; da Fonseca, G. A. B.; Kent, J. Biodiversity Hotspots for Conservation Priorities. *Nature* **2000**, *403*, 853–858.

Pascal, J. P. *Wet Evergreen Forests of the Western Ghats of India*; Inst. fr. Pondichéry, trav. sec. sci. tech., **1988**; Tome 20.

Patnaik, R.; Prasad, V. Neogene Climate, Terrestrial Mammals and Flora of the Indian Subcontinent. *Proc. Indian Natl. Sci. Acad.* **2016**, *82*, 605–615.

Prasad, V.; Garg, R.; Khowaja-Ateequzzaman; Singh, I. B.; Joachimski, M. M. *Apectodinium* Acme and the Palynofacies Characteristics in the Latest Palaeocene-Earliest Eocene of Northeastern India: Biotic Response to Paleocene-Eocene Thermal maxima (PETM) in Low Latitude. *J. Paleontol. Soc. India* **2006**, *51*, 75–91.

Prasad, V.; Farooqui, A.; Tripathi, S. K. M.; Garg, R.; Thakur, B. Evidence of Late Paleocene– Early Eocene Equatorial Rain Forest Refugia in Southern Western Ghats, India. *J. Biosci.* **2009**, *34*, 777–797.

Puri, R.; Barman, P.; Geeta, R. A Phylogenetic Approach Toward the Understanding of Disjunct Distributions of Plant Taxa in Western Ghats and Northeastern India. *Rheedea* **2016**, *26* (2), 99–114.

Raghavan, R.; Dahanukar, N.; Tlusty, M.; Rhyne, A.; Krishnakumar, K.; Molurand, S.; Rosser, A. M. Uncovering an Obscure Trade: Threatened Freshwater Fishes and the Aquarium Pet Trade. *Biol. Conserv.* **2013**, *164*, 158–169.

Randhawa, M. S. Progressive Desiccation of Northern India in Historical Times. *J. Bombay Nat. Hist. Soc.* **1945**, *45*, 558–565.

Reddy, S.; Robin, V. V.; Vishnudas, C. K. Revisionist History: Rewriting the Story of Indian birds. *BMC Series Blog* **2017**. https://blogs.biomedcentral.com/bmcseriesblog/about/

Ripley, D.; Beehler, B. Patterns of Speciation in Indian Birds. *J. Biogeogr.* **1990**, *17*, 639–648.

Robin, V. V.; Sinha, A.; Ramakrishnan, U. Ancient Geographical Gaps and Paleo-Climate Shape the Phylogeography of an Endemic Bird in the Sky Islands of Southern India. *PLoS One* **2010**, *5* (10), e13321. DOI: 10.1371/journal.pone.0013321

Samant, B.; Mohabey, D. M. Palynoflora from Deccan Volcano-Sedimentary Sequence (Cretaceous-Palaeogene transition) of Central India: Implications for Spatio-Temporal Correlation. *J. Biosci.* **2009**, *34*, 811–823.

Silas, E. G. Further Studies Regarding Hora's Satpura Hypothesis. 2. Taxonomic Assessment and Levels of Evolutionary Divergences of Fishes with the so-called Malayan Affinities in Peninsular India. *Proc. Nat. Inst. Sci. India* **1952**, *18* (5), 423–448.

Smith, A. G. Gondwana: Its Shape, Size and Position from Cambrian to Triassic Times. *J. African Earth Sci.* **1999**, *28*, 71–97.

Sparks, J. S. Molecular Phylogeny and Biogeography of the Malagasy and South Asian Cichlids (Teleostei: Perciformes: Cichlidae). *Mol. Phylog. Evol.* **2004**, *30*, 599–614.

Srinivasan, U.; Prashanth N. S. Preferential Routes of Bird Dispersal to the Western Ghats in India: An Explanation for the Avifaunal Peculiarities of the Biligirirangan Hills. *Indian Birds* **2006**, *2* (5), 116–119.

Srivastava, G.; Mehrotra, R. C. Endemism Due to Climate Change: Evidence from *Poeciloneuron* Bedd. (Clusiaceae) Leaf Fossil from Assam, India. *J. Earth Syst. Sci.* **2013**, *122*, 283–288.

Veevers, J. J. In *Phanerozoic Earth History of Australia*; Oxford Monographs on Geology and Geophysics. Clarendon Press: Oxford, **1984**; vol 2, pp 1–418.

Vijayakumar, S. P.; Pyron, R. A.; Dinesh, K. P.; Torsekar, V.; Srikanthan, A. V.; Swamy, P.; Stanley, E. L.; Blackburn, D. C.; Shanker, K. A New Ancient Lineage of Frog (Anura: Nyctibatrachidae: Astrobatrachinae subfam. nov.) Endemic to the Western Ghats of Peninsular India. *Peer. J.* **2019**, *7*, e6457 https://doi.org/10.7717/peerj.6457

Wadia, D. N. In *Geology of India*, 3rd Ed.; MacMillan: London, **1957**.

Wanntrop, L.; Kocyan, A.; Renner, S. S. Wax Plants Disentangled: A Phylogeny of *Hoya* (Marsdenieae, Apocynaceae) Inferred from Nuclear and Chloroplast DNA Sequences. *Mol. Phylogen. Evol.* **2006**, *39*, 722–733.

Wirthmann, A. Tropical Geomorphology with or Without Structural Control. In *Indian General Geomorphology Part-I*, Chap. 3; Dikshit, K. R., Kale, V. S., Kaul, M. N., Eds.; India Geomorphological Diversity; Rawat Publications: Jaipur and New Delhi, **1994**; pp 63–75.

CHAPTER 22

SPATIAL CONSERVATION PRIORITIZATION OF LANDSCAPES IN NILGIRI BIOSPHERE RESERVE, WESTERN GHATS

K. V. SATISH, S. VAZEED PASHA, and C. SUDHAKAR REDDY

Forest Biodiversity and Ecology Division, National Remote Sensing Centre, Indian Space Research Organisation, Hyderabad, Telangana, India

ABSTRACT

The Western Ghats is one of the global biodiversity hotspots, which is threatened due to increased anthropogenic impacts such as deforestation, degradation, fragmentation, fires, and biological invasions. Spatial patterns and distributions of anthropogenic influences are still uncertain, which have implications for conservation plans. Therefore, modeling conservation prioritization using ecological models is critical to minimize the negative impacts on biodiversity. In this study, we developed spatial conservation priorities for the Nilgiri Biosphere Reserve (NBR) in the Western Ghats by integrating extensive field data and remote sensing-based observations. Based on the integrated data, the fragmentation index, disturbance index, and biologically rich areas were generated by the spatial landscape analysis and modeling. The spatially explicit data have indicated varying patterns of biodiversity prioritization across zones, protected areas, and vegetation types. The new knowledge and data that were developed in this study help long-term conservation and management of the NBR.

22.1 INTRODUCTION

Tropical forests are one of the richest biodiversity reserves on Earth (Whitmore, 1998; Morris, 2010). Measuring the biodiversity of such forests has become a difficult task due to the complexity of living organisms and the multiple dimensions of their understanding (Gaston, 2009). Understanding spatial patterns, the distribution of biodiversity, threats, and dynamics of biodiversity is essential to frame effective conservation strategies. Biodiversity

Biodiversity Hotspot of the Western Ghats and Sri Lanka. T. Pullaiah, PhD (Ed.)

has a broad impact on the characteristics of the ecosystem and its services for human well-being and is equally affected by human-induced changes in the ecosystem (Díaz et al., 2006). The structure and dynamics of human communities as well as their interactions with the local environment vary significantly across the globe and this is a challenge for developing a unified approach to biodiversity conservation (Koh and Gardner, 2010). It is therefore important to develop individual landscape management plans with explicit recognition of the socioeconomic, political, and ecological context within which they are situated (Koh and Gardner, 2010). Determination of the extent to which protected areas (PAs) are effective in the conservation of biodiversity or in the achievement of management objectives is therefore critically important for subsequent management measures to be taken.

Deforestation, habitat fragmentation, overexploitation, the spread of invasive species, intense grazing pressure, and fire are major threats to tropical forest biodiversity (Morris, 2010; Weatherspoon and Skinner, 1995) as well as indirect consequences of anthropogenic influence such as climate change. The species-level impacts of habitat fragmentation are well known globally (Newmark, 1991; Ferreira and Laurance, 1997; Benitez-Malvido, 1998; Benítez-Malvido and Martínez-Ramos, 2003; Laurance et al., 1998, 2011; Didham et al., 1998; Benedick et al., 2006). Overexploitation of valuable species (Ragusa-Netto, 2002) may alter the structure of the population of the species, leading to the depletion of such diversity. Dry tropical forests are overexploited (Ragusa-Netto, 2002) rather than other types of forest because they have economically important species (Mahapatra and Tewari, 2005). There is a clear evidence of overexploitation of timber in tropical forests (Asner et al., 2009). Therefore, Protected Areas including biosphere reserves (BRs) have been declared worldwide to minimize the threats and conflicts between development and conservation. Conservation approaches for PAs focus primarily on economic/ecological species, habitats, vegetation types, and landscape units. The concept of the BRs is designed to be flexible and sufficient to meet the local needs and conditions (Force, 1974; Meena et al., 2005). The Indian National Biosphere Reserve Program was launched in 1986 by establishing the first Indian Biosphere Reserve, the Nilgiri Biosphere Reserve (NBR) in the Western Ghats (UNESCO, 2020).

There are no comprehensive data on key aspects of biodiversity (Turner, 2014). It requires an understanding of the elements of biodiversity through a rapidly increasing range of remote sensing and Geographic Information System (GIS) software tools and techniques (Turner, 2014). Remote sensing enables direct observation of wide coverage of landscapes and ecosystems. This can track the evolving drivers of global biodiversity loss (Pettorelli et al., 2014). Geospatial techniques are essential for assessing the status of PAs such as size and condition, species diversity, and other disturbances to the forest (Nagendra et al., 2013). Biodiversity characterization at the landscape level in Meghalaya was carried out by Roy and Tomar (2000) using field integration and IRS LISS III data, while Xiao et al. (2002) carried out forest-type characterization in Northeastern China using multitemporal SPOT-4 VEGETATION data. The application of high-resolution remote sensing data for the characterization of ecosystems had been explored by Wulder et al. (2004). Conservation of the ecosystem at landscape level is a formal decision rather than the protection of individual habitats and species. There is a knowledge gap in the assessment of the natural resources of many PAs around the world. Development of spatial and temporal change databases on the land cover including forests, vegetation types, fragmentation, forest fires, and disturbances may be useful for ecological and conservation purposes. The identification of conservation priority areas within the BRs is therefore an essential and crucial step for the investment of conservation funds. These spatial databases will help to establish strict conservation policies for the effective management of the BRs.

Spatial characterization of landscapes is an important way to handle change (Clark et al., 2004) in forest ecosystems (Imbernon and Branthomme, 2001). Landscape metrics analysis helps to monitor spatial patterns in forest ecosystems (Munsi et al., 2010) that include fragmentation, patchiness, porosity, interspersion, and juxtaposition, and are the key components of biodiversity and ecosystem evaluation. Among these indices, the patch is the fundamental unit of landscape features that differs from the neighboring patches in terms of size, shape, species composition, distribution, and abundance. Increasing isolation of forest patches that threaten species and lead population dynamics (Zambrano amd Salguero-Gómez, 2014). The effects of fragmentation were measured by Reddy et al. (2013) using landscape indices over the bio-geographical regions of India. The dynamic process of forest fragmentation results in an increase in several patches, a decrease in patch size and an increase in edge effect and patch isolation. These effects can lead to changes in the microclimate in the edges of the forest. Hence, facilitate the establishment of invasive alien species within the forest fragments (Reddy et al., 2013) which may affect and modify the species composition within the fragments (Menon and Bawa, 1997). Monitoring of ongoing changes in regional landscapes is necessary to manage biodiversity in PAs. In this study, landscape analysis was carried out using the following methodology proposed by Roy and Tomar (2000), although a few more parameters were included in the model to improve the performance.

22.2 MATERIALS AND METHODS

22.2.1 STUDY AREA

The NBR is the first Biosphere Reserve designated by the Indian Government under the UNESCO's Man and Biosphere Program on September 9, 1986 (Palni et al., 2012). NBR is located between 10°50′ N and 12°16′ N latitude and 76°00′ E to 77°15′ E longitude. NBR is well known for its unique and endangered tropical habitats within the Western Ghats biogeographic zone. It is one of the world's biodiversity hotspots providing luxurious habitats for global endemic biota. There is a wide range of ecosystems and species diversity in this region. It was therefore a natural choice for the country's largest BR. NBR falls within the biogeographical region of the Malabar rain forest area. NBR consists of substantial intact natural vegetation areas, ranging from dry scrublands to wet evergreen forests which contribute to the richness of biodiversity. Altitude and climate gradients support and nourish the different types of vegetation.

22.2.2 PHYTOSOCIOLOGICAL ANALYSIS

22.2.2.1 MEASURING MENHINICK'S INDEX (D)

A total of 228 samples were collected (0.1 ha) and analyzed for species diversity. Trees with a diameter of over 30 cm in breast height were counted and the girth at breast height (GBH) was measured. Species richness is the number of species within a community or area (Menhinick, 1964). These are independent indices of scale. The simplest species richness index is based on the total number of species and the total number of individuals in each sample or habitat and indicates that higher the value greater the species richness (D). It is calculated using the following formula:

$$D = \frac{n}{\sqrt{N}}$$

22.2.2.2 MEASURING SHANNON–WEINER INDEX (H')

The Shannon–Weiner Index is a diversity index with a maximum value of more than one. Total diversity depends on the number of species and the evenness component (distribution of relative abundance). Higher diversity occurs when the number of species and the evenness component are large, that is, low dominance.

This index, based on theory of information, is a measure of uncertainty. The higher the value of H, the greater is the uncertainty or the probability that the next individual is chosen at random from a collection of species containing N individuals will not belong to the same species as the previous one (Smith and Smith, 2001).

$$H' = -\sum_{i=1}^{N} pi \ln pi$$

where pi is the proportion of individuals belonging to the ith species, N is the total number of species in a sample.

22.2.3 REMOTE SENSING AND ANCILLARY DATA

Indian Remote Sensing (IRS) P6 (Resourcesat-1) Linear Imaging and Self Scanning Sensor (LISS-III) data were used to map modify this as 'NBR's vegetation types and land cover. Ancillary data from various sources including Digital Elevation Model (DEM) generated by the Shuttle Radar Topography Mission (SRTM), Survey of India (SOI) toposheets on 1:250,000 and 1:50,000 scales were used for extraction of the road network, and major settlements from satellite imagery. In addition, a large number of small settlements have been digitized from the ground truth and Google Earth images. Locations of on-site observations of invasive species, forest fire, deforestation, degradation including tree cutting, tourism, grazing, firewood collection were collected using handheld Global Positioning System

(GPS), simultaneously, observations of all types of land cover were recorded and used for image classification and accuracy assessment.

22.2.4 VEGETATION TYPES AND LAND COVER MAPPING

Mapping of vegetation types and land use/land cover was carried out on a scale of 1:50,000 using a hybrid classification technique that combines both visual and digital techniques. Forest types were mapped according to the Champion and Seth's forest-type classification system (Champion and Seth, 1968). Dry and wet season satellite data were used to delineate vegetation types by forest physiognomy. Wet season data were used to identify peak leaf growth and similarly dry season data for the identification of deciduous forests at the leaf fall stage. Satellite data on distinct time windows were used to account for the phenological variations of different vegetation types (See Satish et al., 2014 for more detailed methodology of this section).

22.2.5 LANDSCAPE ANALYSIS AND SPATIAL LANDSCAPE MODELING

Spatial Landscape Analysis and Modeling (SPLAM) is a geospatial modeling software developed by the Department of Space (DOS) that was used to perform a multicriteria spatial analysis (Jeganathan and Narula, 2006). The vegetation-type map was the primary input for the derivation of several landscape indices. SPLAM is a program generated that can perform the analysis of porosity, interspersion, fragmentation, juxtaposition, terrain complexity (TC), disturbance index, and biological richness. The generic binary image is the input in this software and the output will also be generated in the same format. Fragmentation, porosity, patchiness, patch density, interspersion, and juxtaposition are the main outputs. In the case of ecosystem uniqueness (EU) based on International Union for Conservation of Nature (IUCN) data, it is calculated based on the species composition of each type of vegetation and their importance, their value in terms of endemic-nature and their medicinal values in the ecosystem. Each species was assigned weights based on the IUCN status and weights for each type of vegetation were averaged according to the EU value for that type of vegetation. Finally, these values were arranged between 0 and 10 before feeding this value to the SPLAM software. The value of biodiversity is based on the potential economic uses of each type of vegetation. Different economic values were assigned to each species based on primary uses such as grazing, medicinal products, firewood, food, fuel, timber, charcoal, and any other uses. Ancillary road and settlement database was used to prepare a proximity buffer map. The contour map with an interval of 20 m was used to generate TC map. In addition, major deforestation sites, frequent fire areas, ground samples of forest degradation, invasive species locations, disturbance sources, and phytosociological data have been integrated into SPLAM to improve the accuracy in the identification of fragmentation, disturbance, and biological richness areas in NBR. It has the flexibility to change the size of variable grid. All of these maps were integrated with the field data on species richness, EU, and biodiversity value to generate a map of areas categorised as per disturbance index and biological richness (Jeganathan and Narula, 2006; Roy et al., 2012).

22.2.5.1   FRAGMENTATION INDEX (FI)

Fragmentation is defined as the number of forest and nonforest type patches per unit area. It helps to identify the number of forest patches within the landscape. For computing forest fragmentation, input needs to be reclassified into two classes, namely, forest and nonforest resulting in a new spatial data layer. An $n \times n$ ($n = 500$ m) user grid cell is a spatial layer with the criteria for the number of forest patches within the grid cell. Iteration is repeated by moving the grid cell across the entire spatial layer. The output layer with patch numbers is derived and associated with this a Look-Up Table (LUT) is generated which keeps the normalized patch data per cell within the range of 0–10. The 30-m spatial resolution of the land cover map was used to characterize the fragmentation levels around the pixel of the forest.

22.2.5.2   BIOTIC DISTURBANCE (BD)

The basic information required here is the disturbance caused by human interaction with the forest and its surroundings. The variable buffering of the radial distance from the point of disturbance shall be affected by imposing a condition that "the greater the distance, the lesser the weightage," in other words, weightage shall be inversely proportional to the distance. The same criteria apply to the line, the point, and the polygon. Field-specific disturbance from the point of disturbance, invasive species, frequent fire locations, and historic deforested locations were considered for buffering.

22.2.5.3   TERRAIN COMPLEXITY (TC)

This parameter has been computed using SRTM DEM. The appropriate contour level of 20 m is used for digitization. During the creation of SRTM DEM, drainage, river, and streams were considered to be at lower and higher controls. The SRTM DEM, which is a spatial layer, was used for computing of grid cell-based variance for the entire spatial coverage in the raster mode. Output data is generated with a LUT representing the standardized (0–10) data elements for the complexity of the terrain.

22.2.5.4   ECOSYSTEM UNIQUENESS (EU)

Floristic representativeness is one of the important parameters for the uniqueness of the ecosystem. This is decided based on native and non-native species found in the region. The uniqueness of the ecosystem was determined based on field data, species (flora and fauna) composition, area size, contiguity, landscape importance, and critical habitat value of the patch. To assign weights, the interpreter would refer field data.

The most representative of this type of vegetation was the region with either indigenous or endemic species but no exotic (introduced either deliberately or accidentally) species (Dhar et al., 1997).

22.2.5.5   DISTURBANCE INDEX (DI)

The *DI* is calculated by adopting linear combinations of defined parameters based on adaptive probabilistic weightages. Details of the method used to calculate different parameters were provided in SPLAM (Roy et al., 2013). The final spatial data will be scaled to a range of 0–100 for final preparation.

$$DI = \sum_{i=1}^{n} (Frag_i \times Wt_i 1 + PO_i \times Wt_i 2 + P_i \times Wt_i 3 + Int_i \times Wt_i 4 + Jux_i \times Wt_i 5 + BD_i \times Wt_i 6)$$

where *DI* is disturbance index, *Frag* is fragmentation index, *PO* is porosity, *Int* is interspersion, *BD* is biotic disturbance, *Jux* is juxtaposition, *P* is patchiness, and *Wt* is weightages.

22.2.5.6   BIOLOGICAL RICHNESS (BR)

Landscape BR is defined as the function of EU, species diversity (H'), biodiversity value (BV), TC, and disturbance index (DI). Main parameters such as *EU* and *H'* are derived from phytosociological data as well as from the list of species in the IUCN category for rare, endangered and threatened (RET) species. The evergreen forest types were evaluated on the terrain and finally, the biological richness (BR) is calculated as follows:

$$BR = \sum_{i=1}^{n} (DI_i \times Wt_i 1 + TC_i \times Wt_i 2 + SR_i \times Wt_i 3 + BV_i \times Wt_i 4 + EU_i \times Wt_i 5)$$

where *BR* is biological richness, *DI* is disturbance index, *TC* is terrain complexity, *SR* is species richness, *BV* is biological values, *EU* is ecosystem uniqueness, and *Wt* is weightages.

The output of the BR spatial layer is scaled (0–100) for the final presentation. A schematic representation of geospatial modeling for the identification of biologically rich areas has been provided in Figure 22.1.

The output scale values (0–100) of FI, DI, and BR were further classified as low, medium, and high for easy handling and presentation of data.

22.2.6   ANALYSIS, SPATIAL MODELING, AND QUERYING

The SPLAM software package has been used to access the database, as described above, and to facilitate the analysis, spatial modeling, and user-based query of the database. Remote sensing data was imported from the image processing platform into the GIS domain after the classification of different forest types. This was one of the most important primary data sources to be used for further analysis and modeling. The software design has kept the provision to process the data in either vector or raster mode, as the requirements are of both types. The entire data handling was taken care of under the software package.

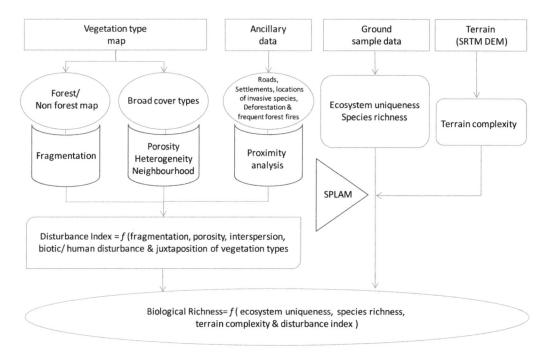

FIGURE 22.1 Schematic representation of the geospatial model for biodiversity characterization at landscape level using SPLAM.

22.3 RESULTS

22.3.1 *VEGETATION TYPE AND LAND COVER*

The vegetation class defined in the NBR is wet evergreen, semi-evergreen, moist deciduous, dry deciduous, riparian (riverine) forest, shola, savannah, reed brakes, scrub, grasslands, and plantations. The vegetation types and land use/land cover of the NBR for the year 2012 are shown in Figure 22.2. The total natural vegetation coverage of the NBR is estimated to be 4878.22 km², accounting for approximately 79.80% of the total geographical area (Figure 22.1). Among the forests, the dominant type was a moist deciduous forest of 1613.00 km² (25.86%) of the total NBR area, followed by a dry deciduous forest (25.18%), a wet evergreen forest (11.91%), and semi-evergreen forest (5.87%). Scrub formations occupy a proportionately significant area, which is approximately 6.37% of the geographical area.

The important vegetation types delineated through remote sensing are presented below.

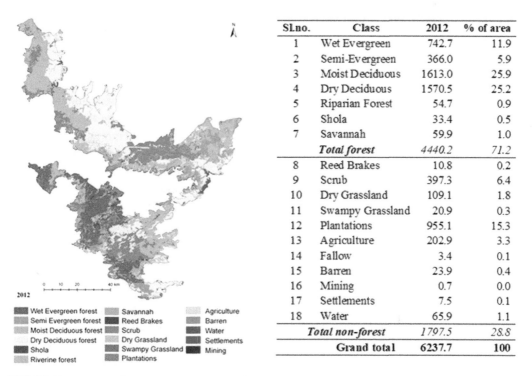

SLno.	Class	2012	% of area
1	Wet Evergreen	742.7	11.9
2	Semi-Evergreen	366.0	5.9
3	Moist Deciduous	1613.0	25.9
4	Dry Deciduous	1570.5	25.2
5	Riparian Forest	54.7	0.9
6	Shola	33.4	0.5
7	Savannah	59.9	1.0
	Total forest	*4440.2*	*71.2*
8	Reed Brakes	10.8	0.2
9	Scrub	397.3	6.4
10	Dry Grassland	109.1	1.8
11	Swampy Grassland	20.9	0.3
12	Plantations	955.1	15.3
13	Agriculture	202.9	3.3
14	Fallow	3.4	0.1
15	Barren	23.9	0.4
16	Mining	0.7	0.0
17	Settlements	7.5	0.1
18	Water	65.9	1.1
	Total non-forest	*1797.5*	*28.8*
	Grand total	**6237.7**	**100**

FIGURE 22.2   Vegetation types and land cover: 2012 (left), and areal extent of vegetation and land cover of NBR (right).

22.3.1.1   *TROPICAL WET EVERGREEN FOREST*

It is a dense forest with evergreen trees, more than 25-m tall. The plants of this forest form a multi-layered vertical structure: the shrubs cover the layer closer to the ground (under storey), followed by the short trees (middle storey) and the tall trees (top storey) and are enriched with climbers, epiphytes, orchids, ferns, and mosses. The luxuriant evergreen forests are distributed in Silent valley national park. It is characterized by species such as *Cullenia excelsa, Elaeocarpus tuberculatus, Palaquium ellipticum, Mesua ferrea, Calophyllum elatum, Cinnamomum zeylanicum, Holigarna arnottiana, Hopea parviflora, Litsea wightiana, Gordonia obtusa, Callicarpa lanata.*

22.3.1.2   *TROPICAL SEMI-EVERGREEN FOREST*

This forest is intermediate between tropical evergreen forests and moist deciduous forests. Dominant trees include both evergreen and deciduous. Semi-evergreen forests are found in high-altitude areas of Wayanad, Silent Valley, Mukurti National Park, and Mudumalai Sanctuary. It consists of *Artocarpus hirsuta, Ailanthus integrifolia, Macaranga peltata,*

Toona ciliata, Bombax ceiba, Pterospermum rubiginosum, Syzygium cumini, Mangifera indica, Radermachera xylocarpa, Palaquium ellipticum, Protium serratum, Schleichera oleosa, Holoptelea integrifolia, Mesua ferrea, and *Pterocarpus marsupium.*

22.3.1.3   MOIST DECIDUOUS FOREST

According to Champion and Seth (1968), these forests are part of the Southern system of moist deciduous forests. It is the most prevalent type of vegetation in the NBR. Moist deciduous forests are distributed in areas where rainfall exceeds 1200 mm per year. Climax moist deciduous forests are mostly found in the western parts of Mudumalai, Wayanad, and Silent Valley. These forests show the prominence of *Terminalia paniculata, Terminalia alata, Madhuca indica, Albizia odoratissima, Pterocarpus marsupium, Lannea coromandelica, Diospyros montana, Terminalia bellirica, Buchanania lanzan, Syzygium cumini,* and *Bridelia retusa.*

22.3.1.4   DRY DECIDUOUS FOREST

These forests are characterized by seasonal leaf shedding, typically found in Nagarahole, Mudumalai, and Bandipur National Parks. These forests cover an area of 1570.53 km^2. Common associations of this type are *Anogeissus latifolia, Terminalia alata, Diospyros mantana, Buchanania lanzan, Lagerstroemia parviflora, Tectona grandis,* and *Phyllanthus emblica.* It occurs mostly at an altitude of less than 400 m. *Lantana camara,* an invasive species, has been identified as an indicator of disturbances in these forests. Its bushes make thickets and render the area unsuitable for most wild animals. It occupies most of the open forest lands in Nagarahole, Mudumalai, and Bandipur.

22.3.1.5   REED BRAKES

The species of Reed bamboo (*Ochlandra travancorica*) forms almost pure formations in degraded evergreen and deciduous forests. These reed brakes are often very dense and form tangled thickets. It is considered to be the succession/degradation stage of the once climax forests. Reed bamboos have invaded degraded wet evergreen forests in the Agasthyamalai Biosphere Reserve (Dutta and Reddy, 2016). This study revealed the progression of temporal invasion and the spatial distribution of spread.

22.3.1.6   RIPARIAN FOREST

The Riparian forest is an ecologically unique type of forest, intermingled in deciduous systems, wherever streams and rivers flow over longer periods than surroundings. It covers an area of 54.74 km^2. The species that make up such forests include *M. indica, Diospyros malabarica, Glochidion zeylanicum, Pongamia pinnata, Mangifera sylvatica, Terminalia arjuna, Oroxylum indicum, Trema orientalis, Memecylon umbellatum*, etc.

22.3.1.7 SHOLA

Shola forests are southern montane wet temperate forests. It is a unique patchy forest confined to sheltered valleys, hollows depressions, and surrounded by grasslands in high altitude areas (>1000–1700 m) of the NBR. It consists of stunted short, bold evergreen trees that are unable to regenerate in open areas due to a lack of tolerance to fire and frost. It is characterized by dominant species including *Calophyllum polyanthum, Actinodaphne hookeri, Litsea stocksii, Cinnamomum sulphuratum, Rhododendron arboreum* subsp. *nilagiricum, Gaultheria frangitissima, Psychotria nilgiriensis, Litsea wightiana, Symplocos foliosa, Mahonia leschenaultii, Neolitsea cassia, Syzygium tamilnadensis, Meliosma simplicifolia, Syzygium calophyllifolium, Gardneria ovata, Myrsine wightiana*. A large portion of these forests is protected by the Mukurthi National Park.

22.3.1.8 SAVANNAH

It is ecologically homogeneous grassland on which woody plants are more are less evenly distributed. In the NBR, savannahs can be found in the forest and grassland transition zone and occupy an area of 59.88 km² (0.96%) of the total geographical area.

22.3.1.9 SCRUB

It covers an area of 397.26 km². It is dominated by shrubs and grasses. Although very few small trees (mostly <5 m in height) are sporadically distributed throughout the scrub. The scrub is the result of an anthropogenic degradation (in some areas) or an edaphic system (due to natural factors such as climate and geology). The most frequent species in the scrub are *Erythroxylum monogynum, Canthium parviflorum, Randia dumetorum, Ziziphus mauritiana, Maytenus emarginata*, and *Opuntia dillenii*.

22.3.1.10 GRASSLANDS

Generally, grasslands are mixed with barren areas, fringes of deciduous forests, and shola. It is distributed over an area of 109.07 km². Swampy grasslands are naturally found in nearby water bodies and these habitats are critically dependent on fluctuations in natural water levels. These grasslands can be found seasonally or permanently. Grasslands are composed of species such as *Apluda mutica, Themeda triandra, Pogostemon erectum, Cyperus triceps, Fimbristylis falcata, Rumex dentatus, Carex* sp., *Scirpus tuberosus, Cyperus bulbosus, Eleocharis*. It covers an area of 20.91 km² (0.34% of the total vegetation area).

22.3.1.11   OTHER NONVEGETATION CLASSES

In the NBR 3.25% of the area is under agriculture (202.90 km²). The other classes mapped together were barren land (0.38%), water (1.06%), and settlement (0.12%), which had a relatively much better distinction and ease of interpretation as far as the accuracy of the current scheme was considered.

22.3.2   LANDSCAPE ANALYSIS

Landscape metrics quantify changes in the geometry of the area, the size of the classes, the fragmentation of the land cover, etc., and thus quantify the relationship between the spatial characteristics of the patches, the class of patches, or the entire landscape with the ecological processes. The study area has large evergreen forest patches surrounded by several small patches of semi-evergreen, moist deciduous, and secondary evergreen forests, that have developed due to the long history of working in these areas. The intactness of these evergreen forests is quantified in terms of fragmentation, porosity, and heterogeneity of the land cover of around 250 m of large evergreen patches. It was found that 42% of these evergreen forests are under medium porosity and fragmentation levels. The analysis of land cover along the 250 m periphery of large evergreen patches also clearly shows the degree of heterogeneity of the land cover. The secondary evergreen and moist deciduous forests constitute a large percentage of the periphery of these forests. Out of the four large patches selected, two patches, approximately 30–40% of the surrounding patch, are covered with disturbing secondary land cover types like reeds, open moist deciduous forests, scrubs, etc. The analysis shows the degree and the type of impact on the evergreen forest. The phytosociological analysis also showed that 25% of the species and individuals belong to the edge category.

22.3.2.1   FRAGMENTATION INDEX

Forest fragmentation of the NBR is presented in Figure 22.3. The study area has been analyzed and the level of segregation is categorized as low, medium, and high fragmentation.

22.3.2.1.1   Distribution of Forest Fragmentation over Zonation

Among the four zones of the NBR, the core and manipulation tourism are the least fragmented and show 81.9% and 87.5%, respectively. A medium level of fragmentation is equally distributed. High fragmentation was found in the restoration zone (28.6% of the total restoration zone area) followed by the manipulation-forestry zone (25.8%) (Table 22.1 and Figure 22.3). Large portion of highly fragmented forests are distributed in manipulation forestry (663 km²) and restoration (240.6 km²) zones. The classified map of the NBR FI was presented in Figure 22.3.

22.3.2.1.2 Distribution of Forest Fragmentation over PAs

Among the six PAs, Wayanad Wildlife Sanctuary is the least fragmented (98.5% of its total forest area) followed by Nagarahole National Park (88.4%) and Mudumalai Wildlife Sanctuary (80.4). Medium-level fragmentation is spread over Nagarahole National Park (9.7%) and Mudumalai Wildlife Sanctuary (8.6%) remaining PAs with an almost similar trend. High levels of fragmentation were found in Mukurthi National Park (78.8%) followed by Bandipur national park (19.5%), Silent Valley National Park (19.3%), and Mudumalai Wildlife Sanctuary (11.0%). Area-wise statistics showed that Bandipur National Park has a larger extent of high fragmentation (152 km^2), followed by Mukurthi National Park (61.9 km^2). Detailed information is provided in tabular and graphical form (Table 22.1 and Figure 22.3).

22.3.2.1.3 Distribution of Fragmentation over Forest Types

Forest fragmentation analysis was carried out for major forest types, PAs, and the NBR zone. Fragmentation analysis statistics were analyzed and categorized into high, medium, and low areas. The wet evergreen forest is the least fragmented with 83.5% in the low fragmentation category followed by the moist deciduous (77.8%), riparian (72.4%), semi-evergreen (69.2%), dry deciduous forests (68.6%) and Shola (24.4%). A medium degree of fragmentation can be seen in semi-evergreen (10.7%) and moist deciduous (10.7%). Shola forest is the most fragmented type of forest with 73.1% of its area. Excluding the Shola forests, dry deciduous forests are highly fragmented due to deforestation activities such as forest roads, frequent fire activities, and fire lines. Large areas (389.3 km^2) of dry deciduous forests are highly fragmented followed by moist deciduous (186.2 km^2) and semi-evergreen forests (74 km^2) (Table 22.1 and Figure 22.3).

22.3.2.2 DISTURBANCE INDEX

The disturbance index map is generated by fragmentation, porosity, juxtaposition, and biotic disturbance. Analysis of the disturbance index revealed that 30–40% of evergreen and semi-evergreen forests are under medium disturbance (Table 22.1 and Figure 22.3). Moist deciduous forests were also found with 40% of disturbance levels. This type of spatial delineation of disturbance levels in conjunction with ground data facilitates the selection of suitable forest patches for further analysis in terms of their endemicity, species diversity, economic potential, and the plan for appropriate conservation methods.

22.3.2.2.1 Distribution of Disturbance over Zonation

The disturbance index was carried out over the NBR zones. The core and manipulation forestry zones remain protected from high disturbance regimes at 42.1% and 43.8%

respectively due to effective conservation and mitigation of the disturbance levels. The medium level of disturbance is equally distributed in core, manipulation forestry, and restoration zones, though, in manipulation tourism zone is most evident with medium disturbance. The manipulation forestry zone is highly disturbed (with 27.6%) followed by restoration (25%) (Table 22.1 and Figure 22.3). The classified map of the disturbance index of the NBR was shown in Figure 22.3.

22.3.2.2.2   Distribution of Forest Disturbance over PAs

The PAs-wise disturbance analysis shows that low disturbance levels were found in Silent Valley National Park with 82.5% of its total forest area followed by Nagarahole National Park with 42.8% followed by Wayanad Wildlife Sanctuary-II (22.4%) and Mukurthi National Park (19.7%). A medium level of disturbance was observed in Mudumalai Wildlife Sanctuary (72.6%) followed by Bandipur National Park (70.1%) and Wayanad Wildlife Sanctuary-II (65.8%). Mukurthi National Park is mostly distributed by fires and grazing pressure and shows a 79.5% disturbance index followed by Wayanad Wildlife Sanctuary-I (32.5%), Bandipur National Park (19.9%) (Table 22.1 and Figure 22.3).

22.3.2.2.3   Distribution of Disturbance over Forest Types

Area statistics for different forest types have shown that low disturbances were found in wet evergreen forests (91.3%) followed by semi-evergreen (56.2%), moist deciduous (45.6%), shola (25.2%), and riparian forests (18.8%).The medium disturbed areas are conspicuous in the riparian (61.1%), dry deciduous forests (56.8%), and moist deciduous forests (39.8%). High disturbances were observed in the Shola-grassland ecosystem (73.4%) followed by dry deciduous (23.1%), riparian (20.1%), and semi-evergreen (19.8%) (Table 22.1 and Figure 22.3).

22.3.2.3   BIOLOGICAL RICHNESS

To delineate biological richness zones, the endemicity, species diversity, the economic potential, and the complexity of the terrain were combined with the disturbance image. Based on the phytosociological analysis and published information, it was found that the percentage of endemic species as well as individuals is high in the evergreen forest. Also, the remaining forest types, a few endemic species are found only with a large number of individuals. Hence, the distribution of endemic species itself indicates the ecological uniqueness and relative intactness of the evergreen forest and disturbance levels in other forest types. In addition, the overall species diversity is high in evergreen forests followed by semi-evergreen and moist deciduous forests. Furthermore, the economic potential estimated as a function of different economic uses is shown to be high in moist deciduous forests.

The degree of variability in the above parameters is combined with the spatially generated disturbance levels to map the biological richness. The biological richness analysis has shown that around 82% of evergreen forests are under high biological richness zones and 18% are under very high biological richness (Table 22.1 and Figure 22.3). These richness zones that are delineated using multilayer analysis should be useful for prioritizing conservation activities.

22.3.2.3.1 *Distribution of Biological Richness Areas over Zonation*

Core and manipulation forestry zones of the reserve were observed at 42.1% and 43.8% respectively, of high biological richness areas. The manipulation tourism zone is observed with 72.4% of the area falling under the medium biodiversity richness class, followed by the core (46.1%) and the restoration zone (35.4%). Low biodiversity richness was found in the manipulation tourism zone (27.6%) followed by restoration (25.0%) (Table 22.1 and Figure 22.3). The classified map of the biological richness of the NBR is presented in Figure 22.3.

22.3.2.3.2 *Distribution of Biological Richness Areas over PAs*

Among the six PAs, Silent Valley National Park has high biological richness (82.5%) followed by Nagarahole National Park (42.8%) and Wayanad Wildlife Sanctuary-II (22.4%), Low biological richness areas were identified in Mukurthi National Park (79.5%) followed by Wayanad Wildlife Sanctuary-I (32.5%) and Bandipur National Park (19.9%) (Table 22.1 and Figure 22.3).

22.3.2.3.3 *Distribution of Biological Richness over Forest Types*

Biodiversity richness value was assigned based on ground sample measurements integrated with landscape spatial properties, including importance index and forest density. The biological richness areas were found over the major forest types, 91.3% of wet evergreen forests are under high biological richness followed by semi-evergreen (56%) and moist deciduous (45.6%). Medium and low biological richness areas were found in dry riparian forests (61.1%) followed by deciduous forests (56.8%). The percentage of low biological richness was observed in the Shola-grassland complex followed by dry deciduous forests (23.1%) and riparian forests (20.1%) (Table 22.1 and Figure 22.3).

22.4 DISCUSSION

22.4.1 *FRAGMENTATION*

The core zone of the biosphere is the least fragmented followed by manipulation tourism. High fragmentation was found in the restoration zone followed by the manipulation forestry zone. The PAs-wise fragmentation analysis shows that the Wayanad Wildlife Sanctuary shows the lowest level of fragmentation followed by the Nagarahole National Park and Mudumalai Wildlife Sanctuary. Medium-level fragmentation is almost similar in all PAs. A high level of

TABLE 22.1 Spatial Distribution and Levels of Fragmentation, Disturbance, Biological Richness Indexes Over, Zonation, PAs, and Forest Types.

	Level of fragmentation (%)						Level of disturbance (%)						Level of biological richness (%)					
	Low		Medium		High		Low		Medium		High		Low		Medium		High	
	Area (km²)	Area (%)	Area (km2)	Area (%)	Area (km2)	Area (%)	Area (km2)	Area (%)	Area (km2)	Area (%)	Area (km2)	Area (%)	Area (km2)	Area (%)	Area (km2)	Area (%)	Area (km2)	Area (%)
Zonation																		
Core	1047.5	81.9	77.2	6	154.7	12.1	866.6	42.1	246.6	46.1	166.3	11.8	150.8	11.8	589.8	46.1	538.8	42.1
Manipulation forestry	1659.8	64.7	242.8	9.5	663	25.8	1083	43.8	709.8	28.7	773.2	27.6	707	27.6	735.7	28.7	1123.3	43.8
Manipulation tourism	278.1	87.5	21.1	6.6	18.5	5.8	216.9	21.4	67.2	72.4	33.7	6.3	19.9	6.3	230	72.4	67.9	21.4
Restoration	526.2	62.5	75.2	8.9	240.6	28.6	407.5	39.6	192.5	35.4	242.2	25	210.5	25	298.1	35.4	333.6	39.6
Protected Areas																		
Nagarahole National Park	469.7	88.4	51.6	9.7	10	1.9	412.9	42.8	91.6	53.7	27	3.4	18.1	3.4	285.6	53.7	227.7	42.8
Wayanad Wildlife Sanctuary-I	54.7	98.5	0.8	1.4	0.1	0.1	16.1	14.9	17	52.6	22.3	32.5	18	32.5	29.1	52.6	8.3	14.9
Wayanad Wildlife Sanctuary-II	139.3	97.6	2.8	1.9	0.6	0.4	94.1	22.4	24.9	65.8	23.5	11.8	16.8	11.8	93.8	65.8	31.9	22.4
Mudumalai Wildlife Sanctuary	258.9	80.4	27.9	8.6	35.5	11	218.6	16.5	56.7	72.6	47.1	10.9	35.1	10.9	234	72.6	53.2	16.5
Bandipur National Park	605.3	77.7	22.2	2.8	152	19.5	470.4	9.9	161.4	70.1	147.3	19.9	155.2	19.9	546.4	70.1	77.5	9.9
Mukurthi National Park	16.1	20.6	0.5	0.7	61.9	78.8	6.1	19.7	25.3	0.8	47.2	79.5	62.5	79.5	0.7	0.8	15.5	19.7
Silent Valley National Park	174.1	71.4	22.8	9.3	47.1	19.3	131.4	82.5	59.3	0.7	53.4	16.9	41.1	16.9	1.6	0.7	201.4	82.5
Forest types																		
Wet evergreen	629.1	83.5	60.4	8	63.5	8.4	476.5	91.3	169.6	0.9	107.1	7.8	58.8	7.8	7.1	0.9	687.3	91.3
Semi-evergreen	254.5	69.2	39.2	10.7	74	20.1	151.9	56.2	100.3	24.1	115.7	19.8	72.7	19.8	88.6	24.1	206.6	56.2
Moist deciduous	1261.6	77.8	173.1	10.7	186.2	11.5	886.7	45.6	412.4	39.8	321.9	14.7	238	14.7	644.6	39.8	738.4	45.6
Dry deciduous	1084.2	68.6	107.1	6.8	389.3	24.6	884.4	20	335.9	56.8	360.5	23.1	365.4	23.1	898.6	56.8	316.9	20
Riparian forest	37.5	72.4	4.1	8	10.2	19.6	22.4	18.8	14.9	61.1	14.4	20.1	10.4	20.1	31.6	61.1	9.7	18.8
Shola	6.7	24.4	0.7	2.5	20.2	73.1	1.4	25.2	8.7	1.4	17.4	73.4	20.2	73.4	0.4	1.4	7	25.2

fragmentation was found in Mukurthi National Park due to scattered patches of shola forest. The Bandipur National Park forests are highly fragmented. The wet evergreen forest is least fragmented, followed by the moist deciduous, riparian semi-evergreen, dry deciduous forests, and Shola. Shola forest is the most fragmented type in fact, this high natural fragmentation and is due to the natural and scattered distribution of forest patches influenced by extreme topographic, climatic and edaphic factors. The study identified that the dry deciduous forest is highly fragmented due to high human disturbances including deforestation driven by structure the forest roads, frequent fires activities, and fire lines. Roy and Tomar's (2000) study identified fragmentation and porosity in the Garo hills of Meghalaya. A study in Aravallis found that the broad-leaved hill forest is less fragmented than other types due to relatively inaccessible areas or under protection (Harikrishna et al., 2014).

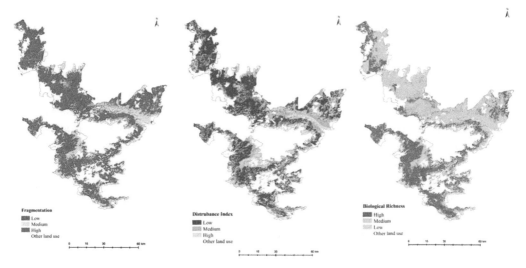

FIGURE 22.3   Spatial distribution of fragmentation, disturbance, and biological richness indexes in NBR.
Source: Authors

22.4.2   *DISTURBANCE REGIMES*

The disturbance index map was produced based on the integration of various landscape indices and anthropogenic disturbances including deforestation and degradation sites, invasive species locations, roads, settlements, forest juxtaposition, and interspersion. The zonation wise analysis showed that there was a high disturbance in the manipulation forestry zone (27.6% of its total area) and the restoration zone (25%). Usually, these zones have a high level of human intervention which has caused a high level of disturbance (i.e., invasions, fire, fragmentation, and deforestation, etc.) in the respective zones. Medium and low disturbances were found in manipulation tourism (72.4%) and core (42.1%), respectively. Analysis over PAs noted low disturbances in the Nagarahole National Park, followed by Mudumalai Wildlife Sanctuary and Silent Valley National Park. Mukurthi National Park and Wayanad Wildlife Sanctuary are the most disturbed. In Mukurthi National Park, disturbances are caused by fires, dams, plantations, and degradation, while in Wayanad Wildlife Sanctuary, disturbances are caused by deforestation due to commercial plantation expansion and plant invasions (e.g., *Senna spectabilis*). The wet evergreen forest is least disturbed followed by dry deciduous and moist deciduous forests. High disturbances are recorded in the dry grasslands of the shola-grassland ecosystems followed by semi-evergreen, riparian, and dry deciduous forests. High disturbance is caused by frequent fire activities in these shola-grassland ecosystems. High disturbances are mainly in the form of habitat degradation and loss due to fuelwood collection and small wood cutting, bamboo cutting, lopping, overgrazing, fire, and spread of invasive alien species. Most of the forest fringes and open forests are invaded by alien species, that is, *Lantana camara*, *Chromolaena odorata*, *Imperata cylindrica*, and *Parthenium hysterophorus*. Roy and Tomar (2000) generated disturbance maps for Meghalaya, the high disturbance levels were found in Garo hills, medium, and low disturbances were found in Kashi and Jaintia hills, respectively. Harikrishna (2014) observed that a low disturbance regime in more than 45% of the forested areas of Aravallis.

22.4.3 BIOLOGICAL RICHNESS

Determination of the biological richness at the landscape level is based on the uniqueness of the ecosystem, the richness of the species, the degree of TC, and the low disturbance index. Core and manipulation forestry zones of the reserve were observed with high biological rich areas, medium biodiversity richness areas were found in the core and restoration zones. Low levels of biodiversity richness areas were observed in the manipulation tourism zone, the human disturbance including the tourism, collection of firewood, non-timber forest products, and prescribed burning is quite common in this zone. Among the six PAs, Silent Valley National Park is dwelling with high biological richness areas followed by Nagarahole National Park and Wayanad Wildlife Sanctuary. Low biological richness areas were observed in Mukurthi National Park. In fact, the Mukurthi has shola forest patches that hold high biological richness just like evergreen forest, but are highly fragmented, of lesser size and grassland (low species diversity) is the predominant land cover in this National Park, and often affected by grazing and fires, and these factors kept under low biological richness class. Six major forest types, four zones and six NBR PAs were found with different levels of diversity, endemicity, and disturbance regimes. Biological richness map was generated as a function of the integrated presence of all the parameters indicating a spatial distribution of high, medium, and low biological richness areas. These biological richness areas can be prioritized for conservation and bioprospecting purposes as the high biological richness areas identified in the study area consist of high levels of species diversity, endemism, and low disturbance. High biological richness areas were found in wet evergreen forests followed by semi-evergreen, moist deciduous, and dry deciduous forests. Medium and low biological richness areas were found in dry deciduous forests and riparian forests. A study conducted by Roy and Tomar (2000) found that the Jaintia and Kashi hills have high biological richness areas compared to the Garo Hills of Meghalaya. A study in Rajasthan found that remarkably high biological richness areas were found in Mt. Abu, Kumbalgarh, Phulwari, Sitamata, and Sariska (Harikrishna et al., 2014).

22.5   SUMMARY AND CONCLUSIONS

Vegetation-type, fragmentation, disturbance, and biological richness maps are the primary input for ecological analyses as they represent mapping units that can be linked to biodiversity and the status of the ecosystem. Conservation prioritization of landscapes is essential for effective management of the BRs. The anthropogenic pressure on the forests of the NBR has decreased due to the management efficiency reflected in the rate of deforestation. The results indicate the remarkable effectiveness of the conservation of forests, after the declaration of the reserve in 1986. However, threat indicators such as forest fires and invasive alien species need to be considered for holistic management of forest resources and biodiversity. Contiguous forests are found to be porous due to the presence of dry grasslands, bamboo brakes, and semi-evergreen forests. Evergreen forests are found in low and medium porous conditions. Species diversity (6.58) was found to be high in evergreen forests. Approximately 85.8% of evergreen forests are found with low and medium disturbance levels. Biological richness maps generated based on disturbance index levels in conjunction with endemicity, naturalness, and species variability facilitate the prioritization of conservation plans.

ACKNOWLEDGMENTS

The work was carried out under national project "Inventorisation and Monitoring of BRs in India using remote sensing and GIS technology," supported by the Ministry of Environment, Forests, and Climate Change, Government of India. The Authors are thankful to Chief Wildlife Wardens and Field Directors, Nilgiri Biosphere Reserve, and the State Forest departments of Karnataka, Kerala, and Tamil Nadu for their permission and facilities to carry out the fieldwork.

CONFLICT OF INTEREST

Authors declared no conflict of interest.

KEYWORDS

- **remote sensing**
- **fragmentation**
- **disturbance**
- **biological richness**
- **modeling**

REFERENCES

Asner, G. P.; Rudel, T. K.; Aide, T. M.; DeFries, R.; Emerson, R. A Contemporary Assessment of Change in Humid Tropical Forests. *Conserv. Biol.* **2009,** *23* (6), 1386–1395.

Benedick, S.; Hill, J. K.; Mustaffa, N.; Chey, V. K.; Maryati, M.; Searle, J. B.; Schilthuizen, M.; Hamer, K. C. Impacts of Rain Forest Fragmentation on Butterflies in Northern Borneo: Species Richness, Turnover and the Value of Small Fragments. *J. Appl. Ecol.* **2006,** *43* (5), 967–977.

Benitez-Malvido, J. Impact of Forest Fragmentation on Seedling Abundance in a Tropical Rain Forest. *Conserv. Biol.* **1998,** *12* (2), 380–389.

Benítez-Malvido, J.; Martínez-Ramos, M. Impact of Forest Fragmentation on Understory Plant Species Richness in Amazonia. *Conserv. Biol.* **2003,** *17* (2), 389–400.

Champion, S. H.; Seth, S. K. *A Revised Survey of the Forest Types of India,* **1968.** Government of India, New Delhi.

Clark, J.; Darlington, J.; Fairclough, G. J. *Using Historic Landscape Characterisation: English Heritage's Review of HLC; Applications 2002-03.* English Heritage and Lancashire County Council, **2004.**

Dhar, U.; Rawal, R. S.; Samant, S. S. Structural Diversity and Representativeness of Forest Vegetation in a Protected Area of Kumaun Himalaya, India: Implications for Conservation. *Biodivers. Conserv.* **1997,** *6* (8), 1045–1062.

Díaz, S.; Fargione, J.; Chapin III, F. S.; Tilman, D. Biodiversity Loss Threatens Human Well-Being. *PLoS Biol.* **2006,** *4* (8), e277.

Didham, R. K.; Hammond, P. M.; Lawton, J. H.; Eggleton, P.; Stork, N. E. Beetle Species Responses to Tropical Forest Fragmentation. *Ecol. Monogr.* **1998,** *68* (3), 295–323.

Dutta, K. Reddy, C. S. Geospatial Analysis of Reed Bamboo (*Ochlandra travancorica*) Invasion in Western Ghats, India. *J. Indian Soc. Remote Sens.* **2016,** *44* (5), 699–711.

Ferreira, L. V.; Laurance, W. F. Effects of Forest Fragmentation on Mortality and Damage of Selected Trees in Central Amazonia. *Conserv. Biol.* **1997,** *11* (3), 797–801.

Force, M. T. *Criteria and Guidelines for the Choice and Establishment of Biosphere Reserves*; UNESCO MAB Report Series, Paris, **1974**; vol 22.

Gaston, K. J. *Geographic Range Limits of Species*; The Royal Society London, 1009.

Harikrishna, P. Reddy, C. S.; Singh, R.; Jha, C. S. Landscape Level Analysis of Disturbance Regimes in Protected Areas of Rajasthan, India. *J. Earth System Sci.* **2014,** *123* (3), 467–478.

Imbernon, J.; Branthomme, A. Characterization of Landscape Patterns of Deforestation in Tropical Rain Forests. *Int. J. Remote Sens.* **2001,** *22* (9), 1753–1765.

Jeganathan, C.; Narula, P. *User guide for Spatial Landscape Modelling (SPLAM) software package in ArcGIS using ArcObjects and Visual basic: Operational Manual*; Indian Institute of Remote Sensing (National Remote Sensing Agency) Department of Space, **2006.**

Koh, L. P.; Gardner, T. A. Conservation in Human-Modified Landscapes. In *Conservation Biology for All*; Oxford University Press, **2010**; pp 236–261.

Laurance, W. F.; Camargo, J. L.; Luizão, R. C.; Laurance, S. G.; Pimm, S. L.; Bruna, E. M.; Stouffer, P. C.; Williamson, G. B.; Benítez-Malvido, J.; Vasconcelos, H. L. The Fate of Amazonian Forest Fragments: A 32-Year Investigation. *Biol. Conserv.* **2011,** *144* (1), 56–67.

Laurance, W. F.; Ferreira, L. V.; Rankin-de Merona, J. M.; Laurance, S. G. Rain Forest Fragmentation and the Dynamics of Amazonian Tree Communities. *Ecology* **1998,** *79* (6), 2032–2040.

Mahapatra, A. K.; Tewari, D. D. Importance of Non-timber Forest Products in the Economic Valuation of Dry Deciduous Forests of India. *Forest Policy Econ.* **2005,** *7* (3), 455–467.

Meena, R. L.; Verma, Y. L.; Korvadiya, V. T.; Pathak, B. J.; Kshatriya, A. R. Kachchh Biosphere Reserve Management (A Plan for Protection, Conservation, Research and Development); Gujarat State Forest Department, Gujarat, **2005.**

Menhinick, E. F. A Comparison of Some Species-Individuals Diversity Indices Applied to Samples of Field Insects. *Ecology* **1964,** *45* (4), 859–861.

Menon, S.; Bawa, K. S. Applications of Geographic Information Systems, Remote-Sensing, and a Landscape Ecology Approach to Biodiversity Conservation in the Western Ghats. *Curr. Sci.* **1997,** *73* (2), 134–145.

Morris, R. J. Anthropogenic Impacts on Tropical Forest Biodiversity: A Network Structure and Ecosystem Functioning Perspective. *Philos. Trans. Royal Soc. B Biol. Sci.* **2010,** *365* (1558), 3709–3718.

Munsi, M.; Areendran, G.; Ghosh, A.; Joshi, P. K. Landscape Characterisation of the Forests of Himalayan Foothills. *J. Indian Soc. Remote Sens.* **2010,** *38* (3), 441–452.

Nagendra, H.; Lucas, R.; Honrado, J. P.; Jongman, R. H.; Tarantino, C.; Adamo, M.; Mairota, P. Remote Sensing for Conservation Monitoring: Assessing Protected Areas, Habitat Extent, Habitat Condition, Species Diversity, and Threats. *Ecol. Indic.* **2013,** *33*, 45–59.

Newmark, W. D. Tropical Forest Fragmentation and the Local Extinction of Understory Birds in the Eastern Usambara Mountains, Tanzania. *Conserv. Biol.* **1991,** *5* (1), 67–78.

Palni, L. M. S.; Rawal, R. S.; Rai, R. K.; Reddy, S. V. *Compendium on Indian Biosphere Reserves: Progression During Two Decades of Conservation*; GB Pant Institute of Himalayan Environment & Development, **2012**.

Pettorelli, N.; Safi, K.; Turner, W. *Satellite Remote Sensing, Biodiversity Research and Conservation of the Future*; The Royal Society, **2014**.

Ragusa-Netto, J. Exploitation of *Erythrina dominguezii* Hassl. (Fabaceae) Nectar by Perching Birds in a Dry Forest in Western Brazil. *Brazilian J. Biol.* **2002,** *62* (4B), 877–883.

Reddy, C. S.; Sreelekshmi, S.; Jha, C. S.; Dadhwal, V. K. National Assessment of Forest Fragmentation in India: Landscape Indices as Measures of the Effects of Fragmentation and Forest Cover Change. *Ecol. Eng.* **2013,** *60*, 453–464.

Roy, P. S.; Kushwaha, S. P. S.; Murthy, M. S. R.; Roy, A.; Kushwaha, D.; Reddy, C. S.; Behera, M. D.; Mathur, V. B.; Padalia, H.; Saran, S. *Biodiversity Characterisation at Landscape Level: National Assessment*; Indian Institute of Remote Sensing: Dehradun, India, **2012**; vol 140.

Roy, P. S.; Kushwaha, S. P. S.; Roy, A.; Karnataka, H.; Saran, S. Biodiversity Characterization at Landscape Level Using Geospatial Model.; *Anais XVI Simpósio Brasileiro de Sensoriamento Remoto–SBSR, Foz Do Iguacu,* PR, Brasil, **2013**; pp 3321–3328.

Roy, P. S.; Tomar, S. Biodiversity Characterization at Landscape Level Using Geospatial Modelling Technique. *Biol. Conserv.* **2000,** *95* (1), 95–109.

Satish, K.V. (2017) Phytodiversity Characterisation at Landscape Level in Nilgiri Biosphere Reserve, Western Ghats Using Remote Sensing and GIS. Unpublished PhD thesis. Andhra University, Visakhapatnam.

Satish, K. V.; Saranya, K. R. L.; Reddy, C. S.; Krishna, P. H.; Jha, C. S.; Rao, P. P. Geospatial Assessment and Monitoring of Historical Forest Cover Changes (1920–2012) in Nilgiri Biosphere Reserve, Western Ghats, India. *Environ. Monit. Assess.* **2014,** *186* (12), 8125–8140.

Smith, R. L.; Smith, T. M. In *Ecology and Field Biology*; Wesley Longman: Inc., San Francisco, CA, 2001.

Turner, W. Sensing Biodiversity. *Science* **2014,** *346* (6207), 301–302.

UNESCO. *Biosphere Reserves-United Nations Educational, Scientific and Cultural Organization.* https://en.unesco.org/biosphere **2020**.

Weatherspoon, C. P.; Skinner, C. N. An Assessment of Factors Associated with Damage to Tree Crowns from the 1987 Wildfires in Northern California. *Forest Sci.* **1995,** *41* (3), 430–451.

Whitmore, T. C. Potential Impact of Climatic Change on Tropical Rain Forest Seedlings and Forest Regeneration. In *Potential Impacts of Climate Change on Tropical Forest Ecosystems;* Springer, **1998**; pp 289–298.

Wulder, M. A.; Hall, R. J.; Coops, N. C.; Franklin, S. E. High Spatial Resolution Remotely Sensed Data for Ecosystem Characterization. *BioScience* **2004,** *54* (6), 511–521.

Xiao, X.; Boles, S.; Liu, J.; Zhuang, D.; Liu, M. Characterization of Forest Types in Northeastern China, Using Multi-Temporal SPOT-4 VEGETATION Sensor Data. *Remote Sens. Environ.* **2002,** *82* (2–3), 335–348.

Zambrano, J.; Salguero-Gómez, R. Forest Fragmentation Alters the Population Dynamics of a Late-Successional Tropical Tree. *Biotropica* **2014,** *46* (5), 556–564.

CHAPTER 23

PHYSIOGRAPHY, CLIMATE, AND HISTORICAL BIOGEOGRAPHY OF SRI LANKA IN MAKING A BIODIVERSITY HOTSPOT

SANDUN J. PERERA[1] and R. H. S. SURANJAN FERNANDO[2]

[1]*Department of Natural Resources, Faculty of Applied Sciences, Sabaragamuwa University of Sri Lanka, Belihuloya, Sri Lanka*

[2]*Centre for Applied Biodiversity Research and Education, Pallekumbura, Doragamuwa, Kandy, Sri Lanka*

ABSTRACT

The island of Sri Lanka, located immediately southeast of the Indian mainland, shares the continental shelf of the India–Sri Lanka (Deccan) plate. The recognition of Sri Lanka and southern India as a globally unique biogeographic entity goes back to Alfred Russel Wallace, who identified the Ceylonese subregion of Oriental region in global zoogeography, which with much refinement of the boundary to include only moist forests have given rise to the Western Ghats and Sri Lanka biodiversity hotspot, a global conservation priority today. This delimitation recognizes the concentration of endemic assemblages of flora and fauna, currently under the threat of being lost owing directly to human impacts. Being a continental island with lesser degree of isolation compared with an oceanic one, together with its topographic and climatic heterogeneity and the eventful geological history, Sri Lanka has assembled a unique biota with an exceptionally high diversity and endemicity for its size, unlike isolated oceanic islands which usually possess highly endemic biotas depauperate in diversity. Sri Lanka experienced species colonization events from the mainland after its separation from India ~50 Ma, initially during the Oligocene followed by dispersal events in the late Miocene, Pliocene, and Pleistocene epochs during glacial periods with lowered sea level (at least five times over the last 500,000 years), followed by isolation-driven speciation during interglacial periods. The northward drifting of the Deccan plate starting deep from global south to the northern hemisphere just above the equator over the last 200 million years have given it so many different elements of biota, which took refuge in Sri Lanka especially during the volcanism of Deccan traps. Remarkable combination of climatic and topographic isolation mechanisms within the island have resulted in insular speciation making endemism values as high as 98% in freshwater crabs and 89% in land snails and amphibians. The remaining fragmented natural habitat patches of the island hence serve as only places for those species to strive on Earth, making this crucible of speciation in southwestern jungles of Sri Lanka a stronghold for biodiversity conservation amidst the challenges placed by the climate change.

23.1   SRI LANKA, AN INTRODUCTION

Being an island of moderate size (65,610 km²), Sri Lanka possesses an exceptionally high biodiversity and endemism for its size. As a continental island in the Indian Ocean, sharing the same continental shelf of the Deccan plate with India, it is located immediately southeast of the mainland between latitudes 5°55' and 9°51' N and longitudes 79°41' and 81°53' E. Sri Lanka is separated from India through the narrow (minimum width of about 53 km) and shallow (less than 100 m deep) Palk Strait. This island located just north of the equator makes the last land mass at its longitude until Antarctica in the south. Sri Lanka harbors a great diversity of Ecosystems, representing several types of global biomes, that is, forests, savanna, thicket or a similar semi-arid thorny scrubland vegetation and grasslands in addition to inland aquatic and coastal vegetation. These harbor many species of flowering plants and vertebrates including a remarkable percentage endemism, together with abundant nonvascular and thalloid flora and an enormous invertebrate fauna are still largely undescribed. Island biotas serve as natural laboratories for biogeographers, although it is a well-established fact that oceanic islands tend to have higher levels of endemism with impoverished biotas. However, the case in Sri Lanka is quite unique, primarily being a continental island with a historical land-bridge connection with the mainland, and hence with a much richer species composition compared with its size, although possessing an exceptionally high percentage endemism among both the flora and fauna. This chapter discusses the making of such a biodiversity and endemicity within this unique island of Sri Lanka.

23.2   SRI LANKA AND THE WESTERN GHATS OF INDIA IN GLOBAL BIOGEOGRAPHY

The first map of global zoogeographic regions developed was based on the then knowledge on taxonomy, distribution patterns, and endemism of mammals by Alfred Russel Wallace, the father of biogeography in 1876. It included a spatial entity centered around Sri Lanka. Wallace (1876) delimited the Ceylonese subregion, south of the Indian subregion, as one of the four subregions within the Oriental region of global biogeography. Ceylonese subregion included the entire island of Sri Lanka, then known as *"Ceylon"* together with the southern Deccan Plateau of India, including the Western Ghats south of the Tapti river, Eastern Ghats south of the Godavari river, and the dry southern coastal plains of India. Although the exact boundaries of the Ceylonese region were questioned and further refined by other biogeographers, the understanding on a unique Sri Lankan biota distributed south of the broader Indian biota remained unchanged. Among such treatments, even before Wallace, Blyth (1871) recognized the Cinghalese subregions including the Western Ghats and the western coast of India together with southern half of Sri Lanka, as a subregion of the Australasian Region. However, northern Sri Lanka belonged to the Indian subregion together with the rest of the India, which he classified under the Ethiopian Region. This recognition of the African elements of Indian biota also adds light to one other important aspect of the origins of broader regional biota also agreed by Wallace, however scarce in the Western Ghats and Sri Lanka (WG&SL) according to Blanford (1876). Hence, Blanford's regionalization cotemporaneous with Wallace recognized Malabar province to include the Western Ghats and southern Sri Lanka, and the Indian province to include northern Sri Lanka and India except Western Ghats, similar to Blyth, however, both provinces belong to the Oriental region. Later, Blanford (1901) divided Sri Lanka into two separate tracts, namely, the northern Ceylon tract related to the Carnatic tract of peninsular India and the southern hill tract related to the Malabar Coast tract. Although Wallace's (1876) mammal-based zoogeographical regions were influenced by early descriptions of global avifaunal regions by Philip Lutley Sclater (Sclater, 1858), Sclater's (1891) subregional delimitations suggested Ceylon to be combined with the Indian peninsula to form a single Indian subregion. Similarly, his son's work based on the similarity of the mammalian assemblages suggested that Wallace's Ceylonese subregion does not warrant such a recognition separated from the Indian subregion (Sclater, 1896; Sclater and Sclater, 1899). However, with the advent of molecular taxonomy, current knowledge suggests otherwise, especially with recent investigations on herpetofauna and other taxa with poor dispersal ability. Such disputed regional affinities and subregional classifications are being resolved in various subsequent biogeographic treatments. While reviewing zoogeographical and phytogeographical regionalizations proposed for nearly a century from Wallace, Udvardy (1975) proposed eight biogeographical realms, in which the entire Indian subcontinent belonged to the Indo-Malayan realm. Among the 28 biogeographical provinces recognized within the Indo-Malayan realm, two provinces represent Sri Lanka; Ceylonese rainforest and Ceylonese monsoonal forest, whereas the southern Deccan Plateau of India was recognized with three provinces: Malabar rainforest, Coromandel, and Deccan Thorn Forest. Ceylonese rainforest and Malabar rainforest represented the tropical humid forest biome, while the Ceylonese monsoonal forest, Coromandel, and Deccan Thorn Forest represented the tropical dry or deciduous forest biome. However, recent conservation biogeographical delimitation of 36 global biodiversity hotspots (Myers et al., 2000) identifies the Western Ghats with the entire island of Sri Lanka as a hotspot, without considering the biomic differences in southern and northern Sri Lanka.

23.3 GLOBAL CONSERVATION PRIORITIES AND BIODIVERSITY HOTSPOTS

Several attempts have been made to set spatial conservation priorities globally toward the allocation of the limited resources and funds for conservation (Wilson et al., 2006). Some notable approaches included the Biodiversity Hotspots, where species- and especially endemic-rich ecosystems under imminent threat are selected (Myers et al., 2000; Mitter-meier et al., 2004); Endemic Bird Areas (Bibby et al., 1992; Stattersfield et al., 1998) and Global 200 Ecoregions were identified based on the representativeness (Olson and Dinerstein, 1998); Tropical Wilderness Areas aimed at maintaining evolutionary process (McCloskey and Spalding, 1989) and Megadiversity Nations, countries that are unusually rich in biodiversity (Mittermeier et al., 1997).

The aforementioned concept of biodiversity hotspots was originally proposed by Myers (1988, 1990) who identified the first 10 of such hotspots. The concept was globally accepted with 25 hotspots, with the seminal paper titled *Biodiversity hotspots for conserva-tion priorities* (Myers et al., 2000). Four years later, it was expanded into 34 hotspots in the book *Hotspots revisited: earths biologically richest and most endangered ecoregions* (Mittermeier et al., 2004), which was further expanded by the works of Williams et al. (2011) and Noss et al. (2015) into 36 such hotspots by today. Despite the numerous limita-tions of the biodiversity hotspots approach (see Ginsberg, 1999; Brummitt and Lughadha, 2003; Ovadia, 2003; Orme et al., 2005; Ceballos and Ehrlich, 2006; Grenyer et al., 2006), it can hardly be denied that the hotspots approach has made a significant impact on global conservation. Conservation International and the MacArthur foundation adopted Myers' hotspots as the guiding approach for their conservation investment, making them receive the largest financial support for any single conservation strategy (Myers, 2003; Myers and Mittermeier, 2003). Biodiversity hotspots have been designated using a biological criterion of floristic endemism, that is, the area must contain at least 0.5% of the world's vascular plant species (1 500 species) as endemics, in addition to a threat criterion where 70% or more of the primary vegetation of the area must have been lost due to human impact (Myers et al., 2000; Mittermeier et al., 2004). This means that animal endemism per se has not been critical for hotspot selection. Nevertheless, vertebrates are being assumed to follow a similar pattern of endemism to higher plants (Myers et al., 2000).

To comply with the endemism criteria of biodiversity hotspot identification, WG&SL harbors 3049 endemic plants (Kumar et al., 2004), although the vertebrates have added more color to the picture with at least 820 species (Gunawardene et al., 2007), while those numbers have considerably increased over the last 15 years of active taxonomic research in both the components of the hotspot, especially on the herpetofauna and freshwater fish. This hotspot harbors a remarkably rich assemblage of endemic plants, reptiles, and amphibians among other biodiversity hotspots; whereas in the threat criteria, only 6.8% of primary vegetation remains anthropogenically untransformed in WG&SL (Myers et al., 2000), while those primary habitats are heavily fragmented (Legg and Jewell, 1995; Naggs and Raheem, 2005; Raheem et al., 2008), placing it among 11 hyper-hot biodiversity hotspots (Myers et al., 2000). However, the main highlight of this chapter is on the southern isolated component of this hotspot, the island of Sri Lanka.

23.4 SPECIES RICHNESS AND ENDEMISM OF SRI LANKA IN A NUTSHELL

As stated above, Sri Lanka has an exceptionally high species richness and endemism for its size. From an early comparative assessment of species diversity per 10,000 km² among countries in Asia (China, India, Indonesia, Myanmar, Thailand, Philippines, Vietnam, Malaysia, and Sri Lanka), Baldwin (1991) reported Sri Lanka to hold the highest density of species richness, based on the biota known by then. Sri Lanka ranked first in average species richness per 10,000 km^2 for flowering plants, amphibians, reptiles, and mammals with a considerable gap respectively with Bangladesh that ranked second for flowering plants, and Malaysia ranked second for the latter two taxa. Even for the bird fauna, Sri Lanka was only marginally second to Malaysia. Those early assessments have recently been supported with much more field data and molecular evidences identify Sri Lanka as a herpetofaunal hotspot (Meegaskumbura et al., 2002; Erdelen, 2012; Perera et al, 2019a) as well as a center of mammalian phylogenetic endemism (Rosauer and Jetz, 2014). The recently published biodiversity profile of Sri Lanka (MoMD&E, 2019) provides updated information on the island's high species richness and endemicity. The overall percentage endemicity of 29% for flowering plants, 51% for vertebrates, and 50% for relatively well-known invertebrate groups is noteworthy. The endemicity levels of 98% for freshwater crabs, 89% for land snails, 80% for millipedes, 78% for scorpions, 68% for damselflies, and 49% for spiders are the highest among invertebrate groups, while some other relatively well-known invertebrate groups, such as termites (26%), dragonflies (24%), ants (14%), bees (14%), and butterflies (13%) are reported with lower degrees of endemicity. Among the vertebrate groups, high endemicity percentages of 89% for amphibians, 68% for reptiles, and 63% for freshwater fish indicate a high degree of isolation-driven speciation among the groups with lesser dispersal ability, while the highly mobile mammals (20%) and birds (15%) show lower endemicity.

Furthermore, the endemic biota of Sri Lanka shows an uneven distribution, with the endemism being largely concentrated in the southwestern wet zone (SWWZ), both in lowland and montane habitats (Gunatilleke et al., 2008). More than 60% and 34% of the Sri Lankan endemic flowering plants are found respectively in the lowland and montane SWWZ (Peeris, 1975; Gunatilleke et al., 2008). Similarly, Kotagama (1989) estimated more than 70% of the vertebrate species endemic to the island are inhabiting the SWWZ. Further, many insular endemics in Sri Lanka are restricted to narrow ranges, while Gunatilleke et al. (2008) report on 108 microendemic species representing both flowering plants and vertebrates, each was recorded only from a single site, again from the lowland wet and hill zones of Sri Lanka, representing flowering plants, mammals, agamid lizards, amphibians, and freshwater fish species. A narrow endemic taxon is largely restricted to a single biogeographic zone and considered here to cover less than one-fourth of the island (see Perera et al., 2018 for more details of narrow endemism), while the micro/point endemism is defined here as being restricted to a single population within a very restricted range, virtually a single habitat patch < 10 km^2 (Caesar et al., 2017); hence, narrow endemism in an area includes the richness of both the micro and narrow endemic species. All in all, Sri Lanka serves as a local center of endemism within the WG&SL biodiversity hotspots, especially in relation to taxa with lower dispersal ability, which have considerably diverged from their Indian sister lineages than we previously thought (Bossuyt et al., 2004). More on the species endemism in Sri Lanka is discussed with examples later.

23.5   CONTRIBUTORS IN MAKING SRI LANKA A BIODIVERSITY HOTSPOT

The extraordinary biodiversity and endemism on the island of Sri Lanka are a result of an interesting combination of various ecological and historical factors. The ecological biogeographic factors can be elaborated into the geographical location of the island, its physiography made by the unique topography, geology and soils, and the internal variation of the island's climate. Sri Lanka's biogeographic history is influenced mainly by its tectonic movements and the islands' affinity to the Indian mainland, as shown by the various biogeographical elements presently found in the islands biota. A number of theories and hypotheses in historical biogeography help solving the puzzle of the makings of Sri Lankan biodiversity.

Factors highlighted above in making the shelf island of Sri Lanka a global biodiversity hotspot are discussed in the following sections. However, they explain only the endemism criteria of the hotspot concept, together with the high species richness on the island, while the threat criteria fulfilled by anthropogenic pressures will not be elaborated in this chapter except for a brief discussion in the concluding remarks.

23.6   ECOLOGICAL BIOGEOGRAPHY OF SRI LANKA

23.6.1   SRI LANKA'S LOCATION IN THE WORLD

Current location of Sri Lanka within the tropical belt just a few degrees north of the equator as a continental island within the intertropical convergence belt causes: (1) high incipient solar radiation and (2) the annual climatic cycles determined by the monsoon. Both the above factors have directly influenced the high species richness in Sri Lanka. Being closer to the equator, the island experiences no seasons and receives ample incident solar radiation throughout the year to support a high biomass, year-round supply of food, and high rates of molecular evolution, hence inevitably a higher diversity of life (Bromham and Cardillo, 2003). Further, being an island, its biota has been biologically isolated allowing unique adaptations and speciation. Sri Lanka, being a continental or a land-bridge island that shares the same continental shelf with India, has allowed many species to colonize it from the mainland and establish new populations, which have further undergone insular speciation, causing higher species endemism. This aspect will be discussed in detail under the historical biogeography of the island. Additionally, Sri Lanka is the southernmost island at its longitude with the vast Indian Ocean south of it until Antarctica, while there is no major land mass in the Bay of Bengal until the Malay Peninsula to the East and in the Arabian Sea until Africa to the West along its latitude, except for a few small stepping stone islands. The monsoonal climate determined by the location of Sri Lanka within the intertropical convergence belt is discussed separately.

23.6.2   GEOLOGY AND SOILS

When the physiography of Sri Lanka is concerned, the geology, soils, and the topography/relief have caused different types of vegetation together with different assemblages of flora and fauna with a rapid turnover of species in a relatively small area. All in all, the physiography of Sri Lanka has triggered in situ speciation of new taxa by isolating populations within fairly narrow ranges, especially in the SWWZ of the island.

Geologically, a vast majority of the Sri Lankan landmass is comprised of Precambrian high-grade metamorphic rocks, while the rest is covered by a few different sedimentary sequences (Cooray, 1984; Chandrajith, 2020). Lithologically, based on the grade of metamorphism and isotope characteristics, Sri Lankan metamorphic terrain can be divided into three divisions: (1) the Highland Complex (central highlands and lowlands in the SWWZ as well as a northeasterly directed belt until Trincomalee), (2) Vijayan Complex (east of the Highland Complex), and Wanni Complex (west of the Highland Complex). Among the sedimentary sequences, limestone underlies the northern and northwestern coastal stretch of the island which has deposited during the Miocene period when the area was part of the continental shelf submerged under the sea, while a small outcrop of Miocene shale and limestone occur in the Southeastern Coast at Minihagalkanda, together with small Jurassic beds in few isolated patches (Tabbowa, Andigama, and Pallama) in the northwest of the island (Cooray, 1984; Chandrajith, 2020). Further, the Precambrian terrain also harbors a few igneous intrusions scattered here and there (Chandrajith, 2020).

According to Cooray (1984), all the Sri Lankan soils except on the Northwestern Coast are resulting from the weathering of the underlying Precambrian crystalline rock, while the soil in the Northwestern Coast and the Jaffna peninsula is calcareous due to the underlying Miocene deposits of sedimentary limestone. Eight major soil groups have been identified in Sri Lanka, corresponding mainly to parent material and to the pattern of rainfall that act as the main agent of rock weathering and soil formation (Cooray, 1984). More recent soil classifications used its morphology, mainly the soil profile and characters of each horizon leading to Great Soil Groups, while subsequent soil taxonomy uses WRB-FAO legend with soil series (Mapa, 2020). Six soil taxonomic orders can be found in Sri Lanka representing 118 soil series, which are grouped into the wet, intermediate, and dry climatic zones and further into up, mid, and low countries, with 33 soil series in the wet zone, 36 in the intermediate zone, and 54 in the dry zone (Mapa, 2020; Dassanayake et al., 2020a, 2020b, 2020c). In general, the red-yellow podzolic soils or the leached lateritic soils dominate the wet zone, while the reddish brown earths or the non-lateritic loamy soils are commonly found in the dry zone. Further, in terms of the altitude, highlands are dominated by reddish brown latosolic soils which are partially laterized or immature brown loams which are clayey, while lower reaches of river valleys are comprised on sandy alluvial soils, and coastal areas are rich in sandy regosol soils.

Divisions of the underlying metamorphic rocks and the resulting differences in soil composition may not have caused a marked difference in vegetation of those areas, resulting in a dearth of edaphically determined vegetation types in Sri Lanka, however, the limestone beds in the northwest have caused a unique vegetation and hence a characteristic biota in those areas.

23.6.3   TOPOGRAPHY AND RELIEF

The island can geomorphologically be considered in three erosional levels or low-relief plains which are referred to as the first peneplain or the coastal/lowland plain, second peneplain or the uplands, and the third peneplain or the highlands (Adams, 1929; Cooray, 1967; Vitanage, 1970; Chandrajith, 2020). The coastal lowland plain comprising mostly a flat and undulating landscape is the largest, covering a vast area especially in the north and east of the island and generally below 100 m in attitude (0–122 m; Cooray, 1967), while upland and highlands show a characteristic ridge and valley topography with a variety of geomorphological features (Chandrajith, 2020). Furthermore, the lowlands possess many eroded remnant hills and rock outcrops, which can sometimes reach up to above 600 m and more abundant in the eastern side (e.g., Maragala/Moneragala, Kokagala, and the Gal Oya hills), while inselbergs, such as Ritigala, Galgiriyakanda, and Doluwekanda occur north of the highlands. The steep rise from about 100–300 m up to the second peneplain starts from the inner edge of the lowland plain. This upland plain spanning from 365 m to 762 m (Cooray, 1967) can be recognized widely in eastern and northern parts of the country, while being narrower in the southern edge of the highlands. The next rise from upland plains to the highlands steeply escalates from about 900–1200 m, with the third peneplain or the highland plain undulating from 1529 m–1828 m, with highest peaks, such as Pidurutalagala (2524 m), Kirigalpotta (2370 m), Totapola (2359 m), Kikiliyamana (2239 m), Samanala Kanda/Sripada Peak/Adam's Peak (2150 m), and Namunukula (2035 m) ascending from there. The third peneplain is characterized by mountain chains and deep valleys, together with plateaus, such as the Maha Eliya or the Horton plains, Elk plains, Moon plains in Nuwara Eliya, Seeta Eliya plains, Agara Pathana, and the Hatton plateau. The central massif is arranged roughly in an anchor-shaped mountain ridges in which the base of the anchor marks the clear southern escarpment of the island from Samanala Kanda in the west through Kirigalpotta to Namunukula peak in the East, also making one of the highest precipices in the island known as the *World's End* with a sheer drop of above 800 m, although the central bar of the anchor are made by Totapola, Kikiliyamana, and Pidurutalagala peaks from south to north, and the northern end is marked by Ramboda-Pussellawa. Additionally, the Lunugala-Madulsima range lies on the eastern margin of the central mountains with two parallel ridges running along the north–south direction with a deep valley, whereas the Dolosbage hill cluster scattered on the western flank of the central mountains runs along the east–west direction.

The two upper peneplains together form three major mountain ranges within the south-central half of the island, namely, the central massif discussed above, and the Rakwana massif southwest of it and the Knuckles massif in the northeast. Sheer precipices similar to the *World's End* in southern flank and almost continuously along the southern boundary exists also in the eastern escarpments in Madulsima of the Lunugala-Madulsima range and in Pitawala Pathana of the Knuckles range. The Knuckles massif, rising to an altitude of about 1500 m with peaks like Gombaniya (1580 m) and Wamarapugala (1661 m) is separated from the central massif through long-term river erosion along Dumbara Mitiyawatha, a ~500 m deep broad valley created by the Mahaweli river. Similarly, the Rakwana massif with the highest peak of Gongala (1300 m) is separated from the central massif by broader Kalu and Walawe river valleys (for more details, see Figure 23.1).

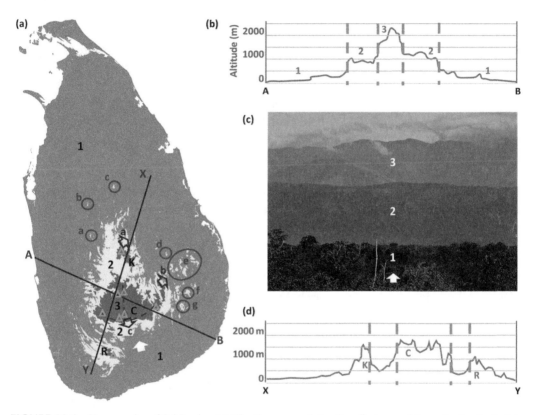

FIGURE 23.1 Topography of Sri Lanka: **(a)** The three peneplains (1 – first peneplain or the lowland plain that ranges between 0 and 122 m asl is marked here by the 300 m contour; 2 – second peneplain that ranges between 305 and 762 m asl is marked here by the 900 m contour; 3 – third peneplain that ranges between 1539 and 1828 m asl is marked above the 900 m contour); inselbergs in the lowland plains are marked by red circles (a – Doluwekanda, b – Galgiriyakanda, c – Ritigala, d – Kokagala, e – Gal Oya hills, f – Govinda Hela, and g – Maragala/Moneragala); three major mountain ranges separated by river valleys (K – Knuckles massif, C – Central massif, and R – Rakwana massif); highest peaks of the third peneplain are shown by light blue open triangles (Gombaniya in K, Gongala in R, and the anchor-shaped formation of ridges in C is shown by blue-dashed line with highest peaks – Samanala Kanda/Sripada Peak/Adam's Peak, Kirigalpotta, and Namunukula from west to east along the base of the anchor and Totapola, Kikiliyamana, and Pidurutalagala on the central ridge from south to north); sheer escarpment precipices at inland edges of the second peneplain (a – *Small World's End* escarpment in Pitawala Pathana of K, b – *Small World's End* escarpment in Madulsima, at the edge of the eastern extension of C, and c – *World's End* escarpment of the southern edge of C); **(b)** Cross-section of the island along A–B showing the three peneplains; **(c)** Kaltota escarpment rising from 1 to 2 and the *World's End* escarpment rising from 2 to 3 when viewed from southeast in the Udawalawe National Park on southern lowland plains as indicated by the white arrow (photo by Sandun J. Perera); **(d)** Cross-section of the island along X–Y showing broad river valleys between the three major mountain massifs (Mahaweli between K and C, and Kalu and Walawe between C and R).

Physiographic characters of the island with peneplains separated by gradual slopes from some directions and by steep escarpment precipices at some localities act as barriers for dispersion, especially of taxa with less mobility (Senanayake et al., 1977), ultimately causing isolation-driven speciation and hence the island's high species diversity and endemism. Recent findings of new species and emerging evidence from molecular studies suggest that the three major mountain ranges have isolated ancestral species within them showing very interesting narrow endemic evolutionary radiations; the ancestral species include the lizard genera of *Ceratophora* and *Cophotis* (Schulte et al., 2002; Samarawickrama et al., 2006; Karunarathna et al., 2020) and the *Pseudophilautus* shrub frogs (Meegaskumbura et al., 2002, 2019; Ellepola et al., 2021) as well as the land snail genus *Corilla* (Raheem et al., 2017). Further, the radiation of *Cnemaspis* day geckos in the island with many being narrow and especially microendemics shows the role played by highlands and isolated inselbergs rising from lowland plains in their evolution (Karunarathna et al., 2019; see Fernando, 2011 for more information on the biodiversity in three selected inselbergs Doluwekanda, Kokagala, and Maragala/Monaragala).

Furthermore, the south-central positioning of the third peneplain results in receiving a heavy rainfall, especially during the southwest monsoon creating a radial system of 103 rivers originating from highlands, also with a ring of waterfalls along its flanks especially along the southern escarpment. A dozen or so of major river basins are isolating their own biotic assemblages, especially of aquatic taxa (Chandrajith, 2020; see Pethiyagoda and Sudasinghe, 2021 for more details).

The origin of three peneplains were initially debated between Adams (1929) and Wadia (1945). Adams described the geomorphology of the island suggesting a slow vertical rising of the landmass exposing the three peneplains to erosion and denudation over successive times in geological history, making the highland peneplain the oldest, while Wadia has proposed the block uplift theory suggesting highlands as the youngest peneplain uplifted along faults. However, recent investigations by Vitanage (1970, 1972) suggest otherwise, explaining a differential up warping and down flexing at different locations over a long period by a series of vertical movements and horizontal thrusting respectively at wrench faults and lineaments. This phenomenon has caused the lowlands, uplands, and the highlands of the island to possess different combinations of slope and height characteristics (See Chandrajith, 2020 for an illustrative description of the orogeny of the geomorphology of Sri Lanka). These claims by Vitanage are consistent with that of Emmel et al. (2012) who suggest that Sri Lanka is one of the most stable shield regions in the post-Gondwanan geological history. However, it should be noted that the orogeny of the physiography and geomorphology of Sri Lanka, especially of its highlands is still being understood (Cooray, 1984; Pethiyagoda and Sudasinghe, 2021).

23.6.4   CLIMATE

Sri Lanka experiences a non-seasonal tropical climate with little variation of daylight hours throughout the year. Being within the intertropical convergence belt, the climate of the island is controlled mainly by the monsoon of south Asia, which is seasonally blowing across Sri

Lanka southwesterly during May–September and northeasterly during December–February (IMD, 2021). The climatic variation within the island is partly influenced by its topography with the south-central location of the highlands, which intercept both the above-mentioned monsoons causing differential patterns of precipitation in the SWWZ of the island and the broader Northeastern dry plains. The moisture laden wind of the southwestern monsoon which sweeps the Indian Ocean before entering the island causes an enormous amount of rainfall making the perhumid SWWZ with an annual average precipitation above 2500 mm, though it reaches above 5000 mm at places. Further, when monsoonal winds retreat in between the two seasons, convection causes two more inter-monsoon rainfall peaks in many parts of the island, the first in March–April and the second in October–November. Due to these inter-monsoonal rains, the wet zone receives a year-round rainfall with no marked dry season. This makes SWWZ of Sri Lanka the only region in Asia outside the Malay Archipelago with no dry season, permitting proper rain forest biota to thrive here. Northeasterlies blown above the Bay of Bengal before entering the island also bring a fair amount of water vapor that will again be intercepted by the highlands and distributed over a vast track of land, making the dry zone receive an annual precipitation less than 1800 mm per year, which is seasonal.

In general agreement, the monsoons partition the country into four major climatic zones, such as the SWWZ, vast dry zone in the north and east, a transitional intermediate zone in between the above two, and a semi-arid zone along rain shadows (see Figure 23.2; also see the map of annual rainfall isohyets in Pethiyagoda and Sudasinghe, 2021). The narrow belt of an intermediate zone is found with an average annual precipitation of 1800–2500 mm y^{-1}, which crosses the western coast between Chilaw and Negombo and the southern coast between Tangalla and Matara. Two semi-arid coastal belts are in the rain shadows of both the monsoons: (1) the Hambantota Coast in the southeast and (2) the Mannar to Jaffna coastline in the northwest, where both the monsoons are not really giving a significant rainfall, however they receive a seasonal inter-monsoon precipitation of about 1000 mm y^{-1}. Although the climatic zones of Sri Lanka are defined primarily based on the annual rainfall, a note on the temperature and relative humidity would support understanding the heterogeneity of habitats even within the same climate zone. The relative humidity in the morning is kept high throughout the island, usually above 70% in supporting a high taxonomic diversity, whereas it can be as high as 90% in the perhumid lowland wet zone (Punyawardena, 2009; Pethiyagoda and Sudasinghe, 2021). According to Punyawardena (2009), the average annual maximum and minimum temperatures vary between 29°C–38°C and 20°C–26°C throughout the dry zone, while the highest temperature is in the northern coastal belt. However, the temperature regime is not that simple in the intermediate and wet zones of Sri Lanka, where they show a high spatial as well as temporal variability providing a diversity of ecological niches for the biodiversity. In the lowland wet and intermediate zones, the average annual maximum and minimum temperatures vary between 29°C–35°C and 20°C–26°C, whereas the values decrease with increasing altitude ranging between 27°C–33°C and 18°C–23°C in the mid-country and 22°C–29°C and 13°C–18°C in the upcountry wet zone.

Being the only area south of the westward extension of the Kangar-Pattani line (Whitmore, 1984) in south Asia, and hence with no marked dry season and very high humidity, SWWZ of Sri Lanka supports rain forests similar to those in Southeast Asia dominated by the family Dipterocarpaceae (Gunatilleke et al., 2017; Pethiyagoda and Sudasinghe, 2021). It is believed that these perhumid conditions were created with the onset of the Asian monsoon in the late Miocene (Rajbhandary et al., 2011). Such aseasonal perhumid rainforest habitats have caused higher narrow endemism with new species adapted to narrower niches both among the flora and fauna, such as the Sri Lankan endemic radiations; the wet crevice dwelling endemic rock frog genus *Nannophrys*, the endemic monophyletic radiations of Platystictid damselflies and of freshwater crabs with restricted distributions and some very specialized species like the rot hole dwelling tree climbing crab *Perbrinckia scansor* (Bahir et al., 2005; Fernando et al., 2007; Bedjanič et al., 2016). Further, the recent climatic fluctuations during the Pleistocene and Holocene causing intermittent aridification events at least in the highlands as indicated by Premathilake and Risberg (2003), caused by fluctuations in the southwest monsoon regime must have influenced narrow endemism in montane cloud forests by isolating humid niches in contracting forest patches, as in the case of Afromontane forests along the southeastern escarpment of Africa (Perera et al., 2011).

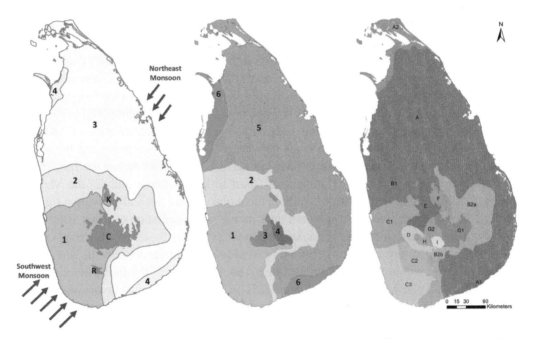

FIGURE 23.2 Climate and vegetation-driven zonation of Sri Lanka: **(a)** Climatic zones generated by interception of two monsoons of Asia by the highlands of Sri Lanka comprised of K – Knuckles massif, C – Central massif, and the R – Rakwana massif: 1 – Wet Zone, 2 – Intermediate Zone, 3 – Dry Zone, and 4 – Semi-Arid Zone. **(b)** Bioclimatic zones (Wijesinghe et al., 1993) combining the climate with topography: (1) Low and mid-country wet zone, (2) Low and mid-country intermediate zone, (3) Montane wet zone, (4) Montane intermediate zone, (5) Dry zone, and (6) Semi-arid zone, and **(c)** Nineteen Floristic regions of Sri Lanka indicating different floristic assemblages made by the habitat heterogeneity, initially adapted from Gaussen et al. (1964) by Ashton and Gunatilleke (1987) and updated by MoMD&E (2019), 16 floristic regions mapped here: A – Dry, A1 – Southern arid, A2 – Northern arid, B1 – Northern intermediate, B2a – Eastern intermediate, B2b – Hill savanna, C1 – Northern wet lowlands, C2 – Sinharaja and Ratnapura, C3 – Southern lowlands and hills, D – Foothills of Adam's peak and Ambagamuwa, E – Kandy and upper Mahaweli, F – Knuckles, G1 – Mountain intermediate, G2 – Mountain wet, H – Adam's peak, I – Horton plains, together with three unmapped narrow floristic regions, Coastal, Wet zone freshwater, and Other southern local floristic regions.

23.6.5 VEGETATION, BIOCLIMATIC ZONES, AND THE BIOGEOGRAPHIC REGIONALIZATION

According to Gunatilleke et al. (2008), the variety of the climate and topography together with the various types of soils found in certain areas have resulted in a surprisingly diverse array of different vegetation types within a relatively smaller area in Sri Lanka. The low country wet zone harbors the most luxuriant rain forests of the island, characterized by tall and multilayered canopies, together with marshy wetlands in the lowlands. However, this landscape has recently been cleared in large scale for human habitations, subsistence agriculture, and especially for export-oriented plantations of rubber, coconut, and lowland tea. What is left today from these rainforests is a fraction of what they were, while been severely fragmented by synanthropic habitats. Hill country wet zone, although being very small in size, harbors the most rugged and biogeographically interesting terrain of the island. The landscape is characterized by hill ranges, high elevation plains, and deep valleys. The vegetation is mainly comprised of montane cloud forests of subalpine nature, interspaced by montane grasslands, while the habitats get increasingly drier toward the eastern section of the hill zone. These habitats have faced the highest level of anthropogenic threat during the recent history primarily by tea-dominated export plantations, tobacco, and monoculture forest plantations. Most of the dry zone is now converted to synanthropic habitat variants such as secondary forests or sparse open forests, as much of these lands were recently brought under paddy cultivation through new irrigation schemes, though some of the irrigation tanks and canals been restored from the ancient hydraulic civilization. All in all, severe levels of deforestation and forest fragmentation during the last two centuries in Sri Lanka have made a profound influence on its endemic biodiversity, ranges of restricted species and their populations as elaborated in Reddy et al. (2017) and Raheem et al. (2008), and are further discussed below in the concluding remarks.

Gunatilleke et al. (2008) and MoMD&E (2019) summarized six major forest types in the island together with nine edaphic and five anthropogenic variants listed in the latter reference. Among them, tropical lowland wet evergreen forests (rain forests), tropical mid-elevation evergreen forests (submontane forests), and tropical montane evergreen forests (cloud forests) are found in the wet zone, respectively in lowlands, midlands, and highlands separated by the 900 m and 1500 m contours (Gunatilleke et al., 2008). Tropical moist mixed-evergreen forests (intermediate forests), tropical dry semi-evergreen forests (monsoon forests), and tropical arid semi-evergreen forests (thorn scrub forests) respectively dominate the intermediate, dry, and semi-arid zones. This island is also blessed with a variety of natural grasslands that were categorized into five types by Gunatilleke et al. (2008) and has been further elaborated into eight types by MoMD&E (2019), together with five edaphic and three anthropogenic variants. Montane upper wet *patana*, montane lower wet *patana*, and humid zone dry *patana* grasslands, as well as upland savannas can be found predominantly in the eastern aspects of mountains of the wet and intermediate zones, whereas summer zone dry *patana* grasslands, lowland savannas, dry grasslands (*damana*), and floodplain grasslands (*villu*) can be found in the intermediate and dry zones, together with *thalawa* grasslands and *kekilla* fernlands in the intermediate and wet zone, and the pasturelands in the dry zone exist as anthropogenic variants (Pemadasa, 1984,

1990; MoMD&E, 2019). Furthermore, a diversity of lentic and lotic inland aquatic ecosystems, coastal, and marine ecosystems and islets, other important natural ecosystems such as caves as well as manmade ecosystems, such as public parks and gardens, home gardens, and agro-ecosystems provide habitats to the diverse biota of Sri Lanka (Gunatilleke et al. 2008; MoMD&E, 2019)

In describing the vegetation of Sri Lanka, Whistler (1944) suggested that this island may be one of those few places in which the vegetation would be detailed so clearly by its physical geography (topography and climate), which was used by Senanayake et al. (1977) in delimiting faunal zones of Sri Lanka, with an emphasis on endemism. When the flora is concerned, Ashton and Gunatilleke's (1987) floristic regions adapted Gaussen et al. (1964). By intersecting climatic zones by the topography and also capturing habitat heterogeneity in the SWWZ, they mapped 15 different floristic regions, which were later updated by national experts on the flora, bringing it up to 19 floristic regions in the island (MoMD&E, 2019; see Figure 23.2). By simplifying the combined effect of the climate and topography on the biota of Sri Lanka, Wijesinghe et al. (1993) mapped six bioclimatic zones: (1) Low and mid country wet zone, (2) Low and mid country intermediate zone, (3) Montane wet zone, (4) Montane intermediate zone, (5) Dry zone, and (6) Semi-arid zone (see Figure 23.2).

Biogeographic regionalization or the mapping of spatial distribution patterns of unique assemblages of biotas in Sri Lanka goes long way back in the history. Studies on the patterns of plant distribution goes back to Trimen who first related those distributions to climate and topography (Pethiyagoda and Sudasinghe, 2017). Trimen (1885, 1886) initiated such phytogeographic studies on the plant endemism and floristic affinities of Ceylon, followed by J. C. Wills who wrote on the floras of hilltops and their narrow endemism (Wills, 1906, 1908, 1911) and proposed his age and area theory (Wills, 1915, 1922), while Ashton and Gunatilleke's (1987) floristic regions of Sri Lanka being the latest along the tradition. Similarly, a number of regionalization attempts have been placed by animal experts. After Blyth's (1871) and Wallace's (1876) initial attempts of identifying the uniqueness of Ceylonese faunal assemblages, Blanford (1876, 1901) and Ripley (1949, 1980) followed a similar tradition. The history of zoogeographic regionalization within Sri Lankan goes back to the avifaunal zones (Legge, 1880; updated by Kotagama, 1989) followed by mammalian zones (Eisenberg and McKay, 1970), and ichthyological zones (Senanayake and Moyle, 1982). The above zones were updated by respective expert groups during the preparation of the *Biodiversity Profile* for *Sri Lanka's Sixth National Report to the Convention on Biological Diversity* (MoMD&E, 2019), while introducing new zonations for amphibians and reptiles as well as for few selected groups of fairly well-known invertebrates, such as freshwater crabs, butterflies, Odonates, and land snails. Except in the case of MoMD&E (2019), all above proposals for faunal zones were established predominantly based on an individual's intuitive discernment, relying on their field explorations, while those in MoMD&E (2019) were still intuitive despite the consensus of many experts on the fauna. Further, almost all above regionalizations were built upon using the climatic zones of the island as a template, not fulfilling the prime objective of bioregionalization to identify patterns of actual distribution of biota. The intuitive approach lacks repeatability and uniform spatial scale while the resulting zones may be biased and too coarse for conservation applications. So far, Sri Lanka lacks numerical bioregionalization attempts to understand the evolutionary histories of areas and to inform conservation planning. However, preliminary numerical biogeographic analyses employing objective and quantitative methods were recently being tested for a few taxonomic groups, when distribution data are available with adequate quality and completeness at an appropriate equal-area grid scale, for flowering plants, birds and reptiles (Ramdhani et al., 2010; Perera et al., 2015; Wijesundara et al., 2015; Perera et al., 2019a, 2019b), although such analyses for Sri Lanka are still in their infancy.

23.7 HISTORICAL BIOGEOGRAPHY OF SRI LANKA

23.7.1 GEOLOGICAL AFFINITY WITH THE INDIAN MAINLAND

Although now been extinct, mammalian mega fauna in the Sri Lankan paleobiodiversity together with the "*Balangoda*" man (*Homo sapiens*) have dispersed to Sri Lanka during the Pleistocene Epoch through the broad land-bridge connecting the island with the Indian mainland (Deraniyagala, 1958; Manamendra-Arachchi et al., 2005; Rodrigo and Manamendra-Arachchi, 2020). However, molecular studies on less mobile animal groups by Bossuyt et al. (2004) revealed unexpectedly high local endemism within Sri Lanka, with very limited genetic interchange although the island had at least intermittently been connected to India until ~11,000 years ago (Vaz, 2000). Nevertheless, the prolonged influence from Indian biota to the Sri Lankan biodiversity is inevitable as the island has always been connected to the mainland throughout the post-Gondwanan times. It is evident by the presence of an unusually large mammal like the Asian Elephant to thrive in an island like Sri Lanka, representing an endemic subspecies (*Elephas maximus maximus*). Even today, the direct distance between Sri Lanka and India is just 53 km at the narrowest point of the Palk Straight, where the continental shelf rarely exceeds 70 m in depth. The depth of sea is even less than 10 m in most locations along the 30 km long Rama's Bridge, a chain of limestone shoals connecting the Mannar island of Sri Lanka and Ramsewaran island of India. Past biotic dispersals from India to Sri Lanka occurred mostly during the intermittent glacial periods, as the sea level has dropped below 70 m at least five occasions during the last 500,000 years, exposing the Palk isthmus wider than 100 km (Rohling et al., 1998; Bossuyt, et al., 2004; Sudasinghe et al., 2018). However, it should also be noted that a considerable amount of Sri Lanka's biota is non-Indian, which is explained below with the tectonic history of Sri Lanka and the elements found in its biota.

23.7.2 TECTONIC HISTORY

Although Sri Lanka forms a part of the Asian continent today, its biota has diverse biogeographic elements including those representing the ancient supercontinent Gondwanaland as well as the Southeast Asia. Those biogeographic elements are discussed next while the history of Sri Lanka's journey starting from the southern polar region of the Earth to its current location, which made it to pick some of its diverse biotic elements is discussed here.

Sri Lanka formed a part of the supercontinent Gondwanaland located in the southern hemisphere of the Earth as far as 500–600 million years ago (Ma) toward late Proterozoic to Paleozoic Era, when Indian (Deccan) plate was sandwiched between the African–Malagasy plate and the Antarctic plate in the Gondwanaland (Yoshida et al., 1992; Meert and Woo, 1997; Chetty, 2017). The Deccan plate consists of peninsular India, south of the present Indo-gangetic plain together with Sri Lanka at its southeastern tip. Geological evidence suggests that the supercontinent Gondwanaland started breaking up around 180–167 Ma, during which the Indian plate parted from other plates over successive rifting (Storey et al., 1995; Chatterjee et al., 2013). Antarctica–Australia started separating from Madagascar–Seychelles–India around 132 Ma, while Sri Lanka, which was connected to Antarctica from its current eastern coastline was separated from the Antarctic plate about 128 Ma, joining the Indian plate. Sri Lanka had been connected to India during northeastward drifting until about 50 Ma (Yoshida et al., 1992; Reeves, 2014; Chatterjee et al., 2013; Pethiyagoda and Sudasinghe, 2021). A Sri Lanka microplate was formed during this separation from Antarctic plate due to a failed rift, which made the microplate to rotate anticlockwise with respect to India until about 120 Ma, bringing it to its current orientation (Ratheesh-Kumar et al., 2020; Pethiyagoda and Sudasinghe, 2021). Madagascar was separated from the African plate around the same time, that is, ~121 Ma and Seychelles–India–Sri Lanka then drifted away from Madagascar ~90 Ma, while the Deccan plate was broken from Seychelles ~65 Ma, during its northward drifting as an island across the Tethys Sea, while sea floor spreading south of it created the Indian Ocean (Rabinowitz et al., 1983; Storey et al., 1995; Storey, 1995). It is noteworthy that the climate of the Indian plate underwent a fast transformation from temperate to tropical during its northward journey, presumably impacting some of the Gondwanan biota it was carrying. Furthermore, the K/T boundary ending the Cretaceous Period around 66 Ma was concurrent to the separation of India–Sri Lanka plate form Seychelles, while an extended period of volcanism was experienced in the northwestern section of India–Sri Lanka plate (Deccan Trap) during the same time. Deccan Trap lava flows have caused much of the Gondwanan biota on the Deccan plate to go extinct, except in the perhumid refugia in southern region of the current WG&SL (Ashton and Gunatilleke, 1987; Gunatilleke et al., 2017; Pethiyagoda and Sudasinghe, 2021). The Deccan plate had its first contacts with the Asian landmass of the northern hemispheric supercontinent Laurasia around 57 Ma in the Cenozoic era (toward the end of Paloecene Epoch), whereas the time of the completion of its hard collision has been disputed. However, it must have happened within the Eocene Period, closing the Tethys Sea, and raising the Himalayan mountain chain and the Tibetan Plateau, also opening up avenues for exchanging of tropical biota between the Asia and India (Aitchison et al., 2007; van Hinsbergen et al., 2011; Pethiyagoda and Sudasinghe, 2021). Much of the biota we know in Sri Lanka today are the results of colonizations during the Cenozoic era. According to Guleria (1992), Morley (2000), Kent et al. (2008); Zachos et al. (2008), Wan et al. (2009), Cohen et al. (2013), and Klaus et al. (2010, 2016), it started with the dispersal of Asian taxa into India corresponding to the tropical perhumid climate in the late Paleocene (66-56 Ma), Eocene (56-34 Ma), and Oligocene (34-23 Ma) up to the middle of Miocene (23-5.3 Ma) until ~15 Ma. Miocene climatic optimum was followed up by the

Miocene aridification together with global cooling, which was later reverted to an optimum climate in late Miocene toward the Pliocene (5.3-2.6 Ma) after the inception of the Asian monsoon. Then the glacial and interglacial cycles of the Ice Age occurred through the Pleistocene (2.6 Ma to 11,700 year ago).

Among the biotic dispersions from Asia into Indian plate after accretion to the Eurasian mainland, which continued all the way south to Sri Lanka, the ancestors of the toad genus *Duttaphrynus* that show an endemic radiation in WG&SL, as well as the frog genus *Lankanectes*, representing a relict and species depauperate deep lineage, the subfamily Lankanectinae endemic to Sri Lanka, have dispersed very early, possibly in Oligocene to early Miocene, causing them to show higher taxonomic level endemism in WG&SL (Dubois and Ohler, 2001; Frost et al., 2006; Van Bocxlaer et al., 2009; Vijayakumar et al., 2019; Pethiyagoda and Sudasinghe, 2021). The ancestors of the shrub frog genus *Pseudophilautus* that dominate the Sri Lankan amphibian fauna have made their journey in the early Oligocene (~31 Ma; Meegaskumbura et al., 2019). Ancestors of a similarly interesting monophyletic radiation of Platystictid damselflies may have dispersed to Sri Lanka around the same time before being isolated in sky islands of the Western Ghats and Southwestern Sri Lanka over the next 20 million years or so (Bedjanič et al., 2016).

Oligocene stands out as a period of a large number of plant and animal dispersals between India and Sri Lanka due to the opening of the Palk Isthmus (Pethiyagoda and Sudasinghe, 2021). However, Sri Lanka was separated from mainland India during the early Miocene due to marine transgression (~20 Ma; Jacob, 1949; Cooray, 1967; Lajmi et al., 2019); hence, the dispersal of taxa to Sri Lanka was restricted. Even when the land-bridge appeared due to lowered sea level during the next glacials, the prolonged aridification together with the global cooling in the Miocene blocked the land corridor through a climatic filter (Pethiyagoda and Sudasinghe, 2021). However, taxa like the geckoes of the genus *Hemidactylus* adapted to the dry climates of India have made their dispersal to Sri Lanka during an occasion the Palk Isthmus emerged during the Miocene, and later diverged into Sri Lankan endemics ~ 13 Ma (Lajmi et al., 2019; Reuter et al., 2021; Pethiyagoda and Sudasinghe, 2021).

23.7.3 BIOGEOGRAPHIC ELEMENTS IN SRI LANKAN BIOTA

The biogeographic history of Sri Lanka has been differently supported by vicariance and dispersal events in different taxa resulting in Sri Lanka to harbor a wide array of biogeographical elements, in addition to the Sri Lankan endemic element and the obvious South Indian element. Affinities of biota with currently discontinuous distributions are not rare in Sri Lanka and they range from species not with their closest relatives in India, but in Andaman Islands, Malaysia, and/or even in Sumatra in the east to the Comoros Islands, Madagascar, and South Africa in the west. They are briefly described here, though, their existence on the island is supported by various biogeographic theories/hypotheses. See Pethiyagoda and Sudasinghe (2021) for a comprehensive review of the historical biogeography of Sri Lanka.

23.7.3.1   GONDWANAN ELEMENT

Most ancient links among the biogeographic links shown by the Sri Lankan biota have come from those that evolved before the break-up of Gondwanaland and underwent vicariance events such as the creation of the Indian Ocean separating the Deccan plate from Africa, Australia, and Antarctica. Some taxa of such type with Gondwanan origins have rafted with the Deccan plate giving rise to today's species with marked geographic disjunctions (Conti et al., 2002), whereas others may have dispersed to Sri Lanka from Gondwanaland through various dispersal paths (Macey et al., 2000).

A classic example of the former category, the lands snail family Acavidae with a Gondwanan origin is presently found in Madagascar, Seychelles, and Sri Lanka is showing a disjunct pattern of distribution with two endemic genera in Sri Lanka, that is, *Acavus* and *Oligospira* (Emberton, 1990; Hausdorf and Perera, 2000). Further, the Gerrhopilid blind snakes represented by two endemic species in Sri Lanka with disjunct distributions in southern India, Madagascar, and Southeast Asia has been found to be a lineage of Gondwanan origin rafted through the Deccan plate and dispersed "out of India" (Vidal et al., 2010). Acrodont lizards belonging to families Chamaeleonidae and Agamidae (Macey et al., 2000), several subfamilies of Ranid frogs (Bossuyt and Milinkovitch, 2001), Cichlid fishes (Sparks, 2004; Karanth, 2006), and strepsirrhine primates (Yoder, 1997; Yoder and Yang, 2004; Karanth, 2006) seem to have followed similar paths "out of India." Gondwanan vicariance has been revealed for several Sri Lankan invertebrates (Pethiyagoda and Sudasinghe, 2021) including pill-millipede family Arthrospaeridae (Wesener et al., 2010; De Zoysa et al., 2016) spider genus *Indoetra* (Kuntner, 2006), and monotypic cascade beetle genus *Scoliopsis* (Toussaint et al., 2016). A similar scenario has been proposed for Heterometrinae forest scorpions with at least two genera in Sri Lanka and southern India (Loria and Prendini, 2020).

Gondwanan origin hypothesis for the rainforest tree family Dipterocarpaceae is supported by fossil evidence, and recent molecular phylogenetic evidence also shows a similar pattern of disjunct distribution today (Dayanandan et al., 1999). As suggested by the basal clades in Africa and South America Dipterocarps must have evolved some 120 Ma in the Gondwanaland and then dispersed through the drifting Indian plate into Asia and diversified in the ever-wet parts of southeast Asia. Monophyletic endemic clades of the genera *Stemonoporus* and *Doona* (*Shorea*) in Sri Lanka are considered independent divergences resulting from the perhumid refugium in the SWWZ, also suggest an unknown wet evergreen corridor linking Malesia with Sri Lanka during the geological history (Ashton and Gunatilleke, 1987; Morley, 2003; Gamage et al., 2006; Dutta et al., 2011; Ghazoul, 2016). Further, Raven and Axelrod (1974) and Jayasooriya et al. (1993) suggest that the mid-elevation montane forests of Sri Lanka harbors Gondwanan plant elements based on a phytosociological analysis. Crypteroniaceae, a plant family with ancient Gondwanan origin of more than 106 Ma have also rafted across Tethys Sea and dispersed "out of India," and are occurring today in Southeast Asia, while *Axinandra zeylanica* remains endemic to Sri Lanka (Conti et al., 2002; Karanth, 2006). The Gondwanan fern genus *Amauropelta*, widely distributed in South America today, with small numbers of species in Africa and Mascarene Islands, has a single Asiatic species, *A. hakgalensis* which is endemic to Sri

Lanka. The fern *Elaphoglossum spatulatum* shows a similar distribution (Ashton and Gunatilleke, 1987; Sladge, 1982). Furthermore, the southern hemispheric distribution of the ancient lineage of scaly tree ferns in the Family Cyatheaceae also shows evidence of a Gondwanan vicariance with seven native species, of which five are endemic to Sri Lanka (Korall and Pryer, 2014; Ranil et al., 2017).

23.7.3.2   AFRICAN ELEMENT

The African element of Sri Lankan biota also represents old lineages that evolved in the African and Madagascan plates has been shown mainly by floristic groups. The only Sri Lankan Cactaceae, *Rhipsalis baccifera* is also distributed widely in South Africa, Mauritius, and Seychelles (Abeywickrama, 1956; Ashton and Gunatilleke, 1987). Two members of the family Orchidaceae in Sri Lanka also show African origin which include *Aerangis hologlottis*, found in Southeast Kenya to Northeast Tanzania, Mozambique, and Sri Lanka, and *Angraecum zeylanicum* is extensively distributed in tropical Africa, Seychelles (Mahé), and in Sri Lanka (Fernando, 2013). Further, Sri Lankan endemic flowering plants in the genus *Exacum* have their center of diversification in Madagascar (Yuan et al., 2005). With a different path out of Africa, the long-distance dispersal of *Begonia* through the Himalaya into Southeast Asia and Sri Lanka marks another genus with African affinity, also providing evidence of the importance of the Asian monsoon in facilitating such dispersals during the late Miocene climate optima (Rajbhandary et al., 2011). When fauna are concerned, the narrowly endemic highland bird *Elaphrornis palliseri* (Sri Lanka Bush Warbler) representing an endemic monotypic genus in Sri Lanka shows an African affinity with *Schoenicola brevirostris* (Fan-tailed Grassbird) in sub-Saharan Africa (Krishan et al., 2020). Similarly, hornbills represented by two species in Sri Lanka show an African origin in Oligocene period and colonizing the Asian forests with diversification in Southeast Asia (Gonzalez, 2012) with a sister species pair of *Ocyceros* gray hornbills; *O. gingalensis* is endemic to Sri Lanka and *O. griseus* is endemic to the Western Ghats.

23.7.3.3   HIMALAYAN AND NORTH ASIAN ELEMENT

Species that show this affinity originally belonged to mainland Asia before the collision of the Indian plate and then dispersed to northern India, and later found their way to Sri Lanka during the optimum climatic conditions. Hence, they obviously are restricted to the highlands and the perhumid wet zone of the island, while some of them are surprisingly absent from Southern India. The genera *Rhododendron* and *Berberis*, which have diversified in the Himalayan region are found with range-restricted species in Sri Lankan highlands (Abeywickrama, 1956; Ashton and Gunatilleke, 1987). Similarly, the genera *Spiranthes* and *Satyrium* represent Himalayan affinities in Sri Lankan orchid flora (Fernando, 2013). Further, many species of the Sri Lankan Pteridophyte and Bryophyte flora have been reported with strong Himalayan and Asian relationships (Sledge, 1982; O'Shea, 2003). Among the fauna, the Crovid genus of blue magpies, *Urocissa*, shows

Himalayan and largely Asian affinities; *U. ornata* is restricted to ever-wet closed-canopy forests in Sri Lanka, while *U. flavirostris* and *U. erythrorhyncha* are found in the Himalayan range together with two other species extending to Taiwan and Hainan of southern China (Kazmierczak and Perlo, 2000).

23.7.3.4 MALAYAN ELEMENT

Biota originated and diversified in the Malayan peninsula have also made their journey into Sri Lanka, either through land after the Indian plate collided with Asia or through trans-oceanic long-distance dispersal through the Bay of Bengal (Ashton and Gunatilleke, 1987). In the meantime, it has recently been suggested through a molecular study that the Indian plate may have tracked along Sumatra and parts of Southeast Asia during its northward drift before its collision with Eurasia, supporting an "into India" dispersal event of some Malayan elements, at least aquatic taxa such as Gecarcinucid crabs (Klaus et al., 2010). However, when the plants are concerned, the orchid genus *Bromheadia* with its center of diversity in Borneo is represented on the island by the single endemic species *B. srilankensis* (Kruizinga et al., 1997; Repetur et al., 1997; Fernando, 2013). Furthermore, there are many Pteridophytes showing Malayan elements in Sri Lanka (Sledge, 1982). Among the fauna, the beautiful freshwater fish genus *Belontia* has only two known species, *B. signata*, which is endemic to Sri Lanka and *B. hasselti*, which is distributed in Java and Sumatra (Senanayake, 1993). Similarly, the evolutionary district sub-fossorial snake genus *Cylindrophis*, representing the monotypic family Cylindrophiidae, has a single endemic species *C. maculatus* in the lowland wet zone of Sri Lanka, while the rest of the species are distributed in the southeast Asian archipelago (de Silva, 1980; Erdelen, 1989; Zaher et al., 2012). Among the birds, the *Phaenicophaeus* Malkoha lineage is represented in Sri Lanka with two species, the *P. pyrrhocephalus* (red-faced malkoha) is endemic to the island and has its center of diversity in the Malayan region.

23.7.3.5 SOUTH INDIAN ELEMENT

Sri Lanka had its direct contact with South India during its entire geological history. Hence, a majority of our biota has South Indian affinities, with many genera and species endemic to South India and Sri Lanka, especially to the WG&SL (Abeywickrama, 1956; Ashton and Gunatilleke, 1987; Erdelen, 1989). Among the flora, flowering plants genera, such as *Farmeria, Kuruna,* and *Taprobania* as well as Orchid genera *Cottonia, Seidenfadeniella,* and *Sirhookera* are restricted to South India and Sri Lanka together with many species of ferns (Sledge, 1982; Kumar and Manilal, 1994; Fernando, 2013; Attigala et al., 2014). Among the ferns, *Alsophila crinita* is restricted to the highlands of the WG&SL representing the Family Cyatheaceae (Korall and Pryer, 2014; Ranil et al., 2017).

According to Erdelen (1989), at least 30% of animal species found in Sri Lanka are shared with South India. For instance, the land snail genus *Corilla* representing 10 species is endemic to Sri Lanka with *C. anax* from the Annamalai hills in southern Western

Ghats (Naggs and Raheem, 2005), while the strepsirrhine primate genus *Loris* includes *L. tardigradus* endemic to the rainforests of Sri Lanka and *L. lydekkerianus* shared its distribution among Sri Lanka and South India. The geological history and the Miocene aridification have isolated humid mountains of south India and Sri Lanka allowing the WG&SL highlands to harbor evolutionary relicts and undergo rapid speciation. All in all, species endemic to the broader WG&SL biodiversity hotspot (except those found only in a single center of endemism) represent the South Indian elements of Sri Lankan biota.

FIGURE 23.3   Himalayan affinity among Sri Lankan endemics: **(a)** The highland endemic *Berberis aristata* representing a Himalayan affinity in Sri Lanka is being fed by Sri Lanka (hill) white-eye *Zosterops ceylonensis* narrowly restricted to the highlands of Sri Lanka, representing the most basal clade of the global white-eye lineage. **(b)** The wet zone endemic Sri Lanka blue magpie *Urocissa ornata*, representing a disjunct Himalayan affinity, preys upon the black-lipped lizard *Calotes nigrilabris* narrowly endemic to the Sri Lanka's central highlands, in the border of the Hackgala Strict Nature Reserve, one of the few protected areas representing the threatened ecosystem of Sri Lankan cloud forests (Figure 23.2b). **(c)** *Rhododendron arboreum* which shows a Himalayan affinity along the highlands of western Ghats is also flowering in the Horton Plains National Park, Sri Lanka (subsp. *zeylanicum*), again restricted to the highlands. (Photos by Sandun J. Perera)

23.7.3.6   SRI LANKAN ENDEMIC ELEMENT

As described above, the historical biogeography of Sri Lanka and the southwest monsoon supported the expansion of lush rain forests in SWWZ of the island, isolating its biota in narrow niches, allowing insular speciation and providing refuge for evolutionary relicts (Gunatilleke et al., 2017). Hence, the high degree of specialization and endemism is common for many groups of the Sri Lankan biota, with an overall percentage endemicity of 29% for flowering plants, 51% for vertebrates, and 50% for known invertebrate groups (MoMD&E, 2019). Some lineages have undergone isolation for periods as long as 30 million years and some others show degrees of endemism as high as 98% (freshwater crabs) even with shorter periods of isolation.

Recent molecular studies, such as those by Bossuyt et al. (2004) and Meegaskumbura et al. (2019), as well as information summarized in Pethiyagoda and Sudasinghe (2021), have emphasized the noteworthy local endemism in Sri Lanka, within the WG&SL biodiversity hotspot, showing that Sri Lanka and South India have been much more isolated from each. The high degree of higher taxonomic level endemicity in Sri Lanka indicates the refugial role played by the island's SWWZ. Hence, the rainforests of the island may still hold many unknown relict species, especially in less explored taxonomic groups as in the case of the relict ant of Sri Lanka *Aneuretus simoni*, representing the monotypic subfamily Aneuretinae. *A. simoni* shows a widespread fossil record of nine extinct species across the Northern Hemisphere in the Tertiary ~50 Ma, but presently restricted to a few patches of lowland rainforests of Sri Lanka (Wilson et al., 1956; Dlussky and Rasnitsyn, 2003; Ward et al., 2010). Among other relict faunal groups are the vertebrates, such as the Ranid frog genus *Lankanectes* with two endemic species, that is, *L. corrugatus* in the lowland and *L. pera* in the montane wet zone, representing an endemic subfamily Lankanectinae (Dubois and Ohler, 2001; Senevirathne, et al., 2018), and the monotypic endemic fish genus *Malpulutta* which is found in the wet zone (Erdelen, 1989; Senanayake, 1993). Recent molecular evidence revealed the role of the montane cloud forests of Sri Lanka in providing refuge to the most basal species of the global white-eye lineage *Zosterops ceylonensis* (Wickramasinghe et al., 2017). Furthermore, the Gondwanan lineage of scaly tree ferns in the Family Cyatheaceae with five out of seven species being endemic in Sri Lanka including relict species such as *Alsophila sinuata*, the only known simple leaf tree ferns in the world found narrowly endemic in a few patches of lowland rainforests (Ranil and Pushpakumara, 2012; Ranil et al., 2017).

As summarized in MoMD&E (2019) and added recently by Srikanthan et al. (2021), Sri Lanka possesses at least 57 plant and animal genera endemic to the island, while 30 (53%) of them are monotypic. Seventeen flowering plant genera are found endemic to Sri Lanka representing families, such as Annonaceae (*Chlorocarpa* and *Phoenicanthus*), Arecaceae (*Loxococcus*), Dilleniaceae (*Schumacheria*), Dipterocarpaceae (*Doona, Stemonoporus*), Euphorbiaceae (*Podadenia*), Gesneriaceae (*Championia*), Malvaceae (*Dicellostyles*), Monimiaceae (*Hortonia*), Orchidaceae (*Adrorhizon*), Poaceae (*Davidsea*), Rubiaceae (*Diyaminauclea, Leucocodon, Nargedia,* and *Scyphostachys*), and Zingiberaceae (*Cyphostigma*), while 12 of them except *Phoenicanthus, Schumacheria, Doona, Stemonoporus, Hortonia,* and *Scyphostachys* are monotypic. Genus *Doona sensu lato* is presently considered a synonym of *Shorea*. However, *Doona sensu stricto* is considered endemic to the Sri Lankan wet zone. Among vertebrate families, the two monotypic genera of mammals representing families Muridae (*Srilankamys*) and Soricidae (*Solisorex*) are endemic to Sri Lanka. When birds are concerned, the monotypic genus *Elaphrornis* of the family Locustellidae, characterized by the relict narrow endemic *E. palliseri* (Sri Lanka Bush Warbler), has recently been placed as a genus endemic to Sri Lanka and still is under taxonomic investigation (Krishan et al., 2020; S. S. Seneviratne *pers. com.* 2021). Reptiles show the highest genus level endemism among vertebrates in the country with eight endemic genera, representing Agamidae (*Ceratophora, Cophotis, Lyriocephalus,* and *Otocryptis*), Colubridae (*Aspidura*), and Scincidae (*Chalcidoseps, Lankascincus,* and *Nessia*), while the genera *Lyriocephalus* and *Chalcidoseps* are found as monotypic. Four genera of amphibians representing four families are found endemic to the island; these include Bufonidae (*Adenomus*), Dicroglossidae (*Nannophrys*), Nyctibatrachidae (*Lankanectes*), and Rhacophoridae (*Taruga*). Similarly, two genera of freshwater fish representing two families are endemic including the family Osphronemidae represented by the monotypic genus *Malpulutta* and the family Cyprinidae represented by the genus *Rasboroides*. Among invertebrate groups with adequate knowledge and taxonomic revolution, millipedes show the highest genus-level endemism with nine endemic genera, followed by freshwater crabs and land snails with five endemic genera each. Six families of millipedes are represented by these endemic genera: Chordeumatida (*Lankasoma*), Cryptodesmidae (*Pocodesmus* and *Singhalocryptus*), Fuhrmannodesmidae (*Lankadesmus*), Paradoxosomatidae (*Pyragrogonus*), Polydesmidae (*Eustaledesmus*), as well as Pyrgodesmidae (*Catapyrgodesmus, Cryptocephalopus,* and *Styloceylonius*), while seven among the above genera except *Lankasoma* and *Singhalocryptus* are monotypic. Two families of freshwater crabs are represented by these endemic genera: Gecarcinucidae (*Clinothelphusa, Mahatha, Pastilla,* and *Perbrinckia*) and Parathelphusidae (*Ceylonthelphusa*); among them, *Clinothelphusa* and *Pastilla* are monotypic genera. Similarly, three families of land snails are represented by these endemic genera: Acavidae (*Acavus* and *Oligospira*), Ariophantidae (*Ratnadvipia* and *Ravana*), and Cyclophoridae (*Aulopoma*) and among which *Ravana* is a monotypic genus. Among Odonates, the family Lestidae (*Sinhalestes*) and Platystictidae (*Ceylonosticta, Platysticta*) are represented by the three endemic genera, among which *Sinhalestes* is monotypic, while the family Formicidae of ants is represented by the endemic and monotypic genus *Aneuretus*.

The island also harbors some unique insular endemic radiations of flora, again, especially in the SWWZ. The Sri Lankan endemic genus *Stemonoporus* shows restricted distribution patterns of about 27 species localized within the rain forests, including a few in highlands marking one of the highest altitudes, a Dipterocarp has reached (Ashton and Gunatilleke, 1987; Greller et al., 1987; Greller and Balasubramanium, 1993; Rubasinghe, 2007; Rubasinghe et al., 2008). The genus *Kuruna*, recently described temperate bamboo lineage endemic to the highlands of Sri Lanka and India with seven species, of which four are endemic to the island (Attigala et al., 2014), the Monimiaceae genus *Hortonia* is endemic to the island with only three species (Rajapakse et al., 2012), and the basal Melastomataceae genus *Memecylon* has 32 species distributed in all climatic zones of Sri Lanka, among which 25 are endemic (Amarasinghe et al., 2021), and are just a few indicators of the evolutionary distinctive Sri Lankan flora. Another group of flowering plants with higher taxonomic level endemism in Sri Lanka includes the subtribe Adrorhizinae of the family Orchidaceae in which the monotypic genus *Adrorhizon* is endemic to the island, while the genus *Sirhookera* with two species is endemic to the WG&SL, while the other genus of the subtribe *Bromheadia* is represented in the island by endemic *B. srilankensis*, while their center of diversity is Borneo (Fernando, 2013). While most of these endemic plants are restricted to the SWWZ of the island, unusual endemic species such as the hemiparasitic *Dendrophthoe ligulata* of Loranthaceae are restricted in the northwestern dry zone of Sri Lanka (Dassanayake, 1987). Sri Lankan endemic flora also include at least 87 microendemic species, such as *Stemonoporus moonii* and *Mesua stylosa,* and both are restricted to seasonally flooded swamp forests known only from a single site in the island at Waturana near Bulathsinhala; *S. reticulatus* is restricted to Kanneliya, *Kuruna densifolia* is restricted to the Horton Plains, *Hortonia ovalifolia* is restricted to the Peak Wilderness range (protected area around Samanala Kanda/Adam's Peak), *Farmeria metzgerioides* s restricted to Hakkinda, and orchids *Oberonia dolabrata* and *O. fornicata* are distributed in Knuckles (see Gunatilleke et al., 2008 for a total list of all microendemics).

Furthermore, the large insular radiation of some 59 endemic species of the direct-developing *Pseudophilautus* shrub frogs in Sri Lanka represents an intriguing scenario, while a majority been narrowly endemic, and the genus being endemic to WG&SL (Meegaskumbura et al., 2002, 2019; Ellepola et al., 2021). The genus demonstrates isolation-driven speciation within the three major mountain ranges of the island represented by sister species in each range (Meegaskumbura and Manamendra-Arachchi, 2005). The island is of particular importance for the unique assemblage of its herpetofauna, among which a few genera of narrowly speciated tetrapod reptiles stand out with a Sri Lankan identity (Erdelen, 2012). Among them is one of the rarest instances of rostral scale evolution of an agamid into a horn-like appendage exhibited by the endemic genus *Ceratophora* (Schulte et al., 2002; Johnston et al., 2012); again with sister species in isolated but adjacent mountain ranges of Rakwana hills, Central mountains, and the Knuckles massif, similar niches are taken by different species (Senanayake, 1993; Pethiyagoda and Manamendra-Arachchi, 1998) also with two isolated wet lowland species (Karunarathna et al., 2020). Furthermore, the narrow endemism among Sri Lankan reptiles may be best explained by the radiation of the day gecko genus *Cnemaspis*, which is still being explored, with all known Sri Lankan members in the genus being endemic to the island, and many are restricted to isolated hills also with a vertical pattern of speciation (Karunarathna et al., 2019, and references therein).

Gecarcinucid freshwater crabs clearly exhibit a similar pattern of local endemism in Sri Lanka (Bossuyt et al., 2004) with a radiation of 50 endemic species, while half of them being narrow endemics, with a sister lineage endemic to the Western Ghats (Bahir et al., 2005; Beenaerts et al., 2010; Pethiyagoda and Sudasinghe, 2021). A similar pattern can also be seen in the near-endemic land snail genus *Corilla* with 10 endemic species in wet and montane zones of the island (Raheem et al., 2017). The remarkable diversity of Platystictid damselflies in Sri Lanka, with an endemic monophyletic subfamily Platystictinae, shows higher taxonomic level endemism with a radiation of 22 endemic species of *Ceylonosticta* and *Platysticta* in narrow ranges within the SWWZ (Dijkstra et al. 2014; Bedjanič et al., 2016). Gunatilleke et al. (2008) also listed 21 microendemic species of vertebrates restricted to single sites such as the freshwater fish *Pethia bandula* in Galapitamada and *Systomus asoka* in Kitulgala, and herpetofauna such as *Pseudophilautus alto* and *Zakerana greenii* in Horton Plains, *Adenomus dasi* and *P. caeruleus* in Peak Wilderness, *Ceratophora erdeleni, Microhyla karunaratnei,* and *P. simba* in Morningside and Sinharaja, and *C. tennentii* and *P. fulvus* in Knuckles. Even among vertebrate groups with lower endemicity such as mammals, the subspecies endemism corresponding to each bioclimatic/elevational

FIGURE 23.4 Fauna representing some endemic genera in Sri Lanka. **(a)** *Perbrinckia punctata*, a narrow endemic in upper montane wet Patana grasslands here in Horton Plains National Park. **(b)** *Acavus phoenix*, **(c)** *Otocryptis wiegmanni*, both restricted to the lowlands to mid-elevations of the wet zone, here in Sinharaja World Heritage Conservation Forest. **(d)** *Taruga eques,* **(e)** *Aspidura trachyprocta*, and **(f)** *Elaphrornis palliseri*, all three species are restricted to cloud forests of the central highlands, here in Horton Plains. (Photos by Sandun J. Perera)

zone shown by endemic species, such as *Macaca sinica, Trachypithecus vetulus,* and *Loris tardigradus* is noteworthy (Dittus, 2013; Gamage et al., 2017).

Accordingly, Sri Lanka has been identified as a hotspot for amphibian endemism (Meegaskumbura et al., 2002), reptile endemism (Erdelen, 2012; Perera et al., 2019a),

Platystictid damselfly endemism (Bedjanič et al., 2016), as well as for the mammalian phylogenetic endemism (Rosauer and Jetz, 2014). In a recent study on the drivers of bird beta diversity in WG&SL, Sreeker et al. (2020) reveal that the divergence of even a mobile group like birds is higher than expected when Sri Lanka and Western Ghats are compared. They suggest dispersal limitation set by the Palk Strait and climate as drivers of this beta diversity respectively at larger and smaller scales, and also recommending Sri Lanka and the Western Ghats to be treated as separate entities in avifaunal conservation. This fact has also been illustrated in the chapter on birds in this book (Dayananda et al., 2022) with seven sister species pairs being endemic to Sri Lanka and the Western Ghats, while five sister species pairs being endemic to within Sri Lanka and the Western and Eastern Ghats together, and one Sri Lankan endemic representing a monotypic genus with its sister genus is found in the Western Ghats.

FIGURE 23.5 **(a)** Sri Lankan rhino-horned lizard, *Ceratophora stoddartii*, narrowly restricted to cloud forests of the central mountain massif **(given in b)**, represents an endemic genus radiated in Sri Lanka with sister species in three major mountain ranges isolated from each other by broad valleys and a few other species isolated in the wet lowlands. **(c)** The world's end escarpment along the southern edge of the central massif acts as a barrier for dispersal of the rhino-horned lizard isolating it above 1500 m asl. (Photos by Sandun J. Perera)

23.7.4 THEORIES TO EXPLAIN SRI LANKAN BIOGEOGRAPHY

A number of historical biogeographic theories support the existence of diverse affinities of the Sri Lankan biota, which may be used in combination to explain their affinities.

Vicariance, the separation of a population by a geographic barrier, such as a mountain or a body of water, resulting in speciation in separated populations, makes a prominent theory among others. Theories of vicariance and plate tectonics together explain much of the ancient Gondwanan and African affinities of Sri Lankan biota, while the theory of long-distance dispersal may support many of the more recent affinities. Hence, dispersal and vicariances have long been used as alternative hypotheses with shifting roles in studying the historical biogeography of areas (Zink et al., 2000). Ancestral species in Gondwanaland must have undergone vicariant speciation due to the formation of the Indian Ocean between the Deccan plate and African–Malagasy Plates (Croizat, 1968). This theory claims the role of plate tectonics in determining biogeographic patterns and can explain the disjunction of distribution in African, Malagasy biotas, and the Indo-Sri Lanka biota, such as the case of the land snail family Acavidae. However, the Malagasy link of the genus *Exacum* in Sri Lanka has suggested that its distribution is the result of long-distance dispersal, and molecular dating suggests a younger divergence than the vicariant event (Yuan et al., 2005). Similarly, the Malayan link in the genus *Bromheadia* is suspected to result from a long-distance dispersal event from Borneo, its center of origin (Repetur et al., 1997).

Attempted explanations of post-breakup dispersal of biota through the hypothesis of ephemeral land brides among Africa, Madagascar, Seychelles, and the Deccan plate are yet to be proven because recent geological evidence does not support the submergence of such islands in the Indian Ocean (Erdelen, 1989; Pethiyagoda and Sudasinghe, 2021); however, the land-bridge theory of past land connection allowing the exchange of fauna and flora among regions has not yet been totally rejected as the Deccan plate together with Madagascar may have provided for migration between present-day Africa, Australia, and Antarctica. At a much finer resolution, Sri Lanka was lodged between Madagascar and the Indian subcontinent doing the same (Raven and Axelrod, 1974; Ashton and Gunatilleke, 1987). Further the theory of fluctuating archipelagos by Deraniyagala (1940, 1946) suggesting the transfer to India–Sri Lanka, of African elements via Seychelles–Madagascar and Malayan elements via Andaman and Nicobar Islands has also been developed on same lines.

The south-to-north rafting of the Deccan plate has ferried Gondwanan and African elements to Asia generally referred to as the "Out of India hypothesis" (Bossuyt and Milinkovitch, 2001; Karanth, 2006; Datta-Roy and Karanth, 2009). However, during the rapid northeastward momentum, the Deccan plate experienced a rapid change from the temperate to a tropical climate, Deccan trap lava flows, and the end-Cretaceous (K/T boundary) mass extinction event, during all of which Sri Lanka has served as a refugium with an aseasonal humid climate (Ashton and Gunatilleke, 1987). The Deccan trap theory (Mani, 1974) explains a massive volcanic eruption in the northeastern Deccan plate possibly around the Narmada basin which may have made a devastative impact on much of the Gondwanan fauna it was carrying, concurrently to the K/T boundary, leaving only a few fossorial or semi-fossorial Gondwanan species to remain in India today (ancient snakehead genus *Aenigmachanna*, and Nasikabatrachid fossorial frogs in India, Gerrhopilid ground snakes and Acavid land snails in Sri Lanka; Pethiyagoda and Sudasinghe, 2021; Naggs and Raheem, 2005). Deccan traps must have also made a vicariant barrier for the dispersal of taxa between the north and south of the Deccan plate explaining the geographical discontinuity of the Sri Lankan and Southern Indian biota with North India (Naggs and Raheem, 2005). This phenomenon has provided an opportunity for testing more historical biogeographic hypotheses on Sri Lanka such as the Relict theory (Erdelen, 1989) or the Continues Range theory (Kurup, 1974), or the Ecological Pocket theory (Mani, 1974). All in all, they explain geographic relics, the living representatives of once widespread taxa which disappeared from other places due to changes in physiographic, climatic, biotic, ecological, or catastrophic events, making disjunct distributions and restricting them to the most optimal niches. Sri Lanka, at least the southern half of the island is believed to have maintained a perhumid climate during the end of the Mesozoic catastrophe, providing a good refuge for at least some taxa. Hence, the present-day Sri Lankan biota, especially that in the southwest, harbors signatures of a number of relict biogeographic elements of regional and global significance (Abeywickrama, 1956; Ashton and Gunatilleke, 1987; Bossuyt et al., 2004).

The glaciation theory explains climatic cycles of alternative cooler (glacial) and warmer (interglacial) periods during the Plio-Pleistocene (Medlicott and Blanford, 1879) causing corresponding fluctuations of the sea level, exposing and submerging a wide land-bridge connecting the Palk bay from Puttalam to Jaffna of Sri Lanka with the southern part of Coromandel coast of India (Voris, 2000). Corresponding connection and isolation of Sri Lanka from the mainland resulted in the island to harbor a uniquely interesting biota, receiving immigration from the mainland during the glacials, and isolation-driven speciations during interglacials (Meegaskumbura et al., 2002; Bossuyt et al., 2004; Biswas and Pawar, 2006; Agarwal et al., 2017; Lajmi et al., 2019; Reuter et al., 2021). Further, it could be the fluctuation of humidity rather than the temperature that affected the migration of biota during those global cooling periods, assuming glacial periods with high humidity (Hora, 1949), also explaining the migration of Himalayan biota into the mountains of southern India and Sri Lanka during cooling and restricting them to refugia in upper elevations during warming (Erdelen, 1989). The biotic exchanges during glacial periods have also been supported by increased humidity during climatic optima that facilitated forest expansions, hence supporting forest dwellers to disperse, amidst the Gulf of Mannar and the northwestern arid belt of Sri Lanka forming a barrier for dispersal during the Miocene aridification (Bossuyt et al., 2004; Biswas and Pawar, 2006; Agarwal and Karanth, 2015; Meegaskumbura et al., 2019).

Similarly, Indo–Malayan and Indo-Chinese elements in southern India and Sri Lanka have been explained through the Satpura hypothesis (Hora, 1949, 1953; Mani, 1974), where dispersal is hypothesized along mountain ranges from eastern Himalayas through Central hills of Assam, Vindhyas range, Kaimur ridge, and the Satpura range of hills to western Ghats, supported by colder and wetter climates along the mountain bridges, compared with those of the plains, during the glacial periods of the Pleistocene (Karanth, 2015; Joshi and Karanth, 2013). This theory, initially developed to explain the distribution patterns of hill-stream fishes by Hora (1949), was later used for other taxa (Ripley, 1949; Moore, 1960).

While the global and regional theories of historical biogeography shed light on understanding the Sri Lankan biota, Wills (1915, 1922) used the data on Sri Lankan plant distributions to develop his "age and area theory" analyzing the high incidence of narrow-range endemism in Sri Lanka (Pethiyagoda and Sudasinghe, 2017). The theory simply postulates that the species that range widely are of ancient origin than those that are narrow ranged. Although the theory did not stand due to biases in underlying data, the "age and area" is still popular in ecological literature (Pethiyagoda and Sudasinghe, 2021)

Adding light to the picture from most recent molecular phylogenetic studies, evidence is now emerging on an "Out of Sri Lanka Hypothesis" of gene exchange through back migrations, amidst the evidence for large-scale colonization of Sri Lanka during the climate optima. Such evidence exists on *Pseudophilautus* shrub frogs (Meegaskumbura et al., 2019), *Duttaphrynus* dwarf toads (Jayawardena et al., 2017), and Gecarcinucid freshwater crabs (Beenaerts et al., 2010), with at least one back migration in latter cases and two in *Pseudophilautus*. While all these back migrations are timed to have happened during the late Miocene rain forest expansion at around 7–8 Ma for shrub frogs and freshwater crabs and even later for dwarf toads, increasing interest in molecular studies involving lesser explored taxa could provide more evidence on similar lines, also for Pleistocene glacials.

However, solving the puzzle of the historical biogeography of Sri Lanka would need several more decades of research with more molecular data and well-resolved and calibrated phylogenies together with a better understanding of the geomorphological evolution with much higher resolution. Some theories and explanations of certain biogeographical affinities of Sri Lanka have been fine-tuned with recent findings, while some are strongly debated and remain inconclusive. For example, the Satpura Hypothesis of Hora (1949, 1953) was further studied and supported by Silas (1952) and contradicted by Daniels (2001), while still being used and corroborated from time to time, for example, Kuttapetty et al. (2014). One such criticism over the Satpura hypothesis is linked to the approximately 300 km wide gap between Garo and Rajmahal mountains in Northeast India (Erdelen, 1989), while Daniels (2001) argued that more recent findings on the taxonomy, ecology, and distribution of freshwater fish fauna provide less support to Hora's theory. Further, the long-believed vicariance of *Chamaeleo zeylanicus* in the Indian peninsula and Sri Lanka, connected to the break-up of Gondwanaland separating the Indian plate from the center of Chamaeleon diversification was recently nullified with molecular evidence of its divergence being younger than the breakup of Malagasy and Indian plates, hence suggesting long-distance dispersal of Chamaeleon through the Indian Ocean (Raxworthy et al., 2002; Tolley et al., 2013). Similarly, the carnivorous plant family Nepenthaceae with its center of diversity in the Malayan peninsular and represented by a single endemic species in Sri Lanka and another in India was considered earlier as a Malayan link of the Sri Lankan flora (Clarke and Lee, 2004; Fernando, 2011), and has recently been unveiled to have a Gondwanan origin and shows an "out of India" dispersal to Southeast Asia with high endemism in Borneo, Sumatra, and the Philippines (Murphy et al., 2020; Biswal et al., 2018). In contrast, *Sibynophis subpunctatus* in Sri Lanka representing the current subfamily Sibynophiinae, earlier believed to have Malagasy affinities (de Silva, 1980) is now resolved with molecular phylogenetic analyses to be of Asian origin (Zaher et al., 2012; Chen et al., 2013). With this backdrop of arguments, the complete historical biogeographic story of making the Sri Lankan biota is yet to be unfolded.

23.8 CONCLUDING REMARKS

This chapter compiles updated information on the remarkable biodiversity and endemism within the continental island of Sri Lanka and the making of it. In doing so, the ecological and historical biogeography of the island has been explored with its theoretical background and examples from Sri Lanka for some of the theories involved. The SWWZ of Sri Lanka which has served as a refugium during paleoenvironmental changes and geological cata-clysms, currently provides the highest level of habitat heterogeneity serving as a speciation hub, also with the potential for continuing to serve as a stronghold of biodiversity amidst the challenges placed by the current climate change. Already subjected to an unprecedented loss of habitats over the past 200 years or so, the remaining fragmented patches of natural habitats ranging from several types of forests, through thicket and savanna to grasslands in Sri Lanka, hence serve as the only places left for those species to strive on Earth. Further-more, Sri Lanka is still in the process of inventorying its innumerable biological diversity, while there are a large number of species, not collected during the past century or so.

Hence, the recognition of WG&SL as a "hyper hot" biodiversity hotspot, with less than 10% of its primary vegetation remaining for its highly endemic biodiversity (Myers et al., 2000) rings a loud alarm for the scientists, policymakers, and the general public of this blessed island. Making things even complicated, Cincotta et al. (2000) reported WG&SL to be the biodiversity hotspot with the highest human population density on the Earth, the value being sevenfold higher compared with the global average. Biologists in Sri Lanka have documented 20 amphibian species extinctions, all of which have been endemic to the island (MoE, 2012) and four confirmed flowering plant species extinctions and 128 possible extinctions, 58 been endemic to Sri Lanka (MoE and DNBG, 2020). Hence, many of the undescribed species, especially of those less explored groups, such as invertebrates and lower plants could be on the brink of extinction by the time they are described or may go extinct beforehand. The National Red List (MoE, 2012) documents 46% of vertebrates, including 28% of bird species as threatened with extinction, while 33% of threatened birds are endemic to Sri Lanka. Among the threatened invertebrates on the island, the most numerous are land snails with 179 species, while the highest threatened percentage of 50% is for freshwater crabs (MoE, 2012). According to the ongoing revision of the National Red List, a considerable percentage of 72%, 75%, and 59% of the Sri Lankan endemics are threatened with extinction, respectively for flowering plants, freshwater fish, and birds (MoE and DNBG, 2020; Goonatilake et al., 2020; MoE, 2021). These assessments of the threat status of biota emphasize the role that humanity has to play in this Anthropocene to decide the future of biodiversity and its immeasurable services, that in turn will indicate the fate of humanity.

ACKNOWLEDGMENTS

We would like to thank Ms. R. H. M. Padma Abeykoon (Director) and Ms. Chanuka Maheshani Kumari (Development Officer-Environment) of the Biodiversity Secretariat of Sri Lanka for steering the initial stage of this work. The first author thanks Emeritus Professor Sarath Kotagama who implanted initial ideas on the biogeography of Sri Lanka during his undergraduate learning, and Professor Şerban Proches who has molded biogeo-graphic thinking in him during doctoral training. We also acknowledge Emeritus Profes-sors Savitri and Nimal Gunatilleke who encouraged both authors in perusing research on the biogeography of the island and also provided insightful comments on early versions of parts of this manuscript as doctoral supervisors of the second author.

KEYWORDS

- **biodiversity hotspot**
- **biogeography**
- **climate**
- **conservation**
- **endemism**
- **island**
- **Sri Lanka**
- **topography**

REFERENCES

Abeywickrama, B. A. The Origin and Affinities of the Flora of Ceylon, *Proceedings of the 11th Annual Session of the Ceylon Association for the Advancement of Science, Colombo,* **1956;** vol 2, pp 99–121.

Adams, F. D. The Geology of Ceylon. *Candian. J. Res.* **1929,** *11,* 425–511.

Agarwal, I.; Biswas, S.; Bauer, A. M.; Greenbaum, E.; Jackman, T. R.; Silva, A. D.; Batuwita, S. Cryptic Species, Taxonomic Inflation, or a Bit of Both? New Species Phenomenon in Sri Lanka as Suggested by a Phylogeny of Dwarf Geckos (Reptilia, Squamata, Gekkonidae, Cnemaspis). *Syst. Biodivers.* **2017,** *15,* 427–439.

Agarwal, I.; Karanth, K. P. A Phylogeny of the only Ground-Dwelling Radiation of *Cyrtodactylus* (Squamata, Gekkonidae): Diversification of Geckoella Across Peninsular India and Sri Lanka. *Mol. Phylogenet. Evol.* **2015,** 82, 193–199.

Aitchison, J. C.; Ali, J. R.; Davis, A. M. When and where did India and Asia Collide? *J. Geophys. Res.* **2007,** *112B.*

Amarasinghe, P.; Barve, N.; Kathriarachchi, H.; Loiselle, B.; Cellinese, N. Niche Dynamics of *Memecylon* in Sri Lanka: Distribution Patterns, Climate Change Effects, and Conservation Priorities. *Ecol. Evol.* **2021,** *11* (24), 18196–18215.

Araujo, M. B.; Thuiller, W.; Williams, P. H.; Reginster, I. Downscaling European Species Atlas Distributions to a Finer Resolution: Implications for Conservation Planning. *Glob. Ecol. Biogeogr.* **2005,** *14,* 17–30.

Ashton, P. S.; Gunatilleke, C. V. S. New Light on the Plant Geography of Ceylon. I. Historical Plant Geography. *J. Biogeogr.* **1987,** *14,* 249–285.

Attigala, L.; Triplett, J. K.; Kathriarachchi, H. S.; Clark, L. G. A New Genus and a Major Temperate Bamboo Lineage of the Arundinarieae (Poaceae: Bambusoideae) from Sri Lanka Based on a Multi-Locus Plastid Phylogeny. *Phytotaxa* **2014,** *174* (4), 187–205.

Bahir, M. M.; Ng, P. K.; Crandall, K.; Pethiyagoda, R. A Conservation Assessment of the Freshwater Crabs of Sri Lanka. *Raffles Bull. Zool.* **2005,** *12,* 121–126.

Baldwin, M. *Natural Resources of Sri Lanka-Conditions and Trends*; Natural Resources Energy and Science Authority of Sri Lanka: Colombo, **1991;** p 280.

Bedjanič, M.; Conniff, K.; Dow, R. A.; Stokvis, F. R.; Verovnik, R.; Tol, J. V. Taxonomy and Molecular Phylogeny of the Platystictidae of Sri Lanka (Insecta: Odonata). *Zootaxa* **2016,** *4182* (1), 1–80.

Beenaerts, N.; Pethiyagoda, R.; Ng, P. K.; Yeo, D. C.; Bex, G. J.; Bahir, M. M.; Artois, T. Phylogenetic Diversity of Sri Lankan Freshwater Crabs and its Implications for Conservation. *Mol. Ecol.* **2010,** *19* (1), 183–196.

Bibby, C. J.; Collar, N. J.; Crosby, M. J.; Heath, M. F.; Imboden, C.; Johnson, T. H.; Long, A. J.; Stattersfield, A. J.; Thirgood, S. J. *Putting Biodiversity on the Map: Priority Areas for Global Conservation*; International Council for Bird Preservation: Cambridge, **1992.**

Biswal, D. K.; Debnath, M.; Konhar, R.; Yanthan, S.; Tandon, P. Phylogeny and Biogeography of Carnivorous Plant Family Nepenthaceae with Reference to the Indian Pitcher Plant *Nepenthes khasiana* Reveals an Indian Subcontinent Origin of *Nepenthes* Colonization in South East Asia During the Miocene Epoch. *Front. Ecol. Evol.* **2018**, *6*, 108.

Biswas, S.; Pawar, S. S. Phylogenetic Tests of Distribution Patterns in South Asia: Towards an Integrative Approach. *J. Biosci.* **2006**, *31*, 95–113.

Blanford, W. T. The African Element in the Fauna of India: A Criticism of Mr. Wallace's Views as Expressed in the 'Geographical Distribution of Animals'. *J. Nat. Hist.* **1876**, *18*, 277–294.

Blanford, W. T. The Distribution of Vertebrate Animals in India, Ceylon, and Burma. *Proc. Roy. Soc. London* **1901**, *67*, 484–492.

Blyth, E. A Suggested New Division of the Earth into Zoological Regions. *Nature* **1871**, *3*, 427–429.

Bossuyt, F.; Meegaskumbura, M.; Beenaerts, N.; Gower, D. J.; Pethiyagoda, R.; Roelants, K.; Mannaert, A.; Wilkinson, M.; Bahir, M. M.; Manamendra-Arachchi, K. Local Endemism within the Western Ghats-Sri Lanka Biodiversity Hotspot. *Science* **2004**, *306*, 479–481.

Bossuyt, F.; Milinkovitch, M. Convergent Adaptive Radiation in Madagascan and Asian Ranid Frogs Reveal Covariation between Larval and Adult Traits. *Proc. Natl. Acad. Sci. USA* **2000**, *97*, 6585–6590

Bromham, L.; Cardillo, M. Testing the Link Between the Latitudinal Gradient in Species Richness and Rates of Molecular Evolution. *J. Evol. Biol.* **2003**, *16* (2), 200–207.

Brummitt, N.; Lughadha, E. N. Biodiversity: Where's Hot and Where's Not. *Conser. Biol.* **2003**, *17*, 1442–1448.

Caesar, M.; Grandcolas, P.; Pellens, R. Outstanding Micro-Endemism in New Caledonia: More than one out of Ten Animal Species have a Very Restricted Distribution Range. *PloS One* **2017**, *12* (7), e0181437.

Ceballos, G.; Ehrlich, P. R. Global Mammal Distributions, Biodiversity Hotspots, and Conservation. *Proc. Natl. Acad. Sci. USA* **2006**, *103*, 19374–19379.

Chandrajith, R. Geology and Geomorphology. In *The Soils of Sri Lanka*; Mapa, R., Ed.; World Soils Book Series; Springer International: Switzerland, 2020; pp 23–34.

Chatterjee, S.; Goswami, A.; Scotese, C. R. The Longest Voyage: Tectonic, Magmatic, and Paleoclimatic Evolution of the Indian Plate During its Northward Flight from Gondwana to Asia. *Gondwana Res.* **2013**, *23*, 238–267.

Chen, X.; Huang, S.; Guo, P.; Colli, G. R.; de Oca, A. N. M.; Vitt, L. J.; Pyron, R. A.; Burbrink, F. T. Understanding the Formation of Ancient Intertropical Disjunct Distributions using Asian and Neotropical Hinged-Teeth Snakes (*Sibynophis* and *Scaphiodontophis*: Serpentes: Colubridae). *Mol. Phylogenet. Evol.* **2013**, *66* (1), 254–261.

Chetty, T. R. K. *Proterozic Orogens of India: A Critical Window to Gondwana*; Elsevier: Amsterdam, 2017; p 426.

Cincotta, R. P.; Wisnewski, J.; Engelman, R. Human Population in the Biodiversity Hotspots. *Nature* **2000**, *404* (6781), 990–992.

Clarke, C.; Lee, C. In *Pitcher Plants of Sarawak*; Natural History Publication: Borneo, 2004.

Cohen, K. M.; Finney, S. C.; Gibbard, P. L.; Fan, J. X. The ICS International Chronostratigraphic Chart. *Episodes* **2013**, *36*, 199–204; Updated version 2020/3.

Conti, E.; Eriksson, T.; Schonenberger, J.; Systsma, K. J.; Baum, D. A. Early Tertiary Out-of-India Dispersal of Crypteroniaceae: Evidence from Phylogeny and Molecular Dating. *Evolution* **2002**, *56* (10), 1931–1942.

Cooray, P. G. *An Introduction to the Geology of Ceylon*; National Museum of Ceylon Publication: Colombo, 1967.

Cooray, P. G. *Geology of Sri Lanka*, 2nd revised ed.; National Museum of Sri Lanka Publication: Colombo, 1984.

Croizat, L. The Biogeography of India: A Note of Some of its Fundamentals, *Proceedings of Symposium on Recent Advances in Tropical Ecology;* Varanasi, India, **1968**; vol II, pp 544–590.

Daniels, R. J. R. Endemic Fishes of the Western Ghats and the Satpura Hypothesis. *Curr. Sci.* **2001**, *81*, 240–244.

Dassanayake A. R.; De Silva G. G. R.; Mapa R. B. Major Soils of the Dry Zone and Their Classification, In *The Soils of Sri Lanka*; Mapa, R., Ed.; World Soils Book Series; Springer International: Switzerland, **2020a**; pp 49–67.

Dassanayake A. R.; Senarath A.; Hettiarachchi L. S. K.; Mapa R. B. Major Soils of the Wet Zone and Their Classification. In *The Soils of Sri Lanka*; Mapa, R., Ed.; World Soils Book Series; Springer International: Switzerland, **2020b**; pp 83–94.

Dassanayake A. R.; Somasiri L. L. W.; Mapa R. B. Major Soils of the Intermediate Soils and Their Classification. In *The Soils of Sri Lanka*; Mapa, R., Ed.; World Soils Book Series; Springer International: Switzerland, **2020c**; pp 69–82.

Dassanayake, M. D. *A Revised Handbook to the Flora of Ceylon*; Oxford and IBH Publishing Co.: New Delhi, **1987**; vol 6.

Datta-Roy, A.; Karanth, K. P. The Out-of-India Hypothesis: What do Molecules Suggest? *J. Biosci.* **2009,** *34* (5), 687–697.

Dayananda, S. K.; Perera, S. J.; Senevirathne, S. S.; Kotagama, S. W. Diversity, Distribution and Biogeography of Sri Lankan Birds, In *Biodiversity of Hotspots of Indian region: Western Ghats-Sri Lanka*; Pullaiah, T., Ed.; Apple Academic Press: Ware Town, **2023**.

Dayanandan, S.; Ashton, P. S.; Williams, S. M.; Primack, R. B. Phylogeny of the Tropical Tree Family Diptero-carpaceae based on Nucleotide Sequences of the Chloroplast rbcL gene. *Amer. J. Bot.* **1999,** *86* (8), 1182–1190.

de Silva, P. H. D. H. *Snake Fauna of Sri Lanka*; National Museums of Sri Lanka: Colombo, **1980**.

De Zoysa, H. K. S.; Nguyen, A. D.; Wickramasinghe, S. Annotated Checklist of Millipedes (Myriapoda: Diplopoda) of Sri Lanka. *Zootaxa* **2016,** *4061* (5), 451–482.

Deraniyagala, P. E. P. Some Post-Gondwana Land Links of Ceylon, *Proceedings of Indian Science Congress* 27th Meeting, **1940**; Part IV, pp 119–120.

Deraniyagala, P. E. P. Some Phases of Evolution of Ceylon, *Proceedings of Ceylon Association of Advancement of Science,* **1946**; section D, pp 69–88.

Deraniyagala, P. E. P. In *The Pleistocene of Ceylon*; Sri Lanka National Museums: Colombo, **1958**.

Dijkstra, K. D. B.; Kalkman, V. J.; Dow, R. A.; Stokvis, F. R.; van Tol, J. Redefining the Damselfly Families: A Comprehensive Molecular Phylogeny of Zygoptera (Odonata). *Syst. Entomol.* **2014,** *39* (1), 68–96.

Dittus, W. P. J. Subspecies of Sri Lankan Mammals as Units of Biodiversity Conservation, with Special Reference to the Primates. *Ceylon J. Sci. (Biol. Sci.)* **2013,** *42*, 1–27.

Dlussky, G. M.; Rasnitsyn, A. P. Ants (Hymenoptera: Formicidae) of Formation Green River and Some Other Middle Eocene Deposits of North America. *Russ. Entomol. J.* **2003,** *11*, 411–436.

Dubois, A.; Ohler, A. A New Genus for an Aquatic Ranid (Amphibia, Anura) from Sri Lanka. *Alytes* **2001,** *19* (2/4), 81–106.

Dutta, S.; Tripathi, S. M.; Mallick, M.; Mathews, R. P.; Greenwood, P. F.; Rao, M. R.; Summons, R. E. Eocene Out-of-India Dispersal of Asian Dipterocarps. *Rev. Palaeobot. Palynol.* **2011,** *166*, 63–68.

Eisenberg, J. F.; McKay, G. M. An Annotated Checklist of the Recent Mammals of Ceylon with Keys to the Species. *Ceylon J. Sci. (Biol. Sci.)* **1970,** *8*, 69–99.

Ellepola, G.; Herath, J.; Manamendra-Arachchi, K.; Wijayathilaka, N.; Senevirathne, G.; Pethiyagoda, R.; Meegaskumbura, M. Molecular Species Delimitation of Shrub Frogs of the Genus *Pseudophilautus* (Anura, Rhacophoridae). *PloS One* **2021,** *16* (10), e0258594.

Emberton, K. C. Acavid Land Snails of Madagascar: Subgeneric Revision Based on Published Data (Gastropoda: Pulmonata: Stylommatophora). *Proc. Acad. Nat. Sci. Philadelphia* **1990,** *142*, 101–117.

Emmel, B.; Lisker, F.; Hewawasam, T. Thermochronological Dating of Brittle Structures in Basement Rocks: A Case Study from the Onshore Passive Margin of SW Sri Lanka. *J. Geophys. Res. Solid Earth* **2012,** *117* (B10).

Erdelen, W. Aspects of the Biogeography of Sri Lanka. *Forschungen auf Ceylon* **1989,** *3*, 73–100.

Erdelen, W. R. Conservation of Biodiversity in a Hotspot: Sri Lanka's Amphibians and Reptiles. *Amphib. Reptile Conserv.* **2012,** *5* (2), 33–51.

Fernando, R. H. S. S. Biodiversity of Ecological Communities and the Biogeography of their Species in Three Isolated Hills in Sri Lanka. PhD Dissertation, University of Peradeniya, Sri Lanka, **2011**.

Fernando, R. H. S. S. Distribution and Habitat Selection of Threatened Orchids of Sri Lanka, *Proceedings of the 11th Asia Pacific Orchid Conference, Okinawa, Japan*, 2013; pp 103–108.

Fernando S. S.; Wickramasingha L. J. M.; Rodirigo R. K. A New Species of Endemic Frog Belonging to Genus *Nannophrys* Gunther, 1869 (Anura: Dicroglossinae) from Sri Lanka. *Zootaxa* **2007,** *68* (1403), 55–68.

Frost, D. R.; Grant, T.; Faivovich, J.; Bain, R. H.; Haas, A.; Haddad, C. F. B.; de Sa, R. O.; Channing, A.; Wilkinson, M.; Donnellan, S. C.; Raxworthy, C. J.; Campbell, J. A.; Blotto, B. L.; Moler, P.; Drewes, R. C.; Nussbaum, R. A.; Lynch, J. D.; Green, D. M.; Wheeler, W. C. The Amphibian Tree of Life. *Bull. Am. Mus. Nat. Hist.* **2006,** *297*, 1–370

Gamage, D. T.; de Silva, M. P.; Inomata, N.; Yamazaki, T.; Szmidt, A. E. Comprehensive Molecular Phylogeny of the Sub-Family Dipterocarpoideae (Dipterocarpaceae) Based on Chloroplast DNA Sequences. *Genes Genet. Syst.* **2006**, *81*, 1–12.

Gamage, S. N.; Groves, C. P.; Marikar, F. M. M. T.; Turner, C. S.; Padmalal, K. U. K. G.; Kotagama, S. W. The Taxonomy, Distribution, and Conservation Status of the Slender Loris (Primates, Lorisidae: Loris) in Sri Lanka. *Primate Conserv.* **2017**, *31*, 83–106.

Gaussen, H.; Legris, P.; Viart, M.; Labroue, L. *International map of the vegetation: Ceylon*; Ceylon Survey Department: Colombo, **1964**.

Ghazoul, J. *Dipterocarp Biology, Ecology, and Conservation*; Oxford University Press: Oxford, 2016.

Ginsberg, J. Global Conservation Priorities. *Conserv. Biol.* **1999**, *13* (1), 5.

Gonzalez, J. C. T. *Origin and Diversification of Hornbills (Bucerotidae)*. PhD Dissertation, University of Oxford, UK, **2012**.

Goonatilake, S. De A.; Fernando, M.; Kotagama, O.; Perera, N; Vidanage, S.; Weerakoon, D.; Daniels, A. G.; Máiz-Tomé, L. *The National Red List of Sri Lanka: Assessment of the Threat Status of the Freshwater Fishes of Sri Lanka 2020*; IUCN, International Union for Conservation of Nature, Sri Lanka and the Biodiversity Secretariat, Ministry of Environment and Wildlife Resources: Sri Lanka, Colombo, **2020**.

Greller, A. M.; Balasubramanium, S. In *Physiognomic, Floristic and Bioclimatological Characterisation of the Major Forest Types of Sri Lanka*, Proceedings of the International and Interdisciplinary Symposium, 'Ecology and Landscape Management in Sri Lanka', March 12–26, 1990; Erdelen, W., Preu, C., Ishwaran, N., Madduma Bandara, C., Eds.; Verlag Josef Margraf: Weikersheim, Germany, **1993**; pp 55–78.

Greller, A. M.; Gunatilleke, I. A. U. N.; Jayasuriya, A. H. M.; Gunatilleke, C. V. S.; Balasubramaniam, S.; Dassanayake, M. D. *Stemonoporus* (Dipterocarpaceae)-Dominated Montane Forests in the Adam's Peak Wilderness, Sri Lanka. *J. Trop. Ecol.* **1987**, *3* (3), 243–253.

Grenyer, R.; Orme, C. D. L.; Jackson, S. F.; Thomas, G. H.; Davies, R. G.; Davies, T. J.; Jones, K. E.; Olson, V. A.; Ridgely, R. S.; Rasmussen, P. C.; Ding, T. S.; Bennett, P. M.; Blackburn, T. M.; Gaston, K. J.; Gittleman, J. L.; Owens, I. P. F. Global Distribution and Conservation of Rare and Threatened Vertebrates. *Nature* **2006**, *444*, 93–96.

Guleria, J. S. Neogene Vegetation of Peninsular India. *Palaeobotanist* **1992**, *40*, 285–331.

Gunatilleke, N.; Gunatilleke, S.; Ashton, P. S. South-West Sri Lanka: A Floristic Refugium in South Asia. *Ceylon J. Sci.* **2017**, *46* (5), 65/78.

Gunatilleke, N.; Pethiyagoda, R.; Gunatilleke, S. Biodiversity of Sri Lanka. *J. Natl Sci. Found. Sri Lanka* **2008**, *36*, 25–62.

Gunawardene, N. R.; Daniels, A. E. D.; Gunatilleke, I. A. U. N.; Gunatilleke, C. V. S.; Karunakaran, P. V.; Nayak, K. G.; Prasad, S.; Puyravaud, P.; Ramesh, B. R.; Subramanian, K. A.; Vasanthy, G. A Brief Overview of the Western Ghats-Sri Lanka Biodiversity Hotspot. *Curr. Sci.* **2007**, *93*, 1567–1572.

Hausdorf, B.; Perera, K. K. Revision of the Genus *Acavus* from Sri Lanka (Gastropoda: Acavidae). *J. Molluscan Stud.* **2000**, *66* (2), 217–231.

Hora, S. L. Satpura Hypothesis of the Distribution of Malayan Fauna and Flora of Peninsular India. *Proc. Natl. Inst. Sci. India* **1949**, *15*, 309–314.

Hora, S. L. The Satpura Hypothesis. *Sci. Prog.* **1953**, *41*, 245–255.

International Union for Conservation of Nature. *The IUCN Red List of Threatened Species. Version 2021–2022 [Online]*. https://www.iucnredlist.org (accessed Nov 06, 2021).

Jacob, K. Land Connections Between Ceylon and Peninsular India. *Proc. Natl. Inst. Sci. India* **1949**, *15* (8), 341–343.

Jayasuriya, A. H. M.; Greller, A. M.; Balasubramaniam, S.; Gunatilleke, C. V. S.; Gunatilleke, I. A. U. N.; Dassanayake, M. D. Phytosociological Studies of Mid-Elevational (Lower Montane) Evergreen Forests in Sri Lanka, *Proceedings of the International and Interdisciplinary Symposium, 'Ecology and Landscape Management in Sri Lanka'*, March 12–26, 1990; Erdelen, W., Preu, C., Ishwaran, N., Madduma Bandara, C., Eds.; Verlag Josef Margraf: Weikersheim, Germany, **1993**; pp 79–94.

Jayawardena, B.; Senevirathne, G.; Wijayathilaka, N.; Ukuwela, K.; Manamendra-Arachchi, K.; Meegaskumbura, M. Species Boundaries, Biogeography and Evolutionarily Significant Units in Dwarf Toads: *Duttaphrynus Scaber* and *D. atukoralei* (Bufonidae: Adenominae). *Ceylon J. Sci.* **2017**, *46* (5), 79–87.

Johnston, G. R.; Lee, M.; Surasinghe, T. D. Morphology and Allometry Suggest Multiple Origins of Rostral Appendages in Sri Lankan Agamid Lizards. *J. Zool.* **2012,** *289,* 1–9.

Joshi, J.; Karanth, P. Did Southern Western Ghats of Peninsular India Serve as Refugia for its Endemic Biota During the Cretaceous Volcanism? *Ecol. Evol.* **2013,** *3* (10), 3275–3282.

Karanth, K. P. Out-of-India Gondwanan Origin of some Tropical Asian Biota. *Curr. Sci.* **2006,** *90,* 789–792.

Karanth, P. An Island Called India: Phylogenetic Patterns Across Multiple Taxonomic Groups Reveal Endemic Radiations. *Curr. Sci.* **2015,** *108,* 1847–1851

Karunarathna, S.; Poyarkov, N. A.; Amarasinghe, C.; Surasinghe, T.; Bushuev, A. V.; Madawala, M.; Gorin, V. A.; De Silva, A. A New Species of the Genus *Ceratophora* Gray, 1835 (Reptilia: Agamidae) from a Lowland Rainforest in Sri Lanka, with Insights on Rostral Appendage Evolution in Sri Lankan Agamid lizards. *Amphib. Reptile Conserv.* **2020,** *14* (3), 103–126.

Karunarathna, S.; Poyarkov, N. A.; De Silva, A.; Madawala, M.; Botejue, M.; Gorin, V. A.; Surasinghe, T.; Gabadage, D.; Ukuwela, K. D.; Bauer, A. M. Integrative Taxonomy Reveals Six New Species of Day Geckos of the Genus *Cnemaspis* Strauch, 1887 (Reptilia: Squamata: Gekkonidae) from Geographically Isolated Hill Forests in Sri Lanka. *Vertebr. Zool.* **2019,** *64,* 247–298.

Kazmierczak, K.; van Perlo, B. *A field guide to the birds of India*; O. M. Book Service: Delhi, **2000.**

Kent, D. V.; Muttoni, G. Equatorial Convergence of India and Early Cenozoic Climate Trends. *Proc. Natl Acad. Sci. USA* **2008,** *105,* 16065–16070.

Klaus, S.; Morley, R. J.; Plath, M.; Zhang, Y. P.; Li, J. T. Biotic Interchange Between the Indian Subcontinent and Mainland Asia Through Time. *Nat. Commun.* **2016,** *7* (1), 1–6.

Klaus, S.; Schubart, C. D.; Streit, B.; Pfenninger, M. When Indian Crabs were not yet Asian - Biogeographic Evidence for Eocene Proximity of India and Southeast Asia. *BMC Evol. Biol.* **2010,** *10* (1), 1–9.

Korall, P.; Pryer, K. M. Global Biogeography of Scaly Tree Ferns (Cyatheaceae): Evidence for Gondwanan Vicariance and Limited Transoceanic Dispersal. *J. Biogeogr.* **2014,** *41,* 402–413.

Kotagama, S. W. Map of the Avifauna Zones of Sri Lanka, In *Strategy for the Preparation of a Biological Diversity Action Plan for Sri Lanka*: Colombo, **1989.**

Krishan, K. T.; Weerakkody, S.; Seneviratne, S. Phylogenetic Affinities of an Endemic Cloud Forest Avian Relict: Sri Lanka Bush Warbler *(Elaphrornis palliseri*), *27ᵗʰ International Forestry Symposium*; University of Sri Jayawardanapura: Colombo, 2020; p 36.

Kruizinga, J.; van Scheindelen, H. J.; De Vogel, E. F. Revision of the Genus *Bromheadia* (Orchidaceae). *Orchid Monographs* **1997,** *8,* 79–118.

Kumar, C. S.; Manilal, K. S. *A Catalogue of Indian Orchids*; Mentor Books & IAAT: Calicut, **1994.**

Kumar, A.; Pethiyagoda, R; Mudappa, D. Western Ghats and Sri Lanka. In *Hotspots Revisited: Earth's Biologically Richest and most Endangered Ecoregions*; Mittermeier, R. A., Robles Gil, P., Hoffmann, M., Pilgrim, J., Brooks, T., Mittermeier, C. G., Lamoreux, J., Da Fonseca, G. A. B., Eds.; CEMEX: Mexico City, 2004; pp 152–157.

Kuntner, M. Phylogenetic Systematics of the Gondwanan Nephilid Spider Lineage Clitaetrinae (Araneae, Nephilidae). *Zool. Scr.* **2006,** *35* (1), 19–62.

Kurup, G. U. Mammals of Assam and the Mammal Geography of India. In *Ecology and Biogeography in India*; Mani, M. S., Ed.; Dr. W Junk Publishers: Hauge, 1974; pp 585–613.

Kuttapetty, M.; Pillai, P. P.; Varghese, R. J.; Seeni, S. Genetic Diversity Analysis in Disjunct Populations of *Rhododendron arboreum* from the Temperate and Tropical Forests of Indian Subcontinent Corroborate Satpura Hypothesis of Species Migration. *Biologia* **2014,** *69* (3), 311–322.

Lajmi, A.; Bansal, R.; Giri, V.; Karanth, P. Phylogeny and Biogeography of the Endemic *Hemidactylus* Geckos of the Indian Subregion Suggest Multiple Dispersals from Peninsular India to Sri Lanka. *Zool. J. Linn. Soc.* **2019,** *186,* 286–301.

Legg, C.; Jewell, N. A 1: 50 000-Scale Forest Map of Sri Lanka: The Basis for a National Forest Geographic System. *The Sri Lanka Forester, Special Issue (Remote Sensing)* **1995,** pp 3–24.

Legge, W. V. *A History of the Birds of Ceylon*; Published by author: London, **1880.**

Loria, S. F.; Prendini, L. Out of India, Thrice: Diversification of Asian Forest Scorpions Reveals Three Colonisations of Southeast Asia. *Sci. Rep.* **2020,** *10,* 22301.

Macey, J. R.; Schulte II, J. A.; Larson, A.; Ananjeva, N. B.; Wang, Y.; Pethiyagoda, R.; Rastegar-Pouyani, N.; Papenfuss, T. Evaluating Trans –Tethys Migration: An Example Using Acrodont Lizard Phylogenetics. *Syst. Biol.* **2000**, *49*, 233–256.

Manamendra-Arachchi, K.; Pethiyagoda, R.; Dissanayake, R.; Meegaskumbura, M. A Second Extinct Big Cat from the Late Quaternary of Sri Lanka. *Raffles Bull. Zool.* **2005**, *12*, 423–434.

Mani, M. S. Biogeogrographical Evolution in India. In *Ecology and Biogeography in India*; Mani, M. S., Ed.; Dr. W Junk Publishers: Hauge, 1974; pp 698–724.

Mapa R. B. Soil Research and Soil Mapping History. In *The Soils of Sri Lanka*; Mapa, R., Ed.; World Soils Book Series; Springer International: Switzerland, 2020; pp 1–12.

McCloskey, J. M.; Spalding, H. A Reconnaissance-Level Inventory of the Amount of Wilderness Remaining in the World. *Ambio* **1989**, *18*, 221–227.

Medlicott, H. B.; Blanford, W. T. *A Manual of the Geology of India*; Calcutta: India, **1879**.

Meegaskumbura, M.; Bossuyt, F.; Pethiyagoda, R.; Manamendra-Arachchi, K.; Bahir, M.; Milinkovitch, M.; Schneider, C. Sri Lanka: An Amphibian Hot Spot. *Science* **2002**, *298*, 379–379.

Meegaskumbura, M.; Manamendra-Arachchi, K. Description of Eight New Species of Shrub Frogs (Ranidae: Rhacophorinae: *Philautus*) from Sri Lanka. *Raffles Bull. Zool.* **2005**, *12*, 305–338.

Meegaskumbura, M.; Senevirathne, G.; Manamendra-Arachchi, K.; Pethiyagoda, R.; Hanken, J.; Schneider, C. J. Diversification of Shrub Frogs (Rhacophoridae, *Pseudophilautus*) in Sri Lanka – Timing and Geographic Context. *Mol. Phylogenet. Evol.* **2019**, *132*, 14–24.

Meert, J. G.; Van Der Voo, R. The Assembly of Gondwana 800-550 Ma. *J. Geodyn.* **1997**, *23*, 223–235.

Mittermeier, R. A.; Gil, P. R.; Hoffmann, M.; Pilgrim, J.; Brooks, T.; Mittermeier, C. G.; Lamoreux, J.; da Fonseca, G. A. B. *Hotspots Revisted: Earth's Biologically Wealthiest and Most Threatened Ecosystems*; CEMEX: Mexico City, **2004**.

Mittermeier, R. A.; Gil, P. R.; Mittermeier, C. G. *Megadiversity: Earth's Biologically Wealthiest Nations*; CEMEX: Mexico City, 1997.

MOE. *The National Red List 2012 of Sri Lanka; Conservation Status of the Fauna and Flora*; Biodiversity Secretariat, Ministry of Environment: Colombo, Sri Lanka, **2012**; p viii+476.

MoE. *The National Red List 2021 – Conservation Status of the Birds of Sri Lanka*; Biodiversity Secretariat, Ministry of Environment: Colombo, Sri Lanka, **2021**.

MoE and DNBG. *The National Red List 2020 – Conservation Status of the Flora of Sri Lanka*; Biodiversity Secretariat, Ministry of Environment and the National Herbarium, Department of National Botanic Gardens: Colombo, Sri Lanka, **2020**; p xviii+254.

MoMD&E. *Biodiversity Profile - Sri Lanka, Sixth National Report to the Convention on Biological Diversity*; Jayakody, S., Wikramanayake, E. D., Fernando, S., Wickramaratne, C., Arachchige, G. M., Akbarally, Z., Eds.; Biodiversity Secretariat, Ministry of Mahaweli Development and Environment: Colombo, Sri Lanka, **2019**; p 211.

Moore, J. C. Squirrel Geography of the Indian Subregion. *Syst. Zool.* **1960**, *9*, 1–19.

Morley, R. J. *Origin and Evolution of Tropical Rain Forests*; John Wiley & Sons: Chichester, **2000**.

Morley, R. J. Interplate Dispersal Paths for Megathermal Angiosperms. *Perspect. Plant Ecol. Evol. Syst.* **2003**, *6*, 5–20.

Murphy, B.; Forest, F.; Barraclough, T.; Rosindell, J.; Bellot, S.; Cowan, R.; Golos, M.; Jebb, M.; Cheek, M. A Phylogenomic Analysis of *Nepenthes* (Nepenthaceae). *Mol. Phylogenet. Evol.* **2020**, *144*, 106668.

Myers, N. Threatened Biotas: Hotspots in Tropical Forests. *Environmentalist* **1988**, *8*, 178–208.

Myers, N. The Biodiversity Challenge: Expanded Hot-Spots Analysis. *Environmentalist* **1990**, *10*, 243–256.

Myers, N. Biodiversity Hotspots Revisited. *Bioscience* **2003**, *53*, 916–917.

Myers, N.; Mittermeier, R. A. Impact and Acceptance of the Hotspots Strategy: Response to Ovadia and to Brummitt and Lughadha. *Conserv. Biol.* **2003**, *17*, 1449–1450.

Myers, N.; Mittermeier, R.; Mittermeier, C.; da Fonseca, G.; Kent, J. Biodiversity Hotspots for Conservation Priorities. *Nature* **2000**, *403*, 853–858.

Naggs, F.; Raheem, D. Sri Lankan Snail Diversity: Faunal Origins and Future prospects. *Rec. West. Aust. Mus.* **2005**, *68*, 11–29.

Noss, R. F.; Platt, W. J.; Sorrie, B. A.; Weakley, A. S.; Means, D. B.; Costanza, J.; Peet, R. K. How Global Biodiversity Hotspots may go Unrecognized: Lessons from the North American Coastal Plain. *Divers. Distrib.* **2015,** *21* (2), 236–244.

O' Shea, B. J. Biogeographical Relationship of the Mosses of Sri Lanka. *J. Hattori. Bot. Lab.* **2003,** *93,* 293–304.

Olson, D. M.; Dinerstein, E. The Global 200: A Representation Approach to Conserving the Earth's Most Biologically Valuable Ecoregions. *Conserv. Biol.* **1998,** *12* (3), 502–515.

Orme, C. D. L.; Davies, R. G.; Burgess, M.; Eigenbrod, F.; Pickup, N.; Olson, V. A.; Webster, A. J.; Ding, T. S.; Rasmussen, P. C.; Ridgely, R. S.; Stattersfield, A. J.; Bennett, P. M.; Blackburn, T. M.; Gaston, K. J.; Owens, I. P. F. Global Hotspots of Species Richness are not Congruent with Endemism or Threat. *Nature* **2005,** *436,* 1016–1019.

Ovadia, O. Ranking Hotspots of Varying Sizes: A Lesson from the Nonlinearity of the Species-Area Relationship. *Conserv. Biol.* **2003,** *17,* 1440–1441.

Peeris, C. V. S. The Ecology of Endemic Tree Species of Sri Lanka in Relation to their Conservation. PhD Dissertation, University of Aberdeen, UK, 1975.

Pemadasa M. A. Grasslands. In *Ecology and Biogeography in Sri Lanka*; Fernando, C. H., Ed.; Dr. W Junk Publishers: Hauge, **1984**; pp 453–492.

Pemadasa, M. A. Tropical Grasslands of Sri Lanka and India. *J. Biogeogr.* **1990,** *17,* 395–400.

Perera, S.; Surasinghe, T.; Somaweera, R.; Ramdhani, S; Karunaratne, S. Hyper-Hotspots Within a Biodiversity Hotspot: Patterns of Reptile Endemism in Sri Lanka. *ATBC, Asia Pacific Chapter Meeting*; Thulhiriya, Sri Lanka, Sept 10–13, 2019; Association for Tropical Biology and Conservation: Colombo, **2019a**; p 63.

Perera, S. J.; Fernando, R. H. S. S.; Gunatilleke, C. V. S.; Gunatilleke, I. A. U. N. Revisiting the Floristic Regions of Sri Lanka, Using Recent Distribution Data for a Numerical Analysis with an Emphasis on the Intermediate Zone. *ATBC, Asia Pacific Chapter Meeting*; Thulhiriya, Sri Lanka, Sept 10–13, 2019; Association for Tropical Biology and Conservation: Colombo, **2019b**; p 132.

Perera, S. J.; Procheṣ, Ş.; Ratnayake-Perera, D.; Ramdhani, S. Vertebrate Endemism in South-Eastern Africa Numerically Redefines a Biodiversity Hotspot. *Zootaxa* **2018,** *4382,* 56–92.

Perera, S. J.; Ramdhani, S.; Procheş, Ş.; Ratnayake-Perera, D. New Biogeographic Insights for Partitioning of Sri Lanka: A Preliminary Attempt of a Numerical Regionalization, *National Symposium on the Biogeography and Biodiversity Conservation in Sri Lanka in a Changing Climate,* Colombo, Sri Lanka, Nov 12–13, 2015; National Science Foundation: Colombo, 2015; p 10.

Perera, S. J.; Ratnayake-Perera, D.; Proches, S. Vertebrate Distributions Indicate a Greater Maputaland-Pondoland-Albany Region of Endemism. *South African J. Sci.* **2011,** *107,* 1–15.

Pethiyagoda, R.; Manamendra-Arachchi, K. A Revision of the Endemic Sri Lankan Agamid Lizard Genus *Ceratophora* Gray, 1835, with Description of Two New Species. *J. South Asian Nat. Hist.* **1998,** *3* (1), 1–50.

Pethiyagoda, R.; Sudasinghe, H. The Development of Sri Lankan Biogeography in the Colonial Period. *Ceylon J. Sci.* **2017,** *46,* 5–18.

Pethiyagoda, R.; Sudasinghe, H. *The Ecology and Biogeography of Sri Lanka: A Context for Freshwater Fishes*; WHT Publications (Private) Limited: Colombo, 2021.

Premathilake, R.; Risberg, J. Late Quaternary Climate History of the Horton Plains, Central Sri Lanka. *Quat. Sci. Rev.* **2003,** *22* (14), 1525–1541.

Punyawardena, B. V. R. Climate of Sri Lanka, *Proceedings of the first national conference on global climate change and its impacts on Agriculture, 'Forestry and Water in the tropics',* **2009**; pp 7–20.

Rabinowitz, P. D.; Coffin, M. F.; Falvey, D. The Separation of Madagascar and Africa. *Science* **1983,** *220,* 67–69.

Raheem, D. C.; Breugelmans, K.; Wade, C. M.; Naggs, F. C.; Backeljau, T. Exploring the Shell-Based Taxonomy of the Sri Lankan Land Snail *Corilla* H. and A. Adams, 1855 (Pulmonata: Corillidae) Using Mitochondrial DNA. *Mol. Phylogenet. Evol.* **2017,** *107,* 609–618.

Raheem, D. C.; Naggs, F.; Preece, R. C.; Mapatuna, Y.; Kariyawasam, L.; Eggleton, P. Structure and Conservation of Sri Lankan Land-Snail Assemblages in Fragmented Lowland Rainforest and Village Home Gardens. *J. Appl. Ecol.* **2008,** *45* (4), 1019–1028.

Rajapakse, S.; Iddamalgoda, P.; Ratnayake, R.; Wijesundara, D. S. A.; Bandara, B. R.; Karunaratne, V. Evaluation of Species Limits of *Hortonia* by DNA Barcoding. *J. Natl. Sci. Found. Sri Lanka* **2012,** *40* (4), 345–349.

Rajbhandary, S.; Hughes, M.; Phutthai, T.; Thomas, D. C.; Shrestha, K. K. Asian Begonia: Out of Africa via the Himalayas. *Gard. Bull. Singapore* **2011**, *63* (1&2), 277–286.

Ramdhani, S.; Perera, S. J.; Proches, S. Introducing a Quarter Degree Grid System for Sri Lanka as a Biogeographical Tool, *3rd International Symposium of SUSL*, Belihuloya, Sri Lanka, Aug 26–28, 2010; Sabaragamuwa University of Sri Lanka: Belihuloya, 2010; p 10.

Ranil, R. H. G.; Pushpakumara, D. K. N. G. Taxonomy and Conservation Status of Pteridophyte Flora of Sri Lanka. In *The National Red List 2012 of Sri Lanka; Conservation Status of the Fauna and Flora*; Weerakoon, D. K., Wijesundara, S., Eds.; Ministry of Environment: Colombo, Sri Lanka, 2012; pp 148–164.

Ranil, R. H. G.; Pushpakumara, D. K. N. G.; Wijesundara, D. S. A.; Bostock, P. D.; Ebihara, A.; Fraser-Jenkins, C. R. Diversity and Distributional Ecology of Tree Ferns of Sri Lanka: A Step Towards Conservation of a Unique Gene Pool. *Ceylon J. Sci.* **2017**, *46* (5), 127–135.

Ratheesh-Kumar, R. T.; Dharmapriya, P. L.; Windley, B. F.; Xiao, W. J.; Jeevan, U. The Tectonic "Umbilical Cord" Linking India and Sri Lanka and the Tale of their Failed Rift. *J. Geophys. Res. Solid Earth* **2020**, *125* (5), e2019JB018225.

Raven, P. H.; Axelrod, D. I. Angiosperm Biogeography and Past Continental Movements. *Ann. Missouri Bot. Gard.* **1974**, *61* (3), 539–673.

Raxworthy, C. J.; Forstner, M. R. J.; Nussbaum, R. A. Chameleon Radiation by Oceanic Dispersal. *Nature* **2002**, *415* (6873), 784–787.

Reddy, C. S.; Manaswini, G.; Jha, C. S.; Diwakar, P. G.; Dadhwal, V. K. Development of National Database on Long-Term Deforestation in Sri Lanka. *J. Indian Soc. Remote Sens.* **2017**, *45* (5), 825–836.

Reeves, C. The Position of Madagascar Within Gondwana and its Movements During Gondwana Dispersal. *J. African Earth Sci.* **2014**, *94*, 45–57.

Repetur, C. P.; Van Welzen, P. C.; De Vogel, E. F. Phylogeny and Historical Biogeography of the Genus *Bromheadia* (Orchidaceae). *Syst. Bot.* **1997**, *22* (3), 465–477.

Reuter, M.; Harzhauser, M.; Piller, W. E.; The Role of Sea-Level and Climate Changes in the Assembly of Sri Lankan Biodiversity: A Perspective from the Miocene Jaffna Limestone. *Gondwana Res.* **2021**, *91*, 152–165.

Ripley, S. D. Avian Relicts and Double Invasions in Peninsular India and Ceylon. *Evolution* **1949**, *3*, 150–159.

Ripley, S. D. Avian Relicts of Sri Lanka. *Spolia Zeylanica* **1980**, *35*, 197–202.

Rodrigo, K.; Manamendra-Arachchi, K. Reconsidering the Palaeo-Environmental Reconstruction of the Wet Zone of Sri Lanka: A Zooarchaeological Perspective. *Int. J. Environ. Ecol. Eng.* **2020**, *14* (10), 284–294.

Rohling, E. J.; Fenton, M. J. J. F.; Jorissen, F. J.; Bertrand, P.; Ganssen, G.; Caulet, J. P. Magnitudes of Sea-Level Lowstands of the Past 500,000 Years. *Nature* **1998**, *394* (6689), 162–165.

Rosauer, D. F.; Jetz, W. Phylogenetic Endemism in Terrestrial Mammals. *Glob. Ecol. Biogeogr.* **2015**, *24* (2), 168–179.

Rubasinghe, S. C. K. Species Limits and Phylogenetics of the Endemic Genus *Stemonoporus* (Dipterocapaceae). MPhil, Dissertation, University of Peradeniya, Sri Lanka, 2007.

Rubasinghe, S. C. K.; Yakandawala, D. M. D.; Wijesundara, D. S. A. Phylogenetics of the Endemic Genus *Stemonoporus* Thw. (Dipterocarpaceae). *J. Natl. Sci. Found. Sri Lanka* **2008**, *36* (4), 281–297.

Samarawickrama, V. A. M. P. K.; Ranawana, K. B.; Rajapaksha, D. R. N. S.; Ananjeva, N. B.; Orlov, N. L.; Ranasinghe, J. M. A. S.; Samarawickrama, V. A. P. A New Species of the Genus *Cophotis* (Squamata: Agamidae) from Sri Lanka. *Russian J. Herpetol.* **2006**, *13* (3), 207–214.

Schulte, J. A.; Macey, J. R.; Pethiyagoda, R.; Larson, A. Rostral Horn Evolution among Agamid Lizards of the Genus *Ceratophora* Endemic to Sri Lanka. *Mol. Phylogenet. Evol.* **2002**, *22*, 111–117.

Sclater, P. L. On the General Geographical Distribution of the Members of the Class Aves. *Zool. J. Linn. Soc.* **1858**, *2*, 130–136.

Sclater, P. L. The Geographical Distribution of Birds; An Address delivered before the Second International Ornithological Congress, *2nd International Ornithological Congress*; Budapest, Hungary, May 1891; Taylor and Francis: London, 1891.

Sclater, W. L. The Geography of Mammals: no. V. The Oriental Region (continued). *Geogr. J.* **1896**, *8*, 378–389.

Sclater, W. L.; Sclater, P. L. *The Geography of Mammals*; K. Paul, Trench, Trübner & Company: London, **1899**.

Senanayake, F. R. The Evolution of the Major Landscape Categories in Sri Lanka and Distribution Patterns of Some Selected Taxa: Ecological Implications, *Proceedings of the International and Interdisciplinary*

Symposium, 'Ecology and Landscape Management in Sri Lanka', March 12–26, 1990; Erdelen, W.; Preu, C.; Ishwaran, N.; Madduma Bandara, C., Eds.; Verlag Josef Margraf: Weikersheim, Germany, 1993; pp 201–219.

Senanayake, F. R.; Moyle, P. B. Conservation of Freshwater Fishes of Sri Lanka. *Biol. Conserv.* **1982,** *22* (3), 181–195.

Senanayake, F. R.; Soulé, M.; Senner, J. W. Habitat Values and Endemicity in the Vanishing Rain Forests of Sri Lanka. *Nature* **1977,** *265,* 351–354.

Senevirathne, G.; Samarawickrama, V. A. M. P. K.; Wijayathilaka, N.; Manamendra-Arachchi, K.; Bowatte, G.; Samarawickrama, D. R. N. S.; Meegaskumbura, M. A New Frog Species from Rapidly Dwindling Cloud Forest Streams of Sri Lanka—*Lankanectes pera* (Anura, Nyctibatrachidae). *Zootaxa* **2018,** *4461* (4), 519–520.

Silas, E. G. Further Studies Regarding Hora's Satpura Hypothesis. 2. Taxonomic Assessment and Levels of Evolutionary Divergences of Fishes with the So-Called Malayan Affinities in Peninsular India. *Proc. Natl. Inst. Sci. India* **1952,** *18* (5), 423–448.

Sledge, W. A. An Annotated Check-List of the Pteridophyta of Ceylon. *Bot. J. Linn. Soc.* **1982,** *84* (1), 1–30.

Sparks, J. S. Molecular Phylogeny and Biogeography of the Malagasy and South Asian Cichlids (Teleostei : Perciformes : Cichlidae). *Mol. Phylogenet. Evol.* **2004,** *30,* 599–614.

Sreekar, R.; Koh, L. P.; Mammides, C.; Corlett, R. T.; Dayananda, S.; Goodale, U. M.; Kotagama, S. W.; Goodale, E. Drivers of Bird Beta Diversity in the Western Ghats–Sri Lanka Biodiversity Hotspot are Scale Dependent: Roles of Land Use, Climate, and Distance. *Oecologia* **2020,** *193,* 801–809.

Srikanthan, A. N.; Adhikari, O. D.; Ganesh, S. R.; Deuti, K.; Kulkarni, V. M.; Gowande, G. G.; Shanker, K. A Molecular and Morphological Study of *Otocryptis* Wagler, 1830 (Squamata: Agamidae) Reveals A New Genus from the Far South of the Western Ghats, Peninsular India. *Zootaxa* **2021,** *5016* (2), 205–228.

Stattersfield, A. J.; Crosby, M. J.; Long, A. J.; Wege, D. C. *Endemic Bird Areas of the World-Priorities for Biodiversity Conservation*; BirdLife International: Cambridge, 1998.

Storey, B. C. The Role of Mantle Plumes in Continental Breakup: Case Histories from Gondwanaland. *Nature* **1995,** *377,* 301–308.

Storey, M.; Mahoney, J. J.; Saunders, A. D.; Duncan, R. A.; Kelley, S. P.; Coffin, M. F. Timing of Hot Spot-Related Volcanism and the Breakup of Madagascar and India. *Science* **1995,** *267* (5199), 852–855.

Sudasinghe, H.; Ranasinghe, R. T.; Goonatilake, S. D. A.; Meegaskumbura, M. A Review of the Genus *Labeo* (Teleostei: Cyprinidae) in Sri Lanka. *Zootaxa* **2018,** *4486* (3), 201–235.

Tolley, K. A.; Townsend, T. M.; Vences, M. Large-Scale Phylogeny of Chameleons Suggests African Origins and Eocene Diversification. *Proc. Roy. Soc. B: Biol. Sci.* **2013,** *280* (1759), 20130184.

Toussaint, E. F.; Fikáček, M.; Short, A. E. India–Madagascar Vicariance Explains Cascade Beetle Biogeography. *Biol. J. Linn. Soc.* **2016,** *118* (4), 982–991.

Trimen, H. Remarks on the Composition, Geographical Affinities, and Origin of the Ceylon Flora. *J. Royal Asiatic Soc. Ceylon Branch* **1885,** *9,* 1–21.

Trimen, H. On the Flora of Ceylon, Especially as Affected by Climate. *J. Bot.* **1886,** *24,* 801–805.

Udvardy, M. D. A Classification of the Biogeographical Provinces of the World; *IUCN Occasional Paper no. 18*; International Union for Conservation of Nature and Natural Resources: Morges, Switzerland, 1975.

Van Bocxlaer, I.; Biju, S. D.; Loader, S. P.; Bossuyt, F. Toad Radiation Reveals Into-India Dispersal as a Source of Endemism in the Western Ghats-Sri Lanka Biodiversity Hotspot. *BMC Evol. Biol.* **2009,** *9* (1), 1–10.

van Hinsbergen, D. J.; Steinberger, B.; Doubrovine, P. V.; Gassmöller, R. Acceleration and Deceleration of India-Asia Convergence since the Cretaceous: Roles of Mantle Plumes and Continental Collision. *J. Geophys. Res. Solid Earth* **2011,** *116* (B6).

Vaz, G. G. Age of Relict Coral Reef from the Continental Shelf off Karaikal, Bay of Bengal: Evidence of Last Glacial Maximum. *Curr. Sci.* **2000,** *78,* 228–229.

Vidal, N.; Marin, J.; Morini, M.; Donnellan, S.; Branch, W. R.; Thomas, R.; Vences, M.; Wynn, A.; Cruaud, C.; Hedges, S. B. Blindsnake Evolutionary Tree Reveals Long History on Gondwana. *Biol. Lett.* **2010,** *6* (4), 558–561.

Vijayakumar, S. P.; Pyron, R. A.; Dinesh, K. P.; Torsekar, V. R.; Srikanthan, A. N.; Swamy, P.; Stanley, E. L.; Blackburn, D. C.; Shanker, K. A New Ancient Lineage of Frog (Anura: Nyctibatrachidae: Astrobatrachinae subfam. nov.) Endemic to the Western Ghats of Peninsular India. *Peer J.* **2019,** *7,* e6457.

Vitanage, P. W. Post Precambrian Uplifts and Regional Neotectonic Movements in Ceylon, *Proceedings of the 24th International Geological Congress;* Montreal, 1972; section 3, pp 642–654.

Vitanage P. W. A Study of the Geomorphology and the Morphotectonics of Ceylon, *Proceedings of the UNESCO/ESCAFE 2nd international seminar on geochemical prospecting methods and techniques;* New York, 1970; pp 391–405.

Voris, H. K. Maps of Pleistocene Sea Levels in Southeast Asia: Shorelines, River Systems and Time Durations. *J. Biogeogr.* **2000,** *27,* 1153–1167.

Wadia, D. N. The Three Superimposed Peneplains of Ceylon: Their Physiography and Geological structure. *Rec. Dept. Mineral. (Ceylon)* **1945,** *1,* 25–32.

Wallace, A.R. *The geographical distribution of animals* (2 vols); Macmillan: London, UK, **1876**.

Wan, S.; Ku¨rschner, W. M.; Clift, P. D.; Li, A.; Li, T. Extreme Weathering/Erosion During the Miocene Climatic Optimum. Evidence from Sedimentrecord in the South China Sea. *Geophys. Res. Lett.* **2009,** *36,* L19706.

Ward, P. S.; Brady, S. G.; Fisher, B. L.; Schultz, T. R.; Phylogeny and Biogeography of Dolichoderine Ants: Effects of Data Partitioning and Relict Taxa on Historical Inference. *Syst. Biol.* **2010,** *59* (3), 342–362.

Wesener, T.; Raupach, M. J.; Sierwald, P. The Origins of the Giant Pill-Millipedes from Madagascar (Diplopoda: Sphaerotheriida: Arthrosphaeridae). *Mol. Phylogenet. Evol.* **2010,** *57,* 1184–1193.

Whistler, H. The Avifaunal Survey of Ceylon Conducted Jointly by the British and Colombo Museums. *Spolia Zeylanica* **1944,** *23,* 119–322.

Whitmore, T. C. A Vegetation map of; Malesia at Scale 1: 5 Million. *J. Biogeography* **1984,** *11,* 461–471.

Wickramasinghe, N.; Robin, V. V.; Ramakrishnan, U.; Seneviratne, S. S. Non-Sister Sri Lankan White-Eyes (Genus *Zosterops*) are a Result of Independent Colonizations. *PLoS One* **2017,** *12* (8), e0181441.

Wijesinghe L. C. A. de S.; Gunatilleke I. A. U. N.; Jayawardana S. D. G.; Kotagama S. W.; Gunatilleke, C. V. S. *Biological Conservation in Sri Lanka: A National Status Report;* IUCN, International Union for Conservation of Nature, Sri Lanka: Colombo, **1993**.

Wijesundara, D. S. A. Areas of Endemism in the Angiosperm Flora of Sri Lanka, *National Symposium on the Biogeography and Biodiversity Conservation in Sri Lanka in a Changing Climate,* Colombo, Sri Lanka, Nov 12–13, 2015; National Science Foundation: Colombo, 2015, p 7.

Williams, K. J.; Ford, A.; Rosauer, D. F.; De Silva, N.; Mittermeier, R.; Bruce, C.; Larsen, F. W.; Margules, C. Forests of East Australia: the 35th biodiversity hotspot. In *Biodiversity Hotspots. Distribution and Protection of Conservation Priority Areas;* Zachos, F. E., Habel, J. C., Eds.; Springer: Berlin, **2011**; pp 295–310.

Willis, J. C. *The flora of Ritigala, an isolated mountain in the North-Central Province of Ceylon; a study in endemism;* Annual Report of the Botanic Gardens: Peradeniya, **1906**; vol 2, pp 271–302.

Willis, J. C. *The Floras of Hill Tops in Ceylon;* Annual Report of the Botanic Gardens: Peradeniya, 1908; vol 4, pp 131–138.

Willis, J. C. *The Flora of Naminakuli-kanda, A Somewhat Isolated Mountain in the Province of Uva;* Annual Report of the Botanic Gardens: Peradeniya, **1911**; vol 5, p 217.

Willis, J. C. The Endemic Flora of Ceylon, with Reference to Geographical Distribution and Evolution in General. *Philos. Trans. R. Soc. Lond. Ser. B* **1915,** *206,* 307–342.

Willis, J. C. *Age and Area: A Study in Geographic Distribution and Origin of Species;* The University Press: Cambridge, 1922.

Wilson, E. O.; Eisner, T.; Wheeler, G. C.; Wheeler, J. *Aneuretus simony* Emery, A Major Link in Ant Evolution. *Bull. Mus. Comp. Zool.* **1956,** *115,* 81–99.

Wilson, K. A.; McBride, M. F.; Bode, M.; Possingham, H. P. Prioritizing Global Conservation Efforts. *Nature* **2006,** *440,* 337–340.

Yoder, A. D. Back to the Future: A Synthesis of Strepsirrhine Systematics. *Evol. Anthropol.* **1997,** *6,* 11–22.

Yoder, A. D.; Yang, Z. Divergent Dates for Malagasy Lemur Estimated from Multiple Gene Loci: Geological and Evolutionary Context. *Mol. Ecol.* **2004,** *13,* 757–773.

Yoshida, M.; Funaki, M.; Vitanage, P. W. Proterozoic to Mesozoic East Gondwana: The Juxtaposition of India, Sri Lanka, and Antarctica. *Tectonics* **1992,** *11,* 381–391.

Yuan, Y.; Wohlhauser, S.; Moller, M.; Klackenberg, J.; Callmander, M. W.; Kupfer, P. Phylogeny and Biogeography of *Exacum* (Gentianaceae): A Disjunctive Distribution in the Indian Ocean Basin Resulted from Long Distance Dispersal and Extensive Radiation. *Syst. Biol.* **2005,** *54* (1), 21–34.

Zachos, J. C.; Dickens, G. R; Zeebe, R. E. An Early Cenozoic Perspective on Greenhouse Warming and Carbon-Cycle Dynamics. *Nature* **2008**, *451*, 279–283.

Zaher, H.; Grazziotin, F. G.; Graboski, R.; Fuentes, R. G.; Sánchez-Martinez, P.; Montingelli, G. G.; Zhang, Y. P.; Murphy, R. W. Phylogenetic Relationships of the Genus *Sibynophis* (Serpentes: Colubroidea). *Pap. Avulsos Zool.* **2012**, *52*, 141–149.

Zink, R.; Blackwell-Rago, R.; Ronquist, F. The Shifting Roles of Dispersal and Vicariance in Biogeography. *Proc. Roy. Soc. B-Biol. Sci.* **2000**, *267*, 497–503.

CHAPTER 24

OVERVIEW OF SRI LANKAN FUNGI AND LICHEN RESEARCH

MAHESH C. A. GALAPPATHTHI[1], SAMANTHA C. KARUNARATHNA[2], CHANDRIKA M. NANAYAKKARA[3], STEVEN L. STEPHENSON[4], LUCAS DAUNER[5], NALIN WIJAYAWARDENE[2], and UDENI JAYALAL[6]

[1]Postgraduate Institute of Science, University of Peradeniya, Peradeniya, Sri Lanka

[2]Center for Yunnan Plateau Biological Resources Protection and Utilization, College of Biological Resource and Food Engineering, Qujing Normal University, Qujing, Yunnan, P.R. China

[3]Department of Plant Sciences, University of Colombo, Colombo, Sri Lanka

[4]Department of Biological Sciences, University of Arkansas, Fayetteville, AR, USA

[5]Kunming Institute of Botany, Chinese Academy of Sciences, Kunming, Yunnan, P.R. China

[6]Department of Natural Resources, Sabaragamuwa University of Sri Lanka, Belihuloya, Sri Lanka

ABSTRACT

Sri Lanka, with a humid tropical climate, is a potentially fungus-rich country. It has been estimated to have 33,000 species of fungi, but little more than 2000 of these have been documented. Compared to limited number studies of microfungi, more research has been directed toward lichens and macrofungi by mycologists in Sri Lanka. Some studies have been carried out on the taxonomy of microfungi, but based on estimates of the size of this group for other regions of the world, many more microfungi remain to be discovered. Very few edible mushrooms are consumed by people in Sri Lanka mainly due to the lack of knowledge, while commercial mushroom cultivation is also limited to a very few examples. This chapter addresses the current status of macrofungi, microfungi (including fungus-like organisms), and lichenological studies in Sri Lanka.

Biodiversity Hotspot of the Western Ghats and Sri Lanka. T. Pullaiah, PhD (Ed.)
© 2024 Apple Academic Press, Inc. Co-published with CRC Press (Taylor & Francis)

24.1   INTRODUCTION

Sri Lanka is a tropical island, with an approximate land cover of 65,000 km², separated from the Indian mainland in the Miocene. The central portion of the island has mountains with elevations up to 2524 m. Sri Lanka, which has a tropical humid climate (Silva, 1988), is estimated to be the home for more than 33,000 species of fungi, but only little more than 2000 of these have been documented. Although Sri Lanka is a potentially fungus-rich country, relatively few studies have been carried out on either fungi or lichens. Moreover, accurate estimates of Sri Lankan fungi are further complicated due to the widely scattered available information and the synonyms used in different studies (Karunarathna et al., 2017).

Houttuyn (1783) described the first fungi (*Peziza ceylonische* and *P. lembosa*) recorded from Sri Lanka. Tom Petch (1908–1947), who worked at the royal botanical gardens in Peradeniya, documented various plant pathogenic and other fungi. In addition, Petch and Bisby (1950) also reported numerous taxa in the genera *Agaricus* (26 species), *Lepiota* (53 species), *Marasmius* (39 species), and *Mycena* (23 species). In this work, they also reported 30 genera of gasteromycetes and 17 genera of polypores recorded from the wet zone of Sri Lanka (Wijesundera, 1986). Coomaraswamy (1979), Coomaraswamy and de Fonseka (1981), and Coomaraswamy (1981) reported the results of three other significant studies in which ca. 100 genera of plant pathogenic fungi, 63 genera of soil fungi, and 53 genera of agarics were reported. Adikaram and Yakandawala (2020) published a checklist of 372 plant pathogenic fungi from Sri Lanka. The checklist provided species names of the pathogens and in most cases, species names of the host plants. In addition to fungi, Jayalal et al. (2020) reported 876 species of lichens of 233 genera and 60 families from Sri Lanka.

The present chapter is divided into three main sections. These are macrofungi, microfungi (including fungus-like organisms), and lichenological studies.

24.2   MICROFUNGI AND FUNGUS-LIKE ORGANISMS—TAXONOMY AND CHEMISTRY

Coomaraswamy (1979) reported several species of plant pathogenic fungi belonging to the Mastigomycotina of the genera such as *Pseudoperonospora*, *Plasmopara*, *Peronospora*, *Albugo*, *Phytophthora*, *Pythium*, and *Synchytrium*. She also reported that all these fungi were widely distributed in the montane zone and wet lowland areas, except for species of *Synchytrium*. Coomaraswamy and de Fonseka (1981) reported several soil-inhabiting fungi belonging to the Zygomycotina, including genera such as *Motriella*, *Cunningemella*, *Zygorrhyncus*, *Mucor*, *Circinella*, *Rhizopus*, *Absidia*, *Helicostylum*, and *Choanephora*, along with genera in the Ascomycotina such as *Sordaria*, *Eupenicillium*, and *Thielavia*. The occurrence of species of *Sclerocystis*, *Glomus*, *Acaulospora*, *Gigaspora*, *Petriellidium*, *Acheatomium*, *Allescharia*, *Chaetomium*, and *Pseudoeurotium* in the soils of rubber tree plantations was reported by Wijesundera (1986). Herath et al. (2017) described the biocontrol properties of Sri Lankan soil fungi. In this study, 22 taxa of chitinolytic and glucanolytic fungi such as *Aspergillus* sp., *Penicillium* sp., *Trichoderma* sp., and *Fusarium* sp. showed biocontrol activity against seven species of pathogenic fungi. Ratnaweera et al. (2018) reported the antibacterial activities of 72 endophytic fungi isolated from six plants belonging to the family Cyperaceae and their data revealed that 66 fungal extracts were active against at least one of the bacterial species they tested. Ferdinandez et al. (2020) carried out a morphological and molecular characterization of *Exserohilum rostratum* and *Exserohilum oryzicola*, which are associated with rice and early barnyard grass in Sri Lanka. Thambugala et al. (2021) reported the isolation of 20 endophytic fungi belonging to the Ascomycetes from the leaves of *Camellia sinensis* in Sri Lanka. Ferdinandez et al. (2021) reported three new species of *Curvularia* associated with cereal crops and weedy grass hosts in Sri Lanka, and also provided their molecular phylogeny and morphological descriptions. Dissanayake et al. (2016) reported the antimicrobial activity of the endophytic fungus *Chaetomium globosum*, isolated from *Nymphaea nouchali* in Sri Lanka. Silva et al. (2021) reported the antimicrobial activities of eight species of fungi associated with walls of caves that were identified based on morphological and molecular analyses. Four of these (*Penicillium cremeogriseum*, *Penicillium panissanguineum*, *Trichoderma yunnanense*, and *Aspergillus bertholletius*) were new records for Sri Lanka.

Arbuscular mycorrhizae represent a symbiotic association between roots of higher plants and fungi in the phylum Glomeromycota (Singh et al., 2010). Weerawardena and Bandara (2018) reported the mycorrhizal fungi associated with *Piper nigrum* plants in reddish brown latesolic soils in Sri Lanka. Studies on post-harvest losses caused by fungi are also very important (Garcia-Solache and Casadevall, 2010). Karunarathna et al. (2012) listed several significant post-harvest pathogens such as *Colletotrichum* spp., *Botryodiplodia theobromae*, *Theilaviopsis paradoxa*, *Penicillium digitatum*, *P. italicum*, *Phytophthora* spp., and *Diaporthe* spp. from the literature that had published during the period from 1989 to 2008. Priyantha et al. (2016) reported 14 fungal pathogens infecting seeds and mentioned that seed germination and seedling vigor are negatively affected by certain seed-borne fungi.

The myxomycetes (also known as myxogastrids or plasmodial slime molds) are another group of fungus-like organisms. Myxomycetes are often common to abundant in forests, but there appear to be few publications' reports of the group from Sri Lanka. The first of these was by Berkeley and Broome (1871), who listed species of myxomycetes among the fungi they reported. Alexopoulos (1963), in his review of what was then known about the distribution of these organisms, listed only a single publication (Petch, 1945) dealing with the myxomycetes of the island. However, there are two earlier publications (Petch, 1909, 1910a) by the same author, both of which reported records of these organisms.

24.2.1 FUNGAL PATHOGENS

Fungal infections of healthy and immunocompetent humans and animals are relatively uncommon when compared to viral and bacterial infections. Among the millions of fungal species in the world, only a small number of species can infect and cause diseases to immunocompromised individuals (Gnat et al., 2021). Atapattu and Samarajeewa (1990) reported 61 fungal isolates from 25 different dried fish. Jayasekera et al. (2015) mentioned that fungal diseases also cause a considerable damage to human health. In this study, they discussed invasive aspergillosis and candidemia as the leading causes for fungal-associated deaths in Sri Lanka. Mucormycosis caused by *Mucor* sp. has been reported among immuno-compromised patients from Sri Lanka, and it is not new to the country (Jayasekara, 2021).

24.3 MACROFUNGAL TAXONOMY AND MUSHROOM CULTIVATION

The colonial periods, first under the Dutch (1658–1796) and then under the British (1796–1948), can be considered as important periods for macrofungal studies in Sri Lanka. Since the end of the colonial period, very few studies have been carried out by native Sri Lankans on the taxonomy, phylogeny, and cultivation of macrofungi in the country.

Houttuyn (1783) was probably the first westerner to write about the macrofungi of Sri Lanka. He was a Dutch naturalist and a systematic botanist (Wijnands et al., 2017). Berkeley and Broome (1871) provided descriptions for 23 species of mushroom belonging to the Entolomataceae (Pegler, 1977). In this publication, Berkeley and Broome (1871) reported 403 species of agarics and 305 were new to science (Berkeley and Broome, 1870, 1871, 1873; Karunarathna et al., 2017). A few years later, Cesati and Beccari (1879) reported few species of agarics. Höhnel (1908, 1909, 1914) and Petch (1905–1925) published several valuable papers on Sri Lankan mushrooms (Karunarathna et al., 2017).

Pegler (1972) published a revision of the genus *Lepiota*. Descriptive morphological features, habitats, and locations of 34 species of *Lepiota*, two species of *Macrolepiota*, and four species of *Leucocoprinus* were considered in this work. Pegler (1977) also carried out a revision of species in the Entolomataceae (Agaricales) from Sri Lanka and India, and he redescribed each species and provided line drawings. Coomaraswamy and Kumarasingham (1988) published a handbook of Sri Lankan macrofungi.

In 2011, *Lentinus giganteus* Berk. was transferred to *Pleurotus* on the basis of morphological and sequence analyses of collections from Sri Lanka and Thailand (Karunarathna et al., 2011). Ediriweera et al. (2015) two new dung fungi records (*Panaeolus sphinctrinus* and *P. foenisecii*) from dry zone forests in Sri Lanka. Tipbromma et al. (2017) reported *Favolaschia auriscalpium, F. manipularis, Tremella fuciformis, Leucocoprinus cretaceus, Boletellus emodensis, Lentinus sajorcaju*, and *Russula* cf. *virescens* as new records for Sri Lanka. Recently, Ediriweera et al. (2021) reported nine new records of polypores from Sri Lanka based on both morphological and phylogenetic data.

Most of the people of Sri Lanka have consumed wild mushroom from ancient times (Gamage and Ohga, 2018). Tipbromma et al. (2017) documented edible wild mushrooms such as *Lentinus squarrosulus, Russula* cf. *virescens, Lentinus sajor-caju, Boletellus emodensis*, and *Tremella fuciformis* in Sri Lanka. Furthermore, Gunasekara et al. (2021) reported the nutritional composition of three species of *Termitomyces* and four other wild edible mushrooms in Sri Lanka, while Adikaram et al. (2020) documented the first report of the edible mushroom *Helvella crispa* from the island.

Commercial mushroom cultivation was introduced to Sri Lanka by the United Nations development program (UNDP) in 1985 (Rajapakse, 2014; Karunarathna et al., 2017). After the establishment of the Sri Lankan export development board, spawn laboratories and mushroom houses were built, and the department of agriculture improved the mushroom cultivation by connecting it with research and developmental activities (Karunarathna et al., 2017). Cultivation of black ear mushrooms (*Auricularia* spp.) and the milk mushrooms (*Callocybe indica* Purkay. & A. Chandra) was first introduced to the country by Udugama and Wickramarathna (1991). Rupasinghe and Nandasena (2006) documented the potential of refused tea as an alternative substrate for mushroom cultivation. Rajapakse et al.

(2010) reported the efficiency of oyster mushroom (*Pleurotus* spp.) cultivation in different compost mixtures. Later, Pathmashini et al. (2008) discussed the efficiency of cultivating *Pleurotus ostreatus* on several types of sawdust media. Rajapakse (2011) introduced a new cultivation technology for *Volvariella volvacea*. Thanuejah et al. (2013) reported the potential of cultivating oyster mushroom in Sri Lanka. Jeewanthi et al. (2017) discussed the growth and yield of *Ganoderma lucidum* (Curtis) P. Karst on different sawdust substrates. The mycochemical composition of *Volvariella volvacea* cultivated on used palm oil bunch was discussed by Nasir and Siva (2019).

24.4   LICHENOLOGICAL STUDIES

G.H.K. Thwaites first collected lichens in 1868, while Leighton (1870) identified 199 species by examining the material the former had collected. Alston (1932) reported 89 species of lichens from Kandy of Sri Lanka. Jayalal et al. (2008) reported four new species of lichens from a survey in Horton Plains national park during 2004–2006. Recently, Weerakoon and her collaborators carried out the most significant work on lichens in Sri Lanka. Weerakoon et al. (2012, 2014) reported six and 13 new lichens respectively, all belonging to the family Graphidaceae. A subsequent study by Weerakoon et al. (2015) described six new lichens belonging to the Graphidaceae from Horton Plains National Park. Interestingly, Weerakoon and Aptroot (2016) reported nine new species and 64 new records from different areas and national parks covering various elevations and climatic zones. In the same year, Weerakoon et al. (2016) reported eight new species and 88 new records from Sri Lanka. The most recent publication by Aptroot and Weerakoon (2018) reported three new species and 10 new records of lichens belonging to the Trypetheliaceae.

24.5   CONCLUSIONS

With increasing human population and the associated development across the landscape, the forest cover and microhabitats of fungi and lichens are shrinking in Sri Lanka, as is the case all over the world. Therefore, some unknown fungi and lichens might become extinct in their original habitats without humans ever knowing their value. The identification of mushrooms, microfungi (including plant pathogens, saprobes, and epiphytes), and lichens is an important part of mycological research in any country, and Sri Lanka is no exception. There is an urgent need for the establishment of both a national collection of fungi and lichens and a network of local and international working groups (Wijayawardene et al., 2021). It is very important to have a central point for mycological activities within the national herbarium, which would coordinate the identification and documentation of all fungi and lichens locally. The maintenance of national herbarium collections and living cultures, the initiation of collaborative efforts to apply molecular-based identification, and updation of mycological knowledge regularly are also emphasized. Finally, we appreciate international and local mycologists for their great work, especially in recent years on the developments of Sri Lankan mycology.

FIGURE 24.1 A photo plate of some microfungi, macrofungi, and lichens reported from Sri Lanka. (a) *Boletellus emodensis*, (b) *Cyptotrama asprata*, (c) *Lentinus sajor-caju*, (d) *Lentinus squarrulosus*, (e) *Ophiocordyceps formicarum*, (f) *Favolaschia manipularis*, (g) Blister blight of tea (caused by *Exobasidium vexans*), (h) *Hypogymnia* sp., and (i) *Megalospora* sp.

ACKNOWLEDGMENT

Samantha C. Karunarathna thanks the CAS President's International Fellowship Initiative (PIFI) for young staff under grant 2020FYC0002 and the National Science Foundation of China (NSFC, project code 31851110759). Appreciation is extended to Michael Pilkington and Ruvini Menaka Karunarathna for providing images of lichens and microfungi from Sri Lanka.

KEYWORDS

- **fungal pathogens**
- **lichens**
- **macrofungi**
- **microfungi**
- **mushroom cultivation**
- **myxomycetes**

REFERENCES

Adikaram, N. K. B.; Yakandawala, D. M. D. A Checklist of Plant Pathogenic Fungi and Oomycota in Sri Lanka. *Ceylon J. Sci.* **2020,** *49* (1), 93–123.

Adikaram, N. K. B.; Yakandawala, D.; Jayasinghe, L. The First Report of *Helvella crispa* (Ascomycota, Pezizales), A Rare Fungal Species in Sri Lanka. *Ceylon J. Sci.* **2020,** *49* (4), 485–489.

Alexopoulos, C. J. The Myxomycetes II. *Bot. Rev.* **1963,** *29,* 1–78.

Alston, A. H. G. *The Kandy Flora.* The Government Press: Colombo, **1932**; pp 78–80.

Aptroot, A.; Weerakoon, G. Three New Species and Ten New Records of Trypetheliaceae (Ascomycota) from Sri Lanka. *Mycologie* **2018,** *39* (3), 373–378.

Atapattu, R.; Samarajeewa, U. Fungi Associated with Dried Fish in Sri Lanka. *Mycopathologia* **1990,** *111,* 55–59. DOI: https://doi.org/10.1007/BF02277304

Berkeley, M. J.; Broome, C. E. On Some Species of the Genus *Agaricus* from Ceylon. *Trans. Linn. Soc. London* **1870,** *27,* 149–152.

Berkeley, M. J.; Broome, C. E. The Fungi of Ceylon (Hymenomycetes from *Agaricus* to *Cantharellus*). *Bot. J. Linn. Soc.* **1871,** *11,* 494–567.

Berkeley, M. J.; Broome, C. E. Enumeration of the Fungi of Ceylon. Part 2. *Bot. J. Linn. Soc.* **1873,** *14,* 29–140.

Cesati, V.; Beccari, O. *Mycetum in itinere Borneensi Lectorum a cl. Od. Beccari.* Typis Regiae Scientiarum Academiae: Neapoli, **1879**; vol 8, pp 1–28.

Coomaraswamy, U. *A Handbook to the Fungi Parasitic on the Plants of Sri Lanka.* MAB-UNESCO Publication No. 4. National Science Council: Sri Lanka, **1979.**

Coomaraswamy U. *A Handbook to the Agarics of Sri Lanka.* MAB-UNESCO Publication No. 5. National Science Council: Sri Lanka, **1981.**

Coomaraswamy, U.; de Fonseka, R. N. *A Handbook to the Soil Fungi of Sri Lanka.* MAB-UNESCO Publication No. 7 Sri Lanka: National Science Council. **1981.**

Coomaraswamy, U.; Kumarasingham, S. *A Handbook to the Macrofungi of Sri Lanka.* Natural Resources, Energy and Science Authority: Sri Lanka, **1988.**

Dissanayake, R. K.; Ratnaweera, P. B.; Williams, D. E.; Wijayarathne, C. D.; Wijesundera, R. L. C.; Andersen, R. J.; de Silva, E. D. Antimicrobial Activities of Endophytic Fungi of the Sri Lankan Aquatic Plant *Nymphaea nouchali* and Chaetoglobosin A and C, Produced by the Endophytic Fungus *Chaetomium globosum*. *Mycology* **2016**, *7* (1), 1–8, DOI: 10.1080/21501203.2015.1136708

Ediriweera, S. S.; Nanayakkara, C. M.; Weerasena, O. V. D. S. J.; Karunarathna, S. C.; Wijesundera, R. L. C.; Piyatissa, M. A. S. U. Morphology and Phylogeny Reveal Nine New Records of Polypores from Dry Zone of Sri Lanka. *Chiang Mai J. Sci.* **2021**, *48* (3), 893–908.

Ediriweera, S.; Wijesundera, R. L. C.; Nanayakkara, C.; Weerasena, J. First Report of *Panaeolus sphinctrinus* and *Panaeolus foenisecii* (Psathyrellaceae, Agaricales) on Elephant Dung from Sri Lanka. *Front. Environ. Microbiol.* **2015**, *1*, 19–23.

Ferdinandez, H. S.; Manamgoda, D. S.; Udayanga, D.; Deshappriya, N.; Munasinghe, M. L. A. M. S. Morphological and Molecular Characterization of Two Graminicolous *Exserohilum* Species Associated with Cultivated Rice and Early Barnyard Grass from Sri Lanka. *Ceylon J. Sci.* **2020**, *49*, 381–387.

Ferdinandez, H. S.; Manamgoda, D. S.; Udayanga, D.; Deshappriya, N.; Munasinghe, M. S.; Castlebury, L. A. Molecular Phylogeny and Morphology Reveal Three Novel Species of *Curvularia* (Pleosporales, Pleosporaceae) Associated with Cereal Crops and Weedy Grass Hosts. *Mycol. Prog.* **2021**, *20*, 431–451.

Gamage, S.; Ohga, S. A Comparative Study of Technological Impact on Mushroom Industry in Sri Lanka: A Review. *Adv. Microbiol.* **2018**, *8*, 665–686. DOI: 10.4236/aim.2018.88045

Garcia-Solache, M. A.; Casadevall, A. Global Warming Will Bring New Fungal Diseases for Mammals. *mBio* **2010**, *1*, e00061- 10. DOI:10.1128/mBio.00061-10

Gnat, S.; Łagowski, D.; Dyląg, M.; Zielinski, J.; Studziński, M.; Nowakiewicz, A. Cold Atmospheric Pressure Plasma (CAPP) as a New Alternative Treatment Method for Onychomycosis Caused by *Trichophyton verrucosum*: In Vitro Studies, *Infection* **2021**, *49* (6), 1233–1240. DOI: 10.1007/s15010-021-01691-w

Gunasekara, N. W.; Nanayakkara, C. M.; Karunarathna, S. C.; Wijesundera, R. L. C. Nutritional Aspects of Three *Termitomyces* and Four Other Wild Edible Mushroom Species from Sri Lanka. *Chiang Mai J. Sci.* **2021**, *48* (5), 1236–1246.

Herath, H. H. M. A. U.; Wijesundera, R. L. C.; Chandrasekharan, N. V.; Wijesundera, W. S. S. Exploration of Sri Lankan Soil Fungi for Biocontrol Properties. *Afr. J. Biotechnol.* **2017**, *16* (20), 1168–1175. DOI: 10.5897/AJB2017.15905

Höhnel, F. Fragmente zur Mycologie V, Nr. 169 bis 181. *Akad. Wiss. Wien Math.-nat. Kl., Band* **1908**, *117*, 1–48.

Höhnel, F. Fragmente zur Mycologie VI, Nr. 182 bis 288. *Akad. Wiss. Wien Math.-nat. Kl., Band* **1909**, *118*, 1–178.

Höhnel F. Fragmente zur Mycologie XVI, Nr. 813 bis 875. *Akad. Wiss. Wien Math.-nat. Kl., Band* **1914**, *123*, 1–107.

Houttuyn, M. *Handleiding tot de plant – en kruidkunde, benevens eene uitvoerige beschrijving der boomen, planten, heesters, kruiden, varens, mossens, bollen grasplanten, volgens het zamenstel van C. Linnaeus.* Bij Lodewyk van Es: Amsterdam, **1783**; vol 14, pp 585–698.

Jayalal, R. G. U.; Wijesundara, D. S. A.; Karunaratne, V. In *Lichenological Studies in Sri Lanka.* 13th International Forestry and Environmental Symposium, University of Sri Jayawardhanapra: Sri Lanka, **2008**; pp 69–70.

Jayasekara, P. *Mucormycosis–Commonly Known as "Black fungus,"* **2021**. https://slmicrobiology.lk/2021/05/10/mucormycosis-commonly-known-as-black-fungus/.

Jayasekera, P. I.; Denning, D. W.; Prerera, P. D.; Fernando, A.; Kudavidanage, S. Is Fungal Disease in Sri Lanka Underestimated? A Comparison of Reported Fungal Infections with Estimated Disease Burden Using Global Data. *Sri Lankan J. Inf. Dis.* **2015**, *5* (2), 73–85.

Karunarathna, S. C.; Mortimer, P. E.; Xu, J.; Hyde, K. D. Review of Research of Mushrooms in Sri Lanka. *Rev. Fitotec. Mexicana* **2017**, *40* (4), 399–403.

Karunarathna, S. C.; Yang, Z. L.; Raspé, O.; Ko, T. W. K.; Vellinga, E. C.; Zhao, R. L.; Bahkali, A. H.; Chukeatirote, E.; Degreef, J.; Callac, P.; Hyde, K. D. *Lentinus giganteus* Revisited: New Collections from Sri Lanka and Thailand. *Mycotaxon* **2011**, *118*, 57–71.

Karunarathna, S. C.; Udayanga, D.; Maharachchikumbura, S. N.; Pilkington, M.; Manamgoda, D. S.; Wijayawardene, D. N. N.; Ariyawansa, H. A.; Bandara, A. R.; Chukeatirote, E.; McKenzie, E. H. C.; Hyde, K.

D. Current Status of Knowledge of Sri Lankan Mycota. *Curr. Res. Environ. Appl. Mycol.* **2012**, *2* (1), 18–29. DOI: 10.5943/cream/2/1/2

Leighton, W. A. The Lichens of Ceylon Collected by G.H.K. Thwaites. *Trans. Linn. Soc. London* **1870**, *27*, 161–185.

Nasir, I. A.; Siva, R. Mycochemical Composition of Straw Paddy Mushroom (*Volvariella volvacea*) Grown on Used Palm Oil Bunch. *J. Sustain. Sci. Manag.* **2019**, *14*, 12–21.

Pathmashini, L.; Arulnandhy, V.; Wijeratnam, R. S. W. Cultivation of Oyster Mushroom (*Pleurotus ostreatus*) on Sawdust. *Ceylon J. Sci.* (*Bio. Sci.*) **2008**, *37* (2), 177–182.

Pegler, D. N. A Revision of the Genus *Lepiota* from Ceylon. *Kew Bull.* **1972**, *27*, 155–222.

Pegler, D. N. A Revision of Entolomataceae (Agaricales) from India and Sri Lanka. *Kew Bull.* **1977**, *32*, 189–220.

Petch, T. The Phalloideae of Ceylon. *Ann. Roy. Bot. Gard., Peradeniya* **1908a**, *4*,139–184.

Petch T. The Genus *Endocalyx* Berkeley and Broome. *Ann. Bot.* **1908b**, *22*, 389–400.

Petch, T. New Ceylon Fungi. *Ann. Roy. Bot. Gard., Peradeniya* **1909**, *4*, 299–307.

Petch, T. Revisions of Ceylon Fungi (Part II). *Ann. Roy. Bot. Gard., Peradeniya* **1910a**, *4*, 373–444.

Petch, T. A List of the Mycetozoa of Ceylon. *Ann. Roy. Bot. Gard., Peradeniya* **1910b**, *4*, 309–371.

Petch, T. Termite Fungi: A Résumé. *Ann. Roy. Bot. Gard., Peradeniya* **1913**, *5*, 303–341.

Petch, T. The Pseudo-Sclerotia of *Lentinus similis* and *Lentinus infundibuliformis*. *Ann. Roy. Bot. Gard., Peradeniya* **1915a**, *6*, 1–18.

Petch, T. Horse-Hair Blights. *Ann. Roy. Bot. Gard., Peradeniya* **1915b**, *6*, 43–68.

Petch, T. A Preliminary List of Ceylon Polypori. *Ann. Roy. Bot. Gard., Peradeniya* **1916a**, *6*, 87–144.

Petch, T. Ceylon Lentini. *Ann. Roy. Bot. Gard., Peradeniya* **1916b**, *6*, 145–152.

Petch, T. Revisions of Ceylon Fungi (Part IV). *Ann. Roy. Bot. Gard., Peradeniya* **1916c**, *6*, 153–183.

Petch, T. Additions to Ceylon Fungi. *Ann. Roy. Bot. Gard., Peradeniya* **1917a**, *6*, 195–256.

Petch, T. Revisions of Ceylon Fungi (Part V). *Ann. Roy. Bot. Gard., Peradeniya* **1917b**, *6*, 307–355.

Petch, T. Revisions of Ceylon Fungi (Part VI). *Ann. Roy. Bot. Gard., Peradeniya* **1919**, *7*, 1–14.

Petch, T. Additions to Ceylon Fungi II. *Ann. Roy. Bot. Gard., Peradeniya* **1922**, *7*, 279–322.

Petch, T. *The Diseases of the Tea Bush*. MacMillan & Co. Ltd.: London, UK, 1923.

Petch, T. Thread Blights. *Ann. Roy. Bot. Gard., Peradeniya* **1924a**, *9*, 1–46.

Petch T. Revisions of Ceylon Fungi (Part. VII). *Ann. Roy. Bot. Gard., Peradeniya* **1924b**, *9*, 119–181.

Petch, T. Ceylon Pink-Spored Agarics. *Ann. Roy. Bot. Gard., Peradeniya* **1924c**, *9*, 201–216.

Petch, T. Agaricaceae Pleuropods Zeylanicae. *Ann. Roy. Bot. Gard., Peradeniya* **1924d**, *9*, 217–227.

Petch, T. Additions to Ceylon Fungi III. *Ann. Roy. Bot. Gard., Peradeniya* **1925**, *9*, 313–328.

Petch, T. Additions to Ceylon Fungi IV. *Ann. Roy. Bot. Gard., Peradeniya* **1926a**, *10*, 131–138.

Petch, T. *Rhacophyllus* B. & Br. *Trans. Br. Mycol. Soc.* **1926b**, *11*, 238–251.

Petch, T. Revisions of Ceylon Fungi VIII. *Ann. Roy. Bot. Gard., Peradeniya* **1927**, *10*, 161–180.

Petch, T. Tropical Root Disease Fungi. *Trans. Br. Mycol. Soc.* **1928**, *13*, 238–253.

Petch, T. Ceylon Fungi, Old and New. *Trans. Br. Mycol. Soc.* **1945**, *2*, 137–147.

Petch, T. A Revision of Ceylon Marasmii. *Trans. Br. Mycol. Soc.* **1947**, *31*, 19–44.

Petch, T.; Bisby, G. R. *The Fungi of Ceylon*. Peradeniya Manual 6. Government Pub. Bureau: Colombo, Sri Lanka, **1950**.

Priyantha, M. G. D. L.; Athukorala, A. R. J.; Jayasinghe, J. A. V. J.; Sato, M.; Takahashi, H. Seed Borne Pathogens Associated with Seed Lots of Major Food Crops of Sri Lanka. *Trop. Agric.* **2016**, *164*, 71–81.

Rajapakse, P. In *New Cultivation Technology for Paddy Straw Mushroom* (*Volvariella volvacea*), Proceedings of the 7th International Conference on Mushroom Biology and Mushroom Products (ICMBMP7), **2011**; pp.446–451.

Rajapakse, J. C. In *Status of Edible and Medicinal Mushroom Research in Sri Lanka*, Proceedings of the 8th International Conference on Mushroom Biology and Mushroom Products (ICMBMP8), vol I & II, New Delhi, India, Nov 19–22, 2014; Singh, M., Ed.; ICAR-Directorate of Mushroom Research: Solan, India, **2014**; pp 417–421.

Rajapakse, J. C.; Rubasingha, P.; Dissanayake, N. N. The Potential of Using Cost-Effective Compost Mixtures for Oyster Mushroom (*Pleurotus* spp.) Cultivation in Sri Lanka. *Trop. Agric. Res. Ext.* **2010**, *10*, 29–32.

Ratnaweera, P. B.; Walgama, R. C.; Jayasundera, K. U.; Herath, S. D.; Abira, S.; Williams, D. E.; Andersen, R. J.; de Silva, E. D. Antibacterial Activities of Endophytic Fungi Isolated from Six Sri Lankan Plants of the family Cyperaceae. *Bangladesh J. Pharmacol.* **2018**, *13*, 264–272. DOI: 10.3329/bjp.v13i3.36716

Rupasinghe, K.; Nandasena, K. Potential Exploitation of Refuse Tea as an Alternative Medium in Mushroom Cultivation. *Trop. Agric. Res.* **2006,** *18* (51), 399–404.

Silva, S. S. D. *Reservoirs of Sri Lanka and Their Fisheries, Food and Agriculture Organization of the United Nations,* **1988;** pp 128. https://www.fao.org/3/T0028E/T0028E00.htm#toc

Silva, E. I. P.; Jayasingha, P.; Senanayake, S.; Dandeniya, A.; Munasinghe, D. H. Microbiological Study in a Gneissic Cave from Sri Lanka, with Special Focus on Potential Antimicrobial Activities. *Int. J. Speleol.* **2021,** *50* (1), 41–51. DOI: 10.5038/1827-806X.50.1.2343

Singh, S. R.; Singh, U.; Chaubey, A. K.; Bhat, M. Mycorrhizal Fungi for Sustainable Agriculture – A Review. *Agric. Rev.* **2010,** *31* (2), 93–104.

Thambugala, K.; Daranagama, D.; Kannangara, S.; Kodituwakku, T. Revealing the Endophytic Mycoflora in Tea (*Camellia sinensis*) Leaves in Sri Lanka: The First Comprehensive Study. *Phytotaxa* **2021,** *514* (3), 247–260.

Thanuejah, S.; Karunakaran, S.; Arulnandhy, V. In *Potential for Oyster Mushroom Cultivation in Sri Lanka–A Review,* 2nd International Symposium on Minor Fruits and Medicinal Plants for Better Life, 2013; pp 198–205.

Tipbromma, S.; Hyde, K. D.; Jeewon, R.; Maharachchikumbura, S. S. N.; Liu, J. K.; Bhat, D. J.; Jones, E. B. G.; McKenzie, E. H. C.; Camporesi, E.; Bulgakov, T. S.; Doilom, M.; et al. Fungal Diversity Notes 491–602: Taxonomic and Phylogenetic Contributions to Fungal Taxa. *Fungal Divers.* **2017,** *83,* 1–261.

Udugama, S.; Wickramaratna K. *Artificial Production of Naturally Occurring Lentinus giganteus* (*Uru Paha*)*, A Sri Lankan Edible Mushroom.* Horticultural Crop Research & Development Institute (HORDI): Gannoruwa, Peradeniya, **1991.**

Weerakoon, G.; Aptroot, A. Nine New Lichen Species and 64 New Records from Sri Lanka, *Phytotaxa* **2016,** *280* (2), 152–162.

Weerakoon, G.; Jayalal, U.; Wijesundara, S.; Karunaratne, V.; Lücking, R. Six New Graphidaceae (Lichenized Ascomycota: Ostropales) from Horton Plains National Park, Sri Lanka. *Nova Hedwigia* **2015,** 101, 1–12 DOI: 10.1127/nova_hedwigia/2015/0241

Weerakoon, G.; Lucking, R.; Lumbsch, H. T. Thirteen New Species of Graphidaceae (Lichenized Ascomycota: Ostropales) from Sri Lanka. *Phytotaxa* **2014,** *189* (1), 331–347.

Weerakoon, G.; Wijeyaratne, C.; Wolsely, P. A. Six New Species of Graphidaceae from Sri Lanka. *Bryologist* **2012,** *115* (1), 74–83.

Weerakoon, G.; Wolseley, P. A.; Arachchige, O.; Cáceres, M. E. D. A.; Jayalal, U.; Aptroot, A. Eight New Lichen Species and 88 New Records from Sri Lanka. *Bryologist* **2016,** *119* (2), 131–142. DOI: 10.1639/ 0007-2745-119.2.131

Weerawardena, T. E.; Bandara, W. M. S. B. Spore Abundance and Morphological Root Modifications of Arbuscular Mycorrhizal Fungi-Infected Black Pepper (*Piper nigrum* L) Plants in Reddish Brown Latesolic Soil of Matale in Sri Lanka. *Sri Lanka J. Food Agric.* **2018,** *4* (2), 1–5.

Wijayawardene, N. N.; Rajakaruna, S.; Jayasekara, S.; Warnakula, L.; Ariyawansa, S.; Fernando, E. Y.; Karunarathna, S. C.; Tibpromma, S.; Ukuwela, K.; Jayalal, U.; Jayasinghe, P.; Sinhalage, D.; Hunupolagama, D.; Jayasinghe, G. G.; Rajapakse, J.; Kumara, W.; Palihawadana, N.; Attanayake, A.; Muthumala, C. K.; Kirk, P. M.; Weerakoon, G.; Udugama, S.; Jayasekara, P.; Dai, D. Q.; Stephenson, S. L.; Nanayakkara, C.; Wijesundara, S. Necessity of National Fungaria and a Culture Collection for Environmental Fungi in Sri Lanka. *Chiang Mai J. Sci.* **2021,** *49* (2), 248–271 (in press).

Wijesundera, R. L. C. Floristic Studies and Distribution of Fungi in Sri Lanka. *Proc. Indian Acad. Sci.* (*Plant Sci.*) **1986,** *96* (5), 375–377.

Wijnands, D. O.; Heniger, J.; Veldkamp, J. F.; Fumeaux, N.; Callmander, M. W. The Botanical Legacy of Martinus Houttuyn (1720–1798) in Geneva. *Candollea* **2017,** *72,* 155–198. DOI: http://dx.doi.org/10.15553/ c2017v721a11

CHAPTER 25

ANGIOSPERMS AND GYMNOSPERMS OF SRI LANKA

SUDHEERA M. W. RANWALA

Department of Plant Sciences, Faculty of Science, University of Colombo, Colombo, Sri Lanka

ABSTRACT

Angiosperm flora of Sri Lanka encompasses 3103 indigenous species from 186 families. Due to the common origin, India and Sri Lanka share many angiosperm species; however, 863 endemic species and 16 endemic angiosperm genera are exclusive to Sri Lanka. The ever-wet rain forests in the south western quarter of the island accommodate nearly 75% of the endemic angiosperms and nearly half of these species are confined to montane areas. The semi-deciduous forests also accommodate diverse angiosperm flora and 50% of the woody endemic plant species. Apart from the forests, *Pathana* grasslands also exhibits distinctive flora in their unique habitats. The diversity of the angiosperms is also reflected by many plant groups; orchids, aquatic plants, insectivorous species, mangroves, and other marine angiosperm species in a remarkable manner. Indigenous gymnosperms are limited to few species of the family Cycadaceae as family Pinaceae is not represented by the wealth of plants native to Sri Lanka.

25.1 INTRODUCTION

Angiosperms of Sri Lanka has been a famous group of plants explored, studied and documented in the European literature during 17th and 19th centuries. The first ever scientific publication, "Catalog of the Indigenous and Exotic Plants Growing in Ceylon," published by Sir Alexander Moon in 1824 cited 761 species recorded from wild, 366 cultivated species, and 164 species new to science (Pethiyagoda, 2007). The series of volumes of "Flora of Ceylon" authored by Sir Henry Trimen (1893–1900) was the masterpiece in plant taxonomic studies nearly a century until the "Revised handbook to the Flora of Ceylon" by Dassanayake et al. (1980–2000) was published.

In 1956, Abeywickrama reported 2855 indigenous angiosperm species belonging to 1065 genera and 171 families and categorized into 6 phyto-geographic elements based

Biodiversity Hotspot of the Western Ghats and Sri Lanka. T. Pullaiah, PhD (Ed.)
© 2024 Apple Academic Press, Inc. Co-published with CRC Press (Taylor & Francis)

on the origin and affinities to other regions, viz. (1) Indo-Ceylon, (2) Malayan, (3) Himalayan, (4) African (moved across India from drier parts such as Africa), (5) Ceylon and (6) pantropic or cosmopolitan related (Figure 25.1).

FIGURE 25.1   Photographs of some native plant species that represent phyto-geographic elements: (A) Indo-Ceylon element- *Michelia nilagirica,* (B) Malayan element- *Dillenia indica,* (C) Himalayan element- *Berberis tinctoria,* (D) African component- *Salvadora persica.*

Photo credit: 25.1 (A) Tharanga Abeywickrama, (B)- (D) S.M.W. Ranwala

Recent discoveries report that 3103 out of 4203 angiosperms are native, 1100 are exotic (MOMD&E, 2016), and 863 are endemic to Sri Lanka (Wijesundara et al., 2020). No endemic angiosperm plant families are reported in Sri Lanka (Abeywickrama, 1955), but 16 endemic genera, namely, *Stemonoporus* (Dipterocarpaceae), *Adrorhizon* (Orchidaceae), *Chlorocarpa* (Achariaceae), *Phoenicanthus* (Annonaceae), *Loxococcus* (Arecaceaea), *Kokoona* (Celestraceaea), *Schumacheria* (Dilleniaceaea), *Championia* (Gesnariaceae), *Dicellostyles* (Malvaceae), *Hortonia* (Monimiaceae), *Davidsea* (Poaceae), *Farmeria* (Podostemaceae), and *Diyaminauclea, Leucocodon, Nargedia* and *Scyphostachys* (Rubiaceae). The 10 largest angiosperm families in Sri Lanka with updated number of species are: Poaceae (248), Fabaceae (217), Orchidaceae (193), Rubiaceae (174), Cyperaceae (169), Acanthaceae (106), Asteraceae (86), Melastomataceae (71), Lamiaceae (70), Malvaceae (69) (Wijesundara et al., 2020). According to the Red List 2020, out of the 186 families evaluated, >50% threatened plant species are reported from 64 families, whereas in 25 families (each represented by <5 species), all the plant species are recognized as threatened. Only 40 families report no threatened species (Wijesundara et al., 2020).

25.2   SOME ANGIOSPERMS OF LOWLAND RAIN FORESTS

Gunatilleke and Gunatilleke (1980) recorded common plant genera in the UNESCO World Heritage Site Sinharaja as *Stemonoporus* (Dipterocarpaceae), *Hortonia* (Monimiaceae), *Doona* (Dipterocarpaceae), *Acrotrema* and *Schumacheria* (Dilleniaceae). Kanneliya Forest Reserve is also rich in Dipterocarps as 07 genera are represented. Moreover, about 10 species including *Stemonoporus kanneliyansis, Vatica affinis, Hopea jucunda* ssp. *jucunda, Agrostistachys hookeri* are confined only to the Kanneliya whereas *Pterospermum canescens, Symplocos cornata, Vanilla moonii* have been found only in the Nakiyadeniya Forest Reserve (Singhakumara, 1995). Waturana, the one and only freshwater swamp forest of the country with an extent of 6.2 ha, is the only site in which *Stemonoporus mooni* and *Mesua stylosa* are found in the wild (Figure 25.2).

FIGURE 25.2   (A) Tender leaves, (B) flower, and (C) twig of *Stemonoporus mooni,* (D) *Mesua stylosa.*
Photo credit: 25.2 (A)- (D) Nalinda Peiris

Members of Dipterocarpaceae, Dilleniaceae, Clusiaceae, Annonaceae, Sapotaceae plant families are abundant in rain forests below 1600 m altitude. Noteworthy features of 58 species of Dipterocarps belonging to 9 genera (*Dipterocarpus, Doona, Shorea, Vatica, Vateria, Hopea, Sunaptea, Balanocarpus,* and *Stemonoporus*) are well described including the natural pattern of the distribution of *Shorea* along with the altitude (Ashton et al., 2018). Seventeen endemic species of Annonaceae, namely, *Desmos zeylanica, Desmos elegans, Uvaria semecarpifolia, Uvaria sphenocarpa, Sageraea thwaitesii, Phoenicanthus coriacea, Phoenicanthus obliqua, Alphonsea hortensis, Polyalthia persicaefolia, Polyalthia moonii, Miliusa zeylanica, Enicosanthum acuminata, Xylopia nigricans, Goniothalamus gardneri, Goniothalamus hookeri, Goniothalamus thomsonii,* and *Goniothalamus salicina,* belonging to 10 genera out of 16 are recorded in areas up to 1500 m altitude (Dassanayake and Fosberg, 1980). Several endemic members of the genus *Semecarpus* of family Anacardiaceae, and 03 genera of the family Dilleniaceae, namely *Dillenia, Tetracera, Schumacheria* are common species found in lowland rain forests up to 900 m elevation (Figure 25.3).

FIGURE 25.3   Some attractive angiosperm species in lowland rain forests. (a) flowers of *Shorea trapezifolia,* (b) pink stipules of *Dipterocarpus zeylanicus,* (c) fruits of *Vateria copallifera,* (d) fruits of *Goniothalamus gardneri,* (e) flower of *Uvaria* sp., (f) poisonous *Semecarpus obovata,* (g) twig of *Schumacheria castaneifolia,* and (h) *Dillenia retusa.*

Photo credit: 25.3 (A), (E) Savitri Gunatilake, (B)- (D), (F)- (H) S.M.W. Ranwala

25.3   SOME ANGIOSPERMS OF SEMI-DECIDUOUS FORESTS

In almost all semi-deciduous forests of the dry zone *Manilkara hexandra* (Sapotaceae), *Drypetis sepiaria* (Euphorbiaceae), *Azadairachta indica* (Neem), *Cassia fistula* (Golden shower), and *Tamarindus indica* (Tamarind) are commonly seen. Euphorbiaceae, Sapotaceae, Rutaceae, Sapindaceae, Ebenaceae, and Myrtaceae are some of the common plant families found (Perera, 2012). In addition, several species belonging to families Asclepiadaceae, Rhamnaceae, Rubiaceae, Sterculiaceae, and Tiliaceae become abundant toward the fringes of the arid zone. Comparatively, a much higher endemism is observed in among the genera *Canthium* (Rubiaceae), *Diplodiscus* (Tiliaceae), *Memecylon* (Melastomataceae) (Wijesundara, 2012). Nevertheless, point endemics are not common in these forests of the dry zone. Three, point endemic species, namely *Wrightia flavidorosea* (Apocynaceae), *Hopea brevipetiolaris* (Dipterocarpaceae), *Oplismenus thwaitesii* (Poaceae) each known only from a single site <100 km² are reported from wetter habitats of forests from the intermediate zone (Gunatilleke et al., 2008). Many valuable and good timber yielding trees also inhabit semi-deciduous forests. These include both, *Diospyros ebanum* (Ebony) and *Diospyros quaesita* (Calamander) from Family Ebenaceae, *Chloroxylon sweitenia* (Rutaceae), *Vitex altissima* (Verbenaceae), *Melia azedarach* (Meliaceae), and *Haldina cordifolia* (Rubiaceae) (Figure 25.4).

FIGURE 25.4   (a) *Manilkara hexandra*, (b) *Drypetis sepiaria* during fruiting season, (c) *Memecylon* sp., (d) *Hydrocarpus venenata*, (e) *Vitex altissima*, (f) *Haldina cordifolia*, (g) *Chloroxylon swietenia*, (h) *Hopea cordifolia* (courtesy-Kew Herbarium, UK).

Photo credit: 25.4 (A)- (G) S.M.W. Ranwala, (H) Kew Herbarium, UK

25.4   SOME ANGIOSPERMS OF MONTANE FORESTS AND WET PATHANA GRASSLANDS

Some of the most common angiosperm families in the montane zone are Clusiaceae, Myrtaceae, Lauraceae, Theaceae, and Symplocaceae. Members of family Meliaceae are totally absent. Although canopy of montane forests do not exceed a height taller than 10–12 m, exceptionally taller tree species are few, including *Calophyllum walkeri* (Clusiaceae), *Magnolia nilagrica* and *Michelia champaka* (Magnoliaceae), *Syzygium rotundifolium* (Myrtaceae) are also reported. Similarly, many members of genera; *Syzygium* (Myrtaceae), especially *S. revolutum, S. rotundifolium, S. hemisphericum, S. zeylanicum*, and *S. sclero-phyllum* occupy the canopy of montane forests. Often *Neolitsea, Litsea, Cinamomum,* and *Eugenea* species (Lauraceae) are equally abundant (Rathnayake et al., 1996). Among the members of the endemic genus *Stemonoporus*, few species such as *Stemonoporus cordifolius, S. rigidus*, and *S. gardneri* are also found in some montane areas occurring co dominantly with *Palaquium rubiginosus* (Greller et al., 1987).

Werner (1995) describes an elevational range between 1500 and 1800 m of the montane zone as an ecotone where some species continue to exist, but others disappear. Theaceae members such as *Gordonia speciosa, Ternstroemia gymnanthera, Semecarpus nigro-viridis* (Anacardiaceae) continue to grow whereas *Actinodaphne ambigua* (Lauraceae), and *Syzygium rotundifolium* is found beyond 1800 m. Around 2000 m altitude, members of Symplocaceae (*Symplocos cordifolia, Symplocos cochinchinensis*, and *Symplocos cordifolia),* *Garcinia echinocarpa* (Clusiaceae), *Gordonia ceylanica,* and few other species which are common between 1500 and 1800 m elevational range are replaced by *Rhododendron arboreum* (Ericaceae), *Euyra* spp. (Theaceae), *Psydrax montanus* and *Rhodomyrtus tomentosa* (Myrtaceae), *Eleocarpus subvillosus* (Eleocarpaceae), *Litsea* spp. and few others.

Apart from the montane forests the pathana grasslands at Horton Plains also provide a unique habitat for many herbaceous angiosperms. The typical grasses that occur in the grassland are *Chrysopogon zeylanicus* and *Arundinella villosa*. The grassland also provides micro habitat for many genera including *Anaphalis* spp. (Asteraceae), *Alchemilla indica* (Rosaceae), *Exacum trinervium* (Gentianaceae), and many more (Figure 25.5).

FIGURE 25.5   (a) Young reddish flush of *Syzygium rotundifolium*, (b) *Rhododendron arboretum* (Ericaceae), (c) *Rhodomyrtus tomemtosa*, (d) *Eugenea marboides*, (e) *Litsea ovalifolia*, (f) *Litsea quinqueflora*, (g) *Syzygium fergusoni* (found only in Knuckles), (h) *Exacum trinervium*, and (i) *Anaphalis* sp.

Photo credit: 25.7 (A)- (I) S.M.W. Ranwala

25.5   ORCHIDS WITH EXCEPTIONAL BEAUTY

The family Orchidaceae is represented by 189 native species, out of which 55 are endemic to the country (Fernando and Ormerod, 2008). Many genera of native orchids belong to subfamilies Epidendroideae and Randoideae. Only one genus from sub family Apostasia is represented in Sri Lanka while members from subfamily Cypripedioideae not reported (Fernando et al., 2003). Many Sri Lankan orchids are of Indo-Ceylon origin, thus the genera; *Cottonia, Ipsea, Seidenfadeniella, Diptocentrum,* and *Sirhookera* can be found only in peninsular India and Sri Lanka. In addition, some native orchid genera in Sri Lanka also show affinity to North Asian/Himalayan orchids. A small number of genera, namely *Apostasia, Bromheadia,* represent the Malayan element of origin, whereas comparatively, many genera including *Angraecum* and *Aerangis* exhibit affinities to orchids of the African element. Many genera such as *Spiranthus, Liparis, Goodyera,* and *Malaxis* exhibit a cosmopolitan affinity and *Adrorhizon* is the only endemic Orchid genus found in Sri Lanka (Fernando, 2013).

Many Sri Lankan orchids are abundant in wet and cooler climates, hence nearly half of the native orchids (54%) and some endemic species such as *Eria tricolor, Phreatia jayaweerae, Zeuxine regia, Arundina minor,* and *Habenaria pterocarpan* are reported from the lower montane zone between 900 and 1500 m altitude. The second highest species diversity is reported from the semi deciduous forests and Savanna grasslands. Despite of the lowest species richness reported from the Dry zone, species such as *Rhynchostylis retusa, Habenaria roxburghii,* and *Nervilia* species are restricted to much drier habitats of the Eastern part of the country (Fernando, 2013).

Although a large number of wild orchids are epiphytes, a few may grow (e.g., *Eria tricolor*) close to the ground level. Saprophytic terrestrial Orchids such as *Epsea speciosa* that grow on litter are mostly seasonal and appear when soil is moist (Fernando et al., 2003). Many epiphytic Orchids prefer moist humid conditions, but some species of *Cymbidium* and *Vanda* tolerate low humid conditions and high intensity of sunshine. Some would prefer to be in extreme shade too (e.g., *Anoectochilus* and *Zeuxine*).

25.6 AQUATIC ANGIOSPERM SPECIES

As for any other groups of plants, several endemic species have been reported among aquatic species of Sri Lanka. *Cryptocoryne* and *Lagenandra* genera (family: Araceae), accommodate the highest number of endemic species (10 and 07, respectively) among water plants (Figure 25.7). The *Cryptocoryne* occurs in the slow running water or seasonally inundated riverine habitats mainly in the wet south–western region, however, few are found in similar habitats of the dry zone of Sri Lanka. All *Lagenandra* species are restricted to the wet zone and occur alongside the river banks (Yakandawela, 2012). *Eriocaulon* is another genus in which nearly half of the species are endemic to Sri Lanka with restricted distribution.

Unlike other water plants members of family Podostemaceae have very unusual habit being thalloid in shape to withstand the water currents in their fast-flowing areas of natural habitat. Out of the seven members, six species occur on rocks in the Mahaweli river whereas *Zeylanidium lichenoides* is reported as confined to the montane region (thought to be Probably Extinct category). *Farmeria metzgerioides* and *Polypleurum elongatum* both endemic to Sri Lanka are considered as Vulnerable, while *Polypleurum stylosum* is listed as Critically Endangered (Yakandawela, 2012). Hence, the members of the family Podostemaceae have been special consideration in conservation plans.

FIGURE 25.6   (a) *Adrorhizon purpurascens*–the monotypic Orchid genus endemic to Sri Lanka, (b) *Vanda testacea* growing in dry zone, (c) *Rhynchostylis retusa,* (d) *Vanda tessellata* var. *kotiya,* (e) *Arundina graminifolia,* (f) *Epsea speciosa,* (g) *Dendrobium crumenatum,* (h) *Oberonia quadrilatera* in their natural habitats.

Phorto credit: 25.6 (A), (C), (D), (F) Sanath B. Herath, (B),(E-F) S.M.W. Ranwala

FIGURE 25.7   (a,b) *Cryptocoryne* spp. (c) *Lagenendra* sp. (d) *Nymphea nouchali.*

Photo credit: 25.7 (A) Nalinda Peiris, (B-D) Tharanga Abeywickrama

Among the native aquatic angiosperms in Sri Lanka, free-floating *Woffia arrhiza* is considered to be one of the smallest vascular plants on the earth. It is a plant with a thalloid habit, about 1 mm (Yakandawela, 2012).

25.7   INSECTIVOROUS ANGIOSPERMS

Among the few carnivorous plants of the world, angiosperms from three plant families, namely Droseraceae, Nepenthecae, and Lentibulariceae are represented in Sri Lanka. The monotypic family represents only one endemic species *Nepenthes distillatoria,* which is found only in the lowland rain forests. The genus *Utricularia* is represented by 15 species including endemic species *U. moniliformis,* all of which are found in semi-aquatic habitats in most parts of the country (Figure 25.8).

FIGURE 25.8   (a) *Nepenthes distillatoria,* (b) *Drosera burmanni,* (c) *Utricularia* sp.

Photo credit: 25.8 (A-C) S.M.W. Ranwala

25.8   GYMNOSPERMS OF SRI LANKA

The only native gymnosperms represented in Sri Lanka are the Cycads. According to Lindstrom and Hill (2007), two members of the Family Cycadaceae have been identified in Sri Lanka, namely *Cycas nathorstii* and *Cycas zeylanica. Cycas nathorstii,* mostly occurring in the northern part in drier habitats has been reported as endemic to Sri Lanka. The latter is said to have highly threatened due to its limited distribution in southern Sri Lanka and, in the Andaman and Nicobar Islands. Recent studies by Mudannayake et al. (2019) on the *Cycas* group using morphological and genomic variation have identified three clusters, namely *C. zeylanica* complex (subsection Rumphiae) and *C. circinalis* and *C. nathorstii* (subsection Cycas). They have further identified that both *C. circinalis* and *C. nathorstii* inhabit mostly the dry zone of Sri Lanka whereas the *C. zeylanica* thrives in the wet zone. Moreover they suggest the occurrence of a fourth group resulted in as a "hybridized group" within the subsection *Cycas.* Further, studies on this subject area are being conducted to resolve the taxonomic identity of *Cycas* spp. in Sri Lanka.

KEYWORDS

- Sri Lanka
- angiosperms
- gymnosperms
- flora
- plants

REFERENCES

Abeywickrama, B. A. The Origin and Affinities of the Flora of Ceylon. *Proc. Annual Sess. Ceylon Assoc. Adv. Sci.* **1955,** *2,* 1–22.

Ashton, M.; Hooper, E. R.; Singhakumara, B.; Ediriweera S. Regeneration Recruitment and Survival in an Asian Tropical Rain Forest: Implications for Sustainable Management. *Eosphere* **2018,** *9* (2), 1–16.

Dassanayake, M. D.; Fosberg, F. R.; Clayton, W. D., Eds. *A Revised Handbook to the Flora of Ceylon,* 1980–2004; vol 1–14.

De Silva, K.; De Silva, M. *A Guide to the Mangrove Flora of Sri Lanka.* WHT Publications: Sri Lanka, **2006.**

Fernando, S. S. In *Distribution and Habitat Selection of Threatened Orchids of Sri Lanka,* Proceedings of the APOC 11 Okinawa International Orchid Show, Okinawa, Japan, Feb 2–11, 2013.

Fernando, S. S.; Ormerod, P. An Annotated Checklist of the Orchids of Sri Lanka. *Rheedea* **2008,** *18* (1), 1–28.

Fernando, M.; Wijesundera, S.; Fernando, S. Orchids of Sri Lanka – A Handrail for Conservationists. IUCN: Sri Lanka, **2003.**

Greller, A.; Gunatilleke I. A. U. N.; Jayasuriya, A.; Gunatilleke, C.; Balasubramaniam, S.; Dassanayake, M. *Stemonoporus* (Dipterocarpaceae)-Dominated Montane Forests in the Adam's Peak Wilderness, Sri Lanka. *J. Trop. Ecol.* **1987,** *3* (3), 243–253.

Gunatilleke, I. A. U. N.; Gunatilleke, C. V. S. The Foristic Composition of Sinharaja – A Rain Forest in Sri Lanka with Special Reference to Endemics. *Sri Lanka For.* **1980,** *14* (3&4), 171–179.

Gunatilleke, N.; Pethiyagoda, R.; Gunatilleke, S. Biodiversity of Sri Lanka. *J. Natl. Sci. Found.* **2008,** 36, 25–61.

Jayatissa, L. Present Status of Mangroves in Sri Lanka. In *The National Red List 2012 of Sri Lanka; Conservation Status of the Fauna and Flora*; Weerakoon, D. K., Wijesundara, S., Eds.; Ministry of Environment: Colombo, Sri Lanka, **2012**; pp 340–345.

Lindstrom, A. J.; Hill. K. D. The Genus *Cycas* (Cycadaceae) in India. *Telopea* **2007,** *11* (4), 463–488.

MOMD&E. *National Biodiversity Strategic Action Plan* 2016–2022. Biodiversity Secretariat, Ministry of Mahaweli Development and Environment: Colombo, Sri Lanka, **2016.**

Mudannayake, A.; Ranaweera, L.; Samaraweera, P.; Sooriyapathirana, S.; Perera A. The Morpho-Genetic and Ecological Niche Analyses Reveal the Existence of Climatically Restricted *Cycas zeylanica* Complex in Sri Lanka. *Sci Rep* **2019,** *9* (1), 16807. DOI: 10.1038/s41598-019-53011-w

Perera, A. Present Status of Dry-Zone Flora in Sri Lanka. In *The National Red List 2012 of Sri Lanka; Conservation Status of the Fauna and Flora*; Weerakoon, D. K., Wijesundara, S., Eds.; Ministry of Environment: Colombo, Sri Lanka, **2012**; pp 165–174.

Pethiyagoda, R. *Pearls, Spices and Green Gold: An Illustrated History of Biodiversity Exploration in Sri Lanka.* Wildlife Heritage Trust of Sri Lanka, **2007.**

Rathnayake, R. M. W.; Solangaarachchi, S. M.; Jayasekara, L. R. In *A Quantitative Study of Pigmy Forest at 2000 m in Hakgala Strict Natural Reserve,* Proceedings of the Second Annual Forestry Symposium 1996: Management and Sustainable Utilization of Forest Resources, Sri Lanka; Amarasekera, H. S., Ranasinghe, D. M. S. H. K., Finlayson, W., Eds.; Department of Forestry and Environmental Science, University of Sri Jayawardenapura: Sri Lanka, **1998.**

Singhakumara, B. M. P. *Ecological Assessment of Kanneliya, Dediyagala-Nakiyadeniya (KDN) Forest Complex.* Department of Forestry and Environmental Science, University of Sri Jayewardenepura: Nugegoda, **1995.**

Trimen, H. *The Handbook to the Flora of Ceylon.* Dalau: London, **1893–1900**; vol 1–5.

Werner, W. L. Biogeography and Ecology of the Upper Montane Rain Forests of Sri Lanka. In *Tropical Montane Cloud Forests;* Lawrance, S. H., James, O. J., Scatena, F. N., Eds.; Springer Verlag: New York, NY, **1995**; pp 343–352.

Wijesundara, S. Present Status of Montane Forests in Sri Lanka. In *The National Red List 2012 of Sri Lanka; Conservation Status of the Fauna and Flora*; Weerakoon, D. K., Wijesundara, S., Eds.; Ministry of Environment: Colombo, Sri Lanka, **2012.**

Wijesundara, S.; Ranasinghe, S.; Jayasinghe, H. Present Status of Angiosperms in Sri Lanka. In *The National Red list 2020- Conservation Status of the Flora of Sri Lanka*; Biodiversity Secretariat of the Ministry of

Environment and the National Herbarium of the Department of the National Botanic Gardens, Peradeniya, **2020**; pp 1–7.

Yakandawela, D. Present Status of Fresh Water Aquatic Flora of Sri Lanka. In *The National Red List 2012 of Sri Lanka; Conservation Status of the Fauna and Flora*; Weerakoon, D. K.; Wijesundara, S., Eds.; Ministry of Environment: Colombo, Sri Lanka, **2012**; pp 186–196.

CHAPTER 26

SRI LANKAN INSECTS WITH AN OVERVIEW OF DIVERSITY AND BIOGEOGRAPHY

AMILA P. SUMANAPALA

Department of Zoology and Environment Sciences, University of Colombo, Colombo, Sri Lanka

ABSTRACT

Systematic work on insect taxonomy and research in Sri Lanka goes back to the 19th Century. Early descriptive taxonomy and recent work have resulted in documenting about 11,500 insect species from Sri Lanka. This chapter provides an overview on the diversity of major insect orders in the country, that is, Coleoptera, Hymenoptera, Hemiptera, Lepidoptera, Diptera, Odonata, Mantodea, Ephemeroptera, Orthoptera, Neuroptera, Phasmatodea, Plecoptera, Dermaptera, Blattodea, Thysanoptera, and Trichoptera. The present knowledge on the biogeography of Sri Lankan insects and distribution zones recognized, are discussed with a general overview on the distribution of insects in the country. Gaps in the current knowledge are discussed providing recommendations for the future work on Sri Lankan insects.

26.1 INTRODUCTION

26.1.1 GEOLOGICAL HISTORY OF SRI LANKA

Sri Lanka is a continental island located near the southern tip of the Indian Peninsula. Geologically, it is a part of the Indian plate and shares a common Gondwanan origin as the Indian sub-continent. About 140 million years ago, Indian plate separated from the rest of the Gondwanaland and started drifting northwards (Kumar et al., 2007). This movement continued over millions of years and it collided with Laurasia resulting in the current position of the subcontinent. Despite being parts of the same plate, recurrent land bridges between the mainland and Sri Lanka formed during glacial periods and were disrupted during interglacials. This allowed recurrent episodes of connectivity and isolation between the two regions. However, despite the periodic land connectivity, the rainforests of south

Biodiversity Hotspot of the Western Ghats and Sri Lanka. T. Pullaiah, PhD (Ed.)

western Sri Lanka have never been in contact with the rainforests of the Western Ghats for the last 7 million years (MoMD&E, 2019). The periodic connectivity of the island with the mainland and the long isolation of the wet forests supported a highly diverse fauna with shared elements with Peninsular India as well as a significant endemism in the biota.

26.1.2   CLIMATE AND GEOGRAPHY

The topography of Sri Lanka varies considerably within the island despite its relatively smaller size (65,610 km^2). Three distinct peneplains are identified as the coastal plains with an altitude between 0 and 30 m, the inland plain with an altitude of 30–300 m, and the highlands with an altitude over 300 m. The highlands are located primarily in the central part of the island and consist of several mountain ranges separated by river valleys.

Precipitation in the country is mainly provided by monsoons and convective rainfalls. The monsoons arrive from the south–west and north–east, and due to the position of the central highlands, the two monsoons provide rainfall to two different areas. The south–western monsoonal winds which come across the Indian Ocean and thus contain more moisture, are the main source of precipitation to the south–western part of the island. This higher precipitation has formed a relatively wetter area in the central and south–western parts of the island, which is recognized as the wet zone of the country and receives an annual precipitation of over 2500 mm. The northern and eastern halves of the island receive relatively lower precipitation and thus are recognized as the dry zone where the annual precipitation is less than 1750 mm. An intermediate climatic zone occurs in between the wet and the dry zones with an annual precipitation between 1750 and 2500 mm.

26.1.3   ECOSYSTEMS

The diverse topographic and climatic characters of Sri Lanka provide a diverse array of ecological conditions, where an equally diverse range of ecosystems have been established. This is evident from the diversity of the floristic zones of Sri Lanka, where vegetation composition and structure change significantly. When major ecosystem types are considered, there are six different forest types (Lowland wet evergreen forests, Mid-elevation evergreen forests, Montane forests, Moist-mixed evergreen forests, Dry-mixed evergreen forests, Arid-mixed evergreen forests); five grassland types (Wet patana, Dry patana, Upland and lowland savannas, Dry [Damana] grasslands, Wet-intermediate [Thalawa] grasslands); four freshwater wetland types (Villus; Rivers and Streams; Reservoirs, tanks, lakes, and ponds; Marshes), three coastal vegetation types (Sand dunes; Beaches; Seagrass, salt marshes, and mangroves), among the natural ecosystems in the country (MoMD&E, 2019).

26.2   DIVERSITY OF SRI LANKAN INSECTS

The vast amount of work published on or include the insect fauna of Sri Lanka has documented a large number of species from the country. The only compilation of the total documented insect diversity of Sri Lanka in recent times was done by Wijesekara and Wijesinghe (2003) who reports a total of 11,144 insect species. With numerous new species discoveries, new country records, and taxonomic revisions in recent years, the total number of insects documented from the country at present should be no less than 11,500. However, the exact species richness of insects at present can only be determined after an extensive compilation of the literature.

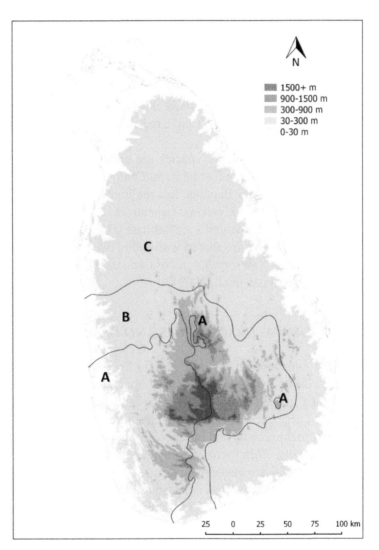

FIGURE 26.1 Map of Sri Lanka. (a) Wet climatic zone, (b) Intermediate climatic zone, (c) Dry climatic zone.

Among insect orders present in Sri Lanka, Coleoptera, Lepidoptera, Hemiptera, Hymenoptera, and Diptera are the most diverse. The following accounts discuss the documented present diversity, leading past taxonomic work, and recent studies (especially, the work published after Wijesekara and Wijesinghe, 2003) in the major insect orders. Some prominent insect groups and families that have received significant taxonomic attention, are also discussed under their relevant order.

Note: The taxonomic knowledge on all insect orders is not equally updated for the Sri Lankan fauna. The present exercise is a result of available and accessible literature. Therefore, there may be certain deviations with the species and endemics in taxa that have not been recently updated.

26.2.1 ORDER EPHEMEROPTERA: MAYFLIES

Mayflies that belong to Order Ephemeroptera are aquatic insects with exclusively aquatic larval stages and very short-winged adult stages. 52 species of mayflies in 10 families are known from Sri Lanka according to the revised checklist of Ephemeroptera in the Indian sub-region (Sivaramakrishnan et al., 2009) and a recent new species addition by Martynov and Palatov (2020).

Among the Sri Lankan mayflies, 41 species are endemic. Several genera including *Kimminsula* and *Megaglena* of Leptophlebiidae, and *Indobaetis* and *Indocleon* of Baetidae are endemic to the country. Family Baetidae has the highest diversity within Sri Lankan Ephemeroptera with 25 species (Sivaramakrishnan et al., 2009) and representatives distributed in all zones in both lentic and lotic habitats (Hubbard and Peters, 1984). Even though most non-endemic species are shared with India, two species, that is, *Teloganodes tristis* of Teloganodidae and *Sparsorythus jacobsoni* of Tricorythidae, are shared with Southeast Asia (Sivaramakrishnan et al., 2009).

26.2.2 ORDER ODONATA: DRAGONFLIES AND DAMSELFLIES

Odonata is one of the most studied insect groups in Sri Lanka. Currently, there are 131 species of Odonata recorded from the country (Kalkman et al., 2020; Sumanapala, 2021). Out of these, 63 belong to the sub-order Zygoptera and the remaining 68 belong to the sub-order Anisoptera. These species represent 12 families and 67 genera. Of the known species, 59 are endemic to the country and further eight species are endemic at the subspecies level. Three genera, namely, *Platysticta* and *Ceylonosticta* of family Platysticitidae and *Sinhalestes* of family Lestidae are also endemic.

The taxonomic interest in Odonata has increased significantly in recent years, especially after the year 2000. A distribution atlas of Sri Lankan Odonata, the first such publication on any faunal taxa in the country, was published by Bedjanič et al. (2014) compiling all known distribution data available. The new millennia saw the discovery of 14 new endemic species and two endemic subspecies.

26.2.2.1 FAMILY PLATYSTICTIDAE

The representatives of Platystictidae in Sri Lanka are classified into the endemic subfamily, Platyastictinae. It includes two endemic genera, that is, *Ceylonosticta* and *Platysticta*, and

26 endemic species. In recent years, Platysticitdae has received major taxonomic attention which resulted in the discovery of 12 new species (Bedjanič, 2010; Bedjanič et al., 2016; Priyadharshana et al., 2016, 2018). Bedjanič et al. (2016) who studied taxonomy and molecular phylogeny of Sri Lankan Platystictidae reports that the lineage of subfamily Platysticitnae is monophyletic, with its origins even preceding the separation of the old and the new worlds.

FIGURE 26.2 Some endemic Sri Lankan odonates. (a) *Sinhalestes orientalis* (Lestidae), (b) *Ceylonosticta tropica* (Platystictidae), (c) *Heliogomphus nietneri* (Gomphidae), (d) *Lyriothemis defonsekai* (Libellulidae). (Photos: Author)

26.2.3 *ORDER ORTHOPTERA: GRASSHOPPERS, CRICKETS, AND KATYDIDS*

According to Cigliano et al. (2021), 317 species of Orthopterans are known from Sri Lanka with 211 of them being endemic. They represent 15 families. Tettiogonidae, Gryllidae, Acrididae, and Tetrigidae are the largest Orthopteran families represented in the country while the family Pamphagidae is represented only by a single species.

26.2.3.1 *FAMILY TETRIGIDAE: PYGMY GRASSHOPPERS*

At present, there are 41 species of tetrigid species known from the country and 30 of these are considered endemic. The monotypic genera *Cingalina*, *Gignotettix*, and *Spadotettix*, and genus *Cladonotus* with four species, are endemic to the country. Several genera including *Euscelimena* and *Gavialidium* are confined to Sri Lanka and India and are shared between the countries (Cigliano et al., 2021).

26.2.3.2   FAMILY TETTIGONIIDAE: KATYDIDS

Katydids of the family Tettigoniidae are the most diverse taxa among Sri Lankan Orthoptera with 75 known species. 63 of these (84%) are endemic to the country. The monotypic genera *Aulocrania, Labugama, Perula, Scytoceroides, Temnophylloides, Vetralla,* and genera *Acrodonta, Brunneriana, Ischnophyllus,* and *Zumala* are endemic. According to the present understanding of the species distribution, many other genera represented in the country are distributed across the oriental region (Cignaliano et al., 2021) while some such as *Brochopeplus* and *Decolya* are confined to the Indian subcontinent. Some genera like *Cratioma, Gonyatopus,* and *Pseudophaneroptera* are shared only between Sri Lanka and parts of Southeast Asia while the genus *Phyllophora* is also shared with the Australian region (Cignaliano et al., 2021). Two Tettigoniidae genera found in the country have clear African affinities. The genus *Diogena*, represented in the country by a single endemic species *Diogena lanka*, is distributed widely in Africa other than southern parts (Massa, 2015) while the genus *Perelinus* with three endemic species in Sri Lanka, is also found in Seychelles islands (Matyot, 1998).

26.2.3.3   FAMILY ACRIDIDAE: GRASSHOPPERS, LOCUSTS

At present, there are 51 species of Acrididae grasshoppers in Sri Lanka. Thirty-two of these species are endemic to the country (Cignaliano et al., 2021).

Among the recorded genera, eight are endemic including the monotypic genera *Cingalia, Dubitacris, Ochlandriphaga, Pseudophlaeoba, Urugalla,* and *Wellawaya*. Of the other genera, many are widely distributed across the continent and sometimes even in the African region. Several genera are restricted to the Indian subcontinent and are shared between Sri Lanka and India (Cignaliano et al., 2021). Among those, the genera *Bambusacris* is restricted to the Western Ghats and Sri Lanka and is represented in each region by a single endemic species. The genus *Zygophlaeoba*, represented in the country by two endemic species, is confined to Sri Lanka and Tamil Nadu (Cignaliano et al., 2021; Shishodia et al., 2010).

26.2.3.4   FAMILY GRYLLIDAE: CRICKETS

Sri Lankan Gryllidae consists of 59 species with 31 endemics (Cignaliano et al., 2021). Among these, five genera, that is, *Coiblemmus, Hemilandreva, Homalogryllus, Scapsipedoides,* and *Poliotrella*, are considered endemic. Of the other genera represented in the country, a majority have a widespread distribution. The monotypic genus *Apterocryncus* has a restricted distribution limited only to South India and Sri Lanka (Shishodia et al., 2010).

Two genera, namely, *Jareta* and *Varitrella* are only distributed in Sri Lanka and South–East Asia (Cignaliano et al., 2021).

FIGURE 26.3   Some Sri Lankan orthopterans. (a) *Cingalia dubia* (Acrididae), (b) *Rakwana ornata* (Pyrgomorphidae), (c) *Euscelimena gavialis* (Tetrigidae), (d) *Orchetypus* sp. (Chorotypidae). (Photos: Author)

26.2.4   ORDER NEUROPTERA: ANTLIONS, OWLFLIES, LACEWINGS, AND MANTIDFLIES

Neuroptera is a relatively less studied group of insects in Sri Lanka with no extensive taxonomic work being conducted in the past. Scattered work across the literature reports the presence of about 45 species in the country. Among these work, Meinander (1982) studied the family Coniopterygidae of Sri Lanka with two new species discoveries. Ohl (2004), who published a catalogue to the Mantidflies of the world, mentions 14 species from the country with 11 of them being reported only from Sri Lanka.

26.2.5   ORDER PHASMATODEA: STICK INSECTS AND LEAF INSECTS

There are 62 valid species of Phasmatodea recorded from the country. Among these, 57 species (92%) are endemic to Sri Lanka. Of the recorded species, two are leaf insects of genera *Cryptophyllium* (Cumming et al., 2021) and *Pulchriphyllium* (Banks et al., 2021) while the rest are stick insects.

26.2.6   ORDER PLECOPTERA: STONEFLIES

Stoneflies of Order Plecoptera are a group of insects that have a relatively low representation within the country. Even though they have been surveyed during country-wide entomological surveys (Kawai, 1975), only 10 species in two genera have been described (Kawai, 1975; Zwick, 1980, 1982). In addition to these, several other possible new species to science have also been recognized through a limited number of materials (Zwick, 1982). It is not clear how many of the known species are endemic to the country. However, considering their distribution patterns within Sri Lanka, it is evident that the majority of species is restricted to the wet zone and highlands and thus might represent endemic species.

26.2.7   ORDER DERMAPTERA: EARWIGS

Earwigs are classified under the order Dermaptera. These are terrestrial insects characterized by the cerci that develop like a pair of prominent claspers. Steinmann (1989) has reported 71 species of Earwigs from Sri Lanka in his World Catalogue of the order and 25 of these species (36%) are endemic to the country (as cited by Wijesekara and Wijesinghe, 2003).

26.2.8   ORDER MANTODEA: MANTISES

Sri Lankan Mantodea fauna consists of 59 documented species in 10 families (Battiston, 2016; Ehrmann, 2002; Patel and Sing, 2016). Though species richness is comparatively less, Sri Lanka is richer in endemism than the Indian region in terms of Mantises (Mukherjee et al., 1995). Sixteen of the recorded species are endemic to the country. Among Sri Lankan mantis fauna, the members of Family Iridopterygidae are noteworthy as five out of six recorded genera, that is, *Hapalopezella, Iridopteryx, Micromantis, Muscimantis,* and *Pezomantis* are monotypic and endemic (Ehrmann, 2002), suggesting an interesting evolutionary history behind them.

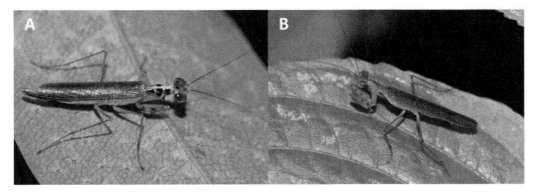

FIGURE 26.4   Some endemic Sri Lankan mantises. (a) *Hapalopazella maculata*, (b) *Iridopteryx irridipennis.* (Photos: Author)

26.2.9 BLATTODEA: TERMITES AND COCKROACHES

26.2.9.1 ISOPTERA: TERMITES

Edirisinghe et al. (2016) provides a comprehensive summary on the termites of Sri Lanka discussing taxonomy, habitats, and feeding habits. According to Edirisinghe et al. (2016), 76 species of termites in 29 genera and four families are known from Sri Lanka. Among these, 33 species are considered endemics. The majority of forest-dwelling species recorded from Central Sri Lanka inhabit the soil. Only three genera of termites in the family Termitidae (*Macrotermes, Odontotermes,* and *Hypotermes*) are known to dwell in earthen mounds.

26.2.9.2 COCKROACHES

The current knowledge on Sri Lankan cockroaches is based on a number of scattered studies in the history (Wijesekara, 2006). Wijesekara and Wijesinghe (2003) reported that there are 66 species of cockroaches known to occur in the country.

26.2.10 ORDER THYSANOPTERA: THRIPS

The first extensive survey on Sri Lankan Thysanopteran fauna, focusing on the diversity, host plants and distribution, was conducted by Tillekaratne et al. (2011). This survey primarily covered the agricultural habitats and reported 72 thrips species from 324 species of host plants. Edirisinghe et al. (2016) provides a comprehensive summary of the studies on Sri Lankan thrips and updated knowledge on them. At present, there are 104 recorded species of thrips in Sri Lanka including seven species that have not been identified to the species level (MoMD&E, 2019).

26.2.11 ORDER HEMIPTERA: BUGS

Hemiptera is the third-largest insect order in Sri Lanka. Wijesekara and Wijesinghe (2003) reported that there are 1783 known Hemiptera (including Homoptera) species in the country. The only comprehensive work on the entire Hemipteran fauna of the country is the seven volumes of the Fauna of British India series by W. L. Distant (1902–1918). Later, several authors have separately worked on different taxa within the order.

26.2.11.1 FAMILY APHIDIDAE: APHIDS

Aphids are small hemipteran insects with economic importance as many species are considered agricultural pests. Wijerathna and Edirisinghe (1999) conducted a general survey of aphids in Sri Lanka sampling close to 1000 plant species. With the findings of this work and literature, Edirisinghe et al. (2016) provides a comprehensive summary of knowledge on Sri Lankan Aphids. At present, 74 species of aphids in 40 genera are known from Sri Lanka. Ten of the known species are considered endemic (Edirisinghe et al., 2016).

26.2.11.2   FAMILY CICADELLIDAE: LEAFHOPPERS

Cicadellid leafhopper fauna of Sri Lanka consists of 258 species in 120 genera (Gnaneswaran, 2012; Viraktamath and Gnaneswaran, 2013). Gnaneswaran (2012) reports on host plants, habitats, and climatic zones inhabited by some Cicadellids primarily based on work conducted in human-dominated and agricultural landscapes.

26.2.11.3   FAMILY LYGAEIDAE: SEED BUGS

Slater (1982) reports 139 Seed bug species known to occur in Sri Lanka based on historical work and collections made during Lund University Expeditions in the country. Of these species, 26% are only known from Sri Lanka. Slater (1982) also notes on the habitats of the species in the Lund University collection.

26.2.11.4   FAMILY CICADIDAE: CICADAS

According to Price et al. (2016) who revised the list of Cicadids in several South Asian countries, there are 23 species of Cicadids in Sri Lanka. They are classified into two subfamilies: Cicadinae and Cicadettinae, and 13 genera. Of the recorded species, 11 are only reported from Sri Lanka.

26.2.11.5   FAMILY DELPHACIDAE: PLANT HOPPERS

Plant hoppers in the family Delphacidae in Sri Lanka have been studied by multiple authors in the past. Fennah (1975) studied the Sri Lankan material collected during the Lund University Expeditions and reported 69 species including 25 new species based on the collection.

26.2.11.6   FAMILY ALEYRODIDAE: WHITEFLIES

The Whiteflies of Aleyrodidae are economically important insects as some species are considered agricultural pests. The revision of Aleyrodidae in Sri Lanka by David and Dubey (2008) reported 58 species of Whiteflies in Sri Lanka, which are classified into 24 genera, with several new records, host plant records, and a new species description.

26.2.11.7   FAMILY FLATIDAE: FLATID PLANT HOPPERS

Pioneering work on Sri Lankan Flatidae has been conducted by Kirby (1891), Melichar (1903), and Distant (1912), (Medler, 2006). Based on the historical collections and collections made during the biosystematic studies of the Insects of Sri Lanka project, Medler has recently revised the Flatidae fauna of the country. According to Medler (2006), there are 30 species of Flatidae classified into 18 genera in Sri Lanka.

26.2.11.8   FAMILY MIRIDAE: MIRID BUGS

Miridae is a diverse family of predominantly phytophagous bugs. A compilation of taxonomic work on the family by Wijesekara and Henry (1999) reports 118 species of 69 genera from the country. A review of subfamily Bryocorinae in Sri Lanka reported a new country record and an unidentified species (Basnagala et al., 2002).

26.2.11.9   FAMILY REDUVIIDAE: ASSASSIN BUGS

Members of Reduviidae are generally referred to as Assassin Bugs and are predatory insects. Sri Lankan Reduviidae is not studied comprehensively in recent times. Based on the historical work and sparse recent work including the discovery of a new cave dwelling, thread-legged assassin bug by Ghate et al. (2018), 146 Reduviidae species are recorded from Sri Lanka. Sixty of these are only known from the country and thus are possibly endemic (Ambrose, 2006; Distant, 1904, 1910; Ghate et al., 2018).

26.2.12   AQUATIC HEMIPTERA

The list of aquatic hemipterans in the country has been compiled in the publication series "A Guide to the Freshwater Fauna of Ceylon." Fernando (1974) reports 117 species of aquatic hemipterans including four species of water scorpions (Nepidae: Nepinae), eight species of water stick-insects (Nepidae: Ranatrinae), two species of giant water-bugs (Belostomatidae), three species of creeping water-bugs (Naucoridae), three species of Helotrephid Backswimmers (Helotrephidae), two species of Pygmy Backswimmers (Pleidae), 11 species of Backswimmers (Notonectidae), 24 species of Water Boatmen (Corrixidae), four species of Water Measurers (Hydrometridae), two species of Pond Skaters (Mesovelidae), 18 species of Riffle bugs (Veliidae), three species of Velvet Water-bugs (Hebridae), 30 species of Water Striders (Gerridae), two species of Shore bugs (Saldidae), and one species of Spiny-legged bugs (Leptopodidae). Among the aquatic Hemiptera of the country, two are open ocean species, that is, *Halobates germanus* and *H. micans*, which are known to occur in the Sri Lankan waters (Andersen and Foster, 1992).

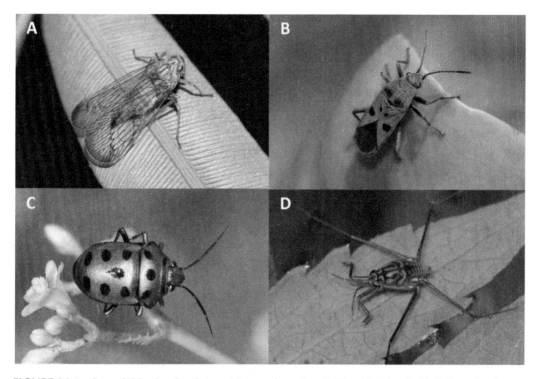

FIGURE 26.5 Some Sri Lankan hemiptera. (a) *Leusaba rufitarsis* (Tropiduchidae), (b) *Spilostethus hospes* (Lygaeidae), (c) *Chrysocoris purpureus* (Scutelleridae), (d) *Metrocoris* sp. (Gerridae). (Photos: Author)

26.2.13 ORDER HYMENOPTERA: ANTS, BEES, SAWFLIES, AND WASPS

Hymenoptera is the fourth most diverse order among Sri Lankan insects (Wijesekara, 2006). This includes two suborders, Symphyta and Apocrita. The suborder Apocrita which includes ants, wasps, and bees is among the most studied insect taxa in the country.

Wijesekara and Wijesinghe (2003) reported that there are 1519 known species of Hymenoptera in Sri Lanka. With recent work on taxa such as Mutillidae (Lelej, 2005), Formicidae (Dias et al., 2020), Apoidae (Karunaratne et al., 2017), and multiple new species additions by several authors including Buhl (2014) and Sureshan (2006, 2007, 2010), this number has increased considerably. As of now, the number of Hymenoptera species known from Sri Lanka is about 1700.

26.2.13.1 FAMILY FORMICIDAE: ANTS

As a result of historical work and recent taxonomic surveys, 301 species of ants are currently known from the country (Dias et al. 2020a,b). They represent 11 subfamilies with Formicinae, Mymicinae, and Ponerinae being the most diverse. Of the recorded species, 68 are endemic to Sri Lanka while 18 are introduced species with established populations. When the distribution of ants is considered, the wet zone of the country has the highest diversity of ants followed by the dry zone while the intermediate zone has the comparatively lowest ant diversity among the climatic zones (Dias et al., 2020a,b). Recent taxonomic work has described three new species from Sri Lanka, that is, *Protanilla schoedli* (Baroni Urbani and De Andrade, 2006), *Tyrannomyrmex legatus* (Alpert, 2013), and *Pristomyrmex sinharaja* (Yamane and Dias, 2016), and Dias et al. (2019) described the worker cast of *P. schoedli*.

26.2.13.2 *SUPERFAMILY APOIDEA: BEES*

Bees are one of the most well-known pollinators which have been studied by multiple authors over the years. At present, there are 159 bee species recorded from Sri Lanka and 22 of these are endemic (MoMD&E, 2019). Recent work on Sri Lankan bees has resulted in many new discoveries including new country records (Karunaratne et al., 2017), new species discoveries (Griswold, 2001; Karunaratne et al., 2005, Engel, 2008), and redis-covery of a species (Silva et al., 2018).

Edirisinghe et al. (2016) who provide an overview of the bee diversity in the country and discuss their nesting habits and foraging behaviors, reports that most of Sri Lankan bees are pollen bees of which large majority are ground-nesting species. Some species of carpenter bees, leaf-cutting bees, and parasitic cuckoo bees are also found among the pollen bees. Five bee species are honey bees which include three *Apis* species and two *Tetragonula* species.

26.2.13.3 *WASPS*

Wasps represent a large group of known Hymenoptera in Sri Lanka classified under a number of Hymenoptera families. Many groups of wasps were studied by K. V. Krombein through his work in the project biosystematic studies on Ceylonese insects during 1978–1998 (Wijesekara, 2006). Their work lead to the publication of 22 volumes in a series titled biosystematic studies of Ceylonese Wasps, covering the taxonomy and natural history of many different wasp taxa. This series of publications covers the Subfamily Amiseginae (Number I); Family Scoliidae (Number II); Species *Paraleptomenes mephitis* (Number III); Subfamily Kudakrumiinae (Number IV); Family Ampulicidae (Number V); Affinities and derivation of the wasp fauna (Number VII); Family Philanthidae (Number VIII); Family Tiphiidae (Number IX); Subfamily Oxybelinae (Number X); Subfamily Amiseginae and Loboscelidiinae (Number XI); Family Sphecidae (Number XII), Subfamily Stizinae (Number XIII); Genus *Carinostigmus* (Number XIV); Subfamily Alyssoninae, Nyssonisae and Gorytinae (Number XV); Genus Gastrocericus (Number XVI); Genus *Bembix* (Number XVII); Genus *Trachepyris* (Number XVIII); Family Eumenidae, Vespidae, Pompilidae and Crabronidae (Number XIX); Genus *Tachysphex* (Number XX); Subfamily Bethylinae and Epylinae (Number XXI) and Genus *Bethsmyrmilla* (Number XXII) of Sri Lanka.

Family Scoliidae is the only wasp family to be fully revised under the work of this publication series. Twenty-five species of Scoliid wasps were reported from the country by Krombein (1978) with several new species descriptions. Majority of these are wide-spread species in the oriental region while six species can be considered endemic based on original findings of Krombein (1978) and the recent revision of Indian Scoliidae by Gupta and Jonathan (2003).

26.2.13.4 FAMILY MUTILLIDAE: VELVET ANTS

Velvet ants belong to the family Mutillidae of order Hymenoptera. Lelej (2005) published a revision of the Mutillidae in the Oriental region including Sri Lankan taxa, verified previous records, and described several new species from the country as well. Based on revisions and previous literature records compiled, there are 87 known species of Velvet ants recorded from Sri Lanka (Lelej, 2005) and 52 of these are endemic (Cascio, 2015; Lelej, 2005). Among the known taxa, genera such as *Lehritilla* and *Serendibiella* are considered endemic to the country (Lelej, 2005).

FIGURE 26.6   Some Sri Lankan wasps. (a) *Ammophila atripes* (Specidae), (b) *Scolia affinis* (Scolidae), (c) *Trogaspidia soror* (Mutillidae), (d) *Eustenogaster eximia* (Vespidae). (Photos: Author)

26.2.14 ORDER COLEOPTERA: BEETLES

Beetles are the most specious Order among insects. Wijesekara and Wijeyasinghe (2003) report that there are at least 3033 beetle species in Sri Lanka, and this is about 27% of all reported insects at the time. Considering the recently described species and new country records, it is likely that the current known number of Coleoptera species in the country is over 3100. Among beetle families are known from Sri Lanka. Carabidae, Scarabaeidae, Chrysomelidae, Staphylinidae, Tenebrionidae, and Curculionidae are some of the most diverse.

26.2.14.1 FAMILY CARABIDAE: GROUND BEETLES

Carabidae is one of the largest beetle families in Sri Lanka. Fowler and Andrews have reported 368 Carabidae species from the country with 34.5% endemics (as cited in Erwin, 1984). Later, Erwin (1984) reports the presence of 525 species in 143 genera primarily based on collections made in entomological expeditions carried out in the 1960s and 1970s. He also reports 10 endemic genera to Sri Lanka. Considering the possibility of some species recorded by Fowler and Andrews not being recorded in his catalogue, Erwin (1984) suggests that the total Carabid fauna in Sri Lanka could exceed 600 species.

From a biogeographic standpoint, Sri Lankan Carabids are primarily related to the Oriental fauna. Based on Fowler and Andrews's checklists, 31% of the species are shared with India and Burma while 28% are shared with the entire Oriental region. Other regions such as the Australian region, Ethiopian region, and the Mediterranean are having only a small percentage of species shared with Sri Lankan fauna (Erwin, 1984).

Among Sri Lankan Carabids, Tiger Beetles of Cicindelinae is probably the most studied group. There are 64 species of Tiger Beetles known from the country and 39 of them are endemic (Dangalle, 2018; Dangalle et al., 2011; Abeywardhana et al., 2020). The ground-dwelling tiger beetles of Tribe Cicindelini are relatively well known and are found in a diverse range of habitats. The arboreal tiger beetles of Tribe Collyridini are relatively a less studied group with 32 species in Sri Lanka (Dangalle, 2018; Abeywardena, 2020). Despite the relatively continuous attention given to the group, new discoveries have been made even in recent years. The rediscovery of *Jansenia laeticolor* which was not reported for over a century (Thotagamuwa et al., 2016) and the first report of *Tricondyla gounellei* in Sri Lanka (Abeywardhana et al., 2020) are such examples.

26.2.14.2 FAMILY CHRYSOMELIDAE: LEAF BEETLES

Chrysomelid leaf beetles are represented by a very diverse fauna in Sri Lanka. A major preliminary compilation of the Chrysomelid fauna in the region was done by Jacoby (1908) and Maulik (1919, 1926, 1936) as contributions to the Fauna of British India series. Collectively, they report about 270 species of leaf beetles to occur in Sri Lanka. During later decades, multiple authors contributed in expanding the taxonomic knowledge on Sri Lankan leaf beetle fauna. Scherer (1969) published a monographic work on the Alticinae of the Indian subcontinent with 82 Sri Lankan species with 64 endemics and one new species described from the country. Medvedev published on several groups of Chysomelids in Sri Lanka during 1970s and 1980s (ZIN RUS, 2021). Mohamedsaid (1997) treated the Sri Lankan Galerucinae collected during the Lund University Expedition in a paper in which he described two new species and report 10 new country records. Among recent work, Kimoto (2003) reports 123 Chrysomelidae species with nine new species descriptions and

14 new country records based on the collections of Akio Otake made between 1973 and 1975. Staines (2015) recorded 41 Hispinae species from Sri Lanka with 20 endemics in the World catalogue of Hispinae.

FIGURE 26.7   Some carabidae beetles. (a) *Cicindela bicolor*, (b) *Omphra hirta*, (c) *Ophionea indica* and Chrysomelidae beetles (d) *Agonita apicipennis*, (e) *Diapromorpha turcica*, (f) *Lema quadripunctata* of Sri Lanka. (Photos: Author)

26.2.14.3   WEEVILS

Weevils are classified into several families under the superfamily Curculionoidea. Many of them possess elongated snouts and some are considered serious agricultural pests. The taxonomic work on Sri Lankan weevils is not that comprehensive. Marshall (1916) reports

23 species from Sri Lanka in his volume of Fauna of British India series. Bark Beetles of Sri Lanka were studied by Schedl (1959) who provided a checklist of 111 Scolytinae species and 13 Platypodinae species (both were considered separate families earlier). With additional new records, new species and taxonomic changes, Schedl (1971) updated the list of Sri Lankan Scolytinae making the number of species 122. The world catalogue of tribe Mecysolobini (Curculionidae: Molytinae) by Andrew and Ramamurthy (2010) reports 22 species from Sri Lanka and the catalogue of tribe Cleonini (Curculionidae: Lixinae) by Meregalli (2017) lists nine species from the country.

26.2.14.4   FIREFLY BEETLES

Thirty-seven species of firefly beetles in the families Lampyridae and Rhagophthalmidae are currently known from Sri Lanka (MoMD&E, 2019; Packova and Kundrata, 2021). Two species, namely, *Harmatelia discalis* and *Harmatelia bilinia* are confirmed as endemics (MoMD&E, 2019). A study done on the Luciolinae firefly beetles in grassland ecosystems across the country indicates that the species *Luciola chinensis* is the most common firefly species in Sri Lanka (Wijekoon et al., 2012). The collection of all Cantharoid beetles including fireflies in the "Sri Lankan firefly collection" housed at the Department of National Museums, Sri Lanka has been systematically revised by Wijekoon et al. (2016).

26.2.14.5   FAMILY SCARABAEIDAE: SCARAB BEETLES

This is one of the most diverse groups of beetles in the country. Arrow (1910, 1917, 1931) treated most of the groups in the family in his volumes of Fauna of British India including the Sri Lankan taxa. Since the publication of the above series, no comprehensive work has been conducted on the entire family but later authors have worked on selected subfamilies and groups.

Taxonomy of Sri Lankan scarabid beetles of Tribe Sericini has been studied extensively in recent years. The monograph on the group produced by Fabrizi and Ahrens (2014) described 43 new species to science. Further, four new species were described by Ranasinghe et al. (2020) with some new faunistic records. At present, the number of known Sericini in Sri Lanka stands at 81 with 71 endemics (87.6%).

Kudavidanage and Lekamge (2012) provides a relatively comprehensive account on the dung beetles of Scarabinae based on their work on diversity, distribution, and the effect of habitat fragmentation and land-use changes on the dung beetle fauna in Sri Lanka. The provisional checklist of Sri Lankan dung beetle fauna comprised of 103 species with 21 endemics (20.4%). The diversity and abundance of dung beetles are higher in the dry zone where most of the large mammals are also distributed. The montane zone has the lowest dung beetle diversity in Sri Lanka (Kudavidanage and Lekamge, 2012). Aphodiinae, a subfamily that consists of small dung beetles, consists of 18 known species in the country (Stebnicka, 1988).

26.2.14.6 FAMILY TENEBRIONIDAE: DARKLING BEETLES

Tenebrionidae is a large group of beetles with many different forms. Sri Lankan Tenebrionidae was revised by Kaszab in 1979 and he reported 304 species from the country. According to Kaszab (1979), 190 of these species are endemic (62.5%) and two-third of the endemics are reported from the Central Province. Therefore, it was suggested that the central highlands of Sri Lanka has facilitated speciation among Tenebrionidae resulting in numerous restricted range endemics. The Tenebrionid fauna in the dry zone, especially in the semi-arid zones are largely shared with Southern India. However, some wingless species are confined to Sri Lanka.

The world catalogue of Amarygmini, Rhysopaussini, and Falsocossyphini lists 31 species from the country with 25 endemics (Bremer and Lillig, 2014). A total of 22 species with two endemics from Sri Lanka are listed in the world catalogue of *Gonocephalum* (Iwan et al., 2010).

26.2.14.7 FAMILY LUCANIDAE: STAG BEETLES

The diversity of Stag Beetles in Sri Lanka is relatively low. Only eight species are currently known. The monotypic genus *Platyfigulus,* with the species *Platyfigulus scorpio,* is the only endemic genus in the country (Bartolozzi and Bomans, 1962). A new species in the genus *Dinonigidius,* which was only known from single species in India prior to this, was described based on material collected in and around Kandy (Paulsen, 2016).

26.2.14.8 LONGICORN BEETLES

Longicorn Beetles are wood-boring insects. Some of these are well-known agricultural pests. According to the recent checklist compiled by Makihara et al. (2008), there are 85 Cerambycidae Longicorn beetles (excluding subfamily Laminae) and one Vesperidae longicorn beetle known from the country. 53 of these (62.4%) are known to occur only in Sri Lanka. Makihara et al. (2008) also report the known host plants of the species recorded. In addition to this, the Cerambycidae species *Acanthophorus serraticornis* was recorded for the first time in the country by Delahaye et al. (2011).

26.2.14.9 FAMILY MELOIDAE: BLISTER BEETLES

Members of Family Meloidae are known as blister beetles because they secrete a blistering agent named canthridin as a defensive mechanism. Sri Lankan blister beetles have been studied by Mohamedsaid (1979). Currently, 16 blister beetle species with three endemics are recognized from the country (Ghoneim, 2013; Mohamedsaid, 1979).

26.2.14.10   FAMILY ANTHICIDAE: ANT-LIKE FLOWER BEETLES

Anthicid collection of the Lund University Ceylon Expedition was studied by Bonadona (1988) who reported 31 species including 14 new species and eight new country records. Six additional new species were described by Bonadana (1989) based on Sri Lankan material. Recent work by Kejval (2011) described a species from Sri Lanka in a revision of Oriental *Notoxus*.

26.2.14.11   FAMILY ELATERIDAE: CLICK BEETLES

Click beetles are a group of less studied Coleoptera in Sri Lanka with no extensive work on the group in the country ever been published. Scattered historic work by multiple authors has described the known Sri Lanka Elaterids. Ohira (1973) studied the Elateridae collection made during the Lund University Ceylon Expedition and reported 31 species including four new species and four new country records. Of the species reported by Ohira (1973), 13 are considered to be distributed only in Sri Lanka according to present knowledge, suggesting considerable endemism among Sri Lankan Elaterid fauna.

26.2.14.12   FAMILY STAPHYLINIDAE: ROVE BEETLES

Rove beetles are distinct taxa of beetles with short elytra. Herman (2001) compiled a catalogue of Staphylinidae of the world excluding four subfamilies that is, Aleocharinae; Paederinae; Pselphinae, Scaphidiinae, and Scydmaeninae, which was considered a separate family by the time. According to the catalogue, 213 species of rove beetles are known from the country with 76 endemics. A species of *Megarthrus* in subfamily Proteininae was described (Cuccodoro and Liu, 2016) subsequent to the publication of the catalogue. About 128 species of Aleocharinae (Hammond, 1975; Kistner, 2012); 86 species of Paederinae (Hammond, 1975; Assing, 2012; de Raugemont, 2014); 44 species of Scaphidiinae (Löbl, 1971); 97 species of Pselphinae (Jeannel, 1961; Li and Yin, 2020; Löbl, 2017; Löbl and Kubratov, 2001), and 198 species of Scydmaeninae (Franz, 1982, 1983; Jaloszyński, 2009, 2020) are known to occur in the country. Considering these, the total number of reported rove beetle species in the country is 767.

26.2.14.13   AQUATIC BEETLES

There are several Coleoptera families which are primarily aquatic in habit. According to the publication, A Guide to the Freshwater Fauna of Ceylon, and its supplements, there are 202 aquatic coleopterans in the country (Mendis and Fernando, 1962; Fernando, 1963, 1964, 1969, 1974; Fernando and Weerawardhena, 2002). Among the major families of aquatic coleopterans, the above list includes 54 species of Diving beetles (Dytiscidae), 62 species of Water scavenger beetles (Hydrophilidae), 18 Whirligig beetles (Gyrinidae), 34 species of Riffle beetles (Elmidae), and 15 species of Marsh beetles (Scirtidae), that have been recorded from the country.

FIGURE 26.8   Some beetles of Sri Lanka. (a) *Platyfigulus scorpio* (Lucanidae), (b) *Indopolemius dimidiatus* (Cantharidae), (c) *Hycleus ceylonicus* (Meloidae), (d) *Gymnopleurus koenigi* (Scarabaeidae), (e) *Sandracottus festivus* (Dytiscidae), (f) *Paederus fuscipes* (Staphylinidae). (Photos: Author)

Among recent work, Freitag (2012) reported the first record of the genus *Ancyronyx* of family Elmidae in the Indian subcontinent with a new species discovery from Sri Lanka. A revision of the genus *Cyphon* (Scirtidae) by Ruta (2007) described a new species from Sri Lanka. Another new aquatic beetle species in the genus *Peschetius* (Dytiscidae) was described by Biström and Bergsen (2015). The world catalogue on Dytiscidae reports 39 species from Sri Lanka (Nilsson and Hájek, 2019) and Noteridae includes seven species from the country according to recent catalogues (Nilsson, 2011; Nilsson and Hájek, 2019).

26.2.14.14 OTHER COLEOPTERA FAMILIES

Based on recent global and regional taxonomic catalogues, the number of species recorded from Sri Lanka in several Coleoptera families is as follows; four species of false skin beetles in Byphillidae (Węgrzynowicz, 2015), four species of cedar beetles in Callirphidae (Hájek, 2011), two species of Megalopodid leaf beetles in Magalopodidae (Rodríguez-Mirón, 2018), four species of bark-gnawing beetles in Trogossitidae (Kolibáč, 2013). Zahradník (2008) described a species of Ptinidae and also notes that there are 21 known species in the family belonging to seven subfamilies.

26.2.15 ORDER TRICHOPTERA: CADDISFLIES

Sri Lankan Trichoptera has been studied in detail by F. Schmid (in 1958) and H. Malicky (Wijesekara and Wijesinghe, 2003). According to Wijesekara and Wijesinghe (2003), there are 188 species of Trichoptera recorded from the country. In recent decades, several new species have been added to the Trichoptera fauna of Sri Lanka. Oláh and Johanson (2008) and Oláh et al. (2007) described five species in the genus *Cheumatopsyche* while Flint (2003) described three species in the genus *Macrostemum*.

26.2.16 ORDER LEPIDOPTERA: BUTTERFLIES AND MOTHS

Lepidoptera is the second largest insect order in Sri Lanka. Majority of these are moths while the butterfly fauna of Sri Lanka consists of 248 species.

26.2.16.1 BUTTERFLIES

The butterflies of Sri Lanka are classified into six families. Family Lycaenidae is the most specious family in the country while the family Riodinidae is only represented by a single species. Among Sri Lankan butterflies, 31 species are currently recognized as endemics (MoMD&E, 2019). Majority of the endemic species are distributed in the wet climatic zone of Sri Lanka with some species like *Spindasis greeni*, *Parantica taprobana*, and *Lethe daretis* being restricted to the montane regions of the zone. The dry climatic zone of the country is generally occupied by more generalist species but one endemic species, *Spindasis nubilis*, is restricted to this zone. Both wet zone and dry zone butterfly fauna are represented in the intermediate climatic zone with the endemic species *Ypthima singala* being restricted to the savanna grasslands of the zone (van der Poorten and van der Poorten, 2018). Compilations of host plants used by Sri Lankan butterflies were published by Jayasinghe et al. (2014, 2021).

FIGURE 26.9   Some endemic Sri Lankan butterflies. (a) *Mycalesis rama*, (b) *Idea iasonia*, (c) *Kallima philarchus*, (d) *Lethe dynsate*. (Photos: Author)

Sri Lankan butterflies share their origin with the butterflies of India. All but one genus of butterflies in the country are also found in the Western Ghats region. The genus *Ideopsis*, which is distributed in Southeast Asia, is absent in India but found in the Southwestern region of Sri Lanka. It has been suggested that the genus has been introduced with the plants brought into the country in ships. Three species of butterflies that were first recorded in the recent past (*Catopsylia scyla*, *Cephrenes trichopepla*, and *Erionota torus*) are thought to be recent introductions through plants (van der Poorten and van der Poorten, 2018) and thus considered exotic species recorded in the country (MoMD&E, 2019).

26.2.16.2   MOTHS

The moths of Sri Lanka consist of approximately 2100 known species in 44 families. Most of the moth diversity in Sri Lanka, especially those of the Microlepidoptera, are relatively lesser-known. Four species of Saturnidae; one species of Callidulidae (Old World butterfly moths) around 300 species of 204 genera in Crambidae (Crambid snout moths); about 300 species of 123 genera in Geometridae (Geometer moths), about 210 species of Noctuidae (Cutworm moths), 57 Sphingidae species (Hawk moths), about 125 species in Tortricidae (Tortricidae leafroller moths), over 300 species in 53 genera of Eupterotidae (monkey moths), more than 150 species in Pyralidae (Pyralid snout moths), about 70 species in 23 genera of Nolidae (Tufted moths), about 25 species of 13 genera in Notodontidae (Prominent moths), about 25 species in 14 genera of Euteliidae (Euteliid moths), about 30 species of Uraniidae (Swallowtail moths), about 15 species of Drepanidae (Hooktip moths), about 20 species in 11 genera of Zygaenidae (Burnet moths), 32 species in 14 genera of Limacodidae (Slug-caterpillar moths), about 20 species in 11 genera of Lasiocampidae (Lappet moths), about 12 species of Cossidae (Carpenter moths), one species of Callidulidae (Old world butterfly moths), about 20 species of Thyrididae (Window-winged moths), about five species in four genera of Sesiidae (Clearwing moths), about 23 species of Psychidae (Bagworm moths); about 10 species of Choreutidae (Metalmark moths) and more than 15 species of Pterophoridae (Plume moths), are known from Sri Lanka (Jayawardana and Jayawardana, 2020).

26.2.17   ORDER DIPTERA: FLIES

Diptera is the fifth largest order of insects in Sri Lanka (Wijesekara and Wijesinghe, 2003) with over 1350 known species being reported from the country.

26.2.17.1   FAMILY ASILIDAE: ROBBERFLIES

Robber flies are aggressive predatory dipterans that feed primarily on insects that are caught in the flight. Work by Joseph and Parui in the oriental (1983) and Indian (1998) regions discuss the Sri Lankan fauna. At present, there are 53 species of Robber flies that have been recorded from the country and 26 of these are considered to be endemic (Geller-Grimm, 1997; Joseph and Parui, 1983, 1998; Scarbrough and Hill, 2000).

26.2.17.2   FAMILY CALLIPHORIDAE: BLOWFLIES

The Blowflies of Sri Lanka have been studied by Kurahashi (2001) based on the collections of Lund University Ceylon Expedition and field surveys of medically important flies in Sri Lanka conducted in 1986, 1987, and 1989. A total of 35 species are reported by Kurahashi (2001).

26.2.17.3   FAMILY STRATIOMYDAE: SOLDIERFLIES

Sri Lankan Soldier flies have been studied by multiple authors in the past. Hauser and Rozkošný (1999) revised the Sri Lankan Stratiomydae fauna with new records, new descriptions, and taxonomic notes on recorded genera. Later, Rozkošný and Hauser (2011) revised the records of the genus *Ptecticus* in the country describing a new species. According to these, a total of 30 Stratiomydae species with 11 endemics are currently known from the country.

26.2.17.4   FAMILY DIOPSIDAE: STALK-EYED FLIES

Feijen and Feijen (2020) revised the stalk-eyed flies (Diopsidae) of Sri Lanka with two new species descriptions. With that, there are eight species of Diopsidae known from the country including two undetermined species in genera *Cyrtodiopsis* and *Diopsis*. All five known species in the genus *Teleopsis* are endemic while the only known species in the genus *Sphyracephala* is shared with South India.

26.2.17.5   FAMILY TABANIDAE: HORSEFLIES

Horseflies are relatively large flies of which the females are facultative hematovores. John F. Burger reviewed the entire Tabanid fauna of Sri Lanka describing multiple new species and providing notes on the Zoogeography. According to Burger (1982), there are 51 known species of Tabanidae in Sri Lanka with 25 species considered to be endemic. Two genera, namely, *Silviomyza* and *Udenocera*, are monotypic and endemic to the country.

26.2.17.6   FAMILY LIMONIIDAE AND TIPULIDAE: CRANEFLIES

Craneflies are flies with characteristically long legs. Limoniids usually keep their wings folded along the abdomen when at rest while Tipulids keep the wings open. According to the Catalogue of the Craneflies of the World (Naturalis Biodiversity Center, 2021), there are 114 species of craneflies in Sri Lanka. These include 88 species of Limoniidae craneflies and 26 species of Tipulidae craneflies. Among these, 59 limoniids (67%) and 12 tipulids (46%) are considered endemic.

26.2.17.7   FAMILY CULICIDAE: MOSQUITOS

Mosquitos have been studied in Sri Lanka in some detail due to their medical and economic importance. Gunathilaka (2018) published the most updated list of mosquitos in Sri Lanka and also gives an overview of the past work. According to Gunathilaka (2018), there are 159 species of mosquitos recorded from the country. Amarasinghe (1995) published an illustrated key to the mosquito genera in the country and Gunathilaka (2017) updated the identification key of Sri Lankan *Anopheles*.

26.2.17.8   FAMILY DROSOPHILIDAE: FRUIT FLIES

Fruit flies of family Drosophilidae in Sri Lanka were studied extensively by Okada (1988) based on the collections of Lund University Ceylon Expedition. He reports the presence of about 50 Drosophilidae species in the country known prior to his work and expands the species list with 30 new species descriptions and 62 new records.

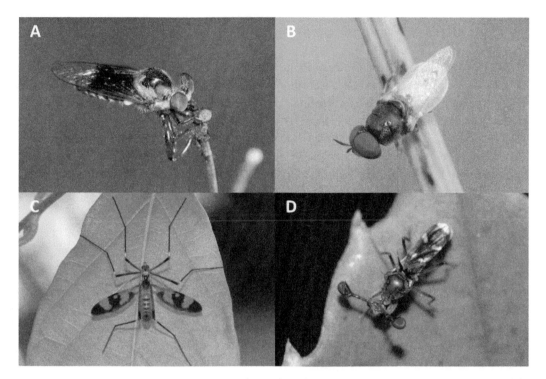

FIGURE 26.10 Some Dipteran in Sri Lanka. (a) *Damalis fulvipes* (Asilidae), (b) *Oplodontha minuta* (Stratiomyidae), (c) *Pselliophora laeta* (Tipulidae), (d) *Teleopsis neglecta* (Diopsidae). (Photos: Author)

26.2.17.9 FAMILY TEPHRITIDAE: TEPHRITID FRUIT FLIES

The Lund university expedition collection of Sri Lankan Tephritidae was studied by Hardy (1971) who reported 21 species in the collection. Tribe Dacini of Tephritidae was extensively studied in Sri Lanka by several authors, and 39 Dacine fruit fly species with 10 endemics are currently known from the country (Leblanc et al., 2018).

26.2.17.10 FAMILY DOLICHOPODIDAE: LONG-LEGGED FLIES

Dolichopodids are small flies with relatively long legs and metallic colors. Among early work on the family, the Indo-Australian Dolichopodidae catalogue reports 29 species from Sri Lanka (Becker, 1922). Recent taxonomic work on Sri Lankan Dolichopodidae based on historical collections has resulted in the description of five species of *Medetera* (Naglis and Bickel, 2012), 10 species of *Griphophanes* (Naglis and Grootaert, 2012), one species each in the genera *Rhaphium* (Naglis and Grootaert, 2011), *Srilankamyia* (Naglis et al., 2011), and *Phoomyia* (Naglis et al., 2013).

26.2.17.11   OTHER DIPTERA FAMILIES

Several World catalogues of Dipteran fauna published in recent years cover the Sri Lankan species in each respective taxon. According to these, there are reports of 33 Bombyliidae (Evenhuis and Greathead, 2015), 16 Spaeroceridae (Marshall et al., 2011), 63 Tachinidae (O'Hara et al., 2020), 1 Rhinophoridae (Cerretti et al., 2020), 20 Cecydomyiidae (Gagńe and Jaschhof, 2014), 30 Ceratopogonidae (Borkent and Dominiak, 2020), and seven Clusiidae (Lonsdale, 2017) species in Sri Lanka.

26.3   DISTRIBUTION AND BIOGEOGRAPHY OF SRI LANKAN INSECTS

26.3.1   DISTRIBUTION PATTERNS AND ZONES

Insects are distributed throughout Sri Lanka in all available terrestrial and freshwater habitats. They also occur in the coastal habitats but only the Sea Skaters of the genus *Halobates* are found in the marine environment.

The distribution patterns and ranges of many Sri Lankan insects are not well known. Some charismatic and relatively well-studied taxa are better known in terms of their distribution patterns while some of the lesser-known taxa are known only from the type locality or a handful of locality records. Based on a large amount of historical and recent work, the butterfly and odonate fauna of the country are the best-known insect groups in terms of the available knowledge on distribution. The recent work through the contribution of multiple researchers resulted in the introduction of butterfly and Odonata zones for Sri Lanka. This preliminary zonation was based on the distribution ranges of species and species composition in different geo-climatic regions of the country (MoMD&E, 2019).

The butterfly and Odonata zones of the country show some general patterns that may be applied to a majority of other Sri Lankan insect taxa. The major climatic zones tend to have different species compositions. Often some generalist species are known to be distributed across the three climatic zones while others are restricted to one or two zones contributing to the differences in the faunal composition of each zone.

The highlands also have a unique species composition. This may be represented by common species distributed across the highlands as it is shown in butterflies, or some specific taxa restricted to certain mountain ranges as evident in Sri Lankan Odonates (MoMD&E, 2019). The major mountain ranges, that is, Piduruthalagala Mountain Range, Peak Wilderness Mountain Range, Namunukula Mountain Range, Knuckles Mountain Range, Rakwana Mountain Range, and other smaller mountain ranges and isolated mountains may have different compositions of species, especially in less mobile and microhabitat specialist insect groups.

The dry zone of Sri Lanka is usually represented by fauna widespread across the zone. However, in Sri Lankan butterflies, some species are largely confined to the north and north–western part of the country and thus a specific butterfly zone (Northern butterfly zone) has been recognized. The relatively arid nature, floral composition (northern arid floristic zone), and proximity to the Indian Peninsula might have facilitated certain insect taxa to be distinct to the zone. This may also be applied to some insects in the south–eastern arid region which is also known to be distinct in its floral composition and recognized as a separate floristic region, that is, southern arid floristic zone (MoMD&E, 2019).

In addition to these, the coastal zone of the country with its diverse coastal formations and habitats can also be important for certain distinct species. Especially, the aquatic insects adapted to brackish water habitats and species that depend on coastal vegetation, are taxa that contribute to the distinct species composition of the coastal zone.

Distribution range maps are not available for the species in most of the insect taxa other than the butterflies and odonates. However, the historical collection data and recent field observations provide a preliminary idea on the range of some commonly recorded species in several major insect groups. It is of utmost importance to understand the distribution of these lesser-known insects to facilitate further studies and conservation implementations where necessary.

26.3.2 *BIOGEOGRAPHIC AFFINITIES*

The biogeographic affinities of Sri Lankan insect fauna have only been studied by several authors with relevance to specific taxonomic groups of interest. Most of these are preliminary work based on descriptive compilations or basic numerical methods.

The biogeographic origin of Sri Lankan butterflies has been discussed in detail by van der Poorten and van der Poorten (2016). Most of the present diversity of Sri Lankan butterflies are of a relatively recent origin. It is hypothesized that the original butterfly fauna which originated from the Gondwanan times, faced extinction due to the volcanic activity in the Indian tectonic plate that took place around 65–70 million years ago. This allowed the later influx of species from surrounding regions to establish in Sri Lanka. The present diversity of butterflies in Sri Lanka is largely of an Oriental origin. Few species of butterflies originating from the Palearctic region and African region, as well as some species with affinities to the Australian region are also present in the country. This is supported by a study determining the center of diversity of each butterfly genera present in Sri Lanka by van der Poorten and van der Poorten following the method used by Kunte (2016).

Other work on different insect taxa including Carabidae (Erwin, 1984), Dermaptera (Brindle, 1977), Ephemeroptera (Hubbard and Peters, 1984), Odonata (Laidlaw, 1951), Tabanidae (Burger, 1982), Tenebrionidae (Kaszab, 1979), also discuss the biogeographic affinities and highlight the affinity of the Sri Lankan insects to the Oriental region.

The recent work by Bedjanič et al. (2016) and Goonesekera et al. (2019) used molecular methods to study the biogeographic history of Sri Lankan Platystictidae (Odonata) and *Mycaleis* (Lepidoptera: Nymphalidae), respectively.

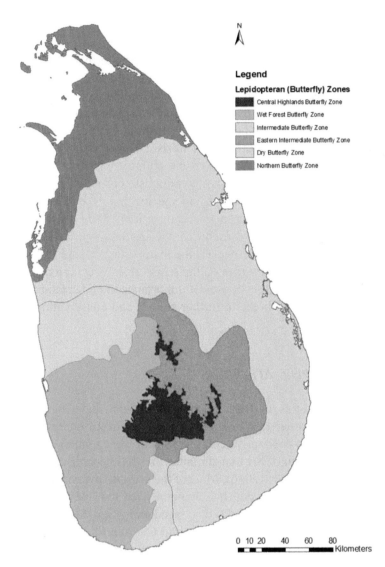

FIGURE 26.11 Butterfly zones of Sri Lanka.

Source: Based on MoMD&E, 2019

26.4 KNOWLEDGE GAPS AND RECOMMENDATIONS

The diversity of many insect orders in Sri Lanka is incompletely known. Comprehensive and updated work covering entire taxa are only available for handful of taxonomic groups such as butterflies and odonates. Some taxonomic groups have been comprehensively studied in the past, but no recent work has updated these and thus the present state of those taxa and the present validity of the taxonomic knowledge are uncertain. For most of the insect taxa, only a little is known beyond the descriptive taxonomic knowledge. Areas such as ecology, phylogeny, and biogeography are still untouched for all but few insect taxa in Sri Lanka.

FIGURE 26.12   Odonata zones of Sri Lanka.
Source: Modified based on MoMD&E, 2019

Systematic entomological surveys are few and scattered in the recent decades inhibiting the advance of insect taxonomic work in the country. Most of the recent species discoveries and taxonomic revisions involving Sri Lankan insects are based on decades old material collected during wide and organized entomological surveys. The absence of similar comprehensive surveys with wider taxonomic focus and limited number of research might slow down the advance in the taxonomic knowledge in the years to come. Most of the entomological surveys carried out in the recent years are limited in their geographical and ecological coverage and thus more focused work are required to provide a comprehensive understanding on the taxa. The slow advance of taxonomic knowledge and relative difficulty to acquire, it inhibits the expansion of many other fields of research on Sri Lankan insects, as taxonomic knowledge provides the baseline for these.

Wijesekara (2006) recognized several major impediments for insect taxonomic work in Sri Lanka, that is, lack of passionate amateur entomologists, lack of well-curated insect collections in the country, lack of well-organized education in systematic entomology and the lack of comprehensive literature collections. In addition to these, lack of funding for biodiversity surveys and basic taxonomic research is also a major limiting factor.

Based on previous work and the understanding on the present state of insect taxonomy and natural history work in Sri Lanka, following recommendations are made.

- Accessibility to collections and literature should be improved. Digitizing previous work and maintaining an online repository can largely improve accessibility to material.
- Capacity of local researchers, students, and interested amateurs should be developed in conducting research in insect taxonomy, ecology, and conservation.
- Citizen science should be promoted and used in collecting observations on insects.
- Country wide systematic insect taxonomic surveys should be conducted with the involvement of national research and academic institutions.
- Knowledge on insects, their diversity and importance should be popularized among the general public through awareness programs, educational materials, field guides, and other media.
- Preliminary conservation assessments of insects should be conducted, at least for the major taxonomic groups.

ACKNOWLEDGMENT

The author wishes to extend his sincere gratitude to Mrs. D. R. Sumanapala for the support given in data compilation, map preparation, and manuscript preparation; Dr. D. P. Wijesinghe for the extensive support extended during the literature survey and valuable comments provided on the manuscript; Prof. D. K. Weerakoon for the guidance and continuous encouragement; Mr. T. R. Ranasinghe for the continuous support in exploring the insect diversity in Sri Lanka; all colleagues for support and companionship.

KEYWORDS

- **insects**
- **biodiversity**
- **taxonomy**
- **natural history**
- **Sri Lanka**

REFERENCES

Abeywardhana, D. L.; Dangalle, C. D.; Mallawarachchi, Y. W. Cicindelinae of Sri Lanka: New Record of the Arboreal Tiger Beetle *Tricondyla gounellei* Horn, 1900. *J Natl Sci Found. Sri Lanka* **2020**, *48* (2), 213–216.

Alpert, G. D. A New Species of *Tyrannomyrmex* (Hymenoptera: Formicidae) from Sri Lanka. *Zootaxa* **2013**, *3721* (3), 286–290.

Amarasinghe, F. P. Illustrated Keys to the Genera of Mosquitoes (Diptera: Cuculidae) in Sri Lanka. *J. Natl. Sci. Council Sri Lanka* **1995**, *23* (4), 183–211.

Ambrose, D. P. A Checklist of Indian Assassin Bugs (Insecta: Hemiptera: Reduviidae) with Taxonomic Status, Distribution and Diagnostic Morphological Characteristics. *Zoos Print J.* **2006**, *21* (9), 2388–2406.

Andersen, N. M.; Foster, W. A. Sea Skaters of India, Sri Lanka, and the Maldives, with a New Species and a Revised Key to Indian Ocean species of *Halobates* and *Asclepios* (Hemiptera, Gerridae*), J. Nat. Hist.* **1992**, *26* (3), 533–553.

Andrew, J. P.; Ramamurthy, V. V. A Checklist of Weevils of the Tribe Mecysolobini (Coleoptera: Curculionidae: Molytinae). *Orient. Insects* **2010**, *44* (1), 271–336.

Assing, V. The *Rugilus* Species of the Palaearctic and Oriental Regions (Coleoptera: Staphylinidae: Paederinae). *Stuttg. Beitr. zur Naturkd. A, Neue Serie* **2012**, *5*, 115–190.

Banks, S.; Cumming, R. T. Li, Y.; Henze, K.; Le Tirant, S.; Bradler, S. A Tree of Leaves: Phylogeny and Historical Biogeography of the Leaf Insects (Phasmatodea: Phylliidae). *Commun. Biol.* **2021**, *4* (932), 1–12.

Baroni Urbani, C.; De Andrade, M. L. A New *Protanilla* Taylor, 1990 (Hymenoptera: Formicidae: Leptanillinae) from Sri Lanka. *Myrmecol. Nachr.* **2006**, *8*, 45–47.

Bartolozzi, L.; Bomans, H. E. The Lucanidae (Coleoptera) of Sri Lanka. Reports from the Lund University Ceylon Expedition in 1962: vol. 3. *Entomol. Scand. Suppl.* **1988**, *30*, 77–92.

Basnagala, S.; Wijesekara, G. A. W.; Wijayagunasekara, H. N. P. Review of the Subfamily Bryocorinae (Heteroptera: Miridae) of Sri Lanka. *Trop. Agric. Res.* **2002**, *14*, 154–164.

Battiston, R. *Mantis religiosa*. The IUCN Red List of Threatened Species **2016**, e.T44793247A44798476. DOI: 10.2305/IUCN.UK.2016-1.RLTS.T44793247A44798476.en

Becker, T. H. Dipterologische Studien. Dolichopodidae der Indo-Australischen Region. *Cap. Zool.* **1922**, *1* (4), 1–247.

Bedjanič, M. Three New *Drepanosticta* Species from Sri Lanka (Zygoptera: Platystictidae). *Odonatologica* **2010**, *39* (3), 195–215.

Bedjanič, M.; Conniff, K.; Dow, R. A.; Stokvis, F. R.; Verovnik, R.; van Tol, J. Taxonomy and Molecular Phylogeny of Platystictidae of Sri Lanka (Insecta: Odonata). *Zootaxa* **2016**, *4182* (1), 1–80.

Bedjanič, M.; van der Poorten, N.; Conniff, K.; Salamun, A. *Dragonfly Fauna of Sri Lanka: Distribution and Biology with Threat Status of its Endemics*; Pensoft: Sofia, **2014**, p 321.

Biström, O.; Bergsten, J. A New Species of *Peschetius* GUIGNOT Described from Sri Lanka (Coleoptera: Dytiscidae). *Koleopterol. Rundsch.* **2015**, *85*, 57–60.

Blackith, R. E. The Tetrigidae (Orthoptera) of Sri Lanka. Reports from the Lund University Ceylon Expedition in 1962, Vol. IV. *Entomol. Scand.* **1988**, (supp 30), 93–109.

Bonadona, P. Anthicidae (Coleoptera) de Sri Lanka. *Entomol. Scand. Suppl.* **1988**, *30*, 55–75.

Bonadona, P. Anthicidae (Coleoptera) Nouveaux des Collections du Muséum d'histoire naturelle de Genève. *Rev. Suisse Zool.* **1989**, *96*, 253–276.

Borkent, A.; Dominiak, P. Catalog of the Biting Midges of the World (Diptera: Ceratopogonidae). *Zootaxa* **2020**, *4787* (1), 1–377.

Bremer, H. J.; Lillig, M. World Catalogue of Amarygmini, Rhysopaussini and Falsocossyphini (Coleoptera; Tenebrionidae). *Mitt. Münch. Entomol. Ges.* **2014**, (supp 104), 3–176.

Brindle, A. Dermaptera from Ceylon. *Rev. Sussie Zool.* **1977**, *84* (2), 453–461.

Burger, J. F. A Review of the Horse Flies (Diptera: Tabanidae) of Sri Lanka. Reports from the Lund University Ceylon Expedition in 1962: Vol. 3. *Entomol. Scand. Suppl.* **1982**, *30*, 81–124.

Cascio, P. L. Worldwide Checklist of the Island Mutillid Wasps (Hymenoptera Mutillidae). *Biodivers. J.* **2015**, *6* (2), 529–592.

Cerretti, P.; Badano, D.; Gisondi, S.; Giudice, G. L.; Pape, T. The World Woodlouse Flies (Diptera, Rhinophoridae). *ZooKeys* **2020**, *903*, 1–130.

Cigliano, M. M.; Braun, H.; Eades, D. C.; Otte, D. Orthoptera Species File, **2021**. http://Orthoptera.SpeciesFile. org (accessed Mar 12, 2021).

Cuccodoro, C.; Liu, Z. *Megarthrus* of Southern India and Sri Lanka, with Notes on Their Phylogenetic and Biogeographical Relationships (Coleoptera: Staphylinidae: Proteininae). *Zootaxa* **2016**, *4097* (4), 530–544.

Cumming, R. C.; Bank, S.; Bresseel, J.; Constant, J. Le Tirant, S.; Dong, Z.; Sonet, G.; Bradler, S. *Cryptophyllium*, the Hidden Leaf Insects–Descriptions of a New Leaf Insect Genus and Thirteen Species from the Former Celebicum Species Group (Phasmatodea, Phylliidae). *ZooKeys* **2021**, *1018*, 1–179.

Dangalle, C. D. The Forgotten Tigers: The Arboreal Tiger Beetles of Sri Lanka. *J. Natl. Sci. Found. Sri Lanka* **2018**, *46* (3), 241–252.

Dangalle, C. D.; Pallewatta, N.; Vogler, A. The Current Occurrence, Habitat and Historical Change in the Distribution Range of an Endemic Tiger Beetle Species *Cicindela (Ifasina) willeyi* Horn (Coleoptera: Cicindelidae) of Sri Lanka. *J. Threat. Taxa* **2011**, *3* (2), 1493–1505.

David, B. V.; Dubey, A. K. Aleyrodid (Hemiptera: Aleyrodidae) fauna of Sri Lanka with Description of a New Species. *Orient. Insects*. **2008**, *42*, 349–358.

Delahaye, N.; Goonatilake, M.; Silva, M. Etude du genre *Acanthophorus* Audinet-Serville, 1832, et première capture d'*A. serraticornis* (Olivier, 1795) au Sri Lanka (Coleoptera, Cerambycidae, Prioninae). *Bull. Soc. Entomol. Fr.* **2011**, *116* (3), 329–335.

de Rougemont, G. A New Species of *Rugilus* (Eurystilicus) from Sri Lanka (Coleoptera, Staphylinidae, Paederinae). *Rev. suisse de zool.* **2014**, *121* (2), 247–248.

Dias, R. K. S.; Guénard, B.; Akbar, S. A.; Economo, E. P.; Udayakantha, W. S.; Wachkoo, A. A. The Ants (Hymenoptera, Formicidae) of Sri Lanka: A Taxonomic Research Summary and Updated Checklist. *ZooKeys* **2020a**, *967*, 1–142.

Dias, R. K. S.; Udayakantha, W. S.; Thotagamuwa, A.; Akbar, S. A. *Tetraponera modesta*, a New Pseudomyrmecine Ant Record (Hymenoptera: Formicidae) for Sri Lanka. *Ukrainska Entomofaunistyka* **2020b**, *11* (2), 23–26.

Dias, R. K. S.; Yamane, S.; Akbar, S. A.; Peiris, A. W. S.; Wachkoo, A. A. Discovery of the Worker Caste of *Protanilla schoedli* Baroni Urbani & De Andrade (Formicidae: Leptanillinae) in Sri Lanka. *Orient. Insects* **2019**, *53* (2), 160–166.

Edirisinghe, J. P.; Karunaratne, W. A. I. P.; Hemachandra, I. I.; Gunawardene, N. R.; Bambaradeniya, C. M. B. An Appraisal of Select Insect Taxa in Sri Lanka. In *Economic and Ecological Significance of Arthropods in Diversified Ecosystems: Sustaining Regulatory Mechanisms*; Chakravarty, A. K., Sridhara, S., Eds.; Springer: Singapore, 2016; pp 81–116.

Ehrmann, R. *Mantodea – Gottesanbeterinnen der Welt;* Verlag Natur und Tier: Münster/Westf, **2002**, p 519.

Engel, M. S. New Species and Records of Ammobatine Bees from Pakistan, Kyrgyzstan, and Sri Lanka (Hymenoptera: Apidae). *Acta Entomol. Slov.* **2008**, *16* (1), 19–36.

Erwin, T. L. Composition and Origin of the Ground Beetles. In *Ecology and Biogeography of Sri Lanka*; Fernando, C. H., Ed.; Martinus Nijhoff Publishers: The Hague, **1984**; pp 371–390.

Evenhuis, N. L.; Greathead, D. J. *World Catalog of Bee Flies (Diptera: Bombyliidae)*, September **2015**. http://hbs.bishopmuseum.org/bombcat/bombcat-revised2015.pdf.

Fabrizi, S.; Ahrens, D. A Monograph of the Sericini of Sri Lanka (Coleoptera: Scarabaeidae). *Bonn zool. Bull. Suppl.* **2014**, *61*, 124.

Feijen, H. R.; Feijen, C. A Revision of the Genus *Teleopsis* Rondani (Diptera, Diopsidae) in Sri Lanka with Descriptions of Two New Species and a Review of the Other Stalk-Eyed Flies from the Island. *ZooKeys* **2020**, *946*, 113–151.

Fennah, R. G. Homoptera: Fulgoroidea Delphacidae from Ceylon. Reports from the Lund University Ceylon Expedition in 1962: Vol. 2. *Entomol. Scand. Suppl.* **1975**, *4*, 79–136.

Fernando, C. H. A Guide to the Freshwater Fauna of Ceylon, Supplement 1. *Bull. Fish. Res. Stn. Ceylon* **1963**, *16*, 29–38.

Fernando, C. H. A Guide to the Freshwater Fauna of Ceylon, Supplement 2. *Bull. Fish. Res. Stn. Ceylon* **1964**, *17*, 177–211.

Fernando, C. H. A Guide to the Freshwater Fauna of Ceylon, Supplement 3. *Bull. Fish. Res. Stn. Ceylon* **1969,** *20,* 13–27.

Fernando, C. H. A Guide to the Freshwater Fauna of Ceylon, Supplement 4. *Bull. Fish. Res. Stn. Ceylon* **1974,** *25,* 27–81.

Fernando, C. H.; Weerawardhena, S. R. *A Guide to the Freshwater Fauna of Sri Lanka, Supplement 5*; **2002,** p 61.

Flint, O. S. The Genus Macrostemum Kolenati (Trichoptera: Hydropsychidae) in Sri Lanka. *Proc. Entomol. Soc. Wash.* **2003,** *105* (4), 816–831.

Franz, H. Coleoptera: Die Scydmaenidae Sri Lankas (mit Ausnahme der Genera *Cephennium* s. lat., *Clidicus* und *Syndicus*). Reports from the Lund University Ceylon Expedition in 1962: Vol. 3. *Entomol. Scand. Suppl.* **1982,** *11,* 125–274.

Franz, H. Scydmaeniden des Ungarischen Naturwissenschaftlichen Museums in Budapest aus Südostasien: Sri Lanka, Thailand und Vietnam. *Folia Entomol. Hung.* **1983,** *44,* 175–187.

Freitag, H. *Ancyronyx jaechi* sp.n. from Sri Lanka, the First Record of the Genus *Ancyronyx* Erichson, 1847 (Insecta: Coleoptera: Elmidae) from the Indian Subcontinent, and a World Checklist of Species. *Zootaxa* **2012,** *3382,* 59–65.

Gagné, R. J.; Jaschhof, M. *A Catalog of the Cecidomyiidae (Diptera) of the World: 3rd Ed., Digital version 2,* **2014**.

Geller-Grimm, F. A New Species of *Damalis* FABRICIUS from Sri Lanka (Diptera, Asilidae). *Stud. Dipterol.* **1997,** *4,* 197–200.

Ghate, H. V.; Kulkarni, S.; Benjamin, S. P. Giant assassin in the Cave: A New Species of the Genus *Myiophanes* from Sri Lanka (Hemiptera: Heteroptera: Reduviidae: Emesinae). *Zootaxa* **2018,** *4524* (2), 237–244.

Ghoneim, K. S. Global Zoogeography and Systematic Approaches of the Blister Beetles (Coleoptera: Meloidae): A Bibliographic Review. *Intern. J. Res. BioSci.* **2013,** *2* (3), 1–45.

Gnaneswaran, R. Provisional Checklist of the Leafhoppers (Hemiptera: Cicadellidae) in Sri Lanka. In *The National Red List 2012 of Sri Lanka; Conservation Status of the Fauna and Flora;* Weerakoon, D. K., Wijesundara, S., Eds.; Ministry of Environment: Colombo, Sri Lanka, **2012;** pp 431–437.

Goonesekera, K.; Lee, P. L. M.; van der Poorten, G.; Ranawaka, G. R. The Phylogenetic History of the Old World Butterfly Subtribe Mycalesina Extended: The *Mycalesis* (Lepidoptera: Nymphalidae) of Sri Lanka. *J. Asia-Pac. Entomol.* **2019,** *22,* 121–133.

Griswold, T. Two New Species of Trap-Nesting Anthidiini (Hymenoptera: Megachilidae) from Sri Lanka. *Proc. Entomol. Soc. Wash.* **2001,** *3* (1–2), 269–273.

Gunathilaka, N. Illustrated Keys to the Adult Female *Anopheles* (Diptera; Culicidae) Mosquitoes of Sri Lanka. *Appl. Entomol. Zool.* **2017,** *52,* 69–77.

Gunathilaka, N. Annotated Checklist and Review of the Mosquito Species (Diptera: Culicidae) in Sri Lanka. *J. Insect Biodivers.* **2018,** *7* (3), 038–050.

Gupta, S. K.; Jonathan, J. K. *Fauna of India and the Adjacent Countries, Hymenoptera: Scoliidae*; Zoological Survey of India: Kolkata, **2003,** p 277.

Hájek, J. World Catalogue of the Family Callirhipidae (Coleoptera: Elateriformia), with Nomenclatural Notes. *Zootaxa* **2011,** *2914,* 1–66.

Hammond, P. M. Coleoptera: Staphylinidae Oxytelini from Ceylon. Reports from the Lund University Ceylon Expedition in **1962**: Vol. 2. *Entomol. Scand. Suppl.* **1975,** *4,* 141–178.

Hardy, D. E. Ditera: Tephritidae from Ceylon. Reports from the Lund University Ceylon Expedition in 1962: Vol. 1. *Entomol. Scand. Suppl.* **1971,** *1,* 287–292.

Hauser, M.; Rozkošný, R. An Annotated List of Stratiomyidae (Diptera) from Sri Lanka with Taxonomic Notes on Some Genera. *Stuttg. Beitr. Naturkd. Ser. A (Biol.)* **1999,** *585,* 1–15.

Herman, L. H. Catalog of the Staphylinidae, Parts I–VII. *Bull. Am. Mus. Nat. Hist.* **2001,** *265,* 1–4218.

Hubbard, M. D.; Peters, W. L. Ephemeroptera of Sri Lanka: An Introduction to the Ecology and Biogeography. In *Ecology and Biogeography of Sri Lanka*; Fernando, C. H., Ed.; Martinus Nijhoff Publishers: The Hague, **1984**; pp 257–274.

Iwan, D.; Ferrer, J.; Ras, M. Catalogue of the World *Gonocephalum* Solier, 1834 (Coleoptera, Tenebrionidae, Opatrini). Part 1. List of the Species and Subspecies. *Ann. Zool.* **2010,** *60* (2), 245–304.

Jacoby, M. *The Fauna of British India Including Burma and Sri Lanka: Coleoptera, Chrysomelidae, Vol. 1*; Taylor and Francis: London, **1908**.

Jałoszyński, P. Revision of *Cephennomicrus* Reitter of Sri Lanka (Coleoptera: Staphylinidae: Scydmaeninae). *Genus* **2009**, *20* (2), 189–197.

Jałoszyński, P. *Clidicus minilankanus* sp. n., with Notes on Remaining Sri Lankan *Clidicus* Species (Coleoptera, Staphylinidae, Scydmaeninae). *Zootaxa* **2020**, *4718* (1), 87–94.

Jayasinghe, H. D., Rajapakha, S. S.; Alwis, C. A Compilation and Analysis of Food Plants Utilization of Sri Lankan Butterfly Larvae (Papilionoidea). *Taprobanica* **2014**, *6* (2), 110–131.

Jayasinghe, H. D.; Rajapakshe, S. S.; Ranasinghe, T. New Additions to the Larval Food Plants of Sri Lankan Butterflies (Insecta: Lepidoptera: Papilionoidea). *J. Threat. Taxa* **2021**, *13* (2), 17731–17740.

Jayawardana, N. C.; Jayawardana, A. *A Handbook to the Moths of Sri Lanka: Vol. 1*; Department of Wildlife Conservation: Sri Lanka, **2020**.

Jeannel, R. Sur les Pselaphides de Ceylan. *Bull. Br. Mus. Nat. Hist. Entomol.* **1961**, *10*, 423–456.

Joseph, A. N. T.; Parui, P. New and Little-Known Indian Asilidae (Diptera) VI. Key to Indian *Ommatius* Wiedemann with Descriptions of Fourteen New Species. *Entomol. Scand.* **1983**, *14*, 85–97.

Joseph, A. N. T.; Parui, P. *Fauna of India and Adjacent Countries–Diptera (Asilidae) (Part I). General Introduction and Tribes Leptogasterini, Laphriini, Atomosini, Stichopogonini and Ommatini*; Zoological Survey of India, **1998**.

Kalkman, V. J.; Babu. R.; Bedjanič, M.; Conniff, K.; Gyeltshen, T.; Khan, M. K.; Subramanian, K. A.; Zia, A.; Orr, A. G. Checklist of the Dragonflies and Damselflies (Insecta: Odonata) of Bangladesh, Bhutan, India, Nepal, Pakistan and Sri Lanka. *Zootaxa* **2020**, *4849* (1), 1–84.

Karunaratne W. A. I. P.; Edirisinghe J. P.; Pauly, A. *An Updated Checklist of Bees of Sri Lanka with New Records. Publication No. 23*; MAB handbook and occasional paper series, National Science Foundation: Sri Lanka, **2005**.

Karunaratne, W. A. I. P.; Edirisinghe, J. P.; Engel, M. S. First Record of a Tear-Drinking Stingless Bee *Lisotrigona cacciae* (Nurse) (Hymenoptera: Apidae: Meliponini), from the Central Hills of Sri Lanka. *J. Natl. Sci. Found. Sri Lanka* **2017**, *45* (1), 79–81.

Kaszab, Z. Faunistik der Tenebrioniden von Sri Lanka (Coleoptera). *Folia Entomol. Hung.* **1979**, *32* (2), 43–128.

Kawai, T. Plecoptera from Ceylon. Reports from the Lund University Ceylon Expedition in 1962: Vol. 2. *Entomol. Scand. Suppl.* **1975**, *4*, 65–78.

Kejval, Z. Taxonomic Revision of the Oriental Species of *Notoxus* (Coleoptera: Anthicidae). *A Acta Entomol. Musei Natl. Pragae.* **2011**, *51* (2), 627–673.

Kimoto, S. The Chrysomelidae (Insecta: Coleoptera) Collected by Dr. Akio Otake, on the Occasion of his Entomological Survey in Sri Lanka from 1973 to 1975. *Bull. Kitakyushu Mus. Nat. Hist. Hum. Hist., Ser. A* **2003**, *1*, 23–43.

Kistner, D. H. A New Species of *Atheta* (Coleoptera: Staphylinidae) from Sri Lanka Found with Termites (Isoptera: Termitidae). *Sociobiology* **2012**, *59* (4), 1563–1570.

Kolibáč, J. Trogossitidae: A Review of the Beetle Family, with a Catalogue and Keys. *ZooKeys* **2013**, *366*, 1–194.

Krombein, K. V. Biosystematic Studies of Ceylonese Wasps, II: A Monograph of the Scoliidae (Hymenoptera: Scolioidea). *Smithsonian Contrib. Zool.* **1978**, *283*, 56.

Kudavidanage, E. P.; Lekamge, D. A Provisional Checklist of Dung Beetles (Coleoptera: Scarabaeidae) in Sri Lanka. In *The National Red List 2012 of Sri Lanka; Conservation Status of the Fauna and Flora*; Weerakoon, D. K., Wijesundara, S., Eds.; Ministry of Environment: Colombo, Sri Lanka. **2012**; pp 438–444.

Kumar, P.; Yuan, X.; Kumar, M. R.; Kind, R.; Li, X.; Chandha, R. K. The Rapid Drift of the Indian Tectonic Plate. *Nature* **2007**, *449*, 894–897.

Kunte, K. Biogeographic Origins and Habitat Use of Butterflies of the Western Ghats. In *Invertebrate Diversity and Conservation in the Western Ghats, India*; Priyadarsanan, D. R., Soubadra, D. M., Subramanian, K. A., Aravind, N. A., Seena, N. K., Eds.; Ashoka Trust for Research in Ecology and the Environment (A TREE): Bengaluru, **2016**; pp 1–21.

Kurahashi, H. The Blow Flies Recorded from Sri Lanka, with Descriptions of Two New Species (Diptera, Calliphoridae). *Jpn. J. Syst. Entomol.* **2001**, *7* (2), 241–254.

Laidlaw, F. F. A Note on the Derivation of the Odonata Fauna of the Island of Ceylon. *Entomological News, Am. Entomol. Soc.* **1951**, *62–63*, 77–83.

Leblanc, L.; Doorenweerd, C.; Jose, M. S.; Sirisena, U. G. A. I.; Hemachandra, K. S.; Rubinoff, D. Description of a New Species of *Dacus* from Sri Lanka, and New Country Distribution Records (Diptera, Tephritidae, Dacinae). *ZooKeys* **2018**, *795*, 105–114.

Lelej, A. S. *Catalogue of the Mutillidae (Hymenoptera) of the Oriental region*; Institute of Biology and Soil Sciences, Russian Academy of Sciences: Far Eastern Branch, **2005**, p 252.

Li, N.; Yin, Z. The First Sri Lankan Species of *Labomimus* Sharp (Coleoptera: Staphylinidae: Pselaphinae). *Zootaxa* **2020**, *4809* (2), 397–400.

Löbl, I. Scaphidiidae von Ceylon (Coleoptera). *Rev. Suisse Zool.* **1971**, *78*, 937–1006.

Löbl, I.; Kurbatov, S. A. The Batrisini of Sri Lanka (Coleoptera: Staphylinidae: Pselaphinae). *Res. Suisse Zool.* **2001**, *108* (3), 559–697.

Löbl, I. On the *Batraxis* Species of Sri Lanka and the Identity of *Maya uzeli* Blattný (Coleoptera: Staphylinidae: Pselaphinae). *Folia Heyrovskyana, Series A* **2017**, *25* (1), 13–19.

Lonsdale, O. World Catalogue of the Druid Flies (Diptera: Schizophora: Clusiidae). *Zootaxa* **2017**, *4333* (1), 1–85.

Makihara, H.; Mannakkara, A.; Fujimura, T.; Ohtake, A. Checklist of Longicorn Coleoptera of Sri Lanka (1) Vesperidae and Cerambycidae Excluding Lamiinae. *Bull. FFPRI* **2008**, *7* (2), 95–110.

Marshall, G. A. K. *The Fauna of British India Including Burma and Sri Lanka: Coleoptera, Rhynchophora – Curculionidae*; Taylor and Francis: London, 1916.

Marshall, S. A.; Roháček, J.; Dong, H.; Buck, M. The State of Sphaeroceridae (Diptera: Acalyptratae): A World Catalog Update Covering the Years 2000–2010, with New Generic Synonymy, New Combinations, and New Distributions. *Acta Entomol. Musei Natl. Pragae.* **2011**, *51* (1), 217–298.

Martynov, A. V.; Palatov, D. M. A New Species of *Indoganodes* Selvakumar, Sivaramakrishnan & Jacobus, 2014 (Ephemeroptera, Teloganodidae) from Sri Lanka. *ZooKeys* **2020**, *969*, 123–135.

Massa, B. New Genera, Species and Records of Phaneropterinae (Orthoptera, Phaneropteridae) from sub-Saharan Africa. *ZooKeys* **2015**, *472*, 77–102.

Matyot, P. The Orthopteroids of the Seychelles: a Threatened Island Fauna. *J. Insect Conserv.* **1998**, *2*, 235–246.

Maulik, S. *The Fauna of British India Including Burma and Sri Lanka: Coleoptera, Chrysomelidae (Hispinae and Cassidinae)*; Taylor and Francis: London, **1919**.

Maulik, S. *The Fauna of British India Including Burma and Sri Lanka: Coleoptera, Chrysomelidae (Chrysomelinae and Halticinae)*; Taylor and Francis: London, **1926**.

Maulik, S. *The Fauna of British India Including Burma and Sri Lanka: Coleoptera, Chrysomelidae (Galerucinae)*; Taylor and Francis: London, **1936**.

Medler, J. T. A Review of the Sri Lankan Flatidae (Homoptera: Fulgoroidea). *Orient. Insects.* **2006**, *40* (1), 231–265.

Meinander, M. The Coniopterygidae of Ceylon (Neuroptera). *Entomol. Scand.* **1982**, *13*, 49–55.

Mendis, A. S.; Fernando, C. H. A Guide to the Freshwater Fauna of Ceylon. *Bull. Fish. Res. Stn. Ceylon* **1962**, *12*, 160.

Meregalli, M. World Catalogue of the Curculionidae: Lixinae: Cleonini. In *International Weevil Community*; Lyal, C. H. C., Ed.; **2017**, http://weevil.info/content/world-catalogue-curculionidae-lixinae-cleonini.

Mohamedsaid, M. S. The Blister Beetles (Meloidae) of Sri Lanka. *Ceylon J. Sci. (Biol. Sci.)* **1979**, *13* (1 & 2), 203–251.

Mohamedsaid, M. S. The Galerucine Beetles of Sri Lanka, with Descriptions of Two New Species (Coleoptera: Chrysomelidae). *Stobaeana* **1997**, *9*, 1–7.

MoMD&E. *Biodiversity Profile – Sri Lanka, Sixth National Report to the Convention on Biological Diversity*; Biodiversity Secretariat, Ministry of Mahaweli Development and Environment: Sri Lanka, **2019**, p 211.

Mukherjee, T. K.; Hazra, A. K.; Ghosh, A. K. The Mantid Fauna of India (Insecta: Mantodea). *Orient. Insects.* **1995**, *29*, 185–358.

Naglis, S.; Bickel, D. J. *Medetera* (Diptera, Dolichopodidae) of Sri Lanka. *Zootaxa* **2012**, *3188*, 55–63.

Naglis, S.; Grootaert, P. A Remarkable New Species of *Rhaphium* Meigen (Diptera, Dolichopodidae) from Sri Lanka. *Zootaxa* **2011**, *2991*, 44–48.

Naglis, S.; Grootaert, P. Review of the Genus *Griphophanes* Grootaert & Meuffels (Diptera, Dolichopodidae), with the Description of Ten New Species from Sri Lanka. *Zootaxa* **2012**, *3525*, 51–64.

Naglis, S.; Grootaert, P.; Brooks, S. E. *Phoomyia*, a New Genus of Dolichopodinae from the Oriental Region (Diptera: Dolichopodidae). *Zootaxa* **2013**, *3666* (1), 83–99.

Naglis, S.; Grootaert, P.; Wei, L. *Srilankamyia*–A New Dolichopodine Genus (Diptera, Dolichopodidae). *Miscellaneous Papers, Centre for Entomological Studies Ankara* **2011**, *155*, 8.

Naturalis Biodiversity Center. *Catalogue of the Craneflies of the World*, **2021**. https://ccw.naturalis.nl/ (accessed Mar 15, 2021).

Nilsson, A. N. *A World Catalogue of the Family Noteridae*, **2011**. http://www.waterbeetles.eu/documents/W_CAT_Noteridae.pdf.

Nilsson, A. N.; Hájek, J. *A World Catalogue of the Family Dytiscidae*, **2019**. http://waterbeetles.eu/documents/W_CAT_Dytiscidae_2019.pdf.

O'Hara, J. E., Henderson, S. J. & Wood, D. M. *Preliminary Checklist of the Tachinidae (Diptera) of the World: Version 2.1, ***2020**, *p 1039*. http://www.nadsdiptera.org/Tach/WorldTachs/Checklist/Worldchecklist.html.

Ohira, H. Coleoptera: Elateridae from Ceylon. Reports from the Lund University Ceylon Expedition in 1962: Vol. 2. *Entomol. Scand. Suppl.* **1973**, *4*, 27–38.

Ohl, M. Annotated Catalog of the Mantispidae of the world (Neuroptera). *Contrib. entomol. Int.* **2004**, *5* (3), 131–262.

Okada, T. Family Drosophilidae (Diptera) from the Lund University Ceylon Expedition in 1962 and Borneo Collections in 1978–1979. Reports from the Lund University Ceylon Expedition in 1962: Vol. 3. *Entomol. Scand. Suppl.* **1988**, *30*, 111–152.

Oláh, J.; Flint, O. S.; Johanson, K. A. *Cheumatopsyche galapitikanda* Species Cluster in Sri Lanka with the Description of Four New Species (Trichoptera: Hydropsychidae). *Braueria* **2007**, *34*, 21–28.

Oláh, J.; Johanson, K. A. Generic Review of Hydropsychinae, with Description of Schmidopsyche, New Genus, 3 New Genus Clusters, 8 New Species Groups, 4 New Species Clades, 12 New Species Clusters and 62 New Species from the Oriental and Afrotropical Regions (Trichoptera: Hydropsychidae). *Zootaxa* **2008**, *1802*, 248.

Packova, G.; Kundrata, R. The Genus *Selasia* Laporte, 1838 (Coleoptera: Elateridae: Agrypninae) in Sri Lanka. *Zootaxa* **2021**, *4926* (2), 285–292.

Park, K. T. Three Hundred and Twenty Newly Described Species of Lecithoceridae (Lepidoptera, Gelechioidea) by K.T. Park Since 1998, with a Tentative Catalogue and Images of Types. *J Asia Pac Biodivers.* **2014**, *7*, e95–e132.

Patel, S.; Sing, R. Updated Checklist and Distribution of Mantidae (Mantodea: Insecta) of the World. *Intern. J. Res. Stud. Zool.* **2016**, *2* (4), 17–54.

Paulsen, M. J. A New Species of *Dinonigidius* de Lisle from Sri Lanka (Coleoptera: Lucanidae). *Insecta Mundi* **2016**, *484*, 1–4.

Polhemus, J. T. Results of the Austrian-Ceylonese Hydrobiological Mission 1970 of the 1st Zoological Institute of the University of Vienna (Austria) and the Department of Zoology of the Vidyalankara University of Ceylon, Kelaniya. Pt. XIX. Aquatic and Semi-Aquatic Hemiptera of Sri Lanka from the Austrian Indo-Pacific Expedition, 1970–1971. *Bull. Fish. Res. Stn. Sri Lanka* **1979**, *29*, 89–113.

Price, B. W.; Allan, E. L.; Marathe, K.; Sarkar, V.; Simon, C.; Kunte, K. The Cicadas (Hemiptera: Cicadidae) of India, Bangladesh, Bhutan, Myanmar, Nepal and Sri Lanka: An Annotated Provisional Catalogue, Regional Checklist and Bibliography. *Biodivers. Data J.* **2016**, *4*, e8051, 1–156.

Puthz, V. Coleoptera: Staphyinidae Steninae von Ceylon. Reports from the Lund University Ceylon Expedition in 1962: Vol. 2. *Entomol. Scand. Suppl.* **1975**, *4*, 1–4.

Ranasinghe, S.; Eberle, J.; Benjamin, S. P.; Ahrens, D. New Species of Sericini from Sri Lanka (Coleoptera, Scarabaeidae). *Eur. J. Taxon.* **2020**, *621*, 1–20.

Rodríguez-Mirón, G. M. Checklist of the Family Megalopodidae Latreille (Coleoptera: Chrysomeloidea). *Zootaxa* **2018**, *4434* (2), 265–302.

Rozkošný, R.; Hauser, M. Additional Records of *Ptecticus* Loew from Sri Lanka, with a New Species and a New Name (Diptera: Stratiomyidae). *Stud. Dipterol.* **2011**, *8* (1), 217–223.

Ruan, Y.; Yang, X.; Konstantinov, A. S.; Prathapan, K. D.; Zhang, M. Revision of the Oriental *Chaetocnema* Species (Coleoptera, Chrysomelidae, Galerucinae, Alticini). *Zootaxa* **2019**, *4699* (1), 1–206.

Ruta, R. Scirtidae of India and Sri Lanka. Part 1. The Chlorizans-Group of *Cyphon* Paykull, 1799 (Insecta: Coleoptera). *Genus* **2007**, *17* (2), 323–340.

Scarbrough, A. G.; Hill, H. N. Ommatiine Robber Flies (Diptera: Asilidae) from Sri Lanka. *Orient. Insects.* **2000,** *34* (1), 341–407.

Schedl, K. E. A Check List of the Scolytidae and Platypodidae (Coleoptera) of Ceylon with Descriptions of New Species and Biological Notes. *Trans. R. Entomol. Soc. Lond.* **1959,** *111* (16), 469–516.

Schedl, K. E. Coleoptera: Scolytidae and Platypodidae from Ceylon. Reports from the Lund University Ceylon Expedition in 1962: Vol. 1. *Entomol. Scand. Suppl.* **1971,** *1*, 274–285.

Scherer, V. G. Die Alticinae Des Indischen Subkontinentes (Coleoptera: Chrysomelidae). *Pac. Insects Monogr.* **1969,** *22*, 1–251.

Shishodia, M. S.; Chandra, K.; Gupta, S. K. An Annotated Checklist of Orthoptera (Insecta) from India. *Records of Zoological Survey of India: Occasional Paper No., 314*, **2010,** p 366.

Silva, T. H. S.; Diyas, G. C. P.; Karunaratne, W. A. I. P.; Edirisinghe, J. P. Rediscovery of *Tetragonula praeterita* after 1860: An Unremarked Common Stingless Bee Endemic to Sri Lanka. *J. Natl. Sci. Found. Sri Lanka* **2018,** *46* (1), 109–113.

Sivaramakrishnan, K. G.; Subramanian, K. A.; Ramamurthy, V. V. Annotated Checklist of Ephemeroptera of the Indian Subregion. *Orient. Insects* **2009,** *43*, 315–339.

Slater, J. A. Hemiptera, Heteroptera: Lygaeidae from Sri Lanka. Reports from the Lund University Ceylon Expedition in 1962: Vol. 3. *Entomol. Scand. Suppl.* **1982,** *11*, 1–27.

Staines, C. L. *Catalog of the Hispines of the World (Coleoptera: Chrysomelidae: Cassidinae)*, **2015**. https://naturalhistory.si.edu/research/entomology/collections-overview/coleoptera/catalog-hispines-world.

Stebnicka, Z. Coleoptera: Scarabaeidae, Aphodiinae from Sri Lanka. Reports from the Lund University Ceylon Expedition in 1962: Vol. 3. *Entomol. Scand. Suppl.* **1988,** *30*, 163–166.

Sumanapala, A. P. *Macromia weerakooni* sp. nov. (Odonata: Anisoptera: Macromiidae), A New Dragonfly Species from Sri Lanka. *Intern. J. Odonatol.* **2021,** *24*, 169–177.

Thotagamuwa, A.; Sumanapala, A.; Dangalle, C.; Pallewatte, N.; Lokupitiya, E. Found After 107 Years: Rediscovery of an Endemic Tiger Beetle *Jansenia laeticolor* W. Horn (Coleoptera: Cicindelidae) in Sri Lanka. *Cicindela* **2016,** *48* (3–4), 77–83.

Tillekaratne, K.; Edirisinghe, J. P.; Gunatilleke, C. V. S.; Karunaratne, W. A. I. P. Survey of Thrips in Sri Lanka: A Checklist of Thrips Species, Their Distribution and Host Plants. *Ceylon J. Sci. (Biol. Sci.)* **2011,** *40*, 89–108.

van der Poorten, G.; Van der Poorten, N. *The Butterfly Fauna of Sri Lanka*; Lepodon Books: Toronto, ON, **2016,** p 418.

van der Poorten, G. M.; van der Poorten, N. E. *Field Guide to the Butterflies of Sri Lanka*; Lepodon Books: Toronto, ON, Canada, 2018, p 250.

Viraktamath, C. A.; Gnaneswaran, R. Review of the Grass Feeding Leafhopper Genus *Gurawa* Distant (Hemiptera: Cicadellidae: Deltocephalinae) from the Indian Subcontinent with Description of Two New Species. *ENTOMON* **2013,** *38* (4), 193–212.

Wahis, R.; Krombein, K. V. A New *Mechaerothrix* Haupt from Sri Lanka with Notes on the Genus (Hymenoptera: Pompilidae: Pepsinae: Ageniellini). *Proc. Entomol. Soc. Wash.* **2000,** *102* (2), 271–279.

Węgrzynowicz, P. Catalogue of Biphyllidae (Coleoptera) of the World. *Ann. Zool.* **2015,** *65* (3), 409–471.

Wijekoon, W. M. C. D.; Wegiriya, H. C. E.; Bogahawatta, C. N. L. Regional Diversity of Fireflies of the Subfamily Luciolinae (Coleoptera: Lampyridae) in Sri Lanka. *Lampyrid* **2012,** *2*, 138–141.

Wijekoon, W. M. C. D.; Wegiriya, H. C. E.; Bogahawatta, C. N. L. Systematic Revision of the Repository Collection of Canthoroidea in the Department of National Museums, Colombo, Sri Lanka (Coleoptera: Cantharidae, Lampyridae, Lycidae, Rhagophthalmidae). *Ceylon J. Sci.* **2016,** *45* (1), 67–74.

Wijerathna, M. A. P.; Edirisinghe, J. P. *A Checklist of Aphids and Their Host Plants from Sri Lanka*. MAB Checklist and Handbook Series. Publication No. 21. National Science Foundation: Sri Lanka, **1999**.

Wijesekara, A. An Overview of the Taxonomic Status of Class Hexapoda (Insecta) in Sri Lanka. In *The Fauna of Sri Lanka: Status of Taxonomy, Research and Conservation*; Bambaradeniya, C. M. B., Ed.; The World Conservation Union: Colombo, Sri Lanka & Government of Sri Lanka, 2006; pp 3–11.

Wijesekara, A.; Wijesinghe, D. P. History of Insect Collection and a Review of Insect Diversity in Sri Lanka. *Ceylon J. Sci. (Biol. Sci.)* **2003,** *31*, 43–59.

Wijesekara, G. A. W.; Henry, T. J. The Taxonomic Status of the Plant Bugs (Hemiptera: Miridae) of Sri Lanka. *Proc. Ann. Symp. Dept. Agri. Sri Lanka* **1999,** *1*, 173–194.

Yamane, S.; Dias, R. K. S. New Species of the *Pristomyrmex profundus* Wang Group from the Oriental Region (Hymenoptera: Formicidae: Myrmicinae). *Eur. Entomol. J.* **2016,** *1*, 188–193.

Yeshwanth, H. M.; Chérot, F.; Henry, T. J. The Isometopinae (Hemiptera: Heteroptera: Miridae) of India and Sri Lanka: A Review of the Subfamily, with Descriptions of Six New Species. *Zootaxa* **2021,** *4903* (2), 151–193.

Zahradník, P. *Lasioderma linnmani* sp. n. (Coleoptera: Bostrichoidea: Ptinidae) from Sri Lanka. *Studies and Reports of District Museum Prague-East Taxonomical Series* **2008,** *4* (1–2), 291–294.

ZIN RUS. *Beetles (Coleoptera) and Coleopterologists: L.N. Medvedev: The Whole List of Scientific Publications*; The Zoological Institute of the Russian Academy of Sciences, **2021**. https://www.zin.ru/animalia/coleoptera/eng/lmedved1.htm (accessed Apr 01, 2021).

Zwick, P. A Revision of the Oriental Stonefly Genus *Phanoperla* (Plecoptera: Perlidae). *Syst. Entomol.* **1982,** *7*, 87–126.

Zwick, P. The Genus *Neoperla* (Plecoptera: Perlidae) from Sri Lanka. *Orient. Insects* **1980,** *14* (2), 263–269.

CHAPTER 27

DIVERSITY OF HONEYBEES IN SRI LANKA[1]

R. W. K. PUNCHIHEWA

ThiSaBi Institute (Bees for Sustainable Development), Homagama, Sri Lanka

ABSTRACT

The importance of honeybees on the island of Sri Lanka comes from their value as opportunistic flower visitors, thus giving an immense service in maintaining the high biodiversity. They play a vital role in the pollination of plants, which is important in maintaining the natural ecosystems while contributing to the agricultural productivity. They are a champion group in utilization-based conservation.

27.1 TAXONOMIC DIVERSITY

It is now recognized that Asia is having 12 (or 14) living species of the genus *Apis* honeybees (Otis and Smith, 2021). These are considered the true honeybees or the stinging honeybees. Of these 12 species, only three species (Apidae: Apinae: Apini) exist in Sri Lanka representing each member of the major divisions of *Apis*; firstly, the cavity nesting honeybees (*Apis cerana*), secondly, the open nesting honeybees, Megapis or the giant honeybees (*Apis dorsata*), and thirdly, Micrapis or the small/dwarf honeybees (*Apis florea*) which are also open nesting (Karunaratne et al., 2005).

In the case of the stingless[2] honeybees (Apidae: Apinae: Meliponini), three species have been identified such as two species of *Tetragonula (T. iridipennis* and *T. praeterita)* and a single species of *Lisotrigona cacciae*. However, *L. cacciae* was rediscovered recently, only from a single site, and its nests have not been discovered up to this day (Karunaratne et al., 2017). Therefore, only the two species of *Tetragonula* are discussed here.

---

[1]For the occasion of Prof Dr Nikolaus Koeniger's 80th birthday and celebrating our collaboration since 1974 with him and his wife Dr Mrs Gudrun Koeniger about honeybees in Sri Lanka.

[2]These honeybees do not have stings to use in self-defense; instead they bite and crawl into eyes, nostrils, ears, etc. In the native language Sinhala, they are called "ear crawlers" (*kaneyiya* කනෙයියා or one that crawls into ears). *Apis* or stinging honeybees use a sting located at the posterior end of the body.

Biodiversity Hotspot of the Western Ghats and Sri Lanka. T. Pullaiah, PhD (Ed.)

Compared to other Asian tropical humid countries such as Borneo, which is considered the center of *Apis* diversity with at least eight species (Koeniger et al., 2010), distantly located Sri Lanka has only three species. Taxonomic diversity is confined to single representation in each of the major divisions. Therefore, diversity of honeybees of Sri Lanka discussed in this chapter focuses on other areas such as their ecological and agroecological significances.

27.2   ECOLOGICAL DIVERSITY

Of the five honeybee species discussed above, the three smallest species does not occur in the regions above 900 m elevation or in the up-country of the island. Sri Lanka, in spite of its relative small size (65,600 km^2) shows well demarcated difference in ecological regions, and broadly one can see several major ecological zones as shown in Table 27.1.

TABLE 27.1   Main Ecological Regions of Sri Lanka (after Panabokke, 1996; Panabokke and Kannangara, 1975).

Local name due elevation features	Local name due to rainfall and monsoonal features
Low country 0 ~ 300 m	**Wet zone**—Annual rainfall >2500 mm benefiting from both northeast and southwest monsoons. >50% of the days of the year are rainy.
Mid country 300 ~ 900 m	**Intermediate zone**—An ecozone lying between wet and dry zones with an annual rainfall of 1550 ~ 2500 mm. 25 ~ 50% of the days of the year receive rain.
Up-country >900 m up to cloud forest range >2000 m	**Dry zone**—Annual rainfall < 1550 mm and major rainfall from northeast monsoon, Oct to Dec. <25% of the days of the year are rainy.

This is essentially a local classification not relevant to universal classification system.
There are no dry zones above elevations of 300 m or in the mid and up-country regions.

Sri Lanka, being a tropical evergreen island, is highly conducive to honeybees as well as to any life form compatible with this environment. However, as shown in Figure 27.1, all honeybees are not equally distributed throughout the island. Sri Lanka has nearly 4300 species of plants (MoMD&E, 2019), of which many are visited by honeybees either for pollen or for nectar, or for both. The majority of the plants may be animal pollinated, as mentioned by Bawa (1990) for Malaysia, a country whose vegetation, in many ways, is comparable to Sri Lanka, and most of them are bee-pollinated.

Honeybees, being opportunistic feeders, may be visiting most of the plants or rather their flowers, sometime or the other and therefore one is compelled to think that there are no shortages of food for them in tropical humid lands like Sri Lanka. Therefore, it could be assumed that apiculture[3] is a lucrative industry here. However, the practical aspect of apiculture does not work in that expected manner.

The nonexistence of relatively small *Apis florea* and even smaller stingless honeybees *Tetragonula* (*T. iridipennis* and *T. praeterita*) in elevations above 900 m or the central hills can be interpreted as due to temperature factor. Average temperature on coastal plains being

---

[3]Apiculture—management and the scientific background for the management of honeybees, usually for honey or wax production or for pollination (Michener, 1974).

28° ± 2°C, and with a temperature drop of 1°C for every 150 m rise in elevation (Panabokke, 1996) when reaching the up-country or the 900 m elevation level, the average temperature would be 22°C. This average temperature seems the determining factor in limiting the smaller honeybees in higher elevations. Considering above facts, it is evident that average ambient temperature is the main factor that determines the distribution of honeybees on the island.

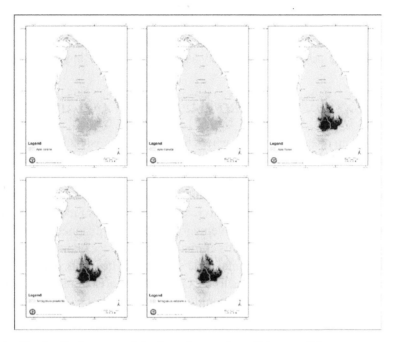

FIGURE 27.1 The distribution maps of five honeybee species which are well-known at present. This image is copied from the sixth national report on biodiversity conservation in Sri Lanka, page 13. This is to highlight the fact that *Apis florea* and both stingless honeybees *Tetragonula (T. iridipennis and T. praeterita)* do not occur in elevations above 900 m or the central hills. The dark areas in the center depict the nonexistence of these three species.

Source: Adapted from *MoMD&E* (2019) Biodiversity Profile—Sri Lanka, p 13 Figure 1.3.

27.3   NESTING SITE DIVERSITY

In Sinhala, the native language of Sri Lanka, the honeybees are broadly divided into two categories such as **stayers** and **movers**.

Stayers (පදිංචිකාරයෝ *Padinchikarayo*, the ones who stay in one place or permanent occupants) are the ones who occupy a particular nesting site for a long period or perennially, and to this category cavity-nesting honeybees are included. Therefore, *Apis cerana* and *Tetragonula (T. iridipennis* and *T. praeterita)* are recognized as stayers.

Movers (යන්නෝ *Yanno*, the one who keeps moving or migrating) are the ones who appear periodically or seasonally, and here open-nesting honeybees are included. Therefore, *Apis dorsata* and *Apis florea* are recognized as movers.

As obvious, this is the simplest way how a land owner recognizes different honeybees who happen to be an integral component of their gardens or farms or natural surroundings.

Depending on the type of their nests, honeybees also could be categorized into two distinct groups, such as:

1. **Open-nesting honeybees** (e.g., Single comb nest honeybees such as giant honeybee of Asia or *Apis dorsata* group and little honeybee of Asia or *Apis florea* group).
2. **Cavity-nesting honeybees** (e.g., Multiple comb nests honeybees such as suspended parallel comb building *Apis cerana* group and prostrate multiple comb building formally *Trigona* renamed currently as *Tetragonula* or stingless honeybee group).

27.3.1   OPEN-NESTING HONEYBEES: APIS FLOREA AND APIS DORSATA

These bees select an open-nesting site where blue sky or sun is visible directly to perform their communication dance, where specific type of body movements are performed among the nest mates, mostly to indicate a location of a food (nectar or pollen) source. A foraging bee, which is successful in finding a food source initially will return to its nest with gathered food and communicate the location to recruit more foragers to gather food. Here, the tempo or the frequency of the body movement indicates the distance from their nest. The small honeybee *Apis florea* performs this communication dance on top of the nest or on top of the honey stores which are horizontal. The giant honeybee *Apis dorsata* performs this communication dance on the vertical surface on a specific area, most open peripheral area of the comb. For more details on dance communication see Frisch (1976), Lindauer (1956, 1957), and Punchihewa et al. (1984).

FIGURE 27.2   A nest of *Apis florea* occupying a twig of a small mango tree (*Mangifera indica*: Anacadiaceae) at a height of about 3 m from ground. Low country wet zone.

FIGURE 27.3 This *Apis dorsata* nesting tree, perhaps the Sri Lanka's largest community nesting site of the giant honeybee, situated near a major roadway, a railway tract, and a residential area in the well-known up-country town Haputale. This banyan tree (*Ficus benghalensis*: Moraceae) bordering a natural forest. Up-country wet zone.

FIGURE 27.4 Generalized areas of an *Apis* comb.

Here, an entire *Apis florea* colony has taken as an example due to its compactness. At this stage, it has come to the reproductive stage where initially the drone phase and then the queen phase, both are visible here. Note the size difference of worker and drone cells. Queen cells are vertical and other cells are horizontal. Incompleteness of bees covering the entire comb and opened queen cells indicates that several swarms[4] have been issued by this colony prior to the photograph. However, they keep covering the brood area well, to maintain proper incubation. As it is shown here and before and even later, Apis honeybees always properly cover their brood areas with workers as opposed to *Tetragonula* honeybees who do not have a similar behavior (Figure 27.6).

---

[4]Swarm—a queen and large group of workers that establish a new colony of honeybees. Natural reproductive procedure of social bees or honeybees.

FIGURE 27.5   This single large *A dorsata* colony built among ground vegetation was only about 2 m above ground and the lower end just 25 cm above ground.

Note the nonuniformity or the broken pattern in the distribution of bees on the nest "left hand side, mid, and lower edge." This area is called the "Entrance area" or functionally flight takeoff and landing area, where returning bees perform the foraging dance to communicate the location of the food source. This area is also called the dancing platform. Flights takeoff, landing, and dance performance all take place in this distinct area of the nest.

27.3.2   CAVITY-NESTING HONEYBEES: TETRAGONULA IRRIDIPENNIS, T. PRAETERITA, AND APIS CERANA

Three species are discussed here such as *Tetragonula irridipennis, T. praeterita* (Apidae: Apinae: Meliponini), and *Apis cerana* (Apidae: Apinae: Apini) who live in cavities available in nature. These honeybees in nature select a cavity with sufficient space that offers them protection from natural factors. Further, inside this dark nest, they are able to perform the "communication dance" to recruit more foragers to gather food. Due to this advantage, they can store food in excess during flowering seasons, over and above their maintenance requirement for future use to tide over dearth periods.

FIGURE 27.6 Cavity-nesting honeybees sharing the same cavity. It is not uncommon to find both these honeybee species nesting together in the same cavity. A plank of wood covering the cavity, where *Tetragonula* occupying was removed to expose their nest. On the other side is an *Apis cerana* nest.

Note the relatively large separating curtains or several thin walls or laminate batumen plates (involucrum) built by the *Tertragonula praeterita* to separate *Apis cerana* on the opposite side of the coconut log cavity. Brood cells (or pots) and food pots are among a network of branched batumen columns. All these batumen structures are rather sticky and perhaps offer protection from any intruders. Batumen is a mixture of bodily wax with plant resins and small plant fragments.

Bold lettered nomenclature according to Michener (1974).

27.3.2.1 DETAILS OF THE NESTS OF TETRAGONULA

FIGURE 27.7 Nest entrance of *Tetragonula praeterita* (Walker) built in a cavity of a stone wall.

Note the rough or non-shiny long entrance tube covered with grit constructed with cerumen or bee glue. The grit consists of tiny pebbles, sand, and plant bark.

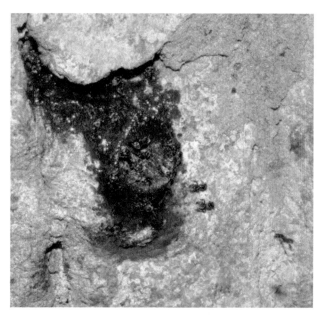

FIGURE 27.8 Nest entrance of *Tetragonula iridipennis* (Smith) build in a cavity in a stone wall.
Note the smooth shiny short entrance tube constructed with cerumen or bee glue. The sticky nature of the short
entrance tube can repel any intruders, mostly ants effectively.

27.3.2.2 DETAILS OF THE NESTS OF APIS CERANA

FIGURE 27.9 A natural nest of *Apis cerana* built inside a thicket of natural vegetation completely covering
it.

Though combs were completely covered by bees, they were moved to a side with the use of smoke for detailed
examination and taking measurements. This nest volume was about 20 L.

FIGURE 27.10 A natural colony of *A. cerana* living in the narrow space of about 150 mm wide in a roof of a house.

On top are asbestos roofing sheets onto which the combs are attached and bottom are wooden planks of the ceiling below which no combs are attached. Both roof and ceiling are partly removed to expose the colony. The measured volume (comb volume) of the colony was about 32 L, however at the time of measurement the live nest has shrunken to about 8 L. The peripheral light-colored combs are honeycombs that get filled during honey-flow periods.

FIGURE 27.11 A comb of *Apis cerana*.

Upper half is honey and pollen store, while the lower half contains the brood. Note the vertical queen cell on the bottom edge. Reproductive female cells or the queen cells and male or drone cells (Figs. 27.4 and 27.12) appear during the reproductive season. Reproductive mechanism of honeybees is called swarming and in commercial beekeeping, a special attention is paid to manage colonies during this period (Punchihewa, 1994; Punchihewa et al., 2000).

FIGURE 27.12   A close-up of *Apis cerana* comb.

Relatively large queen surrounded by her attendants or the "queen court." She is on the light-brown colored drone brood area now in sealed pupa stage of the comb. Compare with the white-colored worker brood area on top. The top seal (or cap) of mature drone brood cell develops a hole in the middle and such an area is clearly seen on the bottom right corner of the photograph. This drone brood cap hole is a unique feature in A. cerana drone pupa cell.

27.4   BEHAVIORAL FEATURES AND THEIR ENVIRONMENTAL SIGNIFICANCE

27.4.1   FORAGING AND FOOD STORAGE BEHAVIORS

In summary, the experiments conducted by Frisch (1976), Koeniger et al. (1982), Lindauer (1956, 1957), and Punchihewa et al. (1985) on dance communication and food gathering behavior also could be interpreted as an adaptation to the floral composition and the flowering pattern in an area concerned. From these experiments, it becomes clear that *Apis dorsata* foraging area is to be over 1 km in radius, *Apis cerana* to be less than 1 km, *Apis florea* to be less than 500 m, and *Tetragonula* to be around or less than 100 m.

The significance of the observations of the native people as **staying** and **moving** honeybees further confirms the ecological significance. Staying honeybees act as resident pollinators and their food storage behavior allows them to fulfill this natural requirement throughout the year without any interruption. Changes or breakdown in continuous flowering causing temporary food shortages for resident pollinators are easily overcome with the food stores in their nests. In the same ecosystem, some plants whose relative densities are high or whose availability is more common cause sudden increase in the food availability for resident honeybees, which is handled by their ability to increase the storage capacity within the intrinsic limits (see Figure 27.10). This phenomenon can be observed when *Apis florea* appear at a particular site, such as a home garden. In tropical humid Asia, it is not surprising to observe dwarf honeybees and *Tetragonula* honeybees even in large cities with some kind of vegetation such as home gardens and small parks. Short-range migratory *Apis florea* has a unique character of some worker bees revisiting the abandoned old nest to recollect some of its wax to build the new nest in the new location, which may be in the close vicinity. However, when flowering becomes even more frequent or superabundant, long-range migratory honeybees come to reside temporarily and as such *A. dorsata* migrates to the area concerned. This phenomenon is operative when one considers floral food (both nectar and pollen) availability in a particular ecological or rather vegetation area, either natural such as natural forest or artificial such as agricultural plantations. Rubber (*Hevea brasiliensis*: Euphobiaceae) plantations and sesame fields are good examples in Sri Lanka. The superabundance of floral food availability is not only confined to natural vegetation or agricultural areas, but even in large cities with ornamental road tree plantings; giant honeybees migrate in the flowering season often to build their nests in tall or prominent buildings, a common site during the flowering season.

In beekeeping, when *Apis cerana* is managed for honey production, they show migratory behavior called "absconding" when food supply becomes lean or during a dearth period. This could easily be overcome by leaving enough honey stores when extracting the honey by the beekeeper or timely supplementary feeding, or changing the foraging range by moving colonies to sites with more nectar availability (Kevan et al., 1995; Punchihewa et al., 1992; Punchihewa, 1994).

Honey water content of different honeybee species is another indicator of the significance of migratory behavior. As given in Table 27.2, migratory or moving honeybees do not evaporate their honeys to a nonfermenting level, as they do not stay at one location for long. In contrast, staying honeybees concentrate their honey to nonfermenting level as they stay perennially at the same location.

TABLE 27.2   Water Content of Honeys of Different Honeybees (Punchihewa's Unpublished Data).

Species	Water content[5]	Comments
Apis cerana	19–21%	Nonfermenting honey, cavity-nesting resident honeybee
Tetragonula	19–20%	Nonfermenting honey, cavity-nesting resident honeybee
Apis florea	27–28%	Fermenting honey, open-nesting, short-range or intrazonal migratory honeybee
Apis dorsata	26–27%	Fermenting honey, open-nesting, long-range or interzonal migratory honeybee

[5]Water content tends to vary depending on relative humidity of different ecoclimatic zones which varies from 95 to 60%. What is given here are the average values of tested honey samples.

27.4.2   FOOD SHARING AND COMPETITION FOR FOOD AMONG HONEYBEES

Food competition among sympatric honeybees in Sri Lanka was investigated properly by Koeniger and Vorwhol (1979) and photographs below are only to enlighten the reader on some simple phenomenon often obvious in tropical evergreen home gardens.

FIGURE 27.13   Both *Apis florea* and *A. dorsata* feeding simultaneously on *Pongamia pinnata* (Fabaceae) flowers.

This is a high resource flower where all five honeybees feast very actively.

FIGURE 27.14   *Apis florea* and *Tetragonula* feeding together on *Pongamia pinnata* (Fabaceae) flowers. Arrows pointing at bees.

FIGURE 27.15   Both *Apis cerana* and *A. florea* feeding together on the same flower cluster (1X) of *Pongamia pinnata* (Fabaceae) flowers.

In this natural observation, it was clear that *Apis* bees cease to visit as the day advances toward noon in the following order, first, *A. dorsata*, then *A. cerana*, and finally, *A. florea*. However, sometimes *A. florea* stayed on till midafternoon, and *Tetragonula* continued throughout the day. It may be interpreted that comparatively, the smallest proboscis of *Tetragonula* allowed them to lick out the tiniest traces of nectar that remained in the hidden floral glands, which the larger bees were unable to extract due to their larger body sizes that restrict access to interior of the flower and larger tongues that are too large to lick the residual traces. In fact, many flowers that are not visited by larger honeybees, that is, *A. dorsata* and *A. cerana* are visited by *Apis florea* quite effectively. Further, some flowers in spite of their impressive displays are not visited by any *Apis* honeybees but actively visited by *Tetragonula* honeybees (Punchihewa's unpublished observations).

27.4.3 MIGRATORY BEHAVIOR

The high diversity of plants and their distribution pattern on the island have revealed 27 vegetation zones (Ashton and Gunatilleke, 1987 and MoMD&E (2019) Biodiversity profile—Sri Lanka).

27.4.3.1   LONG DISTANCE MIGRATORY HABIT OF APIS DORSATA AND ITS IMPLICATIONS

Apis dorsata annually migrates between the up-country high mountain cloud forest (>2000 m) and the low country coastal plains (<300 m). During this annual migration to and fro, they stop by various agroecological regions for them to gather food to rear their young and for the environment to pollinate plants that flower to synchronize their arrival, both agricultural and natural. Here, we examine one impressive example of this. It is most likely that all vegetation zones in Sri Lanka have all if not most of the required "resident" pollinators at disposal when necessary. But there can be occasions where existing resident pollinator population may not be sufficient, when cornucopian flowering takes place. A phenomenon that occurs in two distinctly different and distantly separated two ecosystems is highlighted below.

FIGURE 27.16   Farmers of the southeastern low country dry zone (<300 m) cultivate their sesame fields so that, they come to flower in May–June.

A. dorsata by this time migrate to the lowland plains and they occupy their usual trees to rear brood and benefit from flowering in this ecozone. Both sesame farmers and giant honeybees have mutual benefits from this migration. In the photograph, foreground are large sesame fields in flower and in distant background, central high mountains.

FIGURE 27.17   *Apis dorsata* on sesame flowers (*Sesamum indicum*: Pedaliaceae). Sesame fields are just buzzing from early morning hours till midday with the flower visiting giant honeybees.

Practically one can find an *Apis dorsata* on every single open sesame flower during the peak flowering period.

A similar phenomenon occurs in the cloud forest with which 30 or more species of *Strobilanthes* (*Nelu* නෙළු) make the most extensive protective ground cover.

FIGURE 27.18 Cloud forest during the flowering of *Strobilanthes* sp. (Acanthaceae).
Cloud water harvesting epiphytes are visible on tree trunks through the mist/cloud. Ground covered with flowering *Strobilanthes pulcherrima.*

FIGURE 27.19 *Strobilanthes viscosa* pollinated by *A. dorsata* in the cloud forest.
S. viscosa is another extensive ground cover of the cloud forest

27.4.4   DRONE CONGREGATION AREAS AND MATING BEHAVIOR OF APIS HONEYBEES

It is the common knowledge of inquisitive gardeners and beekeepers, who are familiar with the bee hives in their garden that black-colored rather than stocky looking type of bees fly in and out of the hive quite actively, making their peculiar noise or the "drone buzz" in the late afternoon for a few hours. The "drone buzz" is quite audible to any visitor of *Apis cerana* apiary, and one can easily see flying drones pursuing any object, such as a pebble thrown in the air or any large enough insect such as a butterfly, hornet, dragonfly, etc., who inadvertently happen to cross the space between the trees where the open sky is visible above. Koeniger and Wijayagunasekara (1976) made the first detailed observation on the drone flights of Sri Lankas' *Apis* honeybees.

TABLE 27.3   Drone Flight Periods of Honeybees of Sri Lanka.

Honeybee species	Drone flight period
Apis florea	12:30 ~ 14:30 ± 0:15
Apis cerana	15:00 ~ 17:00 ± 0:15
Apis dorsata	18:00 ~ 18:45 ± 0:15
Tetragonula spp.	Unknown

In controlled experiments, natural queens as well as capsules with queen substance[6] (9 oxo-2-decenoic acid, one of the major ingredients) floated in midair to confirm our results (Punchihewa et al., 1990a, 1990b). In our controlled experiments, we marked mature drone in the hive before the flight period, captured flying drone in drone congregation area (DCA), marked them again, and released to reassess them back in the hives in the vicinity. Finally, these marking, capturing, remarking, and recapturing results confirmed that 90% of the drones in a particular DCA originated from the colonies closest to it. Therefore, with this understanding we have been able to device an appropriate system of breeding and multiplication of *Apis cerana* colonies, an important step in the development of the beekeeping industry in Sri Lanka (Punchihewa, 1992, 1994; Punchihewa et al., 1990a, 1990b).

Even though honeybees are essentially out-breeding or out-crossing, at least for *Apis cerana* in Sri Lanka, a good genetic pool could be maintained by "closed population breeding" by maintaining and constantly replenishing the honeybee colonies with desirable characters in a particular breeding apiary through controlled natural breeding (Punchihewa, 1992).

In *Apis dorsata* colonies, drones **"burst out"** in flight almost at dusk. Even though in most places *A. dorsata* colonies are build high up on tall trees, the flight burst is quite audible and visible from ground, if one happens to be under the tree. Also, it is quite easy to see a mass of drone flying out at the "burst." The *A. dorsata* drone buzz is also quite audible if one happens to walk along nearby tree canopies bordering a well open area, such as open rice fields, meadows, etc.

---

[6]Pheromones are secreted by the mandibular glands of the queen or the reproductive female in a colony of honeybees. These chemicals maintain the colony stability in a major way by coordinating most of the activities of the colony. Queen substance is also used by unmated queens to attract males (or drones) when on their nuptial flight.

FIGURE 27.20   The flying drone assemble in flight among the tree canopies in the near vicinity waiting for the arrival of a mate seeking virgin queen.

Here, a demonstration of such a drone assemblage site is shown by using a caged-queen floated to the drone assemblage space with the use of a hydrogen balloon. These drone assemblage sites are called drone congregation areas or DCA. Drone providing honeybee colonies are underneath the trees in the photograph.

Apis florea drones also vigorously fly with a distinct "drone buzz," in and out of their colonies in the early afternoon hours (see Table 27.3), however up to now, the DCA formation characters have not been elucidated in Sri Lanka and not reported elsewhere.

From our understanding of DCA formation of the two honeybee species discussed above, they always seek some form of a plant canopy for this purpose. The reasons could be a mechanism to prevent bird predation.

KEYWORDS

- **honeybees**
- **pollination**
- **apiculture**
- **conservation**
- **diversity**

REFERENCES

Ashton, P. S.; Gunatilleke, C. V. S. New Light on Plant Geography of Ceylon: 1 Historical Plant geography. *J. Biogeogr.* **1987**, *14* (3), 249–285.

Bawa, K. S. Plant-Pollinator Interactions in Tropical Rain Forests. *Ann. Rev. Ecol. Syst.* **1990**, *21*, 399–422. http://links.jstor.org/sici?sici=0066-4162%281990%2921%3Cght 399%3APIITRF%3E2.0.CO%3B2-.

Frisch, K. von. *The Dance Language and Orientation of Bees*; Belknap Press of Harvard University Press: Cambridge, MA, USA, **1976**.

Karunaratne, W. A. I. P.; Edirisinghe, J. P.; Engel, M. S. First Record of a Tear-Drinking Stingless Bee *Lisotrigona cacciae* (Nurse) (Hymenoptera: Apidae: Meliponini) from the Central Hills of Sri Lanka. *Nat. Sci. Found. Sri Lanka.* **2017**, *45* (1), 79–81. DOI: 10.4038/jnsfsr.v45i1.8042

Karunaratne, W. A. I. P.; Edirisinghe, J. P.; Pauly, A. *An Updated Checklist of Bees of Sri Lanka with New Records*; Publication No. 23. MAB Handbook and Occasional Paper Series. National Science Foundation: Colombo, 2005.

Kevan, P. G.; Punchihewa, R. W. K.; Greco, F. C. Foraging Range of *Apis cerana* and Its Implications for Honey Production and Apiary Management. In *The Asiatic Hive Bee: Apiculture, Biology and Role in Sustainable Development in Tropical and Subtropical Asia;* Kevan, P. G., Ed.; Enviroquest Ltd: Cambridge, ON, Canada, 1995; pp 223–228.

Koeniger, N.; Koeniger, G.; Punchihewa, R. W. K.; Fabritius, Mo.; Fabritius, Mi. Observations and Experiments on Dance Communication. *Apis florea. J. Apic. Res.* **1982**, *21*, 45–52.

Koeniger, N.; Koeniger, G.; Tingek, S. *Honey Bees of Borneo: Exploring the Centre of Apis Diversity*; Natural History Publications (Borneo): Kota Kinabalu, Sabah, Malaysia, **2010**.

Koeniger, N.; Vorwhol, G. Competition for Food Among Four Sympatric Species of Apini in Sri Lanka (*Apis dorsata, A. cerana, A. florea, Trigona iridipennis*). *J. Apic. Res.* **1979**, *21*, 45–52.

Koeniger, N.; Wijayagunasekara, H. N. P. Time of Drone Flight in the Three Asiatic Honeybee Species (*Apis cerana, Apis florea, Apis dorsata*). *J. Apic. Res.* **1976**, *15*, 67–71.

Lindauer, M. Communication Among Honeybees and Stingless Bees in India. *Bee World* **1957**, *38*, 3–14, 34–39.

Lindauer, M. *Über die Verständigung bei indishen Bienen. Z. Vergl. Physiol.* **1956**, *28*, 95–109.

Michener, C. D. *The Social Behavior of Bees*; Belknap Press of Harvard University Press: Cambridge, MA, USA, 1974.

MoMD&E. *Biodiversity Profile – Sri Lanka, Sixth National Report to the Convention on Biological Diversity;* Biodiversity Secretariat, Ministry of Mahaweli Development and Environment: Sri Lanka, **2019**.

Otis, G. W.; Smith, D. R. *Drone Cell Cappings of Asian Cavity-Nesting Honey Bees (Apis* spp.) *Apidologie*; INRAE, DIB and Springer-Verlag, **2021**.

Panabokke, C. R. *Soils and Agro-Ecological Environments of Sri Lanka*; Natural Resources, Energy and Science Authority of Sri Lanka: Colombo, **1996**.

Panabokke, C. R.; Kannangara, R. P. K. Identification and Demarcation of the Agro-Ecological Regions of Sri Lanka. *Proc. Ann. Sess. Sri Lanka Assoc. Adv. Sci.* **1975**, *31* (3), 49.

Punchihewa, R. W. K. *Beekeeping for Honey Production in Sri Lanka: Management of Apis cerana in its Natural Tropical Monsoonal Environment;* Sri Lanka Dept of Agriculture: Peradeniya, **1994**.

Punchihewa, R. W. K. Mating Behaviour of *Apis cerana* is Advantageous for Controlled Natural Breeding. *Apidologie* **1992**, *23*, 348–349.

Punchihewa, R. W. K.; Koeniger, N.; Kevan, P. G. Management of Swarming in *Apis cerana*. In *Asian Bees and Beekeeping: Progress of Research and Development;* Matsuka, M., Verma, L. R., Wongsiri, S., Shesthran, K. K., Pratap, U., Eds.; Oxford & IBH Publishing Co. Ltd.: New Delhi, India, **2000**; pp 114–115.

Punchihewa, R. W. K.; Koeniger, N.; Kevan, P. G.; Gadawski, R. M. Observations on the Dance Communication and Natural Foraging Ranges of *Apis cerana, Apis dorsata* and *Apis florea* in Sri Lanka. *J. Apic. Res.* **1985**, *24* (3), 168–175.

Punchihewa, R. W. K.; Koeniger, N.; Koeniger, G. Congregation of *Apis cerana indica* Fab 1798, Drones in the Canopy of Trees in Sri Lanka. *Apidologie* **1990a**, *21*, 201–208.

Punchihewa, R. W. K.; Koeniger, N.; Koeniger, G. Mating Behaviour of *Apis cerana* in Sri Lanka. In *Social Insects and the Environment;* Veeresh, G. K., Mallik, B., Viraktamath, C. A., Eds.; Proc 11th Congress of IUSSI: Bangalore, India. Oxford & IBH Publishing Co. Ltd.: New Delhi, India, **1990b**; pp 108.

CHAPTER 28

STATUS OF FRESHWATER FISHES OF SRI LANKA

SMPATH DE ALWIS GOONATILAKE

IUCN, International Union for Conservation of Nature, Sri Lanka Office, Pelawatte, Battaramulla, Sri Lanka

IUCN Freshwater fish Specielist Group

ABSTRACT

There are 79 pure freshwater disappearance fish species in Sri Lanka and among them, 54 species are endemic to the island. Another 24 fish species are disappearance both freshwater and brackish or migrating to the saltwater water, and four species are endemic to the island. Meantime another 30 species are exotic species that were introduced to the island on different occasions.

28.1 INTRODUCTION

Sri Lanka is an island situated southernmost tip of the Indian subcontinent. As it is an island, it has diverse natural and manmade freshwater habitats. Among them there are a total of 103 natural river basins, which start from middle highlands and flow throughout the entire island forming a dendritic pattern. These freshwater habitats support a rich aquatic fauna, of which freshwater fish are the most studied taxonomic group (Goonatilake et al., 2020).

28.2 NATURAL HISTORY

The first freshwater fish discovery from Sri Lanka was made by Georges Cuvier and Achille Valenciennes, and followed by P. Bleeker (1960s). George Duncker has explored several localities, including the Gin river basin and made the first checklist of Sri Lankan freshwater fish (Goonatilake, 2012; Goonatilake et al., 2020; Duncker, 1912). In 1878 and

Biodiversity Hotspot of the Western Ghats and Sri Lanka. T. Pullaiah, PhD (Ed.)
© 2024 Apple Academic Press, Inc. Co-published with CRC Press (Taylor & Francis)

1988–1989, Francis Day has published a number of Sri Lankan freshwater fish species through his magnificent works on the fishes of India including Burma and Ceylon (Sri Lanka) also in fish guide of Fauna of British India series (Day, 1878, 1888).

In the mid-20th century, Sri Lankan ichthyologist P.E.P. Deraniyagala has described several new species and produced the first illustrated book on Sri Lankan freshwater fish (Deraniyagala, 1952). His work was followed by A. S. Mendis in 1954, and I. S. R. Munro in 1955 (Goonatilake et al., 2020). During the late 1970s, Ranil Senanayake has laid the foundations to more extensive survey on freshwater fish in Sri Lanka and he has attempted to identify the ichthyological zones of Sri Lanka (Senanayake, 1980; Senanayake and Moyle, 1982). In 1991, Rohan Pethiyagoda has published first color photographic guide on freshwater fishes of Sri Lanka under the Wildlife Heritage Trust flag. This was resulted in the discovery of many new species of freshwater fish (Bailey and Gans, 1998; Kottelat and Pethiyagoda, 1991; Meegaskumbura et al., 2008; Pethiyagoda et al., 2008a, 2008b, 2008c, 2012; Silva et al., 2008, 2011). In 2007, Goonatilake has published first color illustrated field guide in local language.

Between 1991 and 2012, there were a number of nomenclature changers in species belonging to *Puntius, Rasbora,* and *Danio* genera (Pethiyagoda et al., 2012; Silva et al., 2011; Kevin et al., 2010; Miya et al., 2010). Over the past decade, species names of several species were also revised. For example, *Puntius amphibious, Macroganthus aral,* and *Labeo porcellus* were renamed into *Puntius kamalika, Macrognathus pentophthalmos,* and *Labeo lankae* respectively (Silva et al., 2008; Pethiyagoda et al., 2008c; Pethiyagoda, 1994). In 2015, de Silva et al. published a book on freshwater fishes of Sri Lanka including color photographs of all inland recorded fish species (freshwater, Brackish water, and several saltwater species) prior to 2015. It also listed 30 naturally occurring exotic species found in the inland waters of Sri Lanka.

Several Indian fish species names which were applied erroneously to Sri Lankan populations have also been resolved during the last decade. Some of them have now been assigned new names, while some species have evaluated its subspecies name to species level (Batuwita et al., 2015; Conte-Grand et al., 2020; Ng and Pethiyagoda, 2013; Sudasinghe et al., 2016; Sudasinghe and Meegaskumbura, 2016). Meantime some of the endemic species have lost their endemicity and been recognized as Indian species. These are *Esomus thermoicos, Dawkinsia filamentosa* (formerly *Dawkinsia singhala*), and *Systomus sarana* (formally *Sytomus spirulus*) (Sudasinghe et al., 2019a, 2021). Sudasinghe et al. (2019b) phylogenetically recognized Sri Lankan *Amblypharyngodon* species as *Amblypharyngodon grandisquammis* (Jordan and Starks, 1917). Sudasinghe et al. (2018) have reviewed the genus *Labeo* based on the phylogenetic evidence and recognized the all *Labeo* species in Sri Lanka are endemic to the island. Further, their study has indicated that the name *Labeo dussumieri* (Valenciennes, 1842) had been mis-applied to the Sri Lankan population and new name has been given (*Labeo heladiva).*

Puntius layardii (Günther, 1868) and *Puntius tetraspilus* (Günther, 1868) is listed as a species, but its validity has been queried by Pethiyagoda et al. (2008b, 2012) (as it is known only from type specimens). Thus, this species also needs further phylogenetic study.

28.3 FRESHWATER FISHES DIVERSITY

The IUCN Sri Lanka faunal database in 2021 October listed 103 inland water disappearance species ranging from the largest: The Shark catfish (*Wallago attu*), the Long-finned eel (*Anguilla bengalensis*), and the Level-finned eel (*Anguilla bicolor*), to the smallest: Hora dandia (*Horadandia atukorali*), Pallaides rasbora (*Rasboroides pallidus*), and Vateria flower rasbora (*Rasboroides vaterifloris*) and among them 58 species are endemic and another 30 exotic species are also recorded from natural freshwater habitats in the country.

The recorded species belong to 21 families and they can be subdivided into two main categories. Among the recorded species 79 species are purely disappearance within the freshwater waterbodies and among them, 54 species are endemic to the island. Another 24 species (Gobies and Eels) are disappearance both freshwater and brackish or migrating to the saltwater water, and four species are endemic to the island.

28.3.1 BIOGEOGRAPHY OF THE FRESHWATER FISHES

Very few studies were undertaken to understand the freshwater dispersal into the islands or even within the island (Goonatilake et al., 2020). A total number of 40 species representing 19 families have shared the distribution with the Indian peninsula. This relationship was identified by a number of both Sri Lankan and Indian ichthyologists such as Deraniyagala (1952), Pethiyagoda (1991), and Sudasinghe et al. (2018b, 2018d, 2019a, 2019b, 2020c).

In the 1980s, Senanayake has studied the biogeography and ecology of the Sri Lankan freshwater fish fauna (Senanayake, 1980), and based on his findings it has identified four major ichthyological provinces from Sri Lanka (Senanayake and Moyle, 1982). This was done without having phylogenetic knowledge on the freshwater fish species which is not available in that period. In 2018, a new map of ichthyological zones was presented by the Sri Lankan experts during the process of the 6th National Review of the National Biodiversity Strategy Plan. However, it was based on the previous Ichthyological zones proposed by Senanayake and Moyle (1982). After that, there are a number of other species descriptions made by Sudasinghe and his team (Sudasinghe et al., 2019a, 2020c). Including the most recent species descriptions and phylogenetic knowledge, it can be proposed the following Ichthyological zones from Sri Lanka (Map 01). This was based on the geographical isolation of endemicity into the river basin.

28.3.1.1 MAHAWELI ICHTHYOLOGICAL ZONE

The Middle and Upper Mahaweli Ichthyological zones are restricted to the upper head of the Mahaweli basin to the middle-end where the Amban river (main tributary flow from Knuckles mountain region) has connected to main river. However, according to current updated species knowledge, it can be further subdivided into middle Mahaweli and Mahawali headwater zone. A total of 41 freshwater fish species are found within this zone and among them 18 species are endemic. It also has five species restricted to this region. Among the four species are restricted Kcuckles subzone and one species restricted to headwater (Agra) zone.

28.3.1.1.1 Middle Mahaweli Subzone

This is the largest subzone of the Mahaweli Ichthyological zone. This includes a number of major sub-basins (Aban, Heen, Badulu, Loggal, Ma Oya, Hulu, Kothmala) of the Mahaweli river. Its upper boundary starts from the Ginigathhena (651 m Mean Sea Level) to downstream of the Mahaweli main river up to downstream Yakkure (60 m Mean Sea Level) where the Aban river is connected with the main river. This includes one of the main biodiversity hotspots Knuckles mountain range and the majority of the range rested endemic species (*Dawkinsia srilankensis, Labeo fisheri, Systomus martenstyni,* and *Garra philipsi*) are found from the stream network around this Knuckles range.

28.3.1.1.2 Mahaweli Headwater (Agra) Subzone

This includes the Mahaweli headwater including the tributaries of the upper mountain region where the *Devario monticola* has been restricted. It includes the fast-flowing cool streams habitats which have support for very few freshwater species including introduced rainbow trout also occurred.

28.3.1.2 SOUTH -WESTERN ICHTHYOLOGICAL ZONE

This zone is consisting of seven major river basins (Attanagalu, Kelani, Bolgoda, Kalu, Bentara, Gin, and Nilwala) in the southwestern lowland wet zone of Sri Lanka. The northern boundary is the Attanagalu basin and the southern boundary is the Nilwala basin. A total of 83 freshwater fish species were recorded from this region and among them 26 species are endemic. There are 37 species restricted to this zone. This zone supports most of Sri Lanka's endemic freshwater fish species. This zone can be subdivided into the Western wet subzone and the Southern wet subzone, considering the species restriction to each subzone.

28.3.1.2.1 Western Wet Subzone

The western wet subzone covers the streams and rivers of the Aththanagalu, Kelani, and Kalu River basins. There are eight freshwater fish species (*Aplocheilus dayi, Ophisternon bengalense, Pethia bandula, Pethia reval, Rasbora armitagei, Rasboroides vaterifloris, Sicyopterus lagocephalus, Systomus asoka,* and *Stiphodon martenstyni*) restricted to this zone among them five species were endemic to the island. *Pethia bandula* and *Stiphodon martenstynii* are two-point endemic species that is restricted to a small stream at Kelani river and Kalu river basins.

FIGURE 28.1 *Pethia bandula* point endemic species in western wet subzone. Photo courtesy of Goonatilake et al., 2020. © International Union for Conservation of Nature.

FIGURE 28.2 *Pethia cumingi* restricted to southern wet subzone. Photo courtesy of Goonatilake et al., 2020. © International Union for Conservation of Nature.

FIGURE 28.3 *Labeo lankae* restricted to northern dry subzone. Photo courtesy of Goonatilake et al., 2020. © International Union for Conservation of Nature.

FIGURE 28.4 *Labeo fisheri* restricted to Middle Mahaweli subzone. Photo courtesy of Goonatilake et al., 2020. © International Union for Conservation of Nature.

FIGURE 28.5 *Systomus martinstyni* restricted to Middle Mahaweli subzone. Photo courtesy of Goonatilake et al., 2020. © International Union for Conservation of Nature.

FIGURE 28.6 *Dawkinsia srilankensis* restricted to Middle Mahaweli subzone. Photo courtesy of Goonatilake et al., 2020. © International Union for Conservation of Nature.

28.3.1.2.2 *Southern Wet Subzone*

This zone covers the streams of the major rivers Benthara, Gin, and Nilawala River basins. There are seven freshwater fish species (*Aplocheilus werneri, Devario pathirana,*

Macrognathus pentophthalmos, Pethia cumingii, Rasboroides pallidus, and *Schistura scripta*) restricted to this zone and all of them are endemic to the island.

28.3.1.2.3   Ma Oya Subzone

This zone includes the basins of Maha Oya and is an area transitioning from wet to dry climatic zone basins. It supports many of the wet zone endemic species, and *Devario memorialis* is range-restricted to this subzone. This zone has both wet and dry environmental conditions and it has supported several wet zone species. This subzone has limited to one basin and most of the natural habitats have been heavily damaged due to plantations (Tea, coconut, rubber) and human settlements.

28.3.1.3   DRY ICHTHYOLOGICAL ZONE

The dry ichthyological zone is distinguished from the rest of the zones, by having a low number of endemics. It has supported a total of 45 freshwater fish species and eight species (*Channa marulius, Devario memorialis, Labeo lankae, Laubuka hema, Laubuka lankensis, Ompok ceylonensis, Pethia melanomaculata,* and *Rasbora adisi*) are totally restricted to this zone. Among them, only *Channa marulius* is not endemic to the island. Based on the species restriction this region can be further subdivided into four subzones.

28.3.1.3.1   Northern Dry Subzone

The only endemic species that seem to be restricted to this Dry Ichthyological zone is *Labeo lankae. Labeo lankae* was recoded from the entire Dryzone until the early 1980s. However, the current occurrence of *Labeo lankae* is from only the northern part of the Dry Climatic Zone drained by Kala Oya, Malwathu Oya, and Per Aru basins. Other than the above species *Channa marulius* is also restricted to this subzone.

28.3.1.3.2   Southern Dry Subzone

This zone is recognized to be distinct because of the presence of two restricted-range species, *Schistura madhavai,* and *Rasbora naggsi* which are restricted to the Walawa river basin to Kiridi Oya basin. The fish species composition shows some affinity to the South Western Ichthyological Zone.

28.3.1.3.3   Southeast Dry Subzone

This zone is recognized to be distinct because of the presence of two restricted-range species, *Rasbora adisi* and *Laubuka hema* which are restricted to the Manik river basin to

Gal Oya basin. The fish species composition shows some affinity to the South Mahaweli Ichthyological Zone. This the least studied zone in the country and above two species were discovered after 2019.

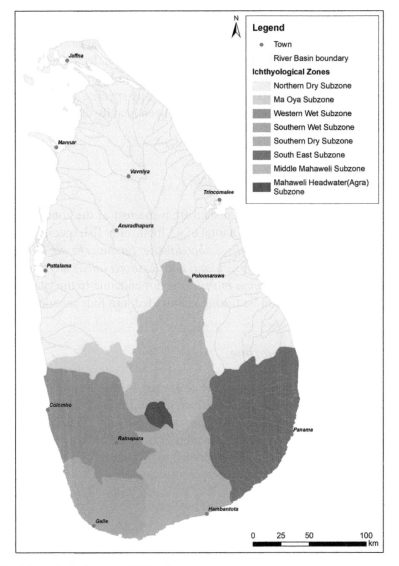

FIGURE 28.7   Ichthyological zones of Sri Lanka.

Source: Modified based on Senanayake and Moyle, 1982.

28.3.2   FRESHWATER FISHES HABITATS

Sri Lanka has a number of aquatic habitats including 103 natural rivers drained across the entire island (MMDE, 2016). These rivers and their tributaries have ultimately formed seasonal or shallow streams. Some of the above rivers have flood plains and some of the rivers form large lake-type shallow estuaries. It also recorded man-made naturalized seasonal or perennial lakes like small, medium, and large reservoirs (12,000 tanks and 13,000 anicuts). Other than the above major habitats number of freshwater marshy habitats are available on the island. Some of them were natural habitats and some of them are man-made seasonal habitats (e.g., paddy fields).

28.4   CONSERVATION OF THE FRESHWATER FISHES

According to the current status approximately 28% of the land area in Sri Lanka has been declared as a protected area and managed by either the Department of Wildlife Conservation or the Forest department (Goonatilake, 2012; Goonatilake et al., 2020). However, most of the endemic species have been restricted to habitats outside the protected area network. These habitats are subject to high human pressure and as a result, the extent and the quality of these habitats are declining rapidly, reducing the carrying capacity of these habitats for freshwater fish (Goonatilake, 2012).

28.4.1   DIRECT AND INDIRECT DIVERS' EFFECT ON FRESHWATER FISH

Sri Lanka is a developing country; there are a number of development projects which directly or indirectly affect the ecology or habitats of the freshwater fish species. Other than the above reasons many anthropogenic activities such as deforestation, gem mining, excessive use of agrochemicals, release of pollutants, and natural hazards such as floods, landslides, and changes in climatic weather patterns directly delicate ecological balance, and threaten the long-term survival ability of many species of freshwater fish in Sri Lanka (Pethiyagoda, 1994). In addition, the introduction of alien invasive fish species, and over-exploitation also led to a significant reduction in the population sizes of native fish.

28.4.2   GLOBAL AND NATIONAL THREAT STATUS

According to the latest (2019) global and national red list assessment, it has assessed 58 endemic species against the red list criteria. The results showed that 12 endemic species were listed as Critically Endangered (CR), 24 range-restricted species were Endangered (EN), and nine species were Vulnerable (VU). A further five species were Near Threatened (NT). Two species that did not have exact distributional data were listed as Data Deficient (DD). The remaining nine species that showed a wide distribution throughout the island and have low threat effects were listed as Least Concern (LC) (Goonatilake et al., 2020). In addition to the above endemic species, 36 native species were assessed against the criteria and of these, only eight species were listed as Threatened (EN: 6 and VU: 2). In addition, five species were listed as Near Threatened and five species as Data Deficient. The remaining 20 species were listed as Least Concern (Goonatilake et al., 2020).

TABLE 28.1   National Threat Status of the Assessed Freshwater Fish Species of Sri Lanka Classified by Family.

Family name	Total number of species	Total number of endemic species	Threat status			NT	DD	LC
			CR	EN	VU			
Aplocheilidae	3	2	0	2	0	0	0	1
Adrianichthyidae	2	0	0	0	0	0	2	0
Anabantidae	1	0	0	0	0	0	0	1
Anguillidae	2	0	0	0	0	2	0	0
Bagridae	4	3	0	1	0	0	0	3
Belonidae	1	0	0	0	1	0	0	0
Channidae	7	3	0	1	1	2	0	0
Cichlidae	2	0	0	0	0	0	0	2
Claridae	1	1	0	0	0	1	0	0
Cobitidae	2	1	0	1	0	0	0	1
Cyprinidae	45	35	7	15	5	2	2	13
Eleotridae	4	0	0	0	0	0	0	2
Gobiidae	11	3	1	4	0	0	3	2
Heteropneustidae	1	0	0	0	0	0	0	1
Mastacembelidae	2	1	1	0	0	0	0	1
Nemacheilidae	4	4	2	1	0	1	0	0
Osphronemidae	3	2	0	1	1	0	0	1
Siluridae	3	2	0	0	1	1	0	1
Synbranchidae	2	1	1	1	0	0	0	0
Syngnathidae	2	0	0	1	0	0	0	0
Toxotida	1	0	0	0	0	0	0	0
Total count	**103**	**58**	**12**	**28**	**9**	**9**	**7**	**29**

CR, critically endangered; DD, data deficient; EN, endangered; LC, least concern; NT, near threatened; VU, vulnerable.

Legal protection: Among the 103 species of freshwater fish species in Sri Lanka, 17 species have been protected by law under the Fauna and Flora Protection Ordinance (FFPO) No.22 of 2009, Schedule VI. In addition, the export of 13 species is prohibited under the Fisheries and Aquatic Resources Act (FARA), No 2 of 1996, and the export of another 13 species is restricted under the same act (Goonatilake et al., 2020).

KEYWORDS

- **Sri Lanka**
- **freshwater fish**
- **natural history**
- **distribution**

REFERENCES

Bailey, R. M.; Gans, C. The New Synbranchid Fishes, *Monopterus roseni* from Peninsular India and *M. desilvai* from Sri Lanka'. *Occas. Pap. Mus. Zool. Univ. Mich.* **1998,** *726,* 1–18.

Batuwita, S.; Maduwage, K.; Sudasinghe, H. Redescription of *Pethia melanomaculata* (Teleostei: Cyprinidae) from Sri Lanka. *Zootaxa* **2015,** *3936* (4), 575–583.

Conte-Grand, C.; Britz, R.; Dahanukar, N.; Raghavan, R.; Pethiyagoda, R.; Tan, H. H.; Hadiaty, R. K.; Yaakob, N. S.; Rüber, L. Barcoding Snakeheads (Teleostei, Channidae) Revisited: Discovering Greater Species Diversity and Resolving Perpetuated Taxonomic Confusions. *PLoS One* **2017,** *12* (9), e0184017. https://doi.org/10.1371/journal.pone.0184017.

Cuvier, G.; Valenciennes, A. In *Histoire Naturelle des Poissons*; P. Bertrand: Paris, **1842;** vol 16, pp 465–487.

Day, F. In *The Fishes of India*; Being a Natural History of the Fishes known to Inhabit the Seas and Fresh Waters of India, Burma, and Ceylon. B. Quaritch: London, **1878;** vol 1& 2.

Day, F. In *The Fishes of India*; Being a Natural History of the Fishes known to Inhabit the Seas and Fresh Waters of India, Burma, and Ceylon. B. Quaritch: London, **1888.**

De Silva, M.; Hapuarachchi, N.; Jayaratne, T. In *Sri Lankan Freshwater Fishes*; Wildlife Conservation Society: Galle, Sri Lanka, **2015.**

Deraniyagala, P. E. P. A New Cyprinoid Fish from Ceylon. *J. Royal Asiatic Society Ceylon* **1943,** *35* (96), 158–159

Deraniyagala, P. E. P. In *A Colored Atlas of Some Vertebrates from Ceylon: Fishes*; Sri Lanka: National Museum: Colombo, Sri Lanka, **1952;** vol 1.

Deraniyagala, P. E. P. *Two New Subspecies and One New Species of Cyprinoid Fishes from Ceylon,* Proceedings of the 12th Annual Sessions of the Ceylon Association for the Advancement of Science, **1956;** vol 1, pp 34–35.

Deraniyagala, P. E. P. In *The Pleistocene of Ceylon*; Ceylon National Museums: Colombo, **1958.**

Duncker, G. Die Susswasserfische Ceylons. *Jahrb. Hamburg Wiss. Anst. Beiheft 2,* **1912,** *29* (2), 241–272.

Goonatilake, S. de A. In *Freshwater Fishes of Sri Lanka* (in Sinhala); Ministry of Environment: Colombo, Sri Lanka, **2007.**

Goonatilake, S. de A. The Taxonomy and Conservation Status of the Freshwater Fishes in Sri Lanka. In *The National Red List 2012 of Sri Lanka; Conservation Status of the Fauna and Flora*; Weerakoon, D. K., Wijesundara, S., Eds.; Ministry of Environment: Colombo, Sri Lanka, **2012;** pp 77–87.

Goonatilake, S. De A.; Fernando, M.; Kotagama, O.; Perera, N.; Vidanage, S.; Weerakoon, D.; Daniels, A. G.; Máiz-Tomé, L. *The National Red List of Sri Lanka: Assessment of the Threat Status of the Freshwater Fishes of Sri Lanka 2020*; International Union for Conservation of Nature, Sri Lanka and the Biodiversity Secretariat, Ministry of Environment and Wildlife Resources, IUCN: Colombo, **2020.**

Kevin, T.; Agnew, M.; Hirt, M.; Sado, T.; Schneider, L.; Freyhof, J.; Sulaiman, Z.; Swartz, E.; Vidthayanon, C.; Miya, M.; Saitoh, K.; Simons, A.; Wood, R.; Mayden, R. Systematics of the Subfamily Danioninae (Teleostei: Cypriniformes: Cyprinidae). *Mol. Phylogenet. Evol.* **2010,** *57,* 189–214.

Meegaskumbura, M.; Silva, A.; Maduwage, K.; Pethiyagoda, R. *Puntius Reval,* A New Barb from Sri Lanka (Teleostei; Cyprinidae). *Ichthyol. Explor. Freshw.* **2008,** *19* (2), 141–152.

Mendis, A. S. Fishes of Ceylon: A Catalogue, Key and Bibliography; *Bulletin of the Fisheries Research Station,* Ceylon, **1954,** 2, 1–222.

Ministry of Mahaweli Development and Environment (MMDE) *National Biodiversity Strategic Action Plan 2016-2022*; Biodiversity Secretariat, Ministry of Mahaweli Development and Environment: Colombo, Sri Lanka, **2016.**

Miya, C. M.; Saitoh, K.; Simons, A. M.; Wood, R. M.; Mayden, R. L. Systematics of the Subfamily Danioninae (Teleostei: Cypriniformes: Cyprinidae). *Mol. Phylogenet. Evol.* **2010,** *57* (1), 189–214.

Munro, I. S. R. *The Marine and Freshwater Fishes of Ceylon*; Department of External Affairs: Canberra, **1955.**

Ng, H. H.; Pethiyagoda, R. *Mystus zeylanicus,* A New Species of Bagrid Catfish from Sri Lanka (Teleostei: Bagridae). *Ichthyol. Explor. Freshw.* **2013,** *24* (2), 161–170.

Pethiyagoda, R. In *Freshwater Fishes of Sri Lanka*; The Wildlife Heritage Trust of Sri Lanka: Colombo, Sri Lanka, **1991.**

Pethiyagoda, R. Threats to The Indigenous Freshwater Fishes of Sri Lanka and Remarks on Their Conservation. In *Ecology and Conservation of Southeast Asian Marine and Freshwater Environments Including Wetlands*; Sasekumar, A., Marshall, N., Macintosh, D. J., Eds.; Reprinted in *Hydrobiologia*, **1994**; vol 285, pp 189–201.

Pethiyagoda, R.; Kottelat, M.; Silva, A.; Maduwage, K.; Meegaskumbura, M. A Review of the Genus *Labuca* in Sri Lanka, with Description of Three New Species (Teleostei: Cyprinidae). *Ichthyol. Explor. Freshw.* **2008a**, *19* (1), 7–26.

Pethiyagoda, R.; Meegaskumbura, M.; Maduwage, K. A Synopsis of the South Asian Fishes Referred to *Puntius* (Pisces: Cyprinidae.). *Ichthyol. Explor. Freshw.* **2012**, *23* (1), 69–95.

Pethiyagoda, R.; Silva, A.; Maduwage, K.; Kariyawasam, L. The Sri Lankan Spiny Eel, *Macrognathus pentophthalmos* (Teleostei: Mastacembelidae), and its Enigmatic Decline. *Zootaxa* **2008c**, *1931* (1), 37–48.

Pethiyagoda, R.; Silva, A.; Maduwage, K.; Meegaskumbura, M. *Puntius kelumi*, A New Species of Cyprinid Fish from Sri Lanka (Teleostei: Cyprinidae)'. *Ichthyol. Explor. Freshw.* **2008b**, *19*, 201–214.

Senanayake, F. R. *The Biogeography and Ecology of the Inland Fishes of Sri Lanka*. PhD Dissertation, University of California, California, USA, **1980**.

Senanayake, F. R.; Moyle, P. B. Conservation of Freshwater Fishes of Sri Lanka. *Biol. Conserv.* **1982**, *22*, 181–195.

Silva, A.; Maduwage, K.; Pethiyagoda, R. *Puntius kamalika*, A New Species of Barb from Sri Lanka (Teleostei: Cyprinidae). *Zootaxa* **2008**, *1824* (1), 55–64.

Silva, A.; Maduwage, K.; Pethiyagoda, R. A Review of the Genus *Rasbora* in Sri Lanka, with Description of Two New Species (Teleostei: Cyprinidae). *Ichthyol. Explor. Freshw.* **2011**, *21*, 27– 50.

Sudasinghe, H.; Dahanukar, N.; Raghavan, R.; Senavirathna, T.; Shewale, D. J.; Paingankar, M. S.; Amarasinghe, A.; Pethiyagoda, R.; Rüber, L.; Meegaskumbura, M. Island Colonization by a 'Rheophilic' Fish: The Phylogeography of *Garra ceylonensis* (Teleostei: Cyprinidae) in Sri Lanka. *Biol. J. Linn. Soc.* **2021**, *132*, 872–893.

Sudasinghe, H.; Meegaskumbura, M. *Ompok argestes*, A New Species of Silurid Catfish Endemic to Sri Lanka (Teleostei: Siluridae). *Zootaxa* **2016**, *4158* (2), 261.

Sudasinghe, H.; Pethiyagoda, R.; Maduwage, K.; Meegaskumbura, M. *Mystus nanus*, A New Striped Catfish from Sri Lanka (Teleostei: Bagridae). *Ichthyol. Explor. Freshw.* **2016**, *27* (2), 163–172.

Sudasinghe, H.; Pethiyagod, R.; Maduwage, K.; Meegaskumbura, M. The Identity of the Sri Lankan *Amblypharyngodon* (Teleostei, Cyprinidae). *ZooKeys* **2019b**, *820*, 25–49.

Sudasinghe, H.; Pethiyagoda, R.; Meegaskumbura, M. A Review of the Genus *Esomus* in Sri Lanka (Teleostei: Cyprinidae). *Ichthyol. Explor. Freshw.* **2019a**, *1106*, 1–18.

Sudasinghe, H.; Pethiyagoda, R.; Meegaskumbura, M. Evolution of Sri Lanka's Giant Danios (Teleostei: Cyprinidae: Devario): Teasing Apart Species in a Recent Diversification. *Mol. Phylogenet. Evol.* **2020c**, *149*, 106853

Sudasinghe, H.; Pethiyagoda, R.; Raghavan, R.; Dahanukar, N.; Rüber, L.; Megaskumbura, M. Diversity, Phylogeny and Biogeography of *Systomus* (Teleostei, Cyprinidae) in Sri Lanka. *Zool. Scr.* **2020d**, *49*, 710–731.

Sudasinghe, H.; Ranasinghe, R. H. T.; Goonatilake, S. de A.; Meegaskumbura, M. A Review of the genus *Labeo* (Teleostei: Cyprinidae) in Sri Lanka. *Zootaxa* **2018**, *4486* (3), 201–235.

CHAPTER 29

DIVERSITY OF AMPHIBIANS OF SRI LANKA

PETER JANZEN

Justus-von-Liebig-Schule, Duisburg, Germany

ABSTRACT

The history of the discovery and study of amphibians is presented. The main research started in the early 1990s and is still ongoing. Many new species have been discovered and made available to science. However, our knowledge of the biology and ecology of many species is still unknown. Many taxonomic findings can only be clarified with collaboration with Indian scientists, as many species of the drier areas of Sri Lanka are closely related or identical to species of India. Completely different are amphibian species distributed only in the humid zone of Sri Lanka. They are separated from India by the dry zone and have evolved separately from Indian influence. The endangerment of Sri Lankan amphibians is discussed.

29.1 AMPHIBIANS OF SRI LANKA

The exploration of Sri Lanka's amphibians started with Albertus Seba's description of caecilians: "Serpens caecilia ceylonica." Ferguson made the first checklist of Sri Lankan herpetofauna, including frog species (1877). The checklist included 41 amphibian species. He sent specimens to Günther in London and Günther named the frog *Ixalus* (today *Pseudophilautus*) *fergusonianus* after him. Günther got specimens from other naturalists like Kelaart and Thwaites and described different new species from Sri Lanka. Altogether most species were described by foreign zoologists, often based on specimens that lacked precise locality data. A problem for recent times is difficult to assign a concrete place of origin if the information is missing. Many holotypes are labeled with "Ceylon." The German brothers Paul Benedikt and Karl Friedrich Sarasin studied the reproduction of the caecilian *Ichthyophis glandulosus*. Kirtisinghe wrote the first treatment of the Sri Lankan amphibian fauna (Kirtisinghe, 1957). He described 33 different frogs and two caecilians in the book. The American Edward H. Taylor described different species of caecilians from Sri Lanka, of which two types are still valid today. Until the 1990s it was difficult to give a name to many Sri Lankan frog species, because the scientific knowledge was very poor. This started to change with Pethiyagoda and his group from Wildlife Heritage Trust (WHT)

Biodiversity Hotspot of the Western Ghats and Sri Lanka. T. Pullaiah, PhD (Ed.)

in 1993. They visited many places in Sri Lanka and tried to discover new species, to describe them and to bring order into the existing chaos. That was a mammoth task. In addition, many high-quality books were published by WHT, one of which was about the amphibians of Sri Lanka (Dutta and Manamendra-Arachchi, 1996). The book gave an overview of the 53 species described and already indicated the description of other species. Pethiyagoda's group were so fascinated by the frogs that they wrote an article, which declared that they would have found 200 undescribed frog species and Sri Lanka would have the greatest density of frog species in the world (Pethiyagoda and Manamendra-Arachchi, 1998). Many species would have a distribution less than 0.5 km^2. The most species were to be assigned to the family Rhacophoridae with 200 species. At that time only 17 rhacophorid species were described from Sri Lanka. The species number relativized soon after this. Meegaskumbura et al. (2002) reduced the number to 140 new species. This number of new species has not been described until today. The number of new species was certainly remarkable and not all possibly new species have been described, but the estimates were clearly wrong. In 2005, the breakthrough came with three publications (Bahir et al., 2005; Manamendra-Arachchi and Pethiayagoda, 2005; Meegaskumbura and Manamendra-Arachchi, 2005). A total of 35 new frog species were described, of which 19 were declared as extinct. Intensive searching had not brought a positive result and it seemed that these frogs were extinct. Many frog species, which were described during colonial time, were not labeled with a detailed place of origin. It was not possible to map it exactly. Probably, the most important person in the Pethiyagoda group was Manamendra-Arachchi. He withdrew and changed to the search for fossils. Pethiyagoda himself left Sri Lanka and went to Australia. What remained was Meegaskumbura, who has a high qualification due to his studies abroad and was able to use molecular genetic methods. He has been the center of Sri Lanka's amphibian research for more than 10 years, increasing the number of extinct frogs to 21 and he described existent new species. Meegaskumbura left Sri Lanka and works in China now. Mendis Wikramasinghe also received great attention. He discovered many new species of frogs and reptiles in Sri Lanka and made articles in magazines and books that show his interest in photography. He rediscovered two species of *Pseudophilautus* and he found *A. kandianus* again. *A. dasi* was later synonymized with *A. kandianus*. At least the number of extinct species is reduced from 21 to 18 (Batuwita et al., 2019). Some frog species are found at new places, not known before. An example for this is *Pseudophilautus femoralis*. It seemed to be restricted to high elevational forests in the Central Hills, but later Peabotuwage et al. (2012) found it at an elevation of 700 m a.s.l. in the Peak Wilderness.

29.2 CAECILIANS (ORDER GYMNOPHIONA)

Caecilians are worm-like amphibians without limbs. The skin is smooth and the tail is reduced and only very short. The body is subdivided into annuli. Gas exchange takes place by lungs and through the skin. Caecilians have small eyes, covered by skin, and they can only distinguish between light and dark. They cannot see contours or moving patterns. Between eyes and nostril is a tentacle located. The caecilan can use this tentacle as a touch organ and they can feel vibration of the ground. All three species are endemic to Sri Lanka.

Ichthyophis glutinosus was first described by Linnaeus in 1758 and this is one of the first researched caecilian. The brothers Pauland Fritz Sarasin collected specimens for science in the Kandy area (1887–1890). All these specimens belong to *Ichthyophis glutinosus*, the most common and widespread caecilian in Sri Lanka. They found females guarding their eggs, too. Kirtisinghe described two caecilian species for Sri Lanka: *I. glutinosus* and *I. monochrous* (1957). *I. monochrous* was described by Bleeker in 1858, but the type locality was Sinkawand, Borneo (Indonesia) (Frost, 2020). Taylor described four more species on the base of collected specimens (1965, 1968) and Nussbaum and Gans reduced the number of species to three after intensive collection by Carl Gans (Nussbaum and Gans, 1980). Gans found specimens in compost heaps, farmland, and wet meadows up to a depth of 50 cm. The soil was wet and loose. Actually, the three Sri Lankan species are recognized: *Ichtyhophis glutinosus, I. orthoplicatus*, and *I. pseudangularis.* Gower et al. (2005) mentioned a possibe fourth cryptic species from Welegama in the Southern Part of Sri Lanka. The specimen was striped and in a molecular analysis turned out to be a sister species of the unstriped *I. orthoplicatus.* So far, no further studies to confirm this potentially new species have been carried out. Only *I. orthoplicatus* has no lateral stripe and the size is about 30 cm. The distribution of *I. glutinosus* comprises from the lowland rainforest of the wet zone up to 1355 m a.s.l. of the montain regions of central Sri Lanka. Is was found in Nilgala Fire Savannah in the intermediate zone too and *I. pseudangularis* was found in the wet zone from Akuressa to Rambukkana up to 1525 m a.s.l. This species is classified as vulnerable and it has been found less often as *I. glutinosus*. Some parts of the distribution of both species overlap; they can be sympatric species. *I. orthoplicatus* is also classified as vulnerable and was found in south central parts of Sri Lanka. The genus *Ichthyophis* dispersed from the upland into lowland rainforest regions and radiated into different species (Gower et al., 2005). All Sri Lankan caecilians are rarely seen during expeditions and it is possible that they have a larger distribution throughout the wet zone. Extensive field work is necessary to increase our knowledge. *I. glutiosus* can be found next to human settlement. It cannot be excluded that it is an anthropophilous species. Further studies are required to investigate the behavior and way of life of Sri Lankan caecilans, especially of *I. orthoplicatus* and *I. pseudangularis*. The way of life of species remains unknown. Studies in captivity are necessary to investigate the reproductive processes. And the exact distribution of all species remains unclear.

29.3 FROGS (ORDER ANURA)

29.3.1 TOADS (FAMILY BUFONIDAE)

There are two genera of toads in Sri Lanka: the slender toads *Adenomus* and the "typical" toads *Duttaphrynus*. All these toads belonged to a cosmopolitan genus *Bufo*. The European *Bufo bufo* has the typical shape for a toad. But this genus *Bufo* was absolutely not monophyletic and Frost et al. (2006) changed the taxonomy. This led to current genera for Sri Lanka. *Duttaphrynus* is a widely distributed genus. The species is distributed from Oman and Iran east to China and Indonesia. Five species of *Duttaphrynus* are described for Sri Lanka. The most widespread is *D. melanostictus* with a range from Pakistan to Sumatra and south to Sri Lanka. And it is invasive in Sulawesi, East Timor, Madagascar, and New Guinea. *D. melanostictus* is a species complex and further investigations will divide the complex into different species. *D. melanostictus* is widely distributed in Sri Lanka in all climate zones up to 1700 m a.s.l. It can be found near human habitation, even close to primary rainforest, but not inside these forests. In such forests it is displaced by *D. kotagamai* and *D. noellerti*. Both species are distributed in the wet zone and only in forests. The knowledge of both species is very low. Peabotuwage et al. (2012) provided new distribution records for *D. kotagamai* from Samanala Nature Reserve near Adam's Peak in the wet zone. The forests must have a closed canopy with wet and cooler climate. Both species avoid human habitation. The original distribution of *D. noellerti* and *D. kotagamai* must have been much larger than today. It became more and more restricted with deforestation in the Sri Lankan wet zone. All the disturbed places offered new habitats for *D. melanostictus*. Dutta and Manamendra-Arachchi (1996) mentioned *D. microtympanum* for southern and central Sri Lanka. They described it as an anthropophilous species avoiding primary forests. Certainly, these specimens were confused with *Duttaphrynus melanostictus*. *D. microtympanum* is endemic for southern parts of India. Dutta and Manamendra-Arachchi (1996) described *Bufo fergusonii*, which is treated as a synonym today. The correct name is *Duttaphrynus scaber*, a small toad (up to 36 mm) limited to the dry zone. This toad is distributed in India, too. Bogert and Senanayake (1966) described with *Duttaphrynus atukoralei* a new species from southeastern Sri Lanka (Yala National Park). This toad is "superficially similar" to *D. scaber* (Dutta and Manamendra-Arachchi, 1996). The form of the parotid glands are a possibilty for distinguishing both species. Genetic analysis delivered the result of one species with three clades. One clade is from northern Sri Lanka and it is shared with Indian toads, a second clade in eastern and southeastern parts of Sri Lanka, which resembles *D. atukoralei* and another clade in southern Sri Lanka. The southern clade is distributed in the wet zone. The author's conclusion is that *D. scaber* evolved in India and spread from southern Sri Lanka eastward and then northward to India. Further investigations will clarify if there are two or only one toad species. The second toad genus in Sri Lanka is *Adenomus*. Two endemic species belong to this genus and make it to endemic. *Adenomus* species are more slender toads and they have no supraorbital ridge. *Adenomus kelaartii* (Fig 29.1) is the more widespread species and can be found in forests of the wet zone up to 1200 m a.s.l.. This is an semi-arboral toad and is often found near streams in closed canopy forests. Haas et al. (1997) reported about breeding success in captivity. The scientific history of the second *Adenomus* is somewhat confusing. Günther described a new species from Sri Lanka and named it *A. kandianus* in 1872. It was thought to have been lost for 137 years and expected as extinct. This species was redecribed by Wikramasinghe in 2009 and the species *A. dasi* was identified as a synonym of *A. kandianus*.

29.3.2   *NARROW-MOUTHED FROGS (FAMILIY MICROHYLIDAE)*

Two genera of narrow-mouthed frogs are available in Sri Lanka: *Microhyla* and *Uperodon*. Narrow-mouthed frogs are often terrestrial or fossorial, but some Sri Lankan species can be found climbing trees. Species of genus *Microhyla* are typical for these frogs. In Sri Lanka they live fossorial and/or terrestrial and are tiny in size up to 25 mm. These species are difficult to find during dry season, but they come out directly with first rain for breeding (explosive breeders). According to the latest findings, the 55 described species of the genus *Microhyla* form two monophyletic clades. The origin of the genus *Microhyla* was in South East Asia. Due to multiple connections of the Indian subcontinent during the continental drift in the Paleocene, there were corresponding possibilities of species immigration to the Indian subcontinent. It is an "out-of-Eurasia" colonization. Actually five species are recognized for Sri Lanka. Three are endemic, only *M. ornata* and *M. rubra* can be found in southern India, too. *M. rubra* and *M. ornata* are commonly found in the dry zone of Sri Lanka up to 460 m a.s.l. Both species breed spontanously after heavy rains and lay eggs in a single clutch.

Microhyla mihintalei (Fig 29.16) can be found in the dry zone next to rivers and stream and on coconut plantations (own record). It is slightly smaller than *M. rubra*. The new species is not conspecific with any of the other known species from the Western Ghats of India and Sri Lanka. *M. zeylanica* is quiet similar to *M. ornata*, but this species is distributed from 1900 m a.s.l. upward in the wet zone. *M. karunaratnei* is an endemic species from the famous Morningside region, eastern part of Sinharaja Forest. *M. karunaratnei* lives in leaf litter next to brooks and breeds in small ponds like gem pits year-round. *Microhyla* species from dry zone illustrate the relationship to the Western Ghats. Species from the wet zone have developed in Sri Lanka and have no direct connection with *Microhyla* from India.

Globular frogs, genus *Uperodon*, are the second genus of narrow-mouthed frogs in Sri Lanka. The genus is restricted to the Indian subcontinent. Two species can be found in Sri Lanka and India. The other four species are endemic in Sri Lanka. *Uperodon systoma* is a fossorial species from the dry and the intermediate zone and difficult to find during dry season and breeds spontanously after heavy rains. *U. systoma* is distributed from Pakistan and Nepal to Sri Lanka. The author found a specimen near Negombo in the wet zone. Some frog species from drier parts seem to benefit from human settlement, which even changes climate conditions. The heavily populated areas in western Sri Lanka have quite different climatic condition as before, when these places were covered by tropical lowland forests. Human settlements are much drier and hotter and offer new conditions for frogs like *Uperodon systoma*. *U. systoma* prefers temporary ponds and puddles for reproduction.

Uperodon taprobanicus lives fossorial and sometimes climbs in trees. This frog was found in all three climatic zones up to 460 m a.s.l.. The habitats cover dry forests, plantations, wetlands, and areas close to human habitations. As an anthropophilous species it is found in gardens, too. The eggs are laid in small temporary ponds and form a single-layer film on the surface. The water can become quiet warm and the concentration of oxygen can be less. In this manner the eggs can get oxgen from air (monolayer). It is sympatric with *U. systoma* and *U. rohani*. The latter is a new species, which was considered as conspecific with the Indian *U. variegatus*. *U. rohani* looks quiet similar to *U. variegatus* and both species have only a small genetic distance. *U. rohani* is widely distributed in lower parts of the country, especially in dry and intermediate zone. This species can exist near human settlements in parts of the wet zone. It can easily be found in bath rooms during dry season, even inside toilets. Three endemic species are described from the wet zone.

The Sri Lankan Globular Frog *U. obscurus* is distributed in the central parts of the country from 100–1220 m a.s.l. in low and midcountry wetzone, central hills, Sinharaja Forest, Knuckles Mountains, and Namunukula. It can be found in tree holes and in leaf litter. *U. obscurus* was easily found in gem pits where males call and the eggs form a surface film in Morningside. *U. obscurus* breeds in phytothelmata and in puddles among tree roots too and artificial ground puddles in gardens. This frog can be found in human settlements if the place provides shaded areas, leaf litter, and tree holes. *U. palmatus* is genetically the next Globular Frog to *U. obscurus*. It has a restricted distribution in the central hills, it replaces *U. obscurus* in higher mountain areas from 1400–2000 m a.s.l.. It can be found in leaf litter and on trees up to 5 m. Tadpoles are found in rock pools next to streams and inside small streams. *Uperodon nagaoi* is a tree hole breeding microhylid frog. This frog was found in other forests in south and central Sri Lanka like Kitulgala and Sinharaja and in forests around Hiyare Forest Reservoir near the city of Galle, where members of a nature club try to support breeding with artifical phytotelmata. *U. nagaoi* should be kept for reproduction in captivity to answer these open questions. *U. nagaoi* is genetically closely related to *U. palmatus* and *U. obscurus*.

Researchers found a possible symbiontic behavior between spiders of the genus *Poecilotheria* and frogs of the genus *Uperodon*. *Uperodon nagaoi* seems to live together with *P. ornata* and *P.* cf. *subfusca* in treeholes. And *Uperodon taprobanicus* was found together with *Poecilotheria fasciata* in Putalam area inside a tree hole. The spiders did not attack the frogs, even when disturbed. Such behavior is described for other frog species, which live in symbiosis with ants or scorpions. Further investigation is necessary to ensure these observations.

29.3.3 ROBUST FROGS (FAMILY NYCTIBATRACHIDAE)

The Sri Lankan Wart Frog *Lankanectes corrugatus* has a long history of name changes. The first name was *Rana corrugata*, Later, it changed to *Dicroglossus* and to *Limnonectes* like many other frogs in Asia. At least the species was placed into the genus *Lankanectes corrugatus* under a new subfamily of frogs the Lancanectinae. Our knowledge about this interesting species is low. Only a very few specimens are available in museums outside Sri Lanka. These conclusions were confirmed by Delorme et al. (2004). They supported the new genus and placed *L. corrugatus* near to *Nyctibatrachus*. And the correctness of the subfamily Lankanectinae was confirmed: "The morphological specializations of this species confirm that it may be the only known representative of an additional major ranid lineage (Lankanectinae)…" (Delorme et al., 2004). Roelants et al. (2004) placed *Lankanectes* as the sister taxon of Nyctibatrachinae and Frost et al. (2006) integrated *Lankanectes* into the family Nayctibatrachidae. The family is endemic for the Indian subcontinent. *L. corrugatus* is an aquatic species found up to 1525 m a.s.l. in western, central, and southern Sri Lanka (Dutta and Manamendra-Arachchi, 1996). It can be found in shallow streams and puddles at shaded places like forests and even in caves (own observation). *L. corrugatus* was many years considered of beeing a monotypic species of an endemic subfamily in Sri Lanka. A second species from the Knuckles Mountains was described. *L. pera* has a restricted distribution and prefers slow flowing streams in primary forests. This makes it to a rare species. *L. pera* is genetically different from *L. corrugatus*. The male specimens are larger (66 and 68.7 mm) than female specimens (42.4–55.8 mm). Female frogs are mostly larger than male, but it is the opposite with some specialized species like *Pyxicephalus adspersus* from Africa. *L. pera* has a large tadpole (up to 45 mm), which lives in clear and clean streams.

29.3.4 FORK-TONGUED FROGS (FAMILY DICROGLOSSIDAE)

Frogs of the family Dicroglossidae were previously counted as a member of Ranidae. Twelve species are recognized for Sri Lanka, belonging to a minimum of five genera. The largest frog in Sri Lanka is *Hoplobatrachus crassus*, females can reach 13 cm. Size makes this species distinctive. *H. crassus* is a semi-aquatic frog and prefers reservoirs, marshlands, rivers, and larger pond in the wet and the dry zone up to 500 m asl. It can be found near human settlements, too. It avoids dense forests. A frog species benefits from human activities and human dispersal over Sri Lanka. Sri Lankas do not eat frogs, but in other Asian countries, *Hoplobatrachus* species can regularly be found on food markets. It is not endemic; it can also be found in India, Bangladesh, and possibly Myanmar (Frost, 2020). Tadpoles of *H. crassus* are aggressive against other tadpoles and kill and ingest these (own observation). As with species of the genus *Minervarya*, *H. crassus* can have a midline on the back. *Euphlyctis hexadactylus* (Fig 29.5) is nearly as large as *H. crassus*. *E. hexadactylus* is entirely aquatic, it even rests at the surface, not on land. It can be seen among floating plants of leaves of water lilies and lotus. *E. hexadactylus* appears on land during the night for foraging. It is distributed in all climatic zones up to 800 m a.s.l. Outside Sri Lanka this frog can be found in India and Bangladesh. Second species of *Euphlyctis* is *E. cyanophlyctis*, one of the most common frogs in Sri Lanka. The name "skipper frog" is related to the escape behavior. The skipper frog can easily be seen at the water's edge. Frogs wait there for food and when disturbed they jump over the water surface like a thrown stone. The size is less than *E. hexadactylus*, with up to 6 cm. *E. cyanophlyctis* is bound to water and prefers standing water bodies. It can be found in all climatic zones up to 1700 m a.s.l., but prefers lower altitudes and it avoids closed canopy places. *E. cyanophlyctis* can be seen in puddles on main roads in Sinharaja Forest near human settlements, but not inside the forest (Kudawa, own observation). The skipper frog is active during day and night and breeding is linked to the rainy season, especially in the dry zone. *E. cyanophlyctis* is a widespread species from Iran to Myanmar and south to Sri Lanka (Frost, 2020) and it builds a species complex. A genetic analysis of different populations from Iran, India, Bangladesh, and Sri Lanka showed that the populations from Sri Lanka form a monophyletic group with the relatively newly described *Euphlyctis mudigere* from southern India. *E. mudigere* is actually a synonym of *E. cyanophlyctis*. Further investigations should be done on both species of *Euphlyctis* in Sri Lanka to clarify their real status. These investigations should be realist in cooperation with Indian herpetologists. A bit complicated is the genus *Minervarya*. The name for frogs of the genus changed several times during the last 20 years, from *Rana* via *Limnonectes* to *Fejervarya* and *Minervarya*. At least three Sri Lankan species belong to

Minervarya. *Minervarya greeni* was originally distributed over most parts of the wet zone up to high altitudes until Manamendra-Arachchi and Gabadage (1996) recognized that it is a complex of two species. The new species, *M. kirtisinghei*, has a greater distribution and can be found in western, central, and southern wet zone and the Knuckles Mountains, too, in altitudes between 150 and 1600 m a.s.l.. Both species are quite similar in morphology and can best differentiated by the distribution areas. *M. greenii* is restricted to higher parts of the central mountains (1700–2100 m a.s.l.) and inhabits pools with vegetation. This montane frog can be seen next to cities, especially in botanical gardens like Hakgala or Nuwara Eliya town. *M. kirtisinghei* is more common and prefers forests, drains near human habitats. The third species *M. limnocharis*. It is a split-off from the species complex *Fejervarya limnocharis*. There have been attempts in the past to clarify the true identity of the Sri Lankan population. The name changed from *Fejervarya* to *Zakerana syhadrensis,* which was later identified as a synonym. Today, the Sri Lanka's specimens belong to *M. agricola*, a specis found in India, too. It is a widespread species in most Indian parts. The Sri Lankan population is widespread all over the country in all climatic zones up to 1400 m a.s.l. It prefers the proximity to water even in human settlements.

Spaherotheca is another genus belonging to Dicroglossidae in Sri Lanka. Two species are proven for Sri Lanka. For long time these species were *S. breviceps* and *S. rolandae*. Dahanukar et al. identified S. *breviceps* as an Indian species (2017), which does not occur in Sri Lanka. *S. rolandae* is distributed in India and Sri Lanka and *S. pluvialis* is now the name for the second Sri Lankan species. *S. pluvialis* is also distributed in southern India. The problem is that we cannot just exchange the names *S. breviceps* against *S. pluvialis*. A new survey of the Sri Lankan *Sphaerotheca* has to be done to confirm the findings of Dahanukar et al. (2017) and the distribution of both species must be investigated. The distribution areas will range over the dry zone. This also includes gardens and agrarian landscapes.

The only endemic genus of Dicroglossidae is *Nannophrys*. These are specialized frogs, which live near flowing waters. Unlike other species in the family, they avoid the dry zone. *N. naeyakai* occurs in the intermediate zone and the other three species live in the wet zone. The wet zone species live next to fast-flowing streams. They hide under boulders during daytime and are active during darkness. *Nannophrys* lay eggs under stones above water level. The tadpoles have a thick round tail, with which they can make small jumps. Tadpole are semiterrestrial and live on rocks in splash water zone and feed on algae from the stone surface. This behavior is quite similar to frogs of the genera *Petropedetes* from Africa, *Indirana* from India, and *Thoropa* from Brazil. *Nannophrys* are not closely related to *Indirana*. Their adaptations to a comparable habitat lead to convergent developments. The sister genus of *Nannophrys* is *Euphlyctis* and the most widespread species is *N. ceylonensis*. This frog prefers fast flowing streams. It is found in the hills of southern, western, and central Sri Lanka from 60–1070 m a.s.l. on rocks in cascades and in wet boulders in lowland and submontane forests. Distribution is discontinuous throughout the wet zone and depends on the availability of streams in forested regions. Eggs and tadpoles are most abundant during the rainy season. They feed on algae, but change to carnivorous food with later developmental stage and changed gut length. Another species is *N. marmorata*. It is mostly restricted to the rock streams of the Knuckles Mountains

(200–1200 m a.s.l.). During daytime, especially in the dry season, the frogs avoid exposure to adverse environmental conditions. *N. marmorata* is difficult to find during the dry season, With onset of rain breeding starts and the frogs lay eggs. *N. naeyakei* is the third species, found at two places in eastern Sri Lanka in the intermediate zone: Kogala (Ampara District) and Yakunattela (Moneragala District, Milgala Fire Savanah) in 200–620 m a.s.l.. Goonewardene et al. (2003) reported about a *Nannophrys* from Nilgala before. *N. naeyakai* is morphologically very similar to *N. marmorata*. *N. naeyakei* lives in seasonal streams with rock crevices. The streams were shallow and formed rock pools. During the wet season the habitat was covered by grasses. With the beginning of the dry season the streams dried out. Bush fires happen frequently during this period. *N. naeyakei* is nocturnal and hides during daytime. The last member of the genus is *N. guentheri*. The status of this species is controversial. It is mostly declared extinct and attempts to prove the existence of this frog failed. The frog was described by Boulenger in 1882 and it is known only from the imprecise type locality of "Ceylon" (Frost, 2020). Bopage et al. (2011) presumed that the description of *N. guentheri* based on misidentification, it cannot be decided here who is right. *N. naeyakei* is endangered because of new human settlements. In the eastern parts the people enter more and more forest and convert it into paddy fields like it is going on in Nilgala Fire Savannah. *N. marmoratus* is rated as critically endangered and there are a few remaining populations left. Tea plantations and forests underplanted with cardamom are widespread in the distribution areas and the use of pesticides is problematic. And local tourists imperil the places with rubbish after stopover at Pitawana Patana. The situation of *N. ceylonensis* is better, because this frog is more widespread in the wet zone and can be found in sanctuaries, too.

29.3.5   *TRUE FROGS (FAMILY RANIDAE)*

The family Ranidae has made fundamental changes over the last two decades. The complete familiy Dicroglossidae formerly belonged to Ranidae. There are three ranid species in Sri Lanka and these changed some of their names, too. First all three species belonged to the genus *Hylarana* until Biju et al. surveyed the *Hylarana* species from the Indian subcontinent and formed four groups on the base of genetic analysis (2014). One group is endemic to Sri Lanka: *Hylarana temporalis* and *H. serendipi*, a new species. *H. gracilis* belongs to another group together with H. *malabaricus*. The name *H. aurantiaca* has exclusively been used for an Indian species from this time on. *H. aurantiaca* is endemic in the Western Ghats. This led to a lower number of endemic species between Western Ghats and Sri Lanka (Biju et al., 2014). Oliver et al. (2015) arranged *Hylarana* species by genetical data again and found that Asian species form a paraphyletic group. For Sri Lanka they placed *Hylarana gracilis* into genus *Hydrophylax*. *Hydrophylax* covers species from Bangladesh, India, Thailand, and Myanmar. *H. gracilis* is widely distributed in dry, intermediate, and wet zone up to 500 m a.s.l.. *Indosylvirana temporalis* and *I. serendipi* are the new names for Sri Lankan *Hylarana temporalis* and *H. serendipi*. *I. temporalis* is a common Sri Lankan species. It was long time misidentified with a species, endemic for the Western Ghats (Biju et al., 2014). *I. temporalis* is restricted to Sri Lanka and can easily be found in forests along streams up to 1800 m a.s.l. *I. serendipi* is endemic for Sri Lanka. At least all three species are endemic for Sri Lanka and belong to the same genera like species from Western Ghats.

29.3.6   FLYING FROGS

Rhacophoridae is a widespread family (Africa and large parts of Asia) and most of Sri Lanka's frog species are among them. All Sri Lankan species belong to the genera *Polypedates*, *Taruga*, and *Pseudophilautus* (Fig 29.2). The genera *Rhacophorus* and *Theloderma* do not apply to Sri Lanka, since the species have been sorted into the three genera mentioned. *Theloderma* does not occur in Western Ghats, but there are some species of *Rhacophorus* in the Western Ghats. The genus *Polypedates* is common in both regions. Three species are named for Sri Lanka. *P. maculatus* is widespread below 3000 m from Nepal, Bhutan, and Sikkim south throughout India, except the northeastern states and Haryana, Rajasthan, Punjab, Bangladesh, and Sri Lanka. The height limit of the distribution is 500 m in Sri Lanka. *P. maculatus* can be found in all climatic zones often in available habitats near human habitation, especially bath rooms. Females are larger than males (70/45 mm). *P. maculatus* is quiet similar to the widespread *P. leucomystax*, which does not occur in Sri Lanka. The eggs are laid in foam nests hanging on leafs, rocks, or artifical constructions above stagnant water. A foam nest contains 210 to 448 eggs and metamorphosis is completed after 50–70 days. The foam protects against solar radiation and predators. Another widespread species is *Polypedates cruciger* (Fig 29.18) the Common Hour-glass Treefrog. It can be found in dry and wet zones up to 1525 m a.s.l., especially near artificial gardens. *P. cruciger* can be found in edges of rainforsts, too (own observation in Morningside). *Polypedates cruciger* was described from Western Ghats. Later these frogs are named a seperate species *Polypedates pseudocruciger*. *P. cruciger* is endemic for Sri Lanka. The females are a bit larger (90 mm) than *P. maculatus*. *P. cruciger* also accepts frogs as food, even larger frogs like *Taruga fastigo* (own observation). Females lay up to 450 eggs in foam nests like *P. maculatus*. A female laid only 100 to 120 eggs in captivity (Herrmann 1993). *P. cruciger* is not picky, when selecting breeding locations. They often use artifical ponds and lay egg on the underside of leaves or on walls. *P. cruciger* tadpoles have a predator-induced plasticity at the time of leaving the foam net. The tadpoles left the foam net for escaping before their development is complete (own observation). It is a behavior well known from *Agalyhnis callidryas*, a treefrog from Central America. *Polypedates ranwellai* is with 40 mm (male) and 65 mm (female) the smallest species of the genus in Sri Lanka and it is only known from a very few places. These places are characterized by high rates of deforestation and anthropogenic activities, which threaten the survival of the species. During the rainy season the males congregate near temporary water holes for breeding. During dry periods it is dificult to find these frogs. Breeding behavior was only observed during night The species is not accessed regarding threat status, but it seems to be a rare species. Threats come from clearing and conversion of forests into human settlements and leisure activities and littering landscape. The genus *Taruga* is endemic to Sri Lanka. There are a few differences in morphology like the calcar at the distal end of the tibia, which is absent in Sri Lankan *Polypedates*. *Taruga* is the sister genus of *Polypedates*. Tadpoles of *T. eques* and *P. cruciger* differ markedly in external morphology. *Taruga* forms an isolated group of frogs. The distribution of all *Taruga* species is restricted to parts of the wet zone, whereas *Polypedates* are mostly widespread in Sri Lanka, even more in the dry zone. *Taruga eques* and *T. fastigo* are restricted to submontane and montane regions. *T. eques* occurs in the Central and the Knuckles Mountains (1202–2135 m a.s.l.) and *T. fastigo* is restricted to the Rakwana Mountains (Morningside, eastern Sinharaja Forest). *Taruga longinasus* is found in the lowland (Kanneliya Forest) and mid-hill regions up to an elevation of about 600 m. *P. longinasus* was originally widespread in the wet zone. Due to the environmental changes caused by human activities and the loss of most rainforests, the distribution area has been significantly reduced. Manamendra-Arachchi and Pethiyagoda described *Taruga fastigo* (Fig 29.13) as a new species and separated eastern populations from *T. longinasus* (2001). *T. fastigo* is endemic for eastern Sinharaja Forest (Morningside). It looks quiet similar to *T. longinasus*. Old foam nests of *T. fastigo* are intensive blue (own observation). All three species of *Taruga* have a hourglass-shaped marking on the back. This can even be seen on older tadpoles.

Pseudophilautus is the most difficult group of frogs in Sri Lanka. Until the book by Dutta and Manamendra-Arachchi (1996) was published, it was difficult to identify any species of the genus in Sri Lanka. The first significant step to improve the situation were the new descriptions by the Pethiyagoda group (Bahir, 2005; Manamendra-Arachchi and Pethiyagoda, 2005; Meegaskumbura and Manamendra-Arachchi, 2005). They named 35 new species 17 of them were considered extinct. Meegaskumbura et al. (2007) described two more extinct species: *P. pardus* and *P. maia*. Both species are known from collections done before 1876. Günther described an egg clutch on the belly of the holotypus of *P. maia* (1876). Meegaskumbura et al. (2007) assumed that the female wanted to press the eggs onto a substratum (leaves) and at that moment, it was killed and prepared. Such behavior is known and described for *P. femoralis*. Three *Pseudophilautus* were redicovered later. *P. stellatus* was found near Adam's Peak in Peak Wilderness Sanctuary and *P. semiruber* was dicovered in the Central Mountains. Some more new species were discovered from Sri Lanka. Menids Wickramasinghe was very efficient in identifying new species. He found eight new species from Peak Wilderness (Wickramasinghe et al., 2013b) and one more from central Sri Lanka between Kandy and Nuwara Eliya (Wickramasinghe et al., 2015). This frog got its name from the tea brand Dilmah: *Pseudophilautus dilmah*. This species is endangered due to the high rate of deforestation and anthropogenic activities. Meegaskumbura et al. (2009) described with *P. tanu* and *P. singu* two new species from southwestern Sri Lanka and later Meegaskumbura et al. (2011) described another two species: *Pseudophilautus schneideri* from Sinharaja Forest (Kudawa) and *P. hankeni* from Knuckles mountains. Batuwita et al. (2019) found *P. conniffae* in southern Sri Lanka (Galle and Matara districts). Altogether 77 species of *Pseudophilautus* have been described from Sri Lanka, 17 of them are considered extinct. All *Pseudophilautus* had a different genus name before. Dutta and Manamendra-Arachchi (1996) used the name *Theloderma schmarda* for *P. schmarda*, *Rhacophorus* for some species like *P. microtympanum*, and *Philautus* for other species like *P. temporalis*. Bossuyt and Dubois (2001) placed the former Sri Lankan *Rhacophorus* into genus *Philautus*. From this time there are no more *Rhacophorus* in Sri Lanka. *Rhacophorus* are distributed from Western Ghats to Indonesia and Philippines. The genus name *Theloderma* also no longer belongs to Sri Lanka. *Theloderma schmarda* is *Pseudophilautus schmarda*. The genus *Theloderma* is distributed from Northeast India to Borneo. At least it is not a genus found in the Western Ghats-Sri Lanka-hotspot. Bossuyt and Dubois (2001) treated the name *Pseudophilautus* as a synonym of *Philautus*. Li et al. (2009) resurrected the genus name *Pseudophilautus* and Yu et al. (2010) presented molecular evidence that *Pseudophilautus* extends from Sri Lanka to the Western Ghats and Pyron et al. (2011) confirmed the monophyly and *Pseudophilautus* as sister taxon to *Raorchestes*. Li et al. (2013) represented a Bayesian tree showing *Pseudophilautus* as sister taxon of *Raorchestes*, including the genus *Kurixalus*. Meegaskumbura et al. (2018) discussed phylogenetics and biogeography of *Pseudophilautus* and suggested that *Raorchestes* (direct development) is likely the sister taxon of *Mercurana* and together the sister taxon of *Pseudophilautus* (direct development). Meegaskumbura et al. (2015a) found the same results, *Pseudophilautus* as sister taxon of *Raorchestes*. They arranged the Sri Lankan *Pseudophilautus* into six clades. One clade contains mountainous species from the central mountain above 800 m a.s.l.. From there they spread into the Knuckles Mountains and the Rakwana Mountains. *P. femoralis* would be the basic species for this radiation. For example, other species of this clade are *P. poppiae* from Rakwana Mountains and *P. mooreorum* from Knuckles Mountains. Both are green-colored species like *P. femoralis*. Meegaskumbura et al. (2018) tried to group more species with these clades, based on genetic comparison. The main event was the arrival of the first ancestor from India by land bridge in the Oligocene. The climate was mild (20°C–25°C) and dry. This stopped with higher sea level and divided Indian populations from Sri Lanka. In Miocene there was a back-migration from Sri Lanka to India. During this period the climate was warm and dry. This stopped again with the separation of Sri Lanka from India. Today, the north of Sri Lanka is hot and dry and contains *Pseudophilautus regius* (Fig 29.7). The southern tip of India has the same climate and the occuring species there is *P. kani*. These should be the descendants of the ancestor species. A main reason for the radiation of the genus in the Central Mountains are the altitudinal belts with different climates. Range-restricted species developed in dependence of the altitude. The source of dispersal among mountains was species adapted to warmer and drier conditions. The mountains operated as species pumps with their diverse habitats. As far as it is known, species of the genus *Pseudophilautus* are direct developers. Meegaskumbura et al. (2015a) researched the reproduction of rhacophorid frogs. Direct developing means without free larval phase. The frogs lay eggs outside water and the development completes within the egg. A froglet will emerge from the egg. It is the same in the sister genus *Raorchestes*. The ancestral basis of reproduction in the family Rhacophoridae was aquatic breeding (e.g., *Buergeria*). The following breeding form was gel nesting. Today, this mode is available in *Theloderma* frogs from southeast Asia. Foam nesting evolved twice, once by ancestors of the genus *Chiromantis* and a second time in the group containing *Polypedates* and *Taruga*. Direct development evolved also twice. One is ancestor of *Philautus* and the second is the ancestor of *Pseudophilautus* and *Raorchestes*. This is another difference between *Philautus* and *Pseudophilautus*. Direct development has evolved from foam nesting species. *Raorchestes* stick their eggs on the upper surface of leaves (Biju, 2003), whereas a few *Pseudophilautus* attach the eggs on the undersides of leaves (Bahir et al., 2005). *P. femoralis* is an example for this. The first description of an unusal breeding behavior of Sri Lankan frogs came from Günther (1876). He gave a description of a female *Pseudophilautus microtympanum*, which laid 20 eggs into soil and guarded the clutch. The eggs were 5–6 mm large. Kirtisinghe (1946) observed the complete development without a free tadpole. Froglets emerged from the eggs. More research was done by Bahir et al. (2005). They bred *Pseudophilautus* at Agra-Bopath Forest Reserve (650 m a.s.l.) in the Central Mountains. The frogs were kept in artificial terraria without artificial light. The food was enhanced by a vitamin powder. Sixteen species were identified as ground nestors, the female digged a cavity, laid eggs in the soil, and closed the cavity again. These species use this mode of reproduction: *Pseudophilautus alto, P. asankai, P. caeruleus, P. femoralis, P. frankenbergi, P. microtympanum, P. schmarda, P. sarasinorum, P. silus, P. sordidus, P. viridis, P. caeruleus, Philautus hallidayi, P. rus, P. zorro,* and *P. decoris.* They laid 6 to 155 eggs per clutch into soil. *Pseudophilautus femoralis* differs from the other species. *P. femoralis* is an arboreal frog and adheres 7 to 22 eggs on the underside of leaves. Development is completed after 37–49 days and brown froglets emerge from the eggs. All species did not show parental care. Bahir et al. (2005) wanted to develop breeding techniques for possible breeding of endangered species. It is believed that *P. poppiae* also adhere her greenish eggs on the underside of leaves. That is a wrong assumption. *P. poppiae* adheres the eggs on the upperside of leaves (Fig 29.11) like *Raorchestes.* Both species *P. femoralis* and *P. poppiae* are restricted to mountainous places, *P. femoralis* from Central Mountains and *P. poppiae* from Rakwana Mountains (Morningside). Both species are green and arboreal, and they are evolutionary close related to each other and to *P. mooreorum* from Knuckles Mountains (Meegaskumbura et al., 2018). It would be very interesting to know if *P. mooreorum* places the egg clutch on the upper or at the underside of the leaf. Karunarathna and Amarasinghe (2007, 2010) reported of an in-situ breeding of *P. regius.* The female digged one cavity and another narrow hole next to the first. The female laid 17 eggs into the narrow hole and closed it. If the two cavities are accidental or not remains unclear. *P. regius* is a species from dry and intermediate zone and can be found in home gardens and bathrooms (Dissanayake and Wellipuli-Arachchi, 2012). Only a very few *Pseudophilautus* species are distributed in the dry and intermediate zone. The wet zone is the preferred climate zone. The distribution in detail remains unclear for some species of the genus. Some are point endemics (very restricted area) like *P. simba* only known from the terra typica in Morningside at 1080 m a.s.l.. Other species have a much larger distribution. For example, *P. popularis,* a frog found in the wet zone up to 1000 m a.s.l. and often at edges of forests, can be found in cities like Kandy. *P. fergusonianus* is another example. It was found in Nilgala in the intermediate zone (Karunarathna et al., 2008), in Knuckles Mountains (Samawickrama et al., 2012), Kandy area, and from Morningside in the Rakwana Mountains. In Morningside the species prefers bathrooms and toilets during daytime. From other species range extensions are reported: *P. cavirostris* (Kandamby and Batuwita, 2001) and *P. sarasinorum* (Peabotuwage et al., 2012). Closer investigations will lead to changes in opinion. *P. cavirostris* and *P. reticulatus* were considered to be residents of closed primary forests, but could also be found in secondary forests (Kumara and Ukuwela, 2009). Some species are very similar to each other, making differenciation difficult in the field, especially when cryptic species are involved. Ellepola et al. (2021) hold out the prospect of 14 previously unidentified new species. These will probably only be identifiable with the help of the exact origin or molecular biological methods. The authors were able to divide the described species of the genus *Pseudophilautus* in Sri Lanka into six clades by molecular biology. There is also the likelihood that some species will be synonymized.

There are also variable species that can be colored or marked differently like *P. asankai*, *P. hoipolloi*, and *P. zorro* (Karunaratne, 2001). So far we only know the type of reproduction from some of the *Pseudophilautus* species; it is assumed that all of them have direct development. That is quiet the same as with *Eleutherodactylus* in the Noetropis. But the exinct species *E. jasperi* gave birth to live young animals (Wake, 1978). At least we have to research the breeding behavior of all *Pseudophilautus* species.

29.4   THREATS TO SRI LANKAN AMPHIBIANS

There are different threats that endanger the continued existence of the amphibians in Sri Lanka. The main factor is loss of habitat and fragmentation of habitats due to population growth (Cincotta et al., 2000). The Western Ghats Sri Lanka hotpot has high population density and the number of people is still growing. In search for food production and possibility of building homes, more and more land will be converted and animals will be brought to the edge of existance. The prominent example is the ongoing human–elephant conflict, but it affects amphibians as well. And "many people in this region treat frogs with almost the same revulsion as snake, despite their ecological importance" (Manamendra-Arachchi, 2000). One good reason for conservation is the feeding of *Aedes* eggs (Senavirathne et al., 2014). The tadpoles of different frog species feed on eggs of Mosquitos, that transmit Dengue, a problematic infection for the country (Tam et al., 2013). In the future climate change will cause problems, especially in mountain regions. The importance for cloud forests should not be underestimated (Werner, 2001). When the temperature rises the animals have to climb up hills in search for new biotope. The weather changed in Nuwara Eilya during the period from 1870 to 2000. The mean temperature increased by 1 degree celcius and the mean rainfall decreased significantly (Schäfer, 1998). This threatens mountainous and cloud forests around Nuwara Eliya, for example Hakgala, Horton Plains, and Pidurutalagala. Introduced and especially invasive animals and plants can be a problem for amphibians, but this is not a main factor in Sri Lanka. Because people in Sri Lanka do not eat frogs, there was no breeding facility with the American Bullfrog *Lithobates catesbeianus*, which is blamed for spreading deadly variants of *Batrachochytrium dendrobatidis* and Ranavirus (Laufer, 2018). Pollution is much more important for Sri Lanka. People throw their garbage into waters, agrochemicals spill into rivers, lakes, and ponds. These are problems for amphibians and for the people as well. Agrochemicals used in tea plantations can be transported by wind into cloud forests, where many species of *Pseudophilautus* live and some of them still await discovery. A much-discussed problem would be *Batrachochytrium dendrobatidis*, a fungus that kills amphibians in different regions around the world. The fungus is verified for Sri Lanka, but no problem has been recorded yet (Rahman et al., 2020). The animal trade and smuggling are often cited as a reason for endangering Sri Lanka's reptiles (Janssen and de Silva, 2019). This is completely meaningless for amphibians. Trade of amphibians exists and also keeping of amphibians in terrariums, especially in Japan, Europe, and North America. This affects almost exclusively poison dart frogs that are already being bred in high numbers in captivity. There is no market for Sri Lankan amphibians and smuggling would not be worthwhile. de Silva (2009) applied for captive breeding and genetic research of endangered or critically endangered amphibian species and Molur (2008) believes that frogs of the genus *Pseudophilautus* are particularly suitable for breeding because of their direct development. Sri Lankan people are not used to eat frogs, but Chinese do it. It can be assumed that local people deliver Chinese in Hambantota harbor and Chinese workers of the Lakvijaya Power Station, Kalpitiya, eat frogs. It is reported that local persons sell snakes to them. That would mainly affect *Hoplobatrachus crassus* and *Euphlyctis cyanophlyctis* (Fig 29.15), two species that are not endangered. It is very important that knowledge about the biology of individual species will increase and not just an inventory of species is set up. This is important for breeding species in ex-situ programmes (Molur, 2008).

TABLE 29.1 Checklist of Amphibians.

Scientific name	IUCN Red List status	Family	Distribution status
Ichthyophis glutinosus	Least concern	Ichthyophidae	Endemic
Ichthyophis orthoplicatus	Vulnerable	Ichthyophidae	Endemic
Ichthyophis pseudangularis	Vulnerable	Ichthyophidae	Endemic
Adenomus kandianus	Critically endangered	Bufonidae	Endemic
Adenomus kelaartii	Endangered	Bufonidae	Endemic
Duttaphrynus atukoralei	Least concern	Bufonidae	Endemic
Duttaphrynus kotagamei	Endangered	Bufonidae	Endemic
Duttaphrynus melanostictus	Least concern	Bufonidae	
Duttaphrynus noellerti	Endangered	Bufonidae	Endemic
Duttaphrynus scaber	Least concern	Bufonidae	
Euphlyctis cyanophlyctis	Least concern	Dicroglossidae	
Euphlyctis hexadactylus	Least concern	Dicroglossidae	
Fejervrya limnocharis	Least concern	Dicroglossidae	
Hoplobatrachus crassus	Least concern	Dicroglossidae	
Minervarya greeni	Endangered	Dicroglossidae	Endemic
Minervarya kirtisinghei	Least concern	Dicroglossidae	Endemic
Nannophrys ceylonensis (Fig 29.3, 29.4)	Least concern	Dicroglossidae	Endemic
Nannophrys guentheri	Extinct	Dicroglossidae	Endemic
Nannophrys marmorata	Critically endangered	Dicroglossidae	Endemic
Nannophrys naevakai	Endangered	Dicroglossidae	Endemic
Sphaerotheca pluvialis	Not assessed	Dicroglossidae	
Sphaerotheca rolandae	Least concern	Dicroglossidae	
Hydrophylax gracilis	Least concern	Ranidae	Endemic

TABLE 29.1 *(Continued)*

Scientific name	IUCN Red List status	Family	Distribution status
Indosylvirana serendipi	Not assessed	Ranidae	Endemic
Indosylvirana temporalis	Near threatened	Ranidae	Endemic
Lankanectes corrugatus	Least concern	Nyctibatrachidae	Endemic
Lankanectes pera	Not assessed	Nyctibatrachidae	Endemic
Microhyla karunaratnei	Critically endangered	Microhylidae	Endemic
Microhyla mihintalei	Not assessed	Microhylidae	Endemic
Microhyla ornate	Least concern	Microhylidae	
Microhyla rubra	Least concern	Microhylidae	
Microhyla zeylanica	Endangered	Microhylidae	Endemic
Uperodon nagaoi	Vulnerable	Microhylidae	Endemic
Uperodon obscurus	Near threatened	Microhylidae	Endemic
Uperodon palmatus	Endangered	Microhylidae	Endemic
Uperodon rohani (Fig 29.8)	Not assessed	Microhylidae	Endemic
Uperodon systoma	Least concern	Microhylidae	
Uperodon taprobanicus	Least concern	Microhylidae	
Polypedates cruciger	Least concern	Rhacophoridae	Endemic
Polpedates maculatus	Least concern	Rhacophoridae	
Polypedates ranwellai	Not assessed	Rhacophoridae	Endemic
Pseudophilautus abundus	Least concern	Rhacophoridae	Endemic
Pseudophilautus adspersus	Extinct	Rhacophoridae	Endemic
Pseudophilautus alto	Endangered	Rhacophoridae	Endemic
Pseudophilautus asankai	Endangered	Rhacophoridae	Endemic
Pseudophilautus auratus	Endangered	Rhacophoridae	Endemic
Pseudophilautus bambaradeniyai	Not assessed	Rhacophoridae	Endemic
Pseudophilautus caeruleus	Endangered	Rhacophoridae	Endemic
Pseudophilautus cavirostris (Fig 29.14)	Endangered	Rhacophoridae	Endemic
Pseudophilautus conniffae	Not assessed	Rhacophoridae	Endemic
Pseudophilautus cuspis	Endangered	Rhacophoridae	Endemic
Pseudophilautus dayawansai	Not assessed	Rhacophoridae	Endemic
Pseudophilautus decoris	Critically endangered	Rhacophoridae	Endemic
Pseudophilautus dilmah	Not assessed	Rhacophoridae	Endemic
Pseudophilautus dimbullae	Extinct	Rhacophoridae	Endemic
Pseudophilautus eximius	Extinct	Rhacophoridae	Endemic
Pseudophilautus extirpo	Extinct	Rhacophoridae	Endemic
Pseudophilautus femoralis	Endangered	Rhacophoridae	Endemic
Pseudophilautus fergusonianus (Fig 29.10)	Least concern	Rhacophoridae	Endemic
Pseudophilautus folicola	Endangered	Rhacophoridae	Endemic

TABLE 29.1 *(Continued)*

Scientific name	IUCN Red List status	Family	Distribution status
Pseudophilautus frankenbergi	Endangered	Rhacophoridae	Endemic
Pseudophilautus fulvus	Endangered	Rhacophoridae	Endemic
Pseudophilautus hallidayi	Vulnerable	Rhacophoridae	Endemic
Pseudophilautus halyi	Extinct	Rhacophoridae	Endemic
Pseudophilautus hankeni	Not assessed	Rhacophoridae	Endemic
Pseudophilautus hoffmani	Endangered	Rhacophoridae	Endemic
Pseudophilautus hoipolloi	Least concern	Rhacophoridae	Endemic
Pseudophilautus hypomelas	Critically endangered	Rhacophoridae	Endemic
Pseudophilautus jagathgunawardanai	Not assessed	Rhacophoridae	Endemic
Pseudophilautus karunarathnai	Not assessed	Rhacophoridae	Endemic
Pseudophilautus leucorhinus	Extinct	Rhacophoridae	Endemic
Pseudophilautus limbus	Critically endangered	Rhacophoridae	Endemic
Pseudophilautus lunatus	Critically endangered	Rhacophoridae	Endemic
Pseudophilautus macropus	Critically endangered	Rhacophoridae	Endemic
Pseudophilautus maia	Extinct	Rhacophoridae	Endemic
Pseudophilautus malcolmsmithi	Extinct	Rhacophoridae	Endemic
Pseudophilautus microtympanum	Endangered	Rhacophoridae	Endemic
Pseudophilautus mittermeieri	Endangered	Rhacophoridae	Endemic
Pseudophilautus mooreorum	Endangered	Rhacophoridae	Endemic
Pseudophilautus nanus	Extinct	Rhacophoridae	Endemic
Pseudophilautus nasutus	Extinct	Rhacophoridae	Endemic
Pseudophilautus nemus	Critically endangered	Rhacophoridae	Endemic
Pseudophilautus newtonjayawardanei	Not assessed	Rhacophoridae	Endemic
Pseudophilautus ocularis	Critically endangered	Rhacophoridae	Endemic
Pseudophilautus oxyrhynchus	Extinct	Rhacophoridae	Endemic
Pseudophilautus papillosus	Critically endangered	Rhacophoridae	Endemic
Pseudophilautus pardus	Extinct	Rhacophoridae	Endemic
Pseudophilautus pleurotaenia	Endangered	Rhacophoridae	Endemic
Pseudophilautus poppiae (Fig 29.11, 29.12)	Endangered	Rhacophoridae	Endemic
Pseudophilautus popularis	Least concern	Rhacophoridae	Endemic
Pseudophilautus procax (Fig 29.17)	Critically endangered	Rhacophoridae	Endemic
Pseudophilautus pruranappu	Not assessed	Rhacophoridae	Endemic
Pseudophilautus regius	Data deficient	Rhacophoridae	Endemic
Pseudophilautus reticularis	Endangered	Rhacophoridae	Endemic
Pseudophilautus rugatus	Extinct	Rhacophoridae	Endemic
Pseudophilautus rus	Near threatened	Rhacophoridae	Endemic
Pseudophilautus samarakoon	Not assessed	Rhacophoridae	Endemic

TABLE 29.1   *(Continued)*

Scientific name	IUCN Red List status	Family	Distribution status
Pseudophilautus sarasinorum	Endangered	Rhacophoridae	Endemic
Pseudophilautus schmarda	Endangered	Rhacophoridae	Endemic
Pseudophilautus schneideri	Not assessed	Rhacophoridae	Endemic
Pseudophilautussemiruber	Data deficient	Rhacophoridae	Endemic
Pseudophilautus silus	Endangered	Rhacophoridae	Endemic
Pseudophilautus silvaticus	Endangered	Rhacophoridae	Endemic
Pseudophilautus simba	Critically endangered	Rhacophoridae	Endemic
Pseudophilautus singu	Endangered	Rhacophoridae	Endemic
Pseudophilautus siriwijesundarai	Not assessed	Rhacophoridae	Endemic
Pseudophilautus sordidus	Near threatened	Rhacophoridae	Endemic
Pseudophilautus steineri	Endangered	Rhacophoridae	Endemic
Pseudophilautus stellatus	Critically endangered	Rhacophoridae	Endemic
Pseudophilautus stictomerus	Near threatened	Rhacophoridae	Endemic
Pseudophilautus stuarti	Endangered	Rhacophoridae	Endemic
Pseudophilautus tanu	Endangered	Rhacophoridae	Endemic
Pseudophilautus temporalis	Extinct	Rhacophoridae	Endemic
Pseudophilautus variabilis	Extinct	Rhacophoridae	Endemic
Pseudophilautus viridis	Endangered	Rhacophoridae	Endemic
Pseudophilautus zal	Extinct	Rhacophoridae	Endemic
Pseudophilautus zimmeri	Extinct	Rhacophoridae	Endemic
Pseudophilautus zorro (Fig 29.9)	Endangered	Rhacophoridae	Endemic
Taruga eques	Endangered	Rhacophoridae	Endemic
Taruga fastigo	Critically endangered	Rhacophoridae	Endemic
Taruga longinasus	Endangered	Rhacophoridae	Endemic

FIGURE 29.1   *Adenomus kelaartii* in amplexus, Kudawa, Sinharaja Forest Reserve, March 2002.

FIGURE 29.2 *Pseudophilautus* spec., Kudawa, Sinharaja Forest Reserve, March 2002.

FIGURE 29.3 *Nannophrys ceylonensis*, Makandawa Forest, Kitulgala, May 2006.

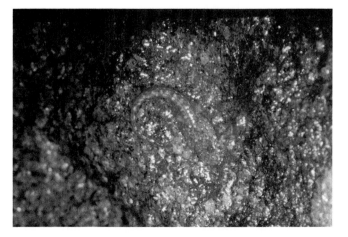

FIGURE 29.4 Tadpole of *Nannophrys ceylonensis*, Makandawa Forest, Kitulgala, May 2006.

FIGURE 29.5   *Euphlyctis hexadactylus* with median line, Tabbowa Wewa, March 2012.

FIGURE 29.6   *Duttaphrynus melanostictus*, male wanting to mate, Puttalam, December 2000.

FIGURE 29.7   *Pseudophilautus regius*, Botanical Garden Peradeniya, Maya 2006.

FIGURE 29.8 *Uperodon rohani*, Putalam, October 2014.

FIGURE 29.9 *Pseudophilautus zorro*, Gonnaruwa forest, March 2006.

FIGURE 29.10 *Pseudophilautus fergusonianus* in lavatory, Suriyakanda, July 2011.

FIGURE 29.11 *Pseudophilautus poppiae*, egg clutch on a leaf, Morningside Estate, July 2010.

FIGURE 29.12 *Pseudophilautus poppiae*, Morningside Estate, July 2011.

FIGURE 29.13 *Taruga fastigo* in amplexus, Suriyakanda, July 2011.

FIGURE 29.14   *Pseudophilautus cavirostris*, Suriyakanda, July 2010.

FIGURE 29.15   *Euphlyctis cyanophlyctis*, Suriyakanda, March 2016.

FIGURE 29.16 *Microhyla mihintalei*, Kalpitiya, April 2013.

FIGURE 29.17 *Pseudophilautus procax* in amplexus, Suriyakanda, July 2011.

29.5 CONCLUSION

Scientific research of the amphibians of Sri Lanka and India has taken place particularly in the past 30 years. The results clearly show that the diversity of species has been under-estimated and that the Western Ghats and Sri Lanka are home to mostly different species. Formerly, the same species have been split into different ones. This was made possible in particular by the results of molecular biological methods. These results notably are related to the families Ranidae and Rhacophoridae. There have not been exchanges of genes between the two regions for very long time (Bossuyt et al., 2001; Gunawardene et al., 2007; Meegaskumbura et al., 2002). The reason for it lies in the repeated physical disconnection of both regions and changes of the climate. With the emergence of the monsoons, southern India and the north of Sri Lanka became drier and became a barrier for species from humid tropical regions. Both areas are therefore climatically separated and genetic exchange was

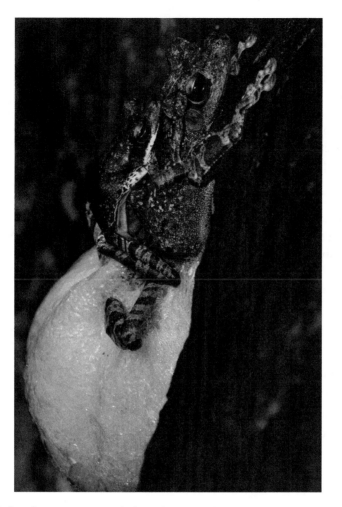

FIGURE 29.18 *Polypedates cruciger*, producing a foam nest, Suriyakanda, March 2016.

impossible (Surasinghe, 2009). The frogs of the Western Ghats and the Sri Lankan wet zone were separated and even within both regions populations became seperated by rivers and changes of local climate and the species number increased over time due to adaptive radiation. We have to consider the Western Ghats-Sri Lanka hotspot as a hotspot with two centers. Both have made a similar development. In both regions, large areas of forest were cleared during the English colonization, especially in higher altitudes. And in both regions the population has grown significantly and this process has not been completed yet. Today, the greatest threat to areas of unspoiled nature comes from population growth. The Western Ghats-Sri Lanka hotspot is one of the most populated hotspots of biodiversity (Cincotta et al., 2000; Erdelen, 2012). In Sri Lanka the wet zone is particularly densely populated (Bahir et al., 2005). Even the species of the dry zone are not completely the same in both countries. Research of Sri Lankan amphibians must accelerate, if we do not want to see more new species as already extinct or we will never discover them. No one knows how many species disappeared unnoticed by the deforestation for coffee and tea. And researching the biology of the species requires a lot of effort. Taxonomy has not a high significance today and is made difficult by species protection laws and amphibian species are often under-sampled in surveys (Pethiyagoda et al., 2007). The removal from nature is strictly regulated and requires a lot of bureaucratic effort. Even the observation is difficult, because walking at night is prohibited in many areas of Sri Lanka. This particularly affects the "normal" people and tourists interested in science. A tourist who photographs frogs at night must expect to be brought to court. On the other hand, there is little or no protection against deforestation, conversion of forests into cultivated areas and dangers like pesticides. Research of the amphibians of Sri Lanka requires international cooperation, especially since it requires technologies that are difficult to obtain in Sri Lanka. This includes in particular the taxonomy based on genomic comparisons. This currently applies to populations of *Euphylctis cyanophlyctis, E. hexadactylus,* and to the genus *Sphaerotheca.* The threats to amphibians in Sri Lanka will further exist independant of their protection.

KEYWORDS

- **Sri Lanka**
- **amphibians**
- **extinction**
- **conservation**
- **history of exploration**

REFERENCES

Bahir, M. M.; Meegaskumbura, M.; Manamendra-Arachchi, K.; Schneider, C. J.; Pethiyagoda, R. Reproduction and Terrestrial Direct Development in Sri Lankan Shrub Frogs (Ranidae: Rhacophoridae: *Philautus*). *Raffles Bull. Zool.* **2005,** *12* (339), 339–350.

Batuwita, S.; De Silva, M.; Udugampala, S. Description of a New Species of *Pseudophilautus* (Amphibia: Rhacophoridae) from Southern Sri Lanka. *J. Threat. Taxa* **2019,** *11*, 13120–13131.

Biju, S. D. Reproductive Mode in the Shrub Frog *Philautus glandulosus* (Jerdon, 1853) (Anura: Rhacophoridae). *Curr. Sci.* **2003,** *84* (3), 283–284.

Biju, S. D.; Garg, S.; Mahony, S.; Wijayathilaka, N.; Senevirathne, G.; Meegaskumbura, M. DNA Barcoding, Phylogeny and Systematics of Golden-Backed Frogs (*Hylarana*, Ranidae) of the Western Ghats-Sri Lanka Biodiversity Hotspot, with the Description of Seven New Species. *Contrib. Zool.* **2014,** *83* (4), 269–335.

Bogert, C. M.; Senanayake, F. R. A New Species of Toad (*Bufo*) Indigenous to Southern Ceylon. *Am. Mus. Novit.* **1966,** *2269,* 1–18.

Bopage, M. M.; Wewalwala, K.; Krvavac, M.; Jovanovic, O.; Safarek, G.; Pushpamal, V. Species Diversity and Threat Status of Amphibians in the Kanneliya Forest, Lowland Sri Lanka. *Salamandra* **2011,** *47* (3), 173–177.

Bossuyt, F.; Dubois, A. A Review of the Frog Genus *Philautus* Gistel, 1848 (Amphibia, Anura, Ranidae, Rhacophorinae). *Zeylanica* **2001,** *6,* 1–112.

Cincotta, R. P.; Wisnewski, J.; Engelmann, R. Human Population in the Biodiversity Hotspot. *Nature* **2000**, *404*, 990–992.

Dahanukar, N.; Suakhe, S.; Padhya, A. Identity of *Sphaerotheca pluvialis* (Jerdon, 1853) and Other Available Names among the Burrowing Frogs (Anura: Dicroglossidae) of South Asia. *J. Threat. Taxa* **2017**, *9* (6), 10269–10285.

Delorme, M.; Dubois, A.; Kosuch, J.; Vences, M. Molecular Phylogenetic Relationships of *Lankanectes corrugatus* from Sri Lanka: Endemism of South Asian Frogs and the Concept of Monophyly in Phylogenetic Studies. *Alytes* **2004**, *22* (1–2), 53–64.

de Silva, A. In *Amphibians of Sri Lanka: A Photographic guide to common Frogs, Toads and Caecilians, 2009*; p 168.

Dissanayake, D. E. S. B.; Weillipuli-Arachchi. Habitat Preferences of the Endemic Shrub Frog *Pseudophilautus regius* (Manamendra-Arachchi & Pethiyagoda 2005) at Mihintale Sanctuary, Sri Lanka. *Amphib. Reptile Conserv.* **2012**, *5* (2), 114–124.

Dutta, S. K.; Manamendra-Arachchi, K. In *Amphibian Fauna of Sri Lanka- a Systematic Review*; WHT, 1996.

Erdelen, W. R. Conservation of Biodiversity in a Hotspot: Sri Lanka's Amphibians and Reptiles. *Amphib. Reptile Conserv.* **2012**, *5* (2), 33–51.

Ferguson, W. In *Reptile Fauna of Ceylon: Letter on a Collection Sent to the Colombo Museum*; Herbert, Ceylon, **1877**.

Frost, D. R. Amphibian Species of the World: An Online Reference. Version 6.1 (*Date of access*) [Online]. https://amphibiansoftheworld.amnh.org/index.php. American Museum of Natural History, New York, USA (accessed Jun 6, 2020).

Frost, D. R.; Grant, T.; Faivovich, J.; Bain, R. H.; Haas, A.; Haddad, C. F. B.; de Sa, O.; Channing, A.; Wilkinson, M.; Donnellan, S. C.; Raxworthy, C. J.; Campbell, J. A.; Blotto, B. L.; Moler, P. E.; Drewes, R. C.; Nussbaum, R. A.; Lynch, J. D.; Green, D. M.; Wheeler, W. C. The Amphibian Tree of Life. *Bull. Am. Mus. Nat. Hist.* **2006**, *297*, 1–370.

Goonewardene, S.; Hawke, Z.; Vanneck, V.; Drion, A.; de Silva, A.; Jayarathne, R.; Perera, J. *Diversity of Nilgala Fire Savannah, Sri Lanka: With Special Reference to its Herpetofauna;* Report of Project Hoona, **2003**.

Gower, D. J.; Bahir, M. M.; Mapatuna, Y.; Pethiyagoda, R.; Raheem, D.; Wilkison, M. Molecular Phylogenetics of Sri Lankan *Ichthyophis* (Amphibia: Gymnophiona: Ichthyophiidae), with Discovery of a Cryptic Species. Contributions to Biodiversity Exploration and Research in Sri Lanka. *Raffles Bull. Zool.* **2005**, *12*, 153–161.

Günther, A. Notes on the Mode of Propagation of Some Ceylonese Treefrogs with Description of Two New Species. *Ann. Mag. Nat. Hist.* **1876**, *4*, XVII.

Haas, W.; Leher, E.; Köhler G. The Tadpole of *Bufo kelaartii* GÜNTHER 1859 from Sri Lanka. *Lyriocephalus* **1997**, *3* (2), 2–6.

Herrmann, H. J. Haltung und Zucht von *Polypedates cruciger cruciger* BLYTH, 1852. *Herpetofauna* **1993**, *15* (85), 31–34.

Janssen, J.; de Silva, A. The Presence of Protected Reptiles from Sri Lanka in International Commercial Trade. *TRAFFIC Bull.* **2019**, *31* (1), 9–15.

Kandamby, D. S.; Batuwita, S. Some Observations on the Distribution of *Rhacophorus cavirostris* (Gunther, 1868) Endemic Tree Frog from Sri Lanka. In *The Amphibia of Sri Lanka: Recent Research. Lyriocephalus Special Issue* **2001**, *4* (1&2), 93–94.

Karunarathna, D. M. S. S.; Abeywardena, U. T. O.; Amarasinghe, A. A. T.; Sirimanna, D. G. R.; Asela, M. D. C. Amphibian Faunal Diversity of Berliya Mukalana Proposed Forest Reserve. *Tigerpaper* **2008**, *35* (2), 12–16.

Karunarathna, D. M. S. S.; Amarasinghe, A. A. T. Observations on the Breeding Behavior of *Philautus regius* Manamendra-Arachchi and Pethiyagoda, 2005 (Amphibia: Ranidae: Rhacophorinae) in Nilgala, Moneragala District in Sri Lanka. *Russian J. Herpetol.* **2007**, *14* (2), 133–136.

Karunarathna, D. M. S. S.; Amarasinghe, A. A. T. Field Observations on the Reproductive Behaviour of *Philautus popularis* Manamendra-Arachchi and Pethiyagoda, 2005 (Amphibia: Rhacophoridae). *Sauria* **2010**, *32* (3), 57–62.

Karunaratne, P. Observations of *Philautus nasutus* Günther, 1886 (Amphibia: Rhacophoridae) Inhabiting the Gannoruwa Forest, Kandy, Sri Lanka. In *The Amphibia of Sri Lanka: Recent Research*; de Silva, A., **2001**; pp 95–100.

Kirtisinghe, P. The Presence in Ceylon of a Frog with Direct Development on Land. *Ceylon J. Sci. B* **1946**, 23, 109–112.

Kirtisinghe, P. The Amphibia of Ceylon; Colombo, Sri Lanka; Privately Printed, **1957**.

Kumara, D. M. N. P.; Ukuwela, K. D. B. A Survey on the Amphibians of Ambagamuwa, a Tropical Wet Midland Area in Sri Lanka. *Herpetol. Notes* **2009**, *2*, 81–85.

Laufer, G. G. *Invasión de rana toro y quitridiomicosis : dos amenazas a la conservación de los anfibios en Uruguay.* Doctor en Ciencias Biológicas–Ecología, Universidad de la República (Uruguay), Facultad de Ciencias–PEDECIBA, **2018**.

Li, J. T.; Che, R. W.; Murphy, R. W.; Zhao, H.; Zhao, E. M.; Rao, D. Q.; Zhang, Y. P. New Insights to the Molecular Phylogenetics and Generic Assessment in the Rhacophoridae (Amphibia: Anura) Based on Five Nuclear and Three Mitochondrial Genes, with Comments on the Evolution of Reproduction. *Mol. Phylogenet. Evol.* **2009**, *53*, 509–522.

Li, J. T.; Li, Y.; Klaus, S.; Rao, D. Q.; Hillis, D. M.; Zhang, Y. P. Diversification of Rhacophorid Frogs Provides Evidence for Accelerated Faunal Exchange between India and Eurasia during the Oligocene. *Proc. Nat. Acad. Sci. U.S.A.* **2013**, *110*, 3441–3446.

Manamendra-Arachchi, K. Know your Frogs. *Sri Lanka Nature* **2000**, *2* (5), 4–16.

Manamendra-Arachchi, K.; Pethiyagoda, R. *Polypedates fastigo*, A New Tree Frog (Ranidae: Rhacophorinae) from Sri Lanka. *J. South Asian Hist.* **2001**, *5* (2), 191–199.

Manamendra-Arachchi, K.; Pethiyagoda, R. The Sri Lankan Shrub-Frogs of the Genus *Philautus* Gistel, 1848 (Ranidae: Rhacophorinae), with Description of 27 New Species. Contributions to Biodiversity Exploration and Research in Sri Lanka. *Raffles Bull. Zool.* **2005**, *12*, 163–303.

Meegaskumbura, M.; Bossuyt, F.; Pethiyagoda, R.; Manamendra-Arachchi, K.; Bahir, M.; Milinkovitch, M. C.; Schneider, C. J. Sri Lanka: An Amphibian Hot Spot. *Science* **2002**, *298*, 379.

Meegaskumbura, M.; Bowate, G.; Manamendra-Arachchi, K.; Meegaskumbura, S. Amphibian Research in Sri Lanka. *Frog Log* **2011**, *98*, 26–29.

Meegaskumbura, M.; Manamendra-Arachchi, K. Description of Eight New Species of Shrubfrogs (Ranidae: Rhacophorinae: *Philautus*) from Sri Lanka. Contributions to Biodiversity Exploration and Research in Sri Lanka. *Raffles Bull. Zool.* **2005**, *12*, 305–338.

Meegaskumbura, M.; Manamendra-Arachchi, K. Two New Species of Shrub Frogs (Rhacophoridae: *Pseudophilautus*) from Sri Lanka. *Zootaxa* **2009**, *2747*, 1–18.

Meegaskumbura, M.; Manamendra-Arachchi, K. Two New Species of Shrub Frogs (Rhacophoridae: *Pseudophilautus*) from Sri Lanka. *Zootaxa* **2011**, *2747*, 1–18.

Meegaskumbura, M.; Manamendra-Arachchi, K.; Pethiyagoda, R. Two New Species of Shrub Frogs (Rhacophoridae: *Philautus*) from the Lowlands of Sri Lanka. *Zootaxa* **2009**, *2122*, 51–68.

Meegaskumbura, M.; Manamendra-Arachchi, K.; Schneider, C. J.; Pethiyagoda, R. New Species amongst Sri Lanka's Extinct Shrub Frogs (Amphibia: Rhacophoridae: *Philautus*). *Zootaxa* **2007**, *1397*, 1–15.

Meegaskumbura, M.; Senevirathne, G.; Biju, S. D.; Garg, S.; Meegaskumbura, S.; Pethiyagoda, R.; Hanken, J.; Schneider, C. J. Patterns of Reproductive-Mode Evolution in Old World Tree Frogs (Anura, Rhacophoridae). *Zoologica Scripta* **2015a**. https://doi.org/10.1111/zsc.12121.

Meegaskumbura, M.; Senevirathne, G.; Manamendra-Arachchi, K.; Pethiyagoda, R.; Hanken, J.; Schneider, C. J. Diversification of Shrub Frogs (Rhacophoridae, *Pseudophilautus*) in Sri Lanka– Timing and Geographic Context. *Mol. Phylogenet. Evolut.* **2019**, *132*, 14–24.

Molur, S. South Asian Amphibians: Taxonomy, Diversity and Conservation Status. In *International Zoo Yearbook*, **2008**; vol 42, pp 143–157.

Nussbaum, R. A.; Gans, C. On the *Ichthyophis* (Amphibia: Gymnophiona) of Sri Lanka. In *Spolia Zeylanica*, **1983**; vol 35, pp 137–154.

Oliver, L.; Prendini, E.; Kraus, F.; Raxworthy, C. J. Systematics and Biogeography of the *Hylarana* Frog (Anura: Ranidae) Radiation Across Tropical Australasia, Southeast Asia, and Africa. *Mol. Phylogenet. Evolut.* **2015**, *90*, 176–192.

Peaboutuwage, I.; Bandara, I. N.; Samarasinghe, D. J. S.; Perera, N.; Madawala, M.; Amarasinghe, C.; Kandambi, H. K. D.; Karunarathna, M. S. S. Range Extension for *Duttaphrynus kotagamai* (Amphibia: Bufonidae) and a Preliminary Checklist of Herpetofauna from the Uda Mäliboda Trail in Samanala Nature Reserve, Sri Lanka. *Amphib. Reptile Conserv.* **2012**, *5*, 52–64.

Pethiyagoda, R.; Gunatilleke, N.; de Silva, M.; Kotagama, S.; Gunatilleke, S.; de Silva, P.; Meegaskumbura, M.; fernando, P.; Ratnayeke, S.; Jayawardene, J.; Raheem, D.; Benjamin, S.; Ilangakoon, A. Science and Biodiversity: The Predicament of Sri Lanka. *Curr. Sci.* **2007,** *92* (4), 426–427.

Pethiyagoda, R.; Manamendra-Arachchi, K.; Evaluating Sri Lanka's Amphibian Diversity. *Occas. Papers Wildlife Heritage Trust* **1998,** *2,* 1–12.

Pyron, R. A.; Wiens, J. J. A Large-Scale Phylogeny of Amphibia Including over 2800 Species, and a Revised Classification of Advanced Frogs, Salamanders, and Caecilians. *Mol. Phylogenet. Evolut.* **2011,** *61,* 543–583.

Rahman, M. M.; Badhon, M. K.; Salaunddin, M.; Rabbe, M. F.; Islam, M. S. Chytrid Infection in Asia: How much do we know amd what else do we ned to know). *Herpetol. J.* **2020,** *20,* 99–111.

Roelants, K.; Jiang, J.; Bosuyt, F. Endemic Ranid (Amphibia: Anura) Genera in Southern Mountain Ranges of the Indian Subcontinent Represent Ancient Frog Lineages: Evidence from the Molecular Data. *Mol. Phylogenet. Evol.* **2004,** *31,* 730–740.

Samarawickrama, V. A. M. P. K.; Samarawickrama, D. R. N. S.; Kumburegama, S. Herpetofauna in the Kaluganga Upper Catchment of the Knuckles Forest Reserve, Sri Lanka. *Amphib. Reptile Conserv.* **2012,** *5* (2), 81–89.

Schäfer, D. Climate Change in Sri Lanka? Statistical Analysi of Longterm Temperature and Rainfall Records. In *Sri Lanka, Past and Present: Archaeology, Geography, Economics–Selected Papers on German Research*; Domrös, M.; Roth, H.; Markgraf Verlag, **1998.**

Senevirathne, G.; Bowatte, G.; Meegaskumbura, M. An Ecological Service of Frogs: Tadpoles Feeding on Dengue Mosquito Eggs. *Frog Log* **2014,** *119* (1), 57–60.

Surasinghe, T. D. Conservation and Distribution Status of Amphibian Fauna in Sri Lanka. *Biodiversity* **2009,** *10* (1), 3–17.

Taylor, E. H. New Asiatic and African Caecilians with Redescriptions of Certain Other Species. *Univ. Kans. Sci. Bull.* **1965,** *46,* 253–302.

Taylor, E. H. In *The Caecilians of the World*; University Kansas Press: Lawrence, **1968**; p 845.

Wake, M. H. The Reproductive Biology of *Eleutherodactylus jasperi* (Amphibia, Anura, Leptodactylidae, with Comments on the Evolution of Live-Bearing Systems. *J. Herpetol.* **1978,** *12* (2), 121–133.

Werner, W. In *Sri Lanka's Magnificient Cloud Forests*; WHT Publications, **2001.**

Wickramasinghe, L. J. M. In *Repertoire. A Pictorial Gateway to Sri Lanka's Nature;* Gunaratne Offset, **2015.**

Wickramasinghe, L. J. M.; Vidanapathirana, D.; Rajeev, M. D. G.; Ariyarathne, S. C.; Chanaka, A. W. A.; Priyantha, L. L. D.; Bandara, I. N.; Wickramasinghe, N. Eight New Species of *Pseudophilautus* (Amphibia: Anura: Rhacophoridae) from Sripada World Heritage Site (Peak Wilderness), a Local Amphibian Hotspot in Sri Lanka. *J. Threatened Taxa* **2013b,** *5,* 3789–3920.

Yu, G. H.; Zhang, M. W.; Yang, J. X. Generic Allocation of Indian and Sri Lankan *Philautus* (Anura: Rhacophoridae) Inferred from 12S and 16S rRNA Genes. *Biochem. Syst. Ecol.* **2010,** *38,* 402–409.

CHAPTER 30

DIVERSITY, DISTRIBUTION, AND BIOGEOGRAPHY OF SRI LANKAN BIRDS

SALINDRA K. DAYANANDA[1,4], SANDUN J. PERERA[2,4],
SAMPATH S. SENEVIRATHNE[3,4], and SARATH W. KOTAGAMA[4]

[1]*Guangxi Key Laboratory of Forest Ecology and Conservation, College of Forestry, Guangxi University, Nanning, China*

[2]*Department of Natural Resources, Faculty of Applied Sciences, Sabaragamuwa University of Sri Lanka, Belihuloya, Sri Lanka*

[3]*Avian Evolution Node, Department of Zoology and Environment Sciences, Faculty of Science, University of Colombo, Sri Lanka*

[4]*Field Ornithology Group of Sri Lanka, Department of Zoology and Environment Science, Faculty of Science, University of Colombo, Sri Lanka*

ABSTRACT

The Western Ghats and Sri Lanka biodiversity hotspot is well established, for which Sri Lanka contributes a relatively small land area (65,610 km²), exceptionally rich in biodiversity and endemism packed in a continental island. The tropical climates, varied rainfall dominated by two monsoons, topographical heterogeneity, and geological history have shaped this biological diversity. Currently, Sri Lanka is home to 479 bird species with 34 country endemics. These bird species belong to 19 orders and 89 families. Among them, 244 (51%) species are breeding residents, 147 (31%) are migratory species, and 88 (18%) are vagrant species. Recent molecular phylogenetic studies have revealed the possibility of adding a few more endemic bird species to Sri Lanka, yet to be further explored. The 66 subspecies endemic to the country provide the guideline to direct future molecular studies with an integrated approach to establishing their taxonomic status. The island can be regionalized to seven well-defined avifaunal zones following the biogeography of resident birds with unique species assemblages in each zone. However, most of the birds are facing elevated levels of threat due to habitat loss, habitat degradation, habitat fragmentation, human-induced land use changes, unplanned development of linear structures, and climate

Biodiversity Hotspot of the Western Ghats and Sri Lanka. T. Pullaiah, PhD (Ed.)
© 2024 Apple Academic Press, Inc. Co-published with CRC Press (Taylor & Francis)

change. To mitigate those threats, Sri Lanka has declared approximately 30% of its land area under varied protected area jurisdictions; identified 70 important bird areas; initiated habitat restoration projects; conducted awareness creation targeting the general public toward biodiversity conservation and sustainability. The need for well-planned monitoring programs on bird population fluctuations; conducting explicit scientific studies, especially on the endemic species; establishment of sound collaboration between government departments, NGOs, and other stockholders, including the private sector, is essential to ensure the long-term conservation of Sri Lankan birds for the future.

30.1 INTRODUCTION

Sri Lanka is a continental island, located south of the peninsular India with a total land area of 65,610 km^2 (between northern latitude 5° 54' and 9° 52' and between eastern longitude 79° 40' and 81° 53'). It holds high density of species richness and habitats with varying climate conditions due to its complex geological past, tropical island nature, and monsoon winds of the Indian ocean (SDSL, 2007; Katz, 1978). The climate in Sri Lanka is mainly governed by the two monsoons, the northeast and southwest, primarily contributing to the delimitation of the wet and dry zones of the island. Being a small island, no place within the country is more than 110 km (70 miles) from the Indian ocean. In addition, its landform of the central mountains and surrounding plains at the south-central part, its varied rainfall throughout the year, etc., all combine to create diverse climatic conditions and habitats that give rise to a rich variety of birds and bird communities to the island (MoMD&E, 2019). And also, the human factor influenced primarily by the Buddhist majority and agriculture has further contributed to its diversity and protection.

At present, the number of birds recorded within Sri Lanka (including the pelagic bird species recorded in Sri Lanka's exclusive economic zone in the ocean) is 478 which include 34 definite endemics. There are another 43 bird species with uncertain records which need careful revision of the reported description; among them, there are restricted species such as the black-and-orange flycatcher *Ficedula nigrorufa* which is endemic to the Western Ghats of India. On average, this number increases by at least one vagrant species per year due to the geographic positioning of the country. Furthermore, the geographic barrier between India and Sri Lanka is the resulting divergence in several bird lineages (Wickramasinghe et al., 2017; Abeyarama and Seneviratne, 2017a). Further, 143 species are shared with mainland India which include 66 endemic subspecies (Kotagama and Ratnavira, 2017), but the endemicity of these subspecies has to be confirmed by further studies. The most recent example of such a split was a genetic study done on *Dinopium* species in Sri Lanka, which resulted in a new endemic species to Sri Lanka, *Dinopium psarodes* Sri Lanka red-backed flameback (Fernando et al., 2016). These 522 species (479 + 43 doubtful records) belong to 19 orders and 89 families (Kotagama and Ratnavira, 2017). The most diverse orders are Passeriformes with 40 families and Charadriformes with 12 families, and more comprehensive and updated information on taxonomy can be found in the section "Current Taxonomy" below.

30.1.1 A BRIEF HISTORY OF BIRD STUDIES IN SRI LANKA

Interest for Sri Lankan birds was reported way back in the past. King Upatissa (368–410 A.D.) is on record as having established "bird-feeding tables" at the royal park of Anuradhapura, in the late fourth century (Kotagama and Ratnavira, 2017). This custom has prevailed for over 500 years. Thus, birds have been observed by even ancient Sri Lankans for aesthetic pleasure. "Ruvan Mala" (means golden flower), a book by King Parakrama Bahu, VI (1412–1467 A.D.) contains a reference to about 24 bird species. The "Vessantara Jataka Kawi" (a Sinhalese poem) has 23, "Saddarmalankaraya" (a Buddhist book in Sinhala) about 17, and "Pujavaliya" (a Buddhist book in Sinhala) about 42 such references of birds. The highest number of names appears in the 12th century, "Vesaturu Da Sannaya"—about 60 birds have been named.

With the arrival of colonial powers in Sri Lanka, firstly the Portuguese in 1505, writers like Bernardim Riberio (1482–1552), Philippus Baldaeus (1632–1671), Eça de Queiroz (1845–1900), and such, during the 16th and 17th centuries had mentioned in their publications about birds. But these were not very descriptive. Among these Europeans, Robert Knox (1641–1720) may be the first European observer to make serious records on the birds (Knox, 1681a, 1681b). In 1743, George Edwards published "Natural History of Uncommon Birds." This may be the first book written specifically on Sri Lankan birds, in which the *Loriculus beryllinus* (Sri Lankan hanging parrot) was described. During 1752–1757, the governor of Sri Lanka, Jan Gideon Loten made extensive collections and drawings of birds. His contribution added numerous species to the island's avifaunal list. Peter Brown, an English naturalist published some drawings in 1766. These drawings were also used by many other naturalists, some of whom were Johann Reinhold Forster (1729–1798).

The first Englishman to record birds in Sri Lanka was R. A. Templeton (1802–1892). Templeton also employed or sought assistance from numerous others to obtain specimens (Pethiyagoda, 2007). Among them, E. L. Layard could be identified as the person who made the biggest contribution. Layard did not confine himself to collecting specimens, but also made many observations which he published in "The Annals and Magazines of Natural History" from 1844–1853, as a series of notes. The list of 183 species compiled by Templeton was increased to 315 by Layard.

Dr. E. F. Kelaart published "Prodromus Faunal Zeylanicae" in which part II was on birds. In 1865, Holdsworth arrived in Sri Lanka and continued the work. He published his work in the "Proceedings of the Zoological Society of London" in 1872. This publication titled "Catalogue of the Birds Found in Ceylon" could be considered as the most complete document on the birds up to that time (Pethiyagoda, 2007). Capt. W. Vincent Legge (1872–1880) spent 8 years in Sri Lanka and has written the most detailed monograph, yet to be published, on Sri Lankan birds. His book "A History of the Birds of Ceylon" had incorporated all the previous details and many new ones. It was not surprising, therefore, that after him till 1925, there were very few additional publications made. The only other publication after 1881 was done by Murray (1890). However, during this long period, numerous articles and publications in journals such as the Royal Asiatic Society Journal, Spolia Zeylanica, Proceedings of Natural History, etc. were written by various persons, mostly British planters living in Sri Lanka. Subsequently in 1920–1930, Stuart Baker published the eight volumes of "Birds of British India including Ceylon and Burma," followed by four volumes on "Nidifications of the Birds of the Indian Empire."

Following the continental interest in birds, the British museum along with the Colombo museum conducted what one could call the first scientific avifaunal survey of Sri Lanka from 1936 to 1939. This work was done by Hugh Whistler and his work was published in "Spolia Zeylanica" vol. 23 (Whistler, 1944). With this work, the total number of species and subspecies was placed at 384, with 22 endemic species and 77 endemic subspecies. This work was followed by Dillon Ripley, who was interested mainly in the "relict" forms and made several contributions (Ripley, 1949). In 1952, W. W. A. Phillips published a "Revised Checklist of the Birds of Ceylon." During this same period, Phillips started a series of papers on "Nests and Eggs of Ceylon Birds" in the Ceylon Journal of Science. These articles appeared till his death in 1982. In 1955, G. M. Henry with his "Guide to the Birds of Ceylon" made a dramatic change in the bird interest in Sri Lanka. More persons became interested in the subject and it became a popular hobby among many, mostly white colonists and their local followers. Scientific studies however were very poorly pursued. Ripley in 1961 published the "Synopsis of the Birds of India and Pakistan together with those of Nepal, Sikkim, Bhutan, and Ceylon." This was the base for the "Handbook of the Birds of India and Pakistan" by Salim Ali and Dillon Ripley in 1968.

By 1975 general distribution patterns and abundance related information of birds were incorporated into the Phillips' "Annotated Checklist of the Birds of Sri Lanka," which was revised again in 1978. Even till 1970s, the interest in birding and bird studies was confined mainly to the English-speaking elite. In 1976, Douglas Ranasinghe brought out his book on "Asirimath Kurulu Lokaya" (meaning "Fascinating world of birds") which helped to infuse interest among many Sinhala-speaking persons. This was followed by the formulations of a standard Sinhala nomenclature in 1978—"Sri Lanka Avifaunal List." This was followed by a reprint of Legge's four volumes in 1983, and the systematization of the Sinhala names on a scientific basis by the publication of the "Systematic Sinhala Nomenclature for the Birds of Sri Lanka" by Perera and Kotagama (1983). The lack of scientific information was however becoming evident by the latter half of 1970s. Except for few accounts on behavior and breeding, most of the works had been concerning the distribution or description of birds. The available status reports were confined to "common," "rare," etc. with no adequate measure of the "commonness" or "rareness." It is not surprising therefore that in the 1978 checklist, there were 36 different abbreviations used to describe the status of abundance, residence, migrants, and breeding of Sri Lankan birds.

An effort to bring about a scientific change in bird studies in Sri Lanka commenced in 1975, with the formation of the field ornithology group of Sri Lanka (FOGSL) at the Department of Zoology, University of Colombo. For the first time, moving away from the traditional "white colonist's pastime," counts of birds and planned research projects were undertaken in Sri Lanka by local and foreign personnel to promote the study of birds as a tool to popularize environmentalism, birds, nature education, and sustainability (Guneratne, 2015). This lays the foundation for a renewed interest and a new chapter of bird studies in Sri Lanka among the broader social spectrum.

30.2 CURRENT TAXONOMY AND EVOLUTIONARY REMARKS

The taxonomy of Asian birds has seen major changes during the last few decades, with new methodologies adopted in ornithology (e.g., Jarvis et al., 2014; Jetz et al., 2012; Tobias et al., 2010; Hackett et al., 2008). Here, we have followed the classification and nomenclature of birds of the world (BOW: www.birdsoftheworld.org) where it has used a single taxonomy, the Clements checklist (Clements et al., 2021). Currently, 522 bird species have been recorded to the political boundary of Sri Lanka, including the oceanic exclusive economic zone (MoMD&E, 2019). These species belong to 23 orders, 89 families, and 266 genera. However, within these recorded birds, 43 species have historical records published, but the accuracy of those records is somewhat uncertain (Because some birds have definite restricted ranges only in Western Ghats and southeast Asia without migratory population); for example, *Cyornis pallidipes* (White-bellied blue flycatcher), where if not carefully observed easily confused with *Eumyias sordidus* (Sri Lanka dusky-blue flycatcher), and *Rhipidura javanica* (Malaysian pied fantail), where melanized mutation of a *Rhipidura aureola* (white-browed fantail) can be easily confused. Hence, need a critical revision of those records. After Warakagoda and Sirivardana (2009) and Kaluthota and Kotagama (2009), we couldn't find a published material about updated taxonomic revision for birds of Sri Lanka.

30.2.1 RECENT FAMILY-LEVEL CHANGES

A total of 11 new families under three orders were newly added to the bird family list of Sri Lanka since 2009. Nine out of those 11 families belong to the order Passeriformes and the remaining two families are divided into Procellariiformes and Accipitriformes.

Few members previously in the Sylviidae family were moved into several new families, the *Schoenicola* genus under Locustellidae family, *Acrocephalus* and *Iduna* genera now under Acrocephalidae family, *Phylloscopus* genus in Phylloscopidae family, and *Hemitesia* genus (previously placed in *Cettia*) now in Scotocercidae family (Fregin et al., 2009). The *Culicicapa* genus previously was under the Muscicapidae family, now considered within the newly declared family Stenostiridae (Pasquet et al., 2002). *Tephrodornis affinis* common woodshrike was previously in the Campephagidae family but now placed in the Vangidae family as endemic species to Sri Lanka (Rasmussen and Anderton, 2005). The genera *Turdoides* and *Garrulax* are now placed in the Leiothrichidae family and the genus *Pellorneum* placed under the family Pellorneidae, where both families were previously placed at Timaliidae family (Alström et al., 2013; Moyle et al., 2012). Once all the old-world barbets and toucans were considered under the Ramphastidae family, but now recognized under the family Megalaimidae (Moyle, 2004). The southern storm-petrels previously placed in family Hydrobatidae belong to genera *Fregetta*, *Oceanites*, and *Pelagodroma* are now treated under family Oceanitidae (Kennedy and Page, 2002). The Osprey previously placed in the family Accipitridae is now considered under separate family Pandionidae (Hackett et al., 2008).

30.2.2 ENDEMIC SPECIES

Legge (1880) reported 47 species endemic to Sri Lanka. The concept of subspecies resulted in a decrease of endemic bird species in Sri Lanka, and as a result, the number of endemic species for the country had remained mostly around 20–33. Based on the biological species concept of Mayr (1942), Wijesinghe (1994) proposed five more species to the mostly accepted list of 20–22 endemics (Wait, 1925; Philips, 1952, 1975; Henry, 1955) making the list jump to 26. Rasmussen and Anderton (2005) proposed several subspecies to elevate as full species, hence the number rose up to 33. Then, Fernando et al. (2016) studies elevated the subspecies of *Dinopium benghalense psarodes* black-rumped flameback to full species level using molecular phylogenetics and named as *D. psarodes* red-backed flameback/Sri Lanka lesser flameback. With this addition, the current number of definite endemic bird species is 34 (Table 30.1). There were some controversies on the species status of *Zoothera imbricata* Sri Lanka scaly thrush with *Zoothera aurea* white's thrush (see species factsheet in BirdLife International, 2021), however most global checklists recognized this taxon as a valid endemic taxon for Sri Lanka (BOW, 2021).

Bird species currently accepted as endemic to Sri Lanka as of December, 2021 include the following 34: *Galloperdix bicalcarata* (Forster, 1781) (Sri Lanka spurfowl); *G. lafayettii* (Lesson, 1831) (Sri Lanka junglefowl); *Columba torringtoniae* (Blyth and Kelaart, 1853) (Sri Lanka woodpigeon); *Treron pompadora* (Gmelin, 1789) (Sri Lanka green pigeon); *Loriculus beryllinus* (Forster, 1781) (Sri Lanka hanging parrot); *Psittacula calthropae* (Blyth, 1849) (Sri Lanka emerald-collared parakeet); *Phaenicophaeus pyrrhocephalus* (Pennant, 1769) (Sri Lanka red-faced malkoha); *Centropus chlororhynchos* (Blyth, 1849) (Sri Lanka green-billed coucal); *Otus thilohoffmanni* (Warakagoda and Rasmussen, 2004) (Sri Lanka Serendib scops owl); *Glaucidium castanotum* (Blyth, 1846) (Sri Lanka chestnut-backed owlet); *Ocyceros gingalensis* (Shaw, 1811) (Sri Lanka gray hornbill); *Psilopogon flavifrons* (Cuvier, 1816) (Sri Lanka yellow-fronted barbet); *P. rubricapillus* (Gmelin, 1788) (Sri Lanka barbet); *Dinopium psarodes* (Lichtenstein, 1793) (Sri Lanka red-backed flameback); *Chrysocolaptes stricklandi* (Layard, 1854) (Sri Lanka Layard's flameback); *Tephrodornis affinis* (Blyth, 1847) (Sri Lanka woodshrike); *Dicrurus lophorinus* (Vieillot, 1817) (Sri Lanka drongo); *Urocissa ornata* (Wagler, 1829) (Sri Lanka blue magpie); *Cecropis hyperythra* (Blyth, 1849) (Sri Lanka swallow); *Pycnonotus melanicterus* (Gmelin, 1789) (Sri Lanka black-capped bulbul); *P. penicillatus* (Blyth, 1851) (Sri Lanka yellow-eared bulbul); *Elaphrornis palliseri* (Blyth, 1851) (Sri Lanka bush warbler); *Pellorneum fuscocapillus* (Blyth, 1849) (Sri Lanka brown-capped babbler); *Pomatorhinus melanurus* (Blyth, 1847) (Sri Lanka scimitar babbler); *Argya rufescens* (Blyth, 1847) (Sri Lanka orange-billed babbler); *Garrulax cinereifrons* (Blyth, 1851) (Sri Lanka ashy-headed laughing thrush); *Zosterops ceylonensis* (Holdsworth, 1872) (Sri Lanka white eye); *Gracula ptilogenys* (Blyth, 1846) (Sri Lanka myna), *Sturnus albofrontatus* (Layard, 1854) (Sri Lanka white-faced starling); *Myophonus blighi* (Holdsworth, 1872) (Sri Lanka whistling thrush); *Geokichla spiloptera* (Blyth, 1847) (Sri Lanka spot-winged thrush); *Zoothera imbricata* (Layard, 1854) (Sri Lanka scaly thrush); *Eumyias sordidus* (Walden, 1870) (Sri Lanka dusky-blue flycatcher/Sri Lankan mountain flycatcher); *Dicaeum vincens* (Sclater, 1872) (Sri Lanka Legge's flowerpecker).

TABLE 30.1   The Endemic Birds of Sri Lanka.

	Scientific name	Common name
01	*Galloperdix bicalcarata* (Forster, 1781)	Sri Lanka Spurfowl
02	*Gallus lafayettii* Lesson, 1831	Sri Lanka Junglefowl
03	*Columba torringtoniae* (Blyth & Kelaart, 1853)	Sri Lanka Woodpigeon
04	*Treron pompadora* (Gmelin, 1789)	Sri Lanka Green-pigeon
05	*Loriculus beryllinus* (Forster, 1781)	Sri Lanka Hanging-parrot
06	*Psittacula calthropae* (Blyth, 1849)	Sri Lanka Emerald-collared Parakeet
07	*Phaenicophaeus pyrrhocephalus* (Pennant, 1769)	Sri Lanka Red-faced Malkoha
08	*Centropus chlororhynchos* Blyth, 1849	Sri Lanka Green-billed Coucal
09	*Otus thilohoffmanni* Warakagoda & Rasmussen, 2004	Sri Lanka Serendib Scops-owl
10	*Glaucidium castanotum* (Blyth, 1846)	Sri Lanka Chestnut-backed Owlet
11	*Ocyceros gingalensis* (Shaw, 1811)	Sri Lanka Grey Hornbill
12	*Psilopogon flavifrons* (Cuvier, 1816)	Sri Lanka Yellow-fronted Barbet
13	*Psilopogon rubricapillus* (Gmelin, 1788)	Sri Lanka Barbet
14	*Dinopium psarodes* (Lichtenstein, 1793)	Sri Lanka Red-backed Flameback/ Sri Lanka Lesser Flameback
15	*Chrysocolaptes stricklandi* (Layard, 1854)	Sri Lanka Greater Flameback/ Sri Lanka Layard's Flameback
16	*Tephrodornis affinis* Blyth, 1847	Sri Lanka Woodshrike
17	*Dicrurus lophorinus* Vieillot, 1817	Sri Lanka Drongo
18	*Urocissa ornata* (Wagler, 1829)	Sri Lanka Blue Magpie
19	*Cecropis hyperythra* (Blyth, 1849)	Sri Lanka Swallow
20	*Pycnonotus melanicterus* (Gmelin, 1789)	Sri Lanka Black-capped Bulbul
21	*Pycnonotus penicillatus* Blyth, 1851	Sri Lanka Yellow-eared Bulbul
22	*Elaphrornis palliseri* (Blyth, 1851)	Sri Lanka Bush-warbler
23	*Pellorneum fuscocapillus* (Blyth, 1849)	Sri Lanka Brown-capped Babbler
24	*Pomatorhinus melanurus* Blyth, 1847	Sri Lanka Scimitar-babbler
25	*Argya rufescens* (Blyth, 1847)	Sri Lanka Orange-billed Babbler
26	*Garrulax cinereifrons* Blyth, 1851	Sri Lanka Ashy-headed Laughingthrush
27	*Zosterops ceylonensis* Holdsworth, 1872	Sri Lanka White-eye
28	*Gracula ptilogenys* Blyth, 1846	Sri Lanka Myna
29	*Sturnus albofrontatus* (Layard, 1854)	Sri Lanka White-faced Starling
30	*Myophonus blighi* (Holdsworth, 1872)	Sri Lanka Whistling-thrush
31	*Geokichla spiloptera* (Blyth, 1847)	Sri Lanka Spot-winged Thrush
32	*Zoothera imbricata* Layard, 1854	Sri Lanka Scaly Thrush
33	*Eumyias sordidus* (Walden, 1870)	Sri Lanka Dusky-blue Flycatcher/ Sri Lankan Mountain Flycatcher
34	*Dicaeum vincens* (Sclater, 1872)	Sri Lanka Legge's Flowerpecker

30.2.3 EVOLUTIONARY REMARKS

The evolutionary distinctiveness of Sri Lankan resident bird species was assessed by Abeyarama and Seneviratne (2017a, 2017b) and found *Batrachostomus moniliger* Ceylon frogmouth (Figure 30.1A) as the most evolutionary distinct (ED) species followed by *Harpactes faciatus* Malabar trogon (Figure 30.1B) and migratory *Pitta brachyura* Indian pitta. These species have oldest branches in the phylogenetic tree which accounts for higher ED value, as closely related species which evolved recently get lower ED values (Abeyarama and Seneviratne, 2017a). When their phylogenetic distinctiveness is considered with the global conservation status, a species gets further highlighted with an evolutionary distinct and globally endangered (EDGE) score. *Otus thilohoffmanni* Serendib scops owl (Figure 30.1C) holds the highest EDGE score among the Sri Lankan birds evolutionary parameters such as ED and EDGE, which are vital to capture the important aspects of evolutionary history of birds and also prioritize the conservation measures when it comes to species-specific conservation activities.

FIGURE 30.1   Three most evolutionary distinct resident bird species in Sri Lanka. (A) *Batrachostomus moniliger* Ceylon frogmouth, (B) *Harpactes faciatus* Malabar trogon, and (C) *Otus thilohoffmanni* Serendib scops owl, the most evolutionary distinct and globally endangered bird species in Sri Lanka. Photo credits: (A) and (B) Gehan Rajeev, (C) Dulan Ranga Vidanapathirana.

30.3 BIOGEOGRAPHY OF SRI LANKAN BIRDS

The biogeographical uniqueness of the avifauna of peninsular India and Sri Lanka was identified more than one and half centuries ago by Alfred Russel Wallace, who developed the first detailed map of global zoogeography (Wallace, 1876), following Sclater's avifaunal regions initially described in 1858 and mapped later in 1891 (Sclater, 1858, 1891). Wallace's regionalization identified the Oriental region (now Indomalayan realm; Udvardy, 1975) among the six global zoogeographical regions, encompassing four subregions including the Ceylonese subregion and south of the Indian subregion (Wallace 1876), as explained in the introductory chapter to Sri Lanka in this book (Perera and Fernando, 2022). Although Sclater (1891) and Sclater and Sclater (1899) contradicted the separation of the Ceylonese and Indian subregions based respectively on the similarities of bird and mammal assemblages known by then, an analysis based on the current knowledge on the biotas of two subregions including their avifauna, and especially their herpetofauna (Bossuyt et al., 2004) would suggest Wallace was right, however, not with the exact boundary he proposed (See Perera and Fernando (2022) for historical details on biogeographic studies in Sri Lanka).

Hence, the century-old literature has treated the Western Ghats and southern Sri Lanka as a true biogeographical entity, from which northern Sri Lanka was eliminated as belonging to an Indian entity, which is in-line with the current knowledge on the avifaunal zones for Sri Lanka (discussed later in this chapter). However, in the demarcation of biodiversity hotspots, particularly considering flowering plant endemism and the threats faced by the biodiversity, Myers et al. (2000) have identified the entire island of Sri Lanka together with Western Ghats of India as a biodiversity hotspot.

The Western Ghats and Sri Lanka harbor about 60 bird species endemic to the hotspot (Rasmussen and Anderton, 2005). This number is comprised of 16 (Bawa et al., 2007) to 22 (Gunawardene et al., 2007) species reported as restricted within the Western Ghats and 34 species endemic to Sri Lanka (MoMD&E, 2019; del Hoyo et al., 2020), together with several species endemic to the entire range of the hotspot, such as *Gracula indica* southern hill myna and *Hypsipetes ganeesa* square-tailed bulbul (MoMD&E, 2019; Bird-Life International, 2021). It is also interesting to note that at least 10 pairs of bird species respectively endemic to the Western Ghats of India and Sri Lanka are closely related or taxonomically sister species to each other. Among the 34 bird species endemic to Sri Lanka, at least 22 species show further restricted ranges within the island, mostly in the per-humid southwestern quarter, characterized by its high habitat heterogeneity, including the south-central rugged highland terrain.

It is noteworthy that both Western Ghats and Sri Lanka have separately been recognized by the BirdLife International as endemic bird areas owing to their high conservation importance for endemic avifauna (Stattersfield et al., 1998). Further, they have separately been identified among the global 200 ecoregions, identified as priority ecoregions for the conservation of a representative sample of the global biodiversity (Olson and Dinerstein, 1998), namely, south Western Ghats moist forests and Sri Lanka moist forests, as spatial conservation priorities representing the tropical and subtropical moist broadleaf forests biome of the Indomalayan realm (Olson et al., 2001). However, more recent investigations, particularly involving molecular phylogenetics indicate Sri Lanka alone to be identified as a unique "local center of endemism" within the biodiversity hotspot, with evidence for insular speciation and diversification from multiple lineages, including those of birds (Jha et al., 2021; Wickramasinghe et al., 2017; Bossuyt et al., 2004). For instance, the "Asian babbler" found in Sri Lanka is a noteworthy group with seven species representing families

Timaliidae, Pellorneidae, and Leiothrichidae; of which four are endemic species and the rest are endemic subspecies, making the entire group endemic to the island.

30.3.1   BIOGEOGRAPHICAL MAKING OF THE SRI LANKAN AVIFAUNA

The avifauna of the island has its intrinsic attributes similar to other fauna and flora. The major factors influencing its biota are the location of Sri Lanka in Asia; it is a continental (land-bridge) island, its internal physiography detailed by the climate, topography or relief, soils, and vegetation; its eventful geological history. Those remain the same for avifauna, with added light from Sri Lanka's close affinity to the Indian mainland, from where many taxa, including birds have colonized the island over the geological history. A combination of the above factors has caused different assemblages of bird species to occupy different avifaunal zones of the island (see chapter 23 on Sri Lanka in this book for more details; Perera and Fernando, 2023).

30.3.1.1   ECOLOGICAL BIOGEOGRAPHY OF BIRDS

The location of Sri Lanka, between northern latitudes 5° and 10° within the intertropical convergence zone, experiencing monsoonal rainfall together with a high incipient solar radiation throughout the year supports a greater avifaunal diversity. The tropical aseasonal climate of the island allows the resident birds to breed several times a year corresponding to the periods of monsoons, and supported by the year-round supply of food (Henry, 1955; Legge, 1983; Kotagama and Ratnavira, 2017). Moreover, being the southernmost island at its longitude, with no landmass south of it in the Indian ocean until the Antarctic continent, Sri Lanka is a destination for the northern hemispheric migratory birds using the central Asian flyway (Kotagama and Ratnavira, 2017; Allport et al., 2021a). Further, being a medium-sized island in the vast Indian ocean, the island inevitably accounts for a considerable richness of pelagic sea birds (Allport et al., 2021b; Panagoda et al., 2020; Seneviratne et al., 2015; Kotagama and de Silva, 2006).

Sri Lanka's climate, detailed in the introductory chapter to Sri Lanka in this book (Perera and Fernando, 2022), was illustrated by Phillips (1929) with three climatic cum faunal zones, with an emphasis on mammals. Later, he mapped the climatic cum avifaunal zones of the island (Phillips, 1952) incorporating the description by Legge (1880) and reproduced with additional data for birds in successive revisions (Phillips, 1975, 1978, 1980). The interception of the two monsoons by the highlands located in the center of the southern half of the island results in heavy rainfall in the southwestern quarter during the southwest monsoon, while comparatively low rainfall sweeps the north and east during the northeast monsoon. In addition to being shaped by the climate, the three marked peneplains in the island's topography have facilitated species to evolve in isolation. The variety of the climate and topography together with the various types of soils found in certain areas have resulted in a surprisingly diverse array of different vegetation types within the relatively smaller land area in this island (see MoMD&E, 2019 and Perera and Fernando, 2022 for a description of the ecosystem diversity in Sri Lanka). Ecological factors explained above have contributed Sri Lanka to be marginally second only to Malaysia in the bird species richness per 10,000 km^2 of land among all Asian countries (Baldwin, 1991). Wijesinghe et al. (1993) combined the climatic and topographic heterogeneity of the island into six different bioclimatic zones. Similarly, the variety of avifaunal assemblages in different parts of the country has made ornithologists of Sri Lanka delimit well-defined avifaunal zones over the history of the islands' ornithological exploration, starting from Legge (1880) (see details later in this chapter).

30.3.1.2 HISTORICAL AFFINITIES OF SRI LANKAN BIRDS

The prolonged geological and tectonic history of the Deccan plate, consisting India and Sri Lanka, has placed many different evolutionary signatures in the Sri Lankan avifauna (Legge, 1983). As explained in the introductory chapter to Sri Lanka in this book (Perera and Fernando, 2023) and references therein, geological evidence suggests that the Deccan plate formed a part of the southern supercontinent Gondwanaland, sandwiched between the African-Malagasy plate and the Antarctic plate. While it parted from other plates over successive rifting, Sri Lanka had been connected to India during its relatively rapid northward drifting over the Tethys sea. Sri Lanka was first separated from mainland India during the early Miocene period due to marine transgression, while the late Miocene aridification and Plio-Pleistocene glacial and interglacial periods resulted in alternative cycles of connection and isolation of the island from the mainland, allowing biotic colonization from the mainland during the glacial and insular speciation in interglacial periods. Biotic dispersal into Sri Lanka was also supported by increased wetness during climatic optima that facilitated forest expansions, hence supporting forest dwellers to disperse, amidst the northwestern arid belt of Sri Lanka and gulf of Mannar forming a barrier for dispersal during the Miocene aridification.

As a result of the tectonic history, Indian avifauna consists of Indo-Chinese, African, and Palearctic affinities, among which Indo-Chinese elements predominate (Ripley, 1961). While the paucity of African/Gondwanan elements in Indian avifauna can be attributed to the plate tectonics and the late divergence of most major bird lineages, the prominent Indo-Chinese elements have been explained through several more recent dispersal routes (Daniels, 2001; Ripley and Beehler, 1990; Srinivasan and Prashanth, 2006) including the Satpura hypothesis (Hora, 1949, 1953), in which dispersal is postulated along mountain ranges from the eastern Himalayas. It is noteworthy to mention that comments on such affinities in Sri Lankan avifauna had already been made by Legge (1880), during a time the knowledge on historical biogeography was far scarce. More recent geological events that isolated the mountains of south India and Sri Lanka allowed the Western Ghats and Sri Lankan highlands to harbor evolutionary relics and to undergo rapid speciation, supported by the expansion of lush rain forests on their western and southern aspects due to heavy rainfall brought in by the southwest monsoon (Jha et al., 2021). The high degree of specialization resulting in endemic lineages within the southwestern wet zone, especially in the highlands is common for many groups of the Sri Lankan biota, including birds (Legge,

1880; De Silva, 1980; Fernando et al., 2016; Wichramsinghe et al., 2017; Abeyrama and Seneviratne, 2017 b).

Ripley (1949) provides evidence for relict avifauna in Sri Lanka with major disjunctions in their distribution. Such discontinuous distributions ranges can be seen among species of which closest relatives are not found in India, but in the eastern Himalayas, Malaysian region, and/or even in Sumatra, Java, and Borneo in the southeast Asia, while such presences in Madagascar or Africa in the west have also been recorded for other taxa (De Silva, 1980; Zaher et al., 2012). Further, Ripley (1980) explains clear historical affinities of Sri Lankan avifauna with south India confirming Legge's (1880) claim on closest relatives of our southwestern wet highland avifauna to be found in the Western Ghats of south India including Nilgiri hills and those of our northwestern semiarid dry zone avifauna in the "Carnatic" dry plains of eastern south India.

Zoogeographical affinities of the endemic birds of Sri Lanka with evidence for Malabar/Western Ghats, peninsular/south Indian, Himalayan, Malayan, and even rarely found African and Afrotropical elements are tabulated by the second author in online supplementary material available at https://www.researchgate.net/publication/359237493_2022_03_15_Zoogeographical_affinities_of_SL_avifauna. This provides a preliminary chorological analysis of the distribution of the ecologically, geographically, and taxonomically closest (or presumed to be closest, when phylogenies are not available) relative in each lineage of all 34 Sri Lankan endemic bird species, based on the phylogenetic super tree of Sri Lankan avifauna presented in Abeyrama and Seneviratne (2017a, 2017b, supported by Wickramasinghe et al., 2017, and Krishan et al., 2020). The analysis documented as:

a. Two monotypic Sri Lankan endemic bird genera, one with a Malabar/Western Ghats affinity (*Sturnornis*, *Elaphrornis*) and the other with an African affinity (*Elaphrornis*)

b. Three Sri Lankan endemic species with a considerable disjunction in their lineage, two of them showing Himalayan affinities (*Pycnonotus penicillatus* and *Urocissa ornata*) and one with Malayan affinity (*Dicaeum vincens*)

c. A total of 20 Sri Lankan endemic species with their closest relative found extralimital to Sri Lanka but not with considerable disjunctions. They include seven species with Malabar/Western Ghats affinity (*Zoothera imbricata*, *Geokichla spiloptera*, *Eumyias sordidus*, *Rubigula melanicterus*, *Psilopogon rubricapillus*, *Tephrodornis affinis*, and *Columba torringtoni*), five species with Western and Eastern Ghats affinity (*Pomatorhinus melanurus*, *Myophonus blighi*, *Treron pompadora*, *Psilopogon flavifrons*, and *Psittacula calthrapae*), five species with peninsular/south Indian affinity (*Gallus lafayetii*, *Galloperdix bicalcarata*, *Argya cinereifrons*, *Argya rufescens*, and *Ocyceros gingalensis*), and three species with Malayan affinity (*Loriculus beryllinus*, *Pellorneum fuscocapillus*, and *Phaenicophaeus pyrrhocephalus*).

d. Nine Sri Lankan endemic species with their closest relative found in Sri Lanka with a wider geographic range extending out of Sri Lanka, showing one species with Malabar/Western Ghats affinity (*Gracula ptilogenys*), two species with peninsular/south Indian affinity (*Glaucidium castanotum* and *Dinopium psarodes*), five species with Malayan affinity (*Otus thilohoffmanni*, *Zosterops ceylonensis*, *Chrysocolaptes stricklandi*, *Centropus chlororhynchos*, and *Dicrurus lophorinus*), and one species with Palaeotropical affinity, except Australia (*Cecropis hyperythra*).

Inferring Sri Lanka's place in the Palaeotropical bird biogeography, the overall result indicated a prominent peninsular/south Indian (35%), Malabar/Western Ghats (26%), and Indo-Malayan (26%) affinities together with uncommon Himalayan (6%), African (3%), and the Palaeotropical (3%) affinities among Sri Lankan endemic avifauna.

30.3.2   AVIFAUNAL ZONES OF SRI LANKA

Regionalization of Sri Lanka based on the distribution of its avifauna dates back to Legge (1880), which was further developed in two recent attempts by Kotagama (1989, 1993) and by consensus of several ornithologists (MoMD&E, 2019). However, all those attempts of establishing avifaunal zones followed the intuitive discernment based on comprehensive field experience, of mainly a single ornithologist at both initial attempt and the opinion from a group of experts at the latter attempt. Although the distributional data on avifauna of the island have now reached a considerable level of completeness, a complete numerical avifaunal regionalization of Sri Lanka still remains unattempted. While the three main steps in the history of avifaunal regionalization in Sri Lanka are presented in the Figure 30.2 and described below, their discrete zones with spatial relationships are presented in Table 30.2. However, compared to less mobile vertebrates and invertebrates, zonal restrictions are poor among birds, due to their high dispersal ability (Erdelen, 1989). Nevertheless, certain avifaunal zones show high endemism at the species level (see Table 30.3 for characteristic and restricted species in each zone). In addition to the three avifaunal zones described above, Senanayake et al. (1977) delimited faunal zones in Sri Lanka, combining the monsoon-driven climatic zones with the three peneplains of the island, with an emphasis on endemism of birds and also supported by data on amphibians and lizards. Senanayake's zonation included, (a) the dry zone of the lowest peneplain, (b) wet zone of the lowest peneplain, (c) wet zone of the second peneplain, and (d) wet zone of the third (highest) peneplain (not mapped here) with three, 10, 18, and 18 species of Sri Lankan endemic birds found within each zone, respectively.

30.3.2.1   AVIFAUNAL ZONES OF LEGGE (1880)

Legge (1880) in his *History of the Birds of Ceylon* identified four intuitive regions of ornithological interest based primarily on his exploration of the avifauna of the island and the knowledge on climatic zones. He called them climatic cum avifaunal zones of Sri Lanka. Those initial avifaunal zones of Legge are listed below with his original descriptions (Legge, 1983; p. xviii).

1. The "damp luxuriant typical Ceylonese region in the Western Province," referring to the lowland southwestern wet zone.
2. The "lofty hill zones of the central highlands forming a perfect mountain zone," which refers to the central mountains possibly including the Knuckles and Rakwana ranges.
3. The "dry forest of the entire north and southeast."
4. The "arid maritime belt of the northwest coast and the Jaffna peninsula with which is grouped the similar belt on the southeast."

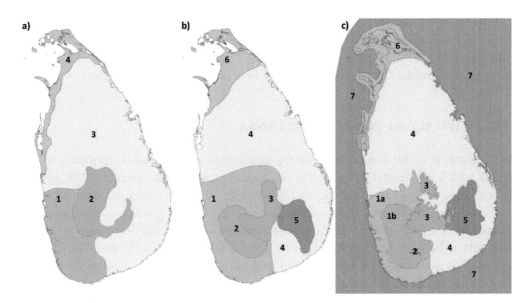

FIGURE 30.2 History of the identification of avifaunal zones of Sri Lanka: **a.** An initial map of the avifaunal zones of Ceylon proposed by Legge (1880) in his intuitive discernment, mapped by Phillips (1952) and further supported by the checklists of birds of Ceylon by Phillips (1975, 1978, 1980), **b.** Avifaunal zones of Sri Lanka revised a century after Legge by Kotagama (1989, 1993), based again on intuition however with much higher data availability, and **c.** Avifaunal zones of Sri Lanka modified through a consultative process involving 14 experts in the field including all coauthors of this chapter during the preparation of the biodiversity profile of Sri Lanka for the sixth national report to the convention on biological diversity (MoMD&E, 2019). See Table 30.2 for the names of avifaunal zones and subzones denoted by numbers.

Source: Modified from MoMD&E, 2019

TABLE 30.2   Avifaunal Zones of Sri Lanka as Developed by Different Authors over the Last 140 Years.

Climatic cum avifaunal zones of Sri Lanka (Legge, 1880; Phillips, 1952)	Avifaunal zones of Sri Lanka (Kotagama, 1989, 1993)	Avifaunal zones of Sri Lanka (MoMD&E, 2019)	
1. Low country wet	1. Low country wet	1. Wet forest	1a. Wet lowlands
	2. Mid country		1b. Wet mid-hills
		2. Rakwana hills	
2. Hill	3. Hill country	3. Highlands	
3. Low country dry	4. Dry zone	4. Dry and intermediate	
	5. Uva	5. Uva	
4. North western coastal	6. Indian	6. Palk bay coastal	
		7. Marine	

Legge's climatic cum avifaunal zones were produced into a published map only by Phillips (1952), namely the low country wet, low country dry, and the central hill zones together with the northwestern coastal zone. Although Legge mentioned the southeastern coast of Sri Lanka to accompany the northwestern coastal zone due to the similarities of the rainfall pattern, Phillips's map only delineates northwest coast and the Jaffna peninsula as an avifaunal zone. This southeastern coastline was not detailed even by Legge, to the same degree he described the northwest, possibly due to the lack of characteristic avifauna of northwest in the southeastern stretch, despite the similarity in climate.

TABLE 30.3 Characteristic and Restricted Species in Each Zone.

Avifaunal zone	Characteristic species	Zone restricted species
Palk Bay Coastal	#*Coturnix coromandelica* (Rain Quail), *Anous stolidus* (Brown Noddy), *Thalasseus bergii* (Great Crested Tern), *Phoenicopterus roseus* (Greater Flamingo)	#*Francolinus pondicerianus* (Grey Francolin), #*Cursorius coromandelicus* (Indian Courser), #*Dicrurus macrocercus* (Black Drongo)
Dry and Intermediate	*Anthracoceros coronatus* (Malabar Pied Hornbill), *Coracina macei* (Large Cuckooshrike), *Ploceus manyar* (Streaked Weaver)	None
Uva	*Phaenicophaeus leschenaultia* (Sirkeer Malkoha)	#*Treron phoenicoptera* (Yellow-footed Green Pigeon), #*Francolinus pictus* (Painted Francolin), #*Perdicula asiatica* (Jungle Bush-quail)
Wet Forest	*Phaenicophaeus pyrrhocephalus* (Sri Lanka Red-faced Malkoha)	*Urocissa ornata* (Sri Lanka Blue-Magpie), *Centropus chlororhynchos* (Sri Lanka Green-billed Coucal), *Dicrurus lophorinus* (Sri Lanka Drongo), *Garrulax cinereifrons* (Sri Lanka Ashy-headed Laughingthrush), *Dicaeum vincens* (Sri Lanka White-throated Flowerpecker)
Rakwana Hills	*Eumyias sordidus* (Sri Lanka Dusky-blue Flycatcher), *Pycnonotus penicillatus* (Sri Lanka Yellow-eared Bulbul)	None
Highlands	*Saxicola caprata* (Pied Bushchat), *Eumyias sordidus* (Sri Lanka Dusky-blue Flycatcher), *Pycnonotus penicillatus* (Sri Lanka Yellow-eared Bulbul)	*Elaphrornis palliseri* (Sri Lanka Warbler), *Myophonus blighi* (Sri Lanka Whistling-thrush)
Marine	Pelagic bird species	None

*Species endemic to Sri Lanka.

#These species, although being restricted to the respective zone, are not endemic to it or either to Sri Lanka.

Henry (1955) in his *"Guide to the Birds of Ceylon"* used an oversimplified faunal zonation simply as the wet zone and dry zone, together with a third "central hill zone," for birds. Nevertheless, the first description of those three climatic cum faunal zones goes back to Phillips (1929), for mammals.

30.3.2.2 AVIFAUNAL ZONES OF KOTAGAMA (1989, 1993)

Avifaunal zones of Sri Lanka were revisited by Kotagama, more than a century after Legge (1880), initially during the documentation of the strategy for the preparation of a biological diversity action plan for Sri Lanka (Kotagama, 1989) and published later as a map of avifaunal zones of Sri Lanka in Kotagama (1993).

Kotagama preferred Legge's avifaunal zones over that of Phillips as the former were recognized not based only on the climate as the latter. Although climate plays a major role in delimiting faunal zones, it is not essentially the only factor determining the actual distribution of the faunal assemblages. While accepting the four avifaunal zones of Legge that were mapped by Phillips (1952), Kotagama (1989, 1993) paid his attention to the peculiarities of the avifauna in the southeast of the island. However, it is not the rain shadow along the southeastern coast, which was proposed by Legge (1880) as being similar to the condition of the northwest, which Kotagama identified as a unique avifaunal zone. Instead, he describes the Uva basin with its peculiar savanna vegetation as a distinct avifaunal zone, owing to the unique (but not endemic) assemblage of species inhabiting there (see Table 30.4). Further, Kotagama (1993) mapped a mid-country transition zone in between the wet lowlands and the hill country in south-western Sri Lanka. This made a total of six avifaunal zones: (a) Indian avifauna in the north, (b) dry zone avifauna distributed in much of the dry plains of Sri Lanka, (c) Uva avifauna that can be found in the low hills of the southeastern sections of the intermediate and dry climatic zones, (d) low country wet avifauna distributed in the lowland rain forests of the southwestern coastal plain of the island, (e) mid-country avifauna found in the submontane rain forests in mid-elevation hills of the southwest, and (f) hill country avifauna in the montane cloud forests of the central highlands of the island. Kotagama's six avifaunal zones are still a product of intuition based on an individual's experience on species distributions, however with a higher degree of data sharing and/or publication during the contemporary development of information technology, compared to Legge.

30.3.2.3 AVIFAUNAL ZONES IN THE BIODIVERSITY PROFILE FOR SIXTH NATIONAL REPORT TO THE CONVENTION ON BIOLOGICAL DIVERSITY

The avifaunal zones of Sri Lanka were modified during the preparation of the biodiversity profile of Sri Lanka for the sixth national report to the convention on biological diversity (MoMD&E, 2019) through a consultative process involving 14 experts in the field including three coauthors of this chapter (SKD, SJP, and SSS), in which those zones of Kotagama (1989, 1993) were used as the basis to which individual field experience of experts was added, specifically considering some restricted distributions of endemic species. However, the seven avifaunal zones and two subzones identified through this consultative process are still a product of intuition based on expert knowledge on the distributions of resident and marine birds. Hence, currently accepted avifaunal zones for the island are, (1) wet forest zone including two avifaunal subzones named (a) wet lowlands subzone and (b) wet mid-hills subzone, (2) Rakwana hill zone, (3) highlands zone, (4) dry and intermediate zone, (5) Uva zone, (6) Palk bay coastal zone, and (7) marine zone (MoMD&E, 2019).

Major amendments made in the current avifaunal zonation to that of Kotagama's included (i) making the mid-country avifaunal zone a subzone within the wet forest avifaunal zone, (ii) addition of the Rakwana hill avifaunal zone, (iii) an extension of the Indian avifaunal zone to include the coastal habitats, and renaming it as the Palk bay coastal avifaunal zone, and (iv) addition of the marine avifaunal zone including both southwest monsoon pelagic avifauna and the northeast monsoon pelagic avifauna, encompassing all pelagic habitats around the island. The boundary modifications were mostly from the updated climatic zone boundaries and a 30-m digital elevation model more accurately defining river basins, the montane and submontane forests in the wet zone (MoMD&E, 2019). The highland avifaunal zone was delimited by the 1000 m contour, also including Haputale and Namunukula hills. The group agreed upon the use of 900 m contours in delimiting Rakwana hills as a unique avifaunal zone recognized during this exercise. Low country and mid-country avifaunal zones of Kotagama were considered in the present classification as subzones of the wet forest avifaunal zone, with wet lowlands up to the foothills of the second peneplain delimited by the 300 m contour, and the wet mid-hills above 300 m altitude with an undulating landscape, to encompass the southwestern corner of the Sabaragamuwa mountain range including Kanneliya-Dediyagala-Nakiyadeniya forest complex, northwest Sinharaja ridge up to Labugama and the foot hill areas of the Rakwana mountain range below 900 m. The distribution of hill savanna habitat was used in delimiting the Uva avifaunal zone ensuring the inclusion of Uva low hills. The current Palk bay coastal avifaunal zone was an expansion from the previous Indian avifaunal zone to encompass the coastal wetlands such as Mannar island, Adam's bridge, Jaffna peninsula, Chundikulam lagoon, and surrounding shallow seas and northern coastal waters identified with immense importance for waterbird aggregation, especially of migratory waders.

30.3.2.3.1   *Wet Forest Avifaunal Zone*

Wet forest avifauna of the island inhabits its dense and prolonged rain forests that acted as a refuge for in situ speciation of endemics. The zone mainly comprises forest birds. The most species-rich mixed species feeding flocks of birds can be seen in these forests comprising many endemic species, occupying the different layers of the forest physiognomy and varied horizontal feeding niches even at the same vertical layer. *Phaenicophaeus pyrrhocephalus* (Sri Lanka red-faced malkoha) is found characteristic to the wet forest avifaunal zone, while *Urocissa ornata* (Sri Lanka blue magpie), *Centropus chlororhynchos* (Sri Lanka green-billed coucal), *Dicrurus lophorinus* (Sri Lanka drongo), *Garrulax cinereifrons* (Sri Lanka ashy-headed laughingthrush), *Argya rufescens* (Sri Lanka orange-billed babbler), and *Dicaeum vincens* (Sri Lanka white-throated flowerpecker) show restricted ranges, however not exclusive within this zone. *Otus thilohoffmanni* (Sri Lanka Serendib scops owl) shows the most restricted distribution within the zone. All the above species are endemic to Sri Lanka. This avifaunal zone captured almost all the area in the wet zone except highlands, and further divided into two subzones as wet lowland avifaunal subzone and wet mid-hill avifaunal subzone. Wet lowland avifaunal subzone is characterized by Dipterocarp—*Shorea*, dominant rain forest tree communities. When birds are concerned, mid-hill avifaunal subzone is characterized by relatively high abundance of some canopy dwelling large-bodied endemic birds such as the *Phaenicophaeus pyrrhocephalus* Sri Lanka red-faced malkoha and *Sturnus albofrontatus* Sri Lanka white-faced starling, while they are seldom found in the wet lowland avifaunal subzone.

30.3.2.3.2   Highland Avifaunal Zones

The land area encircled by the 1000 m contour or the highland avifaunal zone includes the central mountain complex, Namunukula hill range and Haputale hills, together with the Knuckles mountain range. Highland avifauna is characterized by the endemic birds with the most restricted ranges and the avifauna here shows the highest degree of endemism among all avifaunal zones of the island. Both forest and grassland birds are seen while the former shows the highest diversity. *Saxicola caprata* (pied bushchat), the grassland birds together with forest birds such as *Eumyias sordidus* (Sri Lanka dusky-blue flycatcher), *Pycnonotus penicillatus* (Sri Lanka yellow-eared bulbul), and *Zosterops ceylonensis* (Sri Lanka white-eye) are characteristic to the highlands avifaunal zone, while *Elaphrornis palliseri* (Sri Lanka warbler) and *Myophonus blighi* (Sri Lanka whistling thrush) show narrow distribution ranges restricted within the zone. All the above species except the first are endemic to Sri Lanka.

30.3.2.3.3   Rakwana Avifaunal Zones

Eastern section of the Sinharaja rain forest above 900 m in Rakwana hills has been considered here as a separate avifaunal zone. The geographic isolations due to deep valleys and anthropogenic settlements between the central mountain complex and the isolated Rakwana hill ranges are considered here in the identification of two avifaunal zones, since the isolation is more than 25 km by aerial distance. This valley acts as a dispersal barrier to the bird species with weak flight performance such as the *Elaphrornis palliseri* (Sri Lanka bush warbler) making them absent in the Rakwana range. *Eumyias sordidus* (Sri Lanka dull-blue flycatcher) and *Pycnonotus penicillatus* (Sri Lanka yellow-eared bulbul) being endemic to Sri Lanka and other highland birds are characteristic to the Rakwana hills avifaunal zone too, while no bird species are found restricted within this zone.

30.3.2.3.4   Uva Avifaunal Zone

The natural habitats of this avifaunal zone are characterized by a dominant savanna ecosystem mosaic with dry riverine forests and rock outcrops. These environmental features support a combination of floral and faunal communities unique to this avifaunal zone. Three bird species not endemic to Sri Lanka are known to be restricted to this avifaunal zone and all three being naturally rare and difficult to spot. Those are *Francolinus pictus* (Painted francolin), *Treron phoenicoptera* (Yellow-footed green pigeon), and *Perdicula asiatica* (Jungle bush quail), showing restricted ranges within the zone, while *Phaenicophaeus leschenaultia* (Sirkeer malkoha) can be identified as characteristic to the Uva avifaunal zone. Due to low numbers, those three zone-restricted species are critically endangered at the national level even though they are quite widespread in the Indian subcontinent.

30.3.2.3.5 Dry and Intermediate Avifaunal Zone

This zone is the largest avifaunal zone on the island, which occupies > 60% of the country. It spans through a flat landscape with scattered isolated rocky hills covered with dry mixed evergreen forests and transitional moist forests in between wet to dry conditions which can be called as intermediate forests, thorny scrublands, and dry grasslands. Dry and intermediate avifauna is marked mainly by its forest and grassland birds as well as the numerous varieties of storks, herons, egrets, ducks, and other aquatic and wading birds making benefit from the opportunities provided by the ancient and partially restored hydraulic civilization of Sri Lanka, together with an array of waders inhabiting and over-wintering in coastal wetlands. Some bird species characteristic of this avifaunal zone are *Anthracoceros coronatus* (Malabar pied hornbill), *Clamator jacobinus* (Jacobin cuckoo), *Ploceus manyar* (Streaked weaver), *Phaenicophaeus viridirostris* (Blue-faced malkoha), *Glaucidium radiatum* (Jungle owlet), and *Coracina macei* (Large cuckooshrike). None of those characteristic species are endemic to the island, while there are no bird species found strictly restricted within this zone. Furthermore, this avifaunal zone holds the breeding populations of *Charadrius dubius* (Little-ringed plover) and *Charadrius alexandrinus* (Kentish/Hanuman plover), which are known to be the only members in genus *Charadrius* with a breeding population in Sri Lanka.

30.3.2.3.6 Palk Bay and Northern Avifaunal Zone

According to Legge (1880), this zone was called the northwestern coastal belt and the Jaffna peninsula, which is characterized by peculiar thorny vegetation. It is home to a characteristic assemblage of avifauna, which is similar to those in lowland plains of the eastern section of south India. This zone is located in the northernmost coastal area of the island including Mannar island, Rama's bridge, Jaffna peninsula, and coastal belt up to Pulmudai. Most of the land area is dominated by thorny scrublands and salt marshes in this zone. This avifaunal zone holds the only mass nesting sites of pelagic terns in Sri Lanka located in the third island of Rama's bridge, which is now declared as a national park. *Anous stolidus* (brown noddy), *Thalasseus bergii* (great crested tern), *Phoenicopterus roseus* (greater flamingo), and a large number of migratory shorebirds including the *Dromas ardeola* (crab plover), *Limosa lapponica* (bar-tailed grodwit), *Calidris tenuirostris* (great knot), and *Xenus cinereus* (Terek sandpiper) are found characteristic to the Palk bay and northern coastal avifaunal zone. Further, a unique assemblage of birds including *Francolinus pondicerianus* (grey francolin), *Cursorius coromandelicus* (Indian courser), *Dicrurus macrocercus* (black drongo), *Lanius schach caniceps* (long-tailed/rufous-backed shrike), *Streptopelia decaocto* (Eurasian *collared dove),* and *Coturnix coromandelica* (rain quail), which are characteristic of lowland plains of the eastern section of south India are found restricted within this zone, however none of them being endemic to Sri Lanka. Further, the characteristic nature of some species Legge listed for this zone has now been flawed as of current data, especially the *Eremopterix griseus* (ashy-crowned sparrow/finch lark) and the *Euodice malabarica* (Indian silverbill) showing a much wider island-wide distribution.

30.3.2.3.7 Marine Avifaunal Zone

All the marine habitats of pelagic birds are considered as a separate avifaunal zone within the exclusive economic zone of Sri Lanka in the Indian ocean. Pelagic bird species such as shearwaters, petrels, storm petrels, frigatebirds, tropicbirds, jaegers, and pelagic terns such as *Onychoprion fuscatus* (sooty tern) and *Onychoprion anaethetus* (bridled tern) are characteristic to the marine avifaunal zone, while no bird species is found restricted within this zone.

30.3.2.4 AN ATTEMPTED NUMERICAL REGIONALIZATION OF SRI LANKA FOR ITS AVIFAUNA

As explained above, Sri Lanka has been partitioned into several taxa-specific bioregions, among which avifaunal zones have long been established and recently been revised. However, all the above proposals for avifaunal zones were established predominantly based on intuitive discernment, initially based on an individual's field experience and recently with the consensus of many. This approach lacks uniform spatial scale and some of the resulting units may be too coarse for conservation applications. Hence, a preliminary analysis (Perera et al., 2015) was attempted to partition Sri Lanka into numerical avifaunal zoogeographic entities employing an objective and a quantitative method, with a pilot distribution data set for bulbuls of the family Pycnonotidae representing seven species; most being regionally or locally widespread with a single endemic in the lowlands and a single range-restricted endemic species in the highlands. With the high diversity and a considerable percentage of endemism (33%), together with the acceptable level of knowledge on their distribution ranges, bulbuls represent an indicator group to start quantitatively exploring the avian zoogeography of Sri Lanka, and also to see whether they follow similar patterns with plants as described by Ashton and Gunatilleke (1987). These taxa are fairly equally and adequately sampled across the island and have an acceptable level of clarity in their taxonomy. Ramdhani et al. (2010) proposed the use of a quarter degree grid cell (QDGC) system for Sri Lanka, as a standard spatial reference tool to document and express patterns of biological diversity and its distribution, as practiced elsewhere in the world (Larsen et al., 2009). Each $1° \times 1°$ grid cell covering Sri Lanka was divided into four $0.5° \times 0.5°$ half degree grid cells and 16 $0.25° \times 0.25°$ QDGCs (ca. 27×27 km). All QDGCs covering the landmass of Sri Lanka were used as operational geographic units (OGUs) for the analyses. Such a grid system has several analytical advantages and requires minimum assumptions (Perera et al., 2011; Procheş, 2005; Ramdhani et al., 2008); all such grid cells are more or less of equal size, which facilitates analytical procedures where area/scale is important. A species incidence (presence/absence) matrix was populated for selected species in above equal-area OGUs through an extensive review of literature and published maps, supplemented with the authors' field experience. The database was subjected to a numerical bioregionalization with an agglomerative and hierarchical clustering approach using Jaccard index of dissimilarity (Jaccard, 1901) for pairwise comparison of OGUs and the unweighted pair-group method with arithmetic averages (UPGMA) algorithm to define a cluster dendrogram, following Kreft and Jetz (2010) and Proches and Ramdhani (2012). The dendrogram clustered the OGUs with similar avifauna and hence defined hierarchical zoogeographical divisions (provinces and districts), and established relations between those spatial entities. The results of this preliminary regionalization were mapped using ArcMap 10.1 (ESRI, 2012) presenting the avian zoogeographical patterns for the island and also identifying priority regions for conservation.

Results of the said preliminary avifaunal regionalization presented in Figure 30.3, reconfirm the current bioclimatic zones (i.e., the wet zone, dry zone, and the intermediate zone), which are principally determined by the temperature and the annual rainfall, while emphasizing the need to incorporate elevation as a determinant variable for a subordinate hierarchy in regionalization (i.e., montane and the lowland subzones). According to results of the analysis, Sri Lanka is divided into two major avifaunal provinces representing the dry and wet climatic zones, while the subordinate avifaunal districts indicate dissimilarities of avifauna within the provinces, due to the intermediate climates and higher elevations. The dry province for avifauna includes the true dry district and the district with a dry aspect of the intermediate climate. Similarly, the wet province comprises the lowland wet and the montane wet districts in addition to a district with a wet aspect of the intermediate climate. Therefore, the regionalization numerically recovers avifaunal zoogeographical entities not much different from the bioclimatic zones of the island (Wijesinghe et al., 1993). These spatial patterns, together with their species richness and endemism, support the higher conservation value of the southwestern wet zone, especially its montane subsection, thus prioritizing that area for local-scale conservation attempts within the biodiversity hotspot.

Further, the regionalization depicted in Figure 30.3c suggests a transition zone in between the wet and dry avifaunal provinces. The northwestern and southeastern sections of it fall under the wet province, while the northeastern section falls under the dry province indicating the true intermediate nature of those transitional districts. However, the Uva and Palk bay coastal avifaunal zones among those intuitive delimitations are not recovered in this preliminary numerical regionalization, due to lack of species representing such distributions in our database. Similarly, it has not captured the habitat heterogeneity within the wet avifaunal province, which would have been divided into several narrow districts with a complete database with a greater representation of the Sri Lankan avifauna. This study has presented only a preliminary modal analysis for a possible future analysis on a multispecies distribution database representing the entire avifauna of the island, which we believe is currently available at an acceptable level of completeness at the GDGC scale. Such an analysis can inform the conservation management with spatial priority needs for bird conservation. The results demonstrate that most of the intuitive bioregionalization for the country can be numerically confirmed and/or corrected and fine-tuned with such an analysis based on such a comprehensive database at an appropriate scale. One has to be cautious on the spatial scale of the analysis as the incompleteness of data at smaller grain sizes can cause geographical scattering of biogeographic clusters of OGUs making them discontinuous, while increasing the grid size also increases the risk of not identifying natural boundaries of resulting biogeographical entities and missing finer patterns of endemism that are important in conservation planning (Perera et al., 2011, 2018, 2021). However, the preliminary results shown here suggest that the QDGC scale is too coarse for such an analysis to deal with the rapidly changing climatic and topographic heterogeneity of the island (Perera et al., 2015), while the available data, at least for some well-studied taxa, could be complete enough to facilitate a finer grid scale of analysis.

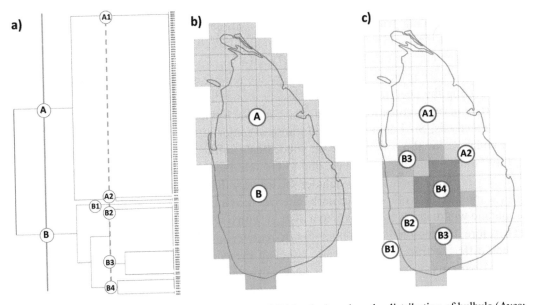

FIGURE 30.3 Numerical avifaunal regionalization of Sri Lanka based on the distribution of bulbuls (Aves: Pycnonotidae), **a)** Agglomerative and hierarchical clustering dendrogram computed based on Jaccard index of dissimilarity and unweighted pair-group method with arithmetic averages (UPGMA), with two phenon lines defining avifaunal entities (solid line for provinces and dashed line for districts), **b)** Proposed avifaunal provinces for Sri Lanka: A—dry avifaunal province and B—wet avifaunal province, and **c)** Proposed avifaunal districts for Sri Lanka: A1—dry avifaunal district, A2—dry intermediate avifaunal district, B1—rejected due to insufficient land cover, B2—lowland wet avifaunal district, B3—wet intermediate avifaunal district, and B4—montane avifaunal district.

Source: Authors

30.4 CONSERVATION STATUS AND THREATS

Being a part of a biodiversity hotspot in the world (Myers et al., 2000), inevitably we have threatened our biodiversity. Ostensibly the magnitude of threats which has a direct impact on birdlife is comparatively low when compared with the other animals. But lack of high-quality and robust studies related to birds, specifically on threats, is a major drawback of Sri Lanka. Even though we have approximately 30% of land area under protected area jurisdiction (MoMD&E, 2019) under the monitoring and management of the department of wildlife conservation and the department of forest conservation, we have observed a decline in bird numbers, a decrease in species richness, and decreased aggregations of waterbirds throughout the country. This observed evidence suggests us that the status is not much satisfactory when it comes to bird conservation (Weerakoon and Gunawardena, 2012). Further, Sri Lanka has declared six Ramsar sites (RSIS, 2021) which are important for wetland and waterbirds, and identified 70 important bird areas/key biodiversity area where most of those sites are protected under law (MoMD&E, 2019).

However, it is important to discuss the observed and realized threats to Sri Lankan avifauna over the last few decades. The last assessment of the conservation status of the birds was done in 2012, which resulted in 18 species for both critically endangered and endangered category, 31 species into vulnerable status, and 35 belong to near threatened. The current assessment which we have done in 2021 yielded an increased number of the birds in threatened categories, which shows us the increased pressure from all the threats we identified (Table 30.4). Here, we discussed most of the identified threats to the avifauna of Sri Lanka, which were also widely applicable for many faunal groups; few birds' specific threats are also discussed where we need immediate attention to mitigate the impact generating from those threats.

TABLE 30.4 The Number of Threatened Birds and Accessed Birds During National-Level Conservation Assessments of 2007, 2012, and 2021 by the Ministry of Environment, Sri Lanka.

	NCS 2007	NCS 2012	NCS 2021*
Critically endangered	12	18	19
Endangered	17	18	48
Vulnerable	24	31	14
Total threatened	**53**	**67**	**81**
Near threatened	NE	35	31
Least concern	NE	138	130
Data deficient	NE	-	2
The total number of species assessed	**53**	**240**	**244**

NCS, National conservation status assessed by biodiversity secretariat, Ministry of environment; NE, Not evaluated.

*In press.

30.4.1 DECIMATION OF TROPICAL RAIN FORESTS AND MONTANE CLOUD FORESTS

Sri Lanka is losing its rain forests in an alarming rate due to agriculture expansions, illegal encroachment, and development of linear infrastructure. The destruction of tropical rain forests and montane cloud forests poses a serious direct extinction risk of many forest species, including birds. Recent satellite remote sensing and GIS technique-mediated study revealed that from 1976–2014, our forest loss was 5.5% (Sudhakar Reddy et al., 2017). Unfortunately, it has occurred in the most diverse climatic zone (the wet zone) and currently, it has suffered from urban expansion, agricultural expansion, as well as unplanned developmental projects. Today, we have to discuss the threat levels and conservation in an already fragmented landscape. It was found that 87% of the forest clearing happened due to small-scale farming and tea plantations being the major drive to the current loss of forest areas (Mattsson et al., 2012).

When it comes to fragmentation (apart from habitat loss), habitat degradation and edge effects are always there to intensify the threats toward biodiversity. Approximately 35% of resident birds highly rely on forested habitats, including 60% of the endemic birds (MOE, 2012). Even though Sri Lanka lacks systematically designed studies on how fragmentation affected birds, the absence of large-bodied, forest species (e.g., *Phaenicophaeus pyrrhocephalus* Sri Lanka red-faced malkoha and *Urocissa ornate* Sri Lanka magpie) in some larger forest fragments such as Yagirala forest reserve (area = 2357.1 ha) indicates the species extirpations due to habitat fragmentation and isolation (Figure 30.4). Further, we have no evidence about the pressure due to habitat loss or habitat degradation faced by the other forest-dwelling endemic species or the population fluctuations. Thus, we highlight the crucial need for properly planned systematic studies on these knowledge gaps linked with deforestation and fragmentation, specifically in the wet zone. After 2009, where the civil unrest has ceased in the north and east, the post-war period up to now clearly resulted from the recolonization of people in the northern dry zone. This has led to a large-scale clearing of northern dry zone forests and this will directly affect large-bodied frugivore species such as *Anthracoceros coronatus* Malabar-pied hornbills, which has been listed as near threatened in the global context.

FIGURE 30.4   (A) *Phaenicophaeus pyrrhocephalus* Sri Lanka red-faced malkoha and (B) *Urocissa ornate* Sri Lanka magpie few of the largest and forest restricted endemic birds threatened with habitat loss and degradation. *Source:* Photo credits, Gehan Rajeev.

30.4.2 HABITAT DEGRADATION

Most of the smaller forest fragments (size < 100 ha) have been invaded by many invasive plant species, hence affecting the native vegetation structures and the species composition. These invasive plants (e.g., *Dillenia suffruticosa*, *Alstonia macropylla*, *Clidemia hirta*, *Prosopis juliflora*, and *Annona glabra*) are spreading fast along the fragment borders and altering the native plant communities. Though some invasive plant species provide food sources for some birds, such as nectarivores and frugivores, we can assume that by altering the native plant species composition they may affect the persistence of food webs in forest ecosystems. But this assumption should be studied to conclude.

Most of the wetlands have been polluted with heavy metals due to excess and uneducated use of insecticides and pesticides. The bioaccumulation causes lethal health hazards to the bird species which utilize the food resources of these wetlands. The salt lagoons of one of the most important Ramsar sites (the Bundala national park) had changed salinity levels, and due to that, we observed once shorebirds abundantly feed mudflats have disappeared, and now the entire lagoon converted to a freshwater pool with many waterbirds. This site was known to hold thousands of *Phoenicopterus roseus* greater flamingos before 1995 and now not a single individual can be seen even during the peak migratory periods, though several individuals nomadically visit to check the food availability.

30.4.3 LAND-USE CHANGES

Currently, there are land-use changes at the local level where farmers and villages are changing their home gardens to multicropping agriculture. The farmers who have agricultural fields along the forest reserves and fragments are doing illegal encroachments to increase their area of the farming field. But the farmers who don't do that intensify the agricultural practices toward monoculture cropping which provides monotonic vegetation. It was found that the abundance of endemic species with small range sizes is higher in the forest than in the agricultural fields in buffer zones near forest reserves (Sreekar et al., 2021). But when it comes to total species richness, agricultural sites with higher vegetation complexities harbor a higher number of species than forests (Hanle et al., 2021), especially in the buffer zones. But these small-scale agricultural lands could act as an ecological trap where the domestic predators (cats and dogs) may also be present in these anthropogenic habitats, thus the actual ecological role of these small-scale farming lands need further investigation to fully understand the actual use of this land-use type by the birds.

Another dominant land-use change that is happening currently, which affects birdlife, is the expansion of prawn farming. There is an emerging trend to clear the mangrove vegetation cover and convert those wetlands to prawn farm industries (Gunawardena and Rowan, 2005). These habitats are important as feeding grounds for many waterbirds as well as a roosting site for many birds. All these changes affect directly and indirectly the long-term survival of many bird species, and if these habitat destructions are going to continue, then most of the waterbird populations will face perilous situations.

30.4.4   UNPLANNED LINEAR INFRASTRUCTURE DEVELOPMENT

The Ministry of environment in Sri Lanka has established clear and strong rules and regulations to mitigate the impacts of developmental projects on the environment through various pre- and post-monitoring procedures (e.g., initial environment assessment, environmental impact assessments, etc.). However, several rapid accelerated projects such as building highways, building wind farms, and building mini-hydro systems seem to have tarnished the environmental screening procedures and still need considerable improvement (Zubair, 2001). We can see some misuses or delaying processes to follow the established procedures to minimize the environmental impacts by the people who engaged in those developmental projects. Continuous pressure building from the environmentalist and nature lovers against those developmental projects indicates the conflicting situation. Malpractices against already established environmental protection have a considerable impact on birdlife as well as other wildlife. Recent development along the critical feeding grounds of migratory birds in the north and northeastern sector of the country highlighted these issues with several wind power projects being implemented. But there are some satisfactory mitigatory actions said to be in place, such as the establishment of an automated radar system to shut down the wind blade rotation when detected the flight of a large flock of birds toward the wind tour. Functionality of these said mitigation system and the longevity of it is highly questionable though.

Projects under developing highway network tend to identify wetlands (marshlands, abandoned paddy fields) as placement of the route because of the efficacy of land clearances are easy. But they tend to fill the wetlands rather than building the highways on a column, which leads to the destruction of the habitat of the waterbirds. Those localize reductions of critical feeding grounds will start a cascade effect on population decline of the waterbirds.

30.4.5   CLIMATE CHANGE

It is predicted that if the current trends of global warming continue, 89% of the land bird species may go extinct by 2100, which are mostly physiologically specialized and restrict range species (Şekercioğlu et al., 2012). It has shown that the climate change may cause significant and gradual upslope shifts, that is, +4.4 m for each 1°C increase in temperature (Pouteau et al., 2018). As a tropical country with higher number of endemic and near-endemic bird species, which are mostly favored by humid microclimatic niches, Sri Lanka is vulnerable to climate change. Even though there are no studies that explicitly discuss the impacts of climate change on Sri Lankan birds, there are a few predictions which showed some susceptibility toward risk of extinction of some bird species, especially on the highland endemics. A recent study conducted in the wet zone of Sri Lanka covering three land-use types with varying elevations revealed that the climate is more important in shaping bird communities at local scale, which exemplified the importance of climate niches (Sreekar et al., 2020). Most of these bird communities comprise restricted-range species such as *Myophonus blighi* Sri Lanka whistling thrush, *Phaenicophaeus pyrrhocephalus* Sri Lanka red-faced malkoha, *Zoothera imbricata* Sri Lanka thrush, and *Sturnornis albofrontus* Sri

Lanka white-faced starling (Sreekar et al., 2021). Further, a climate suitability modeling attempt done for five exclusively highland endemic species, namely *Elaphrornis palliseri* Sri Lanka bush warbler, *Myophonus blighi* Sri Lanka whistling thrush, *Eumyias sordidus* Sri Lanka dusky-blue flycatcher, *Pycnonotus penicillatus* Sri Lanka yellow-eared bulbul, and *Zosterops ceylonensis* Sri Lanka white-eye suggested that by 2050, there won't be any area with suitable climate left for Sri Lanka yellow-eared bulbul, and by 2080, Sri Lankan bush warbler and Sri Lanka dusky-blue flycatcher will face > 80% reduction in their climatic niche (Perera et al., 2016). However, extreme weather events such as heat waves, cold spells, tropical cyclones, severe floods, and landslides (Wickramasinghe et al., 2021) may have adverse impact on some bird species in Sri Lanka which has limited and scattered populations, especially those found in coastal habitats. Climate change predictions suggest decreased northeast monsoon rainfall in the low country dry zone, while the southwest monsoon rainfall is predicted to increase in the wet zone in low, mid, and high altitudes (Alahacoon and Edirisinghe, 2021), the impacts of which on the avifauna is yet to be assessed. Lack of systematically gatshered long-term datasets on ecology of birds has greatly affected any detailed predictions of the impacts of climate change on Sri Lankan birds.

30.4.6   PET TRADE

There are only five species of birds (*Corvus splendens* house crow, *Corvus macrorhynchos* large-billed crow, *Lonchura punctulata* scaly-breasted munia, *Lonchura striata* white-rumped munia, and *Psittacula krameri* rose-ringed parakeet, see FFPO, 1937, section 31, schedule 3), which have not been protected from the fauna and flora protection ordinance (FFPO) of Sri Lanka, the overarching legislative tool for protection of biodiversity in Sri Lanka (FFPO, 1937). Catching, holding or killing any species of bird without a permit even outside the protected area network is illegal under the provisions of FFPO. But there are few cases of selling bird species for pet trade noticed locally. Even though the magnitude of this threat is low, it can seriously affect populations of strictly forest-dwellers, such as Sri Lanka myna and plum-headed parakeet with a population of scattered distribution mostly among the tea estates.

30.4.7   OTHER THREATS

Hunting of birds for meat and collection of eggs, especially large eagles, waterfowls, and members of the Phasianidae family seldom occurs in rural areas. Once abundant Sri Lanka junglefowl, even in suburban areas such as villages and agricultural fields, now see a rapid decline. A frequent occurrence of snares, traps, and direct shooting is reported to the department of wildlife conservation, which indicates that the poaching in Sri Lanka is widespread.

Another neglected but important threat to Sri Lankan birdlife is pollution-related threats such as pesticides and herbicides. The results from Jayaratne et al. (2019) study on heavy metal accumulation in bird feathers (*Egretta garzetta* Little egret) showed the observed accumulation levels already exceeding the known limits. Removal of roosting trees along urban areas can be a considerable threat to the population of egrets and cormorants because most of those roosting colonies are used as nesting sites as well. Therefore, the removal of those roosting sites directs a great impact on waterbird populations, especially with the urban development projects now concern more about the removal of roadside large trees to expand the roads and put walking paths.

Another least concerning threat is the eutrophication of the inland wetland system. A large area of the lowland dry zone of Sri Lanka holds a great network of man-made tank systems, recently identified as one of the globally important agricultural heritage systems by FAO (Santoro et al., 2020). But this agricultural system is now in peril due to unmanaged use of hazardous pesticide, thus resulting in increased virus infections in the bacterial community (Peduzzi and Schiemer, 2004), which are crucial to mediate persistence of the wetland ecosystem. This could be indirectly linked to population declining of waterbirds. The spreading of zoonotic diseases among waterbirds often goes unnoticed but is rarely reported by regional newspaper reporters, which could be caused by *Salmonella*. Because there are some incidents which caused mass deaths of cormorants in north-central province. Thus, proper followed up research on those mass deaths should be implemented to identify the reasons behind those incidents.

Another neglected but important and emerging threat is unethical bird watching and bird photography. In the recent past, many emerging bird photographers and most of the established bird watching tour guides are known to use tape lures to attract birds to them to show to a client or just to photograph. Continuation of such acts may be resulting in permeant displacement of highly territorial birds such as thrushes, owls, nightjars, etc. Because of the misusage of smartphone applications, even some local guides under government conservation-related departments use them to have playbacks activities to show birds to visitors without knowing the consequences of these malpractices. Alteration of vegetation to get a clear shot of individual roosting sites and nesting places is also happening, infrequently done by unethical bird photographers as well as attracting elusive birds toward them to take photographs. This could be exposing the rare and elusive bird to its predators, which could be a human-driven threat to the birdlife. These increasing events of violating ethical wildlife photography subjected the formulation of ethical photography model suggesting a proper ethical framework for wildlife photography in Sri Lanka (Podduwage, 2016).

30.5   CONCLUDING REMARKS

Sri Lanka is a truly remarkable island with densely packed biological diversity, among which birds make the most attracted faunal group not only among scientist but also among the general public, while holding a rich evolutionary and biogeographic history still awaiting a detailed exploration. The historical influence from Buddhism shaped the communities to be compassionate toward most of the wildlife, making Sri Lanka an island with no hunting pressure on birds, compared to the other Asian countries. Institutional structure and the capacity of Sri Lanka for biodiversity conservation including government departments such as the department of wildlife conservation and the forest conservation department; institutes focusing on biodiversity information management such as the biodiversity secretariat in the Ministry of environment; universities with active research culture on ornithology including units such as the field ornithology group of Sri Lanka; bodies such as the wildlife and nature protection society and Ceylon bird club; such other environmental NGOs actively putting their pressure for the need of the conservation of flora and fauna indicates a promising future for the biodiversity of the island promoting its sustainable use for the development. Furthermore, the increasing involvement of conservation management in evidence-based scientific initiatives and the increasing participation of the people of Sri Lanka in citizen science programs such as eBird and iNaturalist are shedding light on a future with more educated nature loving community.

ACKNOWLEDGMENT

We would like to thank Ms. R. H. M. Padma Abeykoon (Director) and Ms. Chanuka Maheshani Kumari (Development Officer-Environment) of the Biodiversity Secretariat for steering the initial stage of this work. We extend our gratitude to Chinthaka Kalutota, Athula Wijesinghe and Samatha Suranjan Fernando for providing necessary literature sources. We thank Dulan Ranga Vidanapathirana and Gehan Rajeev for contributing their photographs of birds and Lasith V. Perera for his support in preparation of maps.

KEYWORDS

- **birds**
- **Sri Lanka**
- **bird taxonomy**
- **bird diversity**
- **biogeography**
- **bird conservation**
- **threats**

REFERENCES

Abeyarama, D.; Seneviratne, S. S. Evolutionary Distinctiveness of Sri Lankan Avifauna. *WILDLANKA J. Dept. Widllife Conserv. Sri Lanka* **2017a,** *5* (1),1–10.

Abeyarama, D. K.; Seneviratne, S. S. Evolutionary Distinctness of Important Bird Areas (IBAs) of Sri Lanka: Do the Species-Rich Wet Zone Forests Safeguard Sri Lanka's Genetic Heritage? *Ceylon J. Sci.***2017b,** *46* (5), 89–99.

Alahacoon, N.; Edirisinghe, M. Spatial Variability of Rainfall Trends in Sri Lanka from 1989 to 2019 as an Indication of Climate Change. *ISPRS Intern. J. Geo-Info.* **2021,** *10*, 84.

Allport, G.; Collinson, M. J.; Shannon, T. J.;Seneviratne, S. S. Eastern Yellow Wagtail *Motacilla tschutschensis* and Western Yellow Wagtail *M. flava* in Sri Lanka with Comments on Their Status in the South Asia Region. *Ibis*. **2021a**.

Allport, G.; Kodikara Arachchi, M. H.; Perera, L. Evidence of a 'Spring' *Stercorarius skua* Passage in West Coast Sri Lanka. *Birding Asia* **2021b**.

Alström, P.; Olsson, U.; Lei, F. A Review of the Recent Advances in the Systematics of the Avian Superfamily Sylvioidea. *Chinese Birds* **2013**, *4*, 99–131.

Ashton, P. S.; Gunatilleke, C. V. S. New Light on the Plant Geography of Ceylon. I. Historical Plant Geography. *J. Biogeogr.* **1987**, *14*, 249–285.

Baldwin, M. *Natural Resources of Sri Lanka-Conditions and Trends Colombo*; Natural Resources Energy and Science Authority of Sri Lanka: Sri Lanka, **1991**; p 280.

Bawa, K.; Das, A.; Krishnaswamy, J.; Karanth, K.; Kumar, N.; Rao, M. *Ecosystem Profile: Western Ghats and Sri Lanka Biodiversity Hotspot*; Western Ghats Region. Conservation International: Arlington, VA, **2007**.

BirdLife International. Data Zone, 2021. http://www.datazone.birdlife.org (accessed on 9.30.2021).

Bossuyt, F.; Meegaskumbura, M.; Beenaerts, N.; Gower, D. J.; Pethiyagoda, R.; Roelants, K.; Mannaert, A.; Wilkinson, M.; Bahir, M. M.; Manamendra-Arachchi, K. Local Endemism Within the Western Ghats-Sri Lanka Biodiversity Hotspot. *Science* **2004**, *306*, 479–481.

Billerman, S. M.; Keeney, B. K.; Rodewald, P. G.; Schulenberg, T.S., Eds. BoW Birds of the World; Cornell Laboratory of Ornithology: Ithaca, NY, **2021**. https://birdsoftheworld.org/bow/home

Clements, J. F.; Schulenberg, T. S.; Iliff, M. J.; Billerman, S. M.; Fredericks, T. A.; Gerbracht, J. A.; Lepage, D.; Sullivan, B. L.; Wood, C. L. The eBird/Clements Checklist of Birds of the World: v2021, **2021**. https://www.birds.cornell.edu/clementschecklist/download/

Daniels, R. J. R. Endemic Fishes of the Western Ghats and the Satpura HYPOTHESIs. *Curr. Sci.* **2001**, *81*, 240–244.

De Silva, R. Avian Relics of Sri Lanka. *Spolia Zeylanica* **1980**, *35*, 197–202.

del Hoyo, J.; Collar, N.; Christie, D. A. Sri Lanka Thrush (*Zoothera imbricata*), Version 1.0. In *Birds of the World*; Billerman, S. M.; Keeney, B. K.; Rodewald, P. G.; Schulenberg, T. S., Eds.; Cornell Lab of Ornithology: Ithaca, 2020.

Edwards, G. *A Natural History of Uncommon Birds: And of Some Other Rare and Undescribed Animals, Quadrupeds, Fishes, Reptiles, Insects, & c*; College of Physicians in Warwick-Lane: London, 1743.

Erdelen, W. Aspects of the Biogeography of Sri Lanka. *Forschungen auf Ceylon* **1989**, *3*, 73–100.

ESRI. *ArcGIS Desktop Software*. Environmental Systems Research Institute, Inc.: Redlands, CA, 2012.

Fernando, S. P.; Irwin, D. E.; Seneviratne, S. S. Phenotypic and Genetic Analysis Support Distinct Species Status of the Red-Backed Woodpecker (Lesser Sri Lanka Flameback: *Dinopium psarodes*) of Sri Lanka. *Auk* **2016**, *133*, 497–511.

FFPO. The Fauna and Flora Protection Ordinance no.2 of 1937 (with Amendments), Department of Wildlife Conservation, Sri Lanka Government, the Democratic Socialist Republic of Sri Lanka, 1937.

Fregin, S.; Haase, M.; Olsson, U.; Alström, P. Multi-Locus Phylogeny of the Family Acrocephalidae (Aves: Passeriformes)—The Traditional Taxonomy Overthrown. *Mol. Phylogen. Evol.* **2009**, *52*, 866–878.

Gjershaug, J.; Diserud, O.; Rasmussen, P.; Warakagoda, D. An Overlooked Threatened Species of Eagle: Legge's Hawk Eagle *Nisaetus kelaarti* (Aves: Accipitriformes). *Zootaxa* **2008**, *1792*, 54–66.

Gunawardena, M.; Rowan, J. S. Economic Valuation of a Mangrove Ecosystem Threatened by Shrimp Aquaculture in Sri Lanka. *Environ. Manag.* **2005**, *36*, 535–550.

Gunawardene, N. R.; Daniels, D. A.; Gunatilleke, I.; Gunatilleke, C.; Karunakaran, P.; Nayak, G. K.; Prasad, S.; Puyravaud, P.; Ramesh, B.; Subramanian, K. A Brief Overview of the Western Ghats–Sri Lanka Biodiversity Hotspot. *Curr. Sci.* **2007**, *93*, 1567–1572.

Guneratne, A. **2015**. A Bird in the Bush: Dillon Ripley, Slim Ali and the Transformation of Ornithology in Sri Lanka. Macalester College DigitalCommons@Macalester College: Faculty Publications, Anthropology Department. http://digitalcommons.macalester.edu/anthfacpub/1

Hackett, S. J.; Kimball, R. T.; Reddy, S.; Bowie, R. C. K.; Braun, E. L.; Braun, M. J.; Chojnowski, J. L.; Cox, W. A.; Han, K.-L.; Harshman, J.; Huddleston, C. J.; Marks, B. D.; Miglia, K. J.; Moore, W. S.; Sheldon, F. H.; Steadman, D. W.; Witt, C. C.; Yuri, T. A Phylogenomic Study of Birds Reveals Their Evolutionary History. *Science* **2008**, *320*, 1763–1768.

Hanle, J.; Singhakumara, B. M. P.; Ashton, M. S. Complex Small-Holder Agriculture in Rainforest Buffer Zone, Sri Lanka, Supports Endemic Birds. *Front. Ecol. Evol.* **2021,** 9.

Henry, G. M. *A Guide to the Birds of Ceylon*; Oxford University Press, **1955.**

Hora, S. L. Satpura Hypothesis of the Distribution of Malayan Fauna and Flora of Peninsular India. *Proc. Natl. Inst. Sci. India* **1949,** *15,* 309–314.

Hora, S. L. The Satpura Hypothesis. *Sci. Progress* **1953,** *41,* 245–255.

Jaccard, P. Distribution de la flore alpine dans le bassin des Dranses et dans quelques régions voisines. *Bull. Soc. Vaudoise Sci. Nat.* **1901.** 37, 241–272.

Jarvis, E. D., Mirarab, S.; Aberer, A. J. et al. Whole-Genome Analyses Resolve Early Branches in the Tree of Life of Modern Birds. *Science* **2014,** *346,* 1320–1331.

Jayaratne, R. L.; Perera, I. C.; Weerakoon, D. K.; Kotagama, S. W. Feathers of Little Egret (*Egretta garzetta*) Fledglings as a Bio Monitoring Tool for Mercury, Arsenic, Cadmium and Lead Pollution in Sri Lanka. *J. Ecotoxicol. Ecobiol.* **2019,** *4,* 103–113.

Jetz, W.; Thomas, G. H.; Joy, J. B.; Hartmann, K.; Mooers, A. O. The Global Diversity of Birds in Space and Time. *Nature* **2012,** *491,* 444–448.

Jha, A.; Seneviratne, S.; Prayag, H. S.; Vasudevan, K. Phylogeny Identifies Multiple Colonisation Events and Miocene Aridification as Drivers of South Asian Bulbul (Passeriformes: Pycnonotidae) Diversification. *Organ. Divers. Evol.* **2021.** https://doi.org/10.1007/s13127-021-00506-y

Kaluthota, C.; Kotagama, S. *Revised Avifaunal List of Sri Lanka.* Occasional Paper No. 2 of the Field Ornithology Group of Sri Lanka, **2009**; 25 pp.

Katz, M. B., Sri Lanka in Gondwanaland and the Evolution of the Indian Ocean. *Geol. Magaz.* **1978,** *115,* 237–244.

Kennedy, M.; Page, R. D. M., Seabird Supertrees: Combining Partial Estimates of Procellariiform Phylogeny. *Auk* **2002,** *119,* 88–108.

Knox, R. *Account of the Captivity of Capt. Robert Knox and Other Englishmen in the Island of Ceylon*; Wolfenden Press: London, **1681a.**

Knox, R. *An Historical Relation of the Island Ceylon, in the East-Indies;* R. Chiswell: London, **1681b.**

Kotagama, S. W. Map of the Avifauna Zones of Sri Lanka. In *Strategy for the Preparation of a Biological Diversity Action Plan for Sri Lanka*; Colombo, **1989.**

Kotagama, S. W. Wildlife Conservation and Development of the South East Dry Zone. In *The Southeast Dry Zone of Sri Lanka*; Agrarian Research and Training Institute: Colombo, **1993.**

Kotagama, S. W.; De Silva, R. I. The Taxonomy and Status of Offshore Birds (Seabirds) of Sri Lanka. In *The Fauna of Sri Lanka: Status of Taxonomy, Research and Conservation*; Bambaradeniya, C. N. B., Ed.; World Conservation Union (IUCN)/Government of Sri Lanka: Colombo, **2006**; pp 288–293.

Kotagama, S. W.; Ratnavira, G. *Birds of Sri Lanka: An Illustrated Guide to the Birds of Sri Lanka*; Field Ornithology Group of Sri Lanka: Colombo, Sri Lanka, **2017.**

Kreft, H.; Jetz, W. A Framework for Delineating Biogeographical Regions Based on Species Distributions. *J. Biogeog.* **2010,** *37,* 2029–2053.

Krishan, K. T.; Weerakkody, S.; Seneviratne, S. Phylogenetic Affinities of an Endemic Cloud Forest Avian Relict: Sri Lanka Bush Warbler (*Elaphrornis palliseri*). *27th International Forestry Symposium.* University of Sri Jayawardanapura: Colombo, **2020**; p 36.

Larsen, R.; Holmern, T.; Prager, S. D.; Maliti, H.; Røskaft, E. Using the Extended Quarter Degree Grid Cell System to Unify Mapping and Sharing of Biodiversity Data. *Afr. J. Ecol.* **2009,** *47,* 382–392.

Legge, W. V. *A History of the Birds of Ceylon*; Published by Author: London, **1880.**

Legge, W. V. *A History of the Birds of Ceylon,* 2nd edn.; Tisara Press, Dutugemunu St.: Dehiwala, Sri Lanka, **1983.**

Mattsson, E.; Persson, U. M.; Ostwald, M.; Nissanka, S. P. REDD+ Readiness Implications for Sri Lanka in Terms of Reducing Deforestation. *J. Environ. Manage.* **2012,** *100,* 29–40.

Mayr, E. *Systematics and the Origin of Species*; Columbia University Press: New York, **1942.** Dover reprint **1964.**

MOE. *The National Red List 2012 of Sri Lanka; Conservation Status of the Fauna and Flora*; Ministry of Environment: Colombo: Sri Lanka, **2012.**

MoMD&E, Biodiversity Profile—Sri Lanka, Sixth National Report to the Convention on Biological Diversity, Jayakody, S.; Wikramanayake, E. D.; Fernando, S.; Wickramaratne, C.; Arachchige, G. M.; Akbarally, Z., Eds.; Biodiversity Secretariat, Ministry of Mahaweli Development and Environment: Sri Lanka, **2019**; p. 211.

Moyle, R. G.Phylogenetics of Barbets (Aves: Piciformes) Based on Nuclear and Mitochondrial DNA Sequence Data. *Mol. Phylogen. Evol.* **2004**, *30*, 187–200.

Moyle, R. G.; Andersen, M. J.; Oliveros, C. H.; Steinheimer, F. D.; Reddy, S. Phylogeny and Biogeography of the Core Babblers (Aves: Timaliidae). *System. Biol.* **2012**, *61*, 631–651.

Murray, J. A. *The Avifauna of the Island of Ceylon;* Kegan,Paul, Trench, Trubner & Co.: London, **1890**.

Myers, N.; Mittermeier, R.; Mittermeier, C.; da Fonseca, G.; Kent, J. Biodiversity Hotspots for Conservation Priorities. *Nature* **2000**, *403*, 853–858.

Olson, D. M.; Dinerstein, E. The Global 200: A Representation Approach to Conserving the Earth's Most Biologically Valuable Ecoregions. *Conservation biology* **1998**, 12, 502–515.

Olson, D. M.; Dinerstein, E.; Wikramanayake, E. D.; Burgess, N. D.; Powell, G. V. N.; Underwood, E. C.; D'amico, J. A.; Itoua, I.; Strand, H. E.; Morrison, J. C.; Loucks, C. J.; Allnutt, T. F.; Ricketts, T. H.; Kura, Y.; Lamoreux, J. F.; Wettengel, W. W.; Hedao, P.; Kassem, K. R. Terrestrial Ecoregions of the World: A New Map of Life on Earth: A New Global Map of Terrestrial Ecoregions Provides an Innovative Tool for Conserving Biodiversity. *BioScience* **2001**, *51*, 933–938.

Panagoda, B. G.; Seneviratne, S. S.; Kotagama, S. WS.;Welikala, D. Sympatric Breeding of Two Endangered Sternula Terns, Saunders's (*S. saundersi*) and Little (*S. albifrons*) Terns, in the Rama's Bridge of Sri Lanka. *BirdingAsia* **2020**, *34*, 76–83.

Pasquet, É.; Cibois, A.; Baillon, F.; Érard, C. What Are African Monarchs (Aves, Passeriformes)? A Phylogenetic Analysis of Mitochondrial Genes. *Comptes Rendus Biol.* **2002**, *325*, 107–118.

Peduzzi, P.; Schiemer, F. Bacteria and Viruses in the Water Column of Tropical Freshwater Reservoirs. *Environ. Microbiol.* **2004**, *6*, 707–715.

Perera, D. G. A.; Kotagama, S. W. *A Systematic Nomenclature for the Birds of Sri Lanka*; Field Ornithology Group of SriLanka: Colombo, **1983**.

Perera, J. N.; Muthuwatta, L. P. Senevirathne, S. Climate Forecasts Suggest a Foggy Future for the Montane Endemics of Sri Lanka. "Siyo Siri" Celebrating 40 Years—FOGSL Rodrigo, M., Ed.; Field Ornithology Group of Sri Lanka: Colombo, Sri Lanka, **2016**; pp 86–89.

Perera, S. J.; Fernando, R. H. S. S. *Physiography, Climate and Historical Biogeography of Sri Lanka in Making a Biodiversity Hotspot*. In *Biodiversity of Hotspots of Indian Region: Western Ghats-Sri Lanka*; Pullaiah, T., Ed., Vol. 3; Apple Academic Press: Ware Town, 2022; pp 405–445.

Perera, S. J.; Herbert, D. G.; Procheş, Ş.; Ramdhani, S. Land Snail Biogeography and Endemism in South-Eastern Africa: Implications for the Maputaland-Pondoland-Albany Biodiversity Hotspot. *PLoS ONE* **2021**, *16*, e0248040.

Perera, S. J.; ProcheŞ, Ş.; Ratnayake-Perera, D., Ramdhani, S. Y. D. Vertebrate Endemism in South-Eastern Africa Numerically Redefines a Biodiversity Hotspot. *Zootaxa* **2018**, *4382*, 56–92.

Perera, S. J.; Ramdhani, S.; Procheş, Ş.; Ratnayake-Perera, D. New Biogeographic Insights for Partitioning of Sri Lanka: A Preliminary Attempt of a Numerical Regionalization. In the National Symposium on the Biogeography and Biodiversity Conservation in Sri Lanka in a Changing Climate; National Science Foundation: Colombo, **2015**.

Perera, S. J.; Ratnayake-Perera, D.; Proches, S. Vertebrate Distributions Indicate a Greater Maputaland-Pondoland-Albany Region of Endemism : Research Article. *SA J. Sci.* **2011**, *107*, 1–15.

Pethiyagoda, R. *Pearls, Spices, and Green Gold: An Illustrated History of Biodiversity Exploration in Sri Lanka*; WHT Publications (Private) Limited: Colombo, Sri Lanka, **2007**.

Phillips, W. W. A. Checklist of the Mammals of Ceylon. *Spolia Zeylanica* **1929**, *15*, 119.

Phillips, W. W. A. *A Revised Checklist of the Birds of Ceylon*; The National Museums of Ceylon: Colombo, **1952**.

Phillips, W. W. A. *Annotated Checklist of the Birds of Ceylon (Sri Lanka)*; Wildlife and Nature Protection Society of Ceylon: Colombo, **1975**.

Phillips, W. W. A. *Annotated Checklist of the Birds of Ceylon* (Sri Lanka), Revised edn.; The Wildlife & Nature Protection Society of Ceylon in association with The Ceylon Bird Club: Colombo, **1978**.

Phillips, W. W. A. *On the Avifauna of Sri Lanka. Spolia Zeylanica* (Centenary Commemoration) Proceedings of the Ceylon Natural History Society, National Meseum of Sri Lanka: Colombo, **1980**; p 35.

Podduwage, D. R. An Ethical Model for the Wildlife Photography of Sri Lanka. *J. Aesthetic Fine Arts* **2016**, *1*, 98–129.

Pouteau, R.; Giambelluca, T. W.; Ah-Peng, C.; Meyer, J.-Y. Will Climate Change Shift the Lower Ecotone of Tropical Montane Cloud Forests Upwards on Islands? *J. Biogeog* **2018**, *45*, 1326–1333.

Proches, S., The World's Biogeographical Regions: Cluster Analyses Based on Bat Distributions. *J. Biogeog.* **2005**, *32*, 607–614.

Proches, S.; Ramdhani, S. The World's Zoogeographical Regions Confirmed by Cross-Taxon Analyses. *BioScience* **2012**, 62, 260–270.

Ramdhani, S.; Barker, N. P.; Baijnath, H., Exploring the Afromontane centre of endemism: *Kniphofia* Moench (Asphodelaceae) as a floristic indicator. *J. Biogeography* **2008**, 35, 2258–2273.

Ramdhani, S.; Perera, S. J.; Proches, S. Introducing a Quarter Degree Grid System for Sri Lanka as a Biogeographical Tool. In *The Third International Symposium of Sabaragamuwa University of Sri Lanka*; Sabaragamuwa University of Sri Lanka: Belihuloya, Sri Lanka, **2010**.

Rasmussen, P. C.; Anderton, J. C. *Birds of South Asia: The Ripley Guide*; Smithsonian Institute and Lynx Edicions, **2005**.

Ripley, S. D. Avian Relicts and Double Invasions in Peninsular India and Ceylon. *Evolution* **1949**, *3*, 150–159.

Ripley, S. D. *A Synopsis of the Birds of India and Pakistan*; Bombay Natural History Society: Bombay, India, **1961**.

Ripley, S. D. Avian Relicts of Sri Lanka. *Spolia Zeylanica* **1980**, *35*, 197–202.

Ripley, S. D.; Beehler, B. M. Patterns of Speciation in Indian Birds. *J. Biogeogr.* **1990**, *17*, 639–648.

RSIS. Annotated List of Wetlands of International Importance Sri Lanka, **2021**. https://www.ramsar.org/wetland/sri-lanka

Santoro, A.; Venturi, M.; Bertani, R.; Agnoletti, M. A Review of the Role of Forests and Agroforestry Systems in the FAO Globally Important Agricultural Heritage Systems (GIAHS) Programme. *Forests* **2020**, *11*, 860.

Sclater, P. L. On the General Geographical Distribution of the Members of the Class Aves. *Zool. J. Linn. Soc.* **1858**, *2*, 130–136.

Sclater, P. L. The Geographical Distribution of Birds; an Address Delivered Before the Second International Ornithological Congress, at Budapest. In International Ornithological Congress; Budapest, **1891**.

Sclater, W. L.; Sclater, P. L. *The Geography of Mammals*; K. Paul, Trench, Trübner & Company, Limited, **1899**.

SDSL. *The National Atlas of Sri Lanka*; 2nd edn.; Survey Department of Sri Lanka, **2007**.

Şekercioğlu, Ç. H.; Primack, R. B.; Wormworth, J. The Effects of Climate Change on Tropical Birds. *Biol. Conserv.* **2012**, *148*, 1–18.

Senanayake, F. R.; SoulÉ, M.; Senner, J. W. Habitat Values and Endemicity in the Vanishing Rain Forests of Sri Lanka. *Nature* **1977**, *265*, 351–354.

Seneviratne, S. S.; Weeratunga, V.; Jayaratne, T. A.; Weerakoon, D. K. First Confirmed Breeding Record of Brown Noddy (*Anous Stolidus*) in Sri Lanka. *Birding Asia* **2015**, *23*, 63–65.

Sreekar, R.; Koh, L. P.; Mammides, C.; Corlett, R. T.; Dayananda, S.; Goodale, U. M.; Kotagama, S. W.; Goodale, E. Drivers of Bird Beta Diversity in the Western Ghats–Sri Lanka Biodiversity Hotspot Are Scale Dependent: Roles of Land Use, Climate, and Distance. *Oecologia* **2020**, *193*, 801–809.

Sreekar, R.; Sam, K.; Dayananda, S. K.; Goodale, U. M.; Kotagama, S. W.; Goodale, E. Endemicity and Land-Use Type Influence the Abundance–Range-Size Relationship of Birds on a Tropical Island. *J. Anim. Ecol.* **2021**, *90*, 460–470.

Srinivasan, U.; Prashanth, N. Preferential Routes of Bird Dispersal to the Western Ghats in India: An Explanation for the Avifaunal Peculiarities of the Biligirirangan Hills. *Indian Birds* **2006**, *2*, 114–119.

Stattersfield, A. J.; Crosby, M. J.; Long, A. J.; Wege, D. C. *Endemic Bird Areas of the World-Priorities for Biodiversity Conservation*; BirdLife International: Cambridge, **1998**.

Sudhakar Reddy, C.; Manaswini, G.; Jha, C. S.; Diwakar, P. G.; Dadhwal, V. K. Development of National Database on Long-Term Deforestation in Sri Lanka. *J. Indian Soc. Remote Sens.* **2017**, *45*, 825–836.

Tobias, J. A.; Seddon, N.; Spottiswoode, C. N.; Pilgrim, J. D.; Fishpool, L. D. C.; Collar, N. J. Quantitative Criteria for Species Delimitation. *Ibis* **2010**, *152*, 724–746.

Udvardy, M. D. *A Classification of the Biogeographical Provinces of the World*; IUCN Occasional Paper No. 18 International Union for Conservation of Nature and Natural Resources, Morges, Switzerland, **1975**.

Wait, W. E. *Manual of the Birds of Ceylon*; Colombo Museum: Colombo, **1925**.

Wallace, A. R. *The Geographical Distribution of Animals*, 2 Vols); Macmillan: London, **1876**.

Warakagoda, D.; Sirivardana, U. The Avifauna of Sri Lanka: An Overview of the Current Status. *Taprobanica* **2009**, *1*, 28–35.

Weerakoon, D. K.; Gunawardena, K. The Taxonomy and Conservation Status of Birds in Sri Lanka. In *The National Red List 2012 of Sri Lanka; Conservation Status of the Fauna and Flora*. Weerakoon, D. K., Wijesundara, S., Eds.; Ministry of Environment: Colombo, Sri Lanka, **2012**; pp 114–117.

Whistler, H. The Avifaunal Survey of Ceylon Conducted Jointly by the British and Colombo Museums. *Spolia Zeylanica* **1944**, *23*, 119–322.

Wickramasinghe, M. R. C. P.; De Silva, R. P.; Dayawansa, N. D. K. Climate Change Vulnerability in Agriculture Sector: An Assessment and Mapping at Divisional Secretariat Level in Sri Lanka. *Earth Syst. Environ.* **2021**, *5*, 725–738.

Wickramasinghe, N.; Robin, V. V.; Ramakrishnan, U.; Seneviratne, S. S. Non-Sister Sri Lankan White-Eyes (Genus *Zosterops*) Are a Result of Independent Colonizations. *PLoS ONE* **2017**, *12* (8), e0181441. https://doi.org/10.1371/journal.pone.0181441

Wijesinghe, D. P. *Checklist of the Birds of Sri Lanka*. Special Publications Series, Issue 2. Ceylon Bird Club Notes, **1994**.

Wijesinghe, L.d.S.; Gunatilleke, I.; Jayawardene, S.; Kotagama, S.; Gunatilleke, C. *Biological Conservation in Sri Lanka: A National Status Report*; Sri Lanka, **1993**.

Zaher, H.; Grazziotin, F. G.; Graboski, R.; Fuentes, R. G.; Sánchez-Martinez, P.; Montingelli, G. G.; Zhang, Y.-P.; Murphy, R. W. Phylogenetic Relationships of the Genus *Sibynophis* (Serpentes: Colubroidea). *Papéis Avulsos de Zoologia* **2012**, *52*, 141–149.

Zubair, L. Challenges for Environmental Impact Assessment in Sri Lanka. *Environ. Impact Assess. Rev.* **2001**, *21*, 469–478.

CHAPTER 31

THREATS AND CONSERVATION OF BIODIVERSITY IN SRI LANKA

SUDHEERA M. W. RANWALA

Department of Plant Sciences, Faculty of Science, University of Colombo, Sri Lanka

ABSTRACT

Sri Lanka experiences a great damage to its biological wealth due to various human impacts. Large collection of records available in the 16th–19th centuries disclose that exploitation of biodiversity of Sri Lanka exploded as a result of the change in the political, economic, social, and cultural environment in the colonial days. In the last century, large extent of natural forests, grasslands, and marshlands were transformed to agricultural land and infrastructure development for increasing human population. As far as the conservation of biodiversity in Sri Lanka is concerned, it is apparent that the influence of Buddhist philosophical thoughts in customary law of land had provided a strong support in protecting biodiversity in ancient times. Similarly, the valuable input by foreign scientists and visitors in developing botanical and zoological literary work, establishment of protected areas and species toward the end of colonial period cannot be overlooked. The country after gaining independence continued and strengthened biodiversity conservation activities by incorporating it into the government policy and becoming a party to many international treaties and agreements.

31.1 THREATS TO BIODIVERSITY OF SRI LANKA: FROM PAST TO PRESENT

Human influences to natural ecosystems in the Island of Sri Lanka have been identified since the hydraulic civilization in the dry zone (500BC to 1200AD), during which forests of the dry zone; Northern, North Central, Eastern, South–Eastern regions of the country were cleared for expansion of agriculture and irrigation. Nevertheless, these events had not been considered as a significant damage to the biodiversity as most of the disturbed land returned back to forests when the historical kingdoms gradually shifted from dry zone to the wet zone during the late medieval period (1200AD to 1500 AD). During the colonial period (1500AD to 1948AD) influences from Portuguese, Dutch, and British brought significant

Biodiversity Hotspot of the Western Ghats and Sri Lanka. T. Pullaiah, PhD (Ed.)
© 2024 Apple Academic Press, Inc. Co-published with CRC Press (Taylor & Francis)

changes in the political, economic, social, and cultural climate of the country and as a result, biological resources had been heavy exploited. When Dutch were administrating the country (1658–1756), the key sources of revenue were the sale of elephants, native timbers such as ebony (*Diospyros ebanum*), calamander (*Diospyros quesita*), and *Nedun* (*Ptericopsis mooniana*). During the British period (1796–1948), about one million acres of montane forests were cleared for coffee, cinchona, tea, coconut, rubber, and vegetable cultivation. Related infrastructure developments drastically changed the landscape and land use pattern. Country's indigenous green repositories became a paradise for sports and hunting for foreign visitors. Forbes in his book "Eleven years in Ceylon" explains how 106 elephants were put down within in few days by four European hunters in 1837. Similarly, Samuel Bakers in his book "The Rifle and the Hound in Ceylon" describes how he inaugurated the sport of hunting Elk with hounds at Horton Plains in the 1850s. During the post-independence period (since 1948), the threat to biodiversity aggravated due to economic growth and increasing demands for biological resources by the increasing population. Nearly, a 4.5-fold increase in the population in the country has been reported over the last 100 years, with an increase in human density from 70 to 305 km^{-2} (Fernando et al., 2015). Increasing demands on land, natural resources and other human needs became directly responsible for complete elimination (loss) of natural habitats, degradation, and fragmentation of ecosystems, while overexploitation of plants and animals from wild, and the introduction of invasive alien species acted as key species related drivers of biodiversity depletion (MOFE,1999).

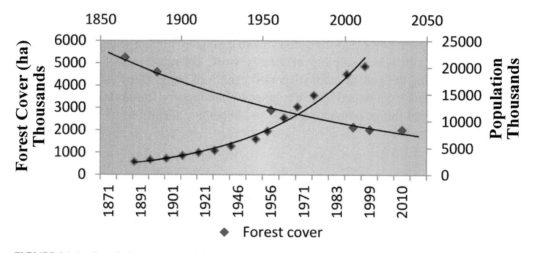

FIGURE 31.1   Population growth and forest cover change in Sri Lanka.

Source: Reprinted from Fernando et al. (2015).

31.2   HABITAT LOSS, DEGRADATION, AND FRAGMENTATION

A drastic drop of forest cover over the 100 years has been observed and the habitat loss, degradation of the environment, and fragmentation of the natural habitats has been limitless.

Shifting cultivation (*chena*) in the dry zone areas has been a traditional practice since the ancient past; however, at present, on many occasions *chena* does not necessarily imply the traditional type of slash-and-burning agriculture mainly because sufficient time is not left for recovery of land after abandonment. Instead, cultivation of cash crops or other land use transformations continue causing a net loss of the forest ecosystem. Records indicate that, the average annual clearing rate of natural forests in the dry zone had been 42,000 ha during 1965–1990. Out of the total area of forest clearance from 1992 to 1996, 87% has been claimed for rain-fed agriculture (Mattsson et al., 2012; Jayasuriya, 2014). Wet zone forests have been transformed for the expansion of plantation and agricultural crops and the loss during 1956–1992 reported to be about 100,000 ha. Forest clearance under permits issued for socioeconomic development had been a key reason for loss and degradation of natural forests. A well-known example is the clearance of forest undergrowth in 2400 ha in Knuckles range for growing of cardamom 1990s (Gunawardene, 2003). Similarly, for passion fruit cultivation, 1200 ha of lowland rain forests in Kalutara district had been cleared impacting the one and only swamp forest, Waturana, where point endemic species *Mesua stylosa* and *Stemonoporus mooni* are naturally found (MMD&E, 2016). Among the examples of habitat loss and fragmentation, large scale exploitation of native trees by mechanized logging in the western part of Sinharaja forest between 1971 and 1977 for plywood industry cannot be forgotten. Similar events have occurred in several areas in Kanneliya, Dediyagala–Naki-yadeniys (KDN) complex in 1950s (Gunatilleke and Gunatilleke, 1983; Jayasuriya, 2014). A considerable area of dry zone forests also has been transformed for irrigation and socio-economic development projects. The accelerated Mahaweli Development Project has also transformed 20,000 ha of dry forests for agriculture, settlements, and related developments. The extent of savanna woodlands in Nilgala valley has been drastically reduced while most of the *Damana* grasslands have disappeared due to urban expansion (Jayasingham, 1988).

The accelerated economic growth, related developments, and urbanization resulted in migration of rural communities to many cities of the western quarter. Due to the change in land use, only about 8% of the lowland rain forests remain in the South–Western region, and they are highly fragmented, degraded and isolated, most of them being <10,000 ha in extent (MOFE,1999). Illegal gem mining in forest lands has also affected the lowland rain forest biodiversity in the western quarter of the Island. In addition to vegetation clearance, it causes changes to the depth of water columns and siltation. Extensive sand mining both in rivers and beaches increase bank erosion, siltation, and saltwater intrusion (CZMP, 2006). In addition, poor slope management in the hill country increase siltation in downstream areas affecting the aquatic biota in low-lying habitats. Earthworks which include large areas cut open and left to erode contribute to increased siltation and erosion irrespective of the location. Diversion of the rivers has also generated problems for aquatic organisms as it alters the quantity of water reaching downstream and speed of water flow (MMD&E, 2016). Among the coastal habitats, extent of Mangroves has been reducing at a high rate since 1990s due to the establishment of prawn culture, expansion of salt pans, and related infrastructure developments.

Threats to biodiversity through habitat loss have been continuously reported from other parts of the country with infrastructure developments including irrigation, hydropower generation, roads and highways, establishment of airports and harbors, establishment of high voltage electricity transmission lines, agricultural, livelihood developments, and resettlement plans. The abandoned areas found in every part of the island are being degraded further due to dumping of waste and frequent fire, causing depletion of the soil fertility, affecting animal behavior (Figure 31.2) (Fernando et al., 2015).

As common to many parts of the world, die back of canopy trees has also become a threat especially to the montane tree flora, viz canopy dominant *Calophyllum* spp. and *Syzygium* spp. (Figure 31.2) in the upper slopes and summits of Totapolakande and Hakgala mountains (Ranasinghe et al., 2009). In some cases, the tree-death is identified due to natural causes as the senescing stage in the forest life cycle, but pollution of atmosphere also had been a cause for tree death in forests (Wijesundera, 2012).

FIGURE 31.2 (A) die back of montane canopy trees in Horton Plains National Park, (B) fragmented landscape in the Dry zone, (C) waste dump in a land adjoining forest has attracted elephants (photographs by author).

Neglecting its important role as flood basins, many wetlands have been reclaimed in the past. The waters have been polluted due to multiple reasons. As a result, forty one (41%) percent of the island's aquatic flora have reported to be threatened (Yakandawela, 2012) and has also posed a severe threat to many aquatic fauna such as eels (e.g., *Monopterus desilvai* and *Monopterus bengalensis*), Fishing Cat (*Prionailurus viverrinus*) and the Otter (*Lutra lutra*) (Weerakoon, 2012).

Nearly one third of all the resident birds in Sri Lanka are forest birds including all the endemic species out of which 60% are restricted to the evergreen forests in the wet zone. Among the isolated populations of three arboreal endemic mammals – the Purple-faced Leaf Monkey (*Semnopithecus vetulus*), the Golden Palm Civet (*Paradoxurus zeylonensis*), and the Red Slender Loris (*Loris tardigradus*), have disappeared from several localities in the South West quarter of the island due to loss of tree cover (Weerakoon, 2012). The reason for low level of endemicity observed in the reptile fauna in the dry zone has been linked to the various disturbances occurred over the hundreds of years. Mortality of reptiles has also been related to man-made forest fires, application of agrochemicals, road kills, non-selective killing of snakes and predation by farm and domestic animals. Displacement of these animals brings into conflict with humans and this is considered as an emerging threat to reptiles in many parts of the country (Wickramasinghe, 2012). Most of the land snails are leaf litter inhabitants, thus clearance of forest floor had been reported detrimental to their survival. In addition, inundation due to large scale development such as dam construction for river diversion (especially the Mahaweli river) and hydroelectric generation often affect land snail populations (Ranawana and Priyadarshina, 2012). The impact on the rare land snail species *Ravana politissima* due to the developments related to Upper Kothmale hydro power project is a good example (IUCN, 2007). Habitat and changes in water quality due to pollution, changes in the speed of water flow are probably the pressing threats for the dragonflies. Land transformations, air pollution, change in species composition, and degradation of suitable habitats due to various human disturbances in many parts of the country are directly responsible for the decreased size of many butterfly populations (van der Poorten and Conniff, 2012).

31.3 OVEREXPLOITATION

Overexploitation of wild species for commercial purposes accelerates their disappearance from natural habitats, impacting biodiversity at species and genetic levels and ultimately degrades the structure and function of natural ecosystems. Population densities of valuable timber species such as *Diospyros ebenum, Chloroxylon swietenia*, and *Manilkara hexandra* in dry forests of the country remain at low levels due to the exploitation of their mature trees long before. Over harvesting of fruits such as *Terminalia bellirica, T. chebula, Phyllanthus emblica* and flower buds of *Careya arborea* especially in savanna ecosystems (Figure 31.3 AI–II, B), and *Dialium ovoideum* and *Manilkara hexandra* fruits from dry forests pose severe threats for their natural regeneration due to the limited propagule mass the next growing season (Perera, 2012). Very often, rattans (*Calamus* spp), Bamboo, and many non-timber woody products such as *Salacia reticulata, Coscinium fenestratum, Tinospora cordifolia*, resins from *Shorea* spp., mushrooms and many herbaceous medicinal species have been overexploited (Kariyawasm, 1996). It has been identified that overexploitation of wood products from natural forests has considerably reduced over the decades, while the over extraction of non-woody products has increased.

Many showy orchids, for example, Vesak orchid (*Dendrobium maccarthiae*), Fox tail (*Rhynchostylis retusa*), and other herbaceous species; *Zeuxine* spp., *Anoectochilus* spp. (Figure 31.3C,D) are common examples for plants illegally collected in large amounts from their wild habitats by ornamental traders (Fernando, 2012). Selective removal of root-balled saplings for landscaping and illegal logging (Figure 31.3E,F) is also becoming a growing threat to maintain the structure and species composition of small forest patches outside the protected area system.

Ornamental aquatic fish and plants for the aquarium industry has also made some species threatened. Many *Cryptocoryne* and *Lagenandra* spp. have become threatened in their natural habitats. Among freshwater fish, 75% of native fish species such as *Puntius titteya, Rasbora vaterifloris* are collected from wild. Among the ornamental marine species, about 300 marine fish and invertebrates had been exported for aquarium trade in the past. Coral mining in the south western border of the country had been a threat for a long time (MMD&E, 2016).

According to Azam et al. (2016), live lobsters, endemic fish and plants, sandal wood, turtle and sea shells, butterflies have been common items that have been illegally collected from natural habitats and attempted to export from Sri Lanka in the past years.

FIGURE 31.3 Examples for some species overexploited, (A-I and A-II) flower buds of *Careya arborea,* (B) *Phyllanthus emblica* fruits, (C) *Dendrobium maccarthiae,* (D) *Zeuxine regia,* (E) illegal logging of *Manilkara hexandra,* (F) root-balled sapling, ready to be removed from forest, (G) lobster (Photographs by author).

31.4   INTRODUCTION OF INVASIVE ALIEN SPECIES

The spread of invasive alien species (IAS) has been identified as a growing threat to Sri Lanka's biodiversity. Based on risk assessments, *Prosopis juliflora, Salvinia molesta,* and *Eichhornia crassipes* plant species, and exotic fish species; *Oncorhynchus mykiss* and *Pterygoplichthys* spp. have been identified as top priority species for control (MMD&E, 2015).

Invasive alien plants with climbing or smothering habit such as Cuscuta (*Cuscuta campestris*) often spread over other plants and prevent them from exposing to sunlight. Woody invaders such as Mesquite (*Prosopis juliflora*), Pond apple (*Annona glabra*), Dillenia (*Dillenia suffruticosa*), and Velvet tree (*Miconia calvescens*) form dense thickets and bring excessive shade to the undergrowth and suppress establishment of seedlings of native species. They often change the nutrient status of soil (Wickramathilake et al., 2013). Spread of *Alstonia macrophylla, Clidemia hirta,* and *Dillenia suffruticosa* in lowland wet evergreen forests and *Miconia calvescens,* and *Clusia rosea* have greatly influenced the mid-elevation evergreen forest habitats and their biota (MOFE, 1999). *Prosopis juliflora* and *Opuntia dillenii* are serious threats to the natural vegetation in Bundala National park as about 60% of the total area is invaded with these two species. Both species have caused major ecosystem changes including the feeding and movements of elephants, and wading birds and butterflies. *Opuntia dillenii* has also displaced the undergrowth fully in many places and suppressed the germination of native species such as *Salvadora persica* and *Cassia auriculata* in such localities. *Lantana camara* has invaded in many parts of the Udawalawe National park drastically reducing food sources available for herbivores including elephants (Perera, 2012). Plant invaders such as *Ageratina riparia, Austroeupatorium inulifolium,* and *Cestrum aurantiacum* in montane forests is facilitated by the gaps created by the die back of canopy species (Rathanyake et al., 1996; Wijesundera, 2012). Their spread has been extended to stream banks and many foot paths in montane forests, hill tops, and grasslands. According to Ekanayake (2016), many sites grown with rare herb *Didymorcarpus humboldtianus* in the Knuckles area has been replaced by *Ageratina riparia*. Several other invasive species including *Miconia calvescens* and *Pennisetum clandestinum* have been observed in the montane region (Wijesundera, 2012). Alien invaders that spread along the river catchments and its tributaries such as *Mimosa pigra* affect the water flow, river bank stabilization, and displace the aquatic biodiversity. Floating aquatic invaders; Water Hyacinth (*Eichhornia crassipes*) and Salvinia (*Salvinia molesta*) form dense mats on water impairing penetration of light into the water. Bellanwila Attidiya sanctuary located in the suburbs of Colombo, is a place where heavy growth and spread of *Eichhornia crassipes* is observed throughout (Yakandawela, 2012).

Spread of invasive alien fish, *Chitala ornata* and *Poecilia reticulata* in the marshes of Bellanwila-Attidiya sanctuary has significantly affected the populations of endemic fish species such as *Esomus thermoicos*, *Clarias brachusoma*, *Aplocheilus dayi*, *Channa orientalis*, and *Puntius singhala*. The presence of *Oreochromis mossambicus* has threatened the native inhabitants of *Labeo porcellus* (*L. lankae*) and *L. dussumieri*. The feeding habits of Clown knife fish (*Chitala ornata*) and the Tank Cleaner (*Pterygoplychthys multiradiatus*), together with other invasive characters contribute them to be superior competitors in inland waters and displace endemic aquatic biota (MOE, 2017).

In addition to the mechanisms explained above, threats from climate change on biodiversity of Sri Lanka have been reviewed (e.g., De Costa, 2011; Iqbal et al., 2014; Miththapala, 2015; Kariyawasam et al., 2019). Although many observations have been reported in literature, only few evidence are available on the specific effects of climate change on species and ecosystems levels.

As reported in other countries, failure of people to understand long-term impacts of their actions, and failure of people to admit the consequences of using inappropriate technology, and failure of economic markets to recognize the true value of biodiversity, failure of government policies to correct for the resultant overuse of biological resources, and the lack of coordination among governmental institutes have been identified as the underlying causes for the depletion of biodiversity of Sri Lanka (MOFE, 1999).

31.5   HISTORICAL RECORDS OF CONSERVATION

History of biodiversity conservation in Sri Lanka dates back to the advent of Buddhism in 236 BE (cir. 250. BCE), after which the kings as upholders of the religion adopted Buddhist philosophical thoughts into the customary law of land. As evident by the records of historical chronicles and inscriptions published in the "Epigraphia Zeylanica," ancient kings reaffirmed that "all life is a unity of which man is only a part" through tradition bound concept of "ahimsa" (meaning that suffering and death should not be caused to any other living being). The compassion and sanctity toward plant and animal life are also reflected by the punishments imposed for those who deliberately or consistently violated the custom (Saparamadu, 2006; Fonseka, 1998). The protected areas declared by the kings included crown land comprising of forests, catchment areas, sanctuaries, and monasteries established at the hill tops. In these areas, destroying habitats, felling of trees, and killing animals for consumption of meat were prohibited. Mahamewna Uyana in Anuradhapua was the first to be named as a wildlife sanctuary and has a continuous record since 240BC (MMD&E, 2016). Mihintale has been made a sanctuary by King Devanampiyatissa in 247BC (MOFE, 1999). Sithulpauwa established by King Kawanthissa (130 BC), and Ritigala established by King Sulathissa (187 BC), later by King Lanjathissa (59BC), and King Sena I (831 AD) are examples for the ancient monasteries which served in a way similar to the modern day forest reserves. Ancient monastic complexes have been declared as Sanctuaries under the Flora and Fauna Protection Ordinance after its enactment in 1937. Records also indicate that the ancient man was well aware of the disastrous impacts of erosion and silting on the reservoirs, thus strict rules had been enforced against clearance of land and cultivation in the catchment areas. Such protected areas have become National Reserves over centuries (Saparamadu, 2006).

This long history in conservation of plant and animal life (see box below) was severely affected when the country was invaded by Magha of Kalinga in 1240, nevertheless Kings who ruled the Kingdoms later were able to abide with the same customary laws to some extent until the Europeans invaded the country in the 16th Century (Fonseka, 1998, MOFE, 1999). Despite the exploitation of resources after the advent of Europeans (1500–1815), the natural ecosystems such as the forests, streams of the country, and jungle life were popular topics for foreign writers such as Robert Knox, R. L. Brohier, Leonard Woolf, Emerson Tennent, and R. L. Spittel. In their books, they deeply unfold their impressionistic studies on wildlife, colorful blossoms of nature, and green shades of the dense forest canopies, which provided resourceful pockets for many botanists, zoologists, and archaeologists to initiate scientific studies (Ranwala, 2014b). Among them was the Dutchman, Paul Herman who was the first to prepare a scientific collection of island's plants in the 1670s, which later enabled Carolus Linnaeus to botanically identify. The governor of the Dutch territories in Sri Lanka, Jan Giden Loten was an ornithologist who contributed enormously to explain the bird life in Sri Lankan forests and provided descriptions to many birds including Loten's sunbird (*Nectarina lotenia* L.).

At the end of 19th century, British botanists Henry Trimen and J. C. Wills were among the first few who revealed the wealth of Flora of Sri Lanka through a series of publications to which they included detailed botanical descriptions of each and every plant they studied. The remaining forest patches of the country also became a popular place of study for distinguished zoologists including Dr. Gardener and A. P. Green, who later revealed taxonomic descriptions on many groups of insects while mammals, birds, and reptiles received their first scientific descriptions by Dr. F. Kelaart (Saparamadu, 2006; Ranwala, 2014b). By this time, the foreigners who arrived with an intention to enjoy hunting in Ceylon forests were met with disappointments and which was a continuous pressure for the British governors to take necessary action on the deprived wildlife and plant resources. The disturbances and damage to the plantations by displaced forest animals due to mass clearing of forests were another serious matter. Recognizing that the flora and fauna are part of the forest produce, legal protection of the wildlife and natural vegetation was initiated by the British for the first time in the country. Drafting the Ordinance No 6 of 1872 to prevent wasteful destruction of sambur, buffaloes, and gamebirds and establishment of closed seasons for 5–6 months with restricted hunting activities was a significant change in the governance of country's natural resources. Steps were taken to protect the forests by controlling the unplanned destruction of forests. Enactment of the Forest Ordinance 10 of 1885 and establishment of the "Forest Department" rooted in governance by initially recognized village forests and forest reserves to regulate uncontrolled destruction of forests. Yala (National Park at present) was proclaimed as the first sanctuary in 1900. Meantime, in 1894, legal backgrounds were made to prohibit the export of horns, hide, and antlers. In addition, legal provisions were made to prevent dynamiting of fish, felling of trees, timber transportation, and many more. At the same time, several scientific descriptions on fauna and flora, forest vegetation, forest survey methodologies, arose through studies by the European Botanists and Zoologists. Enactment of the Plant Protection Ordinance in 1924 became a necessity to initiate the control of accidental introductions of weeds, pests, and diseases injurious to crops and other plants, and subsequent development of quarantine regulations. By consolidating all existing laws on the protection of fauna and flora, the Ordinance No 2 of Fauna and Flora protection was drafted and enacted in 1937. It also categorized protected areas in detail. Since then, this Ordinance remains as a key legal instrument for biodiversity conservation. The Fisheries Ordinance consolidated all laws related to fisheries and fish species. This was a milestone in conserving the marine and aquatic biodiversity of the country. As such, toward of the end of the colonial period conservation initiatives were regained with the intention of recovering the damage incurred to the biological wealth of the country. Hence, the latter part of the colonial time in Sri Lanka (1871–1947) could be considered as a period in which legal background of the country was developed toward conservation of the biological resources of Sri Lanka (Saparamadu, 2006; Ranwala, 2014b; Jayasuriya 2014).

31.6 BIODIVERSITY CONSERVATION IN THE POST-INDEPENDENCE PERIOD

Until the term "biodiversity" was given special recognition and global attention, conservation of the biological resources of the country had been an integral component of environment conservation.

Constitutional Amendments of the government of Sri Lanka and subsequent establishment of the Central Environment Authority and enactment of the National Environment Act No. 47 of 1980 became the key instrument for protecting the country's natural resources. It also provided provisions to declare specific areas of unique biological wealth as Environmental Protection Areas through an order published in the Gazette. Legal instruments to regulate developments in coastal area was made in the same year with the enactment of the Coast Conservation Act No 57 of 1981 and the Coast conservation Department (CCD) became the prime agency for caring the coastal resources, grant permits for developmental activities, prepare management plans for conservation of natural coastal habitats and areas of cultural and recreational value. CCD administrates protected areas designated as Special Management Areas and Fisheries Management Areas. The Marine Pollution Prevention Act enacted in 1981 provided provisions for the prevention, reduction, and control of pollution in Sri Lankan ocean waters. Penal actions against marine pollution and damage to live marine resources have been taken. The Fisheries and Aquatic Resources Act (revised in 1996) became responsible for the integrated management, regulation, conservation and development of fisheries, and aquatic resources in Sri Lanka, along with the declaration of fisheries reserves (MOFE,1999).

Meanwhile, as a response to the World Conservation Strategy (1981), the National Conservation Strategy of Sri Lanka was adopted in 1988 as the center piece of the government's policies to deal with environmental degradation in the country. National Environmental Action Plan was adopted in 1992 (later revised in 1994) focusing on development of many policies directly related to the protection of forests and wildlife, agricultural and plantation resources, and coastal resources and wetlands (MOFE, 1999).

Ratification to the Ramsar Convention 1990 paved the pathway to conserve the wetlands of Sri Lanka. The preservation of the ecological values and functions in the country's wetlands was extended to develop a framework for the international corporation with the Government of Netherlands in the Wetland conservation project (CEA, 1991–1998). The Wetland Conservation Strategy served as a guideline for development of the National Wetland Policy in 2006. Accordingly, the Wetland Management Unit was also established at the Central Environment Authority for an efficient implementation of the wetland protection and management strategies.

Forestry sector was also one of the key areas that received attention throughout the years. However, for many years, the function of the forests was mainly focused on the production of goods (especially timber value) than their services. Nevertheless, toward the end of 1970s, forest conservation was re-gained with the establishment of a network of 36 UNESCO's Man and Biosphere (MAB) reserves within which timber extraction was prohibited (Gunatilleke and Gunatilleke, 1983; Bandaratileke, 1996). In 1995, the National Forestry Sector Master Plan was developed targeting the conservation of all components of the forests including soils, water, along with historical, cultural, religious, and aesthetic values. An amendment was made to the Forest Ordinance in 1995 (No. 23 of 1995) to include a new category of forests called Conservation Forests and 31 conservation forests in the wet zone and Knuckles region were included (Bandaratilake, 1996; Bandaratilake and Fernando, 2010). As a part of the Forestry Sector Development project implemented in 1986, Forest Department, Ministry of Environment, World Conservation Monitoring Center for Food and Agriculture Organization of United Nations, and IUCN designed an optimum protected area system for Sri Lanka's Natural forests through a detailed comprehensive and innovative evaluation of 281 forests, emphasizing their importance in soil and water conservation for watershed protection. The study, commonly referred to as the National Conservation Review (1997) recorded 69,400 records of endemic woody plant species and selected animal groups and related information which contributed immensely for the management of forests. It also identified the wet zone forests as the most important in terms of soil and water conservation and biodiversity as the survey revealed that 79% of the woody plant diversity, 88% of endemic woody plant diversity, 83% of faunal diversity, and 85% of endemic faunal diversity are represented in just eight units of contiguous forests: Bambarabotuwa, forests of the central highlands, Gilimale Eratne, KDN, Sinharaja, all in the wet zone, Knuckles/Wasgomuwa in the intermediate zone, and Ruhuna/Yala in the dry zone (IUCN-WCMC-FAO, 1997). Forest plantation sector was strengthened by growing Teak (*Tectona grandis*), Mahogany (*Swietenia macrophylla* and *S. mahagoni*), Pines (*Pinus caribaea*), and eucalyptus (mainly *Eucalyptus grandis, E. microcorys* and *E.*

robusta) though community-based participation in degraded land mainly in the upcountry of Sri Lanka (Sathurusinghe and Hunt, 2005).

Following ratification of the country to the Convention on Biological Diversity in 1994, relevant changes were made to all policy instruments to accommodate conservation of genetic, species and ecosystem diversities. The strategy for the Preparation of a Biodiversity Action Plan for Sri Lanka (MOFE, 1999) was an obligation to the Convention on Biological Diversity (CBD). Development of the Biodiversity Conservation Action Plan (BCAP) was commenced in 1996 through a wide consultative process, involving government agencies, non-governmental organizations, experts and local communities by the Government with technical assistance from IUCN under "Biodiversity Conservation in Sri Lanka: A Framework for Action in 1998. The addendum (2007) to the Biodiversity Conservation in Sri Lanka: A Framework for Action" together with the chapter reports updated the existing knowledge on gaps and needs in protecting country's biological wealth in a comprehensive way. The preparation of Provincial biodiversity profiles and Action plans was also a follow-up activity of the BCAP (MMD&E, 2016).

Mainstreaming biodiversity conservation into national school and higher education curricula and capacity building programs encouraged the participation of youth in environmental societies, habitat restoration programs, biodiversity assessments, and various other environmental activities jointly conducted with private sector non-governmental, and community-based organizations. Activities organized through the leadership of many professional voluntary organizations developed effective methodologies for knowledge sharing and skill development on conservation of wildlife and forest resources among the youth and younger generation. Corporate social responsibility (CSR) projects established in the private business sector organizations initiated collaborations with many stakeholders to take the message and responsibility of conservation to a wider group of stakeholders.

In situ conservation of biodiversity of Sri Lanka is mainly supported by the network of the protected areas since 1950s. The establishment of Protected Area Management and Wildlife Conservation Project in 2001 identified critically important terrestrial biological resources including biodiversity hotspots, wildlife corridors, and habitat linkages vital for the survival of wildlife in areas outside forests (Jayasuriya et al., 2006). According to 2016 records, the total area declared as protected areas in Sri Lanka is around 2.3 million ha, representing about 35% of the total land of the county. These include six internationally important Ramsar wetland sites (Vankalai Sanctuary, Wilpattu Wetland Cluster, Anawilundawa Tank, Madu ganga, Bundala & Kumana Wetland Cluster), four International Man and Biosphere reserves (Sinharaja, Hurulu, Kanneliya, and Bundala) and two natural world heritage sites (Sinharaja) and part of the Central Highland region (covering Horton Plains, Knuckles, and Peak Wilderness).

In situ conservation of crop wild relatives through enhanced information management and field applications was one of the key projects that explored genetic studies on many crop species. The program identified ecogeographical distribution of 672 species of wild relatives of food crops including wild species of priority crops; rice, piper, green gram/ black gram, banana, cinnamon, okra and peanuts and their *in situ* conservation as a valuable genetic resource (DOA, 2009).

Ex situ conservation of Sri Lanka's biological wealth is mainly supported by the National Zoological Gardens and associated orphanages, Botanic Garden network comprising of six botanic gardens, Plant Genetic Resource Center of the Agriculture Department through seed, field gene banks and *in vitro* collections, and the museums associated with the National Museum. New botanic gardens were also established based on bioclimatic regions, Mirijjawela in the dry zone in 2008 and the other in Seethawakapura, Avissawella in the wet zone in 2016 (MMD&E, 2016).

With the global interest in identifying ecosystem services and launch of the Millennium Ecosystem Assessment in 2005, the contribution of biodiversity and ecosystem services to social well-being in Sri Lanka was incorporated into the government policy by implementing the national strategy for sustainable development through the National Council for Sustainable Development. The National Action Plan for Haritha Lanka Programme entitled Caring for the Environment (2009–2013) covered ten broad missions or thrust areas (NCSD, 2009). Sri Lanka committed to achieve a significant reduction of the rate of biodiversity loss by 2010, through the "2010 Biodiversity Target." Recognizing the need to incorporate Biodiversity and Ecosystem Services (BES) values into decision-making process, "The Economics of Ecosystems and Biodiversity" (TEEB) framework was drafted and the Biodiversity Secretariat of the Ministry of Environment initiated the "Pricing the Biodiversity of the Island" project in 2012. This project resulted in vital recommendations on establishing financing mechanisms through biodiversity valuation and economics. At present, the country has incorporated the targets of Aichie and sustainable development Goals (2015–2030) into their government policies (MMD&E, 2016).

Responding to the principles of CBD, the government of Sri Lanka implemented a strategy to prevent/ minimize adverse impacts of Invasive Alien Species and as a result several national programs were conducted to control and eradicate aquatic and terrestrial plant invaders at Udawalawe, Bundala and Horton Plains National Parks since 2003. In addition, there have been a few small-scale programs conducted by non-governmental agencies to control and eradicate Invasive Alien Species at several locations in Sri Lanka (Ranwala, 2014a). The first mega scale project that focused on several aspects of IAS was conducted from 2011 to 2016, *viz.* "Strengthening Capacity to Control the Introduction and Spread of Alien Invasive Species" supported by the United Nations Development Programmes and Global Environment Facility Funds. As a result, a national IAS Policy was finalized and integrated to the national policy framework, a national IAS Strategy, and Action Plan was finalized and adopted including a fiscal and market-based instruments and financial strategies to support IAS control, a national IAS Control Act was developed, National Invasive Species Specialist Group (NISSG) was established and mandated for advising the government of Sri Lanka. IAS pre-entry and post-entry risk assessment protocols were developed and used for the preparation of priority invasive flora and fauna lists, a website on IAS was developed, in addition, site specific, cost effective, control strategies for four priority invasive alien fauna and four priority invasive alien flora were piloted at selected sites through public–private NGO partnerships, and national IAS Communication Strategy was developed and implemented (MOE, 2010, MOE, 2017).

Understanding that the degradation and deforestation of the world's tropical forests are responsible for about 10% of net global carbon emissions, Sri Lanka became a part of the

UN-REDD (reducing emissions from deforestation and forest degradation) framework for taking remedial action to effectively reduce the rates of emissions UN-REDD was launched in 2008 and became UN-REDD-plus (REDD+) in 2010, to reflect the components; Reducing emissions from deforestation and forest degradation, conservation of forest carbon stocks, sustainable management of forests, and enhancement of forest carbon stocks. The project was successfully completed in 2017 and the National REDD+ Investment Framework and Action Plan was presented 13 Policies and Measures. In addition, National Forest Reference Emission Level/Forest Reference Level was submitted to the UNFCCC and a National Forest Monitoring System was made available (MMD&E, 2016).

Conservation and Management Project (ESCAMP, 2017–2022 targets to improve the management of ecosystems in selected locations in Sri Lanka for conservation and community benefits. Unlike others, this projects management is being continued within landscape level to strengthen the integrity of ecosystems by promoting the involvement of local communities for livelihood enhancement and protection through the sustainable use of existing resources.

Biosafety aspects for biodiversity conservation are also an important aspect that the country has committed since ratification to Cartegena Protocol in 2004 to safeguard the biodiversity against importation of genetically modified organisms (GMOs), safe transfer, and handling of genetically modified living organisms (LMOs). Accordingly, the country drafted the National Biosafety Framework in 2005, National Biosafety Clearing House Mechanism (MONER, 2007), and the National Biosafety Policy in 2011. The implementation of the National Biosafety Framework is being continued (2017–2021) through the National Biosafety Project funded by the Global Environment Facility (NSF, 2020).

Commitments of the state to address threats from climate change were initiated with country's ratification to the UN Framework Convention on Climate Change (UNFCC), Kyoto protocol and Paris agreement. As a response National Climate Change Adaptation Strategy for Sri Lanka (2011–2016) was developed (MERE, 2010) and the National Climate Change Policy was adopted in 2012. The National Adaptation Plan (2016–2025) and Nationally Determined Contributions were developed to identify mitigatory actions and build-up resilience to adapt to climate changes (CCS, 2016, MERE, 2012).

In the year 2016, the National Biodiversity Strategy and Action Plan (NBSAP, 2016–2022) was implemented by the Biodiversity Secretariat, Ministry of Mahaweli Development and Environment, with Technical Assistance from International Union for the Conservation of Nature (Sri Lanka), incorporating 2020 Aichi Biodiversity Targets and Sustainable Development Goals. The NBSAP acts as a guiding policy framework for provincial authorities of Sri Lanka, as well as for the civil society groups and private sector organizations in biodiversity conservation and ecosystems management (MMD&E, 2016).

At present, there are about 30 state institutions and 15 legal instruments directly involved in biodiversity conservation in Sri Lanka. These provide an important framework to support the maintenance of biodiversity into sectoral and cross-sectoral strategies, plans, and programs. Among the government agencies directly contribute to biodiversity conservation, the role of Departments of Wildlife Conservation and Forest Conservation became mandatory services over the years. However, insufficient expertise in particular related areas, poor collaboration among the institutes, inadequate and overlapping policies, inefficient information management system and data sharing mechanism, and adequate financing can be identified as priority areas to be strengthened to further uplift biodiversity conservation in Sri Lanka in the future.

KEYWORDS

- **biodiversity**
- **conservation**
- **threats**
- **Sri Lanka**
- **challenges**
- **protection**

REFERENCES

Azam, A. W. M.; Jayasuriya, K. M. G. G.; Musthafa, M. M.; Marika, F. M. M. T. Sri Lanka Is a Hot Spot for Illegal Transnational Trading of Biodiversity and Wildlife Materials from South Asian Region. *J. Transp. Secur.* **2016,** *9,* 71–85.

Bandaratilake, H. M. *Administration Report of the Conservator of Forests for the Year 1995*; Forest Department: Sri Lanka, **1996.**

Bandaratilake, H. M.; Fernando, M. P. S. *National Forest Policy Review*; Sri Lanka, Forest Department: Sri Lanka, **2010.**

CCS. *National Adaption Plan for Climate Impacts in Sri Lanka*; Climate Change Secretariat, Ministry of Mahaweli Development and Environment, **2016.**

CEA. *Final Report of the Wetland Conservation Project (1991–1998)*; Central Environment Authority and ARCADIS Euroconsult: The Netherlands, **1998.**

CZMP. *Coastal Zone Management Plan.* Coast Conservation Department, Colombo. **2006.**

De Costa, W. A. J. M. A Review of the Possible Impacts of Climate Change on Forests in the Humid Tropics. *J. Natl. Sci. Foundation, Sri Lanka* **2011,** *39* (4), 281–302.

DOA. *In Situ Conservation of Crop Wild Relatives through Information Management and Field Application in Sri Lanka*, Department of Agriculture, Ministry of Environment and Natural Resource, International Plant Genetic Resources Institute (IPGRI), Food and Agriculture Organization (Rome), **2009.**

Ekanayake, S. P. Declining Populations of Some Threatened Plants Associated with Rocky Seepages in the Knuckles Area. *The Climate Change magazine of Sri Lanka.* Climate Change Secretariat, Ministry of Mahaweli Development and Environment: Sri Lanka, **2016.**

Fernando R. H. S. S. Present Status of Family Orchidaceous in Sri Lanka. In *The National Red List 2012 of Sri Lanka; Conservation Status of the Fauna and Flora*; Weerakoon, D. K., Wijesundara, S., Eds.; Ministry of Environment: Colombo, Sri Lanka, **2012;** pp 200–204.

Fernando, S.; Senaratna, A.; Pallewatta, N.; Lokupitiya, E.; Manawadu; Imbulana,U.; De Silva, I.; Ranwala, S. *Assessment of Key Policies and Measures to Address the Drivers of Deforestation and Forest Degradation in Sri Lanka-United Nations Development Programme (UNDP) for the Sri Lanka UN-REDD Programme*, Colombo Science and Technology Cell, Faculty of Science, University of Colombo, **2015.**

Fonseka, L. *Environment Policies of Ancient Kings*; Godage Publishes: Colombo, **1998**.

Gunatilleke I. A. U.; Gunatilleke, C. V. S. Conservation of Natural Forests in Sri Lanka. *The Sri Lanka Forester* **1983**, *16*, 39–56.

Gunawardane H. G. Ecological Implications of Cardamom Cultivation in the High Altitudes of Knuckles Forest Reserve, Central Province, Sri Lanka. *Sri Lanka Forester* **2003**, *26*, 1–9.

Iqbal, M. C. M.; Wijesundera, D. S. A.; Ranwala, S. M. W. Climate Change, Invasive Alien Flora and Concerns for Their Management in Sri Lanka. *Ceylon J. Biol. Sci.* **2014**, *43* (2), 1–15.

IUCN. *Biodiversity Assessment in the Upper Kotmale Hydropower Project—Phase II*; International Union for Conservation: Sri Lanka, **2007**.

IUCN- WCMC-FAO. *Designing an Optimum Protected Areas System for Sri Lanka's Natural Forests. Project Report*, IUCN-The World Conservation Union and the World Conservation Monitoring Centre for the Food and Agriculture Organization (FAO) of the United Nations, **1997**.

Jayasingham, T. Transformation Dynamics of the Damana Grasslands, Sri Lanka: Field Study and a Model Synthesis. *Sri Lanka Forester* **1998**, *23* (1 and 2), 4–17.

Jayasuriya, A. H.; Kitchener, D.; Biradar, C. M. *Portfolio of Strategic Conservation Sites/Protected Area Gap Analysis in Sri Lanka*; EML Consultants: Colombo, Sri Lanka, **2006**.

Jayasuriya, A. H. M. Forests and Plants. In *Sri Lanka's Forests- Nature at Your Service*; De Silva, A. T., Ed.; Sri Lanka Association for the Advancement of Science: Colombo, **2014**; pp 32–48.

Kariyawasam, C. S.; Kumar, L.; Ratnayake, S. Invasive Plant Species Establishment and Range Dynamics in Sri Lanka Under Climate Change. *Entropy* **2019**, *21*, 571. DOI: 10.3390/e21060571

Kariyawasm, D. The Effect of Income, Among the Communities of Sinharaja, on the Extraction of Non-Timber Forest Products: A Preliminary Study. *Sri Lanka Forester* **1996**, *22*, 31–38.

Mattsson E., U. Martin Persson, Madelene Ostwald, S. P. Nissanka, S. P. REDD+ Readiness Implications for Sri Lanka for Reducing Deforestation. *J. Environ. Manag.* **2012**, *100*, 29–40.

MERE. *National Climate Change Adaptation Strategy for Sri Lanka*; Ministry of Environment and Renewable Energy: Battaramulla, Sri Lanka, **2010**.

MERE. *The National Climate Change Policy of Sri Lanka*. Ministry of Environment and Renewable Energy, Battaramulla, Sri Lanka, **2012**.

Miththapala S. Conservation Revisited. *Ceylon J. Sci. (Biol. Sci.)* **2015**, *44* (2), 1–26.

MMD&E. Invasive Alien Species in Sri Lanka: Training Manual for Managers and Policy Makers. Biodiversity Secretariat, Ministry of Mahaweli Development and Environment, **2015**.

MMD&E. *National Biodiversity Strategic Action Plan* 2016–2022. Colombo, Sri Lanka: Biodiversity Secretariat, Ministry of Mahaweli Development and Environment, **2016**.

MOE. *Strengthening Capacity to Control the Introduction and Spread of Invasive Alien Species Project Completion Report*. Biodiversity Secretariat, Ministry of Environment, **2010**.

MOE. *The National Red List 2012 of Sri Lanka; Conservation Status of the Fauna and Flora*; Weerakoon, D. K., Wijesundara, S., Eds.; Ministry of Environment: Colombo, Sri Lanka, **2012**.

MOE. *Terminal Evaluation Report, Strengthening Capacity to Control the Introduction and Spread of Alien Invasive Species in Sri Lanka GEF Project*; Biodiversity Secretariat, Ministry of Environment, 2017.

MOFE. *Biodiversity Conservation in Sri Lanka: A Framework for Action*; Ministry of Forestry and Environment, **1999**.

MONER. *National Biosafety Framework of Sri Lanka*; Ministry of Environment and Natural Resources: Colombo, Sri Lanka. **2005**.

NCSD. *Haritha Lanka Programme*; National Council for Sustainable Development, Presidential Secretariat: Colombo, Sri Lanka, **2009**.

NSF. *Vidurawa*, Vol. 37; National Science Foundation: Sri Lanka, **2020**.

Pethiyagoda, R. *Pearls, Spices and Green Gold: An Illustrated History of Biodiversity Exploration in Sri Lanka*; WHT Publications, **2007**.

Perera, A. Present Status of Dry-Zone Flora in Sri Lanka. In: *The National Red List 2012 of Sri Lanka; Conservation Status of the Fauna and Flora*; Weerakoon, D. K., Wijesundara, S., Eds.; Ministry of Environment: Colombo, Sri Lanka, **2012**; pp 165–174.

Ranasinghe, P. N.; Fernando G. W. A. R.; Wimalasena M. D. N. R.; Siriwardana Y. P. S. Dieback in Tropical Montane Forests of Sri Lanka: Anthropogenic or Natural Phenomenon? *J. Geol. Soc. Sri Lanka* **2009**, *13*, 23–45.

Ranawana, K. B.; Priyadarshana, T. G. M. The Taxonomy and Conservation Status of the Land Snails in Sri Lanka. In *The National Red List 2012 of Sri Lanka; Conservation Status of the Fauna and Flora*; Weerakoon, D. K., Wijesundara, S., Eds.; Ministry of Environment: Colombo, Sri Lanka, **2012**; pp. 65–76.

Ranwala, S. M. W. Stakeholder Participation in Management and Control of Invasive Alien Flora in Sri Lanka. *Proceedings of the National Symposium of Invasive Alien Species, IAS 2014*, organized by University of Kelaniya in collaboration with Biodiversity Secretariat of the Ministry of Environment, 27th November 2014, Sri Lanka Foundation, Colombo, **2014a**; pp 8–21.

Ranwala, S. M. W. Forests and People. In *Forests of Sri Lanka*; *Nature at Your Service*; De Silva, A. T., Ed.; Sri Lanka Association for the Advancement of Science, **2014b**; pp 90–104.

Rathanyake, R. M. W.; Jayasekara, L. R.; Solangaarachchi, S. M. A Quantitative Study of Overstorey Vegetation of an Upper Montane Rain Forest. *Sri Lanka Forester* **1996**, *22*, 43–49.

Saparamadu, S. D. *Sri Lanka: A Wildlife Interlude*, Vol. I; Tisara Prakashakayo: Dehiwala, Sri Lanka, 2006.

Sathurusinghe, A.; Hunt, S. Participatory Forest Resource Management in Sri Lanka—Past, Present and Future Directions. *Proc. 17th Commonwealth Forestry Conference*; Colombo, **2005**.

van der Poorten, G. The Taxonomy and Conservation Status of the Butterflies of Sri Lanka. In *The National Red List 2012 of Sri Lanka; Conservation Status of the Fauna and Flora*; Weerakoon, D. K., Wijesundara, S., Eds.; Ministry of Environment: Colombo, Sri Lanka, **2012**; pp 1–10.

Weerakoon, D. K. The Taxonomy and Conservation Status of Mammals in Sri Lanka. In *The National Red List 2012 of Sri Lanka; Conservation Status of the Fauna and Flora*; Weerakoon, D. K., Wijesundara, S., Eds.; Ministry of Environment: Colombo, Sri Lanka, **2012**; pp 134–137.

Wickramasinghe, L. J. M. The Taxonomy and Conservation Status of the Reptile Fauna in Sri Lanka. In *The National Red List 2012 of Sri Lanka; Conservation Status of the Fauna and Flora*; Weerakoon, D. K., Wijesundara, S., Eds.; Ministry of Environment: Colombo, Sri Lanka, **2012**; pp 99–113.

Wickramathilake, B. A. K.; Weerasinghe, T. K.; Ranwala, S. M. W. Impacts of Woody Invader *Dillenia suffruticosa* (Griff.) Martelli. on Physico-Chemical Properties of Soil and, Below and Above Ground Flora. *J. Trop. Forestry Environ.* **2013**, *3* (2), 66–75.

Wijesundara, S. Present status of montane forests in Sri Lanka. In: *The National Red List 2012 of Sri Lanka; Conservation Status of the Fauna and Flora*; Weerakoon, D. K., Wijesundara, S., Eds.; Ministry of Environment, Colombo, Sri Lanka, **2012**; pp 181–185.

Williams, H. *Ceylon-Pearl of the East*; Surjeet Publications: New Delhi, **2002**.

Yakandawela. D. Present Status of Fresh Water Aquatic Flora of Sri Lanka. In *The National Red List 2012 of Sri Lanka; Conservation Status of the Fauna and Flora*; Weerakoon, D. K., Wijesundara, S., Eds.; Ministry of Environment: Colombo, Sri Lanka, **2012**; pp 186–196.

INDEX

X

Z

9781774913765